WITHDRAWN

FLUVIAL HYDRAULICS

ABOUT THE AUTHORS

Professor at the Swiss Federal Institute of Technology in Lausanne (EPFL) since 1973, **Walter H. Graf** is in charge of its Hydraulic Research Laboratory, where M. S. Altinakar is one of his assistants. His research activities are concentrated on the hydrodynamics of free-surface flow, notably open-channel hydraulics and hydrodynamics of lakes and reservoirs. Previously W. H. Graf had teaching engagements at the University of California-Berkeley, at Cornell University and at Lehigh University in the USA. In 1995, he was named a "Guest Professor" at the prestigious Tsinghua University at Beijing, China. Over the years he served also as a visiting professor in Austria (1972, 1994), Republic of South Africa (1974), China (1984, 1994), Italy (1990, 1993), Japan (1995) and the Soviet Union (1975, 1982, 1989), where he lectured essentially on different topics of Fluvial Hydraulics. Amongst the different distinctions received could be mentioned the "Freeman Fellowship" from the American Society of Civil Engineers in 1972, the "IAHR Lecturer Award" given by the International Association of Hydraulics Research in 1988, the "guest scholarship" from the Kajima Foundation of Tokyo, Japan in 1995 and more recently, in 1997, the K. E. Hilgard Hydraulic Prize, awarded annually by the American Society of Civil Engineers.

Walter H. Graf is the author or editor of four books on *Hydraulics of Sediment Transport* (1971, McGraw Hill, USA; 1984, Water Res. Publ., USA), *Hydrodynamics of Lakes* (1979, Elsevier, NL), *Lake and Reservoir Hydraulics* (1987, Water Res. Publ., USA) and *Hydrodynamique* (1991, Ed. Eyrolles, F.; 1995, Presses polytechniques et universitaires romandes, CH). More recently the two volumes of *Fluvial Hydraulics* (1993, 1996, Presses polytechniques et universitaires romandes, CH) were produced; the last books were done with the collaboration of M. S. Altinakar.

In 1987, the XXII Congress of the International Association of Hydraulic Research (IAHR) was held in Lausanne, under the direction of W. H. Graf with the participation of M. S. Altinakar; together six volumes of proceedings were edited.

In addition to being senior assistant at the Hydraulic Research Laboratory, EPFL, Dr. **M. S. Altinakar** is principal Engineer in the hydraulics department of the Bonnard & Gardel Eng. Consulting firm headquartered in Lausanne. There, M. S. Altinakar was in charge of various large projects in hydraulic, hydropower and municipal engineering.

FLUVIAL HYDRAULICS
Flow and Transport Processes in Channels of Simple Geometry

Walter H. Graf
in collaboration with M. S. Altinakar

Laboratoire de recherches hydrauliques
Ecole polytechnique fédérale
Lausanne, Suisse

JOHN WILEY & SONS
Chichester · New York · Weinheim · Brisbane · Singapore · Toronto

Copyright © 1998 by Presses polytechniques et universitaires romandes,
Lausanne, Suisse

Published 1998 by John Wiley & Sons Ltd,
Baffins Lane, Chichester,
West Sussex PO19 1UD, England

National   01243 779777
International   (+44) 1243 779777
e-mail (for orders and customer service enquiries): cs-books@wiley.co.uk
Visit our Home Page on http://www.wiley.co.uk
or http://www.wiley.com

This book, translated by the authors, was originally published in French entitled Hydraulique Fluviale. Volume 1: Ecoulement permanent uniforme et non uniforme. Copyright © 1993 Presses polytechniques et universitaires romandes, Lausanne, Suisse. ISBN: 2-88074-261-7. Volume 2: Ecoulement non permanent et phénomènes de transport. Copyright © 1996 Presses polytechniques et universitaires romandes, Lausanne, Suisse. ISBN: 2-88074-300-1.

All rights reserved. No part of this publication may be reproduced, stored in a retrieval system, or transmitted, in any form or by any means, electronic, mechanical, photocopying, recording, scanning or otherwise, except under the terms of the Copyright, Designs and Patents Act 1988 or under the terms of a licence issued by the Copyright Licensing Agency, 90 Tottenham Court Road, London UK W1P 9HE, without the permission in writing of the publisher and copyright holder.

Other Wiley Editorial Offices

John Wiley & Sons, Inc., 605 Third Avenue,
New York, NY 10158-0012, USA

WILEY-VCH Verlag GmbH, Pappelallee 3,
D-69469 Weinheim, Germany

Jacaranda Wiley Ltd, 33 Park Road, Milton,
Queensland 4064, Australia

John Wiley & Sons (Asia) Pte Ltd, 2 Clementi Loop #02-01,
Jin Xing Distripark, Singapore 129809

John Wiley & Sons (Canada) Ltd, 22 Worcester Road,
Rexdale, Ontario M9W 1L1, Canada

British Library Cataloguing in Publication Data

A catalogue record for this book is available from the British Library

ISBN 0-471-97714-4

Printed from camera-ready-copy supplied by the authors
Printed and bound in Great Britain by Bookcraft (Bath) Ltd

This book is printed on acid-free paper responsibly manufactured from sustainable forestry, in which at least two trees are planted for each one used for paper production.

Foreword

In 1963, over a third of a century ago, a member of the Water Resources Division of the United States Geological Survey advised Cornell University in Ithaca, New York, to obtain from me an external assessment of the proposal to establish at Cornell a major interdisciplinary programme of teaching and research into various aspects of water. My visit enabled me to meet and discuss with a group of outstandingly able and enterprising academics, including the then Assistant Professor Walter H. Graf, a range of initiatives on which I was able to report wholly favourably, and I have considered it a privilege to watch the successful progress of that initiative since.

Five years later, in 1968, the Mexican government invited the distinguished Hans Albert Einstein (who had been Professor Graf's PhD adviser at the University of California in Berkeley) and me to assist the National Autonomous University of Mexico with the establishment of a postgraduate course in hydraulics and hydrology. One weekend, Professor H.A. Einstein talked about his late father, the great Albert Einstein, who had advised his son to choose research topics in which success was not impossible: in the son's field of interest, the mechanics of fluids, the two "impossible" areas were turbulence and sediment transport. The son drily commented that he had naturally chosen one of these, sediment transport, in order to prove his father wrong!

Professor Graf's own important contribution to these "impossible" topics are well documented, for example in his vastly successful 1971 book *Hydraulics of Sediment Transport* to which the late Hunter Rouse in his magisterial historical survey *Hydraulics in the United States 1776–1976* pays due tribute as the pioneer.

The mutual admiration between Rouse and Graf may not be well known, nor that both were Freeman Scholars, Rouse in 1929–31 and Graf in 1973–74.

After a gap of many years, we resumed contact when he moved to the Swiss Federal Institute of Technology in Lausanne to occupy one of the world's great chairs of fluid mechanics and hydraulic engineering and proved himself one of the pillars of the international community engaged in research into hydraulics. At the same time as proving himself a master of the English, French and German languages, as well as of his academic discipline, he demonstrated to his students, by his example, the importance of the visual arts, music, opera and even mythology!

About fifteen years ago, when lecturing at the Dresden University of Technology in East Germany, I was told by the Head of Department, Professor G. Bollrich, that Professor Graf had been an earlier Western contributor to their series of guest lectures and had captivated his students by the clarity and simplicity of his presentation of quite advanced material and by his friendly encouragement of even the most junior members of his audience.

These qualities illuminate the present book.

It covers a wide range of topics listed by the authors in their Preface. To me this seems a wider range of topics than provided by other books. The authors present them with great skill and special attention to the learning needs of students and are particularly good at finding apt illustrations and examples.

Indeed, the authors have chosen what, in the light of our growing pre-occupation with the sustainable future of the civilised world, seems to me the most important range of subjects, from fundamental science to engineering application, all connected with the supply, control and treatment of water and protection from excess water. In the developing world, water science and engineering will be even more valuable and important.

Here they are set in the context of a systematic and comprehensive teaching volume with a sound foundation in mechanics which is clearly intended to encourage the reader, whether a student new to the field or a practitioner seeking to refresh his knowledge, to *think* and to *understand* as well as to acquire information.

The logical structure of the work is well outlined in the Preface. Having had to read many textbooks on fluid mechanics and hydraulic engineering, as a student or adviser to authors and publishers or as a member of teaching and examining bodies, I am pleased to note the many useful as well as profound features of the text and I am confident that it will be a helpful friend to the serious reader.

Professor Emeritus Peter O. Wolf, FEng
London, May 1997

Preface

> *Wie Hexlein von Bock zu Bock*
> *stürzt junges Wasser von Brock zu Brock,*
> *umtost sie wiegsam weich*
> *in nassen Armen liebesreich.*
> . . .
> *Und seidig fein auch grob und linnen*
> *springt's Wasser nieder, niemals auf -*
> *dem Meere zu; denn urvergesslich*
> *nimmt dieses alle Sünder auf.*
>
> Christa
> 1939–1998

Water is a source of life and of energy. Well before antiquity, man understood the necessity to master the available water resources, of which watercourses are an important part. The fluvial resources have been used for human needs, for agriculture, for transportation, for energy production, for flood and pollution management or as boundaries between regions. To the economic exploitation should be added the ecological preoccupation related to the fluvial systems. Fluvial hydraulics, the study of the physical comportment of our natural and artificial watercourses, was and is of major importance to master our environment. To bring together the present knowledge of the hydraulic aspects of watercourses is now the principal goal of our book.

In this book, entitled in short *Fluvial Hydraulics*, we present, in nine chapters of unequal length, topics on *flow and transport processes in channels of simple geometry*. The subjects treated are essentially limited to the study of undirectional flow. Such a restriction – implying that a watercourse is taken as an (artificial) channel – may be considered as strong, but it imposed itself in order to make the book not too voluminous. Emphasis is put onto the hydraulic concepts of the subject. There might be regrets that such topics as hydrology, geomorphology or engineering of rivers did not receive adequate attention; also this became a necessary and personal choice.

An overall *Introduction* (chap. 1) and theoretical *Hydrodynamic Considerations* (chap. 2) make up the first part of the book. In the second part the flow in channels is treated as being *Uniform Flow* (chap. 3), *Non-uniform Flow* (chap. 4) and *Unsteady Flow* (chap. 5). The third part is devoted to some transport phenomena, such as *Transport of Sediments* (chap. 6), *Local Scour* (chap. 9), *Turbidity Currents* (chap. 7) and *Transport and Mixing of Matter* (chap. 8). A brief introduction preceding each chapter should familiarise the reader with the topics to be treated within the chapter. In support and as an illustration of the main text numerous exercises have been elaborated in great detail. Since each chapter is written almost self-contained – making reference to other chapters, but not demanding necessarily to read these – some repetitions became unavoidable.

The style of the book is such that it should readily be usable as a textbook for students and also as a reference book for engineers. For those who want more information and details there is a selective list of references at the end of the book.

The motivation to write this book stems to a large degree from a long teaching and research engagement with topics related to fluvial hydraulics. However, equally important was the fact that no recent and comprehensive book on this subject was available in the French language. Produced by *Presses polytechniques et universitaires romandes*, Lausanne, the French edition was presented in two volumes in 1993 and 1996. Seemingly this book has been well received, not only by students (in upper-class and graduate courses), but also by practising engineers. Almost all the reviewers of the book asked for a translation into English, which is herewith available, soon to be followed by one into Chinese. It remains our hope that this book and the style in which it is written will continue to be appreciated, all over the world, by those interested in this subject.

It is a pleasant duty to express thanks to all our immediate collaborators over the years, who have contributed in different ways to the creation of this book. Our special thanks go to Prof. Peter O. Wolf from London, who has written the *Foreword*, witnessing thus his professional and personal confraternity.

Walter H. Graf
March, 1998

Contents

1. **Introduction** 1
   1.1 Channels — 3
   1.2 Flow in Channels — 6
   1.3 Distribution of Velocity — 10
   1.4 Distribution of Pressure — 12

2. **Hydrodynamic Considerations** — 17
   2.1 Equation of Continuity — 19
   2.2 Equation of Energy — 21
   2.3 Specific Energy — 25
   2.4 Gravity Waves — 31
   2.5 Hydrodynamic Equations — 36
   2.6 Distribution of Velocity — 50

3. **Uniform Flow** 69
   3.1 Hydrodynamic Equations — 71
   3.2 Coefficient of Friction — 74
   3.3 Discharge Calculation, Fixed Bed — 88
   3.4 Discharge Calculation, Mobile Bed — 92
   3.5 Flow in Curves — 100
   3.6 Instability at Surface — 105
   3.7 Exercises (5 + 25) — 109

4. **Non-uniform Flow** 133
   4.1 Gradually Varied Flow — 135
   4.2 Forms of Water Surface — 141
   4.3 Computation of Water Surface — 149
   4.4 Rapidly Varied Flow — 163
   4.5 Transitions — 179
   4.6 Lateral Inflow — 191
   4.7 Exercises (6 + 29) — 194

5. **Unsteady Flow** 251
   5.1 Hydrodynamic Equations — 253
   5.2 Methods of Solution — 258
   5.3 Kinematic Wave — 270
   5.4 Diffusive Wave — 276
   5.5 Flood Wave — 280
   5.6 Translatory Waves — 282
   5.7 Exercises (5 + 12) — 294

6. **Transport of Sediments** 351
   6.1 Generalities — 353
   6.2 Hydrodynamic Equations — 358
   6.3 Bed-load Transport — 371
   6.4 Suspended-load Transport — 384
   6.5 Total-load Transport — 393
   6.6 Exercises (6 + 8) — 400

7. **Turbidity Currents** 467
   7.1 Generalities — 469
   7.2 Hydrodynamic Equations — 474
   7.3 Interface-profile Curves — 479
   7.4 Entrainment Coefficients — 482
   7.5 Front of Current — 485
   7.6 Distribution of Velocity and of Concentration — 489
   7.7 Exercises (1 + 5) — 491

8. **Transport and Mixing of Matter** 517
   8.1 Theoretical Considerations — 519
   8.2 Diffusion — 532
   8.3 Convection-diffusion in Laminar Regime — 539
   8.4 Convection-diffusion in Turbulent Regime — 542
   8.5 Exercises (7 + 18) — 573

9. **Local Scour** 611
   9.1 General Remarks — 613
   9.2 Pier Scour — 614
   9.3 Abutment Scour — 628
   9.4 Constriction Scour — 633
   9.5 Hydraulic Structures Scour — 638
   9.6 Exercises (5 + 9) — 645

List of Symbols 663

Bibliography 669

Index 678

1. INTRODUCTION

In this book on *fluvial hydraulics* — here taken to be synonymous to open-channel hydraulics — we shall treat the flow and flow-related phenomena in artificial and natural channels with a free surface subjected to the atmospheric pressure.

This chapter of introduction begins with a presentation of the different types of channels as well as with the corresponding flow regimes. Subsequently, the notions of the distribution of velocity and of pressure are exposed.

A list of references as well as a list of symbols shall be presented in the final pages of this volume.

TABLE OF CONTENTS

1.1   CHANNELS
         1.1.1   Kinds of Channels
         1.1.2   Geometry of Channels

1.2   FLOW IN CHANNELS
         1.2.1   Types of Flow
         1.2.2   Flow Regimes

1.3   DISTRIBUTION OF VELOCITY

1.4   DISTRIBUTION OF PRESSURE
         1.4.1   Uniform Current
         1.4.2   Curvilinear Current

1.1. CHANNELS

1° A channel is a transport system where water flows and where the free surface is subject to atmospheric pressure.

2° The hydraulic study of a channel often confronts the engineer with a question of the form :
> for a given longitudinal bed slope, a certain discharge must be conveyed; the form and the dimensions of the channel are to be determined.

1.1.1 Kinds of Channels (see Fig. 1.1)

1° Two categories of channels are to be distinguished :

   i) natural channels,
   ii) artificial channels.

2° *Natural* channels are watercourses, which exist naturally on (or under) the earth, such as gullies, brooks, torrents, rivers, streams and estuaries.

The geometric and hydraulic properties of such channels are generally rather irregular. The application of the hydraulic theory gives only approximate results, since numerous assumptions have to be made.

3° *Artificial* channels are watercourses developed by men on (or under) the earth, such as open channels (navigation channels, power canals, irrigation and drainage channels) or closed channels where flow does not fill the entire section (hydraulic tunnels, aqueducts, drains, sewage canals).

The geometric and hydraulic properties of such channels are generally rather regular. The application of the hydraulic theory gives reasonably realistic results.

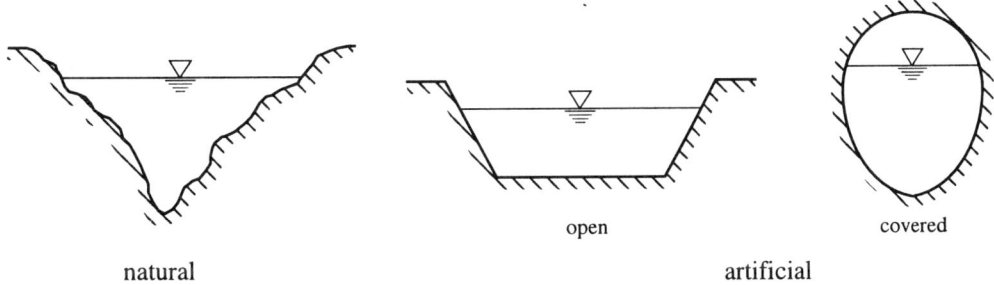

Fig. 1.1   Kinds of channels.

1.1.2 Geometry of Channels (see Fig. 1.2)

1° The (transversal) section of a channel is a section in the cross-sectional plane being normal to the direction of flow.

The section, or better the *wetted surface,* A, is the portion of the cross section occupied by the liquid.

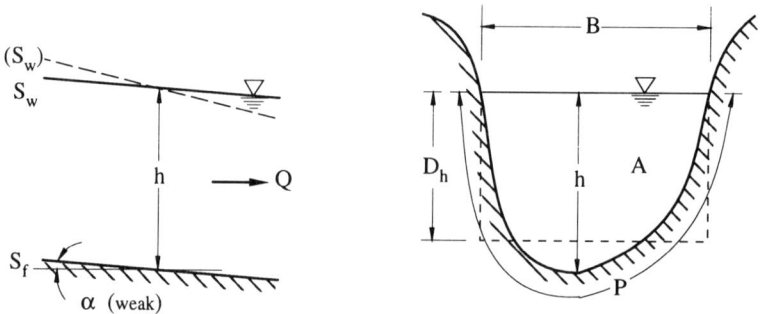

Fig. 1.2 Geometric elements of a channel section.

2° A channel, whose section does not vary and whose longitudinal slope and roughness remains constant — however the flow depth may vary — is called a *prismatic* channel; otherwise the channel is a non-prismatic one.

3° The geometric elements of a section or wetted surface, A, are the following :

i) The *wetted perimeter,* P, of the channel, being formed by the length of the line of contact between the wetted surface and the bed and the side walls, but does not include the free-water surface.

ii) The *hydraulic radius,* R_h, being the ratio of wetted surface, A, to its wetted perimeter, P, or :

$$R_h = \frac{A}{P} \tag{1.1}$$

being often used as a length of reference.

iii) The (top) *width,* B, of the channel being the width at the free surface.

iv) The *hydraulic depth,* D_h, of the channel being defined by :

$$D_h = \frac{A}{B} \tag{1.2}$$

v) The *flow depth,* h, or the water height — if not defined otherwise — is considered to be the maximum depth.

INTRODUCTION

Table 1.1 Geometric elements for different sections of channels.

	Rectangle	Trapezoid	Triangle	Circle	Parabola
Section A	bh	$(b+mh)h$	mh^2	$\frac{1}{8}(\theta - \sin\theta)D^2$	$\frac{2}{3}Bh$
Wetted perimeter P	$b+2h$	$b+2h\sqrt{1+m^2}$	$2h\sqrt{1+m^2}$	$\frac{1}{2}\theta D$	$B + \frac{8}{3}\frac{h^2}{B}$ *
Hydraulic radius R_h	$\dfrac{bh}{b+2h}$	$\dfrac{(b+mh)h}{b+2h\sqrt{1+m^2}}$	$\dfrac{mh}{2\sqrt{1+m^2}}$	$\dfrac{1}{4}\left[1 - \dfrac{\sin\theta}{\theta}\right]D$	$\dfrac{2B^2 h}{3B^2 + 8h^2}$ *
Width B	b	$b+2mh$	$2mh$	$(\sin\theta/2)D$ or $2\sqrt{h(D-h)}$	$\dfrac{3}{2}\dfrac{A}{h}$
Hydraulic depth D_h	h	$\dfrac{(b+mh)h}{b+2mh}$	$\dfrac{1}{2}h$	$\left[\dfrac{\theta - \sin\theta}{\sin\theta/2}\right]\dfrac{D}{8}$	$\dfrac{2}{3}h$

* Valid for $0 < \xi \leq 1$, with $\xi = 4h/B$. If $\xi > 1$: $P = (B/2)\left[\sqrt{1+\xi^2} + 1/\xi \ln\left(\xi + \sqrt{1+\xi^2}\right)\right]$

4° Formulas for the geometric elements for five different types of channel sections (see *Chow*, 1959, p. 21) are given in Table 1.1. A natural watercourse might have a rather irregular geometric form, but often it can be rather well approximated by a trapezoidal or parabolic section.

5° Besides the geometric elements, the longitudinal slopes are also to be considered, namely the :

   i) slope of the bed (bottom or floor), S_f,

   ii) slope of the water surface (piezometric), S_w.

The value of the bottom slope depends essentially on the topography of the terrain; it is generally weak, thus may be expressed by : $S_f = \text{tg }\alpha \cong \sin\alpha$.

6° The wetted perimeter, P, can be composed of a fixed or immobile bed (concrete, rock) or of a mobile bed (granulates of sediments).

1.2 FLOW IN CHANNELS

1° Flow in natural or artificial channels is flow with a free surface, being the surface of separation between air and water; the pressure is there equal to the *atmospheric pressure*.

2° Flow in open channels is essentially due to the inclination (slope) of the bed, while flow in closed conduits (see *Graf & Altinakar*, 1991, chap. PP.1), is due to a difference in the charge between the sections.

1.2.1 Types of Flow

1° A classification of open-channel flow may be done according to the change of the flow depth, h or D_h, with respect to time and space :

$$D_h = f(t, x)$$

2° *Time variation* (see Fig. 1.3)

Flow is *steady* (stationary or permanent) if the average velocity of flow, U, and point velocity, u, but also the flow depth, h or D_h, remain invariable with time, in magnitude and in direction. Consequently the discharge remains constant :

$$U S = Q \qquad (1.3)$$

between the different sections of the channel (see sect. 2.1 and eq. 2.6), supposing there is not lateral inflow or outflow.

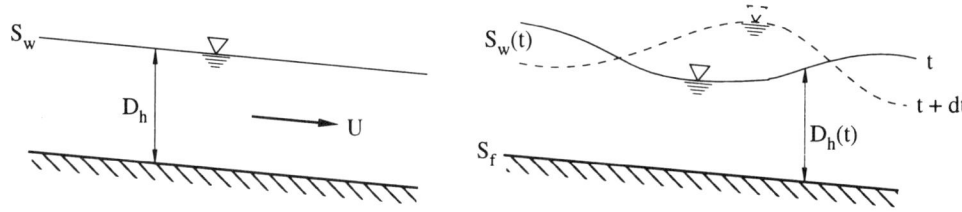

Fig. 1.3 Scheme of steady and unsteady flows.

Flow is *unsteady* if the flow depth, $D_h(t)$, as well as the other parameters vary with time. Consequently, the discharge is no more constant (see sect. 2.1 and eq. 2.1).

Strictly speaking, open-channel flow is rarely steady. However, the temporal variations are often sufficiently slow and the flow may be assumed to be steady and this at least for relatively short time intervals.

INTRODUCTION

3° *Space variation* (see Fig. 1.4)

Flow is *uniform* if the flow depth, D_h, as well as the other parameters, remain unchanged at every section of the channel. The line of the bottom slope is thus parallel to the one of the free-water surface, or $S_f \equiv S_w$.

Flow is *non-uniform* or *varied* if the depth, $D_h(x)$, as well as the other parameters, vary along the length of the channel. The bottom slope is thus different from the slope of the water surface, or $S_f \neq S_w$.

Non-uniform flow can be steady or unsteady.

Varied flow can be accelerated, $dU/dx > 0$, or decelerated, $dU/dx < 0$, depending on the variation of velocity in the direction of flow.

If flow is a *gradually varied* one, the depth, $D_h(x) \cong D_h$, as well as the other parameters, vary slowly from one section to another. Over a small length of the channel, one may assume that the flow is quasi-uniform and the velocity, U, remains essentially constant.

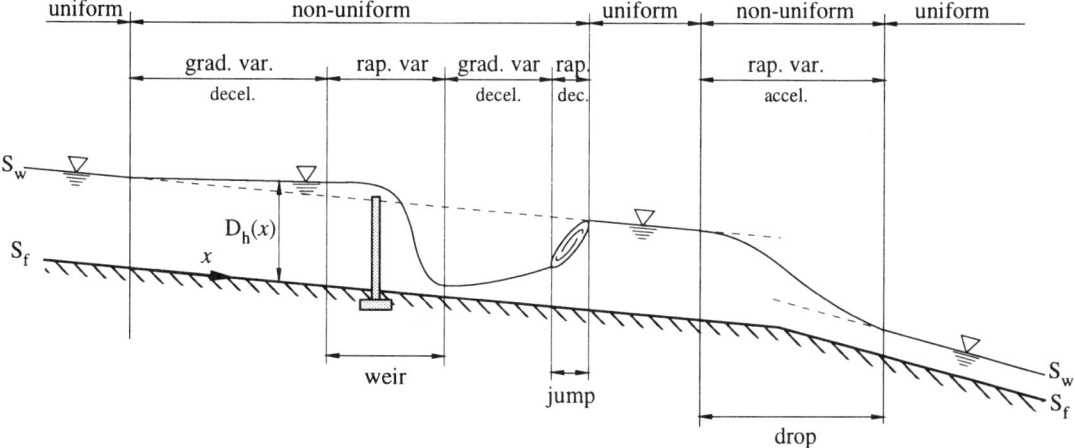

Fig. 1.4 Scheme of steady, uniform and non-uniform flows.

If flow is a *rapidly varied* one, the depth, $D_h(x)$, as well as the other parameters, change abruptly over a comparatively short distance, sometimes with a discontinuity. This happens generally in the neighbourhood of a singularity, such as at a weir or at a change of channel width, but also at an hydraulic jump or an hydraulic drop.

4° The kinds of flow one encounters in fluvial hydraulics (see Fig. 1.3 and Fig. 1.4) can be summarised as follows :

i) Steady flow $\begin{cases} \text{uniform} \\ \text{non-uniform} \begin{cases} \text{gradually} \\ \text{rapidly} \end{cases} \end{cases}$

ii) Unsteady flow $\begin{cases} \text{uniform (}rare\text{)} \\ \text{non-uniform} \begin{cases} \text{gradually} \\ \text{rapidly} \end{cases} \end{cases}$

1.2.2 Flow Regimes

1° The physics of open-channel flow is governed basically by the interplay of the :

- inertia forces,
- gravity forces,
- friction (viscosity and roughness) forces.

2° The (reduced) equations of motion (see *Graf & Altinakar*, 1991, sect. FR.7.2) involve the following dimensionless numbers :

i) the *Froude number,* being the ratio of gravity to inertia forces, or :

$$\frac{\rho g}{\rho U_c^2 / L_c} = \frac{gL_c}{U_c^2} = Fr^{-2} \quad \text{and} \quad Fr = \frac{U_c}{\sqrt{gL_c}} \quad (1.4)$$

ii) the *Reynolds number,* being the ratio of friction to inertia forces, or :

$$\frac{\mu (U_c / L_c^2)}{\rho U_c^2 / L_c} = \frac{\nu}{U_c L_c} = Re^{-1} \quad \text{and} \quad Re = \frac{U_c L_c}{\nu} \quad (1.5)$$

Added to these two numbers is still :

iii) the *relative roughness,* being the ratio of the roughness height, k_s, to a characteristic length, or :

$$\frac{k_s}{L_c} \quad (1.6)$$

INTRODUCTION

U_c and L_c are characteristic velocity and length; one takes often $U_c = U$ and $L_c = R_h$ or $L_c = D_h$.

In the hydraulics of open-channel flow, one generally defines these dimensionless numbers as:

$$Fr = \frac{U}{\sqrt{gD_h}} \quad ; \quad Re = \frac{4R_hU}{\nu} \quad \text{or} \quad Re' = \frac{R_hU}{\nu} \quad ; \quad \frac{k_s}{D_h} \quad (1.7)$$

3° The Reynolds number is used to classify the flow (see *Graf & Altinakar*, 1991, chap. FR.3) as follows:

- laminar flow $Re' < 500$
- turbulent flow $Re' > 2000$
- transition flow $500 < Re' < 2000$

From numerous experiments with different artificial channels (see *Chow*, 1959, p. 10) it results that flow is turbulent if the Reynolds number, Re', reaches a value of 2000.

In general, flow in open channels is a turbulent and often rough flow.

4° The Froude number is used to classify the flow (see sect. 2.3.3) as follows:

- subcritical (fluvial) flow $Fr < 1$
- supercritical (torrential) flow $Fr > 1$
- critical flow $Fr \equiv Fr_c = 1$

In general, flow in open channels can be of the three types.

5° Consequently, the combined effect of the Reynolds number, Re', and the Froude number, Fr, gives the following four regimes of flow:

- subcritical-laminar $Fr < 1$, $Re' < 500$
- subcritical-turbulent $Fr < 1$, $Re' > 2000$
- supercritical-laminar $Fr > 1$, $Re' < 500$
- supercritical-turbulent $Fr > 1$, $Re' > 2000$

A relationship depth/velocity, taken from the experiments by *Robertson et Rouse*, is given in Fig. 1.5; it is valid for very wide rectangular channels.

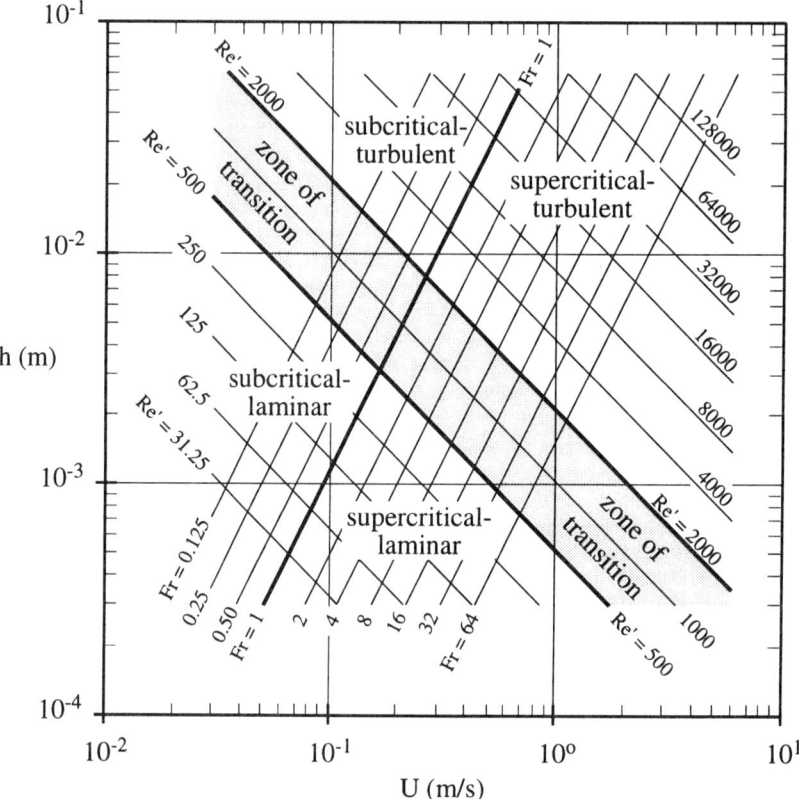

Fig. 1.5 The four regimes of open-channel flow.

1.3 DISTRIBUTION OF VELOCITY

1° In flow along a wall (the bottom of a channel), a distribution of velocity (see *Graf & Altinakar*, 1991, chap. FR. 6) is encountered. Being zero at the wall, the point velocity, u , increases rapidly towards the free surface; its maximum value is often found slightly below this free surface. The velocity profile is approximately logarithmic.

2° Steady flow depends in general on the three variables, x, y and z ; this is called *three-dimensional* flow. For a rectangular channel with a bed and vertical side walls, a schematic distribution of the point velocity, $u(x,y,z)$, is given in Fig. 1.6.

If such a channel has a large width, B — large in comparison with the depth, B > 5 h — flow is considered *two-dimensional*, with the exception of a small distance close to the vertical side walls.

INTRODUCTION

Hydraulic calculations are considerably simplified, if one assumes the flow to be *one-dimensional*. The average velocity, U(x), in a vertical or in a section, is expressed by :

$$U = \frac{1}{h} \int_0^h u(z)\, dz \quad \text{or} \quad U = \frac{1}{A} \int_0^B \int_0^h u(z)\, dz\, dy \qquad (1.8)$$

3° In open channels of simple geometry, one encounters generally turbulent flow where the point velocity, u(x, z), differs little from the average velocity, U(x). In the steady state, such an hypothesis allows to consider the flow as one-dimensional.

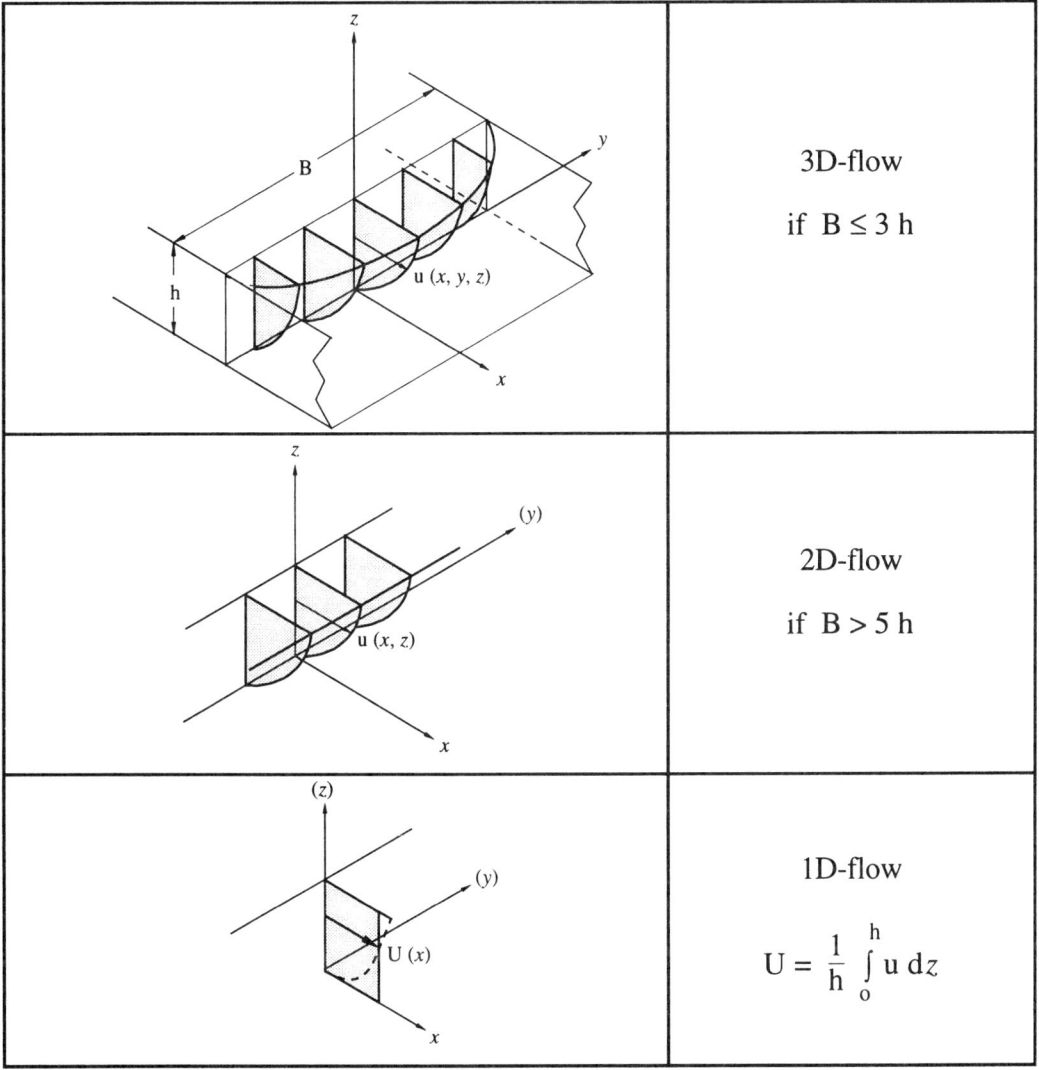

Fig. 1.6 Distribution of velocity.

4° For a determination of the average velocity, U, in a given section, the following approximate relations can be used (see Fig. 1.7) :

$$U \cong (0.8 \text{ à } 0.9) u_s \quad \text{(formula of Prony)}$$
$$U \cong 0.5 (u_{0.2} + u_{0.8}) \quad \text{(formula of USGS)} \quad (1.9)$$
$$U \cong u_{0.4}$$

where $u_{0.2}$, $u_{0.8}$, $u_{0.4}$ and u_s are the point velocities at given positions.

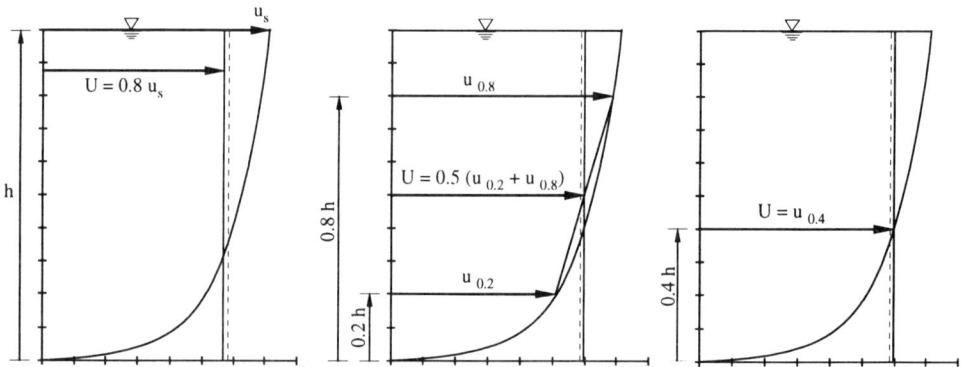

Fig. 1.7 Average velocity.

1.4 DISTRIBUTION OF PRESSURE

1° The equation of steady motion for an incompressible fluid (see *Graf & Altinakar*, 1991, p. 132), written for the normal component, $n(\equiv z)$, is :

$$U \frac{U}{r} = -\frac{1}{\rho} \frac{\partial}{\partial n} (p + \gamma z') \quad (1.10)$$

where (U^2/r) is the centrifugal acceleration of a mass-fluid, which displaces itself on a curved line, r (see Fig. 1.8).

2° Assuming that U and r remain reasonably constant and after integration of eq. 1.10, one obtains :

$$(p + \gamma z') = -\rho \int_0^z \frac{U^2}{r} dn + \text{Cte} = -\rho \frac{U^2}{r} z + \text{Cte} \quad (1.10a)$$

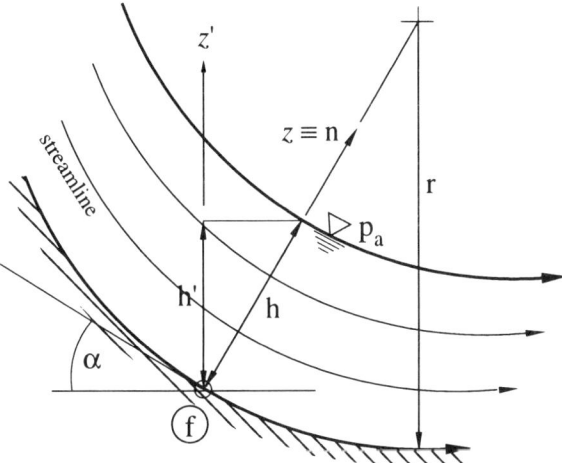

Fig. 1.8 Flow over a concave bottom.

3° Taking a point on the bottom of the channel and another one on the free surface, one respectively writes :

for $z = 0$ ($z' = 0$) : $p = p_f$ where : $p_f = $ Cte

for $z = h$ ($z' = h'$) : $p = p_a$ where : $p_a + \gamma h' = -\rho \dfrac{U^2}{r} h + $ Cte

An expression for the relative (with respect to the atmospheric pressure) pressure on the bottom of the channel, is given by :

$$p_f = \gamma h' + \rho \dfrac{U^2}{r} h + \cancel{p_a}^{= 0} \qquad (1.11)$$

having an hydrostatic and an accelerating contribution.

1.4.1 Uniform Current

1° For uniform flow, when the average velocity, U, remains constant and the streamlines are reasonable rectilinear (with $r \to \infty$), the distribution of pressure is *hydrostatic* in a section, normal to the bottom (see Fig. 1.9). Thus one may write, taking $z \equiv n$ (eq. 1.10), the following :

$$0 = \dfrac{\partial}{\partial z}(\gamma z' + p) \qquad (1.12)$$

2° An expression for the pressure, relative to the bottom, can now be given as:

$$p_f = +\gamma h' \tag{1.13}$$

which gives:

$$\left(\frac{p}{\gamma}\right)_f = h \cos\alpha \tag{1.14}$$

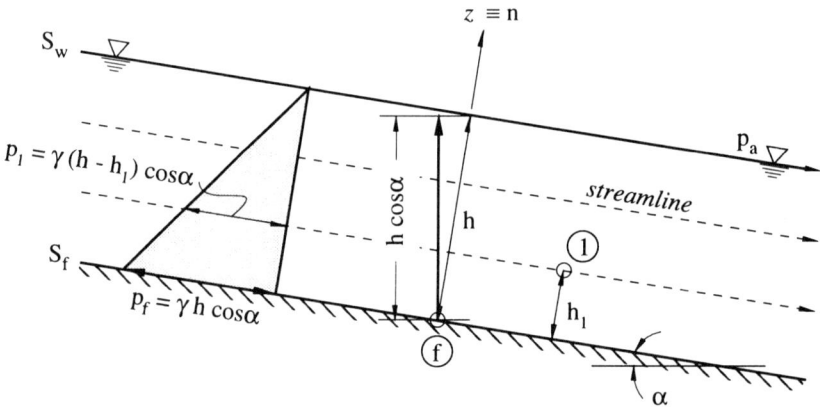

Fig. 1.9 Flow with a uniform current.

3° For the usually encountered open channels, the inclination, α, is rather weak, namely $\alpha < 6°$ or $J_f < 0.1$, implying that $\cos\alpha \simeq 1$. Consequently eq. 1.14 reduces to:

$$\left(\frac{p}{\gamma}\right)_f = h \tag{1.15}$$

where h is the flow depth in the channel.

1.4.2 Curvilinear Current

1° For flow, being (slightly) non-uniform, thus having a curvilinear current of converging or diverging type, there exists an acceleration component caused by the inertia forces. As done above, one writes:

$$\frac{U^2}{r} = -\frac{1}{\rho}\frac{\partial}{\partial n}(p + \gamma z') \tag{1.10}$$

INTRODUCTION

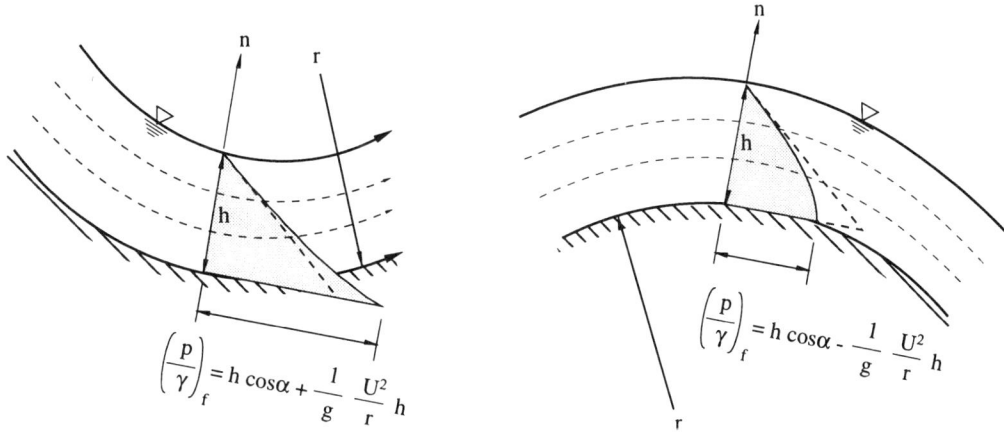

Fig. 1.10 Flow over a concave and a convex bottom.

and the expression for the pressure relative to the bottom is given by :

$$p_f = +\gamma h' \pm \rho \frac{U^2}{r} h \qquad (1.11)$$

being (+) for a concave and (−) for a convex bottom.

Subsequently one obtains :

$$\left(\frac{p}{\gamma}\right)_f = h \cos \alpha \pm \frac{1}{g} \frac{U^2}{r} h \qquad (1.16)$$

2° The distribution of pressure is no more hydrostatic (see Fig. 1.10). For an external concave current, the centrifugal force increases the pressure; while for a convex current, this force decreases the pressure. In the latter case, the pressure could get below the atmospheric pressure, thus causing separation of flow on the channel bed.

2. HYDRODYNAMIC CONSIDERATIONS

Some fundamental notions of hydrodynamics, being the basis of open-channel hydraulics, will be exposed in this chapter.

The equations of continuity and of energy will be developed for the general case. Subsequently, the specific energy, a concept useful for the understanding of different problems, will be introduced. Elementary knowledge of gravity waves is presented.

Finally the hydrodynamic equations are developed, as well as their applications to uniform and non-uniform flow. Experimental results, being a support to the theory, are presented, such as the distribution of velocity, the characteristics of turbulence and also the friction coefficients.

TABLE OF CONTENTS

2.1 EQUATION OF CONTINUITY

2.2 EQUATION OF ENERGY

2.3 SPECIFIC ENERGY
  2.3.1 Specific-energy Curve
  2.3.2 Discharge Curve
  2.3.3 Critical Depth

2.4 GRAVITY WAVES
  2.4.1 Wave Celerity
  2.4.2 Wave Equation
  2.4.3 Flow with a Wave

2.5 HYDRODYNAMIC EQUATIONS
  2.5.1 Equations of Motion
  2.5.2 Uniform Flow
  2.5.3 Non-uniform Flow

2.6 DISTRIBUTION OF VELOCITY
  2.6.1 Laminar Flow
  2.6.2 Turbulent, smooth Flow
  2.6.3 Turbulent, rough Flow
  2.6.4 Turbulence Characteristics

2.1 EQUATION OF CONTINUITY

1° The equation of continuity, one of the basic equations of fluid mechanics, is an expression of the conservation of mass.

The variation of a mass fluid, contained in a given volume during a certain time, must equal the sum of the mass fluid which enters, diminished by the one which leaves.

2° Studied will be a flow of an incompressible fluid, being steady, uniform and almost rectilinear, in an open channel with a free surface and having a weak bed slope (see Fig. 2.1). Considered will be two channel sections; Q will be the entering discharge.

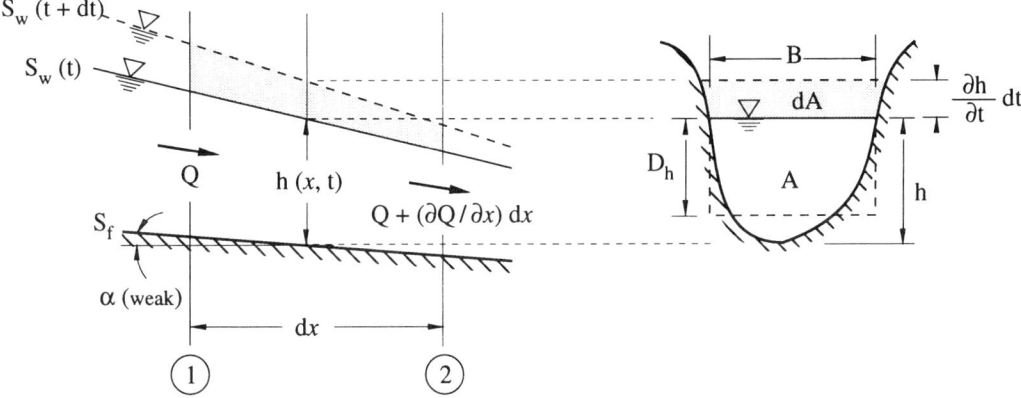

Fig. 2.1 Scheme for the equation of continuity.

The volume, entering by the first section is Qdt; the volume leaving by the second section, being at a distance, dx, from the first one, is $[Q + (\partial Q/\partial x)dx]dt$. The variation of the volume between these two sections during the time, dt, is consequently:

$$-\left(\frac{\partial Q}{\partial x}\right) dx\, dt$$

This variation of the volume is the result of a modification of the free surface, $\partial h/\partial t$, between the two sections during the time, dt; it is expressed by:

$$(B dx)\, \frac{\partial h}{\partial t}\, dt$$

where B(h) is the width of the channel at the free surface and h(x,t) is the flow depth.

Assuming the fluid incompressible, the above two expressions are made equal (see *Chow*, 1959, p. 525) and one obtains:

$$\frac{\partial Q}{\partial x} + \frac{\partial A}{\partial t} = 0 \tag{2.1}$$

where $dA = Bdh$.

3° For a given section, the following relation can be given :

$$Q = UA \tag{2.2}$$

where U is the average velocity in the section, A. Thus eq. 2.1 can be expressed as :

$$\frac{\partial(UA)}{\partial x} + B\frac{\partial h}{\partial t} = A\frac{\partial U}{\partial x} + U\frac{\partial A}{\partial x} + B\frac{\partial h}{\partial t} = 0 \tag{2.3}$$

Using the definition of the hydraulic depth, $D_h = A/B$, one can also write :

$$D_h\frac{\partial U}{\partial x} + U\frac{\partial D_h}{\partial x} + \frac{\partial h}{\partial t} = 0 \tag{2.4}$$

The above equations represent different forms of the equation of continuity, valid for prismatic channels (see sect. 5.1.1).

4° For a rectangular channel, eq. 2.3 is given by :

$$\frac{\partial q}{\partial x} + \frac{\partial h}{\partial t} = h\frac{\partial U}{\partial x} + U\frac{\partial h}{\partial x} + \frac{\partial h}{\partial t} = 0 \tag{2.5}$$

where $q = Q/B$ is the unit discharge.

5° For steady flow, $\partial A/\partial t = 0$, the equation of continuity, eq. 2.1, reduces to :

$$\frac{dQ}{dx} = 0 \tag{2.6}$$

6° If a supplementary discharge leaves (or enters) the channel between the two sections, eq. 2.1 can be adapted, such as :

$$\frac{\partial Q}{\partial x} + \frac{\partial A}{\partial t} \genfrac{}{}{0pt}{}{+}{(-)} q_\ell = 0 \tag{2.7}$$

where q_ℓ is the supplementary discharge per unit length.

2.2 EQUATION OF ENERGY

1° The equation of energy is an expression of the first principle of thermodynamics.

2° The energy for an element of incompressible fluid, written in homogeneous quantities of length (see Fig. 2.2) — here as the height of a liquid with specific weight $\gamma = \rho g$ — in almost rectilinear flow, taken with respect to the plane of reference (PdR), is given by :

$$\frac{u^2}{2g} + \frac{p}{\gamma} + z_P = \frac{p_t}{\gamma} = \text{Cte} \qquad (2.8)$$

The different terms represent :

$\dfrac{u^2}{2g}$ the velocity head

$\dfrac{p}{\gamma}$ the pressure head

z_P the elevation (position) of a point, P

$\dfrac{p_t}{\gamma} = H$ the (mechanical) energy head or the total head

$\dfrac{p}{\gamma} + z_P = \dfrac{p^*}{\gamma}$ the piezometric head

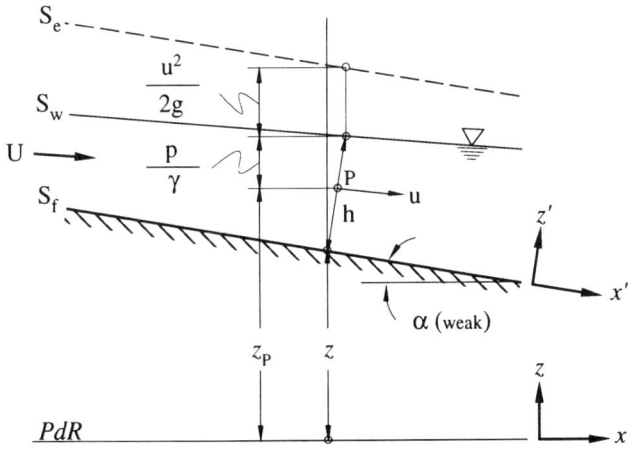

Fig. 2.2 Scheme for the equation of energy in a cross section.

3° The following assumptions shall be applied:

    i)   The piezometric head, $p*/\gamma$, is supposed to be constant over a normal to the bed, implying that the distribution of pressure is hydrostatic.

    ii)   By considering that z gives the elevation of the bed, the slope (weak) of the channel, S_f, is given by:

$$S_f = \operatorname{tg} \alpha = -\frac{dz}{dx} \cong \sin \alpha$$

    iii)   If h is the flow depth, the pressure head at the bed of the channel (see eq. 1.14) is:

$$\left(\frac{p}{\gamma}\right)_f = h \cos \alpha$$

For weak slopes, $\alpha < 6°$, where $S_f < 0.1$, one may take $\cos \alpha \cong 1$. The system of the coordinates, xz, is thus almost identical with the one of the coordinates, $x'z'$, (see Fig. 2.2).

    iv)   In a perfect fluid, each fluid element moves with the same velocity, which is the average velocity in the section, U.

Making use of these reasonable assumptions, the total head in a section is now given by:

$$\frac{U^2}{2g} + h + z = H \tag{2.9}$$

The flow is here considered to be one-dimensional and rectilinear.

The equation of energy, eq. 2.9, is a manifestation of the principle of energy if the liquid is *perfect*. From one to another section, each of the three terms in eq. 2.9 can take a different value, but the sum, H, remains constant.

4° For flow of a *real* fluid with a free surface, being unsteady and non-uniform (gradually varied), the difference of the total head between two sections, separated by a distance, dx, (see Fig. 2.3) is given as:

$$\alpha_e \frac{U^2}{2g} + h + z = \left[\alpha_e \frac{U^2}{2g} + d\left(\alpha_e \frac{U^2}{2g}\right)\right] + [h + dh] + [z + dz] +$$

$$+ \frac{1}{g}\frac{\partial U}{\partial t} dx + \frac{1}{g}\frac{\tau_o}{\rho}\frac{dP}{dA} dx \tag{2.10}$$

HYDRODYNAMIC CONSIDERATIONS

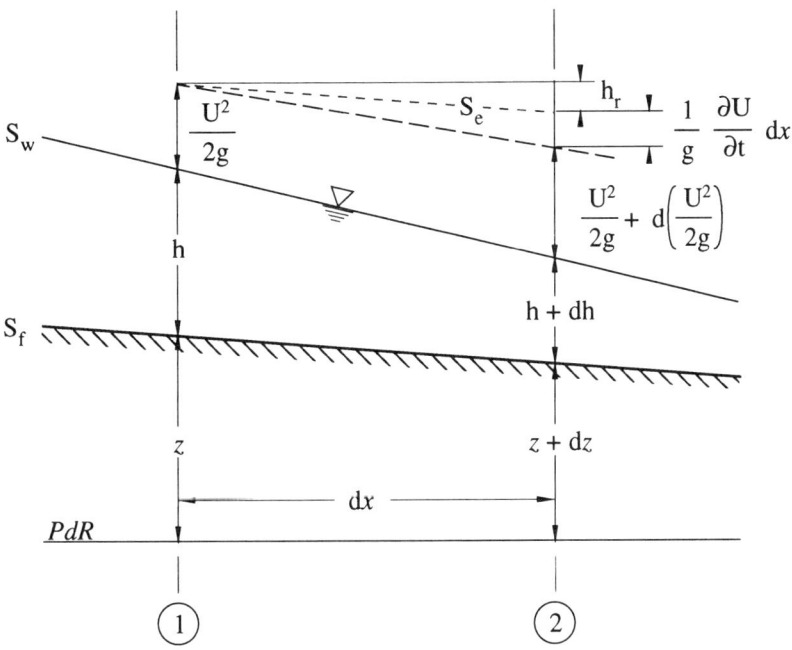

Fig. 2.3 Scheme of the equation of energy, between two sections.

i) $\dfrac{1}{g}\dfrac{\partial U}{\partial t}dx$ is the term of energy due to acceleration in the flow x-direction (see *Graf & Altinakar*, 1991, p. 137).

ii) $\dfrac{1}{g}\dfrac{\tau_o}{\rho}\dfrac{dP}{dA}dx = h_r$ is the term of energy or head loss due to friction (see *Graf & Altinakar*, 1991, p. 138);

The friction forces provoke a dissipation of mechanical (into thermal) energy. dP is the perimeter of an elementary surface, dA, and τ_o is the shear stress due to the frictional forces acting on the surface, $dPdx$. This term, representing the effect of friction, is usually written as h_r.

iii) The kinetic energy correction coefficient, α_e, results from the distribution of the velocity in the section (see Fig. 1.6). Its numerical values (see *Chow*, 1959, p. 28) notably for turbulent flow are very close to unity. In most common cases, the velocity head can thus be taken as:

$$\alpha_e \dfrac{U^2}{2g} \approx \dfrac{U^2}{2g}$$

where U is the average velocity in the section.

The equation of energy, eq. 2.10, can thus be given as :

$$d\left(\frac{U^2}{2g} + h + z\right) = -h_r - \frac{1}{g}\frac{\partial U}{\partial t}dx \qquad (2.11)$$

Dividing by the distance, dx, and using partial differentials, one gets :

$$\underbrace{\underbrace{\underbrace{\frac{1}{g}\frac{\partial U}{\partial t} + \frac{U}{g}\frac{\partial U}{\partial x} + \frac{\partial h}{\partial x} - S_f}_{\text{steady, uniform}}}_{\text{steady, non-uniform}}}_{\text{unsteady, non-uniform}} = -S_e \qquad (2.12)$$

where $h_r = S_e\, dx$ and $S_f = -(dz/dx)$; S_e is the energy slope.

Eq. 2.12 is the dynamic equation for unsteady and non-uniform flow.

The head loss, h_r, must be evaluated with a formula such as the one of Weisbach-Darcy, eq. 3.10, of Chézy, eq. 3.11, or of other experimenters. Such relations are only valid for steady, uniform flow; however — for lack of better information — they are also used (see *Chow*, 1959, p. 217) for unsteady and non-uniform flow.

5° The equation of continuity, eq. 2.5, and the dynamic equation of motion, eq. 2.12, form together the *equations of Barré de Saint-Venant* (see *Chow* 1959, p. 528). Despite the various simplifications made to obtain the equations of Saint-Venant, their solutions are often rather complicated. In some physical cases, which are simple but still realistic, explicit solutions are possible.

6° For flow, which is steady but non-uniform, eq. 2.12 reduces to :

$$\frac{U}{g}\frac{\partial U}{\partial x} + \frac{\partial h}{\partial x} - S_f = -S_e \qquad (2.13)$$

7° For flow, which is steady and uniform, eq. 2.12 reduces to :

$$S_f = S_e \qquad (2.13a)$$

The bed slope, S_f, the energy slope, S_e, and the piezometric slope of the water surface, S_w, are identical. The average velocity, U, and the flow depth, h, are constant; the equation of continuity is given with eq. 2.6.

HYDRODYNAMIC CONSIDERATIONS

8° The dynamic equation of motion, eq. 2.12, can also be obtained by applying the momentum equation. The resulting equation is almost the same (see *Chow*, 1959, p. 51 and *Henderson*, 1966, p. 9), i.e. : eq. 2.12.

2.3   SPECIFIC ENERGY

1° Up till now, the total head, H , in a given cross section was defined with respect to an arbitrary horizontal plane (see Fig. 2.2); for a weak bed slope one writes :

$$\frac{U^2}{2g} + h + z = H \tag{2.9}$$

If the plane of reference is now placed into the bed slope, S_f, a fraction of the total head, called the *specific energy*, H_s , is defined (see *Bakmeteff*, 1932, chap. 4); one writes now (see Fig. 2.4) :

$$\frac{U^2}{2g} + h = H_s \tag{2.14}$$

Using the equation of continuity, Q = UA, one obtains :

$$\frac{Q^2/A^2}{2g} + h = H_s \tag{2.14a}$$

2° The notion of the specific energy is often very useful; it helps to understand and to solve different problems of free-surface flow.

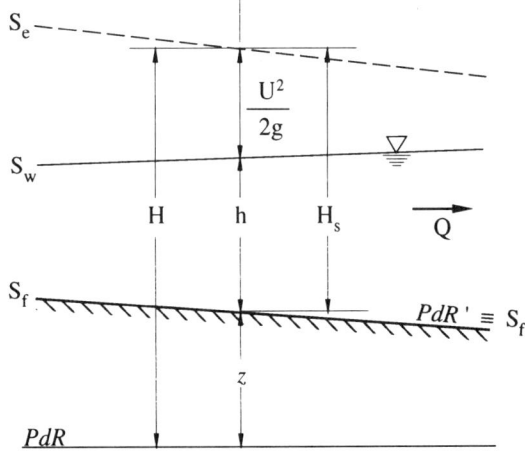

Fig. 2.4 Definition of the total head, H, and the specific energy, H_s .

3° For a given section in the channel, the area of flow, A, is a function of the flow depth, h, and eq. 2.14a gives a relation of the following form:

$$H_s = f(Q, h)$$

which allows a study of the variation of:

i)   h   with H_s,   for   Q = Cte
ii) h   with   Q,   for H_s = Cte.

2.3.1 Specific-energy Curve

1° Eq. 2.14a gives the evolution of the specific energy, H_s, as a function of the flow depth, h, for a given discharge, Q = UA.

This curve (see Fig. 2.5) has two asymptotes:

i)   for   h = 0,   a horizontal asymptote,
ii) for   h = ∞,   the line h = H_s is the other asymptote.

In addition, the curve has a minimum, H_{s_c}, for:

$$\frac{dH_s}{dh} = -\frac{Q^2}{gA^3}\frac{dA}{dh} + 1 = 0 \tag{2.15}$$

Since dA/dh is equal to the width of the channel, B, at the free surface and by using the definition of the hydraulic depth, D_h = A/B, one obtains:

$$\frac{Q^2}{g}\frac{B}{A^3} = \frac{U^2}{gD_h} = 1 \tag{2.15a}$$

2° For a channel with a rectangular cross section, one has D_h = h. The flow depth, h, which corresponds to the minimal specific energy, H_{s_c}, is called *critical depth*, h_c.

3° Following the curve, given with Fig. 2.5, one notices that for a given discharge Q = Cte, and for an arbitrary value of specific energy, H_s, — for the case when flow can take place — there are always two solutions for the flow depth, h_1 and h_2. They are called the corresponding (alternate) depths; one of which, h_1, is smaller and the other one, h_2, is larger than the critical depth, h_c. Both of these depths are indications of different regimes of flow, thus:

h < h_c *supercritical* (torrential) regime

h > h_c *subcritical* (fluvial) regime

h ≡ h_c *critical* regime

HYDRODYNAMIC CONSIDERATIONS 27

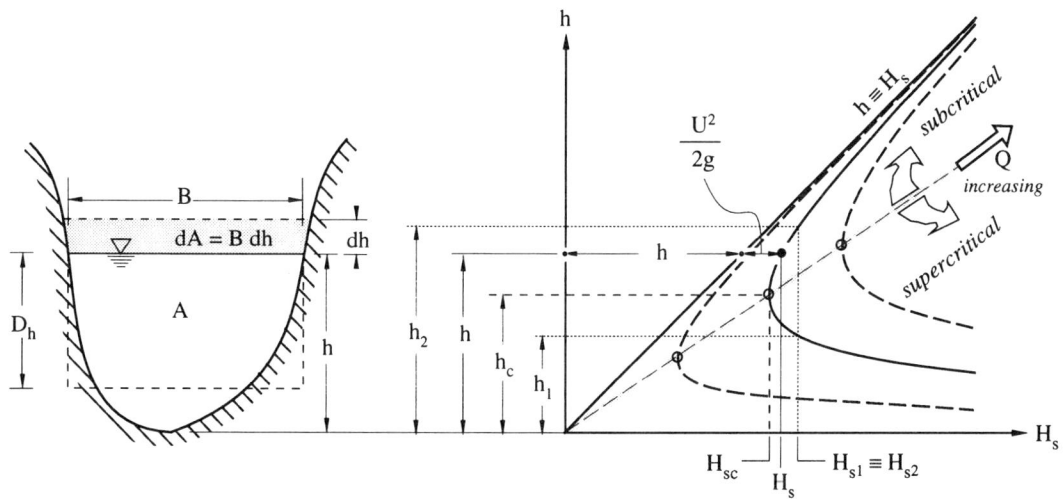

Fig. 2.5 Specific-energy curve, $H_s = f(h)$, for $Q = $ Cte.

Each curve (see Fig. 2.5) has thus two branches. Consequently, a steady flow in a channel can exist in two different ways, both having the same specific energy, H_s :

i)   in supercritical regime, where the flow depth is small and the velocity large,
ii)   in subcritical regime, where the flow depth is large and the velocity small.

4°   For a variation of the discharge, Q, the corresponding curves have the same form ; they follow each other for an increase in the discharge, starting at the origin, (see Fig. 2.5).

2.3.2 Discharge Curve

1°   Eq. 2.14a gives also the evolution of the discharge, Q, as a function of the flow depth, h, for a given specific energy, H_s , such as :

$$Q = A \sqrt{2g \, (H_s - h)} \qquad (2.16)$$

From this curve (see Fig. 2.6), one obtains :

i)   for $h = 0$, $Q = 0$
ii)   for $h = H_s$, $Q = 0$.

2°   In addition, the curve has a maximum value, Q_{max} , for :

$$\frac{dQ}{dh} = \frac{2g \, (H_s - h) \, (dA/dh) - A g}{\left[2g \, (H_s - h)\right]^{1/2}} = 0$$

Taking $dA/dh = B$ and $D_h = A/B$, one may write :

$$\frac{dQ}{dh} = \frac{gB\left[2(H_s - h) - D_h\right]}{\left[2g(H_s - h)\right]^{1/2}} = 0 \tag{2.17}$$

This derivative is zero, if :

$$2(H_s - h) - D_h = 0 \tag{2.18}$$

The values, h and D_h, which correspond to the maximum discharge, Q_{max}, represent the *critical depth*, h_c et D_{h_c}. For flows smaller than Q_{max}, one finds again the two different flow regimes (see Fig. 2.6 and also Fig. 2.5).

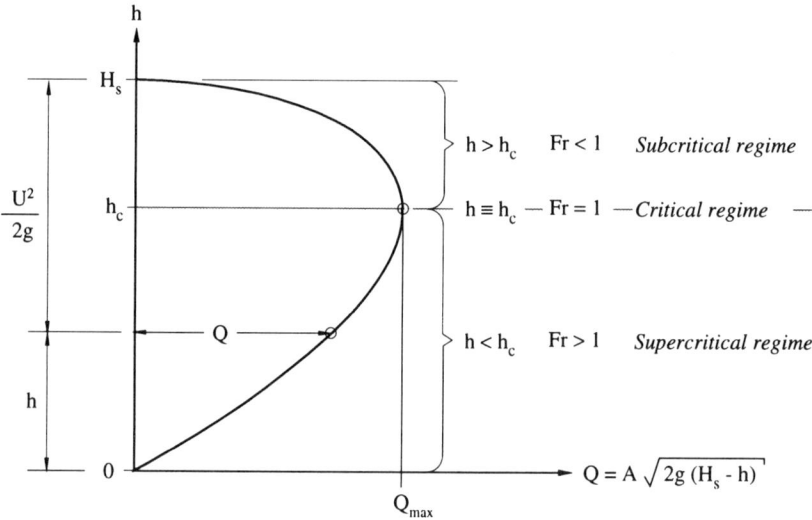

Fig. 2.6 Discharge curve, $Q = f(h)$, for H_s = Cte.

3° For a channel with a rectangular cross section, $D_h \equiv h$, eq. 2.18 becomes :

$$2(H_s - h) - h = 0$$

from where one obtains the critical depth ($h \equiv h_c$ and $H_s \equiv H_{s_c}$) :

$$h_c = \frac{2}{3} H_{s_c} \tag{2.19}$$

For a channel with a triangular or parabolic cross section, one obtains respectively :

$$h_c = \frac{4}{5} H_{s_c} \quad \text{and} \quad h_c = \frac{3}{4} H_{s_c} \tag{2.19a, b}$$

2.3.3 Critical Depth

1° The critical depth, h_c, in a channel is the flow depth at which :

    i)   the specific energy is minimal, H_{s_c}, for a given discharge (see Fig. 2.5),
    ii)   the discharge is maximal, Q_{max}, for a given specific energy (see Fig. 2.6).

2° It follows, that eq. 2.18 can be written as :

$$2(H_{s_c} - h_c) = D_{h_c}$$

and that, using eq. 2.16, the maximum discharge, Q_{max}, is given by :

$$Q_{max} = A\sqrt{gD_{h_c}} \qquad (2.20)$$

The average velocity, which corresponds to the critical hydraulic depth, D_{h_c}, is :

$$U_c = \sqrt{gD_{h_c}} \quad \text{or} \quad \frac{U_c^2}{2g} = \frac{D_{h_c}}{2} \qquad (2.21)$$

In critical regime, the velocity head is thus equal to half of the hydraulic depth.

3° Eq. 2.21 or eq. 2.15a could also be expressed as :

$$\frac{U_c}{\sqrt{gD_{h_c}}} = 1 \qquad (2.22)$$

which is precisely the definition of the Froude number (see eq. 1.4) in critical regime ; here the Froude number, Fr, is equal to unity :

$$Fr_c = 1 \qquad (2.22a)$$

Note that the Froude number, $Fr = U/\sqrt{gD_h}$, is the ratio of inertia to gravity forces per unit volume (see *Graf & Altinakar*, 1991, sect. FR. 7.3). Consequently, the Froude number classifies also the different flow regimes, such as :

    $Fr > 1$ *supercritical* regime $U > U_c$

    $Fr < 1$ *subcritical* regime $U < U_c$

    $Fr = 1$ *critical* regime $U \equiv U_c$

4° The critical velocity, U_c, is given by :

$$U_c = \sqrt{gD_{h_c}} = c \tag{2.21a}$$

This is equal to the celerity, c, of the propagation of (superficial) infinitesimal gravity waves in a channel of hydraulic depth, D_{h_c} (see eq. 2.27 for the general definition).

5° The critical depth for a rectangular channel, $D_h \equiv h$, has been given by :

$$h_c = \frac{2}{3} H_{s_c} \tag{2.19}$$

or equally by :

$$(H_{s_c} - h_c) = \frac{h_c}{2} = \frac{U_c^2}{2g}$$

Using the definition of the unit discharge, q = Uh, one obtains :

$$\frac{h_c}{2} = \frac{q^2}{2gh_c^2} \quad \text{or} \quad h_c = \sqrt[3]{\frac{q^2}{g}} \tag{2.23}$$

The maximum unit discharge, q, which may exist in a channel of rectangular section is equal to :

$$q = \sqrt{gh_c^3} = \sqrt{g\left(\frac{2}{3}H_{s_c}\right)^3} \tag{2.24}$$

6° Experience shows that flow at critical depth, h_c, is often *unstable*, presenting itself by a fluctuating water surface. This is rather evident when observing Fig. 2.5 : even small variations of energy close to the critical value, H_{s_c}, cause large variations in the flow depth, h.

7° According to eq. 2.20 and eq. 2.24, the critical hydraulic depth, D_{h_c}, or the critical flow depth, h_c, depend only on the discharge. Thus it is inviting to use this information for metering flow in open channels :

Here, two examples are given :

   i) *Free Overfall* : Flow in an horizontal channel (or a broad-crested weir) discharges freely into the atmosphere; the critical section is found rather close to the brink (see sect. 4.4.2).

HYDRODYNAMIC CONSIDERATIONS 31

    ii)   *Venturi Canal* : An adequate reduction in the cross section of the channel is provided, where the critical regime (see sect. 4.4.2) takes place.

8°   Flow goes through the critical depth, if the fluvial regime passes to the torrential one. Critical depth is also observed, if the fluvial regime is terminated by a free overfall.

2.4 GRAVITY WAVES

Flow in open channels, which is variable in time, is accompanied by gravity waves at the water surface.

2.4.1 Wave Celerity

1°   Considered will be a periodic, simple wave, representing the propagation of an irrotational motion as satisfied by the equation of Laplace; the pressure at the free surface is constant and the wave amplitudes are small. A channel of rectangular cross section with uniform flow depth is filled with stagnant water; there is thus no flow.

2°   The two-dimensional and progressive wave in the $^-x^+$ direction, will be given (see Fig. 2.7) by a periodic displacement of the free surface as a function of time, t, (see *Kinsman*, 1965, p. 117), such as:

$$\eta(x, t) = A \cos(2\pi x/L - 2\pi t/T) \tag{2.25}$$

where A is the amplitude, being half of the wave height, H = 2A; L is the wave length and T is the wave period. The wave celerity is defined by:

$$c = \frac{L}{T} \tag{2.25a}$$

3°   The hydrodynamic theory for waves of *small amplitude* (see *Lamb*, 1945, pp. 254 and 366, or *Kinsman*, 1965, p. 125), i.e.: H/L << 1 and H/h << 1, gives for the apparent velocity of propagation, also called the *celerity* of a perturbation:

$$c^2 = \frac{gL}{2\pi} \tanh\left(\frac{2\pi h}{L}\right) \tag{2.26}$$

where h is the water depth. Note that the celerity does not depend on the wave height, H.

4° This expression, eq. 2.26, reduces :

   i) for *short* waves or waves of large depth, if L/h < 1, to :

   $$c^2 = \frac{gL}{2\pi} \qquad (2.26a)$$

   ii) for *long* waves or waves of small depth, if L/h >> 1, to :

   $$c^2 = gh \qquad (2.27)$$

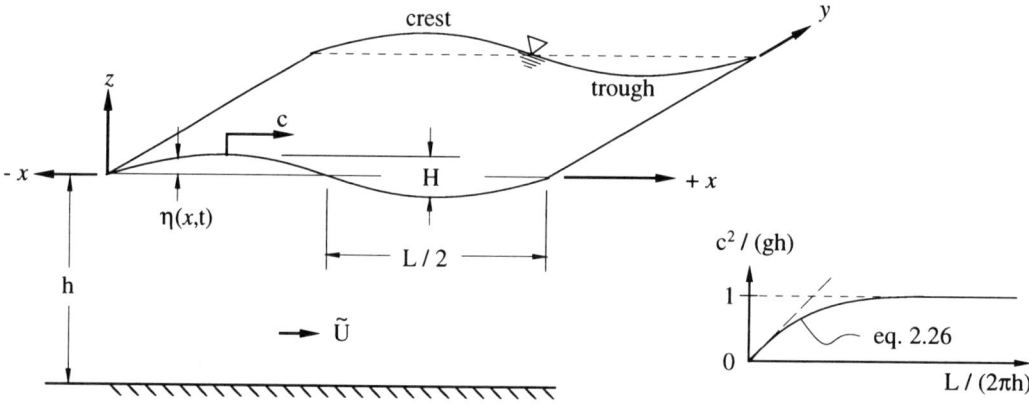

Fig. 2.7 Scheme of a surface wave.

5° If the long wave, where L/h >> 1, is not of small amplitude, thus H/h ≅ 1 , the wave celerity (see eq. 2.27) is given (see *Lamb*, 1945, p. 262) by :

$$c^2 = gh\left(1 + \frac{3}{2}\frac{A}{h}\right) \qquad (2.28)$$

or also (see *Lamb*, 1945, p. 424) by :

$$c^2 = g(h + A) \qquad (2.28a)$$

This last relation was experimentally obtained for a solitary wave.

6° The two signs, which are possible for the celerity, eq. 2.27 or eq. 2.28, show well that the wave can propagate in the direction of x^+ or x^- (see Fig. 2.7).

7° The relation, eq. 2.27, for the celerity, c , of *long* waves can also be obtained by application of the equation of continuity and of energy.

HYDRODYNAMIC CONSIDERATIONS

i) Consider the unsteady flow (see Fig. 2.8a) of a simple wave having an amplitude, $A \equiv \eta$. U is the liquid velocity in the section of the crest. By following the wave — one thus imagines the wave stays immobile — the flow becomes a steady flow (see Fig. 2.8b).

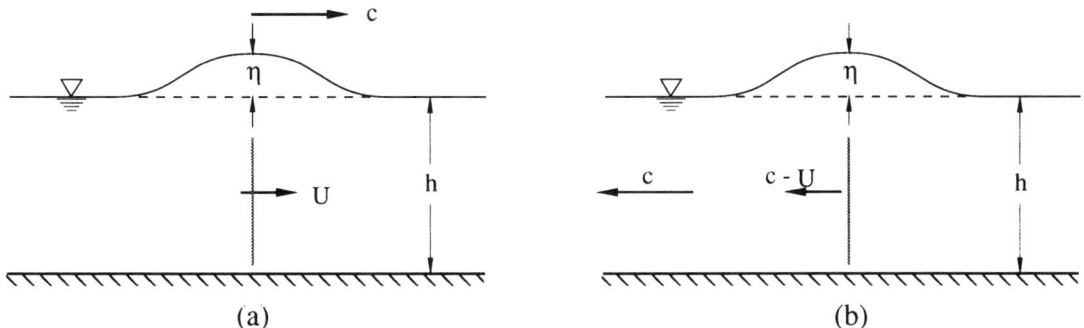

Fig. 2.8 Propagation of a wave.

ii) The equation of continuity reads now :

$$c\,h = (c - U)(h + \eta)$$

If $\eta \ll h$, the wave is thus infinitesimal and one may write:

$$U = c\,(\eta / h) \tag{2.29}$$

iii) The equation of energy reads :

$$h + \frac{c^2}{2g} = (h + \eta) + \frac{(c - U)^2}{2g} \tag{2.30}$$

or written otherwise :

$$\eta = \frac{(c\,U)}{g}\left(1 - \frac{U}{2c}\right) \tag{2.30a}$$

Neglecting the term $(U/2c)$ when compared to unity, one may write:

$$\eta = \frac{c\,U}{g} \tag{2.30b}$$

iv) Through substitution, the eqs. 2.29 and 2.30b give :

$$c^2 = gh \tag{2.27}$$

This is the celerity of wave having a small amplitude, η.

2.4.2 Wave Equation

1° In order to derive the wave equation, the equation of continuity and of motion will be applied to a situation of a wave of small amplitude, which propagates (see Fig. 2.7) in a stagnant liquid.

2° The equation of continuity, eq. 2.5, is expressed as :

$$(h+\eta)\frac{\partial \tilde{U}}{\partial x} + \tilde{U}\frac{\partial (h+\eta)}{\partial x} + \frac{\partial (h+\eta)}{\partial t} = 0$$

where $\tilde{U}$ is the velocity produced by the wave and averaged over the depth. By assuming that depth variation, $\partial h/\partial x = 0$ and $\partial h/\partial t = 0$, are negligible, one may write :

$$(h+\eta)\frac{\partial \tilde{U}}{\partial x} + \tilde{U}\frac{\partial \eta}{\partial x} + \frac{\partial \eta}{\partial t} = 0.$$

If the wave is of small amplitude, $\eta/h \ll 1$, and assuming that $\partial \eta/\partial x \ll 1$, one obtains :

$$h\frac{\partial \tilde{U}}{\partial x} + \frac{\partial \eta}{\partial t} = 0 \qquad (2.31)$$

3° The dynamic equation, eq. 2.12, is expressed as :

$$\frac{1}{g}\frac{\partial \tilde{U}}{\partial t} + \frac{\tilde{U}}{g}\frac{\partial \tilde{U}}{\partial x} + \frac{\partial (h+\eta)}{\partial x} - (S_f - S_e) = 0$$

While the last term shall be omitted, the second term is considered to be small compared to the first one. Since the depth variation is negligible, the above equation simplifies to :

$$\frac{1}{g}\frac{\partial \tilde{U}}{\partial t} + \frac{\partial \eta}{\partial x} = 0 \qquad (2.32)$$

4° One sees immediately that these equations, eq. 2.32 and eq. 2.31, give the following relationship :

$$\frac{\partial^2 \tilde{U}}{\partial t^2} = gh\frac{\partial^2 \tilde{U}}{\partial x^2} \qquad \text{or} \qquad \frac{\partial^2 \eta}{\partial t^2} = gh\frac{\partial^2 \eta}{\partial x^2} \qquad (2.33)$$

HYDRODYNAMIC CONSIDERATIONS 35

This is the classic equation for a progressive wave (see *Lamb*, 1945, p. 255), where $c^2 = gh$ is the celerity of a long wave, previously presented with eq. 2.27. A general solution to it was given with eq. 2.25.

2.4.3 Flow with a Wave

1° It was shown that the celerity, c, with which a gravity wave, being a long one and of small amplitude, propagates in a channel of rectangular section, is given with the relation of eq. 2.27. For a channel of an arbitrary section, one writes :

$$c = \pm \sqrt{gD_h} \qquad (2.27a)$$

where D_h is the hydraulic depth.

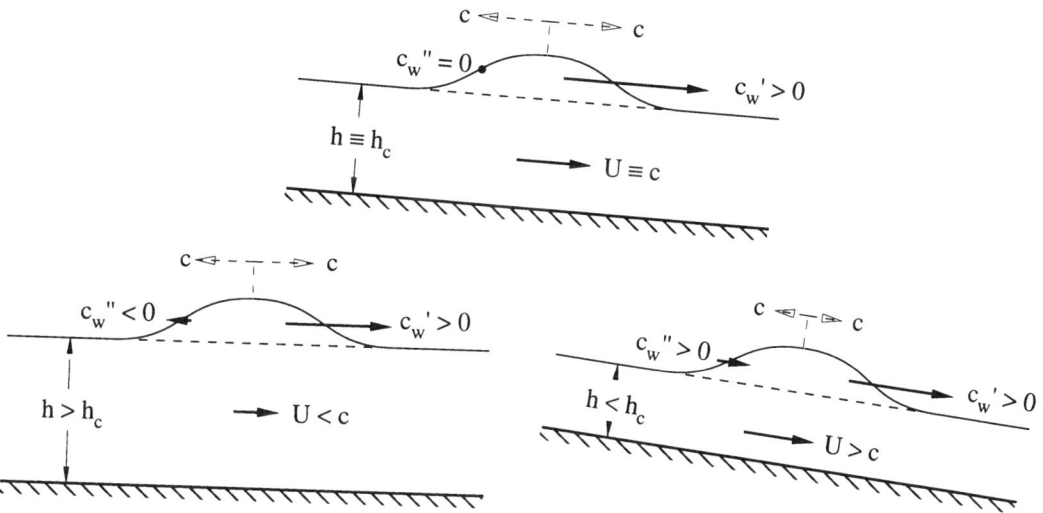

Fig. 2.9 Flow with a wave.

2° This relation, eq. 2.27a, was established for a channel where the liquid was stagnant. However the relation stays valid for the case where the liquid is in motion; the wave superposes itself upon the flow in the channel. Consequently, the *absolute celerity*, c_w, of the wave for a channel having an average velocity, U, can be expressed as:

$$c_w = U \pm \sqrt{gD_h} \qquad (2.34a)$$

and for a channel of rectangular section :

$$c_w = U \pm \sqrt{gh} = U \pm c \qquad (2.34)$$

3° The absolute celerity, c_w, being the velocity with respect to the bed, has evidently two values :

$$c_w' = U + c \quad , \quad c_w'' = U - c \qquad (2.34b)$$

Thus one may distinguish two plus one cases (see Fig. 2.9) :

i) $U < c$, where the wave with celerity, c_w', propagates downstream and where the wave with celerity, c_w'', propagates upstream; the flow regime is fluvial.

ii) $U > c$, where the wave with celerity, c_w', propagates downstream and where the wave with celerity, c_w'', propagates downstream as well; the flow regime is torrential.

iii) At a flow depth, at which the current velocity, U, and the wave celerity, c, are the same, thus :

$$U \equiv c = \sqrt{gh_c}$$

the flow is in critical regime (see sect. 2.3.3); h_c being the critical depth.

4° Flow with a gravity wave, which is long but not of small amplitude, will be treated later (see sect. 5.6).

2.5 HYDRODYNAMIC EQUATIONS

2.5.1 Equations of Motion

1° For flow (see Fig. 2.10), which is two-dimensional, plane, $\vec{V}(\bar{u}, 0, \bar{w})$, and turbulent, the equations of motion and of continuity (see *Graf & Altinakar*, 1991, p. 275 or *Rotta*, 1972, p. 129) can be written as :

$$\frac{\partial \bar{u}}{\partial t} + \bar{u}\frac{\partial \bar{u}}{\partial x} + \bar{w}\frac{\partial \bar{u}}{\partial z} = -\frac{1}{\rho}\frac{\partial \overline{p^*}}{\partial x} + \nu\left(\frac{\partial^2 \bar{u}}{\partial x^2} + \frac{\partial^2 \bar{u}}{\partial z^2}\right) - \left[\frac{\partial}{\partial x}(\overline{u'^2}) + \frac{\partial}{\partial z}(\overline{u'w'})\right]$$

$$\frac{\partial \bar{w}}{\partial t} + \bar{u}\frac{\partial \bar{w}}{\partial x} + \bar{w}\frac{\partial \bar{w}}{\partial z} = -\frac{1}{\rho}\frac{\partial \overline{p^*}}{\partial z} + \nu\left(\frac{\partial^2 \bar{w}}{\partial x^2} + \frac{\partial^2 \bar{w}}{\partial z^2}\right) - \left[\frac{\partial}{\partial x}(\overline{u'w'}) + \frac{\partial}{\partial z}(\overline{w'^2})\right]$$

(2.35)

HYDRODYNAMIC CONSIDERATIONS

$$\frac{\partial \overline{u}}{\partial x} + \frac{\partial \overline{w}}{\partial z} = 0 \qquad (2.35a)$$

- $\overline{u}$ and $\overline{w}$ are the average point velocities in the x and z-direction;
- u' and w' are the velocities due to fluctuations;
- $\rho \overline{u'^2}$, $\rho \overline{u'w'}$, etc. are the supplementary (or Reynolds) stresses due to the turbulence;
- $\overline{p^*}$ is the average (point) driving pressure.

These equations, eqs. 2.35, are known as the *Reynolds equations*. In the absence of turbulence they reduce to the *Navier-Stokes* equations, valid notably for laminar flow.

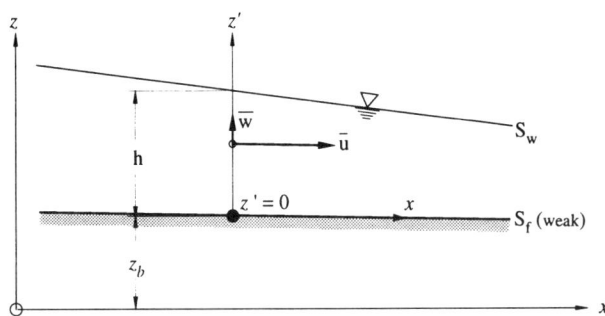

Fig. 2.10 Scheme for the equations of motion.

2° For free-surface flow on weak slopes, $S_f \ll 1$, the terms of the driving pressure, $\overline{p^*}(x,z) = \overline{p}(x,z) + g\rho z$, are expressed as:

$$\frac{\partial \overline{p^*}}{\partial x} = \frac{\partial \overline{p}}{\partial x} - \rho g\, S_f$$

$$\frac{\partial \overline{p^*}}{\partial z} = \frac{\partial \overline{p}}{\partial z} + \rho g \qquad (2.36)$$

where the bed slope is defined as: $S_f = -\dfrac{\partial z_b}{\partial x}$.

The bed of the channel is defined by z_b, but in the following it will be used without an index, thus written as $S_f = -\dfrac{\partial z}{\partial x} = -\dfrac{dz}{dx}$.

3°   Steady free-surface flow may be considered as being flow at high Reynolds numbers. It is thus possible to use the approximations which are developed for boundary-layer flow (see *Graf & Altinakar*, 1991, sect. CL.1).

By considering the order of magnitude of each term in eqs. 2.35 and 2.35a, and by keeping only the terms of the highest order (see *Graf & Altinakar*, 1991, p. 351, or *Rotta*, 1972, p. 130), one has :

$$\bar{u}\frac{\partial \bar{u}}{\partial x} + \bar{w}\frac{\partial \bar{u}}{\partial z} = -\frac{1}{\rho}\frac{\partial \overline{p^*}}{\partial x} + \nu\frac{\partial^2 \bar{u}}{\partial z^2} - \frac{\partial}{\partial x}\overline{(u'^2)} - \frac{\partial}{\partial z}\overline{(u'w')}$$

(2.37)

$$0 = -\frac{1}{\rho}\frac{\partial \overline{p^*}}{\partial z} - \frac{\partial}{\partial z}\overline{(w'^2)}$$

$$\frac{\partial \bar{u}}{\partial x} + \frac{\partial \bar{w}}{\partial z} = 0$$

(2.35a)

4°   In eqs. 2.37, the second one can be integrated and written as :

$$0 = -\frac{1}{\rho}\overline{p^*}(x, z') + \frac{1}{\rho}\overline{p^*}_f(x, z'=0) - \overline{w'^2}(z') + \overline{w'^2}_f(z'=0)$$

$\overline{p^*}_f$ being the driving pressure at the bed, $z' = 0$; due to the no-slip condition one takes $\overline{w'^2}_f = 0$. Thus one may write :

$$\overline{p^*}(x, z') = \overline{p^*}_f(x, 0) - \rho\overline{w'^2}(z')$$

(2.38)

or also :

$$\bar{p} + \rho g z' = \bar{p}_f + \rho g 0 - \rho \overline{w'^2}$$

With $\bar{p}_f = \rho g h$, an expression for the pressure is obtained :

$$\bar{p} = \rho g (h - z') - \rho \overline{w'^2}$$

(2.38a)

The driving pressure is consequently not constant over the flow depth, but will be slightly modified by the Reynolds stress.

The derivation of eq. 2.38 with respect to x, gives :

$$\frac{\partial \overline{p^*}}{\partial x} = \frac{\partial \overline{p^*}_f}{\partial x} - \rho\frac{\partial}{\partial x}\overline{(w'^2)}$$

(2.39)

HYDRODYNAMIC CONSIDERATIONS

where the last term is often neglected; using eq. 2.36, it can be written as:

$$\frac{\partial \overline{p^*}}{\partial x} \cong \frac{\partial \overline{p^*}_f}{\partial x} = \frac{\partial \overline{p}_f}{\partial x} - g\rho S_f = g\rho \left(\frac{\partial h}{\partial x} + \frac{\partial z_b}{\partial x} \right) \qquad (2.40)$$

5° Upon substitution of eq. 2.39 and eq. 2.40 into the first of eqs. 2.37 one gets:

$$\overline{u} \frac{\partial \overline{u}}{\partial x} + \overline{w} \frac{\partial \overline{u}}{\partial z} = -g \left(\frac{\partial h}{\partial x} + \frac{\partial z_b}{\partial x} \right) + \frac{\partial}{\partial z} \left(\nu \frac{\partial \overline{u}}{\partial z} - \overline{u'w'} \right) +$$

$$+ \frac{\partial}{\partial x} \left(\overline{w'^2} - \overline{u'^2} \right) \qquad (2.41)$$

The last term, which is due to the normal Reynolds stress, is also often neglected (see *Rotta*, 1972, p. 130). If one defines the total tangential stress by:

$$\tau_{zx} = \rho \left(\nu \frac{\partial \overline{u}}{\partial z} - \overline{u'w'} \right) \qquad (2.42)$$

one can write eq. 2.41 as:

$$\overline{u} \frac{\partial \overline{u}}{\partial x} + \overline{w} \frac{\partial \overline{u}}{\partial z} = -g \left(\frac{\partial h}{\partial x} + \frac{\partial z_b}{\partial x} \right) + \frac{1}{\rho} \frac{\partial \tau_{zx}}{\partial z} \qquad (2.41a)$$

This equation is also valid for laminar flow, where:

$$\tau_{zx} = \mu \frac{\partial \overline{u}}{\partial z} \qquad (2.42a)$$

6° Free-surface flow, being unsteady, can thus be represented (see *Grishanin*, 1969, p. 59) by a system of equations such as:

$$\frac{\partial \overline{u}}{\partial t} + \overline{u} \frac{\partial \overline{u}}{\partial x} + \overline{w} \frac{\partial \overline{u}}{\partial z} = -g \left(\frac{\partial h}{\partial x} + \frac{\partial z_b}{\partial x} \right) + \frac{1}{\rho} \frac{\partial \tau_{zx}}{\partial z} \qquad (2.41b)$$

$$\overline{p} = \rho g (h - z') - \rho \overline{w'^2} \qquad (2.38a)$$

$$\frac{\partial \overline{u}}{\partial x} + \frac{\partial \overline{w}}{\partial z} = 0 \qquad (2.35a)$$

Note, that the pressure is not quite hydrostatic.

2.5.2 Uniform Flow

1° It will be assumed that flow is steady, unidirectional, $\vec{V}(\bar{u}, 0, 0)$, and uniform on the average.

The equations of motion, eq. 2.41a and eq. 2.38a, and of continuity, eq. 2.35a, reduce to :

$$0 = + g S_f + \frac{1}{\rho} \frac{\partial \tau_{zx}}{\partial z} \tag{2.43}$$

$$\bar{p} = \rho g (h - z') - \rho \overline{w'^2} \tag{2.38a}$$

$$\frac{\partial \bar{u}}{\partial x} = 0 \tag{2.44}$$

where the total tangential stress is expressed by :

$$\tau_{zx} = \mu \frac{\partial \bar{u}}{\partial z} - \rho \overline{u'w'} \tag{2.42}$$

2° After integration over the flow depth, h , one obtains :

i) the equation of motion, eq. 2.43, as being :

$$0 = g S_f \int_0^h dz + \frac{1}{\rho} \int_0^h \frac{\partial}{\partial z} \tau_{zx} \, dz$$

$$0 = + g S_f (h - 0) + \frac{1}{\rho} (0 - \tau_o)$$

and consequently one has :

$$\tau_o = \rho g S_f h \tag{2.45}$$

with τ_o as the stress due to friction, called the wall or bed-shear stress; the ratio, $\tau_o/\rho g h$, gives the energy slope, being now expressed by :

$$S_f = S_e \tag{2.45a}$$

HYDRODYNAMIC CONSIDERATIONS

In eq. 2.43 one notes that the longitudinal pressure gradient, namely the longitudinal component of the gravity (see eq. 2.36), provides the driving force of a uniform flow; the tangential stress (see eq. 2.42) is the dissipating force.

ii)   the equation of continuity, eq. 2.44, as being :

$$\int_0^h \frac{\partial \bar{u}}{\partial x} dz = \frac{\partial}{\partial x} \int_0^h \bar{u}\, dz - \bar{u}_h \frac{\partial h}{\partial x} = 0$$

$$\frac{\partial}{\partial x}(Uh) = 0 \tag{2.46}$$

where $Uh = q$ is the unit discharge and U the average velocity; $\bar{u}_h$ is the velocity at the water surface.

For steady flow, one writes :

$$\frac{\partial q}{\partial x} = U \frac{\partial h}{\partial x} + h \frac{\partial U}{\partial x} = 0 \tag{2.46a}$$

The equation of motion, eq. 2.45a (see eq. 2.13a), and of continuity, eq. 2.46a (see eq. 2.6), in their integral form, form together the simplified equations of Saint-Venant for a steady and uniform flow (see sect. 2.2).

3°   To obtain the distribution (see Fig. 2.11) of the total shear stress, $\tau_{zx}(z)$, the following equation must be integrated over the flow depth :

$$\frac{\partial \tau_{zx}}{\partial z} = \frac{\partial \bar{p^*}}{\partial x} = -\rho g\, S_f \tag{2.43a}$$

As boundary conditions serve :

$z' \cong 0 \;\; (z'/h = 0.05) \quad \Rightarrow \quad \tau_{zx} = \tau_o$

$z' = h \quad \Rightarrow \quad \tau_{zx} = 0$

thus the following is obtained :

$$\tau_{zx}(z') = \rho g\, S_f\, (h - z') \tag{2.47}$$

or written in dimensionless form :

$$\frac{\tau_{zx}(z')}{\tau_o} = \left(\frac{h - z'}{h}\right) \tag{2.47a}$$

This gives (see *Monin et Yaglom*, 1971, p. 268) a linear (triangular) distribution, being valid for turbulent flow with eq. 2.42, and for laminar flow with eq. 2.42a.

Nevertheless, for very small distances from the bed, $z'/h \lesssim 0.05$, the shear-stress distribution may be considered to be constant (see *Graf & Altinakar, 1991*, sect. 6.1) :

$$\frac{\tau_{zx}(z')}{\tau_o} = 1$$

The zone very close to the bed, $z'/h \lesssim 0.20$, where the shear stress is constant, is called the *inner region* (see *Hinze*, 1975, p. 503 and *Monin et Yaglom*, 1971, p. 311) where the total shear-stress variation becomes negligible.

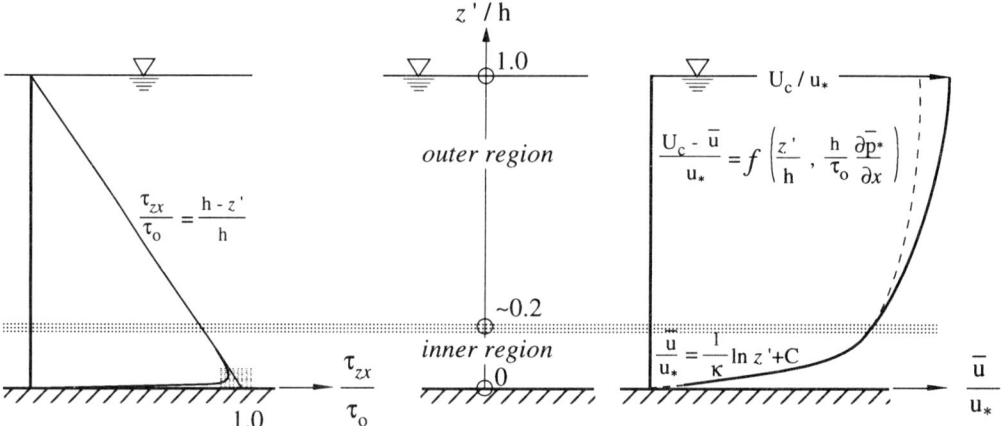

Fig. 2.11 Scheme of the distribution of shear stress, $\tau_{zx}(z)$, and of velocity, $\bar{u}(z)$; for uniform flow.

4° The system of equations, eq. 2.43, eq. 2.38a and eq. 2.44, cannot be used to obtain the distribution of the velocity, $\bar{u}(z)$, (see Fig. 2.11) since the Reynolds stresses are not known. Semi-empirical methods have to be exploited.

i) In the *inner region*, the pressure gradient, $(\partial \overline{p^*}/\partial x)$, — being very weak in uniform flow — in eq. 2.43a may be neglected (see *Monin et Yaglom*, 1971, p. 268); one may write :

$$\frac{\partial \tau_{zx}}{\partial z} = 0 \qquad (2.43b)$$

and consequently :

$$\tau_{zx} = \mu \frac{\partial \bar{u}}{\partial z} - \rho \overline{u'w'} = \text{Cte} \qquad (2.42)$$

To find an expression for this equation, eq. 2.42, one may use the semi-empirical method of the mixing length, $l = \kappa z'$, proposed by *Prandtl* (see *Graf & Altinakar*, 1991, p. 280), such as :

$$\tau_{zx} = \mu \frac{d\bar{u}}{dz} + \rho \kappa^2 z'^2 \left(\frac{d\bar{u}}{dz}\right)^2 \qquad (2.48)$$

where κ is Karman's universal constant.

Outside a very thin region — the viscous region — situated very close to the bed, the shear stress due to viscosity can be neglected, thus one has :

$$\tau_{zx} = \rho \kappa^2 z'^2 \left(\frac{d\bar{u}}{dz}\right)^2 \qquad (2.49)$$

Since the shear stress, τ_{zx}, remains constant in the vertical — more or less correct in the inner region — and equal to the wall-shear stress, τ_o, one may write :

$$\tau_{zx} \equiv \tau_o = \rho \kappa^2 z'^2 \left(\frac{d\bar{u}}{dz}\right)^2$$

After separation of variables, the following differential equation is obtained :

$$d\bar{u} = \frac{\sqrt{\tau_o/\rho}}{\kappa} \frac{dz}{z'}$$

which, upon integration, renders :

$$\frac{\bar{u}}{u_*} = \frac{1}{\kappa} \ln z' + C \qquad (2.50)$$

where $u_* = \sqrt{\tau_o/\rho}$ is the friction velocity. The value of the integration constant, C, must be determined experimentally; in this way the type of the surface (bed), being smooth or rough, will enter.

This *logarithmic law (of the wall)*, eq. 2.50, is only valid in the inner region, $z'/h \lesssim 0.2$, where the shear stress remains constant and the influence of a possible pressure gradient can be neglected (see *Rotta*, 1972, p. 153). The logarithmic law is universal (see *Monin et Yaglom*, 1971, p. 311), being the same for boundary-layer flow, as well as for flow in pipes and in open channels.

In the *inner* region, $z'/h \lesssim 0.2$, the velocity, $\bar{u}(z)$, whose variation is considerable, depends also on the wall-shear stress, on the fluid properties, on the type of the wall (bed) and on the distance from the wall; thus :

$$\bar{u} = f(\tau_o, \rho, \mu, k_s, z').$$

ii) In the *outer region*, $0.2 \lesssim z'/h \lesssim 1.0$, the velocity, $\bar{u}(z)$, whose variation is weak, depends also on the maximum velocity, on the flow depth and the driving pressure (see eq. 2.40), but does not depend on the viscosity or on the type of the wall (bed); thus :

$$\bar{u} = f(U_c, h, \frac{\partial \bar{p}^*}{\partial x} ; \tau_o, \rho, z').$$

In the outer region, a good agreement with experimental data is not possible, since $\tau_{zx} \neq \tau_o$. The logarithmic law, eq. 2.50, must be modified with a function which depends on the flow depth, h, and notably on an dimensionless pressure gradient, $(h / \tau_o) (\partial \bar{p}^* / \partial x)$. The velocity distribution is given here by the *law of velocity defect* (see *White*, 1974, p. 477), such as :

$$\frac{U_c - \bar{u}}{u_*} = f(\frac{z'}{h}, \frac{h}{\tau_o} \frac{\partial \bar{p}^*}{\partial x})$$

Amongst the different relations available (see *Hinze*, 1975, p. 630 et p. 697), the one of Coles shall here be used :

$$\frac{U_c - \bar{u}}{u_*} = \frac{1}{\kappa} \ln (\frac{\delta}{z'}) + \frac{\Pi}{\kappa} (2 - \tilde{\omega}) \qquad (2.51)$$

where the function, known as wake function, is defined by :

$$\tilde{\omega} = 2 \sin^2 (\frac{\pi}{2} \frac{z'}{\delta})$$

The wake parameter of Coles, Π, depends notably on the gradient of the longitudinal pressure :

$$\Pi = f(\beta)$$

where

$$\beta = \frac{h}{\tau_o} \frac{\partial \overline{p^*}}{\partial x} \qquad (2.52)$$

Since this parameter, Π, must remain constant for flow in equilibrium, the β-value is an equilibrium parameter (see *White*, 1974, p. 477). The value of $(2\Pi/\kappa)$ represents the deviation from the logarithmic part of eq. 2.51 for $(z'/\delta) = 1$ (see *Graf & Altinakar*, 1991, p. 288).

The height, $z' = \delta \; (\leq h)$, is the position in the flow section (see Fig. 2.13) where the maximum velocity, U_c, is measured; if U_c is on the water surface, the flow depth, $\delta \equiv h$, is to be taken.

This empirical relation, eq. 2.51, whose validity is made evident by experiments, is valid in the outer region as well as in the inner one, but not in the viscous region. However this relation is valid for both smooth and rough surfaces.

Nevertheless, as a first (and often good) approximation, the logarithmic law can often be applied over the entire flow depth, h (see *Monin et Yaglom*, 1971, p. 298); this is especially so if the flow is uniform having a weak pressure gradient.

iii) The distribution of velocity — called universal since it is independent of the Reynolds number — as given with eq. 2.50 and eq. 2.51, is complete but also complex. For practical purpose, one may also use a simple empirical relation (see *Graf & Altinakar*, 1991, p. 289) of the type :

$$\frac{\overline{u}}{\overline{u}_R} = \left(\frac{z'}{z_R}\right)^q$$

where $\overline{u}_R(z_R)$ is a reference velocity, which was for example previously measured. The variation of $1/10 < q < 1/6$ depends on the Reynolds number; often $q = 1/7$ is taken.

2.5.3 Non-uniform Flow

1° It will be assumed that flow is steady, two-dimensional, $\vec{V}(\overline{u}, 0, \overline{w})$, and gradually varied (non-uniform).

The equations of motion, eq. 2.41a and eq. 2.38a, and of continuity, eq. 2.35a, reduce to :

$$\overline{u}\frac{\partial \overline{u}}{\partial x} + \overline{w}\frac{\partial \overline{u}}{\partial z} = -g\left(\frac{\partial h}{\partial x} - S_f\right) + \frac{1}{\rho}\frac{\partial \tau_{zx}}{\partial z} \qquad (2.53)$$

$$\overline{p} = \rho g(h - z') - \rho \overline{w'^2} \qquad (2.38a)$$

$$\frac{\partial \overline{u}}{\partial x} + \frac{\partial \overline{w}}{\partial z} = 0 \qquad (2.35a)$$

Flow, which is (gradually) non-uniform, may be considered unidirectional if the variation of flow depth, $\partial h/\partial x$, is weak (see *Grishanin*, 1969, p. 59 - 62).

2° After integration over the flow depth, h, (see *Grishanin*, 1969) one obtains :

i) the equation of motion, eq. 2.53, as being :

$$\int_0^h \overline{u}\frac{\partial \overline{u}}{\partial x}dz + \int_0^h \overline{w}\frac{\partial \overline{u}}{\partial z}dz = -g\int_0^h \frac{\partial h}{\partial x}dz + gS_f\int_0^h dz + \frac{1}{\rho}\int_0^h \frac{\partial \tau_{zx}}{\partial z}dz$$

$$\beta_u Uh\frac{\partial U}{\partial x} + U^2 h\frac{\partial \beta_u}{\partial x} = -gh\frac{\partial h}{\partial x} + gS_f h + \frac{1}{\rho}(-\tau_0)$$

where $\beta_u = (1/U^2 h)\int_0^h \overline{u}^2 dz$ is the correction coefficient (of Boussinesq) of the velocity distribution, which is usually taken as $\beta_u = Cte \cong 1$ for turbulent flow. Assuming that the energy slope is given by $S_e = \tau_0/\rho gh$, one may now write :

$$\frac{1}{g}U\frac{\partial U}{\partial x} + \frac{\partial h}{\partial x} - S_f = -S_e \qquad (2.54)$$

HYDRODYNAMIC CONSIDERATIONS

ii) the equation of continuity, eq. 2.35a, as being :

$$\int_0^h \frac{\partial \overline{u}}{\partial x} dz + \int_0^h \frac{\partial \overline{w}}{\partial z} dz = \left(\frac{\partial}{\partial x} \int_0^h \overline{u}\, dz - \overline{u}_h \frac{\partial h}{\partial x} \right) + \overline{u}_h \frac{\partial h}{\partial x} = 0$$

$$\frac{\partial}{\partial x}(Uh) = 0 \tag{2.46}$$

For steady, non-uniform (but also uniform) flow, one writes :

$$\frac{dq}{dx} = U \frac{\partial h}{\partial x} + h \frac{\partial U}{\partial x} = 0 \tag{2.46a}$$

Note, that the term, $\overline{u}_h (\partial h/\partial x) = \overline{w}_h$, gives the equation of the streamline at the water surface.

The equation of motion, eq. 2.54 (see eq. 2.13) and of continuity, eq. 2.46a (see eq. 2.6), form together the simplified equations of Saint-Venant for a steady and non-uniform flow (see sect. 2.2).

3° Furthermore, one may postulate :

i)
$$\frac{\partial h}{\partial x} - S_f = \frac{\partial h}{\partial x} + \frac{\partial z_b}{\partial x} = \frac{1}{\rho g} \frac{\partial \overline{p^*}}{\partial x} \tag{2.40}$$

ii) using the equation of continuity, eq. 2.46a :

$$\frac{\partial U}{\partial x} = - \frac{U}{h} \frac{\partial h}{\partial x}$$

iii) using the definition of head loss (see point 2.2, 4°) :

$$S_e = \frac{\tau_o}{\rho g\, h}$$

In this way, the equation of motion, eq. 2.54, can be expressed as:

$$\rho U^2 \frac{\partial h}{\partial x} = \tau_o \left(1 + \frac{h}{\tau_o} \frac{\partial \overline{p^*}}{\partial x} \right) \tag{2.54a}$$

This equation may be compared with the Karman equation (see *Graf & Altinakar*, 1991, sect. CL.4 : eq. CL.25a) for boundary-layer flow.

The dimensionless longitudinal gradient of the driving pressure :

$$\beta = \frac{h}{\tau_o} \frac{\partial \overline{p}^*}{\partial x} \qquad (2.52)$$

which stands for the ratio of the forces due to the driving pressure and due to friction, defines the equilibrium parameter. This parameter can be used to classify non-uniform flow, being flow with a pressure gradient. Note, however, that also uniform flow (see eq. 2.43a) is flow with a (weak) pressure gradient, where $\beta = -1$.

4° To obtain the distribution (see Fig. 2.12) of velocity, $\overline{u}(z')$, one should distinguish two regions, as was also done with uniform flow.

   i) In the *inner region*, one finds that the logarithmic law of the wall :

$$\frac{\overline{u}}{u_*} = \frac{1}{\kappa} \ln z' + C \qquad (2.50)$$

remains valid (see *White*, 1974, p. 473), and this as long as the driving-pressure gradient is weak (see eq. 2.40), being either positive or negative, $\pm (\partial \overline{p}^*/\partial x)$. One can explain (see *Tennekes* et *Lumley*, 1972, p. 185) this, by assuming the inertia term in eq. 2.53 is negligible and that the term of the driving pressure is weak compared to the term of the Reynolds stress; a zone of quasi-constant stress is thus delimited.

Nevertheless the integration constant, C, may depend upon the pressure gradient (see *Tennekes* et *Lumley*, 1972, p. 186). The thickness of the inner region, $z'/h \leq 0.2$, will now depend on the pressure gradient (see *White*, 1974, p. 473). The inner region can disappear in decelerating flow with strong positive pressure gradients, when flow separation occurs.

   ii) In the *outer region*, one finds (experimentally) that the law of velocity defect, the one of Coles :

$$\frac{U_c - \overline{u}}{u_*} = \frac{1}{\kappa} \ln \left(\frac{\delta}{z'}\right) + \frac{\Pi}{\kappa} (2 - \tilde{\omega}) \qquad (2.51)$$

remains valid. Depending on the gradient of the driving pressure (see eq. 2.40), one has to make the following distinction (see Fig. 2.12) :

   - a positive (unfavourable) pressure gradient, $\partial \overline{p}^*/\partial x > 0$, being always accompanied by a decrease of the average velocity in the direction of the flow (*deceleration*); the velocity profiles get less uniformly distributed;

HYDRODYNAMIC CONSIDERATIONS

- a negative (favourable) pressure gradient, $\partial \overline{p^*}/\partial x < 0$, being usually accompanied by an increase of the average velocity in the direction of flow (*acceleration*); the velocity profiles get more uniformly distributed.

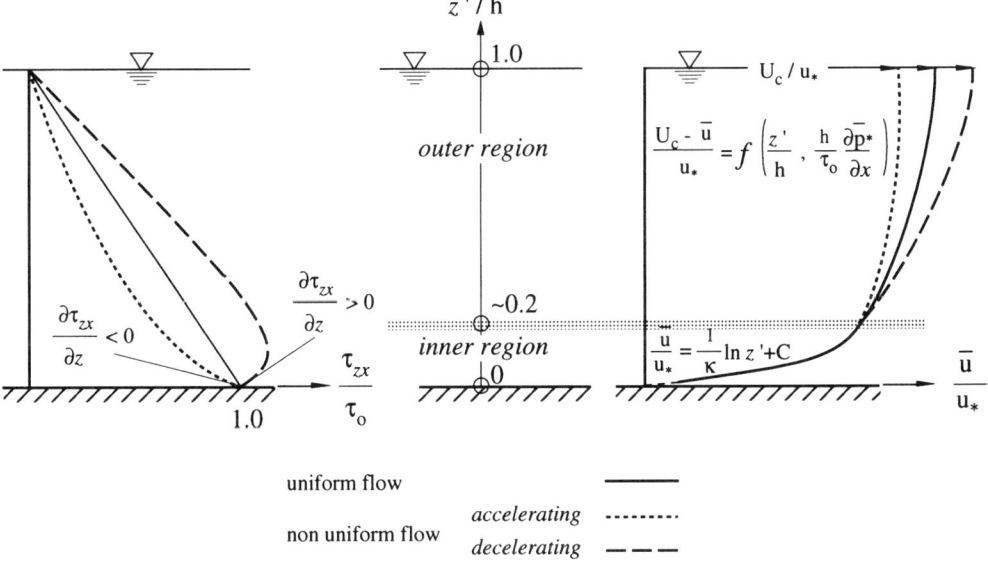

Fig. 2.12 Scheme of the distribution of the shear stress, $\tau_{zx}(z')$, and of the velocity, $\bar{u}(z')$, in non-uniform flow.

5° To obtain the distribution (see Fig. 2.12) of the total shear stress, $\tau_{zx}(z)$, the equation of motion, eq. 2.53, must be integrated (see *Rotta*, 1972, p. 240). The boundary conditions are the following:

$z' = 0$ $\Rightarrow$ $\tau_{zx} = \tau_o$

$z' = h$ $\Rightarrow$ $\tau_{zx} = 0$

Consequently, one obtains:

i) Close to the wall, $z' \ll h$, where the non-slip conditions, $\bar{u} = 0$ and $\bar{w} = 0$, are valid, eq. 2.53 is written as:

$$\frac{\partial \tau_{zx}}{\partial z} = \rho g \left(\frac{\partial h}{\partial x} + \frac{\partial z_b}{\partial x} \right) = \frac{\partial \overline{p^*}}{\partial x} \qquad (2.53a)$$

which subsequently gives (see *White*, 1974, p. 474):

$$\tau_{zx} = \tau_o + \left(\frac{\partial \overline{p^*}}{\partial x}\right) z$$

Depending on the pressure gradient, one has :

- for a positive pressure gradient, $\partial \overline{p^*}/\partial x > 0$, where the flow is *decelerating* :

$$\frac{\partial \tau_{zx}}{\partial z} > 0 \quad \text{and} \quad \tau_{zx} > \tau_o$$

the total stress has its maximum value, $\tau_{zx} \equiv \tau_{max}$, at a certain distance from the wall;

- for a negative pressure gradient, $\partial \overline{p^*}/\partial x < 0$, where the flow is generally *accelerating* :

$$\frac{\partial \tau_{zx}}{\partial z} < 0 \quad \text{and} \quad \tau_{zx} < \tau_o$$

the total stress has its maximum value, $\tau_{zx} \equiv \tau_{max}$, at the wall.

ii) Far from the wall, beyond the point where $\tau_{zx} \equiv \tau_{max}$, the distribution of total shear stress is monotone (see *Rotta*, 1972, p.240) and this up to the water surface, where $\tau_{zx} = 0$.

2.6 DISTRIBUTION OF VELOCITY

1° The experimental results, to support the theory developed in chap. 2.5, will now be presented.

It is taken that the flow of a real and incompressible fluid is completely developed along (the bed of) the channel. Assumed will be that the flow is two-dimensional, but unidirectional in the *x*-direction, being steady and uniform or non-uniform.

2° A direct consequence of a real-fluid flow is the manifestation of the (point) velocity profile, $u(z')$, where the z' is the distance measured from the bed of the channel.

By integration of the velocity profile the average velocity, U, across the flow section is obtained.

3° Between the dimensionless average velocity, U/u_*, and friction coefficient, *f*, there exists (see *Graf & Altinakar*, 1991, p. 433) the following relationship (see eq. 3.8) :

HYDRODYNAMIC CONSIDERATIONS

$$\frac{U}{u_*} = \sqrt{\frac{8}{f}} \qquad (2.55)$$

where $u_* = \sqrt{\tau_0/\rho}$ is the friction velocity.

4° A summary of the velocity distribution, $u(z')$, of the average velocity, U, and of the friction coefficient, f, for uniform flow, both laminar and turbulent, is given in Table 2.1.

2.6.1 Laminar Flow

1° Uniform, steady and laminar flow in a channel of a large width, $R_h \equiv h$, has been studied in great detail (see *Graf & Altinakar*, 1991, p. 257); it is a special case of the *Couette* flow.

2° The distribution of the velocity, $u(z')$, for two-dimensional flow (see Fig. 1.6) is given by a parabolic relation :

$$\frac{u(z')}{u_*} = \frac{1}{2\mu\, u_*} (-\gamma \frac{dh}{dx})\, (2hz' - z'^2) \qquad (2.56)$$

where $h = h + z_b$ and (dh/dx) is the slope of the water surface, given herewith as $S_w \equiv S_f = -(dz_b/dx)$. Using the friction velocity, given as $u_*^2 = gh\, S_f$ (see eq. 3.7), this equation, eq. 2.56, becomes :

$$\frac{u(z')}{u_*} = (\frac{u_* z'}{\nu})\, (1 - \frac{z'}{2h}) \qquad (2.56a)$$

3° The average velocity, U, in the flow section, A, is given (see *Graf & Altinakar*, 1991, p. 257) by :

$$\frac{U}{u_*} = \frac{1}{u_*} \frac{g}{3\nu} S_f h^2 = \frac{1}{3} (\frac{u_* h}{\nu}) \qquad (2.57)$$

a relationship which expresses a proportionality between the average velocity, U, and the bed slope, S_f.

4° Flow is considered to be laminar, if the Reynolds number is :

$$Re' = \frac{Uh}{\nu} \leq 500 \quad \text{or} \quad Re = \frac{4Uh}{\nu} \leq 2000$$

As long as flow stays laminar, the roughness of the bed of the channel is of no consequence.

5° The friction coefficient, f, is obtained by combining eq. 2.55 and eq. 2.57; that is :

$$\sqrt{\frac{8}{f}} = \frac{U}{u_*} = \frac{1}{3}\left(\frac{u_* h}{\nu}\right)\frac{U}{U} = \frac{1}{3}\text{Re'}\sqrt{\frac{f}{8}}$$

or written also as :

$$f = \frac{24}{\text{Re'}} \quad \text{or} \quad f = \frac{6}{\text{Re}} \tag{2.58}$$

The coefficient, $B_1 = 24$, is valid notably in two-dimensional flow, when the width of the channel is large and having an aspect ratio of $B/h > 5$. For channels which are less large, $B/h < 5$, or for channels being not rectangular, this coefficient may be smaller, $14 < B_1 < 24$ (see *Chow*, 1959, p. 11).

2.6.2 Turbulent, smooth Flow

1° The universal velocity distribution for turbulent, smooth flow was developed using the concept of the mixing length (see point 2.5.2, 4°, or *Graf & Altinakar*, 1991, pp. 280 - 289).

2° The distribution of the velocity, $u(z')$, — one shall take now $u \equiv \bar{u}$ for the point-average velocity, or the bar will no more be used — is logarithmic (see eq. 2.50); it is given by :

$$\frac{u(z')}{u_*} = \frac{1}{\kappa}\ln\left(\frac{z' u_*}{\nu}\right) + B_s \tag{2.59}$$

The numerical constants, obtained from numerous experiments with *uniform* flow (see *Reynolds*, 1974, p. 187) are :

$$\kappa = 0.4 \quad ; \quad B_s = 5\,(\pm 25\%)$$

3° For *non-uniform* flows, the numerical constants are only slightly different (see *Reynolds*, 1974, p. 187 and *Cardoso* et al., 1989).

4° This relation, eq. 2.59, is only valid close to the surface (bed), delimited by :

$$35 \leq \frac{z' u_*}{\nu} \leq 200 \quad \text{or} \quad \frac{z'}{h} \leq 0.2$$

HYDRODYNAMIC CONSIDERATIONS

but experiments have shown good agreement over the entire flow depth, h. The region delimited by $(z'/h) \leq 0.2$, is the inner region (see Fig. 2.11) where the shear stress remains essentially constant.

5° Upon integration of eq. 2.59, one obtains an expression for the average velocity :

$$\frac{U}{u_*} = \frac{1}{\kappa} \ln\left(\frac{R_h u_*}{\nu}\right) + \overline{B}_s \qquad (2.60)$$

The constant of integration obtained from numerous experiments (see *Keulegan*, 1938) is given as :

$$\overline{B}_s = 3.5$$

but it depends slightly on the geometry of the cross section and on the Froude number (see *Chow*, 1959, p. 205).

6° The friction coefficient, f, can be now obtained in combining eq. 2.55 with eq. 2.60; this gives :

$$\sqrt{\frac{8}{f}} = \frac{U}{u_*} = \frac{1}{\kappa} \ln\left(\frac{R_h u_*}{\nu}\right) + \overline{B}_s$$

and for $\overline{B}_s = 3.5$, one has :

$$\sqrt{\frac{1}{f}} = 2.03 \log\left(Re' \sqrt{f}\right) + 0.32 \qquad (2.61)$$

or putting $Re' = Re/4$:

$$\sqrt{\frac{1}{f}} = 2.03 \log\left(Re \sqrt{f}\right) - 0.88 \cong 2 \log\left(\frac{Re}{3} \sqrt{f}\right) \qquad (2.61a)$$

The above relations, eq. 2.61, are valid for turbulent flow, $Re' > 500$, in a channel having smooth walls, $(u_* k_s/\nu) < 5$.

2.6.3 Turbulent, rough Flow

1° The universal velocity distribution for turbulent, rough flow was developed, using the concept of the mixing length (see point 2.5.2, 4°, or *Graf & Altinakar*, 1991, pp. 280-289).

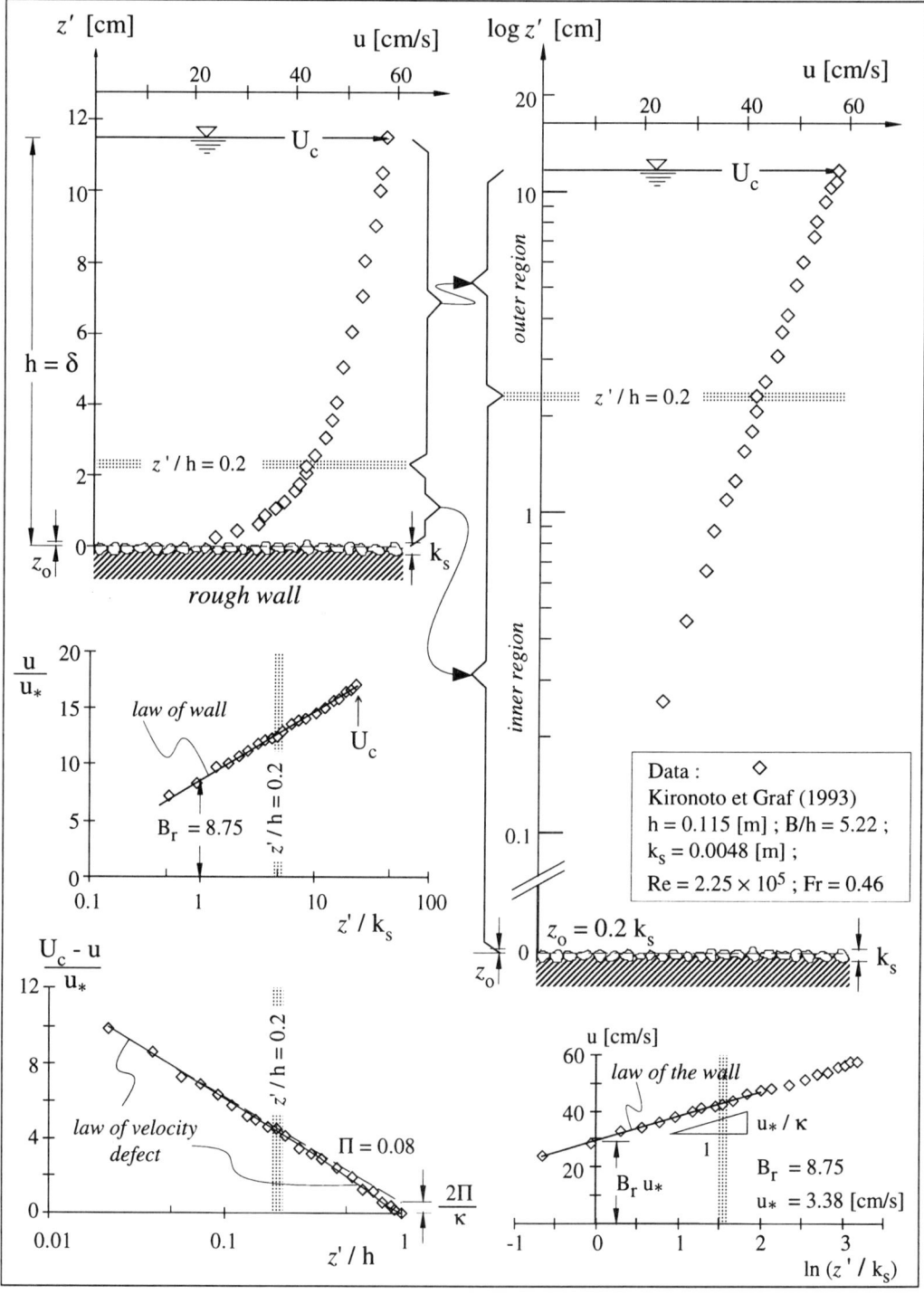

Fig. 2.13 Velocity profile, u(z'); *uniform*, rough flow.

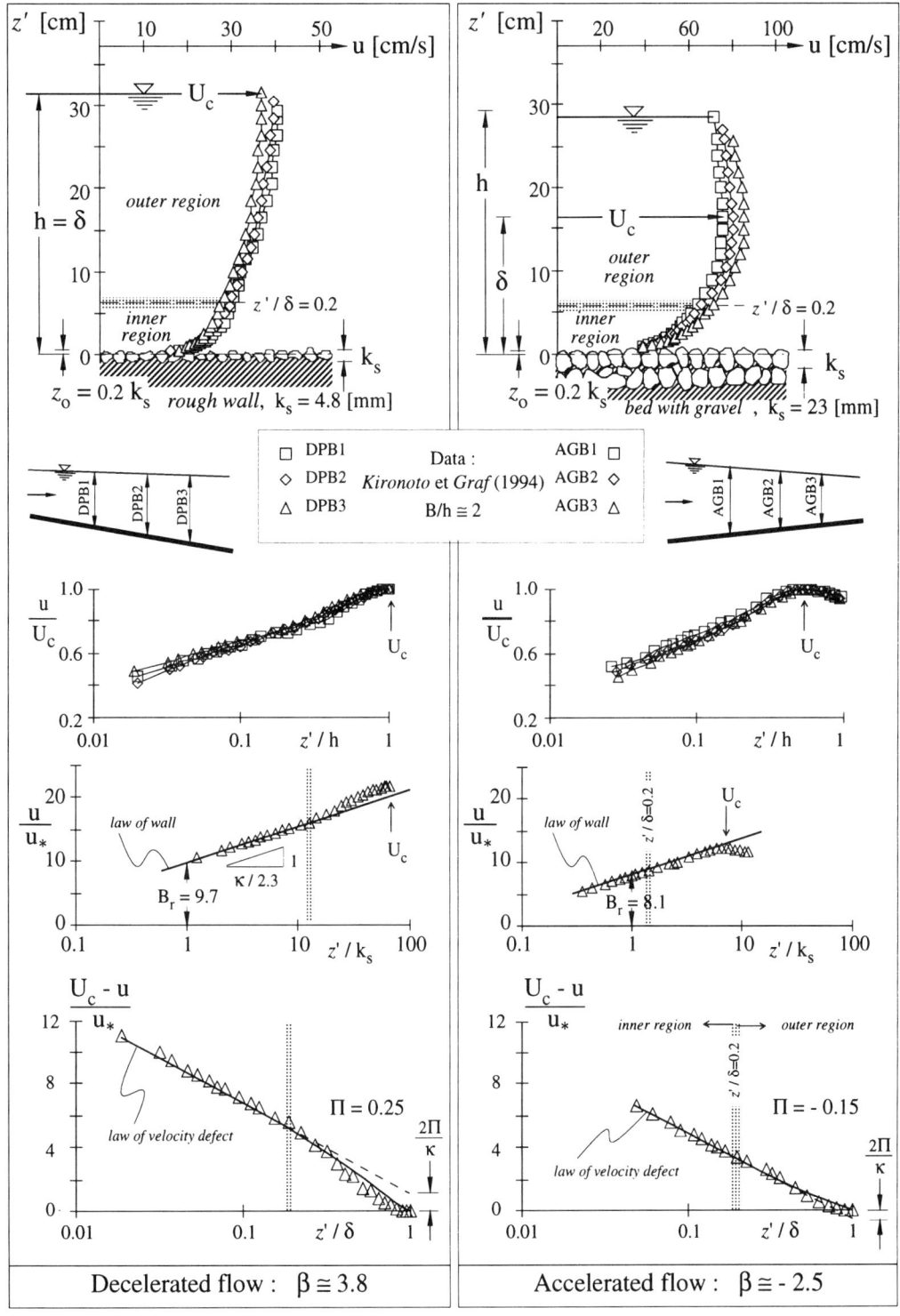

Fig. 2.14 Velocity profile, u(z'); *non-uniform*, rough flow.

2° As an example, the distribution of the velocity — measured in a laboratory flume by *Kironoto et Graf* (1993 et 1994) — is given with Fig. 2.13 for uniform and with Fig. 2.14 for non-uniform flow. Different coordinates are here used :

- u vs z' : is the original (measured) profile;

- u vs log z' : shows the logarithmic form of the original profile;

- $\dfrac{u}{U_c}$ vs $\dfrac{z'}{h}$: shows the auto-similarity, necessary for non-uniform flow in equilibrium;

- u vs $\ln(\dfrac{z'}{k_s})$: used to determine the values of u_* and B_r, if $\kappa = 0.4$ is imposed;

- $\dfrac{u}{u_*}$ vs $\dfrac{z'}{k_s}$: gives a dimensionless representation of the law of wall;

- $\dfrac{U_c - u}{u_*}$ vs $\dfrac{z'}{\delta}$: gives the dimensionless representation of the law of velocity deficit.

3° The distribution of the velocity, u(z') (see Fig. 2.13), is logarithmic (see eq. 2.50); it is given by :

$$\frac{u(z')}{u_*} = \frac{1}{\kappa} \ln\left(\frac{z'}{k_s}\right) + B_r \qquad (2.62)$$

k_s being the equivalent or standard uniform roughness (see *Graf & Altinakar*, 1991, p. 287 et sect. 3.2.1). The numerical constants, which are obtained from numerous experiments (see *Reynolds*, 1974, p. 187 and *Kironoto et Graf*, 1993) for *uniform* flow, are given as :

$$\kappa = 0.4 \quad ; \quad B_r = 8.5\,(\pm 15\%)$$

The vertical distance, z', is measured from a level which passes slightly below the peaks of the roughness (see Fig. 2.13); in general one takes $z_o \cong -0.2\,k_s$ (see *Graf*, 1991 or *Hinze*, 1975, p. 637). The relation of eq. 2.62, just as the one of eq. 2.59, is actually only valid within the inner region, $z'/h \leq 0.2$ (see Fig. 2.11 and Fig. 2.13), but an extension into the outer region is often possible.

4° For *non-uniform* flow, the constant, B_r, is slightly different (see *Kironoto et Graf, 1994*): it is larger in decelerating flow and smaller in accelerating flow (see Fig. 2.14). The same tendency is observed for *unsteady* flow (see *Tu et Graf, 1992*).

5° After integration of eq. 2.62, one obtains the following expression for the average velocity :

$$\frac{U}{u_*} = \frac{1}{\kappa} \ln\left(\frac{R_h}{k_s}\right) + \overline{B_r} \qquad (2.63)$$

The constant of integration, obtained from different experiments (see *Keulegan, 1938, p. 722*) is :

$$\overline{B_r} = 6.25$$

being almost independent of the geometrical form of the channel. For flow with large Froude numbers, $Fr > 1$, the value of $\overline{B_r}$ diminishes (see *Chow, 1959, p. 205*).

6° The friction coefficient, f, obtained by combining eq. 2.55 with eq. 2.63, is written as :

$$\sqrt{\frac{8}{f}} = \frac{U}{u_*} = \frac{1}{\kappa} \ln\left(\frac{R_h}{k_s}\right) + \overline{B_r}$$

substitution of $\overline{B_r} = 6.25$, gives :

$$\sqrt{\frac{1}{f}} = 2.03 \log\left(\frac{R_h}{k_s}\right) + 2.2 \qquad (2.64)$$

This relation is valid for turbulent flow, $Re' > 2 \cdot 10^4$, in channels with completely rough surfaces, $(k_s u_*/\nu) > 70$.

7° Between smooth surface flow, delimited by $(k_s u_*/\nu) < 5$, and rough surface flow, delimited by $(k_s u_*/\nu) > 70$, there exists the transition region, where the experiments of Nikuradse (see *Graf & Altinakar, 1991, p. 427*) are used to make the connection.

8° The friction coefficient, f, for flow over smooth, transition and rough surfaces, is given by the relation of Colebrook et White (see *Graf & Altinakar, 1991, p. 436*) which was adapted for channels by *Silberman et al.* (1963, p. 104), such as :

$$\sqrt{\frac{1}{f}} = -2.0 \log \left(\frac{k_s/R_h}{a_f} + \frac{b_f}{\text{Re}\sqrt{f}} \right) \qquad (2.65)$$

where $12 < a_f < 15$ and $0 < b_f < 6$, being established for sections of different geometrical shapes and $\text{Re} = 4\,R_h U/\nu$. For very wide channels it is recommended to take : $a_f = 12$ and $b_f = 3.4$. If $k_s = 0$, the relation reduces to eq. 2.61a, valid for smooth surfaces; if $\text{Re} \to \infty$, it reduces to eq. 2.64, valid for completely rough surfaces.

9° Outside the inner region (see Fig. 2.11) and up to the entire flow depth, h, is the outer region, delimited by $0.2 < (z'/h) < 1.0$. In this region, the flow is conditioned by the maximum velocity, $u = U_c$, as well as by a possibly existing longitudinal pressure gradient. The distribution of the velocity deviates slightly from the logarithmic law. It is approximately given (see *Graf*, 1991 et *Hinze*, 1975) by a law of velocity deficit :

$$\frac{U_c - u}{u_*} = 9.6 \left(1 - \frac{z'}{h} \right)^2 \qquad (2.66)$$

being valid for turbulent flow, both for smooth and rough surfaces.

Nevertheless, the logarithmic law given by eq. 2.59 and eq. 2.62 can be used over the entire flow depth, if one does not desire a too high precision.

10° The distribution of the velocity over the entire flow depth — with exception of the viscous region —, thus in the zone of $0.01 < (z'/h) < 1.00$ (see Fig. 2.11 and Fig. 2.13), is given by the following law of velocity deficit (see point 2.5.2, 4°) :

$$\frac{U_c - u}{u_*} = \frac{1}{\kappa} \ln \left(\frac{\delta}{z'} \right) + \frac{\Pi}{\kappa} (2 - \tilde{\omega}) \qquad (2.51)$$

where Π is the wake parameter of Coles, which depends notably on the longitudinal pressure gradient, β, (see eq. 2.52). For *uniform* flow over smooth and rough surfaces (see *Kironoto* et *Graf*, 1993) in a channel having a weak pressure gradient, namely the bottom slope, one takes :

$$\Pi \cong 0.2$$

having a variation of $-0.1 \lesssim \Pi \lesssim 0.3$.

For flow (boundary-layer) without pressure gradient, $(\partial \overline{p^*} / \partial x) = 0$, one takes $\Pi \cong 0.55$ (see *Hinze*, 1975, p. 697).

Table 2.1

Summary of velocity profile, u, of average velocity, U, and of friction coefficient, f, for steady, *uniform* flow in channels.

$$\frac{u}{u_*} = \frac{1}{2\mu u_*}(\gamma\, S_f)(2hz' - z'^2)$$

$$\frac{U}{u_*} = \frac{1}{3}\left(\frac{h\, u_*}{\nu}\right)$$

$$f = 24/Re'$$

↑ LAMINAR
――――― Flow ―――――――――――――――――― $Re' = \dfrac{Uh}{\nu} \cong 500$
↓ TURBULENT

smooth
$\dfrac{u_* k_s}{\nu} < 5$

$$\frac{u}{u_*} = \frac{1}{\kappa}\ln\left(\frac{z' u_*}{\nu}\right) + 5.5$$

$$\frac{U}{u_*} = \frac{1}{\kappa}\ln\left(\frac{R_h u_*}{\nu}\right) + 3.5$$

$$\frac{1}{\sqrt{f}} = 2\log(Re'\sqrt{f}) + 0.32$$

smooth
transition
rough

$$\frac{U_c - u}{u_*} = \frac{1}{\kappa}\ln\left(\frac{\delta}{z'}\right) + \frac{\Pi}{\kappa}(2 - \tilde{\omega})$$

$$\frac{1}{\sqrt{f}} = -2\log\left(\frac{k_s/R_h}{a_f} + \frac{b_f}{4Re'\sqrt{f}}\right)$$

rough
$\dfrac{u_* k_s}{\nu} > 70$

$$\frac{u}{u_*} = \frac{1}{\kappa}\ln\left(\frac{z'}{k_s}\right) + 8.5$$

$$\frac{U}{u_*} = \frac{1}{\kappa}\ln\left(\frac{R_h}{k_s}\right) + 6.25$$

$$\frac{1}{\sqrt{f}} = 2\log\left(\frac{R_h}{k_s}\right) + 2.2$$

11° For *non-uniform* flow (see Fig. 2.12 and Fig. 2.14), eq. 2.51 remains still valid, but the wake parameter, Π, has no more a constant value. Such flows must remain in equilibrium, namely the velocity profile must stay auto-similar. An empirical relationship of the type :

$$\Pi = f(\beta)$$

is proposed, where β is the equilibrium parameter :

$$\beta = \frac{h}{\tau_o} \frac{\partial \overline{p^*}}{\partial x} \tag{2.52}$$

which characterizes the longitudinal pressure gradient. Depending on this parameter, β, one finds (see *Kironoto* et *Graf*, 1994) that :

$\beta < -1$: flow is *accelerating*, and the wake parameter is $-1.0 < \Pi < 0.2$;

$\beta > -1$: flow is *decelerating*, and the wake parameter is $\Pi > 0.2$;

$\beta = -1$: flow is *uniform* having a weak pressure gradient, and the wake parameter is $\Pi \cong 0.2$.

The same tendency was observed (see *Tu* et *Graf*, 1992) for *unsteady* flow.

12° For two-dimensional flow, the maximum velocity, U_c, occurs at the water surface, $\delta \equiv h$. For three-dimensional flow, the maximum velocity, U_c, may occur below the water surface, $\delta < h$ (see Fig. 2.15); secondary flow is evident. To parametrise this, one may use a ratio of $(h - \delta)/h$ (see Fig. 2.15). The aspect ratio of $B/h \cong 5$ is the limiting value.

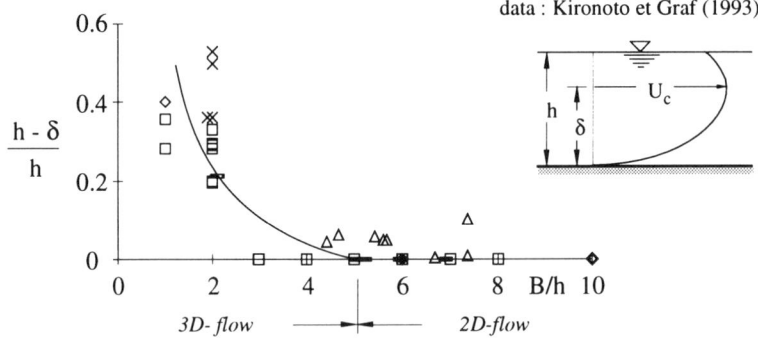

Fig. 2.15 The position of the maximum velocity, U_c ;
in two- and three- dimensional flow.

2.6.4 Turbulence Characteristics

1° Flow at large Reynolds numbers ceases to be laminar; it becomes turbulent. In each point of the flow, the instantaneous velocity, u_i and w_i, is subject to variation in direction and in intensity. The velocity varies around a mean value defined by :

$$u_i = u + u' \quad , \quad w_i = w + w'$$

The fluctuation components, u' and w', are by definition weak as compared to the respective mean values, u and w.

2° The equations of motion for laminar flow — the Navier-Stokes equation (see *Graf & Altinakar*, 1991, sect. FR.1) — are modified by the supplementary stresses due to the turbulence; these are the Reynolds equations, eq. 2.35, (see *Graf & Altinakar*, 1991, sect. FR.5). The supplementary stresses have the form of :

$$\rho \overline{u'^2} \quad , \quad \rho \overline{w'^2} \quad , \quad \rho \overline{u'w'} \quad , \quad \text{etc.}$$

Semi-empirical methods are used to express these stresses.

The characterization of the structure of turbulence is based here on experiments done in channels; some of these will be presented.

3° *Intensity of turbulence*

The temporal mean value of the velocity fluctuations are by definition zero :

$$\overline{u'} = 0 \quad , \quad \overline{w'} = 0$$

This is however not the case for the mean quadratic values, $\overline{u'^2}$. The RMS-value (*Root-Mean-Square*), $\sqrt{\overline{u'^2}}$ and $\sqrt{\overline{w'^2}}$, is commonly used.

The ratio of the RMS-value and the friction velocity (see *Graf & Altinakar*, 1992, p. 267) is used to define the intensity of turbulence :

$$\frac{\sqrt{\overline{u'^2}}}{u_*} = f(z) \quad , \quad \frac{\sqrt{\overline{w'^2}}}{u_*} = f(z) \qquad (2.67)$$

which varies in space.

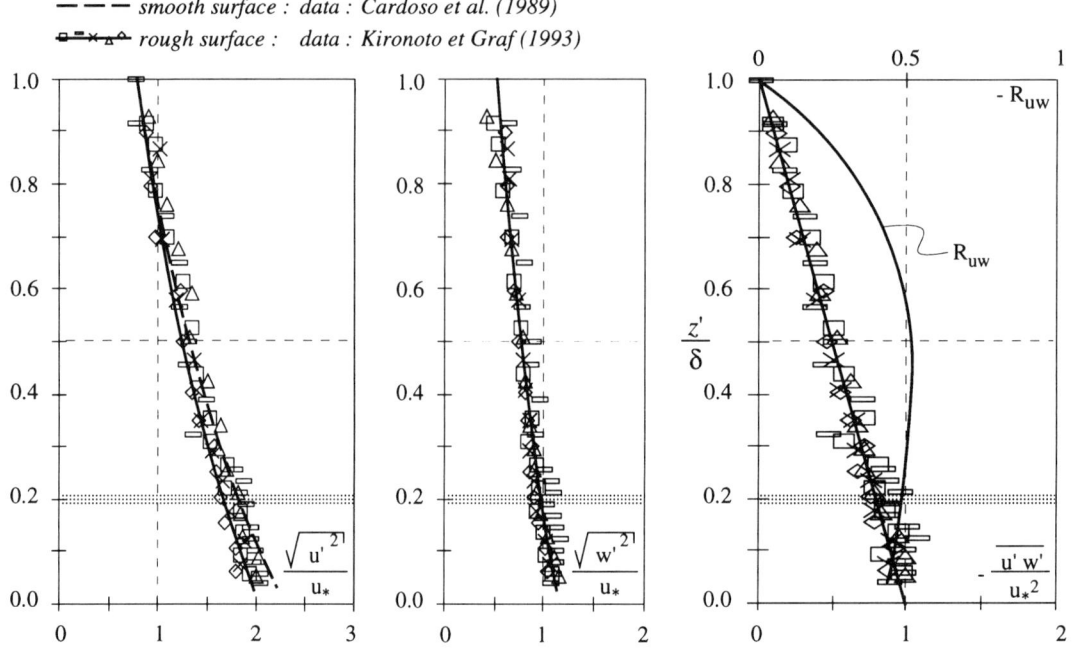

Fig. 2.16 Distribution of the normal and tangential stress of Reynolds; for *uniform* flow.

The vertical distribution of the normal stress, expressed as turbulence intensity, is given in Fig. 2.16 for *uniform* flow; the measurements were performed in the center of the channel having an aspect ratio of $2.1 < B/h < 6.9$. One notices that :

i)   for channels with smooth (see *Cardoso* et al., 1989) and rough surface (see *Kironoto* et *Graf*, 1993), the distribution are reasonably the same;

ii)   close to the bed (surface), in the inner region, one has :

$$\sqrt{\overline{u'^2}} \cong 1.8\, u_* \quad ; \quad \sqrt{\overline{w'^2}} \cong 1.0\, u_*$$

iii)   up to the flow depth, the distribution stays monotone; at the surface one has :

$$\sqrt{\overline{u'^2}} \cong \sqrt{\overline{w'^2}} \cong 0.6\, u_*$$

and the turbulence becomes isotropic.

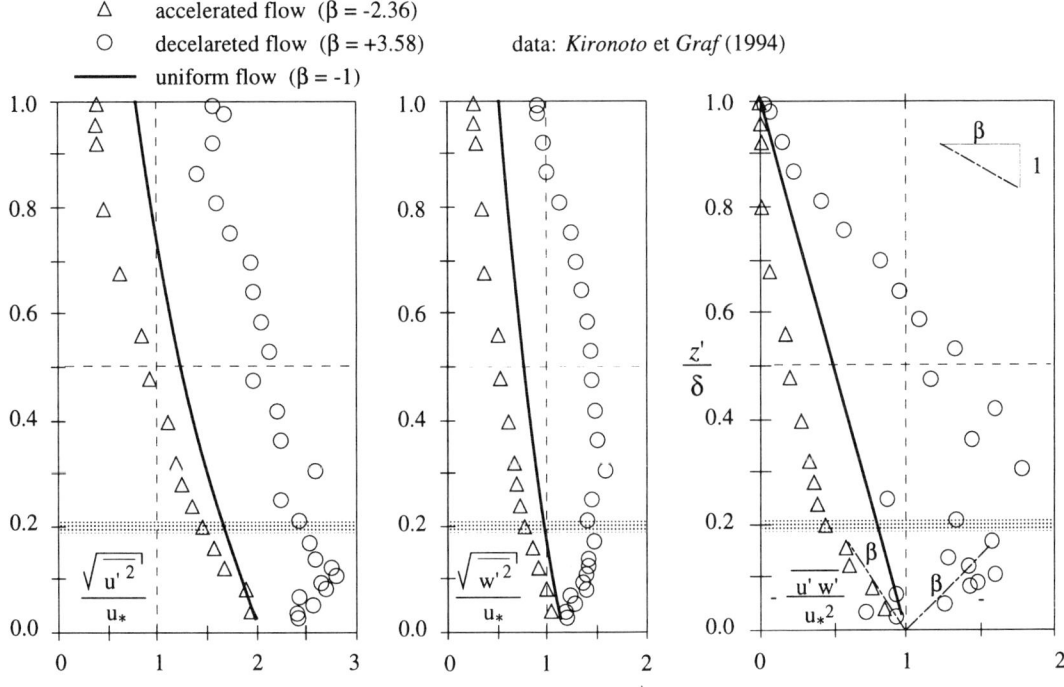

Fig. 2.17 Distribution of the normal and tangential stress of Reynolds; for *non-uniform* flow.

For *non-uniform* flow in equilibrium, the distribution of the normal stress is given in Fig. 2.17; the measurements (see *Kironoto* et *Graf*, 1994) are performed in the center of the channel having an aspect ratio of $B/h \cong 2$. One notices that :

i) for *accelerating* flow, $\beta < -1$, the turbulence intensity is smaller than for uniform flow, $\beta = -1$. The maximum value is at the bed and diminishes towards the water surface. Consequently, in accelerating flow the turbulence is suppressed (see *Hinze*, 1959, p. 66);

ii) for *decelerating* flow, $\beta > -1$, the turbulence intensity is larger than for uniform flow, $\beta = -1$. The maximum value is above the bed and diminishes towards the water surface. Consequently, in decelerating flow the turbulence is enhanced.

Similar experimental observations have also been communicated (see *Bradshaw*, 1978, p. 68) for boundary-layer flow with pressure gradients.

4°   *Reynolds stress*

The total tangential stress, eq. 2.42, are well represented by the supplementary stress, notably for flow at large Reynolds numbers; one writes :

$$\tau_{zx} = -\rho \overline{u'w'} \qquad (2.42b)$$

The vertical distribution of the supplementary stress, in short the Reynolds stress, is given in Fig. 2.16 for experiments with *uniform* flow. One finds — as one has already seen in point 2.5.2, 3° — that :

i)   the distribution is linear, or :

$$\frac{\tau_{zx}}{\tau_o} = \left(\frac{\delta - z'}{\delta}\right) \qquad (2.47b)$$

ii)   with the boundary conditions being :

$$z' \cong 0 \qquad \tau_{zx} \cong \tau_o$$
$$z' \cong \delta \qquad \tau_{zx} = 0$$

For *non-uniform* flow in equilibrium, the distribution of the Reynolds stress is given with Fig. 2.17. One finds — as one has already seen in point 2.5.3, 5° — that :

i)   in *accelerating* flow, $\beta < -1$, the Reynolds stress, whose distribution is concave, diminishes;

ii)   in *decelerating* flow, $\beta > -1$, the Reynolds stress, whose distribution is convex, increases;

iii)   consequently, the energy dissipation, namely the head loss, is larger in decelerating than in accelerating flow;

iv)   close to the bed, the gradient to the curve of distribution is given by :

$$\frac{\partial \tau_{zx}}{\partial z} = \frac{\partial \overline{p^*}}{\partial x} \approx \beta \left(\frac{\tau_o}{h}\right) \qquad (2.53b)$$

which is in agreement with arguments advanced in point 2.5.3, 5° (see Fig. 2.12).

Similar observations have also been done for unsteady flow with a free surface (see *Tu et Graf, 1992*) as well as for boundary-layer flow having pressure gradients (see *Bradshaw, 1978*, p. 68).

A dependence between the velocity fluctuations, u' et w', is often given with a correlation coefficient :

$$-R_{uw} = \frac{\overline{u'w'}}{\sqrt{\overline{u'^2}} \sqrt{\overline{w'^2}}} \qquad (2.68)$$

The distribution of this coefficient for *uniform* flow — for *non-uniform* flow it is similar — is given with Fig. 2.16. One sees that :

i) over a large fraction of the flow depth, $0.1 < z'/\delta < 0.6$, the fluctuation are reasonably well correlated : $R_{uw} \cong -0.5$;

ii) this correlation diminishes close to the bed and close to the water surface.

5° *Mixing length*

The Reynolds stress, eq. 2.42b, can also be expressed by :

$$\tau_{zx} = \rho\, l^2 \left(\frac{\partial \bar{u}}{\partial z}\right)^2 = \rho\, \nu_t \left(\frac{\partial \bar{u}}{\partial z}\right) \qquad (2.49)$$

where l, known as Prandtl's mixing length, is the distance over which the fluid mass displaces itself. ν_t is the mixing coefficient of Boussinesq which has dimensions of the kinematic viscosity, but its value is very large, such as $\nu_t \gg \nu$.

The vertical distribution of the mixing length, l, in *uniform* flow — it is similar in *non-uniform* flow — is given with Fig. 2.18; one finds that :

i) close to the surface (bed), in the inner region, $z'/\delta \lesssim 0.2$, it is given as :

$l = \kappa z'$ where $\kappa = 0.4$

ii) in a large part of the outer region, $0.5 \lesssim z'/\delta \lesssim 1.0$, it is given as :

$l/\delta \cong 0.12$

but the data show a large spread.

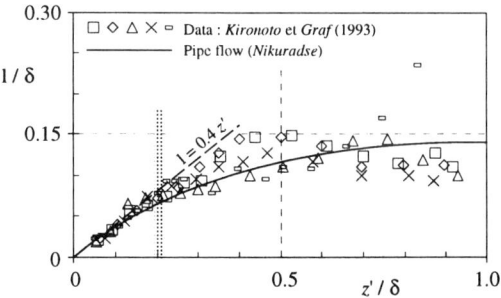

Fig. 2.18 Distribution of the mixing length.

6° *Energy spectrum*

The spectrum of kinetic energy (see *Graf & Altinakar*, 1991, p. 273) provides important information about the turbulence of the flow, namely about the energy distribution of the eddies having different sizes and frequencies.

The spectral function, E(n), gives the turbulent kinetic energy, $\overline{u'^2}(n)$ or $\overline{w'^2}(n)$, for a range of frequencies, dn ; or :

$$\overline{u'^2} = \int_0^\infty E(n)\, dn \qquad (2.69)$$

or written in normalized form, $F(n) = E(n)/\overline{u'^2}$, as :

$$\int_0^\infty F(n)\, dn = 1 \qquad (2.70)$$

F(n) has units of time, [t] , and n has units of frequency, $[t^{-1}]$.

The function, F(n) , — known as the *turbulence spectrum* — is a representation of the way the energy is distributed with the frequency, n. By a transformation — use is made of Taylor's hypothesis of frozen turbulence — of the frequency, n , into a wave number, $(2\pi/u)\, n = k$, the wave-number spectrum is obtained :

$$\int_0^\infty F(k)\, dk = 1 \qquad (2.70a)$$

F(k) has units of length, [m], and k has units of length, $[m^{-1}]$.

With Fig. 2.19 is shown a one-dimensional frequency spectrum for longitudinal, u', and vertical, w', velocity fluctuations for uniform flow in an open channel with a free surface, at different levels, z'. These spectra are rather typical for *uniform* flow over smooth and rough surfaces, having $10^4 < Re < 10^6$ (see *Kironoto et Graf*, 1993) and also for *non-uniform* flow (see *Kironoto et Graf*, 1994).

The energy spectrum is in general rather wide and is usually delimited in three zones (see Fig. 2.19b):

i) The turbulence structure of the largest eddies has no universality; it is anisotropic and depends largely on the flow conditions, thus on the flow Reynolds number. The *macro scale* (see *Reynolds*, 1974, p. 79) of the turbulence, $k_1 = \Lambda^{-1}$, is the upper limit.

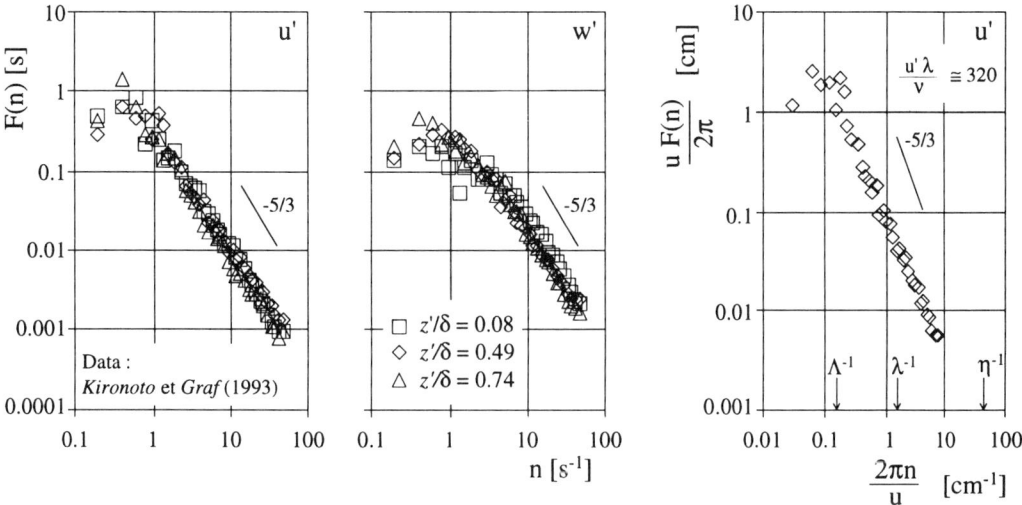

Fig. 2.19 Turbulence spectrum for different levels, z'/δ; in uniform flow.

ii) The turbulence structure of the very smallest eddies, where viscous dissipation dominates, has universality; the lower limit (see *Reynolds, 1974*, p. 99) is given by the Kolmogoroff scale, $k_3 = \eta^{-1} = (\nu^3/\varepsilon)^{1/4}$, where ν is the kinematic viscosity and ε is the dissipated turbulent energy, expressed by :

$$\varepsilon = 15\,\mu\,(\overline{u'^2}/\lambda^2).$$

iii) In the *inertial zone*, if the flow Reynolds number is high, the large eddies disintegrate into smaller eddies and so on (in cascades), and all this by action of the inertia forces. The resulting energy spectrum has universality, thus being independent of the Reynolds number. This part of the spectrum is described by *Kolmogoroff's* hypothesis, which shows by way of a dimensional analysis, that the spectral function, $F(k)$, can be given by a law of the type (see *Reynolds, 1974*, p. 99) :

$$F(k) \propto \varepsilon^{2/3}\,k^{-5/3}$$

In this zone, $k_1 < k \ll k_3$, the turbulence is quasi isotropic and the spectrum is in equilibrium: the *micro-scale* (see *Reynolds, 1974*, p. 79) of the turbulence, $k_2 = \lambda^{-1}$, falls into this zone.

3. UNIFORM FLOW

Flow in a channel is considered as uniform and steady, if the flow depth remains invariable in the flow direction as well as in time. In fluvial hydraulics, uniform flow is taken as the base (reference) for all other considerations, and this despite the fact that truly uniform flow is rarely encountered in reality.

In this chapter, the equations of continuity and of motion will be developed. Subsequently are presented the different relationships for the determination of the coefficients of friction for fixed and mobile channel beds. Knowledge about this coefficient is paramount in all kinds of problems of fluvial hydraulics. The calculation of the discharge for flow over a fixed as well as a mobile bed is elaborated. Elementary knowledge about flow in curves as well as instabilities at the free-water surface will be exposed.

TABLE OF CONTENTS

3.1 HYDRODYNAMIC EQUATIONS
    3.1.1 Notion of Uniformity
    3.1.2 Equation of Continuity
    3.1.3 Equation of Motion

3.2 COEFFICIENT OF FRICTION
    3.2.1 Coefficient of Weisbach-Darcy
    3.2.2 Coefficient of Chézy
    3.2.3 Coefficient of Manning
    3.2.4 Composite Roughness
    3.2.5 Bed Forms
    3.2.6 Coefficient of Friction, mobile Bed

3.3 DISCHARGE CALCULATION, FIXED BED
    3.3.1 Conveyance
    3.3.2 Normal Depth
    3.3.3 Composite Section
    3.3.4 Section of maximum Discharge

3.4 DISCHARGE CALCULATION, MOBILE BED
    3.4.1 Sedimentation Velocity
    3.4.2 Critical Velocity
    3.4.3 Distribution of Shear Stress
    3.4.4 Stable Section

3.5 FLOW IN CURVES
    3.5.1 Super-elevation
    3.5.2 Supercritical Flow
    3.5.3 Head Loss

3.6 INSTABILITY AT SURFACE
    3.6.1 Roll Waves
    3.6.2 Air Entrainment

3.7 EXERCISES
    3.7.1 Problems, solved
    3.7.2 Problems, unsolved

UNIFORM FLOW

3.1 HYDRODYNAMIC EQUATIONS

3.1.1 Notion of Uniformity

1° Flow is considered as uniform and steady (see sect. 1.2.1) if the flow depth, h or D_h, as well as other hydraulic parameters such as the average velocity, the discharge, the roughness and the channel slope, remain invariable in different cross sections of the channel along the axis of flow. The streamlines are rectilinear and parallel and the vertical pressure distribution is hydrostatic. The slope of the bed, S_f, of the water surface, S_w, and of the energy-grade line, S_e, are the same.

2° Truly uniform flow is rather rare in natural, but also in artificial channels. Uniform flow is only possible in very long prismatic channels and this far from the upstream and downstream boundary conditions (see Fig. 3.1).

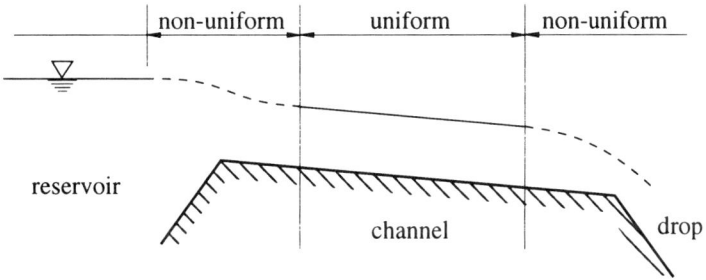

Fig. 3.1 Uniform flow between boundary conditions.

3° Despite the fact that uniform flow occurs rarely, this type of flow is usually taken as the standard (reference) flow for any theoretical and experimental study of other types of flow, but notably for the understanding of the flow resistance.

3.1.2 Equation of Continuity

1° As long as flow is uniform and steady, the cross section of the flow, A, remains the same in direction, x, and in time, t. The equation of continuity (see sect. 2.1) was given as :

$$\frac{\partial (UA)}{\partial x} + \frac{\partial A}{\partial t} = 0 \qquad (2.1)$$

but becomes now :

$$\frac{\partial (UA)}{\partial x} = 0 \qquad (3.1)$$

where Q = UA is the discharge and U is the average velocity.

2° Consequently, the discharge remains constant :

$$Q = Cte \tag{3.2}$$

Between two cross sections (see Fig. 3.2), one has :

$$A_1 U_1 = Q = A_2 U_2 \tag{3.2a}$$

and with $U_1 = U_2$ and $A_1 = A_2$, one writes : $Q = UA$.

3.1.3 Equation of Motion

1° Consider a prismatic channel (see Fig. 3.2). The liquid in motion provokes a friction force at the wetted perimeter :

$$F_F = \tau_o\, P\, dx$$

by an action of the longitudinal component of the gravity force :

$$F_G = \gamma A\, dx\, \sin \alpha = W \sin \alpha$$

In uniform flow, there exists an equilibrium between these forces :

$$\tau_o P\, dx = \gamma A\, dx\, \sin \alpha \tag{3.3}$$

Consequently, one obtains an expression for :

$$\tau_o = \gamma \frac{A}{P} \sin \alpha \tag{3.4}$$

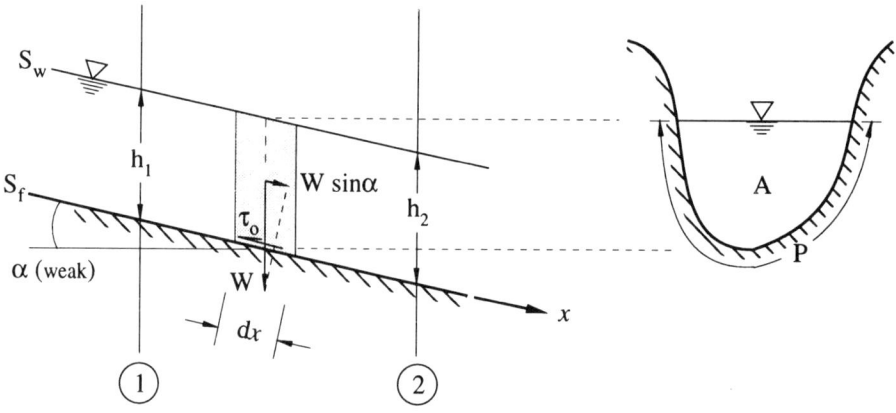

Fig. 3.2 Scheme of uniform flow.

UNIFORM FLOW

The quotient of the wetted cross section, A, and its wetted perimeter, P, defines the hydraulic radius, R_h. The angle, α, is usually very small; thus one may write $\sin \alpha \cong \tg \alpha = S_f$. Above relation, eq. 3.4, now reads :

$$\tau_o = \gamma R_h S_f \tag{3.5}$$

where τ_o is the tension due to the friction forces, called the shear stress, which acts on the wetted surface (wall and bed). Note that eq. 3.5 can be obtained directly from eq. 2.10 or eq. 2.12, by considering the uniformity of the flow.

2° In hydrodynamics, one defines :

$$\tau_o/\rho = u_*^2 \tag{3.6}$$

where u_* is the friction velocity. Thus one can also write :

$$u_* = \sqrt{g R_h S_f} \tag{3.7}$$

Instead of the shear stress, $\tau_o = \rho u_*^2$, one may also use the definition of the *friction coefficient* (see *Graf & Altinakar*, 1991, p. 433) which is given by :

$$f = \frac{\tau_o}{\rho U^2/8} = 8\left(\frac{u_*}{U}\right)^2 \tag{3.8}$$

3° Upon substitution of eq. 3.8 into eq. 3.5, one obtains:

$$(f/8)\rho U^2 = \tau_o = \rho g R_h S_f$$

or, written otherwise :

$$S_f = f \frac{1}{4R_h} \frac{U^2}{2g} \tag{3.9}$$

This relation is known as the equation of *Weisbach-Darcy* (see *Graf & Altinakar*, 1991, sect. FR. 2.1 and sect. PP. 2); it reveals itself as very useful for flow in pipes. The coefficient, f, of friction (head loss) depends on the Reynolds number and the relative roughness, but also on the form of the cross section.

The equation of Weisbach-Darcy can also be written as :

$$U = \sqrt{8g/f} \sqrt{R_h S_f} \tag{3.10}$$

an expression which is frequently given in the form of :

$$U = C \sqrt{R_h S_f} \qquad (3.11)$$

This is called the relationship of *Chézy*, where C is the resistance coefficient of Chézy.

4° In uniform regime (see eq. 3.10 and eq. 3.11), the flow depth, h, which corresponds to the hydraulic radius, R_h, is defined as being the *normal flow depth*, $h \equiv h_n$.

5° Different formulae have been elaborated over the years to render expressions for the friction (resistance) coefficients. Herewith some more common formulae will be presented, namely :

    i)   coefficient of Weisbach-Darcy (see sect. 3.2.1),

    ii)   coefficient of Chézy (see sect. 3.2.2),

    iii)   coefficient of Manning-Strickler (see sect. 3.2.3),

    iv)   coefficient of friction for mobile bed (see sect. 3.2.6).

3.2 COEFFICIENT OF FRICTION

1° It will certainly be useful, to express the friction coefficient, f, for laminar and turbulent flow with the equation of Weisbach-Darcy, eq. 3.10. However, the channel data presently available contain major limitations since channels should be of circular cross section and the roughness should be the standard one.

2° The relation of Chézy, eq. 3.11, is also rather useful as long as the flow in the channel is truly turbulent; this is often the case.

3° These two approaches, eq. 3.10 and eq. 3.11, give satisfactory results, notably for practical problems, if applied correctly and respecting their possible limitations. The ASCE (see *Silberman* et al., 1963) recommended however the use of the equation of Weisbach-Darcy.

The precision which is obtained with these formulae, eq. 3.10 or eq. 3.11, is nevertheless strongly dependent upon the choice of the friction coefficient, f or C.

4° Artificial and particularly natural channels have all types of form of the cross section. No parameter exists which would well take care of the variability in form; the use of the hydraulic radius is often not sufficient.

5° An estimation of the friction coefficient for a fixed or immobile bed is already difficult; but still more difficult will be an estimation for a mobile bed.

UNIFORM FLOW

3.2.1 Coefficient of Weisbach-Darcy

1° In above equations, eqs. 3.9 and 3.10, the definition of friction coefficient, f, is analogous to the one given for circular pipes. For pipes having an industrial roughness, a universal formulation is given (see *Graf & Altinakar*, 1991, sect. PP. 2) by :
   i)   the diagram of Moody-Stanton or
   ii) the relation of Colebrook-White, for turbulent flow.

2° For cross sections, which are geometrically close to circular sections, one may readily use the experiments performed on pipes. However, some modifications are necessary; the hydraulic radius (see *Graf & Altinakar*, 1991, p. 439) should be written as follows :

$$4R_h = 4\frac{A}{P} \qquad (3.12)$$

Thus $4R_h$ becomes the characteristic length, which is to be used in the definition of the Reynolds number, the relative roughness and the equation of Weisbach-Darcy, respectively :

$$Re = \frac{4R_h U}{\nu} \quad ; \quad \frac{k_s}{4R_h} \quad ; \quad S_f = f\frac{1}{4R_h}\frac{U^2}{2g}$$

For the roughness, k_s, in artificial channels, the equivalent roughness, established for industrial pipes, may be taken.

3° The use of the diagram of Moody-Stanton (see *Graf & Altinakar*, 1991, Fig. PP.9) with $Re = 4R_h U/\nu$ and $k_s/4R_h$ gives values for f for laminar and turbulent flow. Subsequently, one obtains the average velocity, U, using eq. 3.10, or the bed slope, $S_f = S_w$, of the channel, using eq. 3.9.

4° Instead of using the diagram of Moody-Stanton, one may also take the semi-empirical *relation of Colebrook-White* (see *Graf & Altinakar*, 1991, p. 436), valid only for turbulent flow, which is written for channels as follows (see eq. 2.65) :

$$\sqrt{\frac{1}{f}} = -2\log\left(\frac{k_s/R_h}{a_f} + \frac{b_f}{Re\sqrt{f}}\right) \qquad (3.13)$$

with $12 < a_f < 15$ and $0 < b_f < 6$, established for different kinds of cross sections, as well as for different types of roughnesses (see *Silberman et al.*, 1963, p. 104).

The equivalent roughnesses, k_s, established for industrial pipes, but considered valid also for artificial channels, are given in Table 3.1. A more complete tabulation is given by *Wallisch* (1990, pp. 235-250).

For channels or watercourses whose bed is made up of a granulate, one generally takes $k_s \cong d_{50}$; d_{50} being the diameter equal to 50% of grains in the granulometric curve.

The importance of the form of the cross section can be somehow taken care of by a factor, which multiplies the hydraulic radius R_h; thus (ϕR_h) and (ϕRe) replace R_h and Re in eq. 3.13. One takes (see *Ghetti*, 1981):

- for a rectangular (B = 2 h) section $\phi = 0.95$
- for a large trapezoidal section $\phi = 0.80$
- for a triangular (equilateral) section $\phi = 1.25$

5° However, it must be pointed out that the results obtained with the diagram of Moody-Stanton or the *relation of Colebrook-White* will only be good approximations.

For channels, being very large and rectangular or very different from circular sections, above methods are less applicable.

Table 3.1 Equivalent roughness for industrial pipes.

Types of wall		Uniform equivalent roughness k_s [mm]
glass, copper, brass		< 0.001
lead		0.025
pipes, steel	new old	0.03 à 0.1 0.4
wrought iron	new old coated	0.25 1.0 à 1.5 0.1
concrete	smooth rough	0.3 à 0.8 < 3.0
wood		1.0 à 2.5
riveted steel		0.9 à 9
stone, worked rough		8 à 15
rock		90 à 600

UNIFORM FLOW 77

6° Natural or artificial channels are usually of large dimensions. Consequently, the Reynolds number, Re = $4R_h U/\nu$, and the roughness, k_s, have also large values. This implies that the turbulent flow is often also a rough one; the value of the friction coefficient, f, remains constant and is no more dependent on the Reynolds number.

This is a justification for using the relation of Chézy, eq. 3.10 or eq. 3.11, where the coefficient of Chézy depends only on the relative roughness, $C = f(k_s/R_h)$, (see eq. 3.13); thus one may write :

$$C = \sqrt{8g} \left(\frac{1}{\sqrt{f}}\right) = \sqrt{8g} \left[2 \cdot \log\left(\frac{a_f}{k_s/R_h}\right)\right] \qquad (3.13a)$$

Subsequently, taking $a_f = 12.7$, one obtains (see also eq. 2.63) :

$$\sqrt{\frac{8}{f}} = 5.6 \log\left(\frac{R_h}{k_s}\right) + 6.25 \qquad (3.13b)$$

For a roughness due to large grains, $R_h/d_{50} \leq 10$, one should take (see *Graf* et al., 1987) :

$$\sqrt{\frac{8}{f}} = 5.75 \log\left(\frac{R_h}{d_{50}}\right) + 3.25 \qquad (3.13c)$$

where $k_s \equiv d_{50}$, with d_{50} as the median grain diameter.

7° In rough channels of large width, $R_h \approx h$, the friction coefficient, f, can be obtained making in-situ measurements of point velocities (see *Graf*, 1966) and assuming a logarithmic distribution (see sect. 2.63) :

$$\frac{u}{u_*} = 5.75 \log\frac{30 z}{k_s}$$

The point velocities, $u_{0.2}$ and $u_{0.8}$, at two elevations, $z' = 0.2 h$ and $z' = 0.8 h$, situated on the same vertical (see Fig. 1.7), are given by :

$$u_{0.8} = 5.75 \, u_* \log (24h/k_s) \qquad\qquad u_{0.2} = 5.75 \, u_* \log (6h/k_s)$$

By elimination of u_* in these two relations and putting $(u_{0.8}/u_{0.2}) = \zeta$, one gets :

$$\log\frac{h}{k_s} = \frac{0.78\zeta - 1.38}{1 - \zeta} \qquad (3.19)$$

The average velocity, U, for a turbulent rough flow (see eq. 2.63) was given by:

$$\frac{U}{u_*} = 5.75 \log \frac{h}{k_s} + 6.25$$

A substitution of eq. 3.19 into this equation – making use of the definition of eq. 3.8 – renders:

$$\frac{U}{u_*} = \frac{1.78 \, (\zeta + 0.95)}{(\zeta - 1)} \tag{3.20}$$

Subsequently one obtains for the coefficient of friction:

$$\sqrt{\frac{1}{f}} = \frac{1.78}{\sqrt{8}} \frac{(\zeta + 0.95)}{(\zeta - 1)} \tag{3.21}$$

and also (see eq. 3.13a):

$$C = \sqrt{8g} \sqrt{\frac{1}{f}} = 1.78 \sqrt{g} \, \frac{(\zeta + 0.95)}{(\zeta - 1)} \tag{3.21a}$$

The coefficients of friction, f or C, are thus obtained in an experimental way for very wide channels, where $h \cong D_h = R_h$, using the hypothesis of a logarithmic velocity distribution.

3.2.2 Coefficient of Chézy

1° For turbulent, rough flow the *formula of Chézy*:

$$U = C \sqrt{R_h S_f} \tag{3.11}$$

can be used. However, it cannot be used for laminar or turbulent smooth flow.

The coefficient of Chézy, $C \, [m^{1/2}/s]$, is a dimensional expression; the numerical values use as unity the meter [m] and the second [s].

Different formulae, being all of empirical nature, have been advanced for the determination of the coefficient of Chézy, C; all of which make use of the hydraulic radius, R_h.

Table 3.2 Coefficients of roughness of Manning, of Strickler and of Kutter.

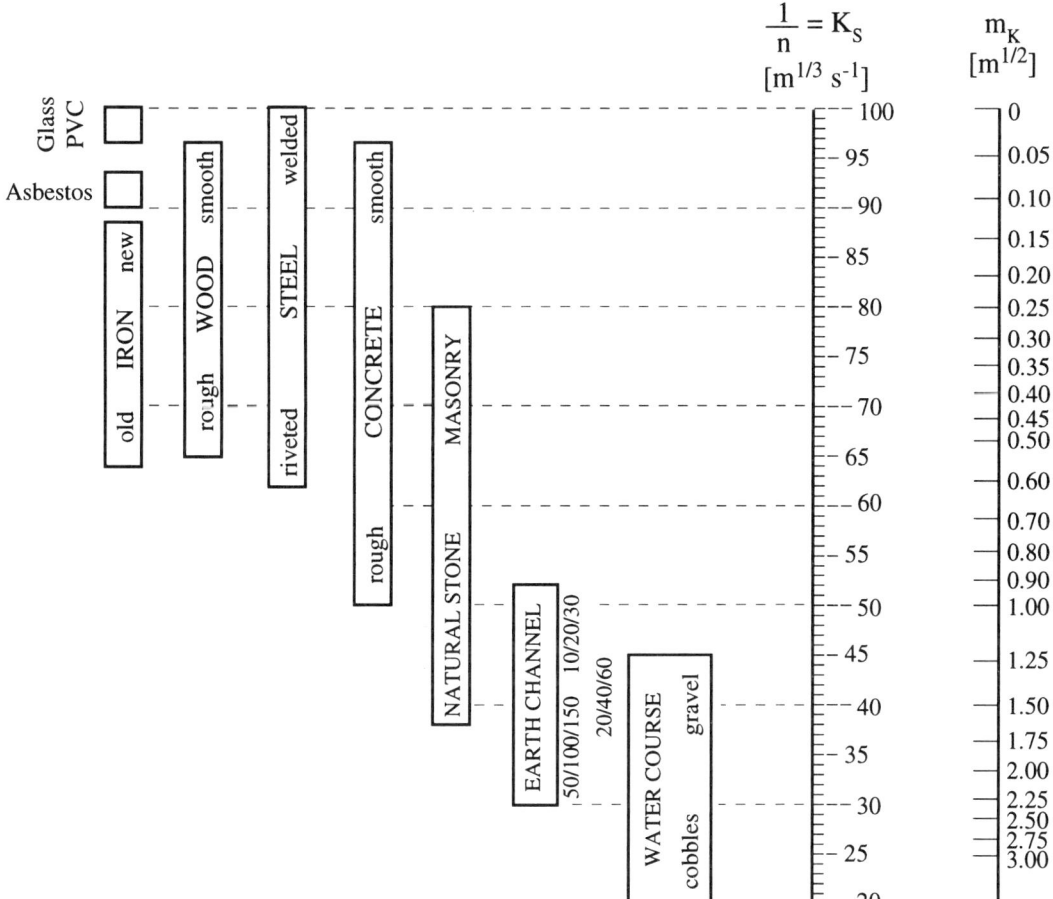

2° The *formula of Bazin* considers C as being a function of the hydraulic radius, R_h [m], and of a coefficient, m_B [$m^{1/2}$], which characterises the roughness of the walls and the bed. Established with data from small artificial channels, this relation reads :

$$C = \frac{87}{1 + (m_B/\sqrt{R_h})} \quad (3.14)$$

The coefficient of Bazin varies from $m_B = 0.06$, for a smooth bed, to $m_B = 1.75$, for a bed made up of stones or covered with vegetation.

3° The *(simplified) formula of Kutter,* established with data from artificial channels as well as from larger rivers, has a similar form, being :

$$C = \frac{100}{1 + (m_K/\sqrt{R_h})} \quad (3.15)$$

where m_K [m$^{1/2}$] is the coefficient of Kutter. Some values for m_K are given in Table 3.2.

4° In the practice, one prefers presently the exponential relations and one uses commonly the *formula of Manning-Strickler* in the form of :

$$U = K_s R_h^{2/3} S_f^{1/2} \qquad (3.16)$$

with

$$C = K_s R_h^{1/6} = \frac{1}{n} R_h^{1/6} \qquad (3.17)$$

Here K_s [m$^{1/3}$s^{-1}] is the coefficient of *Strickler* and n [m$^{-1/3}$s^{1}] is the coefficient of *Manning*. Above relation, eq. 3.16, was elaborated using numerous measurements, performed in both natural and artificial channels. The values of n and K_s are given in Table 3.2. More detailed tables are available (see *Wallisch* 1990, pp. 252-267).

5° There exist other exponential relations, such as :

i) *formula of Forchheimer* : $\quad C = \frac{1}{n} R_h^{1/5}$

ii) *formula of Pavloski* (see *Grishanin*, 1990, p. 45): $\quad C = \frac{1}{n} R_h^q$

for $\quad R_h \le 1$ [m] : $\quad q = 1.5 \sqrt{n}$
$\quad R_h > 1$ [m] : $\quad q = 1.3 \sqrt{n}$

3.2.3 Coefficient of Manning

1° The most popular formula is presently the one of *Manning-Strickler*, often called shortly the *formula of Manning* :

$$U = \frac{1}{n} R_h^{2/3} S_f^{1/2} \qquad (3.16)$$

This is a rather simple relationship, but it must be used only for turbulent, rough flow, thus for flow at large Reynolds numbers. In such a case, the coefficient of Manning, n , stays constant for a given roughness, while the coefficient of Chézy, C , depends (see eq. 3.17) on the relative roughness, $(R_h^{1/6}/n)$.

2° Complete tabulations of the coefficient of Manning, n , have been presented by *Crause* (1951, p. 38), *Chow* (1959, pp. 110-113) and *Graf* (1984, pp. 306-309). Furthermore, *Chow* (1959, pp. 115-123) and *Barnes* (1967) provide photos of different natural and artificial channels as a visual support, to facilitate the choice of the coefficient of Manning in the range of $0.012 < n < 0.15$.

Indicative values are summarised in Table 3.2.

It must be pointed out that the values of the coefficient of Manning are the same both in the metric and in the English system. In the latter case, one has to use the following relation :

$$C = \frac{1.48}{n} R_h^{1/6} \qquad (3.17a)$$

3° For watercourses, where the bed and walls are made up of a non-cohesive granulate, the *formula of Strickler* (see *Strickler*, 1923, pp. 11-15) may be used :

$$K_s = \frac{21.1}{d_{50}^{1/6}} \qquad \text{or} \qquad K_s = \frac{26}{d_{90}^{1/6}} \qquad (3.18)$$

where d_{50} or d_{90} [m] are the diameters, being equal to 50% or 90% of the grains in the granulometric curve.

4° The influence of vegetation on the coefficient of friction is extensively treated by *Chow* (1959, pp. 179-184) and *Wallisch* (1990, p. 229).

3.2.4 Composite Roughness

1° The coefficients of friction, f, n and C , are valid as long as the entire wetted perimeter has the same roughness; thus the wetted section is homogeneous.

2° In sections where the wetted perimeter is *not homogeneous*, the bed and the side walls have different roughnesses (see Fig. 3.3); thus it becomes necessary to compute an equivalent coefficient of friction.

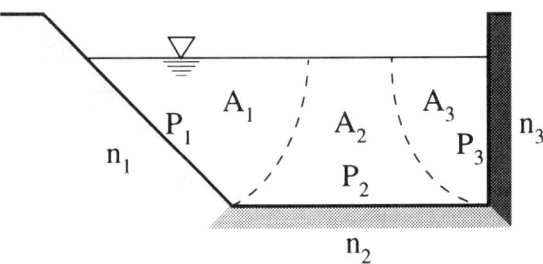

Fig. 3.3 Section of composite roughness.

3° According to *Einstein* (see *Chow*, 1959, p. 136), one divides – in a reasonable way – the wetted surface, A , in N parts, each one having its wetted perimeter, P_1 , P_2 ......P_N , and its coefficient of friction, n_1 , n_2 .....n_N . Furthermore, one assumes that the average velocity of each particular section, A_1 , A_2 .....A_N, is the same and thus also the same as the average velocity of the entire section, $U_1 = U_2 = \ldots\ldots\ldots = U_N \equiv U$.

4° Using, for example, the formula of Manning, eq. 3.16, on writes :

$$U = \frac{1}{n}\left(\frac{A}{P}\right)^{2/3} S_f^{1/2} = \frac{1}{n_1}\left(\frac{A_1}{P_1}\right)^{2/3} S_f^{1/2} = \ldots\ldots\ldots = \frac{1}{n_N}\left(\frac{A_N}{P_N}\right)^{2/3} S_f^{1/2}$$

If one assumes that $A^{2/3} = \sum_1^N A_N^{2/3}$, the equivalent coefficient of friction for a composite roughness can be computed, being :

$$n = \left[\frac{\sum_1^N \left(P_N \, n_N^{3/2}\right)}{P}\right]^{2/3} \tag{3.22}$$

3.2.5 Bed Forms

1° Natural, but also artificial channels may have a *mobile bed*, defined as being a channel bed composed of solid particles (non-cohesive granulate, alluviums), which displace themselves by the action of the flow. The bed may become covered with *bed forms*, commonly called *dunes* (see Fig. 3.4). These solid particles are characterised by the density, ρ_s, the median diameter, $d \equiv d_{50}$, and their granulometric distribution.

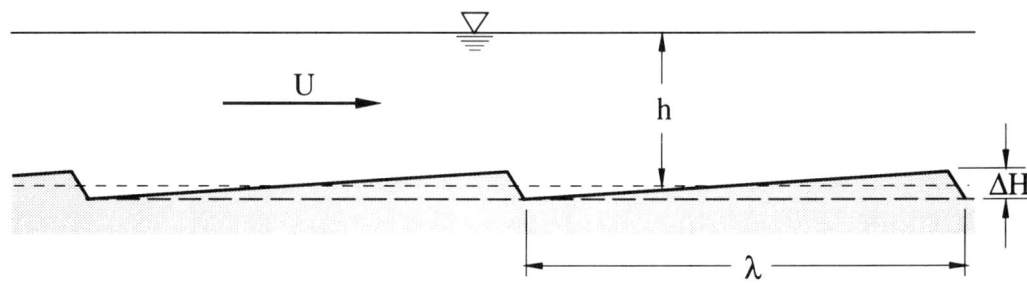

Fig. 3.4 Scheme of a channel bed with a series of dunes.

2° A mobile bed presents successively various aspects, which correspond to the different types of bed deformations. These are usually classified into three regimes, by using the Froude number, Fr (see Fig. 3.5) :

  i) Fr < 1 : The bed remains rather flat, and this till the velocity becomes critical (see sect. 3.4.2) and the sediment (solid) transport begins. Consequently, *mini-dunes* or *ripples* appear, followed by the *dunes* of growing dune length, λ.

UNIFORM FLOW   83

Regime	Transport of sediments	Bed form
Fr < 1	no	flat
	yes	mini-dune
	yes	dune
Fr ≅ 1	yes	flat
Fr > 1	yes	anti-dune

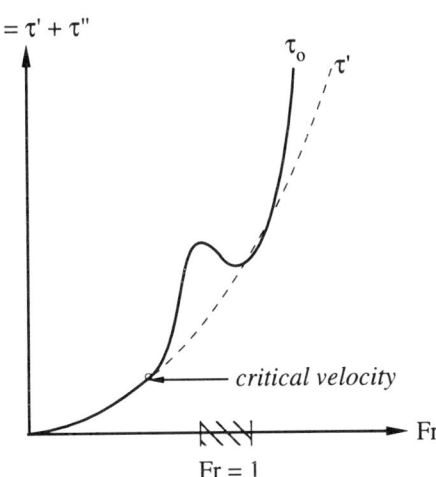

Fig. 3.5 Regime of flow over a mobile bed.

ii) Fr ≈ 1 : As the flow velocity increases, these dunes, already rather long, are washed out and tend to disappear. In this state of *transition*, the bed is once more a flat one.

iii) Fr > 1 : With a further increase of the flow velocity, another kind of dunes appear, commonly called *anti-dunes*, which, contrary to the dunes, travel usually into the upstream direction. The water surface becomes wavy and the sediment transport is very strong.

3° The geometry of a dune (idealised, since sometimes they are not well apparent) is approximated by a triangular form of length, λ, and of height, ΔH (see Fig. 3.4).

Indicatives relations (see *Graf*, 1984, p. 283), made dimensionless by the flow depth, are given as :

$$\frac{\Delta H}{h} < \frac{1}{6} \quad ; \quad \frac{\lambda}{h} \approx 5 \qquad (3.23)$$

4° The presence of bed forms will cause an increase in the flow resistance. For the calculation of the total shear stress on the bed, τ_o, one assumes (see *Graf*, 1984, p. 303) that the contribution of the roughness due to the particles, τ', and the one due to the bed forms, τ'', is additive, namely :

$$\tau_o = \tau' + \tau'' \qquad (3.24)$$

or (see eq. 3.5) :

$$\gamma R_h S_f = \gamma (R_h' + R_h'') S_f$$

where R_h' and R_h'' are the hydraulic radius due to the particle roughness and to the bed forms, respectively.

Using the definition of the friction velocity and of the coefficient of friction, eq. 3.6 and eq. 3.8, one writes :

$$u_*^2 = (u_*')^2 + (u_*'')^2$$
$$f = f' + f''$$
(3.25)

but also :

$$n = n' + n'' \quad ; \quad C = C' + C''$$
(3.26)

5° The total shear stress, τ_o, (see eq. 3.24) varies as a function of the Froude number, Fr . This variation is schematically shown in Fig. 3.5.

3.2.6 Coefficient of Friction, mobile Bed

1° A quantification of friction coefficient for flow over a mobile bed has, up to now, not been very successful over a large range of flow parameters.

2° There exist methods where one determines directly the *entire* coefficient of friction, f or n.

There exist other methods, where one calculates the coefficient of friction due to the grain roughness, f' or n', using the formulae presented above (see sects. 3.2.1 to 3.2.3). Subsequently one determines the coefficient of friction due to the bed forms, f'' or n", using other types of formulae.

3° A selection from the different existing formulae for a direct calculation is given in the following :

  i) The determination of the entire coefficient of friction can be done using an exponential relation of the Chézy type (see eq. 3.11) :

$$U = K_T R_h^x S_f^y$$
(3.27)

Sugio (1972) studied extensively watercourses, having $0.1 < d_{50}$ [mm] < 130, and artificial channels, having $0.2 < d_{50}$ [mm] < 7.0 ; proposed was :

$$U = K_T R_h^{0.54} S_f^{0.27}$$
(3.27a)

UNIFORM FLOW

It should be noted that the exponent of $y = 0.27$ is very different from the one used in the relation of Chézy or of Manning, where $y = 0.50$ for channels with fixed beds.

The values for K_T are : $K_1 = 54$ for mini-dunes
$K_2 = 80$ for dunes
$K_3 = 110$ for the upper regime
$K_4 = 43$ for rivers with meanders.

This relation, eq. 3.27a, is simple to use and compares itself favourably with other formulae advanced for channels in *regime*, namely in equilibrium (see *Graf*, 1984, chap. 10), which are also presented by *Sugio* (1972, p. 24).

ii) The following relationship, presented by *Grishanin* (1990, p. 59), expresses the coefficient of Chézy, C , used in eq. 3.11, as being :

$$C = 5.25 \left(\frac{Ug}{\sqrt[3]{gv}}\right)^{1/2} \left(\frac{D_h}{B}\right)^{1/6} \qquad (3.28)$$

It was established for Russian rivers, having $0.1 < d_{50}\,[\text{mm}] < 0.44$ and $3 \times 10^{-6} < S_f < 2.2 \times 10^{-4}$.

iii) Yet another relationship, presented by *Grishanin* (1990, p. 69) and obtained from different (35) Russian rivers, was given as :

$$U = \frac{1}{M_G^2} \left(\frac{g}{B}\right)^{1/2} D_h \qquad (3.29)$$

where M_G is a local non-dimensional invariant, $M_G = 0.91 \pm 0.12$, for channel beds of sand.

iv) Using a large series of artificial (laboratory) channels as well as natural watercourses, having grain diameters of $0.11 < d_{50}\,[\text{mm}] < 1.35$ and bed slopes of $3 \times 10^{-6} < S_f < 3.7 \times 10^{-2}$, the following relation (see eq. 3.24) was proposed by *Brownlie* (1983, p. 975) :

$$\tau_* = \frac{\tau_0}{d_{50}(\gamma_s-\gamma)} = w\,(q_*S_f)^x\,S_f^y\,\sigma^z\left(\frac{\rho}{\rho_s-\rho}\right) \qquad (3.30)$$

where $q_* = q/\sqrt{gd_{50}^3}$; q is the unit discharge, σ the standard deviation of the grains in the granulometric distribution and $\gamma_s = \rho_s g$ is the specific weight of the granulate. The coefficients were obtained by a statistical analysis, being :

- for channels with mobile beds, having mini-dunes and dunes :

  $w = 0.37$, $x = 0.65$, $y = 0.09$, $z = 0.11$

- for channels with mobile beds, being flat (see Fig. 3.5) or having anti-dunes :

  $w = 0.28$, $x = 0.62$, $y = 0.09$, $z = 0.08$

4° Different empirical relationships have been elaborated for the calculation of the friction velocity and of the coefficient of friction, u_*'' and f'', being due to bed forms. Here are given two relations :

i) The relation proposed by *Einstein-Barbarossa* is given usually in graphical form (see Fig. 3.6), where a large spread is evident. Used were many observations from American rivers, having $0.19 < d_{35}$ [mm] < 4.3 and $1.49 \times 10^{-4} < S_w < 1.72 \times 10^{-3}$. This relationship is expressed (see *Graf*, 1984, p. 310) by :

$$\frac{U}{u_*''} = f\left(\frac{\rho_s - \rho}{\rho} \frac{d_{35}}{R_h' S_f}\right) = f(\psi') \qquad (3.31)$$

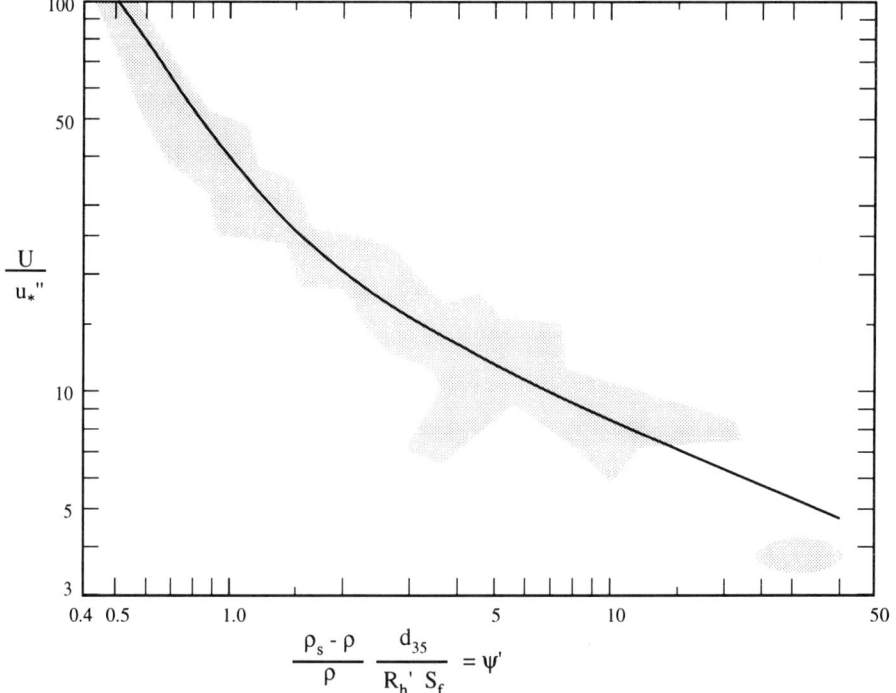

Fig. 3.6 Friction velocity, u_*'', due to bed forms for mobile bed ; after Einstein-Barbarossa.

ii) The relationship proposed by *Alam-Kennedy* is given as :

$$f'' = f\left(\frac{R_h}{d_{50}}, \frac{U}{\sqrt{gd_{50}}}\right) \tag{3.32}$$

Also this one is usually (see *Yalin*, 1972, p. 280) given in a graphical form (see Fig. 3.7), where a large spread (not shown) is evident. A great many data from artificial (laboratory) channels, having $0.04 < d_{50}[\text{mm}] < 0.54$, and American rivers, having $0.08 < d_{50}[\text{mm}] < 0.45$, have been used.

In this figure, Fig. 3.7, the relation of Einstein-Barbarossa corresponds to the region where the lines of the values of $U/\sqrt{g\,d_{50}}$ stay reasonably horizontal, namely for $R_h/d_{50} > 3 \times 10^3$.

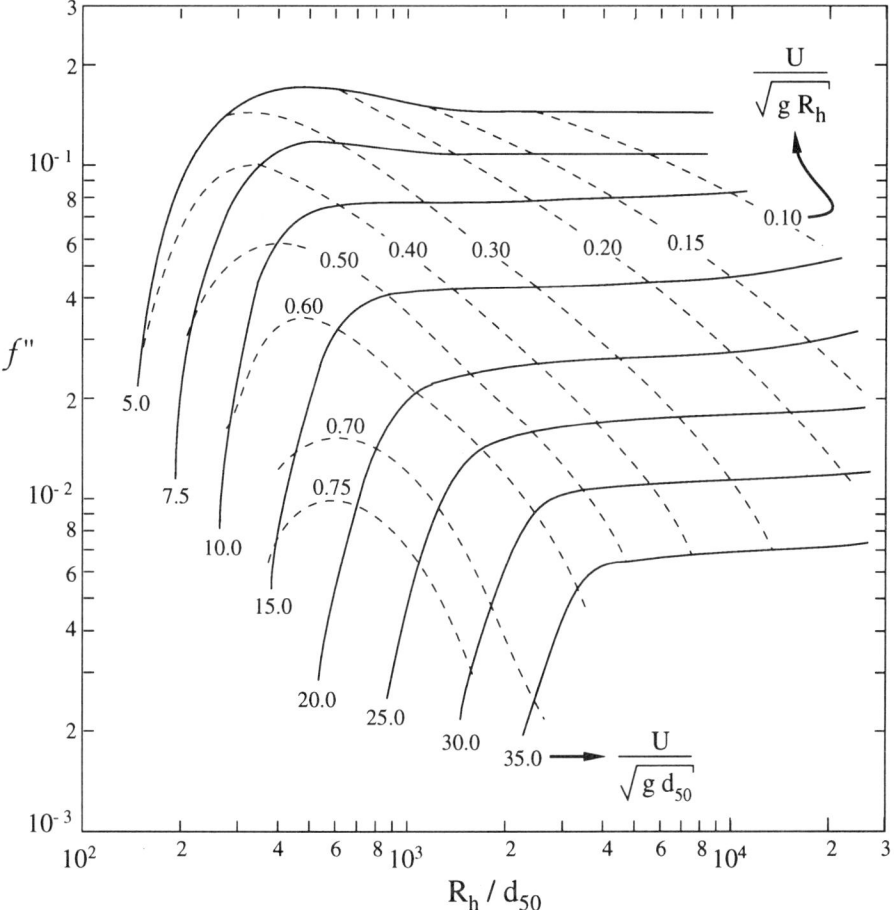

Fig. 3.7 Coefficient of friction, f'', due to bed forms for mobile bed; after Alam-Kennedy.

iii) Some more relations have been presented in *Graf* (1984, pp. 303-320) and *Raudkivi* (1976, chap. 6).

3.3 DISCHARGE CALCULATION, FIXED BED

1° The study of uniform flow in a watercourse is a common task for the hydraulic engineer.

2° Determination of the discharge, Q, in a channel with a fixed bed requires the knowledge of the channel geometry, of the roughness coefficient and of the bed slope.

3° Assumed will be that the walls (bed and side walls) of the channel are fixed or immobile, thus not subject to erosion.

3.3.1 Conveyance

1° The discharge, Q, at uniform flow, given by eq. 3.2a, using the corresponding velocity, given by eq. 3.16, can be expressed as:

$$Q = UA = \frac{1}{n} R_h^{2/3} S_f^{1/2} A \qquad (3.33)$$

2° The values of the wetted section, A, and of the hydraulic radius, R_h, are determined and given by the flow depth, h. Furthermore, the nature of the wall roughness, n, is taken to be known. The following expression can be formed:

$$K(h) = \frac{1}{n} R_h^{2/3} A \qquad (3.34)$$

known as the *conveyance* of the channel (see *Bakhmeteff*, 1932, p. 13), being only a function of the flow depth, $h \equiv h_n$. This depth is known as the *normal depth*, h_n, for the given discharge, Q. Thus, above expression, eq. 3.33, yields:

$$Q = K(h) \sqrt{S_f} \qquad (3.35)$$

or

$$Q/\sqrt{S_f} = f(h) \qquad (3.35a)$$

For a given form (shape) of the section, this relation can be obtained and plotted point by point (see Fig. 3.8). One can readily calculate the conveyance for geometrically simple sections; for complex ones, a graphical solution is necessary.

The normal depth, h_n, increases with the discharge, Q. For identical channels, but having different slopes, S_f, the normal depth increases if the bed slope decreases.

3° The conveyance, K, characterises the channel; it represents a measure of the capacity of water transport through the cross section.

UNIFORM FLOW 89

The curve of the normal depths (see Fig. 3.8) will be found to be rather useful in solving different kinds of problems : if two of the three parameters, h_n, Q and S_f, (see eq. 3.35a) are known, the third one can be found; à priori, the roughness of the walls is taken to be known.

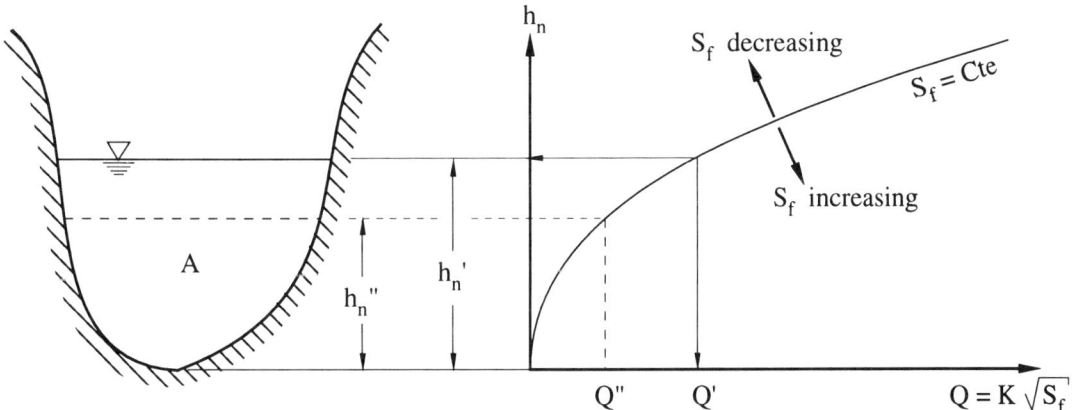

Fig. 3.8 Curve of conveyance or of normal depth.

3.3.2 Normal Depth

1° The normal depth, h_n, (see eq. 3.36) is the flow depth at uniform flow of discharge, Q, at a given bed slope, S_f. (All geometric elements of the cross section, which correspond to the normal depth, h_n, are known as normal elements, such as : R_{h_n}, A_n or P_n.)

2° The normal depth of a channel of a given geometry is calculated using the relation for the discharge :

$$Q = UA = \frac{1}{n} R_h^{2/3} S_f^{1/2} A \qquad (3.33)$$

This relation shows that uniform flow is only possible in a channel whose bed slope is descending, $S_f > 0$. In a horizontal channel, $S_f = 0$, the normal depth would be infinite.

3° For a natural watercourse and for rectangular channels whose width, B, is very large (see Fig. 3.9), one takes $R_h \approx h$ as the hydraulic radius. The relation of the discharge, eq. 3.33, can now be written as :

$$Q = UA = (C h^{1/2} S_f^{1/2}) (h B) \qquad (3.33a)$$

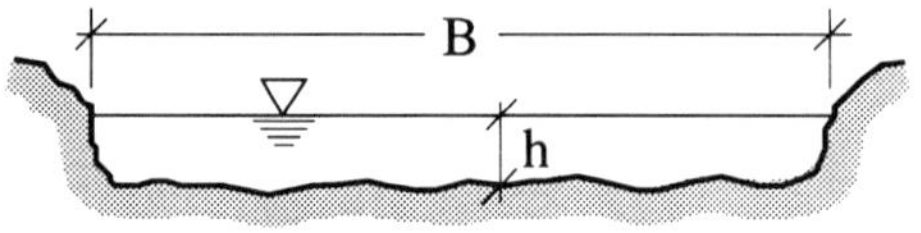

$$h = D_h$$
$$B \gg h$$
$$R_h = \frac{B\,h}{B + 2h} \approx \frac{B\,h}{B} \approx h$$

Fig. 3.9 Section of a channel, having a large width.

For the normal or uniform depth, $h \equiv h_n$, one obtains:

$$h_n = \left(\frac{q^2}{C^2 S_f}\right)^{1/3} \qquad \text{where } q = Q/B \qquad (3.36)$$

3.3.3 Composite Section

1° A cross section of a channel can be composed of different subsections (see Fig. 3.10), of which each one can have a different roughness and a different bed slope.

This is frequently the case during floods, when the flow leaves the channel and enters into the overflow section of the channel.

2° Such a case can be approximately treated by applying the formula of discharge for each subsection:

$$Q = Q_c + Q_o = \frac{1}{n_c} A_c R_{h_c}^{2/3} \left(\frac{\Delta h}{L_c}\right)^{1/2} + \frac{1}{n_o} A_o R_{h_o}^{2/3} \left(\frac{\Delta h}{L_o}\right)^{1/2} \qquad (3.37)$$

Note that the wetted perimeters, P_c and P_o, should be calculated for the lines of contact between water and bed.

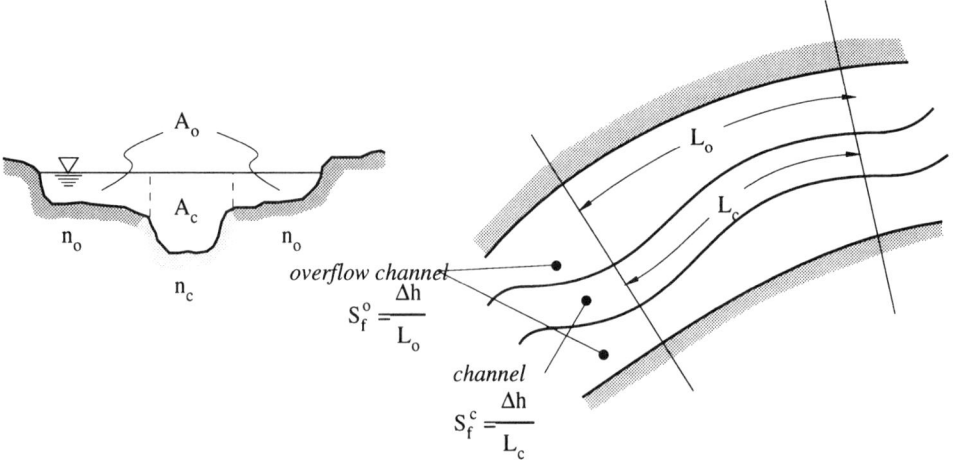

Fig. 3.10 Composite section.

UNIFORM FLOW

3.3.4 Section of maximum Discharge

1° The construction of a channel with a given slope, S_f, and a given roughness, n, which should convey a certain discharge, Q, will be less expensive if the cross section, A, is the smallest possible.

2° Take the formula of discharge :

$$Q = UA = \frac{1}{n} R_h^{2/3} S_f^{1/2} A \qquad (3.33)$$

where for $(S_f^{1/2}/n) = $ Cte one writes :

$$Q = \text{Cte} \, (A^{5/3} P^{-2/3})$$

For a wetted cross section, A, being constant, the above expressions show that the discharge will be maximal, $Q \Rightarrow Q_{max}$, if the hydraulic radius is maximal, $R_h \Rightarrow R_{hmax}$; thus if the wetted perimeter is minimal, $P \Rightarrow P_{min}$.

3° Amongst all geometrical forms possible, the cross section of a *semi-circular* form will give a P_{min} for a given constant cross section, A. This is given (see Fig. 3.11) by :

$$A = \frac{\pi r^2}{2} \quad , \qquad P = \pi r \quad , \qquad R_h = \frac{r}{2} = \frac{h}{2}$$

The semi-circular form can however be only realised (constructed) with artificial channels, made of metal, concrete or wood.

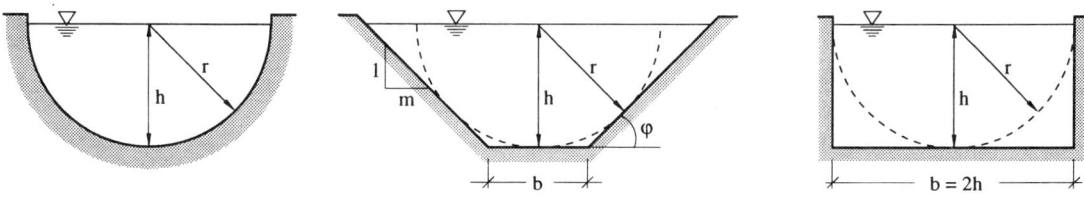

Fig. 3.11 Sections of maximum discharge.

4° For channels in an alluvium, one should also take into account the angle of repose, φ, as well as various constraints due to construction. Consequently, a *trapezoidal* form (see *Crausse*, 1951, p. 51) may be the most reasonable one (see Fig. 3.11) ; where one defines the wetted section, A, and the perimeter, P, such as (see Table 1.1) :

$$A = h\,(b + mh) \quad , \qquad P = b + 2h\sqrt{1 + m^2} \qquad \text{where } m = \text{ctg } \varphi.$$

Subsequently, one takes dA as being zero, since the section, A, remains constant :

$$dA = h\,db + (b + 2\,mh)\,dh = 0$$

If one puts the wetted perimeter, P, as being minimal, this yields :

$$dP = db + 2\sqrt{1 + m^2}\,dh = 0$$

By elimination of db and dh in above equations, one gets :

$$b = 2h\,(\sqrt{1 + m^2} - m)$$

This value, b, can be put in above relations for A and for P, and one obtains an expression for the hydraulic radius, or :

$$R_h = h/2$$

which remains independent from the angle of repose, φ.

5° It should be remarked that for m = 0 the trapeze becomes a rectangle (see Fig. 3.11) such as :

$$b = 2h$$
$$R_h = h/2$$

For a *rectangular* channel, where b ≡ B, the ratio width/depth must be (B/h) = 2.

3.4 DISCHARGE CALCULATION, MOBILE BED

1° Artificial and natural channels, whose flow moves in an alluvium, composed of a (non-cohesive) granulate, are channels of *mobile bed*. The discharge will be calculated using the coefficient of roughness for a mobile bed (see sect. 3.2.6).

2° In such channels, the velocity (in the vicinity of the bed) should :

   i)   not be superior to a certain critical value, otherwise there is a risk of erosion of the solid particles on the bed : this is the permissible maximum velocity or *velocity of erosion*, usually also called the *critical velocity;*

   ii) not be inferior to a certain critical value, otherwise there is a risk of deposition or sedimentation of the solid particles which are possibly suspended in the flow : this is the permissible minimum velocity or *velocity of sedimentation*.

UNIFORM FLOW

3° The flow velocity, U, to be selected for a 'good functioning' of the channel, must lie between the velocity of erosion, $U_E \equiv U_{cr}$, and the one of sedimentation, U_D :

$U_D < U < U_E$.

4° It is evident in Fig. 3.12, that the two velocities, U_D and U_E, will have distinctly different values.

3.4.1 Sedimentation Velocity

1° The allowable minimum velocity or velocity of sedimentation, U_D, is the minimum velocity which is necessary to transport the flow containing solid particles in suspension.

2° Recommended (see *Chow*, 1959, p. 158 and *Crausse*, 1951, p. 16) is to take the following approximate values :

$0.25 < U_D \text{ [m/s]} < 0.9$

depending on fine or very coarse material.

3° The diagram (see Fig. 3.12) which was established by *Hjulstrom* (see *Graf*, 1984, p. 88) delimits the zone of sedimentation as a function of the diameter of the (monodispersed) granulate.

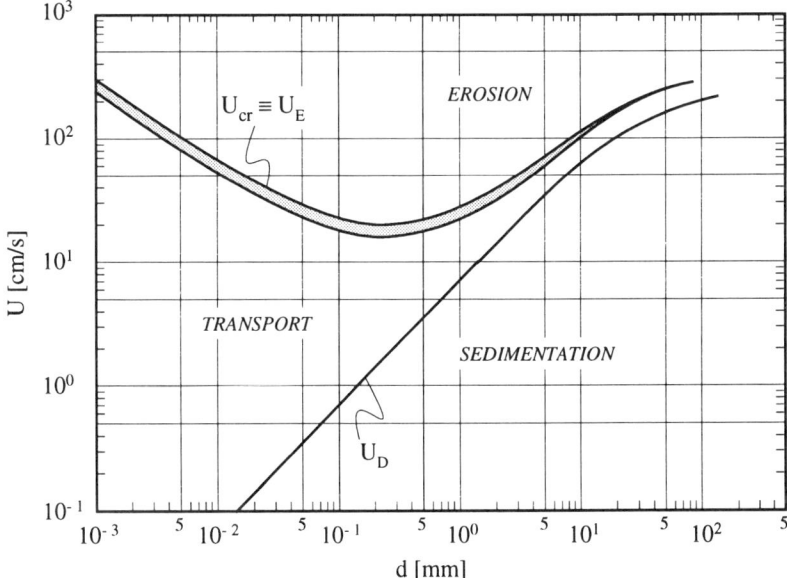

Fig. 3.12 Velocity of sedimentation and of erosion, U_D et U_{cr}, for a uniform granulate, after Hjulstrom.

4° In an experimental study with granulates of $0.49 < d_{50}\,[\mathrm{mm}] < 3.02$, *Graf* et *Pazis* (1977) expressed the critical values by the shear stress, τ_o. It was found that

$$\tau_{o_D} < \tau_{o_E}$$

The difference was however shown to be negligible for this range of granulates; one may thus readily take $\tau_{o_D} \cong \tau_{o_E}$.

3.4.2 Critical Velocity

1° There will be erosion of the bed (and the walls) when one exceeds a certain critical value, expressed with :

  i) the average critical velocity, U_{cr}, or the critical velocity, $u_{b_{cr}}$, at or close to the bed,

  ii) the critical shear stress, $\tau_{o_{cr}}$.

2° From an hydraulic view point, it is more reasonable to use the shear stress, τ_o, as a criterion of erosion.

The shear stress was earlier defined as :

$$\tau_o = \gamma\, R_h\, S_f \qquad (3.5)$$

and the average velocity as :

$$U = C\sqrt{R_h\, S_f} \qquad (3.11)$$

This gives the following ratio, showing the relation between the shear stress, τ_o, and the velocity, U, or :

$$\frac{U}{\sqrt{\tau_o/\rho}} = \frac{C}{\sqrt{g}}$$

In fluvial hydraulics, one uses rather often (see *Graf*, 1984, p. 91) a dimensionless form of the shear stress, τ_*, or :

$$\frac{\tau_o}{(\gamma_s-\gamma)d} \equiv \tau_* = \frac{\gamma\, R_h\, S_f}{(\gamma_s-\gamma)d} \qquad (3.38)$$

where d is the diameter of the granulate (to be specified); γ_s and γ are the specific weight of the granulate and of water respectively. With this relation, one compares the flow parameters, R_h and S_f, with the granulometric parameters, d and (s_s-1).

3° Amongst the different formulae, which one finds in the literature (see *Graf*, 1984, chap. 6), only three will be presented herewith, namely the ones proposed by *Hjulstrom*, by *Neill* and by *Shields*.

4° In an analysis of available data from (monodispersed) uniform granulates, *Hjulstrom* used the average flow velocity, U, instead of the velocity close to the bed, u_b, by assuming that $u_b = 0.4U$. On Fig. 3.12, one can see the limiting zone, where erosion is encountered. This diagram of :

$$U_{cr} = f(d)$$

shows that fine sand ($d \cong 0.1$ mm) is rather easily eroded; the strong resistance to erosion for silt ($d \cong 0.01$ mm) is attributed to the cohesion between the particles.

5° For erosion of a bed composed of a uniform granulate of large diameter, *Neill* proposes the following relation :

$$\frac{\rho U_{cr}^2}{gd(\rho_s - \rho)} = 2.5 \left(\frac{d}{D_h}\right)^{-0.2} \qquad (3.39)$$

being valid for $0.01 < (d/D_h) < 1.0$.

6° Relying on some concepts of the hydrodynamics, *Shields* developed a relation between the dimensionless shear stress, τ_* (see eq. 3.38), and the friction/particle Reynolds number, $Re_* = u_* d/\nu$, such as :

$$\tau_* \equiv \frac{\tau_o}{(\gamma_s - \gamma) d} = f\left(\frac{u_* d}{\nu}\right) \qquad (3.40)$$

where $u_* = \sqrt{\tau_o/\rho}$. *Shields* has determined the form of this relation, using experimental data. An average curve (see *Graf*, 1984, p. 96), reasonably well defined (despite an important scattering), characterises the begin of erosion, expressed by τ_{*cr}. For the particle diameter, one takes usually $d \equiv d_{50}$. It is to be seen (see Fig. 3.13) that these critical values fall roughly in the range of :

$$0.03 < \tau_{*cr} < 0.06$$

The determination of τ_{*cr} is done using the above relation, eq. 3.40, by successive approximations. It must be underlined that the criterion of *Shields* is of great importance for the hydraulic engineer.

7° Since a direct use of the relation of *Shields*, eq. 3.40, is not a simple one, *Yalin* (1972, p. 82) has proposed an interesting combination of terms, such as :

$$\frac{Re_*^2}{\tau_*} = \frac{d^3 g}{v^2} \frac{(\rho_s - \rho)}{\rho}$$

Rather than using the Reynolds number, Re_*, it is now proposed to use a dimensionless diameter of the granulate, given by :

$$d_* = d \left(\frac{\rho_s - \rho}{\rho} \frac{g}{v^2} \right)^{1/3}$$

Consequently, above relation, eq. 3.40, can be expressed as :

$$\tau_* = f(d_*) \tag{3.40a}$$

which is given with Fig. 3.13; one usually takes $d = d_{50}$.

If the properties of the fluid, ρ and v, and of the granulate, ρ_s and d, are known, one can readily determine the corresponding value of τ_{*cr} and subsequently of $\tau_{o_{cr}}$.

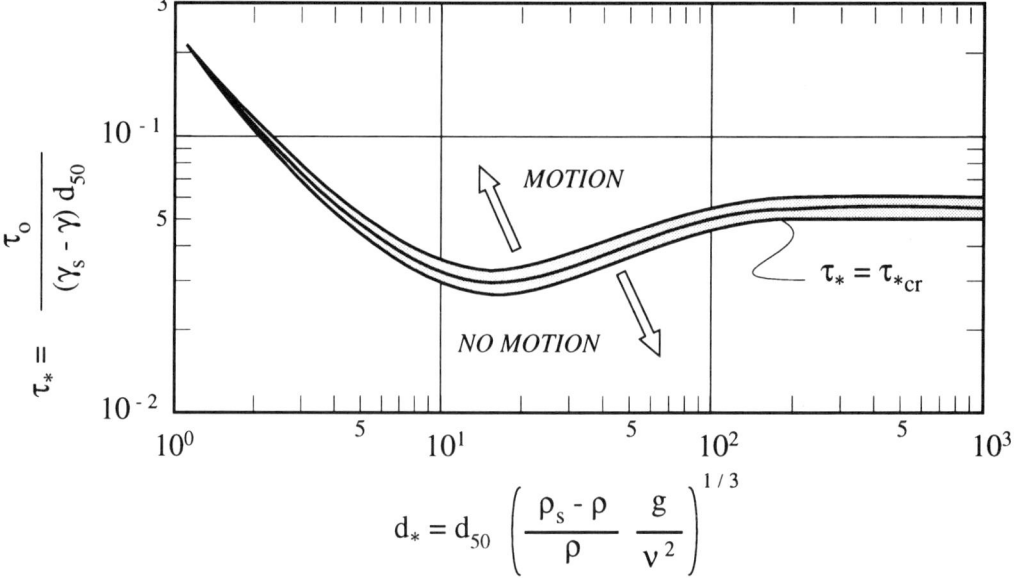

Fig. 3.13 Dimensionless shear stress, τ_*, as a function of the dimensionless diameter, d_*, after Shields-Yalin.

8° For cohesive material, the determination of the critical values, U_{cr} or τ_{*cr}, represents a difficult task; the specialised literature (see *Graf*, 1984, chap. 12, and *Raudkivi*, 1976, chap. 9) should be consulted.

3.4.3 Distribution of Shear Stress

1° The shear stress, τ_o, is given by :

$$\tau_o = \gamma S \frac{1}{P} S_f = \gamma R_h S_f \tag{3.5}$$

For a channel of large width (see Fig. 3.9), when $R_h \equiv h$, one writes :

$$\tau_o = \gamma h S_f \tag{3.5a}$$

2° However, it must be remarked that the shear stress, τ_o, is distributed over the wetted perimeter, P. A typical distribution for a trapezoidal channel (see *Chow*, 1959, p. 169) is given with Fig. 3.14.

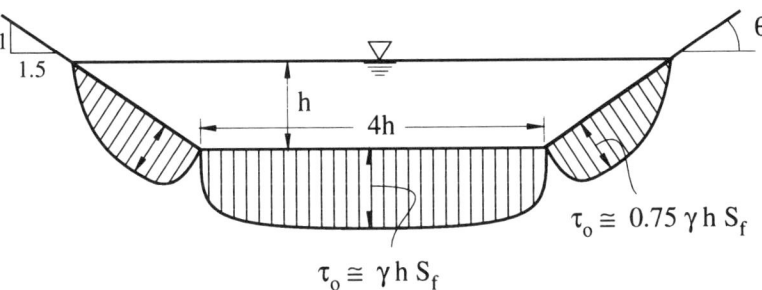

Fig. 3.14 Distribution of the shear stress in a trapezoidal channel.

3° An expression for the shear stress on the channel side walls, $(\tau_{o_{cr}})^w$, was proposed by *Forchheimer* and subsequently by *Lane* (see *Graf*, 1984, p. 116), being of the following form :

$$(\tau_{o_{cr}})^w = \tau_{o_{cr}} \left[\cos\theta \, (1 - tg^2\theta / tg^2\varphi)^{1/2} \right] \tag{3.41}$$

$\tau_{o_{cr}}$ is the critical shear stress on the bed – given for example with Fig. 3.13 –, θ is the inclination of the side wall(s), and φ is the angle of repose. The latter depends on the granulometry and on the cohesion (see *Graf*, 1984, p. 115); it varies such as $20° < \varphi < 40°$. Evidently : $(\tau_{o_{cr}})^w < \tau_{o_{cr}}$, and for stable side walls : $\theta < \varphi$.

3.4.4 Stable Section

1° A *stable* cross section of a channel with a mobile bed, thus erodible, is a section where there is no erosion over the entire wetted perimeter, P.

2° An *ideal stable* cross section – with a maximal discharge and a minimal wetted perimeter – can be calculated with a method advanced by *Glover* et *Lane* (see *Graf*, 1984, p.119). The form of such a section (see Fig. 3.15) can be determined as follows :

   i) Assumed will be that the angle of the side wall at the water surface is identical to the angle of repose, $\theta_s \equiv \varphi$.

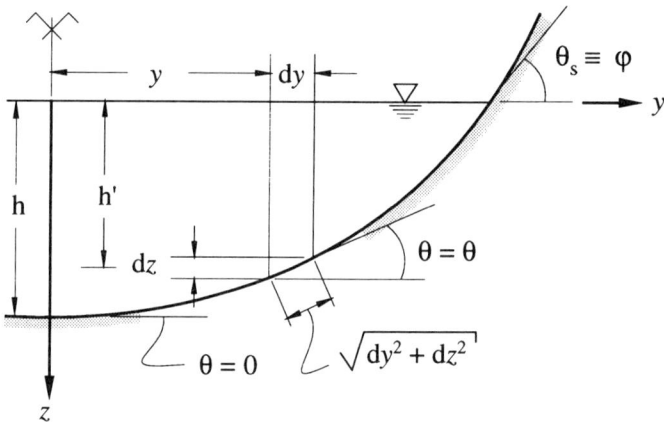

Fig. 3.15 Ideal stable section.

   ii) The shear stress on an element on the bed situated at the side wall is :

   $$(\tau_o^w) = \gamma h' S_f (dy / \sqrt{dy^2 + dz^2}) = \gamma h' S_f \cos\theta$$

   iii) Subsequently, one assumes that this shear stress, (τ_o^w), be critical, $(\tau_{o_{cr}})^w$, over the entire wetted perimeter ; using now the expression of eq. 3.41, one writes :

   $$\gamma h' S_f \cos\theta \Leftarrow (\tau_o^w) \equiv (\tau_{o_{cr}})^w \Rightarrow \gamma h S_f \left[\cos\theta(1-tg^2\theta/tg^2\varphi)^{1/2}\right]$$

   where h is the maximum water depth situated at $y = 0$.

   iv) After mathematical manipulations and taking $(dz/dy) = tg\theta$, one obtains :

   $$\left(\frac{dz}{dy}\right)^2 + \left(\frac{h'}{h}\right)^2 tg^2\varphi - tg^2\varphi = 0.$$

v) The solution of this differential equation is :

$$h' = h \cos\left(\frac{tg\varphi}{h} y\right) \qquad (3.42)$$

which gives the geometry of the ideal stable cross section, being *sinusoidal*.

vi) The other hydraulic parameters of such a section are deduced as being :

$$A = 2h^2 / tg\varphi$$

$$B = \pi h / tg\varphi \qquad (3.42a)$$

$$U = \frac{1}{n} S_f^{1/2} (h \cos\varphi / E)^{2/3}$$

with $h = \tau_o / \gamma S_f$ and $E(\sin\varphi)$ being an elliptic integral, approximated by

$$E \approx (\pi/2)(1 - \frac{1}{4}\sin^2\varphi).$$

vii) The discharge, Q_i , which *can* be conveyed through this *ideal stable* section, is evidently given by $Q_i = UA$.

3° If the discharge, Q , which *must* be conveyed through such a section is different from the ideal discharge , Q_i , thus $Q \neq Q_i$, a corrective calculation must be done :

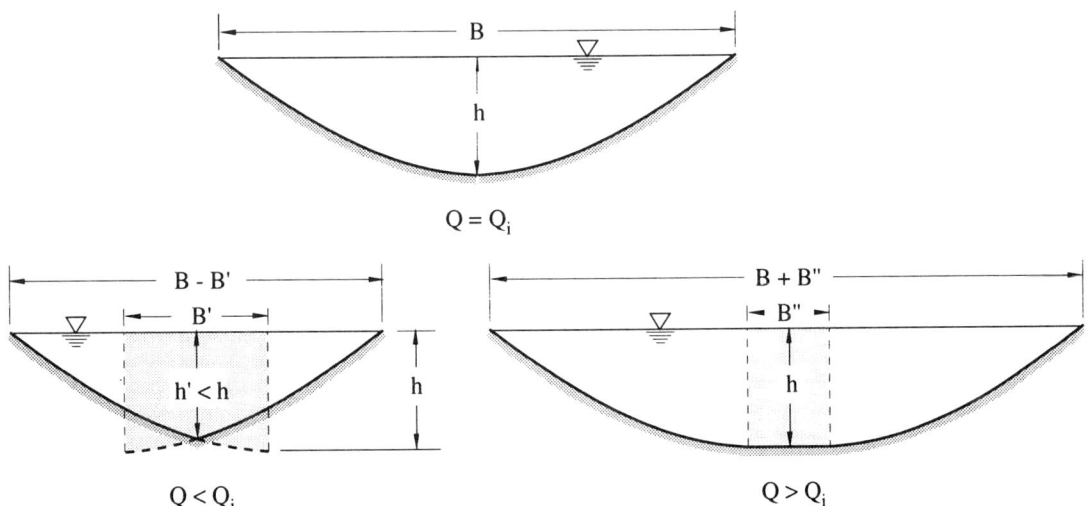

Fig. 3.16 Ideal stable section for different widths.

i) For $Q < Q_i$, the width, B, must be reduced by B', to be computed with:

$$B' = B(1 - \sqrt{Q/Q_i})$$

ii) For $Q > Q_i$, the width, B, must be increased by B", to be computed with:

$$B'' = n(Q - Q_i) / (h^{5/3} S_f^{1/2})$$

The effect of this change in width on the geometry of the channel section is shown in Fig. 3.16 (see *Chow*, 1958, p. 177).

3.5 FLOW IN CURVES

1° A curve or bend, positioned in a rectangular channel, causes a change in the flow direction.

2° If the discharge, Q, remains constant along the curve, the flow velocity, U, as well as the wetted section, A, remain also constant. The sectional distribution of the flow depth, h(y), will be responsible for a transversal water slope and a super-elevation, Δz, at the outside of the curve.

3° The distribution of velocity in the curve can be approximated by the one of a free vortex (see *Graf & Altinakar*, 1991, p. 196). The velocity has a maximum at the inside of the curve (see Fig. 3.17).

3.5.1 Super-elevation

1° In a curve the streamlines will no longer stay parallel and the flow becomes three-dimensional. This is a complex physical phenomenon, for which an adequate analysis seems difficult.

2° The method proposed by *Kozeny* (1953, p. 223) puts forward the following arguments and this for turbulent flow (see Fig. 3.17):

i) Assumed is that the head loss, h_r^c, – following a streamline, s, – can be expressed as:

$$S_e = \frac{h_r^c}{L_c} = \lambda u_s^2$$

UNIFORM FLOW

where λ is a factor of proportionality and $L_c = \alpha r$ is the length of the curve; α being the angle and r the radius of the curve. Thus one may write:

$$u_s = \sqrt{\frac{h_r^c}{\lambda \alpha}} \cdot \sqrt{\frac{1}{r}} = \frac{\kappa}{\sqrt{r}}$$

ii) For $r = r_o$, one takes $u_s = u_a \equiv U$; this implies that the axial velocity, u_a, is identical to the average velocity, U, of the cross section. One obtains now $u_a \sqrt{r_o} = \kappa$ and an expression for the distribution of the velocity, such as:

$$\frac{u_s}{u_a} = \sqrt{\frac{r_o}{r}}$$

iii) If r_2 and r_1 are the outside and inside radius, the corresponding velocities are given by:

$$u_2 = u_a \sqrt{r_o/r_2} \quad \text{and} \quad u_1 = u_a \sqrt{r_o/r_1}$$

iv) The super-elevation (see Fig. 3.17) can now be calculated as being:

$$\Delta z = \frac{u_1^2}{2g} - \frac{u_2^2}{2g} = \frac{u_a^2}{2g}\left(\frac{r_o}{r_1} - \frac{r_o}{r_2}\right)$$

Since $B = (r_2 - r_1)$ is the width of the curve, one can also write:

$$\Delta z = \frac{B\, r_o}{r_1\, r_2} \frac{U^2}{2g} \tag{3.43}$$

If the channel width, B, is small compared to the radius of the curve, r_o, one gets the simplified expression of:

$$\Delta z = \frac{B}{r_o} \frac{U^2}{2g} \tag{3.43a}$$

v) The transversal water profile is convex; one can write:

$$\Delta z_2 = \frac{U^2}{2g}\left(1 - \frac{r_o}{r_2}\right) \quad \text{and} \quad \Delta z_1 = \frac{U^2}{2g}\left(\frac{r_o}{r_1} - 1\right)$$

The super-elevation, $\Delta z = \Delta z_2 + \Delta z_1$, given with eq. 3.43 has its maximal value, $\Delta z = \Delta z_{max}$, usually observed for fluvial flow, Fr < 1, at the entrance of the curve and for supercritical flow, Fr > 1, at the exit of the curve.

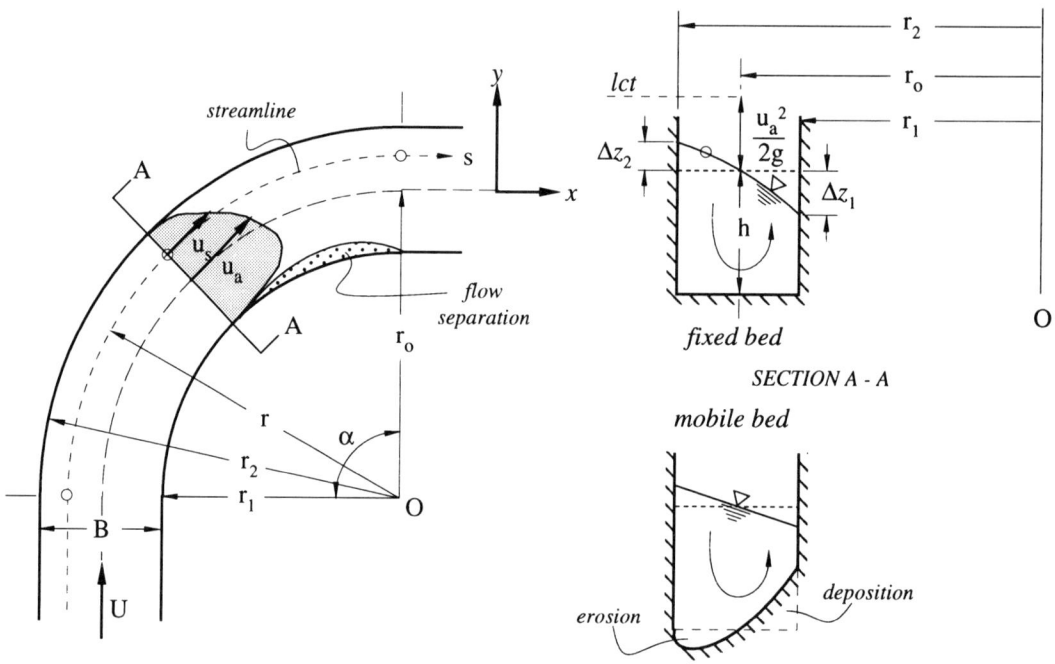

Fig. 3.17 Flow in a curve.

3° One can also define a coefficient of super-elevation by :

$$K^c = \Delta z / (U^2 / 2g) \qquad (3.44)$$

which, taking eq. 3.43, is : $K^c = B r_o / (r_1 r_2)$.

Apmann (1973, p. 73) proposed the following empirical expression, after analysing flow in curves in artificial and natural channels, or :

$$K^c = \frac{5}{4} \text{tgh}\left(\frac{r_o \alpha}{B}\right) \ln\left(\frac{r_2}{r_1}\right) \qquad (3.44a)$$

4° The super-elevation, Δz, can readily be used for the determination of the discharge (see *Apmann*, 1973, p. 70) :

$$Q = A \sqrt{2g \, \Delta z / K^c}$$

This relation is of great use in the determination of flood discharges, which usually leave traces (marks) at their largest occurring flow depth, thus Δz.

UNIFORM FLOW

5° The super-elevation at the outside of the curve (see Fig. 3.17) causes a vertical downward current, which comes back to the water surface at the inside of the curve. Such a secondary current will superpose itself on the primary flow and result in helicoidal flow over the entire reach of the curve.

6° If the outside side wall is of mobile material, erosion will take place; on the inside there will be deposition (see Fig. 3.17).

3.5.2 Supercritical Flow

1° Gravity waves (see sect. 2.4.3) will establish themselves in a curve (see Fig. 3.18), notably if the flow is supercritical, Fr > 1 (see *Ippen*, 1950, p. 563). For a rectangular channel, the celerity of a gravity wave is given by :

$$c^2 = gh \tag{2.27a}$$

2° At the entrance of the curve at the points A and A', one observes under an angle, β:

i) positive perturbations (waves) along the line ABD,

ii) negative perturbations (waves) along the line A'BC,

ii) but no perturbations appear in the zone ABA'.

This angle, β, is approximately defined as being :

$$\sin \beta = \frac{c}{U} = \frac{\sqrt{gh}}{U} = \frac{1}{Fr} \tag{3.45}$$

where Fr is Froude number of the flow upstream of the curve.

3° Consequently, the flow depth varies :

i) increasingly along the line AC, having a maximum at C,

ii) decreasingly along the line A'D, having a minimum at D.

The maximum (+) or minimum (-) flow depth can be calculated (see *Ippen* 1950, pp. 551 and 564) as being :

$$h_{min}^{max} = h\, Fr^2 \sin^2 (\beta \pm \theta / 2) \tag{3.46}$$

The central angle, θ, is determined using geometrical considerations, or :

$$\text{tg } \theta = \frac{B}{(r_o + B/2) \text{ tg } \beta} \tag{3.46a}$$

4° These maxima/minima are then reflected from one to the other side in the curve, such as is indicated in Fig. 3.18. The next maxima/minima appear after an interval of 2θ. This swinging, referred to as *cross waves*, can continue well beyond the end of the curve.

5° The maximum super-elevation, $(\Delta z' + \Delta z)$, due to the gravity waves, can be twice the super-elevation, Δz, obtained with eq. 3.43. It is calculated by :

$$\Delta z' = \frac{B}{r_o} \frac{U^2}{2g} \tag{3.47}$$

The water surface and the resulting transversal slopes are shown schematically with Fig. 3.18.

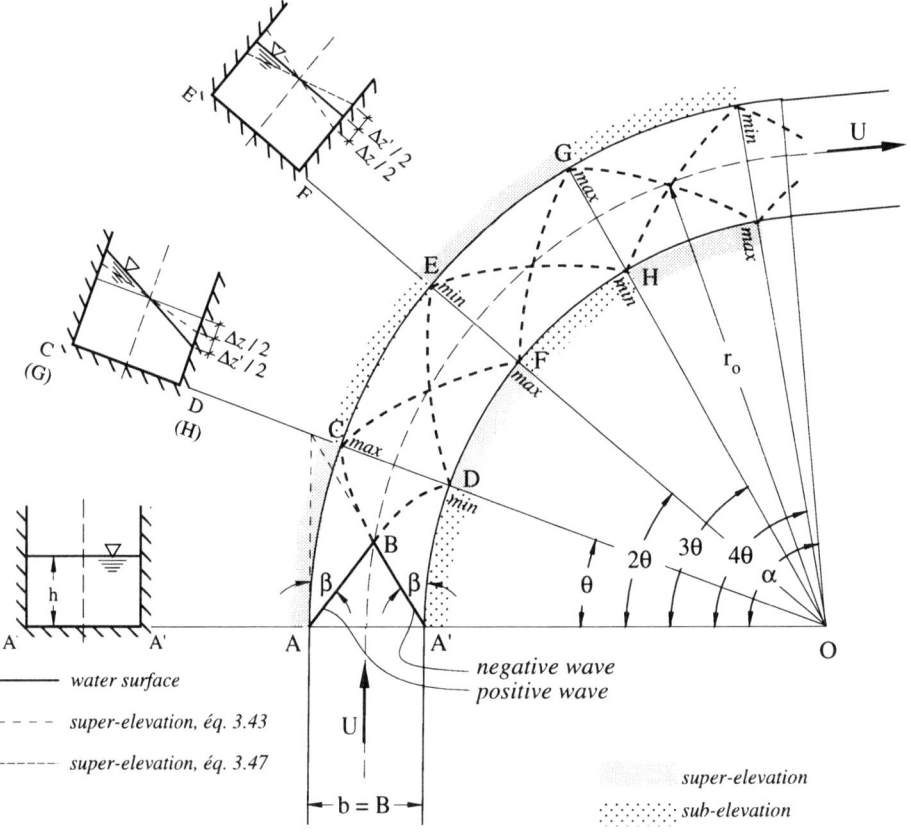

Fig. 3.18 Supercritical flow in a curve.

UNIFORM FLOW

6° If the super-elevations get very large, methods are available to suppress it (see *Naudascher*, 1987, p.206), like the construction of transition curves or the installation of steps on the channel bed.

3.5.3 Head Loss

1° In flow over a curve, one encounters not only a head loss due to friction, h_r, but also one due to the curvilinear flow, h_r^c (see sect. 3.5.1).

2° This additional head loss is usually expressed by :

$$h_r^c = \zeta_c \frac{U^2}{2g} \qquad (3.48)$$

where ζ_c is a coefficient which depends on :

$$\zeta_c = f(\text{Fr, Re}, r_0/B, h/B, \alpha)$$

α being the angle of the curve, Fr and Re are the number of Froude and of Reynolds, respectively. According to numerous experiments, one takes (see *Chow*, 1958, p. 443) :

$$0.1 \leq \zeta_c \leq 1.1$$

where the larger values of ζ_c are for curves of $r_0/B = 0.5$.

3.6 INSTABILITY AT SURFACE

1° If the channel slope is very high and/or if the flow is supercritical, the water surface can become unstable. The normal flow depth, h_n, must now be considered as an average value.

2° Such an instability is characterised by :

    i)   a series of gravity waves of small flow depth, eq. 2.27, called *roll waves*, progressing downstream, and

    ii)   a breaking of these waves, causing an *air entrainment*.

3.6.1 Roll Waves

1° An instability at the water surface is evidenced by the formation of roll waves. The uniform steady flow becomes locally an unsteady one.

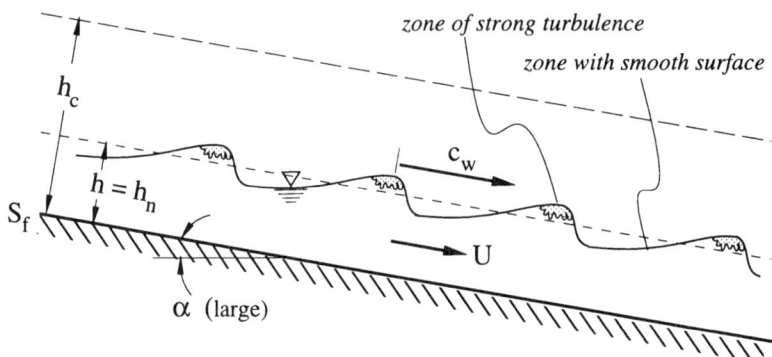

Fig. 3.19 Roll waves.

2° The roll waves are superposed on the uniform flow (see Fig. 3.19). They displace themselves towards the downstream – increasing in height and then collapsing – with an absolute celerity, c_w, being larger than the flow velocity, U, or :

$$c_w = U + \sqrt{gh} \quad > \quad U$$

3° There is no simple criteria available to determine the geometrical dimensions of this type of waves. Their height can, however, attain dimensions of the order of magnitude of the prevailing flow depth (see *French*, 1986, p. 625).

The crests of the roll waves are zones of strong turbulence, while the rest of the waves remains remarkably smooth.

4° Some theoretical considerations for a determination of the (in)stability of uniform flow are presented in *Liggett* (1975, chap. 6).

5° However, there exist useful practical criteria to determine the instability, which is responsible for the creation of roll waves. The geometry of the channel and the type of flow are taken in consideration.

   i) Taken is the Froude number, $Fr = U / \sqrt{gh}$; the flow of a large channel is unstable (see *Albertson* et al., 1960, p.355), if :

   $Fr > 2$ for turbulent, rough flow,
   $Fr \geq 1.5$ for turbulent, smooth flow,
   $Fr \geq 0.5$ for laminar flow.

ii) Taken is the number of Vedernikov (see *Chow*, 1959, p. 210), defined as :

$$Ve = x f_g Fr \tag{3.49}$$

where x is an exponent of the hydraulic radius, R_h, in eq. 3.11 (x = 2 for laminar flow; x = 2/3 or x = 1/2 for turbulent flow, taking the relation of Manning or of Chézy, respectively). f_g is a shape factor given by :

$$f_g = 1 - R_h \frac{dP}{dS}$$

being $f_g = 1$ for very wide channels and $f_g = 0$ for very narrow channels. If :

$$Ve < 1$$

flow can remain stable and any wave at the surface will be depressed. If :

$$Ve \geq 1$$

stable flow is impossible; unsteady flow will prevail and existing waves will amplify and form the roll waves.

6° Roll waves form themselves not only in uniform flow, but are also encountered in non-uniform flow.

3.6.2 Air Entrainment

1° For large channel slopes, S_f, – such as exist also on the downstream face of a weir – the flow is usually supercritical and gravity waves appear at the water surface. These waves will break and entrain air into the water. The turbulence will diffuse (mix) the air bubbles across the entire flow depth; and water droplets will escape into the air.

2° In flow of such an air-water mixture, it becomes a bit difficult to define the flow depth; the water surface is often covered by *white water*.

3° The schematic distribution of the concentration of air :

$$C(z) = \frac{\text{volume (air)}}{\text{volume (air + water)}}$$

is given with Fig. 3.20. Two regions are to be distinguished : bubbles in the water and droplets in the air.

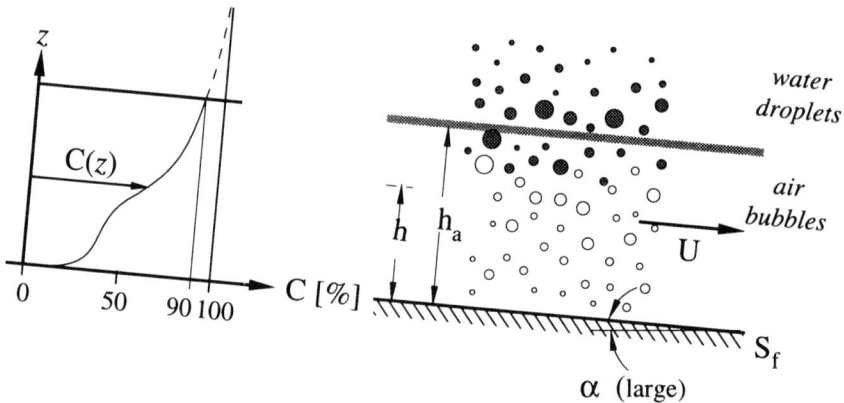

Fig. 3.20 Flow with air entrainment.

4° The equivalent flow depth of water (without the air) is defined by :

$$h = \int_0^\infty (1-C) \, dz \qquad (3.50)$$

and the average velocity of water by :

$U = q/h$

where q is the unit discharge of the water.

The depth of the mixture, which is the depth where the concentration is equal to $C = 90\%$, is given (see *Wood*, 1985, p.21) by :

$$h_a = \frac{h}{(1 - \overline{C})} \qquad (3.51)$$

where $\overline{C}$ is the average concentration in the cross section, given (see *Henderson*, 1966, p. 185) by the relation :

$$\overline{C} = 0.7 \log (S_f / q^{1/5}) + 0.9 \qquad (3.52)$$

which was obtained from experimental studies performed by *Straub* et *Anderson* for $0.14 < q[m^2/s] < 0.93$. Usually (see *Wood*, 1985, p.21) one takes :

$\overline{C} = 0.14$ for slopes of $S_f = 7.5°$
$\overline{C} = 0.71$ for slopes of $S_f = 75°$

5° The air entrainment, or the aspiration of air on the downstream face of a weir, begins at a point where the boundary layer is completely developed, $\delta \equiv h$ (see *Wood*, 1985, p. 18).

UNIFORM FLOW

3.7 EXERCISES

3.7.1 Problems, solved

Ex. 3.A

A channel is to be built of medium-quality concrete to convey a discharge of $Q = 80$ [m³/s]. The channel should have a trapezoidal cross section with a bottom width of $b = 5$ [m] and side slopes of $m = 3$. The channel slope will follow the geomorphology of the terrain which is determined to be $S_f = 0.1$ %. It is admitted that the flow is uniform and the water has a temperature of $T = 10$ [°C].

i) Calculate the flow depth using both the coefficient of Manning and the coefficient of Weisbach-Darcy.

ii) Verify whether the flow is laminar or turbulent and subcritical or supercritical.

SOLUTION :

The geometrical characteristics of a trapezoidal cross section are (see Table 1.1) :

wetted surface : $A = (b + mh) h = (5 + 3h) h$

hydraulic radius : $R_h = \dfrac{(b + mh) h}{b + 2h\sqrt{1+m^2}} = \dfrac{(5 + 3h) h}{5 + 2h\sqrt{1+3^2}}$

i) a) *Calculation of normal flow depth using the coefficient of Manning:*

If one uses the Manning-Strickler formula, eq. 3.16, to express the average flow velocity, the discharge in the channel is given by :

$$Q = UA = \dfrac{1}{n} R_h^{2/3} S_f^{1/2} A$$

The coefficient of Manning is obtained from the Table 3.2 :

For a medium-quality concrete, one can take : $n = \dfrac{1}{K_s} \cong \dfrac{1}{70} = 0.0143$ [m$^{-1/3}$s]

By introducing the expressions for A and R_h, as well as the values of n and S_f into the equation for the discharge, Q, one obtains :

$$Q = 80 = 70 \left(\dfrac{(5 + 3h) h}{5 + 2h\sqrt{10}} \right)^{2/3} \sqrt{0.001}\ (5 + 3h) h$$

This equation can be solved by trial-and-error :

h [m]	Q [m³/s]
2.20	69
2.50	91
2.36	80

The *normal depth* is therefore : $h = h_n = 2.36$ [m]

b) *Calculation of normal flow depth using the coefficient of Weisbach-Darcy* :

The average flow velocity in a channel is given by :

$$U = \sqrt{8g/f}\ \sqrt{R_h\ S_f} \tag{3.10}$$

The discharge is therefore : $Q = U\,A = \sqrt{\dfrac{8g}{f}}\ \sqrt{R_h\ S_f}\ A$

The coefficient of Weisbach-Darcy, f, can be obtained (either by using the Moody-Stanton diagram or) by using the Colebrook-White formula :

$$\sqrt{\dfrac{1}{f}} = -2\log\left(\dfrac{k_s/R_h}{a_f} + \dfrac{b_f}{Re\,\sqrt{f}}\right) \tag{3.13}$$

For a trapezoidal channel one generally uses : $a_f = 12$ and $b_f = 2.5$. However this equation can only be solved by trail-and-error.

The uniform equivalent roughness, k_s, is obtained from the Table 3.1 :
 For a medium-quality concrete, one can take : $k_s = 0.001$ [m] .

It is evident that the velocity, U', calculated with eq. 3.10 for an *estimated* normal depth, h_n , should also satisfy the relationship $U = Q/A$, where A is the wetted surface corresponding to h_n. Using this fact, a trial-and-error type calculation can be devised for calculating the normal depth. The flowchart of this trial-and-error calculation is given hereafter. As it can be seen the algorithm is based on two nested calculation loops. The internal loop calculates the friction coefficient, f, whereas the external loop calculates the normal depth, h_n .

This procedure can be programmed on a microcomputer using a spreadsheet program. The calculation sheet presented below simulates the sequence of calculations depicted in the flowchart. The order of execution of the nested loops is shown in the leftmost column. The detailed explanations of the other columns, numbered from 1 to 10, are given in the table.

The iteration starts with an estimation of $h_n = 2.20$ [m] in the external loop. The internal loop is here executed three times for this same value of h_n. Although the estimated and the calculated values of the Weisbach-Darcy friction coefficient are equal: $f = f' = 0.01419$, at the end of the third iteration, the velocities, U and U', are still different : $\Delta U \neq 0$. Consequently, the calculations must continue by next taking $h_n = 2.50$ [m] and then $h_n = 2.35$ [m]. After the first execution of the internal loop with $h_n = 2.35$ [m] and $f = 0.02$, one finds already that $f' = 0.01402$ and $\Delta U = 0.00$. The iterations can be stopped here without waiting for the convergence of the value of f. The following two calculation lines only help to refine the value of f.

The *normal depth* is therefore : $\quad h = h_n = 2.35$ [m]

UNIFORM FLOW

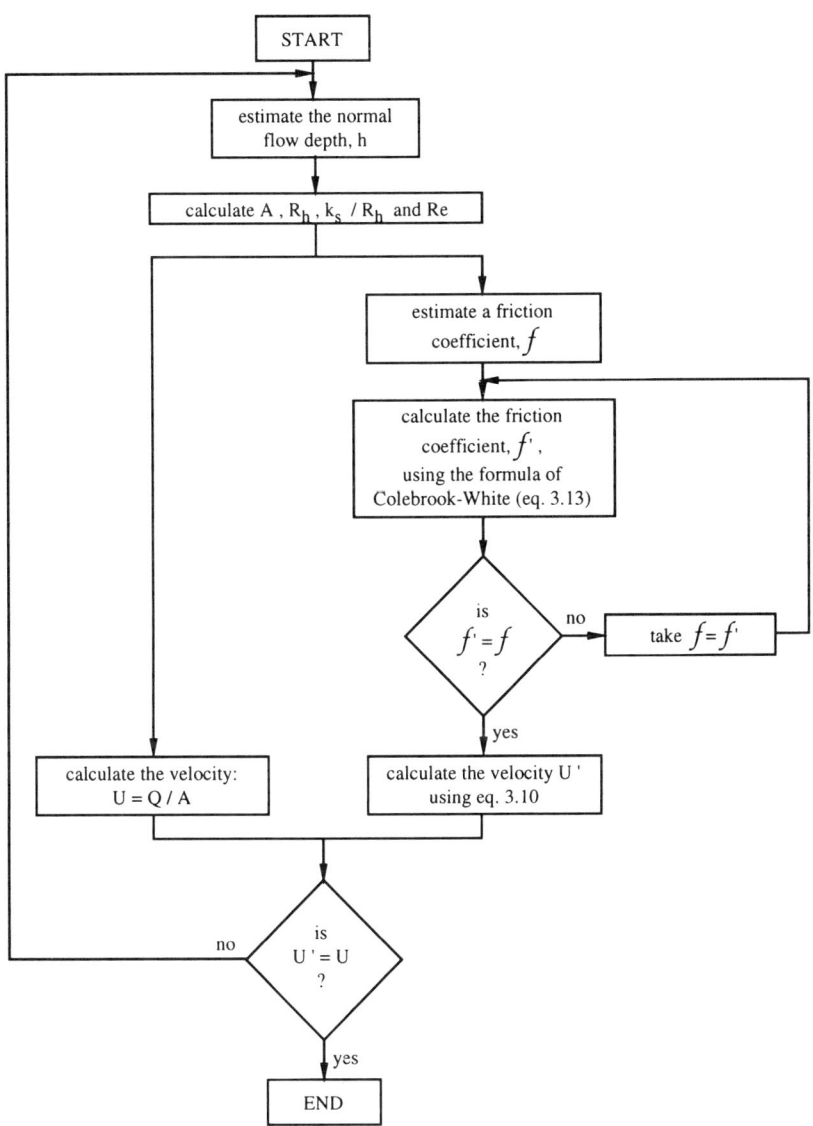

ii) For $h_n = 2.36$ [m], $R_h = 1.43$ [m] and $U = 2.81$ [m/s], one has:

Reynolds number: $\quad Re' = \dfrac{R_h\, U}{\nu} = \dfrac{(1.43)\,(2.81)}{1.31 \times 10^{-6}} = 3.1 \times 10^6 > 2000$

The flow is therefore *turbulent*.

Froude number: $\quad Fr = \dfrac{U}{\sqrt{g h_n}} = \dfrac{2.81}{\sqrt{(9.81)\,(2.36)}} = 0.58 < 1$

The flow is therefore *fluvial*.

Calculation sheet for determining the normal flow depth
by using the coefficient of Weisbach-Darcy,
calculated using the formula of Colebrook-White (eq. 3.13)

$Q = 80$ [m³/s] | $S_f = 0.001$ [-] | | $a_f = 12$ [-]
$b = 5.0$ [m] | $k_s = 0.001$ [m] | $m = 3.0$ [-] | $b_f = 2.5$ [-]
For $T = 10$ [°C] | $\nu = 1.31 \times 10^{-6}$ [m²/s] | (*Graf & Altinakar*, 1991, Table I.3)

iteration number for h_n and f		1 h_n [m]	2 A [m²]	3 R_h [m]	4 U [m/s]	5 k_s/R_h [-]	6 Re [-]	7 f [-]	8 f' [-]	9 U' [m/s]	10 ΔU [m/s]
1	1	2.20	25.52	1.35	3.13	7.41×10⁻⁴	1.29×10⁷	0.02000	0.01417		
	2							0.01417	0.01419		
	3							0.01419	0.01419	2.73	0.40
2	1	2.50	31.25	1.50	2.56	6.66×10⁻⁴	1.18×10⁷	0.02000	0.01388		
	2							0.01388	0.01389		
	3							0.01389	0.01389	2.91	-0.35
3	1	2.35	28.32	1.43	2.83	7.01×10⁻⁴	1.23×10⁷	0.02000	0.01402	2.82	0.00
	2							0.01402	0.01403	2.82	0.00
	3							0.01403	0.01403	2.82	0.00

col.	symbol	explanations	expression
1	h_n	*estimated* normal depth	h
2	A	wetted surface	$(b + mh)h$
3	R_h	hydraulic radius	$\dfrac{(b + mh)h}{b + 2h\sqrt{1+m^2}}$
4	U	average velocity :	Q/A
5	k_s / R_h	relative roughness	k_s / R_h
6	Re	Reynolds number	$4 U R_h / \nu$
7	f	*estimated* friction coefficient	
8	f	friction coefficient calculated using eq. 3.13 : $\left[-2\log\left(\dfrac{k_s/R_h}{a_f} + \dfrac{b_f}{Re\sqrt{f}}\right)\right]^{-2}$	
9	U'	average velocity obtained using eq. 3.10 :	$\sqrt{8g/f'}\ \sqrt{R_h S_f}$
10	ΔU	difference between the velocities	$U - U'$

UNIFORM FLOW 113

> **Ex. 3.B**
>
> The river *Happy* has a variable discharge in the range of $10 < Q \ [m^3/s] < 1000$. In the city of Ste-Justice, the width of the bed is $b = 90 \ [m]$ and the non erodible banks have a slope of 1:1. A bridge crosses the river (without causing any obstruction to the flow) and it is planned to install a measuring gauge at the mid-span of the bridge. A grain-size analysis of the bed material yielded: $s_s = 2.65 \ [-]$ and $d_{50} = 0.32 \ [mm]$, $d_{35} = 0.29 \ [mm]$ and $d_{90} = 0.48 \ [mm]$. The water temperature is $T = 14 \ [°C]$. A survey of the river bed showed that the bed slope is $S_f = 0.0005 \ [-]$.
>
> *i)* Determine the stage-discharge curve, $Q = f(h_n)$, by assuming that the flow is turbulent rough.
>
> *ii)* At what depth will erosion and deposition begin to occur?

SOLUTION :

For water at $T = 14 \ [°C]$ (interpolating the values in Table I.3, in *Graf & Altinakar*, 1991), one has : $\rho = 999.1 \ [kg/m^3]$ and $\nu = 1.186 \times 10^{-6} \ [m^2/s]$.

Using the definition of the specific density (see *Graf & Altinakar*, 1991, p.9), one writes:
$s_s = \rho_s / \rho_{eau} \Rightarrow \rho_s = \rho_{eau} \ s_s = 1000 \times 2.65 = 2650 \ [kg/m^3]$
where $\rho_{eau} = 1000 \ [kg/m^3]$ is the density of water at $p_a = 1 \ [atm]$ and $T = 3.98 [°C]$.

i) The bed (and banks) of a river are generally mobile, composed of erodible granular material. If it is desired to obtain the hydraulic radius, R_h, or the bed shear stress, τ_o, it becomes necessary to make a distinction (see eq. 3.24) between the contribution of the grain roughness, R_h' or τ', and the one of the bed forms, R_h'' or τ''.

The calculations can readily be programmed on a microcomputer using a spreadsheet program. The tabular computation sheet prepared in this way is presented below. Each line of this table represents the computation of the discharge, Q, and some other useful parameters for a given water depth, h. The detailed explanations concerning all the columns are given on the bottom of the table. A brief description of the computation sequence is presented hereafter.

The calculations on a line start by *assuming* a value for the hydraulic radius due to the grain roughness, R_h'. The values of R_h' should be chosen in a way to cover the whole range of the possible discharges, $10 < Q \ [m^3/s] < 1000$. The calculations in the columns 2 and 3 are straightforward. It is interesting to note that, according to the method of Einstein-Barbarossa, the average velocity, U, is calculated using only R_h'. The hydraulic radius due to bed forms, R_h'', calculated in the columns 4 to 7, influences only the flow depth. It is to be noted that the value of U/u_*'' is obtained from Fig. 3.6 with the value of ψ'.

Computation sheet for determining the stage-discharge curve ...									
$b = 90$ [m] $\quad$ $m = 1$ $\quad$ $S_f = 0.0005$ [-]			$T = 14$ [°C] $\quad$ $\rho = 999.1$ [kg/m³] $\quad$ $\nu = 1.186 \times 10^{-6}$ [m²/s]				$\rho_s = 2650$ [kg/m³]		
1	2	3	4	5	6	7	8	9	10
R_h'	u_*'	U	ψ'	$\dfrac{U}{u_*''}$	u_*''	R_h''	R_h	u_*	h
[m]	[m/s]	[m/s]	[-]	[-]	[m/s]	[m]	[m]	[m/s]	[m]
0.02	0.01	0.16	47.92	4.5	0.04	0.26	0.28	0.04	0.29
0.05	0.02	0.29	19.17	6.6	0.04	0.39	0.44	0.05	0.44
0.10	0.02	0.45	9.58	8.7	0.05	0.55	0.65	0.06	0.65
0.15	0.03	0.58	6.39	10.4	0.06	0.63	0.78	0.06	0.79
0.20	0.03	0.69	4.79	11.9	0.06	0.68	0.88	0.07	0.89
0.40	0.04	1.05	2.40	18.3	0.06	0.67	1.07	0.07	1.09
0.60	0.05	1.33	1.60	23.4	0.06	0.66	1.26	0.08	1.29
0.80	0.06	1.58	1.20	32.4	0.05	0.49	1.29	0.08	1.32
1.00	0.07	1.81	0.96	42.6	0.04	0.37	1.37	0.08	1.41
1.25	0.08	2.06	0.77	56.2	0.04	0.27	1.52	0.09	1.57
1.50	0.09	2.30	0.64	73.1	0.03	0.20	1.70	0.09	1.76
2.00	0.10	2.72	0.48	107.2	0.03	0.13	2.13	0.10	2.23
2.50	0.11	3.11	0.38	163.0	0.02	0.07	2.57	0.11	2.71
3.00	0.12	3.46	0.32	6844.0	0.00	0.00	3.00	0.12	3.19

col.	symbol	explanations	expression
1	R_h'	hydraulic radius due to grain roughness (*assumed* value)	
2	u_*'	friction velocity due to grain roughness, eq. 3.24,	$\sqrt{g R_h' S_f}$
3	U	average velocity in the cross section	$u_*' \sqrt{8/f'}$
		with (see eq. 3.13b): $\sqrt{8/f'} = 5.6 \log(R_h'/k_s) + 6.25$	
4	ψ'	parameter of Einstein-Barbarossa, eq. 3.31,	$\dfrac{\rho_s - \rho}{\rho} \dfrac{d_{35}}{R_h' S_f}$
5	$\dfrac{U}{u_*''}$	ratio of velocities corresponding to ψ' (see eq. 3.31 and Fig. 3.6)	
6	u_*''	friction velocity due to bed forms	$U / (U/u_*'')$
7	R_h''	hydraulic radius due to bed forms	$(u_*'')^2 / (g S_f)$
8	R_h	total hydraulic radius, eq. 3.24,	$R_h' + R_h''$
9	u_*	friction velocity, eq. 3.7,	$\sqrt{g R_h S_f}$
10	h	flow depth (see Table 1.1)	

UNIFORM FLOW

... using the method of *Einstein-Barbarossa*				
$U_{cr} = 0.2$ [m/s] (see Fig. 3.12)	⇐	$k_s = \begin{array}{l} d_{35} = 0.00029 \text{ [m]} \\ d_{50} = 0.00032 \text{ [m]} \\ d_{90} = 0.00048 \text{ [m]} \end{array}$ ⇒ $d_* = 7.21$ [-]	⇒	$\tau_{*cr} = 0.04$ [-] (see Fig. 3.13)

11	12	13	14	15	16	17	18	19	20
A	P	Q	$\dfrac{U}{\sqrt{gh}}$	$\dfrac{R_h}{d_{50}}$	$\dfrac{U}{\sqrt{gR_h}}$	$\dfrac{U}{\sqrt{g\,d_{50}}}$	τ_o	τ_*	Notes
[m²]	[m]	[m³/s]	[-]	[-]	[-]	[-]	[N/m²]	[-]	
25.83	90.81	4.2	0.10	889	0.10	2.88	1.39	0.27	¥
40.20	91.26	11.7	0.14	1377	0.14	5.18	2.16	0.42	† ¥
59.25	91.85	26.5	0.18	2016	0.18	7.99	3.16	0.61	† ¥
71.92	92.24	41.4	0.21	2437	0.21	10.27	3.82	0.74	† ¥
81.34	92.53	55.8	0.23	2747	0.23	12.25	4.31	0.83	† ¥
99.28	93.08	103.7	0.32	3333	0.32	18.65	5.23	1.01	† ¥
118.03	93.66	157.4	0.37	3938	0.38	23.80	6.18	1.19	† ¥
120.78	93.74	191.3	0.44	4026	0.45	28.26	6.31	1.22	† ¥
128.53	93.98	232.4	0.49	4274	0.49	32.28	6.70	1.29	† ¥
144.00	94.45	297.3	0.53	4765	0.53	36.84	7.47	1.44	† ¥
161.62	94.98	371.6	0.55	5317	0.56	41.04	8.34	1.61	† ¥
205.28	96.30	559.3	0.58	6662	0.60	48.63	10.45	2.02	† ¥
251.41	97.67	780.9	0.60	8044	0.62	55.44	12.61	2.43	† ¥
297.05	99.02	1026.7	0.62	9375	0.64	61.69	14.70	2.84	† ¥

col.	symbol	explanations	expression
11	A	wetted surface (see Table 1.1)	$(b + mh)\,h$
12	P	wetted perimeter (see Table 1.1)	$b + 2h\sqrt{1 + m^2}$
13	Q	discharge (see eq. 3.2a)	$U\,A$
14	$U/\sqrt{gh}$	Froude number using the flow depth	Fr
15	R_h / d_{50}	relative depth	
16	$U/\sqrt{gR_h}$	Froude number using the hydraulic radius	
17	$U/\sqrt{g\,d_{50}}$	parameter proposed by Alam-Kennedy (see eq. 3.32 and Fig. 3.7)	
18	τ_o	total bed shear stress, eq. 3.6,	$\rho\,u_*^2$
19	τ_*	dimensionless bed shear stress, eq. 3.38,	$\dfrac{\tau_o}{(\gamma_s - \gamma)\,d_{50}}$
20	†	$U > U_{cr}$ ⇒ erosion according to Hjulstrom's criteria (see Fig. 3.12)	
	¥	$\tau_* > \tau_{*cr}$ ⇒ motion according to Shields' criteria (see Fig. 3.13)	

The friction coefficient due to bed forms, f'', can be evaluated using either the method of Einstein-Barbarossa, or the method of Alam-Kennedy (see sect. 3.2.6). The second method is more general but, it necessitates an iterative solution; the calculations are also more elaborate. The method of Einstein-Barbarossa, on the other hand, relies on a straightforward and simple calculation and it will be used for solving the present problem. However, as was already mentioned (see sect. 3.2.6), according to Alam-Kennedy (see Fig. 3.7), the relationship of Einstein-Barbarossa is valid in the region where the friction coefficient due to the bed forms, f'', does not depend on the relative depth, R_h/d_{50}. At the end of the calculation, a verification must be made to check that the calculated f''-value lies in that region.

After calculating the total hydraulic radius, $R_h = R_h' + R_h''$, the flow depth, h, can be obtained using the geometrical relationships. For a trapezoidal cross section (see Table 1.1), the problem is reduced to finding the positive square root of the following quadratic equation :

$$m h^2 + (b - 2 R_h \sqrt{1 + m^2}) h - b R_h = 0$$

The calculations for the remaining columns are explained in the computation sheet. The Froude numbers in column 14 show that the flow is subcritical for all flow depths. Using the values in columns 15 to 17 it can now be checked on Fig 3.7 that all the points fall into the region where the value of f'' is independent from R_h/d_{50}.

The stage-discharge curve, $Q = f(h_n)$, as well as the variation of other useful parameters, U, P, A, R_h', R_h'', R_h, are plotted on the following figure. On this figure, it is interesting to observe the evolution of the curves corresponding to R_h' and R_h''.

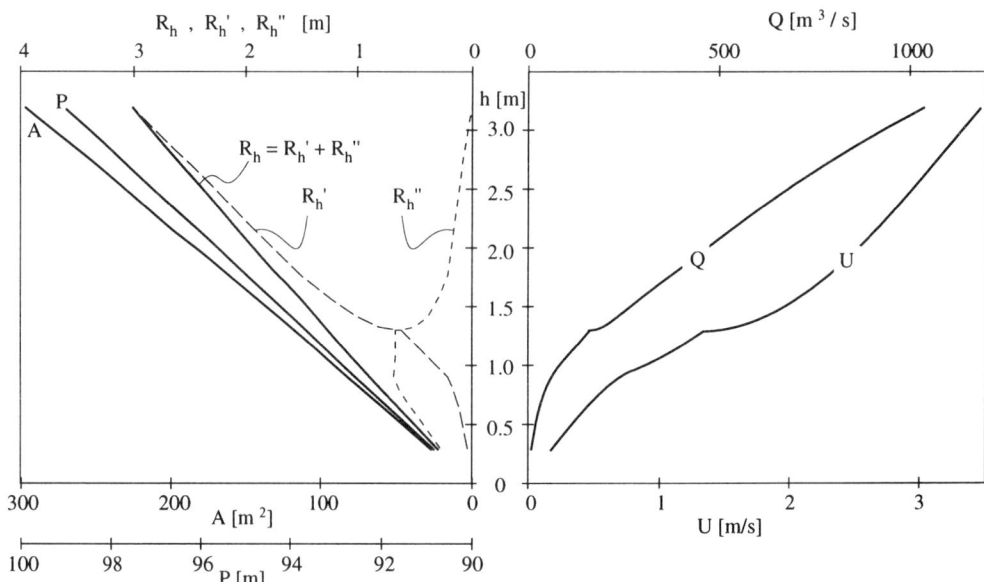

UNIFORM FLOW

ii) a) According to the *criteria of Hjulstrom*, for a given grain size, the critical velocities for the erosion and the sedimentation can be obtained from Fig. 3.12 :

for $d_{50} = 0.00032$ [m] $\Rightarrow$ $U_E = U_{cr} \cong 0.2$ [m/s] and $U_D \cong 0.03$ [m/s]

As it can be seen in the column 20 of the computation sheet, for the average velocities calculated in column 3, one has :

always $\qquad\qquad\qquad\qquad$ $U > U_D$: sediment transport takes place,
always(except for h = 0.29 [m]) $U > U_E$: erosion of the bed should be expected.

It is to be noted that the erosion of the bed *does not mean* the formation of a scour hole in the river bed. Since the sediment transported from the upstream compensates globally the erosion, one should rather talk about a *transport of sediments*.

b) According to the *criteria of Shields*, the initiation of erosion should be verified using the total bed-shear stress, τ_* (see eq. 3.40), whose critical value, τ_{*cr}, for a given dimensionless grain diameter, d_*, is obtained from Fig. 3.13; namely:

$d_{50} = 0.00032$ [m] $\Rightarrow$ $d_* = d_{50} \left(\dfrac{\rho_s - \rho}{\rho} \dfrac{g}{v^2} \right)^{1/3} = 7.21$ [-] $\Rightarrow$ $\tau_{*cr} = 0.04$ [-]

As it can be seen in the column 20 of the computation sheet, the calculated dimensionless total bed-shear stress values (column 19) are :

always $\qquad\qquad\qquad\qquad$ $\tau_* > \tau_{*cr}$: erosion of the bed should be expected.

In order to compute the flow conditions corresponding to $\tau_* = \tau_{*cr}$, one has to consider the values of $R_h < 0.28$ [m]. The flow depth at which the erosion starts can be calculated as being $h_{cr} = 0.05$ [m], which is too shallow for a river of this importance.

Ex. 3.C

A channel excavated in earth should convey a water discharge of $Q = 57$ [m³/s] at an average temperature of $T = 14$ [°C]. The bed slope, $S_f = 0.001$ [-], is given; it will be assumed that the banks will have side slopes of 1.5 horizontal for 1 vertical. A grain-size analysis yielded : $d_{50} = 37$ [mm], $\varphi = 37°$, $s_s = 2.65$ [-] and $n = 0.02$ [m$^{-1/3}$s]. What should be the dimensions of this channel, if no erosion is allowed either at the bottom or on the banks ?

SOLUTION :

For water at $T = 14$ [°C] (interpolating the values in Table I.3, in *Graf & Altinakar*, 1991), one has : $\rho = 999.1$ [kg/m³] $\qquad$ and $\qquad$ $v = 1.186 \times 10^{-6}$ [m²/s].

Using the definition of the specific density (see *Graf & Altinakar*, 1991, p.9), one writes:
$$s_s = \rho_s / \rho_{eau} \quad \Rightarrow \quad \rho_s = \rho_{eau} \, s_s = 1000 \times 2.65 = 2650 \; [kg/m^3]$$
where $\rho_{eau} = 1000 \; [kg/m^3]$ is the density of water at $p_a = 1$ [atm] and $T = 3.98[°C]$.

The stability of the banks requires that the side slopes, θ, should be *smaller* than the angle of repose, $\varphi = 37°$ (see sect. 3.4.3) :

$$tg\theta = 1/1.5 = 0.667 \quad \Rightarrow \quad \theta = 33.7° < 37° \quad \Rightarrow \quad \text{the banks are thus stable.}$$

The critical bed-shear stress on the banks can be calculated using eq. 3.41 :

$$(\tau_{o_{cr}})^w = \tau_{o_{cr}} \left[\cos\theta \left(1 - \frac{tg^2\theta}{tg^2\varphi} \right)^{1/2} \right] = \tau_{o_{cr}} \left[\cos(33.7) \left(1 - \frac{tg^2(33.7)}{tg^2(37)} \right)^{1/2} \right] = 0.39 \, \tau_{o_{cr}}$$

To determine the critical shear stress, the Shields criteria will now be used (see sect. 3.4.2).

The dimensionless grain diameter corresponding to $d_{50} = 0.037$ [m] is :

$$d_* = d_{50} \left(\frac{\rho_s - \rho}{\rho} \frac{g}{\nu^2} \right)^{1/3} = 0.037 \left(\frac{2650 - 999.1}{999.1} \frac{9.81}{(1.186 \times 10^{-6})^2} \right)^{1/3} = 835 \; [-]$$

According to Shields' criteria, the critical value of the shear stress at the bed, τ_{*cr}, is obtained from Fig. 3.13. In the present case, this value falls in the region where $d_* > 2 \times 10^2 \; [-]$, having a constant value of $\tau_{*cr} \cong 0.055 \; [-] \cong$ Cte.

The critical shear stress for the *bed* is (see eq. 3.38) :

$$\tau_{o_{cr}} = \tau_{*cr} \, g \, (\rho_s - \rho) \, d_{50} = 0.055 \times 9.81 \times (2650 - 999.1) \times 0.037 \cong 33 \; [N/m^2]$$

On the *banks*, however, the critical value is already reached for :

$$(\tau_{o_{cr}})^w = 0.39 \, \tau_{o_{cr}} = 0.39 \times 33 = 12.9 \; [N/m^2]$$

The flow depth should be chosen not to exceed these critical values. By taking the critical value of shear stress at the bed as the design criteria and by assuming a wide channel the flow depth can readily be calculated using eq. 3.5a :

$$h = \frac{\tau_{o_{cr}}}{\rho g \, S_f} = \frac{33}{999.1 \times 9.81 \times 0.001} = 3.37 \; [m]$$

UNIFORM FLOW

According to Fig. 3.14, for a wide channel with a trapezoidal cross section having side slopes of m = 1.5, the maximum value of the shear stress on the banks is : $\tau_o = 0.75\, \rho g\, h\, S_f$. By taking this value as the critical value, one concludes that the flow depth should not exceed :

$$h = \frac{(\tau_{o_{cr}})^w}{0.75\, \rho g\, S_f} = \frac{12.9}{0.75 \times 999.1 \times 9.81 \times 0.001} = 1.75\ [m]$$

One should therefore choose h = 1.75 [m] as the *maximum flow depth*.

Now the channel width, b , should be determined such that for a discharge of Q = 57 [m³/s] the uniform flow can be established at a flow depth of h = 1.75 [m]. The discharge in a channel is calculated using eq. 3.33 :

$$Q = U A = \frac{1}{n} R_h^{2/3} S_f^{1/2} A$$

where U represents the average flow velocity. Given that both the hydraulic radius, R_h , and the wetted surface, A , depend on the channel width, b, which is what we are trying to calculate, the above relationship can not be solved directly for b. Therefore, a trial-and-error calculation should be carried out by varying b until the desired discharge is obtained. The computation sheet for the trial-and-error calculation is presented below.

Computation sheet for determining the channel width, b , by trial-and-error.					
m = 1.5 [-] S_f = 0.001 h = 1.75 [m] n = 0.02 [m$^{-1/3}$s]					
b [m]	A = h (b + mh) [m²]	P = b + 2h $\sqrt{1 + m^2}$ [m]	R_h = A / P [m]	U (eq. 3.16) [m/s]	Q = UA [m³/s]
5.00	13.34	11.31	1.18	1.77	23.56
10.00	22.09	16.31	1.35	1.94	42.77
14.00	29.09	20.31	1.43	2.01	58.46
13.63	28.45	19.94	1.43	2.00	57.00

The channel width should therefore be : b = 13.63 [m]. In calculating the shear stress at the bed the channel was assumed to be a wide one. Given that b > h, this hypothesis is now justified.

The uniform flow velocity is U = 2.0 [m/s] (see the above computation sheet). This velocity can be compared with the critical velocity for the erosion according to Hjulstrom. As it can be read on Fig. 3.12, the erosion velocity for d_{50} = 0.037 [m] is $U_{cr} \cong 2$ [m/s] approximately. Given that the uniform flow depth was selected considering the critical shear stress on the banks, the flow is at the limit of the beginning of erosion in accordance with the previous calculations. To have a better security against the erosion it is necessary to choose a flow depth smaller than the one corresponding to $(\tau_{o_{cr}})^w$.

120   FLUVIAL HYDRAULICS

Ex. 3.D

An artificial channel is going to be constructed in a mountainous region. The slope of the channel imposed by the terrain is $S_f = 0.01$ [-]. This channel should convey a discharge of $Q = 30$ [m³/s] at a temperature of $T = 14$ [°C], *without causing any erosion*. The grain-size analysis has shown that the granular material is non cohesive with $d_{50} = 50$ [mm] and $s_s = 2.65$ [-]. The angle of repose of this material is $\varphi = 37°$. The Manning coefficient is estimated to be $n = 0.025$ [m$^{-1/3}$s^1].

i)   What should be the dimensions of this channel which should have a rectangular section with the sides made of wooden boards ? The use of two different approaches is suggested : the critical velocity, U_{cr}, and the critical shear stress, $\tau_{o_{cr}}$.

ii) What will be the dimensions of the channel, if it were to be constructed entirely in its bed material having an ideal stable cross section ?

iii) Compare the dimensions of the channel, obtained using the different methods.

SOLUTION :

For water at $T = 14$ [°C] (interpolating the values in Table I.3, in *Graf & Altinakar*, 1991), one has : $\rho = 999.1$ [kg/m³]   and $\nu = 1.186 \times 10^{-6}$ [m²/s].

Using the definition of the specific density (see *Graf & Altinakar*, 1991, p.9), one writes:
$s_s = \rho_s / \rho_{eau} \quad \Rightarrow \quad \rho_s = \rho_{eau} \, s_s = 1000 \times 2.65 = 2650$ [kg/m³]
where $\rho_{eau} = 1000$ [kg/m³] is the density of water at $p_a = 1$ [atm] and $T = 3.98$ [°C].

i)   a) *Design of the channel using the critical velocity criteria :*

The critical velocity, U_{cr}, according to Hjulstrom (see Fig. 3.12) is :

$d_{50} = 0.05$ [m] $\Rightarrow$ $U_{cr} = 2.5$ [m/s]

The hydraulic radius corresponding to this velocity can be calculated using the formula of Manning-Strickler :

$$U \equiv U_{cr} = \frac{1}{n} R_h^{2/3} S_f^{1/2} \qquad (3.16)$$

With $n = 0.025$ [m$^{-1/3}$s^1] and $S_f = 0.01$ [-], one finds :

$$R_h = \left(\frac{U \, n}{S_f^{1/2}} \right)^{3/2} = \left(\frac{2.5 \times 0.025}{0.01^{1/2}} \right)^{3/2} = 0.494 \text{ [m]}$$

UNIFORM FLOW   121

The wetted surface is (see eq. 1.3)   : $A = Q/U = 30/2.5 = 12 \ [m^2]$

The wetted perimeter is (see eq. 1.1) : $P = A/R_h = 12/0.494 = 24.3 \ [m]$

Given that (see Table 1.1) : $A = bh$ and $P = b + 2h$,

the *dimensions* of the rectangular channel are : $b = 23.1 \ [m]$ and $h = 0.52 \ [m]$

b) *Design of the channel using the critical shear-stress criteria:*

The dimensionless grain diameter corresponding to $d_{50} = 0.05 \ [m]$ is (see sect. 3.4.2) :

$$d_* = d_{50} \left(\frac{\rho_s - \rho}{\rho} \frac{g}{\nu^2} \right)^{1/3} = 0.05 \left(\frac{2650 - 999.1}{999.1} \frac{9.81}{(1.186 \times 10^{-6})^2} \right)^{1/3} = 1129 \ [-]$$

Since d_* is known, the critical shear stress at the bed, τ_{*cr}, according to the Shields criteria, can be obtained from Fig. 3.13. In the present case, the calculated value of d_* falls outside of the limits of this figure. Nevertheless, it can be assumed that for $d_* > 2 \times 10^2$ [-] the critical shear stress has a constant value, being $\tau_{*cr} \cong 0.055 \ [-] \cong$ Cte.

Given the definition of τ_* (see eq. 3.40) : $\tau_{o_{cr}} = \tau_{*cr} (\gamma_s - \gamma) d = \tau_{*cr} g (\rho_s - \rho) d_{50}$

The bed-shear stress is given by eq. 3.5 : $\tau_{o_{cr}} = \gamma R_h S_f = g \rho R_h S_f$

Combining these two expressions, the hydraulic radius can be calculated :

$$R_h = \frac{\tau_{*cr} d_{50}}{S_f} \frac{(\rho_s - \rho)}{\rho} = \frac{0.055 \times 0.05}{0.01} \frac{(2650 - 999.1)}{999.1} = 0.45 \ [m]$$

The velocity corresponding to this hydraulic radius can now be calculated using the formula of Manning-Strickler, eq. 3.16. With $n = 0.025 \ [m^{-1/3} s^1]$ and $S_f = 0.01$ [-], one has :

$$U \equiv U_{cr} = \frac{1}{n} R_h^{2/3} S_f^{1/2} = \frac{1}{0.025} (0.45)^{2/3} (0.01)^{1/2} = 2.35 \ [m/s]$$

The remaining part of the calculations are the same as in the first case (see above) :

$A = Q/U = 30/2.35 = 12.8 \ [m^2]$ and $P = A/R_h = 12.8/0.45 = 28.4 \ [m]$

The *dimensions* of the rectangular channel are : $b = 27.5 \ [m]$ and $h = 0.47 \ [m]$

ii) The critical dimensionless shear stress at the bed, τ_{*cr}, according to the Shields criteria has already been obtained above :

$$d_{50} = 0.05 \text{ [m]} \quad \Rightarrow \quad d_* = 1129 \text{ [-]} \quad \Rightarrow \quad \tau_{*cr} \cong 0.055 \text{ [-]}$$

By using the definition of τ_* (see eq. 3.40), one finds :

$$\tau_{o_{cr}} = \tau_{*cr} \, g \, (\rho_s - \rho) \, d_{50} = 0.055 \times 9.81 \times (2650 - 999.1) \times 0.05 = 44.5 \text{ [N/m}^2\text{]}$$

The maximum depth in the middle of ideal section will be (see sect. 3.4.4) :

$$h = \frac{\tau_{o_{cr}}}{\gamma \, S_f} = \frac{\tau_{o_{cr}}}{g \, \rho \, S_f} = \frac{44.5}{9.81 \times 999.1 \times 0.01} = 0.45 \text{ [m]}$$

The expression for the ideal section is given by eq. 3.42 :

$$h' = h \cos\left(\frac{\text{tg}\varphi}{h} y\right) = 0.45 \cos\left(\frac{\text{tg}(37)}{0.45} y\right) \quad \Rightarrow \quad h' = 0.45 \cos(1.675 \, y)$$

Other characteristics of the section are calculated using eqs. 3.42a :

$$A = 2 h^2 / \text{tg}\varphi = 2 \times (0.45)^2 / \text{tg}(37) = 0.54 \text{ [m}^2\text{]}$$

$$B = \pi h / \text{tg}\varphi = \pi \times 0.45 / \text{tg}(37) = 1.88 \text{ [m]}$$

Approximating the elliptic integral by :

$$E \approx (\pi/2) \left(1 - \frac{1}{4} \sin^2\varphi\right) = (\pi/2) \left(1 - \frac{\sin^2(37)}{4}\right) = 1.429 \text{ [-]}$$

one obtains :

$$U = \frac{1}{n} S_f^{1/2} \left(\frac{h \cos\varphi}{E}\right)^{2/3} = \frac{1}{0.025} (0.01)^{1/2} \left(\frac{0.45 \times \cos(37)}{1.429}\right)^{2/3} = 1.59 \text{ [m/s]}$$

The discharge, Q_i, which *can* be conveyed through this ideally stable section is :

$$Q_i = UA = 1.59 \times 0.54 = 0.86 \text{ [m}^3\text{/s]}$$

This discharge is considerably less than the *required* discharge of $Q = 30$ [m^3/s]. Therefore the wetted surface should be increased by adding a rectangular central part which has a flow depth of $h = 0.45$ [m] and a width of :

$$B'' = n \frac{(Q - Q_i)}{h^{5/3} J_f^{1/2}} = 0.025 \frac{(30 - 0.86)}{(0.45)^{5/3} (0.01)^{1/2}} = 27.57 \text{ [m]}$$

The total width of the channel will thus be : $B + B'' = 1.88 + 27.57 = 29.45$ [m]

This section is drawn below :

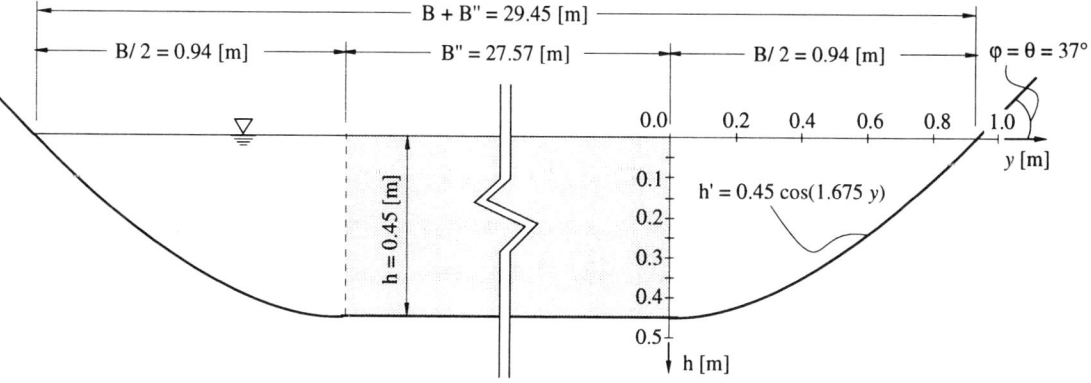

iii) The figure below shows the superposition of the sections, which were determined using the three different methods. It is interesting to note the differences in the dimensions of channels designed according to the erosion criteria and the one designed according to the criteria of an ideal stable section. Each of the three methods rely on different assumptions and a perfect agreement of the results should not be expected.

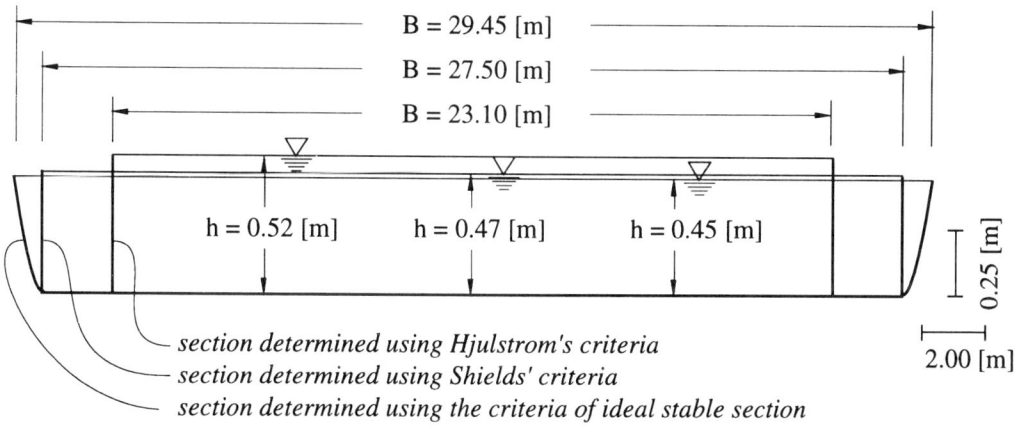

Ex. 3.E

In a riveted steel channel, the uniform flow is established at a depth equal to 70% of the critical depth. This channel has a rectangular cross section with a width of b = 9 [m]. An average velocity of U = 12 [m/s] has been assumed for its design.

i) What bed slope should the channel have for this flow ?
ii) What flow regime should one expect ?
iii) What is the shear stress at the bottom of the canal ?
iv) Verify if there is air entrainment into the flow and if so determine its influence on the flow depth.
v) After a long straight reach, the channel makes a $\alpha = 60°$ curve with a radius of curvature of $r_o = 100$ [m] . How much super-elevation should one expect ? Will cross waves develop in the curved reach ?

SOLUTION :

i) Assuming a turbulent rough, uniform flow regime the average velocity in the channel can be calculated using the formula of Manning-Strickler :

$$U = K_s R_h^{2/3} S_f^{1/2} \tag{3.16}$$

from which one can obtain the bed slope : $S_f = \left[U / (K_s R_h^{2/3}) \right]^2$

The average velocity, U, of the flow is specified. From Table 3.2 the friction coefficient is estimated as being $K_s \cong 65$ [m$^{1/3}$s^{-1}] .

For a rectangular channel (see also Table 1.1), one has : $R_h = \dfrac{b\, h_n}{b + 2h_n}$

The width of the channel, b, is given, but the normal depth, $h = h_n$, has to be calculated using the following relationship: $h_n = 0.7\, h_c$. The critical flow depth, h_c, for a rectangular channel is given by :

$$\dfrac{h_c}{2} = \dfrac{q^2}{2g h_c^2} \quad \text{or} \quad h_c = \sqrt[3]{\dfrac{q^2}{g}} \tag{2.23}$$

where the unit discharge, q, for the uniform flow is :

$q = Q / b = (U\, A) / b = (U\, bh_n) / b = U h_n = U\, (0.7\, h_c)$

One can therefore write :

$$h_c = \sqrt[3]{\dfrac{q^2}{g}} = \sqrt[3]{\dfrac{(0.7\, U h_c)^2}{9.81}} = \sqrt[3]{\dfrac{[(0.7)\,(12)\, h_c]^2}{9.81}} = 1.93\, h_c^{2/3}$$

which yields : $h_c = 7.19$ [m] and $h_n = 0.7\,(7.19) = 5.03$ [m]

UNIFORM FLOW 125

The discharge for uniform flow, $h = h_n$, is :

$Q = (h\ b)\ U = (5.03 \times 9.0)\ 12 = 543\ [m^3/s]$

One can now compute : $\quad R_h = \dfrac{b\ h_n}{b + 2h_n} = \dfrac{9\ (5.03)}{9 + 2\ (5.03)} = 2.38\ [m]$

and the bed slope : $\quad S_f = \left[\dfrac{U}{K_s\ R_h^{2/3}}\right]^2 = \left[\dfrac{12}{(65)\ (2.38)^{2/3}}\right]^2 = 0.0107\ [-]$.

For such a uniform flow, the bed slope should be : $\quad S_f = 0.0107\ [-]$.

ii) Given that : $\quad h_c > h_n \quad \Rightarrow \quad$ the flow is supercritical, namely : $\text{Fr} > 1$.

For a water temperature of $T = 20\ [°C]$, the viscosity is $\nu = 1.004 \times 10^{-6}\ [m^2/s]$ (see *Graf & Altinakar*, 1991, Table I.3). The Reynolds number is then (see eq. 1.7) :

$\text{Re}' = \dfrac{R_h\ U}{\nu} = \dfrac{(2.38)\ (12)}{1.004 \times 10^{-6}} = 2.8 \times 10^7 > 2000 \quad \Rightarrow \quad$ the flow is turbulent.

Moreover, on the Moody-Stanton diagram (see *Graf & Altinakar*, 1991, p.438), it can be verified that the flow is in the region of turbulent rough flow, where the friction coefficient is independent of the Reynolds number.

The flow regime in this channel is therefore *supercritical and turbulent rough*.

iii) The bed-shear stress can be calculated using eq. 3.5. By taking $\rho = 998.2\ [kg/m^3]$ for water at $20\ [°C]$ (see *Graf & Altinakar*, 1991, Table I.3) one has :

$\tau_o = \rho\ u_*^2 = \gamma\ R_h\ S_f = \rho g\ R_h\ S_f = 998.2 \times 9.81 \times 2.38 \times 0.0107 = 249.4\ [N/m^2]$

iv) The Froude number for the flow is :

$\text{Fr} = \dfrac{U}{\sqrt{g\ h_n}} = \dfrac{12}{\sqrt{9.81 \times 5.03}} = 1.71\ [-]$

For a turbulent rough flow the surface instabilities are expected only for $\text{Fr} > 2$. Given that in the present case $\text{Fr} < 2$, there will be no roll-wave forming on the free-water surface.

Since the channel slope is weak, it can also be surmised that there will not be any air entrainment. This can be verified using the following equation to compute the mean air concentration in the cross section :

$\overline{C} = 0.7\ \log\ (\sin\alpha\ /\ q^{1/5}) + 0.9 \qquad (3.52)$

The unit discharge is : $\quad q = Q\ /\ b = U\ h_n = 12 \times 5.03 \cong 60.4\ [m^2/s]$

$\overline{C} = 0.7\ \log\ (0.0107\ /\ 60.4^{1/5}) + 0.9 = -0.73 < 0$

The calculated value of the mean air concentration is negative, which is an impossibility. It can therefore be concluded that no air entrainment takes place across the water surface.

v) The Froude number, Fr = 1.71, calculated above indicates that the flow is supercritical. The cross waves will form in the curved reach. The angle formed between the upstream tangent and the positive and negative waves is given by eq. 3.45 :

$$\sin \beta = \frac{1}{Fr} = \frac{1}{1.71} \quad \Rightarrow \quad \beta = 35.8°$$

By using now data for the curve, $r_o = 100$ [m] and $B = b = 9$ [m], the central angle between a successive maximum and minimum can be calculated using eq. 3.46a :

$$tg\,\theta = \frac{B}{(r_o + B/2)\,tg\,\beta} = \frac{9}{(100 + 9/2)\,tg(35.8)} = 0.119 \quad \Rightarrow \quad \theta = 6.8°$$

Knowing these values, β and θ, one can then calculate the maximum and minimum water depths on the concave and convex side walls of the channel, respectively by using eq. 3.46 :

$$h_{max} = h_n\,Fr^2\,\sin^2(\beta + \theta/2) = 5.03 \times (1.71)^2 \times \sin^2(35.8 + 6.8/2) = 5.88\ [m]$$

$$h_{min} = h_n\,Fr^2\,\sin^2(\beta - \theta/2) = 5.03 \times (1.71)^2 \times \sin^2(35.8 - 6.8/2) = 4.22\ [m]$$

The super-elevation is: $\quad h_{max} - h_{min} = 5.88 - 4.22 = 1.66\ [m]$

It is interesting to compare this value with the one obtained using eqs. 3.43a and 3.47 (see Fig. 3.18) :

$$\Delta z + \Delta z' = 2\,\frac{B}{r_o}\,\frac{U^2}{2g} = 2\,\frac{9}{100}\,\frac{12^2}{2 \times 9.81} = 1.32\ [m] < 1.66\ [m]$$

One can see that the first relationship yields a value larger than the second one. For a better safety a super-elevation of 1.66 [m] can be assumed.

The super-elevation with respect to the normal depth, $h_n = 5.03$ [m], is :

$$h_{max} - h_n = 5.88 - 5.03 = 0.85\ [m]$$

This value will be used in dimensioning the channel cross section in the curved reach.

The formation of cross waves in the curved reach and the downstream straight reach as well as the channel cross sections at different central angles are represented in the following figure. It is important to emphasize that this figure represents a rather simplified image of the reality. A more realistic representation of the water surface can be calculated using the method of characteristics which will be studied in sect. 5.2.2.

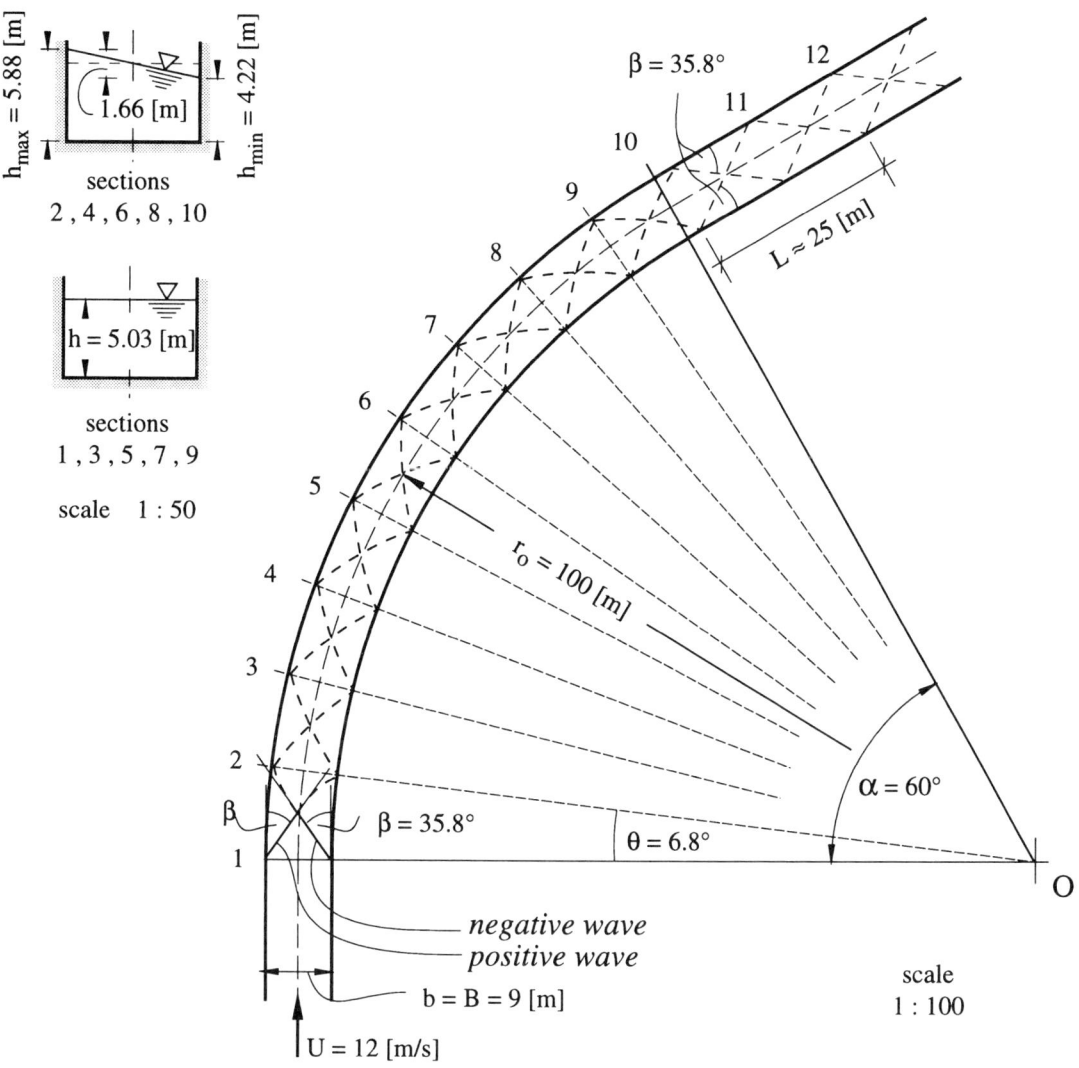

The situation downstream of the curved reach is difficult to predict by a simple method. At the end of the curved reach, at the point of passage from a curved channel to a straight channel, new positive and negative waves are created. According to the ratio between the total angle of the curve, α, and the central angle between a successive maxima and minima, θ, these new waves can be *in* phase or *out* of phase with the existing waves in the curved reach. In some cases a cross-wave pattern of considerable amplitude continues to exist in the downstream channel for some distance before being gradually attenuated by the friction. In the present case, given that the water surface is inclined at the end of the curved reach, it can be expected that the cross waves do continue in the downstream reach. The distance between two maximum (or two minimum) will be in the order of :
$L = 2B / \text{tg}\beta = (2 \times 9) / \text{tg}(35.8) \cong 25$ [m]. This situation is schematically represented on the figure.

3.7.2 Problems, unsolved

Ex. 3.1
The width at the bed of a trapezoidal channel is b = 2.30 [m] and the side slopes are at an angle of 50° with the horizontal plane. The bed drops by 150 [cm] over a distance of 1.2 [km]. Determine the average velocity and the discharge for a flow depth of h = 1.60 [m] and a coefficient of friction after Bazin of $m_B = 0.46$ [$m^{1/2}$]. Calculate also the shear stress on the bed.

Ex. 3.2
Calculate the discharge for the channel as given below. The bed slope was measured, being $S_f = 0.09\%$. Consider the flow as steady and uniform.

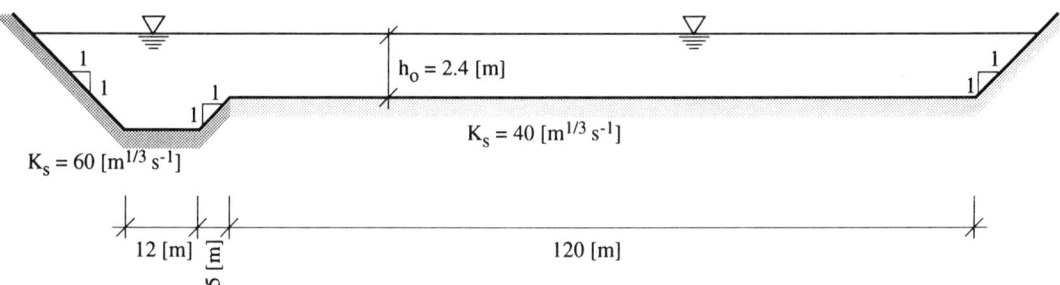

Ex. 3.3
The circular pipe of a sewage system has a coefficient of Manning of n = 0.015 [$m^{-1/3}s$] and is put at an inclination of $S_f = 0.0002$. When the flow depth attains 0.9 of its diameter, the pipe must still transport a discharge of Q = 2.5 [m^3/s]. What should be the diameter? Try to make the same calculations using the coefficient of Weisbach-Darcy.

Ex. 3.4
Water at a temperature of 20 [°C] flows in a large rectangular channel, whose coefficient of Manning was determined previously as being n = 0.011 [$m^{-1/3}s$]. The channel slope is $S_f = 0.0004$ and the flow depth is put at h_n = 1.20 [m]. Calculate the value of the corresponding coefficient of Chezy, C, using the formula of Strickler and the logarithmic expression of $U = 2.5 u_* \ln(41.2 R_h/\delta)$; compare these two resulting values. The viscous sublayer is given by $\delta = 11.6 \, \nu/u_*$.

Ex. 3.5
A semi-circular canal is made of smooth metal, with $n_1 = 0.012$ [$m^{-1/3}s$]; the diameter is D = 2 [m] and the bed slope is $S_f = 0.005$. What diameter should the canal have if it will be reconstructed of corrugated metal, with $n_2 = 0.022$ [$m^{-1/3}s$]?

UNIFORM FLOW   129

Ex. 3.6
A trapezoidal channel must convey a discharge of Q = 5.25 [m³/s] at a flow velocity of U = 1 [m/s]. This channel is made of brickwork (dry, rough stones); the lateral walls are inclined at 45° and the channel slope is S_f = 0.0005. Determine the flow depth and the width of the channel.

Ex. 3.7
A stream should be channelised between the sections 15.2 [km] and 17.5 [km] with a circular pipe having a diameter of 1.5 [m]; the flow drops 4.6 [m] between these two sections. This pipe should transport a maximum discharge of Q = 0.8 [m³/s] at a flow depth of 1/3 of its diameter. What will be the coefficient of Manning, n, for this flow and what pipe material should be selected ?

Ex. 3.8
A rectangular channel being B = 3.6 [m] wide conveys a discharge of Q = 5.50 [m³/s]. What will be the critical flow depth and the corresponding velocity ? For what slope, S_f, will the velocity be critical, if a coefficient of Manning of n = 0.02 [m$^{-1/3}$s] is assumed ?

Ex. 3.9
Find the critical flow depth in a trapezoidal canal, whose width at the bottom is b = 4.0 [m] and whose side slopes are inclined by 1 : 2. The discharge is given as Q = 90 [m³/s].

Ex. 3.10
In a trapezoidal channel the bottom width is b = 6.10 [m] and the side slopes are inclined at m = 2. The elevation of the bed at km 35.0 is 681.30 [m] and the one at km 48.7 is 659.38 [m]. Determine the flow depth for a discharge of Q = 11.33 [m³/s]. A coefficient of Manning of n = 0.025 [m$^{-1/3}$s] was previously determined. Subsequently determine the flow regime, using the average depth as the characteristic length.

Ex. 3.11
A trapezoidal canal must convey a discharge of 16.70 [m³/s] at a flow depth of 1.05 [m] over a distance of 5 [km]. The side slopes are 2 horizontal to 1 vertical; the total drop is given as 8.5 [m]. What will be the bottom width, b, if the flow velocity is supposed to be critical (for the characteristic length use the ratio of the transversal section to the surface width). Give the corresponding coefficient of Manning.

Ex. 3.12
A rectangular channel having a width of B = 6.5 [m] conveys a water discharge of Q = 18 [m³/s]. Establish the specific-energy curve, $H_s = f(h)$, in the range of 0 < h [m] < 8. What is the critical depth ? What would be the specific energy, H_s, if the flow depth is h = 2h_{cr}. What will be the flow depth of uniform, supercritical flow having the *same* specific energy, H_s ?

Ex. 3.13
A rectangular channel, made of smooth concrete and having a width of B = 20 [m], conveys a discharge of Q = 200 [m³/s] at a specific energy of H_s = 3.75 [m]. Determine the flow depth, h_n, and the bed slope, S_f, for uniform, steady flow. Is this flow a supercritical one ?

Ex. 3.14
A trapezoidal channel, made of rather rough concrete and having a side slope of m = 2, is envisioned to transport a discharge of Q = 17 [m³/s] at an average velocity of U = 1.2 [m/s]. Determine the width at the bed, the flow depth, and the bed slope for the hydraulically optimal cross section.

Ex. 3.15
A trapezoidal channel – the coefficient of Strickler is estimated to be K_s = 40 [m$^{1/3}$s^{-1}] – should be designed having a cross section of maximum discharge. The bottom width is b = 2 [m] and the side slopes are of m = 3. The average velocity is fixed at U = 1.98 [m/s]. What will be the geometry of this section, the discharge, Q, and the channel slope, S_f ?

Ex. 3.16
A drainage channel on a highway, running on a slope of S_f = 0.0001, has a triangular section whose side slopes are m = 4 and m = 2. The coefficient of Manning is n = 0.02 [m$^{-1/3}$s]. If flow in the channel is uniform, what will be the flow depth for a discharge of Q = 0.1 [m³/s] ? By how much can the wetted surface be reduced if the channel is made semi-circular ?

Ex. 3.17
A channel of a trapezoidal cross section is built in an alluvium whose granulate is d_{50} = 1 [mm]. The flow depth is h_n = 3 [m] and the width at the bed is b = 4 [m]. The bed slope is S_f = 0.001 and the side slopes of worked stone are 45° inclined. What will be the corresponding velocity, U, and the discharge, Q. Check if the bed will be subject of erosion. May one expect the formation of dunes ?

Ex. 3.18
A very large canal in an alluvium, whose granulate of quartz is d_{50} = 1 [mm], has a bed slope of S_f = 10^{-4}. At what flow depth, h, will erosion commence ? What is the velocity which corresponds to this critical condition ?

Ex. 3.19
One envisions the construction of a non-erodible canal having an ideal, stable cross section. The bed slope is S_f = 10^{-3}. The analysis of the granulometry gave : ρ_s/ρ = 2.65, d_{50} = 6.5 [mm], and an angle of repose of φ = 30°. Establish the design dimensions for the following discharges : Q_1 = 1.5 [m³/s] and Q_2 = 4 [m³/s].

UNIFORM FLOW

Ex. 3.20
Calculate the profile of a trapezoidal channel for a discharge of $Q = 12$ [m³/s] on a bed slope of $S_f = 0.0016$. This channel should be excavated in an alluvium of large gravel. For the side slopes it is recommended to take m = 2.

Ex. 3.21
In the town of Ste-Justice, the construction of a road along the river Happy is envisioned. This project can be realised if the river's width is reduced to 67.5 [m]. All other parameters are the same as the ones in Ex. 3.B.

i) Establish the rating curve, $Q = f(h_n)$, assuming the flow is rough and turbulent.
ii) At which flow depth should one expect the commencement of erosion and deposition ?
iii) Compare these results with the ones of Ex. 3.B. Remark on the consequence of such a reduction in the river width.

Ex. 3.22
The downstream slope of the back of a large weir is 30°; it is long enough such that its flow can be considered uniform. The coefficient of Manning is assumed to be n = 0.0149 [m$^{-1/3}$s]. The width of the weir is B = 7 [m] and the flood discharge is set at $Q = 0.3$ [m³/s]. Determine if one may expect air-entrainment and calculate the flow depth of the mixture as well as the pressure on the back of the weir.

Ex. 3.23
On a slope of 20°, a canal in concrete was projected to evacuate a unit discharge of $q = 35$ [m²/s]. Calculate the flow depth and the pressure on the floor. Will air-entrainment take place ?

Ex. 3.24
A rectangular canal made of wood has a width of B = 8 [m] and should evacuate the flood discharge. The flow depth is h = 1.0 [m] and the flow velocity should be U = 10 [m/s]. Determine the radius of the curve which should be foreseen, under the condition that the maximum flow depth does not exceed a height of $h_{max} = 2.0$ [m].

Ex. 3.25
A rectangular channel, having a bed of sand and side walls of concrete, should be designed to have a cross section of maximum discharge :

i) Knowing that the flow in the channel has a depth of h = 1.0 [m] and runs at a Froude number of Fr = 0.7, what will be the discharge ?
ii) A change in the flow direction of α = 30° should be envisioned. Determine the radius of the curve for the case that the maximum super-elevation is not larger than 15% of the normal depth. What is the head loss in this curve ?

4. NON-UNIFORM FLOW

Steady flow in a prismatic channel is non-uniform or varied, if the flow depth as well as other hydraulic parameters vary from one cross section to another along the length of the channel. In the most general case, the resulting water surface and the streamlines are curvilinear.

The case where the curvature can be neglected will be first investigated : this is the gradually varied flow. The differential equation describing the water-surface profiles are developed; the different types of profiles as well as the methods of computation of these profiles will be exposed.

The case where the curvature is of importance will be presented subsequently : this is the rapidly varied flow, encountered at weirs and hydraulic jumps. This type of flow often takes place over short distances, contrary to gradually varied flow which may extend over very long distances.

Also presented will be the flow in or through transitions, when the cross section or direction of the flow changes suddenly. A few remarks on spatially varied flow will close the chapter.

TABLE OF CONTENTS

4.1 GRADUALLY VARIED FLOW
    4.1.1 Simplified Equations of Saint-Venant
    4.1.2 Equations of Water Surface
    4.1.3 Critical Slope

4.2 FORMS OF WATER SURFACE
    4.2.1 Channels on mild Slope
    4.2.2 Channels on steep Slope
    4.2.3 Channels on critical Slope
    4.2.4 Channels on horizontal Slope
    4.2.5 Channels on adverse Slope

4.3 COMPUTATION OF WATER SURFACE
    4.3.1 Method of successive Approximations
    4.3.2 Method of direct Integration
    4.3.3 Method of graphical Integration

4.4 RAPIDLY VARIED FLOW
    4.4.1 Weirs, Spillways
    4.4.2 Hydraulic Drop
    4.4.3 Underflow Gates
    4.4.4 Hydraulic Jump

4.5 TRANSITIONS
    4.5.1 Channel with variable Bed Floor
    4.5.2 Channel of variable Width
    4.5.3 Oblique Jump

4.6 LATERAL INFLOW

4.7 EXERCISES
    4.7.1 Problems, solved
    4.7.2 Problems, unsolved

NON-UNIFORM FLOW

4.1 GRADUALLY VARIED FLOW

4.1.1 Simplified Equations of Saint-Venant

1° Considered will be (see Fig. 4.1) a prismatic channel with a free-water surface, where the flow is steady but non-uniform (gradually varied) and where the streamlines are almost rectilinear and parallel. Assumed will be a bed slope, S_f, fixed and stationary, being rather weak; the discharge of the incompressible fluid is given by $Q = UA$, with $U(x)$ as the velocity averaged over the entire wetted section, $A(x)$. There is no lateral discharge.

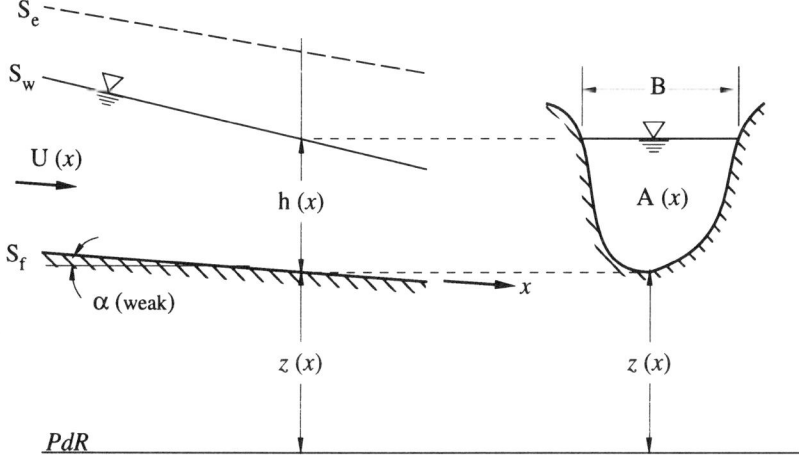

Fig. 4.1 Scheme of steady and non-uniform flow on a fixed bed slope, $z(x)$.

2° The *equation of continuity*, developed in sect. 2.1, is given (see Fig. 2.1) by :

$$\frac{\partial Q}{\partial x} = A\frac{\partial U}{\partial x} + U\frac{\partial A}{\partial x} = 0 \tag{4.1}$$

It would be more correct, notably for non-prismatic channels, to express the second term by :

$$U\frac{\partial A}{\partial x} = U\left(\frac{\partial A}{\partial h}\frac{\partial h}{\partial x} + \frac{\partial A}{\partial x}\bigg|_{h = Cte}\right) \tag{4.1a}$$

since $A = f[\,h(x), x\,]$, and with $\partial A/\partial h = B$.

For a rectangular prismatic channel, above equation, eq. 4.1, yields :

$$h\frac{\partial U}{\partial x} + U\frac{\partial h}{\partial x} = 0 \qquad\qquad B = Cte \tag{4.2}$$

3° The *equation of motion* for steady and gradually varied flow, developed in sect. 2.2, is given (see Fig. 2.3) by :

$$\frac{U}{g}\frac{\partial U}{\partial x} + \frac{\partial h}{\partial x} + \frac{\partial z}{\partial x} = -S_e \qquad (4.3)$$

with $S_f = -(\partial z/\partial x)$ as the slope of the channel bed, $h_r = S_e\,\partial x$ as the head loss and S_e as the slope of the energy-grade line.

As a first (and usual) approximation, one assumes that the energy slope, S_e, can be expressed with a relation established for steady uniform flow (see sect. 3.2) of the type of Weisbach-Darcy, or :

$$S_e = f\,\frac{1}{4R_h}\,\frac{U^2}{2g} \qquad (3.9)$$

or of the type of Chézy :

$$S_e = \frac{8g}{C^2}\,\frac{1}{4R_h}\,\frac{U^2}{2g}$$

Thus the equation of motion, eq. 4.3, can also be written as follows :

$$\frac{1}{g}\frac{\partial}{\partial x}\left(\frac{Q^2}{A}\right) + A\frac{\partial h}{\partial x} + A\frac{\partial z}{\partial x} = -A\,S_e \qquad (4.3a)$$

4° The different terms in the above equation, eq. 4.3, can be considered as representations of slopes. In the first term, the slope is due to a longitudinal variation of the velocity. The second and third term define the slope of the water surface, S_w, while the forth term stands for the energy slope, S_e.

4.1.2 Equations of Water Surface

1° In the study of gradually varied flow the problem consists in the development of the position, x, and the form, $h(x)$, of the profile of the water surface for a given discharge, Q, and a given geometrical form of the cross section, A. A very long channel will be considered.

2° Such a study can be done, using the equation of energy (see Fig. 2.3), or :

$$\frac{U^2}{2g} + h + z = H \qquad (2.9)$$

and the equation of continuity :

$$Q = UA \qquad (2.2)$$

NON-UNIFORM FLOW

3° Differentiating this equation, eq. 2.9, with respect to x, yields:

$$\frac{d}{dx}\frac{(Q/A)^2}{2g} + \frac{dh}{dx} + \frac{dz}{dx} = \frac{dH}{dx} \qquad (4.4)$$

However, taking the following definitions and using the relation of Chézy:

$$\frac{dz}{dx} = -S_f \quad ; \quad \frac{dH}{dx} = -S_e = -\frac{(Q/A)^2}{C^2 R_h}$$

one writes:

$$\frac{d}{dx}\frac{(Q/A)^2}{2g} + \frac{dh}{dx} - S_f = -S_e = -\frac{(Q/A)^2}{C^2 R_h} \qquad (4.5)$$

For a given channel, the arguments, R_h, A and C, are all functions of x and of h, while the bed slope, S_f, is a function of x. This equation, eq. 4.5, is a differential equation of the first order; it allows to determine the flow depth, $h(x)$, as a function of the distance, x, for a given discharge, Q.

Note that eq. 4.5, as well as eq. 4.3a, is a simplified form of the equation of Saint-Venant, being valid for both prismatic and non-prismatic channels.

4° If the channel is prismatic, the cross section, A, is only a function of the flow depth, h; consequently, the first term in eq. 4.5 can be expressed by:

$$\frac{d}{dx}\left(\frac{Q^2}{2gA^2}\right) = \frac{Q^2}{2g}\left(-\frac{2}{A^3}\frac{dA}{dx}\right) = -\frac{Q^2}{gA^3}\left(\frac{dA}{dh}\frac{dh}{dx}\right) \qquad (4.6)$$

which, when using the definition of $dA/dh = B$, renders:

$$\frac{d}{dx}\left(\frac{Q^2}{2gA^2}\right) = -\frac{Q^2}{gA^3} B \frac{dh}{dx} \qquad (4.6a)$$

5° The differential equation, eq. 4.5, can now be written as:

$$-\frac{(Q/A)^2}{gA} B \frac{dh}{dx} + \frac{dh}{dx} - S_f = -\frac{(Q/A)^2}{C^2 R_h} \qquad (4.7)$$

but it is usually written — describing the gradual variation of the flow depth, dh/dx in the longitudinal direction — in the form of:

$$\frac{dh}{dx} = S_f \frac{1 - \dfrac{(Q/A)^2}{C^2 R_h S_f}}{1 - \dfrac{(Q/A)^2}{gA/B}} \qquad (4.8)$$

The equation for gradually varied flow, valid for prismatic channels, allows to draw the line(s) of the free-water surface for the different possible cases ; all this can be done without integrating the above equation.

The above differential equation, eq. 4.8, defines the inclination (slope) with respect to the channel bed and not the slope of the water surface; h is the flow depth and not the elevation, h + z.

6° Interesting remarks concerning this equation, eq. 4.8, follow :

i) The derivative is zero, dh/dx = 0, if the nominator becomes :

$$(Q/A)^2 = C^2 R_h S_f \quad \Rightarrow \quad U = C\sqrt{R_h S_f} \quad (3.11)$$

There is uniform flow; this defines the *normal flow depth*, $h \equiv h_n$ (see sect. 3.3.2). The water surface and bed surface are parallel. Thus for:

$\dfrac{dh}{dx} = 0$ the flow depth remains constant,

$\dfrac{dh}{dx} > 0$ the flow depth increases,

$\dfrac{dh}{dx} < 0$ the flow depth decreases.

ii) The derivative is infinite, dh/dx = ∞, if the denominator becomes :

$$(Q/A)^2 = gA/B \quad \Rightarrow \quad \frac{U^2}{gD_H} = Fr_c^2 = 1 \quad (2.22)$$

There is critical flow; this defines the *critical flow depth*, $h \equiv h_c$ (see sect. 2.3.3). The water surface is perpendicular to the bed surface; this is a theoretical rather than a practical statement.

iii) The normal and the critical flow depth are equal, $h_n \equiv h_c$, if :

$$C^2 R_h S_f = gA/B$$

There is normal flow, being also critical flow. This is used for a definition (see sect. 4.1.3) of the *critical slope*, or :

$$S_f \equiv S_c = \frac{gA}{C^2 B R_h} \quad (4.9)$$

NON-UNIFORM FLOW

7° The equation for gradually varied flow, eq. 4.8, can also be written using the notion of the conveyance (see sect. 3.3.1):

$$K(h) = \frac{1}{n} R_h^{2/3} A = C R_h^{1/2} A \qquad (3.34)$$

For uniform flow, this relation takes the following form:

$$K_n(h) = Q/\sqrt{S_f} \qquad (3.35)$$

The expression in the denominator of above equation, eq. 4.8, now yields:

$$\frac{Q^2 B}{g A^3} = \frac{Q^2}{(C^2 A^2 R_h S_f)} \frac{(C^2 A^2 R_h S_f)}{(g A^3/B)} = \frac{(Q/\sqrt{S_f})^2}{(CAR_h^{1/2})^2} \frac{S_f}{(gA/C^2 B R_h)}$$

$$= \frac{K_n^2}{K^2} \cdot \frac{S_f}{S_c}$$

where S_c is the critical slope, given with eq. 4.9.

Using the above three relations, eq. 4.8 can be written (see *Bakhmeteff*, 1932, p. 52) in the following form:

$$\frac{dh}{dx} = S_f \frac{1 - \left(\frac{K_n}{K}\right)^2}{1 - \left(\frac{K_n}{K}\right)^2 \frac{S_f}{S_c}} \qquad (4.8b)$$

8° Yet there is another form in which this equation, eq. 4.8, can be expressed, notably when the channel is considered as being wide and rectangular.

  i) The normal depth (see sect. 3.3.2) is given by:

$$h_n^3 = \frac{q^2}{C^2 S_f} \qquad (3.36)$$

  ii) The critical depth (see sect. 2.3.3) is given by:

$$h_c^3 = \frac{q^2}{g} \qquad (2.24)$$

where $q = Q/B$, $h \cong D_h$, $B \gg h$, $R_h \cong h$

iii) Using now these two expressions, eq. 3.36 and eq. 2.24, and assuming that the Chézy coefficient, C, is not dependent on the flow depth, h, the above eq. 4.8 becomes :

$$\frac{dh}{dx} = S_f \frac{1 - (h_n/h)^3}{1 - (h_c/h)^3} \qquad (4.8a)$$

This relation is known as the *equation of Bresse* (see *Jaeger*, 1954, p. 74).

If one were to use the Manning coefficient, n, also independent of the flow depth, above eq. 4.8a reads :

$$\frac{dh}{dx} = S_f \frac{1 - (h_n/h)^{10/3}}{1 - (h_c/h)^3} \qquad (4.8c)$$

9° The equation of gradually varied flow, eq. 4.8, was established under the assumption that the streamlines are quasi-rectilinear and parallel. Boussinesq (see *Flamant*, 1923, p. 231 and *Jaeger*, 1954, p. 127) presented a more general equation, including the curvature of the streamlines.

Flow over a gradually varied channel bed was investigated by Massé (see *Jaeger*, 1954, p. 136 and *Chow*, 1959, p. 237) for the prismatic and non-prismatic channels; solutions were obtained using the method of singular points.

4.1.3 Critical Slope

1° The critical slope of a channel, whatever be its cross sectional form, can be expressed by :

$$S_c = \frac{gA}{C^2 BR_h} \qquad (4.9)$$

It is a function of the flow depth, h, which defines the slope in a way that this very depth is the normal depth, h_n, as well as the critical depth, h_c, independently of the discharge.

2° For a wide and rectangular channel, the critical slope is :

$$S_c = \frac{g}{C^2} \qquad (4.9a)$$

3° The notion of the critical slope plays an important role as reference parameter in the study of gradually varied flow.

NON-UNIFORM FLOW

4° If the bed slope — for a given discharge, Q , and roughness coefficient, C — is smaller than the critical slope, the normal depth will be larger than the critical depth, or :

$$S_f < S_c \quad ; \quad h_n > h_c .$$

The channel is said, being of *mild slope*.

The uniform flow, corresponding to this normal depth, will be *subcritical* or fluvial (see sect. 2.3.1 and Fig. 2.5).

5° If the bed slope is larger than the critical slope, the normal depth will be smaller than the critical depth, or :

$$S_f > S_c \quad ; \quad h_n < h_c .$$

The channel is said, being of *steep slope*.

The uniform flow, corresponding to this normal depth, will be *supercritical* or torrential (see sect. 2.3.1 and Fig. 2.5).

4.2 FORMS OF WATER SURFACE

1° A summary of the different (possible) forms (curves or profiles) of the water surface encountered in gradually varied flow and described by eq. 4.8 is given with Fig. 4.2, as a function of the bed slope, S_f , and of the discharge, Q , (see *Bakhmeteff*, 1932, chap. 7).

Each of these curves is not obtained to give an explicit solution, but rather to give a preliminary image of the variation of flow depth, h , with the longitudinal distance, x ; this allows to get acquainted with the hydraulic phenomenon.

2° The boundary conditions to the problems are the following ones :

   i) The curves are asymptotic tangents at the line of the normal depth, $h \equiv h_n$.

   ii) The curves are orthogonal to the line of the critical depth, $h \equiv h_c$.

   iii) For a growing flow depth, $h \to \infty$, the curves will become asymptotically horizontal lines, $dh/dx \to S_f$, where $S_f = - dz/dx$.

3° The flow immediately downstream and upstream of the critical depth, h_c , becomes curvilinear and rapidly varied (see sect. 4.1.1) ; the preceding equation of flow, eq. 4.8, is no more applicable (the dotted curves which are drawn in Fig. 4.2 are merely indicative).

If the flow depth changes suddenly across the line of the critical depth, a discontinuity in the water surface is encountered. For increasing depths, an *hydraulic jump* will occur ; for decreasing depths, a *hydraulic drop* will form (see sect. 4.4).

4° Taking into account above boundary conditions to the equation of flow, eq. 4.8, one may now draw the water-surface profiles for the different (possible) cases as are encountered in open-channel flow.

5° The classification of the water-surface profiles (see Fig. 4.2) is usually done according to the bed slope, S_f. The five following cases are distinguished :

	$S_f < S_c$	channels on *mild* slope	: M
$\mathbf{S_f > 0}$	$S_f > S_c$	channels on *steep* slope	: S
	$S_f = S_c$	channels on *critical* slope	: C
$\mathbf{S_f = 0}$		channels on *horizontal* slope	: H
$\mathbf{S_f < 0}$		channels on *adverse* slope	: A

6° Each curve of a water-surface profile is composed of different branches (or zones). Together they do not represent a real single line of the water surface but are a summary of different possible cases. One real continuous water-surface profile is usually given by one *single* branch. If the water surface is made up of many branches, these may belong to different classes of the water-surface profiles.

7° Above all, it is of major importance, for each curve and its branches, that the following be respected :

i) Subcritical flow, Fr < 1 , is controlled by a singularity situated *downstream* (hydraulic drop, weir, etc.). The computations must be done towards the upstream.

ii) Supercritical flow, Fr > 1 , is controlled by a singularity situated *upstream* (hydraulic drops, gate, etc.). The computation must be done towards the downstream.

Note, that in the examples given in Fig. 4.2a, the *point of control*, which is this singularity, has been always indicated.

8° In what follows, we shall present first the channels on descending slopes, $S_f > 0$, then the ones on horizontal slopes, $S_f = 0$, and finally the ones on adverse slopes, $S_f < 0$. All this is summarised on Fig. 4.2, with selected but typical examples given on Fig. 4.2a.

NON-UNIFORM FLOW

4.2.1 Channels on mild Slope

1° The forms of the water surface in a descending channel, $S_f > 0$, on a mild slope of:

$S_f < S_c$; $h_n > h_c$

are summarised in Fig. 4.2. This curve has three branches : M1, M2 and M3.

2° *Branch M1* :

$h > h_n > h_c$; $Fr < 1$; dh/dx is positive.

The curve comes from upstream, where it approaches asymptotically the normal depth, h_n ; it goes increasingly downstream towards a horizontal tangent.

This branch M1 is encountered, for example :
- upstream of a dam or a weir,
- upstream of a pier,
- at junctures of certain bed slopes.

This curve, which is often encountered in fluvial hydraulics, is *the* backwater curve.

3° *Branch M2* :

$h_n > h > h_c$; $Fr < 1$; dh/dx is negative.

Upstream, this curve approaches asymptotically the normal depth, h_n ; it goes decreasingly downstream towards the critical depth, h_c.

This branch M2 is encountered, for example :
- upstream of an increase in the bed slope,
- upstream of an hydraulic drop.

4° *Branch M3* :

$h_n > h_c > h$; $Fr > 1$; dh/dx is positive.

This curve comes from upstream; it goes increasingly towards the critical depth, h_c, where it terminates with a hydraulic jump (see sect. 4.4.4).

Conditions Eq. 4.8a	$\frac{h_n}{h}$	Sign num.	$\frac{h_c}{h}$	Sign den.	Sign dh/dx	Change of flow depth	Name	Profiles *vertical scale exaggerated*
$S_f > 0$ $S_f < S_c$ $h_n > h_c$	< 1	+	< 1	+	+	increase	M1	
	< 1	+	> 1	−		not possible		
	> 1	−	< 1	+	−	decrease	M2	
	> 1	−	> 1	−	+	increase	M3	
$S_f > 0$ $S_f > S_c$ $h_n < h_c$	< 1	+	< 1	+	+	increase	S1	
	< 1	+	> 1	−	−	decrease	S2	
	> 1	−	> 1	−	+	increase	S3	
$S_f > 0$ $S_f = S_c$ $h_n = h_c$	< 1	+	< 1	+	+	increase	C1	
	> 1	−	> 1	−	+	increase	C3	
$S_f = 0$ $h_n = \infty$		−	< 1	+	−	decrease	H2	
		−	> 1	−	+	increase	H3	
$S_f < 0$ $h_n < 0$	< 1	−	< 1	+	−	decrease	A2	
	< 1	−	> 1	−	+	increase	A3	

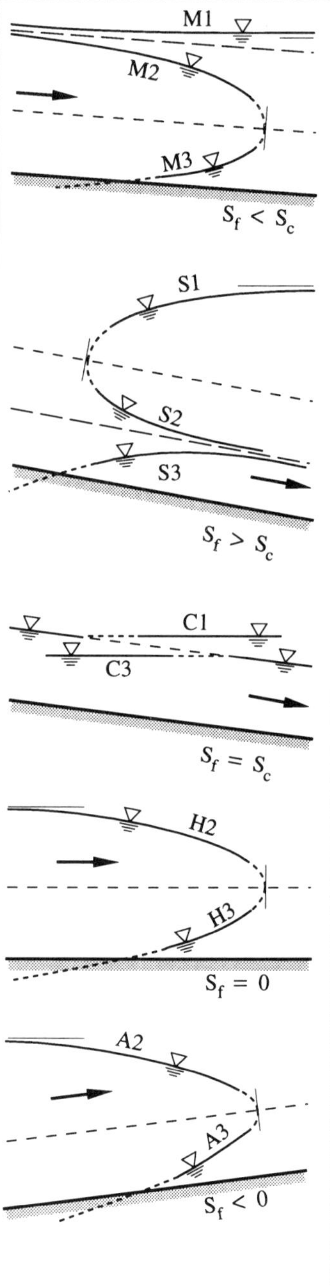

Fig. 4.2 Water-surface profiles for gradually varied flow.
▽ water surface ; — — — normal depth ; − − − − critical depth

NON-UNIFORM FLOW

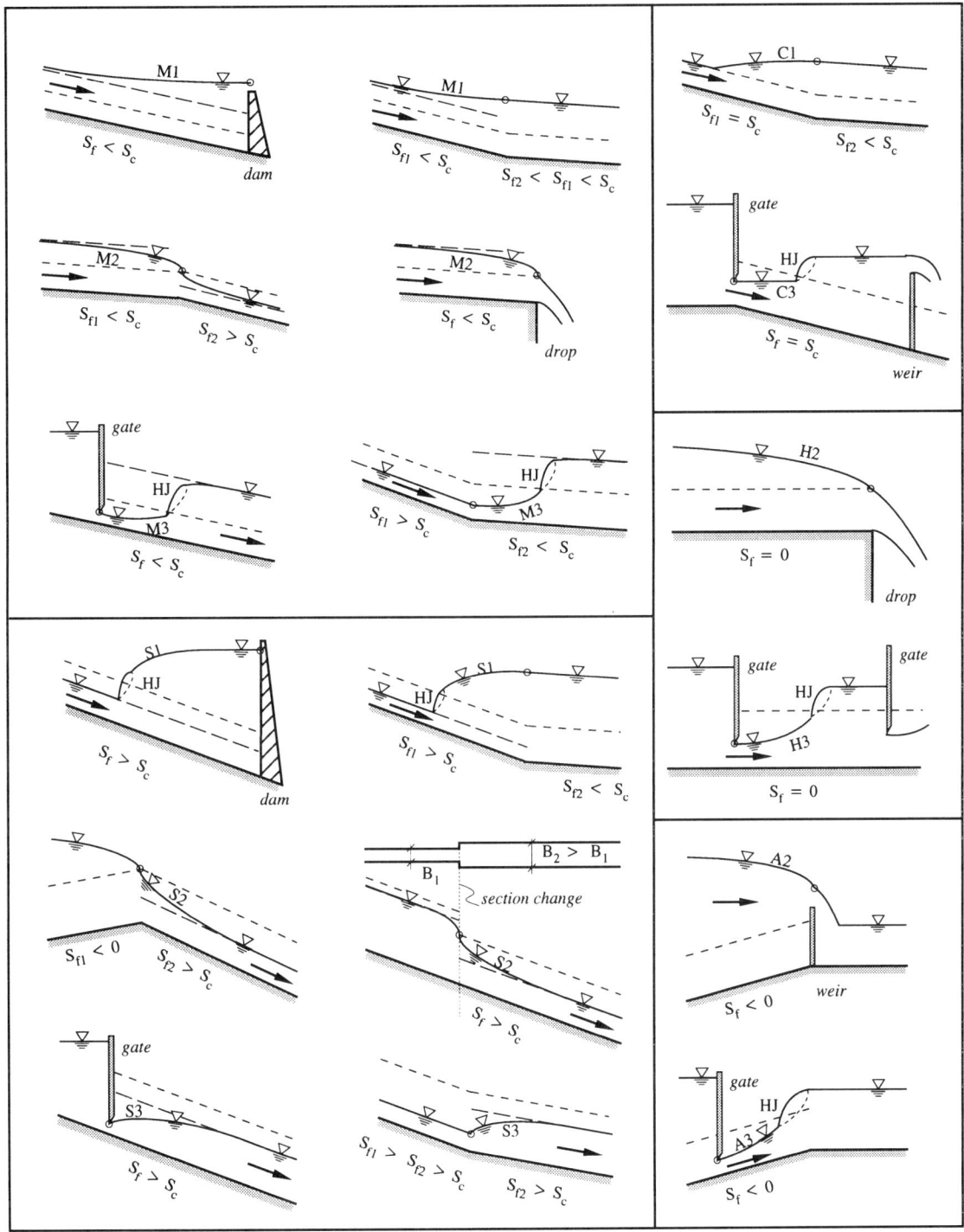

Fig. 4.2a Some examples of flow.
○ control section ; HJ hydraulic jump

This branch M3 is encountered, for example :
- when a supercritical flow enters a mild channel,
- after a change in bed slope from steep to mild.

This curve is often followed by the formation of a hydraulic jump.

4.2.2 Channels on steep Slope

1°   The forms of the water surface in descending channels, $S_f > 0$, on a steep slope of:

$$S_f > S_c \quad ; \quad h_n < h_c$$

are summarised in Fig. 4.2. This curve has three branches : S1, S2 and S3.

2°   *Branch S1 :*

$$h > h_c > h_n \quad ; \quad Fr < 1 \quad ; \quad dh/dx \text{ is positive.}$$

The curve begins at the critical depth, h_c, of a hydraulic jump and terminates increasingly as a tangent to a horizontal line.

The branch S1 is encountered, for example :
- upstream of a dam or a weir,
- at a juncture of certain bed slopes.

This curve is often preceded by an hydraulic jump.

3°   *Branch S2 :*

$$h_c > h > h_n \quad ; \quad Fr > 1 \quad ; \quad dh/dx \text{ is negative.}$$

This curve takes place in the transition between critical depth and uniform flow; it is usually very short.

This branch S2 is encountered, for example :
- downstream of a sudden increase in bed slope,
- downstream of an enlargement of the cross section.

4°   *Branch S3 :*

$$h_c > h_n > h \quad ; \quad Fr > 1 \quad ; \quad dh/dx \text{ is positive.}$$

NON-UNIFORM FLOW 147

This curve takes place in the transition between supercritical flow and uniform flow, which it approaches increasingly as a tangent.

This branch S3 is encountered, for example :
- downstream of a gate, when the flow is below the normal depth,
- when the bed slope is reduced.

4.2.3 Channels on critical Slope

1° The forms of the water surface in channels on a critical slope of :

$$S_f \equiv S_c \quad ; \quad h_n \equiv h_c$$

are summarised in Fig. 4.2. This curve has two branches : C1 and C3.

Profiles of this case represents the transition condition between the preceding cases, M and S; the intermediate branch, C2, disappears, since it presents the case of uniform, critical flow.

2° For $h_n \equiv h_c$, the differential equation, 4.8a , becomes :

$$\frac{dh}{dx} = S_f \tag{4.10}$$

The representative curve is thus horizontal.

Flow at critical depth is usually unstable; undulations begin to appear on the water surface (see sect. 2.3.3).

3° *Branch C1 :*

$$h > h_n \ (\equiv h_c) \quad ; \quad Fr < 1 \quad ; \quad dh/dx \text{ is positive.}$$

The curve is horizontal, when the Chézy formula is used.

This branch C1 is encountered, for example :
- at a juncture of certain bed slopes;
- upstream of a dam (weir).

4° *Branch C3 :*

$$h < h_n \ (\equiv h_c) \quad ; \quad Fr > 1 \quad ; \quad dh/dx \text{ is positive.}$$

The curve is also horizontal.

This branch C3 is encountered, for example :
- when the bed slope is reduced to a critical one,
- downstream of a sluice gate, when the flow is below the normal depth.

4.2.4 Channels on horizontal Slope

1° The forms of the water surface in channels on an horizontal slope of :

$$S_f = 0 \quad ; \quad h_n = \infty$$

are summarised in Fig. 4.2. This curve has two branches : H2 and H3.

2° The normal depth, h_n, is meaningless, becoming infinite for a horizontal slope (see sect. 3.3.2). Consequently, the branch H1 is not established.

3° Profiles of this case, H2 and H3, represent the limiting case of the branches, M2 and M3, when the channel bed becomes horizontal.

The branch H2 is encountered, for example :
- at an hydraulic drop.

The branch H3 is encountered, for example :
- when supercritical flow enters into a horizontal channel.

4.2.5 Channels on adverse Slope

1° The forms of the water surface in channels on an adverse slope of :

$$S_f < 0$$

are summarised in Fig. 4.2. This curve has two branches : A2 and A3.

2° The normal depth, h_n, is meaningless; uniform flow cannot establish itself. The branch A1 is thus not possible.

3° The two branches, A2 and A3, being of parabolic shape, are similar to the ones of horizontal slopes, H2 and H3; they occur only infrequently.

The branch A2 is encountered, for example :
- at a juncture of certain bed slopes.

The branch A3 is encountered, for example :
- when supercritical flow enters into a critical channel.

NON-UNIFORM FLOW   149

4.3 COMPUTATION OF WATER SURFACE

1° Previously the equation, which describes the water surface (see sect. 4.1.2), was established ; it allows to sketch the different forms of the water-surface profile (see sect. 4.2), but this only in a general way.

2° Integration of this equation, eq. 4.8, becomes necessary to compute algebraically and to draw exactly the form of the water-surface profile. The following three methods can be used:
   i)   the method of successive approximations (sect. 4.3.1),
   ii) the method of direct integration (sect. 4.3.2),
   iii) the method of graphical integration (sect. 4.3.3).

3° Each of these methods will give a water-surface profile, but a constant of integration must be supplied. This constant may not be chosen in an arbitrary way; one is thus obliged to find a point (section) which has a physical reality. This will then be the control (reference) point, namely the cross section of control. The *control point* is usually calculated at an hydraulic singularity, which is responsible for the establishment of gradually varied flow; it might be a hydraulic drop, a weir or a gate. Calculations must begin at the control section and proceed in the direction in which the control operates.

4.3.1 Method of successive Approximations

1° The equation of motion shall be used :

$$\frac{d}{dx}\frac{(Q/A)^2}{2g} + \frac{dh}{dx} - S_f = -S_e = -\frac{(Q/A)^2}{C^2 R_h} \tag{4.5}$$

When all is multiplied by dx, this yields :

$$dh = (S_f - \frac{Q^2}{C^2 A^2 R_h})\, dx - \frac{Q^2}{2g}\, d\left(\frac{1}{A^2}\right) \tag{4.11}$$

and taking finite differences, one writes :

$$(h_{i+1} - h_i) = (S_f - \frac{Q^2}{\overline{C}^2\, \overline{A}^2 \overline{R}_h})(x_{i+1} - x_i) - \frac{Q^2}{2g}\left(\frac{1}{A_{i+1}^2} - \frac{1}{A_i^2}\right) \tag{4.12}$$

The flow depth, h_i, takes place at a distance (abscissa), x_i, while a close-by flow depth, h_{i+1}, takes place at a close-by distance, x_{i+1} (see Fig. 4.3).

The values, $\overline{C}$, $\overline{A}$ and $\overline{R}_h$, correspond to the supposed average flow depth of :

$$\overline{h} = \frac{h_{i+1} + h_i}{2}$$

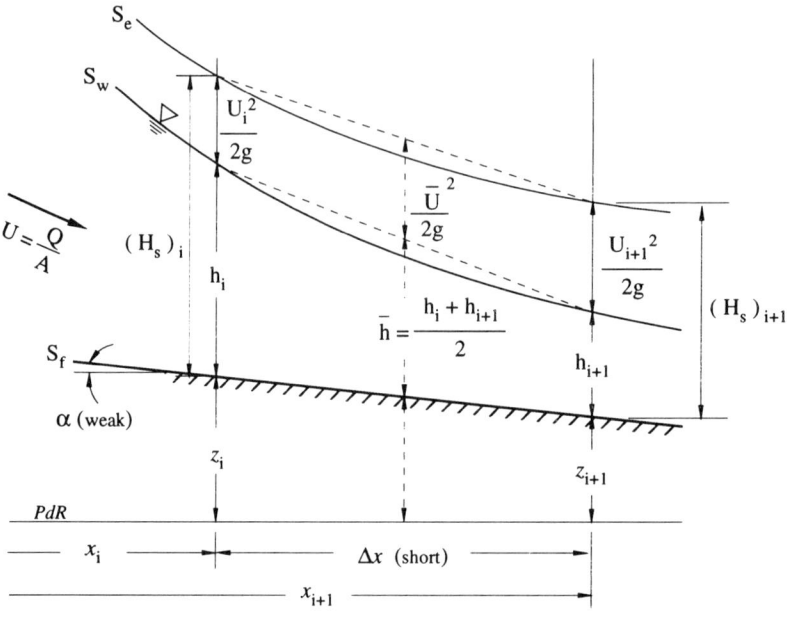

Fig. 4.3 Scheme of a non-uniform flow between two cross sections.

The above equation, eq. 4.12, can also be written (see *Silber*, 1968, p.26) as :

$$(h + z)_{i+1} - (h + z)_i = (- \frac{Q^2}{\overline{C^2 \, A^2 R_h}}) (x_{i+1} - x_i) - \frac{Q^2}{2g} (\frac{1}{A_{i+1}^2} - \frac{1}{A_i^2}) \qquad (4.12a)$$

The two forms of the equation of motion, eqs. 4.12 or 4.12a, are valid for prismatic and non-prismatic channels, thus also for natural watercourses.

For a prismatic channel, above equation, eq. 4.5, can be also written as :

$$\frac{d}{dx} (H_s) = S_f - S_e \qquad (4.13)$$

where $H_s = (h + U^2/2g)$ is the specific energy, eq. 2.14.

Taking now finite differences, one may write :

$$(H_s)_{i+1} - (H_s)_i = (S_f - \frac{\overline{U}^2}{\overline{C^2 R_h}}) (x_{i+1} - x_i) \qquad (4.14)$$

where $\overline{U}$ corresponds to the supposed average flow depth.

NON-UNIFORM FLOW

2° The above developed equations can be used in the following way to draw the water-surface profile :

   i) For a reach of a small distance, $\Delta x = (x_{i+1} - x_i)$, arbitrarily taken, one traces the variation of the flow depth, $\Delta h = (h_{i+1} - h_i)$; this is the method of reaches or the *standard* step method; or

   ii) For a small difference in flow depth, $\Delta h = (h_{i+1} - h_i)$, arbitrarily taken, one traces the distance, $\Delta x = (x_{i+1} - x_i)$, between these two flow depths; this is the method of depths variation or the *direct* step method.

   iii) Before starting the computation with either of the above methods, one has to establish *the control point(s)*, where a known relationship between the flow depth and the discharge exists. A control point may be found at the entrance or exit of the channel, at a weir, a gate or a hydraulic drop.

   iv) Computations proceed towards the upstream for subcritical flow, $Fr < 1$, and towards the downstream for supercritical flow, $Fr > 1$.

   v) The methods of successive approximations are in general rather time-consuming, but are often rather precise.

3° *Method of reaches (Δx is fixed)* :

This method can be used for the solution of the equation of motion in form of eq. 4.12 and eq. 4.14.

   i) One supposes (chooses) to know the flow depth, h_i, at an abscissa, x_i.

   One *wants* to know the flow depth, h_{i+1}, at a close-by abscissa, x_{i+1}.

   One chooses a first value, h'_{i+1}, where A'_{i+1}.

   One calculates the values, $\overline{C}$, $\overline{A}$ and $\overline{R_h}$, or $\overline{U}$, which correspond to the value of the average flow depth, $\overline{h} = (h_i + h_{i+1})/2$.

   One puts these values into the difference equation, eq. 4.12 or eq. 4.14, where S_f is the bed slope of the reach. Thus one obtains a value for the flow depth, h_{i+1}, which will probably be different from the chosen value, h'_{i+1}.

   One then restarts with successive approximations until the value, h_{i+1}, given by this equation becomes equal to the last chosen value, h'_{i+1}, h''_{i+1}, h'''_{i+1} ....

   One goes now to the next reach, etc.

ii) The (implicit) method of reaches – also referred to as the *standard step method* – is in general valid for prismatic, eq. 4.14, as well as for non-prismatic channels, eq. 4.12, where the cross section may change from one reach to another. However, it should be remembered that this method is lengthy and complicated (see *Henderson*, 1966, p. 136 and *Chow*, 1959, p. 265). A more efficient way of using this method was proposed by *Prasad* (see *Ranga-Raju*, 1981, p.145), making use of numerical integration.

iii) For *natural watercourses*, it is usually less important to evaluate the variation of the flow depth, h. The evaluation of the position of the water surface with respect to the horizontal datum, $h = h + z$, is more often of interest since the channel bed may vary continuously. The equation of motion, eq. 4.12a, shall thus be written as follows:

$$(h + z + \frac{Q^2}{2g\,A^2})_{i+1} - (h + z + \frac{Q^2}{2g\,A^2})_i =$$

$$- \frac{Q^2}{\overline{C^2 A^2 R_h}} (x_{i+1} - x_i) - K_{ss} \frac{Q^2}{2g} \left| \frac{1}{A_{i+1}^2} - \frac{1}{A_i^2} \right| \qquad (4.12b)$$

where K_{ss} is a head-loss coefficient for a singularity as caused by a change between two consecutive sections, A_i and A_{i+1}, and/or by other possible irregularities. One takes (see *Silber*, 1968, p. 27) usually:

for accelerating flow : $K_{ss} = 0$

for decelerating flow : $0 < K_{ss} < 1$

By using the above equation, eq. 4.12b, the computations will be long and complicated; the use of the computer will greatly help. This method can be simplified when the variations in the kinetic energy are assumed to be negligible; while the calculations will be less long, they will also be less precise. For computation techniques and procedures for flow in watercourses, extensive information may be found in the books (see *Chow*, 1959, pp. 274-292, *Crausse*, 1951, pp. 171-183 and *Henderson*, 1966, pp. 140-155).

4° *Method of depths variation (Δh is fixed)* :

This method of *direct* steps can be used for the solution of the equation of motion in form of eq. 4.12 and eq. 4.14.

i) One supposes (chooses) to know the flow depth, h_i, at an abscissa, x_i.

One *wants* to know the abscissa, x_{i+1}, for a close-by flow depth, h_{i+1}.

One chooses a value, h_{i+1}, being only slightly different from the given value, h_i. (To avoid any possible inaccuracy, the variation of the flow depth, h_{i+1} and h_i, should be really very small.)

NON-UNIFORM FLOW   153

> One calculates the abscissa, x_{i+1}, using eq. 4.12 or eq. 4.14.
>
> One goes now to the next reach, etc.
>
> ii) It should be noted that this (explicit) method is an efficient one, being less lengthy and complicated than the method of reaches.

5° The use of numerical techniques for the method of successive approximations is elaborated in *Jansen* et al. (1979, p. 257). For different practical aspects such as river networks, the literature should be consulted (see *Jansen* et al., 1979, p. 255, *Chow*, 1959, p. 274, and *Henderson*, 1966, p. 140).

4.3.2 Method of direct Integration

1° The differential equation of motion, eq. 4.8, can generally not be integrated by elementary means. However, different methods have been advanced, all of which use functions to represent the variables.

2° In its general form, the equation of motion was written as :

$$\frac{dh}{dx} = S_f \frac{1 - \left(\frac{K_n}{K}\right)^2}{1 - \left(\frac{K_n}{K}\right)^2 \frac{S_f}{S_c}} \tag{4.8b}$$

where K(h) is the conveyance (see eq. 3.34).

Computing and tracing of the water-surface profiles require that this differential expression, eq. 4.8b, be integrated. Known are the discharge, Q, the bed slope, S_f, and the coefficient of friction, n or C. The variables are the abscissa, x, and the corresponding flow depth, h(x).

The right-hand side of eq. 4.8b is thus evidently a function of the flow depth, h ; the following relation can be postulated :

$$dx = f(h) \, dh \tag{4.15}$$

being a differential equation, relating the two variables. Integrating between two sections, x_i and x_{i+1}, yields :

$$(x_i - x_{i+1}) = \int_{h_{i+1}}^{h_i} f(h) \, dh \tag{4.16}$$

A resolution of this integral, being rarely a simple expression, is a difficult task. Integration becomes possible for the case of a channel having a wide rectangular or a parabolic cross section; this is known as the method of *Bresse* or of *Tolkmitt*, respectively. For channels of some other cross sectional shapes, other methods have been developed by *Bakhmeteff* (1932) and by *Chow* (1959).

The integration of eq. 4.16, as well as the now presented methods, are *direct* ones; successive values of x are independent of each other. This implies that one may proceed with the calculations from one section to another, whatever be the distance between these sections (see Fig. 4.4). However, this may not be done with the method of successive approximations (see sect. 4.3.1), where one must proceed using small distances. Herein lies a main advantage of the method of direct integration.

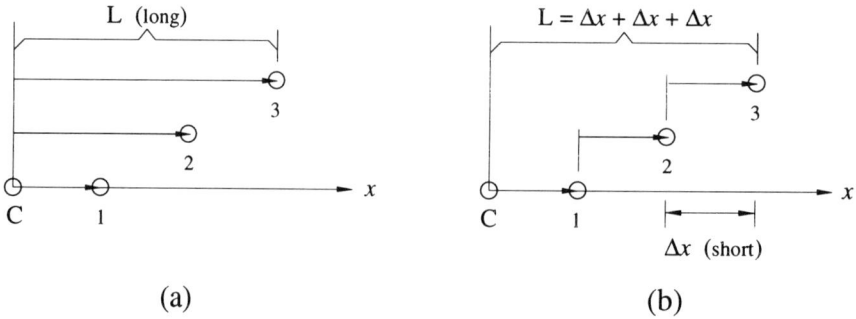

Fig. 4.4 Scheme for calculation of the water surface :
a) with direct integration and b) with successive approximations.

3°   *Method of Bresse* :

i) For a wide rectangular channel – with $h \cong R_h$, $q = Q/B$ and $B = A/h$ – , the normal depth is defined by :

$$h^3_n = \frac{q^2}{C^2 S_f} \qquad (3.36)$$

and the critical depth by :

$$h^3_c = \frac{q^2}{g} \qquad (2.24)$$

The conveyance of the channel is given by :

$$K^2 = C^2 R_h A^2 = C^2 h^3 B^2 \qquad (3.34)$$

Consequently, one may write :

$$\left(\frac{h_c}{h_n}\right)^3 = \frac{C^2 S_f}{g} \quad \text{and} \quad \left(\frac{K_n}{K}\right)^2 = \left(\frac{h_n}{h}\right)^3 \qquad (4.17)$$

NON-UNIFORM FLOW

Since the coefficient of Chézy, C, is taken constant for all flow depths, the critical slope is given by :

$$S_c = \frac{g}{C^2} = S_f \left(\frac{h_n}{h_c}\right)^3 \tag{4.9a}$$

ii) The differential equation, eq. 4.8b, now yields :

$$\frac{dh}{dx} = S_f \frac{1 - \left(\frac{h_n}{h}\right)^3}{1 - \left(\frac{h_n}{h}\right)^3 \frac{S_f}{S_c}} \tag{4.18}$$

and, using the relation for the critical slope, eq. 4.9a, one gets :

$$\frac{dh}{dx} = S_f \frac{1 - \left(\frac{h_n}{h}\right)^3}{1 - \left(\frac{h_c}{h}\right)^3} \tag{4.8a}$$

From here, the hurried reader could directly go to eq. 4.25.

This relation, given already before (see sect. 4.1.2), is the *equation of Bresse* (1860), being valid for wide rectangular channels.

Upon putting : $h/h_n = \eta$ and $dh = h_n \, d\eta$

above equation, eq. 4.8a, can be written as :

$$dx = \frac{1}{S_f} \left[1 + \frac{1 - (h_c/h_n)^3}{\eta^3 - 1} \right] h_n \, d\eta \tag{4.19}$$

iii) By integrating this relation between two sections of abscissas, x_i and x_{i+1}, and of flow depth, h_i and h_{i+1}, one obtains :

$$(x_i - x_{i+1}) = \frac{h_n}{S_f} \left\{ (\eta_i - \eta_{i+1}) - \right.$$

$$\left. - \left[1 - \left(\frac{h_c}{h_n}\right)^3\right] \left[\Phi(\eta_i) - \Phi(\eta_{i+1})\right] \right\} \tag{4.20}$$

with the integral of Bresse (see *Flamant*, 1923, p. 242) defined as :

$$\Phi(\eta,3) = -\int \frac{d\eta}{\eta^3 - 1} = \frac{1}{6} \log_e \left[\frac{\eta^2 + \eta + 1}{(\eta - 1)^2}\right] - \frac{1}{\sqrt{3}} \operatorname{arc\,ctg} \left(\frac{2\eta + 1}{\sqrt{3}}\right)$$

The expression of this function, $\Phi(\eta,3)$, is given in Table 4.1 for $N = 3$; N is the hydraulic exponent in eq. 4.17.

iv) The method of Bresse is simple and fast, but it is limited to channels whose bed width is large compared to the flow depth. It is advised to use the average flow depth, $D_h \cong R_h$, rather than the maximum flow depth.

4° Method of Bakhmeteff :

i) While the method of Bresse is limited to channels of wide rectangular sections, the one of Bakhmeteff is a method which can be used for channels of a more general form of the cross section.

ii) For any type of channel, the conveyance, given by :

$$K(h) = f(h) \qquad (3.35a)$$

shall be expressed with an exponential relation, such as :

$$K^2 = C^2 R_h A^2 = Cte \cdot h^N$$

This allows to formulate the ratio of :

$$\left(\frac{K}{K_n}\right)^2 = \frac{C^2 R_h A^2}{(C^2 R_h A^2)_n} = \left(\frac{h}{h_n}\right)^N \qquad (4.21)$$

The exponent, N, (see *Bakhmeteff*, 1932, p. 84) is the *hydraulic exponent*, to be determined for each form of the cross section of the channel. Some indicative values for different forms are given as follows :

- rectangular form $(B \ll h)$ $N = 2.0$
                    $(B = 2h)$ $N = 2.5$
                    $(B = \infty)$ $N = 3.0$
- trapezoidal form $3.0 < N < 4.0$
- triangular form $5.3 < N < 5.5$
- parabolic form $N = 4.0$

Note that the parabolic form, studied by *Tolkmitt*, can readily be used as an approximation to the form of a natural channel or a watercourse.

iii) One assumes that the ratio (see eq. 4.9) of :

$$\frac{S_f}{S_c} = \beta = S_f \frac{C^2 B R_h}{gA}$$

NON-UNIFORM FLOW

varies little with the flow depth, h ; consequently, this ratio can be considered as being constant in a channel reach.

iv) The differential equation, eq. 4.8b, reads now :

$$\frac{dh}{dx} = S_f \frac{1 - \left(\frac{h_n}{h}\right)^N}{1 - \beta \left(\frac{h_n}{h}\right)^N} \quad (4.22)$$

This relation is the *equation of Bakhmeteff* (1912), being a generalisation of the equation of Bresse, eq. 4.18 ; it is valid for any form of the cross section, being defined by the hydraulic exponent, eq. 4.21.

Upon putting : $h/h_n = \eta$ and $dh = h_n \, d\eta$

above relation, eq. 4.22, can be written (see *Bakhmeteff*, 1932, p. 87) as :

$$dx = \frac{1}{S_f} \left[1 + \frac{1 - \beta}{\eta^N - 1} \right] h_n d\eta \quad (4.23)$$

v) Integrating this relation between two sections of abscissas, x_i and x_{i+1}, and assuming that β and N are constants, one obtains :

$$(x_i - x_{i+1}) = \frac{h_n}{S_f} \left\{ (\eta_i - \eta_{i+1}) - (1 - \beta) \left[\Phi(\eta_i) - \Phi(\eta_{i+1}) \right] \right\} \quad (4.24)$$

with the integral of Bakhmeteff, defined as :

$$\Phi(\eta, N) = - \int \frac{d\eta}{\eta^N - 1} \quad (4.24a)$$

Bakhmeteff (1932, pp. 308-313) has provided tables, giving values of $\Phi(\eta, N)$ for different common values of N (see Table 4.1). For the special case where N = 3, the hydraulic exponent is the one used in the equation of Bresse, eq. 4.8a, where $\Phi(\eta, N) \equiv \Phi(\eta, 3)$.

vi) *Bakhmeteff* (1932, p. 96) proposed also a simplified method, where the kinetic-energy variation is neglected. This implies that $\beta = 0$ in eq. 4.24. This simplified method is however only to be used with success for water-surface profiles of the branch M1.

vii) This method as well as its application to many different practical situations is elaborated in the book by *Bakhmeteff* (1932).

5° *Method of Chow* :

i) In the preceding methods, it was assumed that the ratio given by :

$$\beta = \frac{S_f}{S_c} = \frac{C^2 B R_h}{gA}$$

remains essentially constant; the friction coefficient, C or n, remains thus also constant and independent from the flow depth, h.

The method of *Chow* (1959) goes beyond these restrictions; it is not necessary to assume that β stays constant.

ii) The differential equation, eq. 4.8a, shall be generalised in the following way :

$$\frac{dh}{dx} = S_f \frac{1 - (\frac{h_n}{h})^N}{1 - (\frac{h_c}{h})^M} \qquad (4.25)$$

N is the hydraulic exponent used in the expression for the conveyance, eq. 3.35a ; it will depend on the form of the cross section and on the chosen friction coefficient. M is the second hydraulic exponent. For any type of cross section, it was shown (see *Chow* 1959, p. 66 and p. 131) that :

$$N(h) = \frac{2h}{3A} \left(5B - 2R_h \frac{dP}{dh} \right)$$

$$M(h) = \frac{h}{A} \left(3B - \frac{A}{B} \frac{dB}{dh} \right) \qquad (4.26)$$

Both hydraulic exponents are thus functions of the flow depth, h ; they vary in the range of : $2.0 < N < 5.3$ et $3 < M < 4.8$.

Upon putting : $h/h_n = \eta$ and $dh = h_n \, d\eta$

above relation, eq. 4.25, can be written as :

$$dx = \frac{1}{S_f} \left[1 - (\frac{1}{1-\eta^N}) + (\frac{h_c}{h_n})^M (\frac{\eta^{N-M}}{1-\eta^N}) \right] h_n \, d\eta \qquad (4.27)$$

By taking $N = M = 3$, this equation, eq. 4.27, becomes the equation of Bresse, eq. 4.19, being limited to wide rectangular channels.

NON-UNIFORM FLOW

iii) Integrating this relation between two sections of abscissas, x_i and x_{i+1}, one obtains:

$$(x_i - x_{i+1}) = \frac{h_n}{S_f} \left\{ (\eta_i - \eta_{i+1}) - \int_0^\eta \left(\frac{1}{1-\eta^N}\right) d\eta + \left(\frac{h_c}{h_n}\right)^M \int_0^\eta \left(\frac{\eta^{N-M}}{1-\eta^N}\right) d\eta \right\} \quad (4.28)$$

The first integral is:

$$\Phi(\eta,N) = \int \left(\frac{1}{1-\eta^N}\right) d\eta = - \int \left(\frac{1}{\eta^N-1}\right) d\eta \quad (4.24a)$$

being thus identical to the one in the equation of Bakhmeteff, eq. 4.24.

The second integral can be transformed (see *Chow*, 1959, p. 255) into:

$$\int_0^\eta \left(\frac{\eta^{N-M}}{1-\eta^N}\right) d\eta = \frac{J}{N} \int_0^\zeta \left(\frac{1}{1-\zeta^J}\right) d\zeta = \frac{J}{N} \Phi(\zeta,J) \quad (4.28a)$$

where $\zeta = \eta^{N/J}$ and $J = N/(N-M+1)$. The function, $\Phi(\zeta,J)$, is like the previous function, $\Phi(\eta,N)$, except that the variables, η and N, are replaced by the variables, ζ and J; consequently, the same tabulation, given with Table 4.1, can be used for the two functions, eqs. 4.24a and 4.28a.

iv) Using the definition of these two functions, the above equation, eq. 4.28, can now be written as:

$$(x_i - x_{i+1}) = \frac{h_n}{S_f} \left\{ (\eta_i - \eta_{i+1}) - \left[\Phi(\eta_i,N) - \Phi(\eta_{i+1},N)\right] + \left(\frac{h_c}{h_n}\right)^M \frac{J}{N} \left[\Phi(\zeta_i, J) - \Phi(\zeta_{i+1}, J)\right] \right\} \quad (4.29)$$

v) This method as well as its application to different practical problems is elaborated in the book of *Chow*, 1959, pp. 252-262.

Table 4.1 Functions for gradually varied flow.

$$\Phi(\eta, N) = - \int_0^\eta \frac{d\eta}{\eta^N - 1}$$

The constant of integration is adjusted for $\Phi(0,N) = 0$ and $\Phi(\infty,N) = 0$

η \ N	2.8	3.0	3.2	3.6	4.0	5.0
0.10	0.100	0.100	0.100	0.100	0.100	0.100
0.20	0.201	0.200	0.200	0.200	0.200	0.200
0.30	0.303	0.302	0.302	0.301	0.300	0.300
0.40	0.408	0.407	0.405	0.403	0.402	0.401
0.44	0.452	0.450	0.448	0.445	0.443	0.441
0.48	0.497	0.494	0.492	0.488	0.485	0.482
0.52	0.544	0.540	0.536	0.531	0.528	0.523
0.56	0.593	0.587	0.583	0.576	0.572	0.565
0.58	0.618	0.612	0.607	0.599	0.594	0.587
0.60	0.644	0.637	0.631	0.623	0.617	0.608
0.61	0.657	0.650	0.644	0.635	0.628	0.619
0.62	0.671	0.663	0.657	0.647	0.640	0.630
0.63	0.684	0.676	0.669	0.659	0.652	0.641
0.64	0.698	0.690	0.683	0.672	0.664	0.652
0.65	0.712	0.703	0.696	0.684	0.676	0.663
0.66	0.727	0.717	0.709	0.697	0.688	0.675
0.67	0.742	0.731	0.723	0.710	0.701	0.686
0.68	0.757	0.746	0.737	0.723	0.713	0.698
0.69	0.772	0.761	0.751	0.737	0.726	0.710
0.70	0.787	0.776	0.766	0.750	0.739	0.722
0.71	0.804	0.791	0.781	0.764	0.752	0.734
0.72	0.820	0.807	0.796	0.779	0.766	0.746
0.73	0.837	0.823	0.811	0.793	0.780	0.759
0.74	0.854	0.840	0.827	0.808	0.794	0.771
0.75	0.872	0.857	0.844	0.823	0.808	0.784
0.76	0.890	0.874	0.861	0.839	0.823	0.798
0.77	0.909	0.892	0.878	0.855	0.838	0.811
0.78	0.929	0.911	0.896	0.872	0.854	0.825
0.79	0.949	0.930	0.914	0.889	0.870	0.839
0.80	0.970	0.950	0.934	0.907	0.887	0.854
0.81	0.992	0.971	0.954	0.925	0.904	0.869
0.82	1.015	0.993	0.974	0.945	0.922	0.885
0.83	1.039	1.016	0.996	0.965	0.940	0.901
0.84	1.064	1.040	1.019	0.985	0.960	0.918
0.85	1.091	1.065	1.043	1.007	0.980	0.935
0.86	1.119	1.092	1.068	1.031	1.002	0.954
0.87	1.149	1.120	1.095	1.055	1.025	0.973
0.88	1.181	1.151	1.124	1.081	1.049	0.994
0.89	1.216	1.183	1.155	1.110	1.075	1.015
0.90	1.253	1.218	1.189	1.140	1.103	1.039
0.91	1.294	1.257	1.225	1.173	1.133	1.064
0.92	1.340	1.300	1.266	1.210	1.166	1.092
0.93	1.391	1.348	1.311	1.251	1.204	1.123
0.94	1.449	1.403	1.363	1.297	1.246	1.158
0.95	1.518	1.467	1.423	1.352	1.296	1.199
0.96	1.601	1.545	1.497	1.417	1.355	1.248
0.97	1.707	1.644	1.590	1.501	1.431	1.310
0.98	1.855	1.783	1.720	1.617	1.536	1.395
0.99	2.106	2.017	1.940	1.814	1.714	1.537
0.995	2.355	2.250	2.159	2.008	1.889	1.678

η \ N	2.8	3.0	3.2	3.6	4.0	5.0
1.005	1.818	1.649	1.506	1.279	1.107	0.817
1.01	1.572	1.419	1.291	1.089	0.936	0.681
1.02	1.327	1.191	1.078	0.900	0.766	0.546
1.03	1.186	1.060	0.955	0.790	0.668	0.469
1.04	1.086	0.967	0.868	0.714	0.600	0.415
1.05	1.010	0.896	0.802	0.656	0.548	0.374
1.06	0.948	0.838	0.748	0.608	0.506	0.342
1.07	0.896	0.790	0.703	0.569	0.471	0.315
1.08	0.851	0.749	0.665	0.535	0.441	0.292
1.09	0.812	0.713	0.631	0.506	0.415	0.272
1.10	0.777	0.681	0.601	0.480	0.392	0.254
1.11	0.746	0.652	0.575	0.457	0.372	0.239
1.12	0.718	0.626	0.551	0.436	0.354	0.225
1.13	0.692	0.602	0.529	0.417	0.337	0.212
1.14	0.669	0.581	0.509	0.400	0.322	0.201
1.15	0.647	0.561	0.490	0.384	0.308	0.191
1.16	0.627	0.542	0.473	0.369	0.295	0.181
1.17	0.608	0.525	0.458	0.356	0.283	0.173
1.18	0.591	0.509	0.443	0.343	0.272	0.165
1.19	0.574	0.494	0.429	0.331	0.262	0.157
1.20	0.559	0.480	0.416	0.320	0.252	0.150
1.22	0.531	0.454	0.392	0.299	0.235	0.138
1.24	0.505	0.431	0.371	0.281	0.219	0.127
1.26	0.482	0.410	0.351	0.265	0.205	0.117
1.28	0.461	0.391	0.334	0.250	0.193	0.108
1.30	0.442	0.373	0.318	0.237	0.181	0.100
1.32	0.424	0.357	0.304	0.225	0.171	0.093
1.34	0.408	0.342	0.290	0.214	0.162	0.087
1.36	0.393	0.329	0.278	0.204	0.153	0.081
1.38	0.378	0.316	0.266	0.194	0.145	0.076
1.40	0.365	0.304	0.256	0.185	0.138	0.071
1.42	0.353	0.293	0.246	0.177	0.131	0.067
1.44	0.341	0.282	0.236	0.169	0.125	0.063
1.46	0.330	0.273	0.227	0.162	0.119	0.059
1.48	0.320	0.263	0.219	0.156	0.113	0.056
1.50	0.310	0.255	0.211	0.149	0.108	0.053
1.60	0.269	0.218	0.179	0.123	0.087	0.040
1.70	0.236	0.189	0.153	0.103	0.072	0.031
1.80	0.209	0.166	0.133	0.088	0.060	0.024
1.90	0.188	0.147	0.117	0.076	0.050	0.020
2.00	0.169	0.132	0.104	0.066	0.043	0.016
2.20	0.141	0.107	0.083	0.051	0.032	0.011
2.40	0.119	0.089	0.068	0.040	0.024	0.008
2.60	0.102	0.076	0.057	0.033	0.019	0.005
2.80	0.089	0.065	0.048	0.027	0.015	0.004
3.00	0.078	0.056	0.041	0.022	0.012	0.003
3.50	0.059	0.041	0.029	0.015	0.008	0.002
4.00	0.046	0.031	0.022	0.010	0.005	0.001
5.00	0.031	0.020	0.013	0.006	0.003	0.000
10.00	0.009	0.005	0.003	0.001	0.000	0.000

NON-UNIFORM FLOW   161

4.3.3 Method of graphical Integration

1° The equation of motion for a prismatic channel of an arbitrary cross section, written in its general form, was given as :

$$\frac{dh}{dx} = S_f \frac{1 - \frac{(Q/A)^2}{C^2 R_h S_f}}{1 - \frac{(Q/A)^2}{gA/B}} \tag{4.8}$$

This equation, written in a simplified form, yields :

$$dx = f(h) \, dh \tag{4.15}$$

The function, $f((h)$, is of a form which is difficult to integrate. The parameters, $A(h)$, $R_h(h)$, $B(h)$ and $C(h)$, can not be expressed as simply analytical functions of the flow depth, h.

2° Different methods using an analytical integration have been presented previously (see sect. 4.3.2). However, there exist also several methods using a graphical integration of above equations, eq. 4.8 or eq. 4.15.

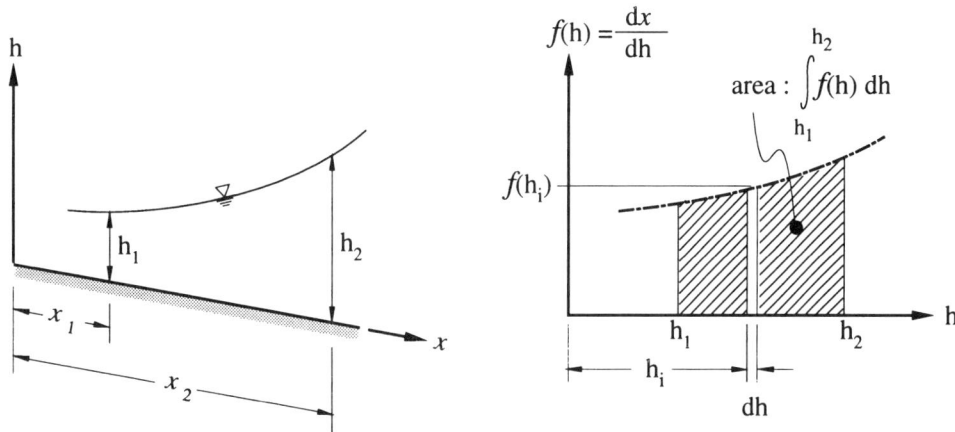

Fig. 4.5 Scheme of the method of graphical integration.

3° *Direct method* :

   i)   One calculates or establishes the function, $f(h_i)$, for different values of the flow depth, h_i.

   One traces (see Fig. 4.5) this curve of $f(h)$ as a function of h.

The area within this curve, $f(h)$, the axis of h and the two lines of the abscissa, h_1 and h_2, gives the distance, (x_2-x_1), which separates the two sections; one writes :

$$(x_2-x_1) = \int_{x_1}^{x_2} dx = \int_{h_1}^{h_2} f(h)\, dh \qquad (4.16)$$

ii) The configuration of these curves – giving the water-surface profiles for mild slopes, M, and for steep slopes, S – is schematically shown in Fig. 4.6 (see also Fig. 4.2).

iii) This method is rather rigorous, but could become rather time-consuming when an entire water-surface profile has to be evaluated. The method is also useful and valid for non-prismatic channels (see *Chow*, 1959, p. 249).

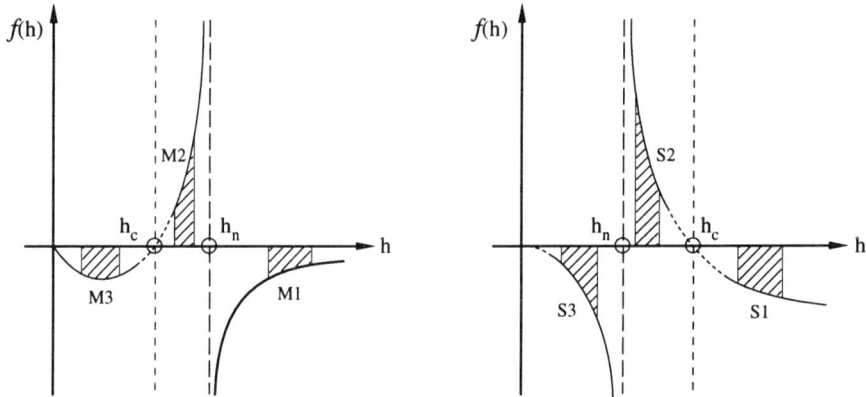

Fig. 4.6 Forms of the water surface; schematic representation of the function, $f(h)$, of eq. 4.15.

4° There exists still other graphical methods; two of which should be mentioned : the method of Raytchine et Châtelain, elaborated in *Carlier*, 1972, p. 380, and the method of Silber, summarised also in *Carlier*, 1972, p. 382.

5° The method of *Silber* (1968) introduces two dimensionless variables, namely a reduced flow depth and discharge :

$$h^* = \frac{h}{H_s} \qquad \text{and} \qquad q^* = \frac{Q}{BH_s\sqrt{2gH_s}}$$

In this method, the concept of the specific energy, H_s (see sect. 2.3) is used.

The corresponding curve, $q^* = f(h^*)$, constitutes a universal characteristic for gradually varied flow; this technique can be rather useful for different problems in fluvial hydraulics.

4.4 RAPIDLY VARIED FLOW

1° Upstream and downstream of the critical depth, h_c, the flow is a rapidly varied one (see Fig. 4.2); all this takes place over a reach of short length. In passing a section of critical depth, the type of flow changes : from supercritical to subcritical or vice versa.

2° The simplifying conditions, which were taken in the derivation of the equation of gradually varied flow, eq. 4.8, are no more valid. The curvature of the flow can no longer be ignored; consequently the pressure distribution is not an hydrostatic one (see sect. 1.4).

3° The passage of the critical depth, h_c, is usually accompanied by a sudden change in flow depth; examples will be studied herewith, namely :

   i) a flow depth *increase* : formation of an hydraulic jump (see sect. 4.4.4).

   ii) a flow depth *decrease* : flow over a spillway (see sect. 4.4.1), formation of an hydraulic drop (see sect. 4.4.2) or flow under a sluice gate (see sect. 4.4.3).

4° Knowledge about rapidly varied flow is often useful for the determination of the *control point(s)*, a necessary information for the computation of gradually varied flow (see sect. 4.3).

4.4.1 Weirs, Spillways

1° A weir or a spillway – a weir is the simple form of a spillway – (see Fig. 4.7) is a device or structure, which can be used to mesure and/or control the discharge, Q, in a channel. A device of this kind consists of a vertical plate, thin or thick, having a certain height, H_D ; it is mounted usually vertically at an angle to the flow and it obstructs more or less the cross section of the channel. The flow passes *over* the weir towards the downstream.

2° Despite the complexity of the flow over a weir – streamlines converge – , it is possible to establish an expression for the computation of the discharge, Q, (see *Graf & Altinakar*, 1991, pp. 183-186).

Applying the equation of energy between sections, sufficiently upstream and at the weir itself, the following expression for the discharge is obtained :

$$Q = L_D K_D \frac{2}{3} \sqrt{2g} \left[\left(H + \frac{U_1^2}{2g} \right)^{3/2} - \left(\frac{U_1^2}{2g} \right)^{3/2} \right] \quad (4.30)$$

where L_D is the effective length of the weir, H is the height (head) of the water surface above the weir, being measured sufficiently upstream from the weir and U_1

is the average velocity in the approach section, measured at a flow depth of $H + H_D$. K_D is the dimensionless discharge coefficient of the weir.

3° The relation, $Q = f(H)$, given with eq. 4.30, can also be written as :

$$Q = L_D K_D' \frac{2}{3} \sqrt{2g} \, H_T^{3/2} \qquad (4.31)$$

where $H_T = H + (U_1^2/2g)$. It can also be given by :

$$Q = L_D K_D \frac{2}{3} \sqrt{2g} \, H^{3/2} = L_D K_D (\frac{2}{3} H) \sqrt{2gH} \qquad (4.32)$$

if the term of the kinetic energy, $U_1^2/2g$, is neglected, notably for the case of a small approach velocity, U_1.

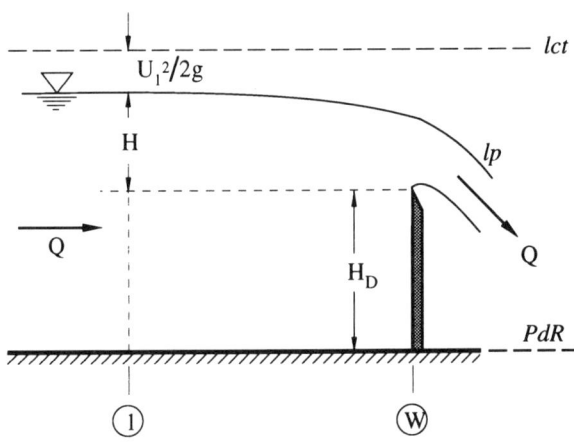

Fig. 4.7 Hydraulic scheme of a weir.

4° The discharge coefficient, K_D or K_D', depends on :

- the curvature and the contraction of the streamlines over the weir,
- the effects of viscosity and turbulence, as well as of surface tension,
- the approach velocity, U_1, itself dependent on the ratio of H_D/H,
- the geometry of the weir and its hydraulic roughness.

In hydraulic engineering, only the influence of the geometry of the weir is considered; other influences are usually ignored.

The coefficient, K_D, for all the different types of weirs, should – if possible – be established by a calibration. Many different forms of weirs have been extensively investigated; their coefficients can be found in norms as well as in the literature.

NON-UNIFORM FLOW 165

5° The following different types of weirs will be briefly presented :
   - sharp-crested weirs,
   - spillways,
   - mobile weirs,
   - broad-crested weirs (see sect. 4.4.2).

6° *Sharp-crested weirs* (see Fig. 4.8)

   i) This type of weir consists of a thin vertical plate, set perpendicular to the flow. The upper part of the plate has a sharp-edged crest; in this way it cuts clearly the flow, forming a nappe or jet.

   Such a weir can be rectangular (with or without lateral contraction), trapezoidal or circular (see *Carlier*, 1972, and *French*, 1985, p. 339).

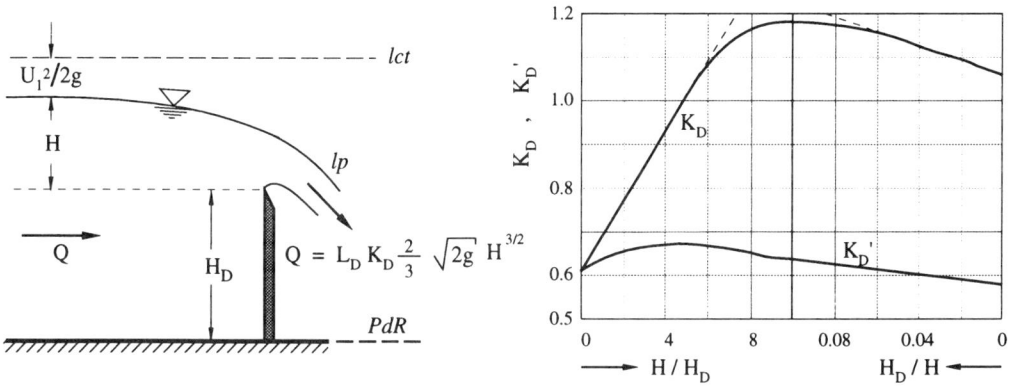

Fig. 4.8 Sharp-crested weir.

   ii) For a weir of *rectangular* section, the discharge, Q , can be calculated with the use of eq. 4.32. The coefficient of discharge, K_D , depends on the height of the plate, H_D , and on the upstream water depth, $H_D + H$. Empirical relations are available; here are given the ones proposed by *Rehbock* and *Böss* (see *Naudascher*, 1987, p. 103), respectively :

$$K_D = 0.61 + 0.08 \, (H/H_D) \qquad (H/H_D) < 6$$
$$K_D = 1.06 \, (1 + H_D/H)^{1.5} \qquad (H_D/H) < 0.06$$
(4.33)

   Between these two relations, there exists a transition at the maximum value of $K_D \cong 1.2$ for $(H/H_D) \cong 10$. This point may be considered as the delimitation between flow over a weir $(H/H_D < 10)$ and flow over a sill $(H/H_D > 10)$.

iii) The use of the coefficient of discharge, K'_D, together with its equation, eq. 4.31, is recommended (see *Naudascher*, 1987, p. 104), since it presents itself (see Fig. 4.8) with a small variation, or :

$$0.58 < K'_D < 0.67.$$

iv) The geometry of the nappe, as well as the distribution of the pressure within the nappe, have been investigated (see *Rouse*, 1938, p. 316 and *Chow*, 1959, p. 361) for different relative heights, (H_D/H).

v) For a weir with a *submerged* nappe, the empirical relation of *Villemonte* (see *Ranga-Raju*, 1981, p. 220) can be used for the computation of the discharge :

$$Q' = Q \left[1 - (H'/H)^{3/2} \right]^{0.385}$$

where Q is the discharge of the free (non-submerged) nappe, given with eq. 4.31; H' is the height above the crest, measured downstream.

vi) Different types of sharp-crested or thin-plated weirs have been investigated (see *Carlier*, 1972, pp. 203-206) such as :

- *inclined* weirs, where the plate is inclined to the vertical;

- *oblique* weirs, where the plate is inclined to the longitudinal axes of the channel;

- *lateral* weirs, where a lateral wall of the channel serves as the weir (see *Chow*, 1959, p. 327).

7° *Spillways* (see Fig. 4.9)

i) The form of a (overflow) spillway is usually determined by the shape of the lower streamline of the nappe over a sharp-crested weir.

Numerous are the forms of a spillway, including notably the ones of *Scimeni* and of *Craeger* (see *Hoffmann*, 1977). Herewith will be presented the standard profile developed in the USA by the Waterways Experiment Station (WES), having a vertical upstream face. For the *design* of such a profile, one uses the design discharge, Q^*, thus a design head, H^*. If the actual discharge, Q, varies, the following should be expected :

- if $H < H^*$: the lower part of the nappe is more curved, resulting in an over-pressure on the back of the spillway;

- if $H > H^*$: the lower part of the nappe is less curved, resulting in a flow separation and an under-pressure on the back of the spillway; there is a risk of introducing cavitation and instabilities.

NON-UNIFORM FLOW

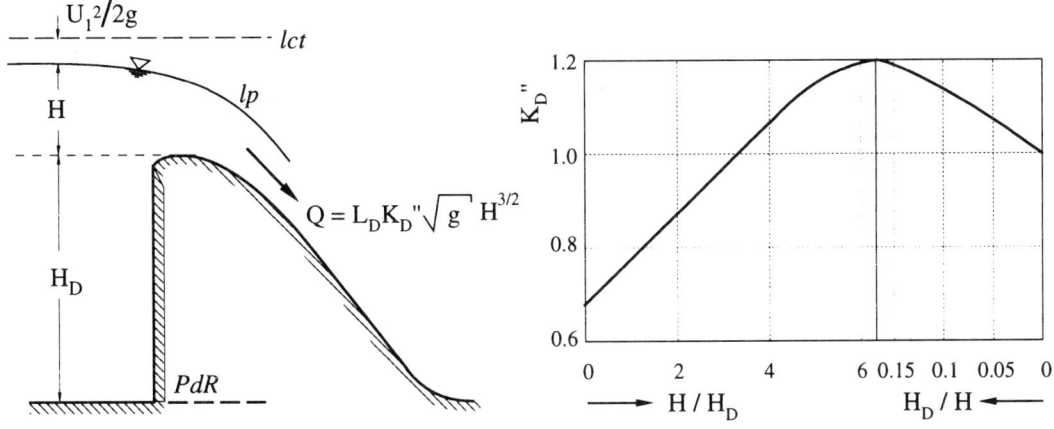

Fig. 4.9 Scheme of spillway and its coefficient.

ii) For a spillway, the discharge, Q, is usually calculated using eq. 4.32 as follows :

$$Q = L_D K''_D \sqrt{g} H^{3/2}$$

This discharge coefficients, K''_D, is given (see *Naudascher* 1987, p. 109) in the Fig. 4.9. The pressure on the back of the spillway stays atmospheric; neither over- nor under-pressure is encountered.

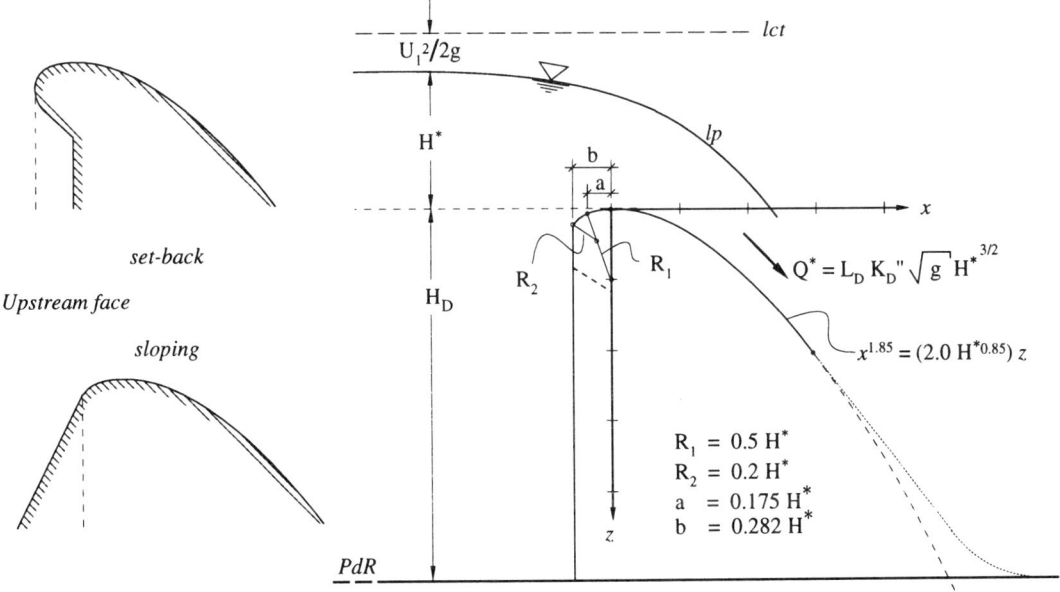

Fig. 4.10 Spillway : the WES profile and other profiles.

8° *Mobile spillways*

i) A spillway – which is a fixed installation – can be equipped, but also partially or totally replaced, by a *mobile spillway*, also called gated spillway or gate (see *Bouvard*, 1984). The mobile element allows, by an opening/closure operation, a better regulation of the passing discharge.

There exist different types of mobile spillways (see Fig. 4.11), such as drum gates, radial gates or combined sluice gates.

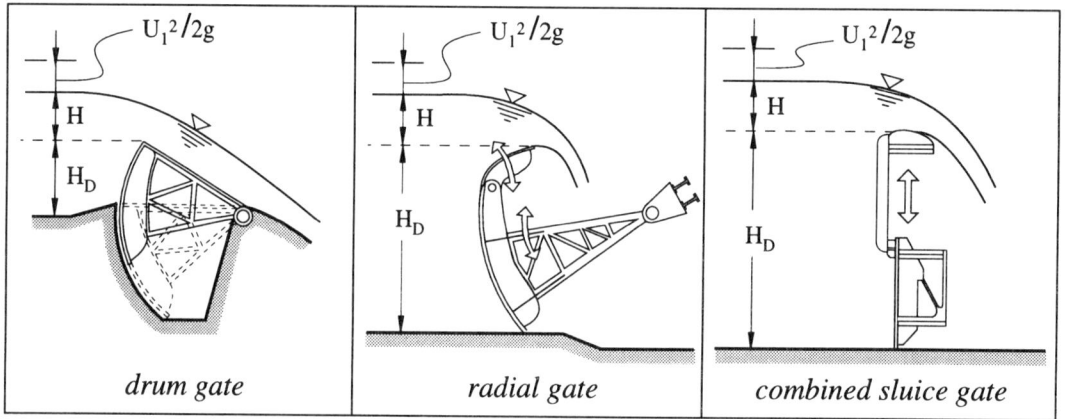

Fig. 4.11 Mobile spillways.

ii) For mobile spillways, the discharge, Q, is calculated using above equations, eq. 4.32 or eq. 4.31. The discharge coefficient, K_D or K'_D, should be obtained by calibration, either in situ or on a model.

iii) In the books of *Chow* (1959), *Naudascher* (1987), *Ippen* (1950), *Sinniger* et *Hager* (1989) and *Hoffmann* (1977), one finds further informations, concerning the different geometries of fixed and mobile spillways, the flow with submerged and non-submerged nappes, as well as the pressure distribution and the air entrainment on the back of the spillway.

4.4.2 Hydraulic Drop

1° In a channel on a weak, $S_f < S_c$, or horizontal, $S_f = 0$, bed slope, the flow is supposed to discharge freely into the atmosphere (see Fig. 4.12); this is an hydraulic drop or free overfall. The water-surface profile is of the type, M2 or H2, respectively; the flow is subcritical.

The specific energy, H_{s_c}, will be minimal at the hydraulic drop, where the critical depth, h_c, is observed ; this is the smallest possible flow depth for a given discharge. The existence of a section with critical depth is an information on maximum value of the discharge for a given specific energy (see sect. 2.3.3).

A decrease in the specific energy, H_s, is accompanied by a decrease in the flow depth, h.

NON-UNIFORM FLOW 169

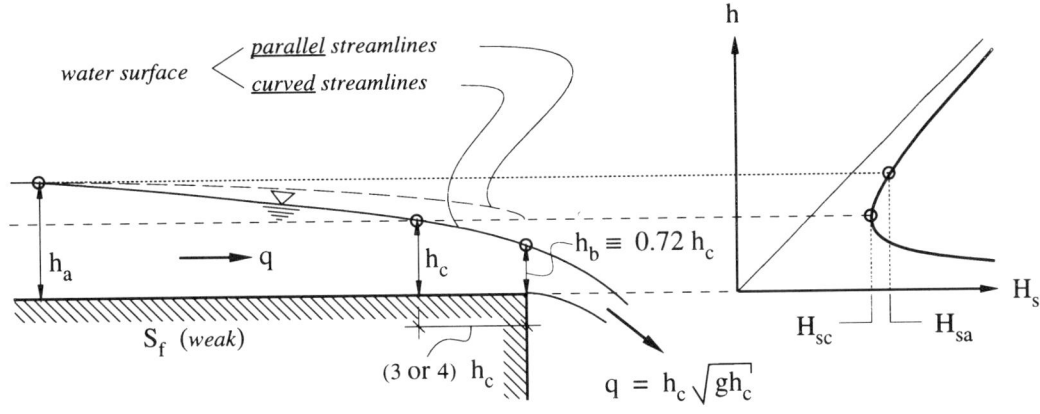

Fig. 4.12 Hydraulic drop.

2° For a rectangular channel, the critical depth is given by :

$$h_c = \frac{2}{3} H_{s_c} \qquad (2.19)$$

and the corresponding average velocity, by :

$$U_c^2 = gh_c \qquad (2.21)$$

With the definition of the unit discharge, q = Uh , one obtains :

$$q = h_c \sqrt{gh_c} = \sqrt{g \left(\frac{2}{3} H_{s_c}\right)^3} \qquad (2.24)$$

which in turn yields :

$$h_c = \sqrt[3]{q^2/g} \qquad (2.23)$$

Channels, having non-rectangular cross sections, have been reviewed in the literature (see *Ranga-Raju*, 1981, chap. 8).

3° However, it must be now remarked that the critical depth, h_c , as defined by eq. 2.23, is only an approximation. The relation is only valid in rectilinear flow, which is not the case in the vicinity of an hydraulic drop. For curvilinear flow, the experimental data (see *Rouse*, 1938, p. 325 and *Bauer et Graf*, 1971) have shown that the critical depth, h_c , appears slightly upstream of the drop (see Fig. 4.12); the depth at the section of the drop is approximately given as :

$$h_b = 0.72\, h_c = 0.72 \sqrt[3]{q^2/g} \qquad (4.34)$$

For channels on weak slopes, S_f , of non-rectangular sections, the numerical value in eq. 4.34 should be taken as being closer to 0.75 (see *Rajaratnam et al.*, 1964).

Nevertheless, in many pratical problems, one may assume that the depth at the drop, h_b, is the same as the critical depth, h_c, such as is defined in eq. 2.23.

4° *Broad-crested weir* (see Fig. 4.13)

i) Take a weir having a broad wall (sill), $B_D > 3H$, being horizontal; this weir can be assimilated to a short rectangular channel with an hydraulic drop. If the flow is a free (non-submerged) one and the streamlines are parallel, the critical depth, h_c, will be attained at the drop. Since the streamlines are curved, the critical depth, h_c, given by eq. 2.23, should be measured slightly upstream of the drop (see Fig. 4.12).

ii) The discharge can be computed by :

$$Q = L_D q = L_D K^+_D h_c \sqrt{gh_c} = L_D K^+_D \sqrt{g \left(\tfrac{2}{3} H_{s_c}\right)^3} \qquad (2.24a)$$

where L_D is the effective length of the weir. K^+_D is the weir coefficient and depends on the geometric characteristics, H_D and B_D, and the roughness of the weir ; it is given (see *Naudascher*, 1987, p. 117) as $0.88 < K^+_D < 0.98$.

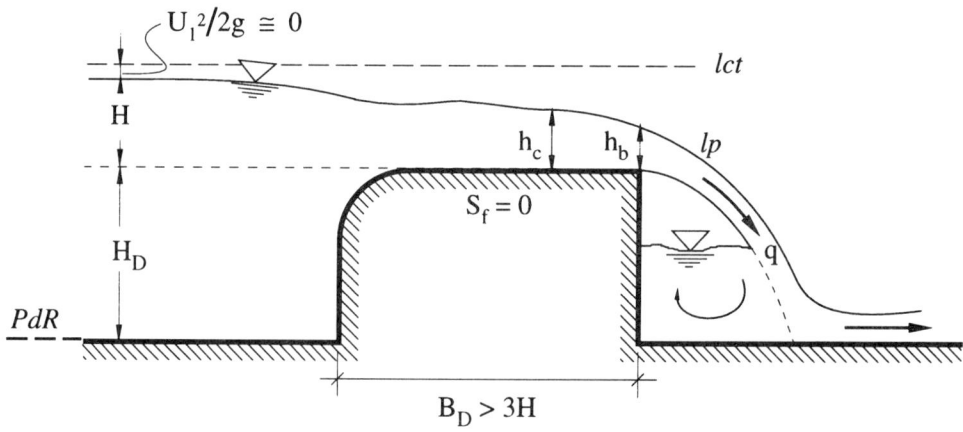

Fig. 4.13 Broad-crested weir.

5° *Measuring flume* (see Fig. 4.14)

i) A measuring flume at critical regime – also called Venturi canal (see *Jaeger*, 1954) or Parshall flume (see *Chow*, 1959) – consists of a vertical and/or horizontal constriction, also called throat, in a (rectangular) channel, being followed by a progressive enlargement.

NON-UNIFORM FLOW

For a correct functioning, the geometry of the measuring flume must allow the formation of the critical depth, h_c, somewhere in the constricted section. Flow, which is subcritical, changes locally into supercritical flow; an hydraulic jump forms itself downstream of the throat. The flow is thus a free (non-submerged) one and the discharge does not depend on the downstream flow depth, h_3. To make sure that the flow does not become submerged, it is recommended to change (increase) locally the floor of the flume.

ii) The discharge, Q, is calculated with the following expression :

$$Q = B_2 \sqrt{g \, h_c^3} \qquad (2.24)$$

with B_2 corresponding to the width at the throat.

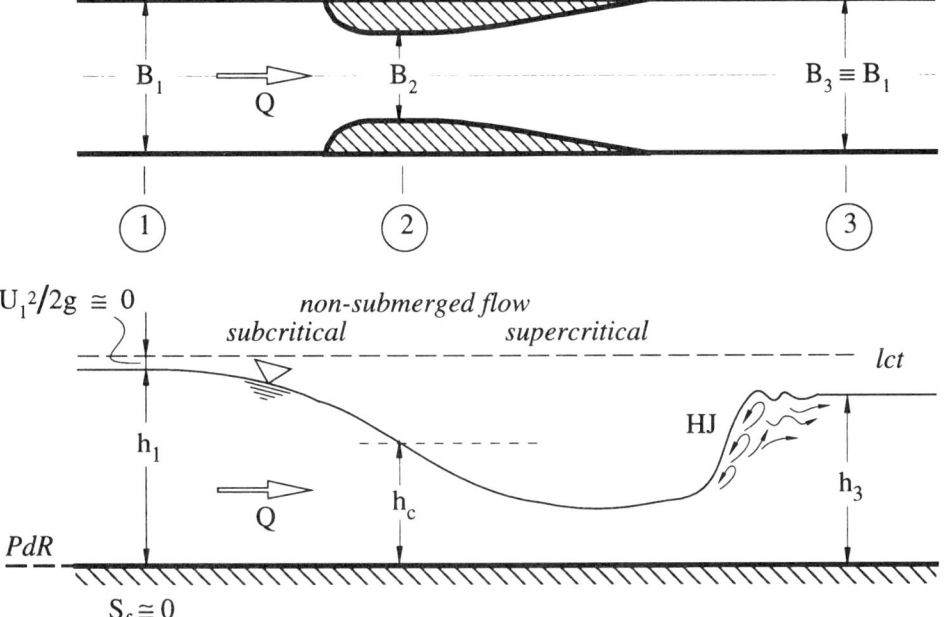

Fig. 4.14 Measuring flume at critical regime.

iii) Since the exact location of the critical depth, h_c, is not known, it is advised to measure the upstream flow depth, h_1, rather than the critical one, h_c.

The specific energy in both of these sections is the same, $H_{s_1} = H_{s_c}$. If the kinetic energy, $U_1^2/2g$, can be neglected, above equation, eq. 2.19, can be written as :

$$h_1 \cong H_{s_1} = H_{s_c} = \frac{3}{2} h_c \qquad \text{or} \qquad h_c = \frac{2}{3} h_1 \qquad (4.35)$$

iv) Consequently, the discharge relation, eq. 2.24, yields :

$$Q = B_2 K_j \sqrt{g \left(\frac{2}{3} h_1\right)^3} \qquad (4.36)$$

This allows the determination of the discharge, Q, in a measuring canal at critical regime, with an unique measurement of the upstream flow depth, h_1. The coefficient, K_j, depends on the geometry of the measuring flume and, to a smaller extend, on the upstream flow depth. For ratios of $B_1/B_2 \geq 3$, this coefficient is slightly below unity. However, for good discharge measurements, it is advisable to calibrate the flume (using for example, micro-propellers or pitot tubes).

4.4.3 Underflow Gates (see Fig. 4.15)

1° An underflow gate is an installation, which is used for measuring and/or controlling the discharge, Q, flowing in a channel or leaving a reservoir. Such a device consists essentially of a vertical wall having an opening of a given surface area, A_o, and being positioned either at or close to the channel bed; water passes across towards the downstream.

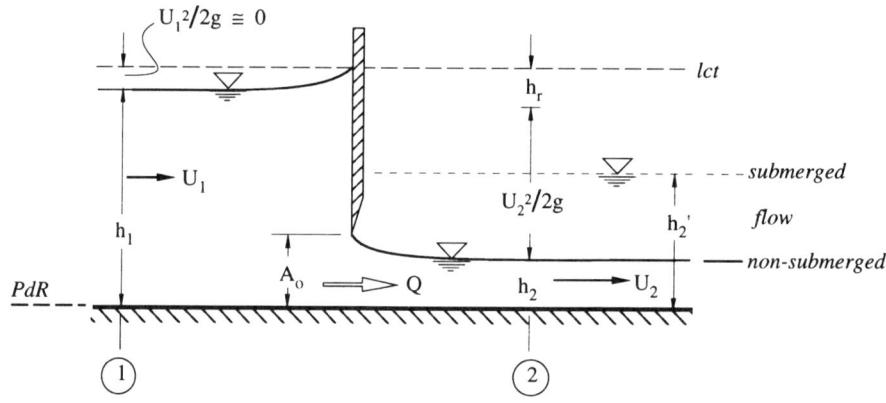

Fig. 4.15 Underflow gate.

2° For flow, being submerged or non-submerged (free), across (or under) an underflow gate – the streamlines will be contracted –, it is possible to establish an expression for the discharge, Q (see *Graf & Altinakar*, 1991, pp. 179-182). Applying the equation of energy between two cross sections before and after the gate, the following expression is obtained :

$$Q = A_o K_v \sqrt{2g (h_1 - h_2)} \qquad (4.37)$$

written also as :

$$Q = A_o K_v \sqrt{2g (h_1)}$$

where $(h_1 - h_2)$ is the difference in the flow depths, h_1 and h_2.

NON-UNIFORM FLOW

3° K_v is the gate coefficient, given by :

$$K'_v = C_q \cdot C_c \frac{1}{\sqrt{1 - C_c^2 (A_o/A_1)^2}}$$

but also by :

$$K_v = C_q \cdot C_c \frac{1}{\sqrt{1 + C_c (A_o/A_1)}} \qquad (4.38)$$

C_q is the discharge coefficient, which takes care of the effects of viscosity and turbulence; it varies such as : $0.95 < C_q < 0.99$. $C_c = (A_2/A_o)$ is the contraction coefficient – the effect of the vena contracta is parametrized –, which depends on the geometry of the opening; it varies such as : $0.5 < C_c < 1.0$. Consequently, there is a large variation of the gate coefficient, K_v , (see *Naudascher*, 1987, pp. 89 and 101), being $K_v < 0.61$; it will depend on the geometric characteristics of the type of the gate and its opening, but also on the type of flow, being submerged (drowned) or non-submerged (free). If an arbitrary-shaped gate is used for measuring the discharge, it is highly recommended to perform a calibration, possibly in situ.

4° An underflow gate is usually installed as a mobile element, namely its opening section, A_o , is adapted to specific needs. In this way, one succeeds in varying the discharge, Q , to be calculated with eq. 4.37; note however, that the gate coefficient, K_v, will also be undergoing these variations.

5° There exist many different types of underflow gates (see Fig. 4.16), such as sluice gates, radial gates or rolling gates.

6° In the books (see *Henderson*, 1966, *Ippen*, 1950, p. 536 and *Naudascher*, 1987, pp. 82-102), one finds further informations concerning different types and geometries of underflow gates as well as their operations.

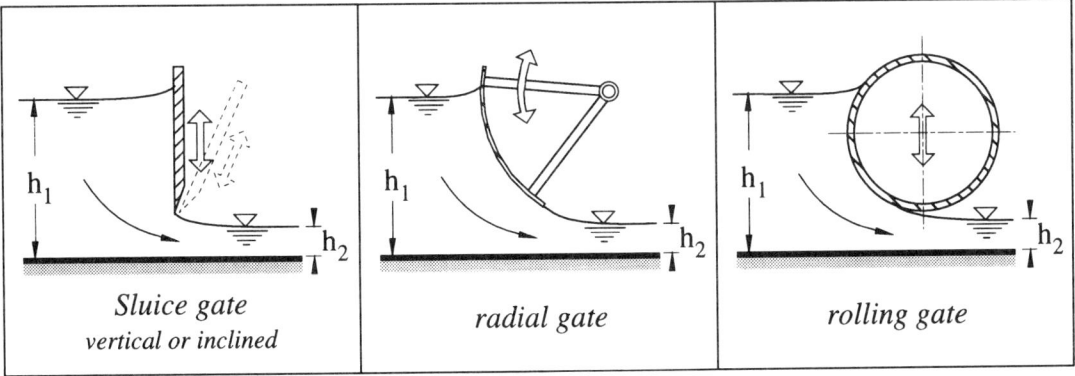

Fig. 4.16 Underflow gates.

4.4.4 Hydraulic Jump

1° The hydraulic jump, or shortly called the jump (see Fig. 4.17):

- is encountered, when flow passes abruptly from an (upstream) supercritical regime, $Fr_1 > 1$, to a (downstream) subcritical one, $Fr_2 < 1$;
- is evidenced as a sudden super-elevation (discontinuity of flow) in the water level, $(h_2 - h_1)$, over a short reach;
- remains at a fixed position – being a stationnary wave – in the channel, if the flow is steady;
- is accompanied by a very turbulent motion with instabilities at the water surface (undulations and air entrainment);
- is often encountered in form of an aerated wave, braking up in rollers;
- causes a strong dissipation of mechanical energy, $h_{HJ} = H_{s_2} - H_{s_1}$.

Some examples, where and when an hydraulic jump forms, are given in Fig. 4.2a.

2° The two depths, h_2 and h_1, are also called the *conjugate* (or sequent) *depths*; they envelop the jump. The *height* of a jump is given by the difference of the conjugate depths, $(h_2 - h_1)$.

3° The *theory* of the hydraulic jump has as objective to render a relation between these conjugate depths, h_2 and h_1, for a given discharge in the channel.

The equation of energy, eq. 2.10, is not very useful to develop such a relation; the head loss, which cannot be estimated a priori, is very large. However, the momentum equation, eq. 4.39, where the external forces get involved, can be successfully used.

i) Considered shall be a prismatic channel with two cross sections, A_1 and A_2, enveloping the jump. The distance between these sections should be sufficiently long to consider the pressure as hydrostatic, but should be sufficiently short to consider the friction forces, F_F, on the bed as negligible.

ii) The momentum equation for steady flow expresses that the quantity of motion leaving across the surface of a fluid volume be equal to the sum of the applied forces (see *Graf & Altinakar*, 1991, p. 145).

Along the longitudinal direction of the channel (see Fig. 4.17), the momentum equation reads:

$$\Sigma F_x = F_{P_1} - F_{P_2} + W\sin\alpha - F_F = \rho Q (U_2 - U_1) \qquad (4.39)$$

where F_{P_2} and F_{P_1} are the pressure forces applied at the sections of the conjugate depths, $W\sin\alpha$ is the gravity force and F_F is the friction force on the bed. These two last forces (together) are usually neglected.

NON-UNIFORM FLOW

iii) For an *horizontal* channel with a rectangular section, introducing :

$$A_1 = h_1 B, A_2 = h_2 B \quad ; \quad q = Q/B \quad ; \quad F_{P_2} = \gamma(h_2/2)A_2, F_{P_1} = \gamma(h_1/2)A_1$$

above equation, eq. 4.39, can be (see *Bakhmeteff*, 1932, p. 240) written as :

$$\frac{q^2}{gh_1} + \frac{h_1^2}{2} = \frac{q^2}{gh_2} + \frac{h_2^2}{2} \tag{4.40}$$

iv) This implies that a specific force (momentum), defined as :

$$M = \frac{q^2}{gh} + \frac{h^2}{2}$$

takes the same value at the conjugate depths of the jump, namely $M_1 = M_2$. A graphical resolution (see Fig. 4.17) is possible; the curves, $M = f(h)$, are traced point-by-point for the different discharges; each curve has two branches and passes through the critical depth, h_c, by a minimum value. Obviously each vertical cuts the curve at the two conjugate depths, h_1 and h_2.

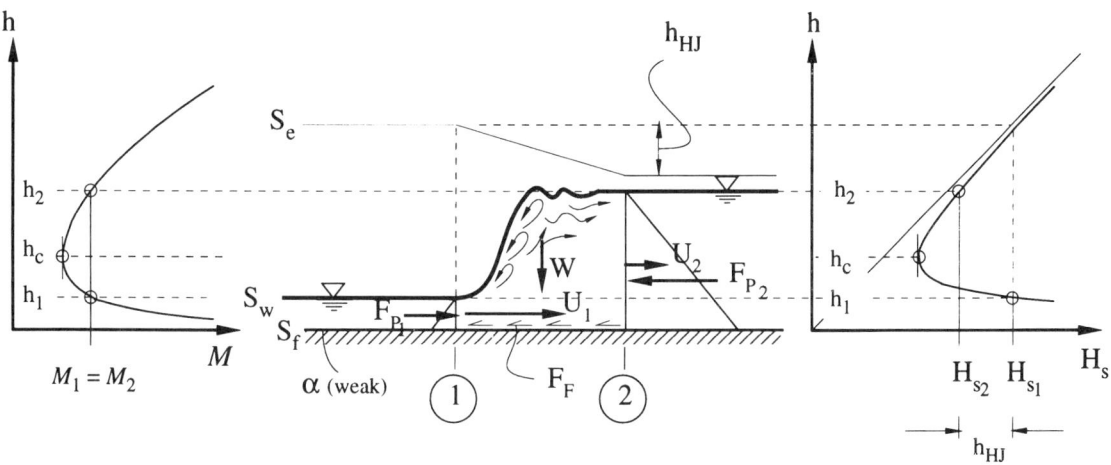

Fig. 4.17 Scheme of an hydraulic jump.

v) By applying the equation of continuity, $q = U_1 h_1 = U_2 h_2$, the above equation, eq. 4.40, yields :

$$\frac{1}{2}(h_2^2 - h_1^2) = \frac{U_1^2}{g} h_1 \left(1 - \frac{h_1}{h_2}\right)$$

which, upon dividing by ($h_2 - h_1$), renders :

$$h_2^2 + h_2 h_1 - 2h_1 \frac{U_1^2}{g} = 0 \tag{4.40a}$$

Of the two roots of this equation, the one which is positive is of interest, or :

$$h_2 = -\frac{1}{2} h_1 \overset{+}{(-)} \sqrt{\frac{h_1^2}{4} + 2h_1 \frac{U_1^2}{g}} \tag{4.41}$$

Usually this is expressed in a dimensionless form, given by :

$$\frac{h_2}{h_1} = \frac{1}{2} \left(\sqrt{1 + 8 \, Fr_1^2} - 1 \right) \tag{4.41a}$$

or:

$$\frac{h_1}{h_2} = \frac{1}{2} \left(\sqrt{1 + 8 \, Fr_2^2} - 1 \right) \tag{4.41b}$$

where the Froude numbers are :

$$Fr_1^2 = \frac{U_1^2}{gh_1} \quad \text{and} \quad Fr_2^2 = \frac{U_2^2}{gh_2}$$

The last two relations – they are known as the *equation of Bélanger* (see *Resch* et *Leutheusser*, 1972) – allow for a given discharge, Q, the calculation of one of the flow depths, h_1 or h_2, if the other flow depth, h_2 or h_1, is known. One can also verify that the critical depth, $h_1 = h_2 = h_c$, satisfies these equations, eqs. 4.41a and 4.41b.

vi) For an hydraulic jump on an *inclined bed*, the gravity force may not be neglected (see eq. 4.39). The ratio of the conjugate depths, h_2/h_1, is calculated (see *Ranga Raju*, 1981, p. 183) with :

$$\frac{h_2}{h_1} = \frac{1}{2} \left(\sqrt{1 + 8 \, \chi_{RH} \, Fr_1^2} - 1 \right)$$

The empirical coefficient given by Rajaratnam is $\chi_{RH} = (10^{0.027} \, \alpha)$; the angle of the bed is expressed in degrees.

vii) Hydraulic jumps in channels of non-rectangular sections have been also studied and are summarized by *Silvester* (1964).

4° The hydraulic jump is an efficient (mechanical) energy dissipator ; this represents an irreversible mechanism. The turbulent characteristics of the flow across the hydraulic jump are rather complex; they depend to a large extend on the upstream conditions of the supercritical flow (see *Resch* et *Leutheusser*, 1972).

NON-UNIFORM FLOW 177

i) For a rectangular channel, the *head loss*, h_{HJ}, across the hydraulic jump can be calculated as being the change in the specific energy (see Fig. 4.17), which is given by :

$$h_{HJ} = H_{s_1} - H_{s_2} = (h_1 + \frac{U_1^2}{2g}) - (h_2 + \frac{U_2^2}{2g})$$

Taking into account the above equation, eq. 4.40a, one may write :

$$h_{HJ} = \frac{(h_2 - h_1)^3}{4 h_1 h_2} \tag{4.42}$$

ii) The efficiency of the hydraulic jump is defined by the ratio of the received potential energy and the lost kinetic energy, or :

$$\eta = \frac{h_2 - h_1}{U_1^2/2g - U_2^2/2g} = \frac{4 h_1 h_2}{(h_1 + h_2)^2} \tag{4.43}$$

If there is a large difference in the water level, η will be small; if this difference is small, η will approach unity.

5° Several distinct types of hydraulic jumps (see Fig. 4.18) are encountered, depending on the Froude number of the upstream supercritical flow (see *Chow*, 1959, p. 395).

i) An *ordinary* jump will have a large height, occuring usually for $h_1/h_c < 0.61$. Observed is the formation of one or more rollers on the water surface; the horizontal axis of these rollers is above the diverging flow. If the upstream Froude number is $1.7 < Fr_1 < 2.5$, the jump is a weak one; if this Froude number is $Fr_1 > 4.5$, the jump is a strong one, being rather stationary.

ii) An *undular* jump will have a small height – the conjugate depths, h_1 and h_2, are close to the critical depth, h_c –, occuring usually for $h_1/h_c > 0.67$. Rollers on the water surface do not form, but the surface is an undulating one. The Froude number is small, $Fr_1 < 1.7$.

6° The *length* of the hydraulic jump, L_{HJ}, is the distance between the two terminal cross sections (see Fig. 4.18). The length is rather well defined for an ordinary jump; this is not the case for an undular jump.

The length for an ordinary jump in a rectangular channel can only be given by empirical relations, such as :

$$5 < (\frac{L_{HJ}}{h_2 - h_1}) < 7 \tag{4.44}$$

Other experiments (see *Henderson*, 1966, p. 219) have shown, that the length can also be taken as :

$$L_{HJ} \approx 6.1 \, h_2 \qquad (4.45)$$

valid for $4.5 < Fr_1 < 13$. Beyond this range of the Froude number, Fr_1, the length is usually smaller. For jumps in non-rectangular channels, the literature should be consulted (see *Silvester*, 1964).

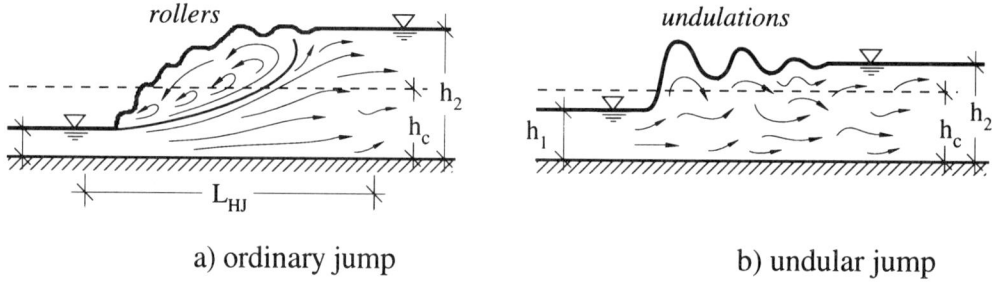

a) ordinary jump b) undular jump

Fig. 4.18 Types of hydraulic jump.

7° If the flow regime passes from a supercritical, $Fr_1 > 1$, to a subcritical one, $Fr_2 < 1$, an hydraulic jump is formed. In order to determine the *position* of the jump (see Fig. 4.19), it is necessary to draw the water-surface profiles upstream and downstream of the jump. The jump will take place at this cross section, where the flow depths are the conjugate ones, h_1 and h_2, calculated with eq. 4.41. [Take the case where the position of the jump is controlled by a downstream hydraulic condition : if the downstream water depth increases, the jump will move upstream; if the water depth decreases, the jump will move downstream.]

To obtain the position of the jump, the following graphical resolution (see Fig. 4.19) is recommended (taken is a channel on a weak slope) :

- draw the upstream and downstream water-surface profiles,
- draw the conjugate curve (see eq. 4.41) of the upstream profile,
- the jump takes place at the intersection of the downstream profile and the conjugate curve,
- a jump of zero length, L_{HJ}, takes now place between the points A' and Z',
- determine now the length, L_{HJ}, of the jump, using eq. 4.45,
- position the jump in a way that the line, Z"Z , parallel to the channel bed has a distance equal to the length of the jump,
- the hydraulic jump of the length, L_{HJ}, establishes itself between the points A and Z.

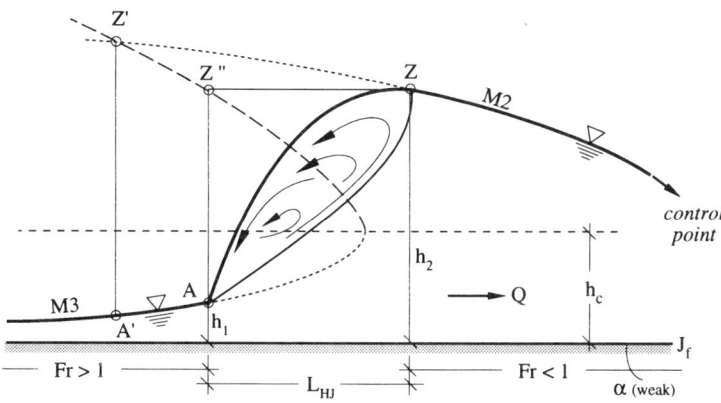

Fig. 4.19 Positioning of hydraulic jump.

8° In hydraulic engineering, the hydraulic jump is often *used*, as an energy dissipator. The existing upstream kinetic energy is considerably reduced by the creation of a jump (see Fig. 4.17) and this over a rather short distance; the jump causes an important head loss and raises the water depth at the downstream side.

For flow with large velocity, $Fr_1 > 1$, a channel with a mobile bed will be subjected to erosion (see sect. 3.4.2); consequently, a protection of the bed with an artificially fixed (paved) floor, called *stilling basin*, will be necessary. The length of such a basin should contain all jumps, namely jumps for all possible discharges; the outgoing velocity should be sufficiently small, $Fr_2 < 1$.

In order to control and/or provoke an hydraulic jump, sills (dentated) or steps may have to be installed on the bed.

Different configurations of stilling basins, as well as their hydraulic performance, are discussed in the literature (see *Chow*, 1959, chap. 15, *Hoffmann*, 1977, chap. 9, *Sinniger* et *Hager*, 1989, chaps. 16 and 17, and *Naudascher*, 1987, chap. 2).

The hydraulic jump may also be used as a *mixer* (of chemicals), by using the fact that the strong turbulent action in the rollers mixes the water with the air or with another liquid.

4.5 TRANSITIONS

1° A sudden transition is a short passage from one type of uniform flow to another one. Transitions will induce rapidly varied flow over a relatively short distance (see sect. 4.4) in a non-prismatic channel, where the cross section is variable. The flow does not necessarily pass across the critical depth.

2° A transition is encountered, if the cross section of the channel suddenly changes :

  i)   vertically : as an abrupt rise or drop in the channel floor (see sect. 4.5.1),

  ii) horizontally : as an expansion or contraction in the channel width (see sect. 4.5.2 and sect. 4.5.3),

  iii) vertically as well as horizontally.

3° A transition can serve as a control section; in this way it may be used for the calculation of the water-surface profile in gradually varied flow (see sect. 4.3).

4° If the transition is a short one, being sufficiently progressive and reasonably streamlined, the encountered head loss, h_r , can often be neglected.

4.5.1 Channel with variable Bed Floor

1° Take a channel of rectangular cross section with a constant width, B , on a very weak channel slope, S_f . At a certain location, an abrupt rise or drop, $\pm \Delta z$, in the bed floor is encountered (see Fig. 4.20).

2° Neglecting the head loss, h_r (see *Rouse*, 1938, p. 285), the equation of energy, eq. 2.9 , for the two sections, A_1 and A_2 , situated before and after the rise (+) or the drop (-) , can be written as :

$$h_1 + \frac{q^2}{2g\,h_1^2} = h_2 + \frac{q^2}{2g\,h_2^2} \pm \Delta z \tag{4.46}$$

The equation of continuity defines the respective velocities :

$$q = U_1 h_1 = U_2 h_2 \quad ; \quad B = Cte$$

For a given unit discharge, q , and a given change in the channel floor, $\pm \Delta z$, the above equation, eq. 4.46, allows to find the downstream flow depth, h_2 , if the upstream flow depth, h_1 , is known, and vice versa.

This equation, eq. 4.46, is a cubic one. The question now is, how many real solutions exist and are *hydraulically* possible. For an answer, it is advantageous to utilise the concept of the specific energy (see sect. 2.3).

3° Consequently, one poses :

$$H_{s_2}\,(H_{s_3}) = H_{s_1} \mp \Delta z \tag{4.47}$$

where $H_s = U^2/2g + h$ is the specific energy, whose curve for a constant discharge, q, is given with Fig. 4.20. A graphical solution to eq. 4.46 is to be followed :

i) The energy-grade line (lct) stays constant and horizontal.

ii) Upstream of the floor change, at a given specific energy, H_{s_1}, two flow depths, h_1 and h_1', are possible : one, $h_1 > h_c$, in the fluvial regime, and one, $h_1' < h_c$, in the torrential regime.

iii) Downstream of the floor change, the specific energy, H_{s_2} (H_{s_3}) is changed by the height of the floor change, Δz ; it has now a value of H_{s_2} (H_{s_3}) = $H_{s_1} \mp \Delta z$. On the curve, $H_s = f(h)$, one notices a shift of Δz towards the left for a rise and towards the right for a drop in the channel floor.

iv) For an abrupt rise (see Fig. 4.20a), the flow depth decreases, $h_2 < h_1$, in the fluvial regime and increases, $h_2' > h_1'$, in the torrential regime. For an abrupt drop (see Fig. 4.20b), the inverse is encountered.

v) For a rise, a flow is only possible, if the rise height, $+\Delta z$, does not decrease the specific energy, H_{s_2}, below its critical value, H_{s_c} ; it is therefore necessary that $H_{s_2} \geq H_{s_c}$. If the specific energy, $H_{s_2} = H_{s_c}$, becomes the critical one, the flow after the rise will be in the critical regime, $Fr_2 \equiv 1$; in such a case the rise in the channel floor can serve as a control section (see Fig. 4.13).

In subcritical flow, the upstream flow depth will increase and a water-surface profile will form itself (see Ex.4.E, point iv); in supercritical flow, an hydraulic jump may form itself.

vi) A minimum specific energy, $H_{s_1}^m$, at the upstream of the rise is necessary for the flow to pass over the rise, $+\Delta z$; it is given by :

$$H_{s_1}^m = H_{s_c} + \Delta z \quad \text{where} \quad H_{s_c} \equiv H_{s_2}.$$

4° If the head loss, h_r, caused by a variable channel floor, is no more negligible, it becomes necessary to obtain a relation for $h_2 = f(h_1)$, using the momentum equation (see *Jaeger*, 1954, pp. 162-171).

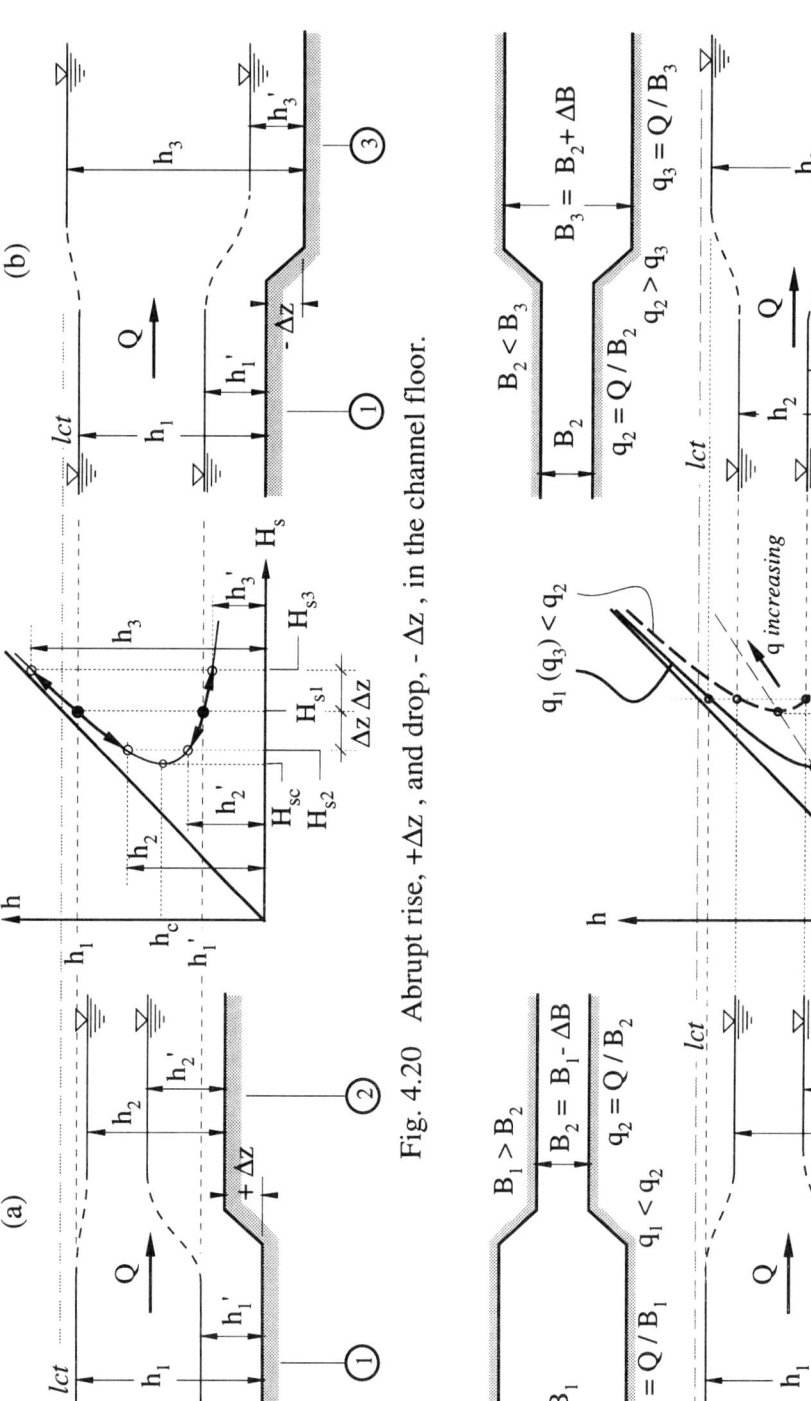

Fig. 4.20 Abrupt rise, $+\Delta z$, and drop, $-\Delta z$, in the channel floor.

Fig. 4.21 Contraction, $-\Delta B$, and expansion, $+\Delta B$, of the channel width.

NON-UNIFORM FLOW

4.5.2 Channel of variable Width

1° Take a channel of rectangular cross section on a very weak and constant channel slope, S_f. The width of the channel, B, no more constant, is changed being either expanded, $+\Delta B$, or contracted, $-\Delta B$.

2° If one assumes that this change in width, $\pm \Delta B$, is achieved progressively over a short distance, the head losses, h_r, can be neglected. This problem can be studied in the way the previous one (see sect. 4.5.1) was developed, namely by considering the equation of specific energy (see sect. 2.3).

3° The equation of continuity for a constant discharge, Q, reads:

$$q_1 B_1 (q_3 B_3) = Q = q_2 B_2$$

However, for a channel of variable width, the unit discharge, q, will be variable.

The specific energy will be maintained throughout the transition, or:

$$H_{s_1} (H_{s_3}) = H_{s_2} \qquad (4.48)$$

A graphical solution to eq. 4.48 is to be followed:

i) The curves (see Fig. 4.21) of the specific energy, H_s, for the two unit discharges, q_1 (q_3) and q_2, are drawn; each one has its own critical depth, h_{c_1} (h_{c_3}) and h_{c_2}.

ii) In the fluvial regime: a contraction causes an increase of the unit discharge, $q_2 > q_1$, and consequently a decrease in the flow depth, $h_2 < h_1$; an expansion causes a decrease of the unit discharge, $q_2 > q_3$, and consequently an increase in the flow depth, $h_3 > h_2$.

iii) In the torrential regime: a contraction causes an increase in the flow depth, $h_2' > h_1'$; an expansion causes a decrease in the flow depth, $h_3' < h_2'$.

iv) One may also ask, as to when the unit discharge is at a maximum, $q_m = Q/B_m$, namely when the upstream, subcritical and supercritical, flow becomes a critical flow, $h_m \equiv h_c$. The reduction in width, $\Delta B = B_1 - B_m$, can thus be calculated (see eq. 2.24); for a rectangular cross section, one obtains:

$$B_m^{-1} \equiv B_c^{-1} = \frac{1}{Q} \sqrt{g\, h_c^3}$$

This concept has been useful in the design of measuring flumes (see Fig. 4.14), where the critical depth establishes itself at the throat of the canal.

v) If the reduction in the width, $B_2 < B_c$, is an important one and the discharge, Q, remains constant, a water-surface profile curve may form itself (see *Ranga Raju*, 1981, p. 121).

4° Should the head loss, h_r, due to the variation in width, be no more negligible, it becomes necessary to obtain a relationship between the respective depths, h_1 and h_2 (see *Jaeger*, 1954, p. 171-180). The head loss, neglected in the preceding study, can be given (see *Crausse*, 1951, p. 229) by :

$$h_r = K_t \frac{U_2^2 - U_1^2}{2g}$$

where $0.05 < K_t < 0.1$ for a progressive contraction and $K_t \approx 0.5$ for an angular one; and where $0.1 < K_t < 0.2$ for an expansion and $K_t \approx 1.0$ for an angular one.

5° A contraction, namely a reduction in width, will also be encountered, when flow passes across piers or other obstacles (see *Crausse*, 1951, p. 233). Due to such obstacles (see Fig. 4.22), the width available for the flow is reduced to $(B_1-\Delta B)$.

In the fluvial regime, the piers will cause a decrease in the flow depth; in the torrential regime, an increase will be evident (see Fig. 4.21).

Note, that this reduced width, $(B_1-\Delta B)$, must be corrected with a contraction coefficient, μ, which is slightly smaller than unity.

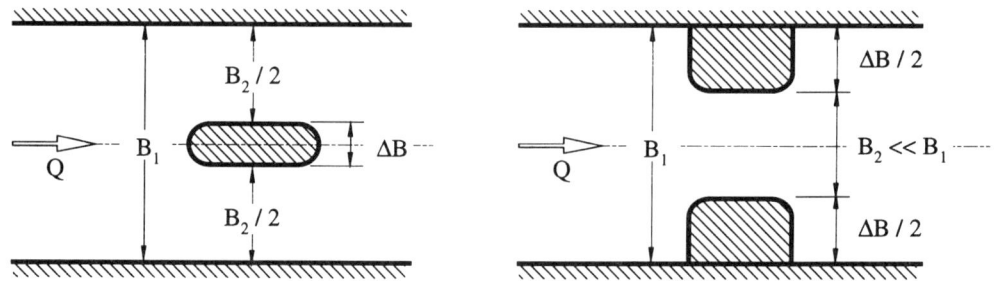

Fig. 4.22 Channel of variable width : pier and constriction.

6° A rather large reduction in width is also encountered during the passage of flow across any type of constriction (see Fig. 4.22), placed into the channel.

The hydraulics of this phenomenon is rather complex; only experimental studies are available. Design recommendations for different types of constrictions exist in the literature (see *French*, 1985, pp. 394-426).

NON-UNIFORM FLOW   185

4.5.3 Oblique Jump

1° At high-velocity flow, Fr > 1, a small variation in flow depth, h, causes usually a large variation in kinetic energy, $U^2/2g$ (see Fig. 2.5).

A transition – such as a change in channel width (see sect. 4.5.2) or in flow direction (see sect. 3.5.2) – will provoke an abrupt variation in the flow depth, h ; at the free-water surface, stationary and stable gravity waves (see sect. 2.4.3) will appear.

2° The formation of these fronts of (standing) *waves*, which are also called *oblique (hydraulic) jumps*, can be explained using the following hydraulic considerations (see *Ippen*, 1949, sect. D.14).

3° Take an upstream flow, being supercritical (see Fig. 4.23), whose Froude number is given by :

$$Fr_1 = \frac{U_1}{\sqrt{gh_1}} = \frac{U_1}{c_1} > 1 \tag{1.4}$$

where U_1 is the average velocity and $c_1 = \sqrt{gh_1}$ is the celerity of the gravity wave (see eq. 2.27).

i) The upstream flow undergoes a directional deflection – caused by a vertical channel wall – , whose angle of deflection is θ. An oblique wave front is formed under an angle of β. Nevertheless, downstream, $h_2 U_2$, and upstream, $h_1 U_1$, of this wave the flow can be considered to be uniform and one-dimensional.

ii) In a channel of unit width, for the direction *normal* to the wave front, the equation of continuity is :

$$h_1 U_1^n = h_2 U_2^n \tag{4.49}$$

and the momentum equation – friction on the bottom is neglected – is :

$$\Sigma F_n = \gamma \frac{h_1^2}{2} - \gamma \frac{h_2^2}{2} = \rho q (U_2^n - U_1^n) \tag{4.50}$$

The corresponding Froude numbers are given as :

$$Fr_1^n = \frac{U_1^n}{\sqrt{gh_1}} \quad \text{and} \quad Fr_2^n = \frac{U_2^n}{\sqrt{gh_2}} \tag{4.51}$$

For the direction *tangential* to the wave front, no change in momentum will take place, or :

$$\Sigma F_t = 0 = \rho q (U_2^t - U_1^t) \tag{4.52}$$

thus, it is apparent that :

$$U_2^t = U_1^t \tag{4.52a}$$

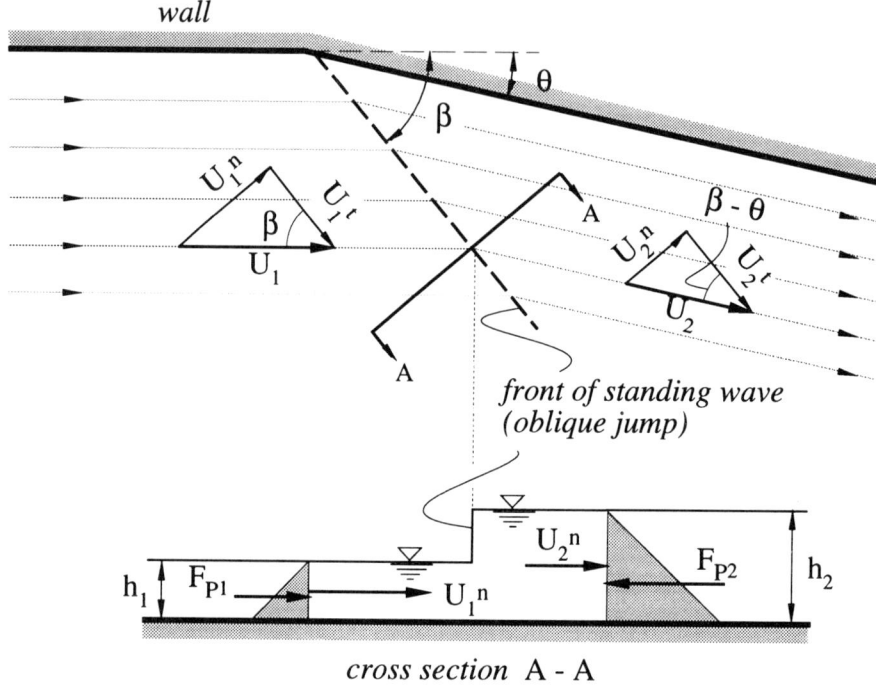

Fig. 4.23 Scheme of an oblique jump.

iii) It follows from the geometry (see Fig. 4.23) that :

$$U_1^n = U_1 \sin \beta \quad \text{and} \quad U_2^n = U_2 \sin (\beta - \theta)$$
$$U_1^t = U_1^n / \text{tg } \beta \quad \text{and} \quad U_2^t = U_2^n / \text{tg } (\beta - \theta)$$
(4.53)

iv) Combining the equation of continuity, eq. 4.49, and the momentum equation, eq. 4.50, yields the depth ratio across the wave front of :

$$\frac{h_2}{h_1} = \frac{1}{2} (\sqrt{1 + 8 \left(Fr_1^n \right)^2} - 1)$$
(4.54)

where the Froude number is defined as :

$$Fr_1^n = \frac{U_1^n}{\sqrt{gh_1}} = \frac{U_1 \sin \beta}{\sqrt{gh_1}} = Fr_1 \sin \beta$$
(4.51a)

The angle, β, of the wave front can be expressed by :

$$\sin \beta = \frac{1}{Fr_1} \sqrt{\frac{1}{2} \frac{h_2}{h_1} (\frac{h_2}{h_1} + 1)}$$

NON-UNIFORM FLOW

Note, that for small variations of the flow depth, $h_1 \approx h_2$, – thus for gradual transitions –, one gets:

$$\sin \beta \approx \frac{1}{Fr_1} = \frac{c_1}{U_1} \tag{3.45}$$

It is also interesting to compare eq. 4.54 with eq. 4.41a; the latter relation was obtained for a normal hydraulic jump, where the wave front is normal to the direction of flow, $\beta = 90°$ or $\sin \beta = 1$.

v) Using the geometric relations, eqs. 4.53, and the equation of continuity, eq. 4.49, the depth change across the wave front can also be expressed by:

$$\frac{h_2}{h_1} = \frac{U_1^n}{U_2^n} = \frac{tg\,\beta}{tg\,(\beta-\theta)} \tag{4.55}$$

vi) Combining the relation of eq. 4.54 and of eq. 4.55, renders (see *Ippen*, 1949, p. 553) an implicit expression for the angle, β, of the wave front, such as:

$$tg\,\theta = \frac{tg\beta\,(\sqrt{1 + 8\left(Fr_1^n\right)^2} - 3)}{2\,tg^2\beta + \sqrt{1 + 8\,(Fr_1^n)^2} - 1} \tag{4.56}$$

This relation between the wall angle, θ, and the front angle, β, given here with Fig. 4.24, was experimentally verified. It is however only valid for positive angles of deflection, θ, thus for contracting channels only. The following is to be noticed:

- for all values of upstream Froude numbers, Fr_1, there exists a maximum value for the angle of deflection, θ_{max};

- for all values below $\theta < \theta_{max}$, two values for β are possible; however, since the flow is subcritical when $Fr_2 < 1$, only values of β for $Fr_2 > 1$ are of practical interest.

As was said, above relation, eq. 4.56, is only valid if the upstream flow converges towards the downstream; for diverging flows, an approximate method (see *Ippen*, 1949, p. 555 and p. 559) must be employed.

4° If a supercritical flow is deviated, a front of a standing wave (oblique jump) is formed; one can predict its height, $h_2 - h_1$, with eq. 4.54 and its direction, β, with eq. 4.56. These waves will propagate towards the downstream; they will be reflected (see Fig. 4.25) at the opposite channel wall(s) and/or be intersected by other waves (see Fig. 4.26).

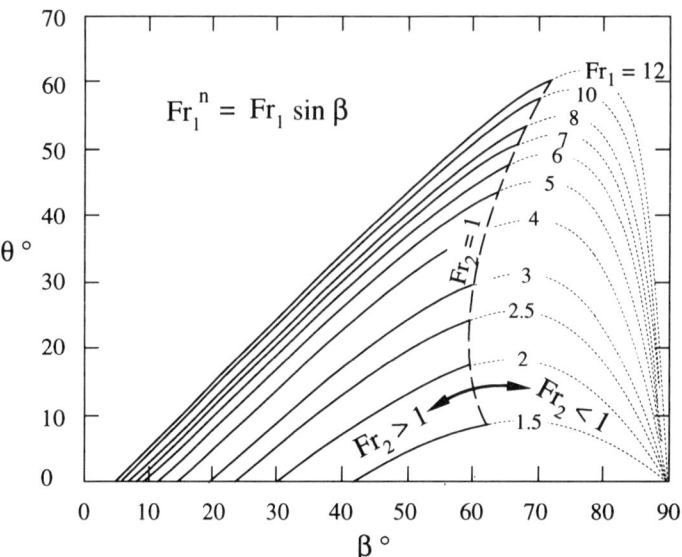

Fig. 4.24 Plot of the relation, eq. 4.56, for the angle, β, of the wave front of an oblique jump.

5° The knowledge about the oblique jump can well be put to use for the study of supercritical flow in rectangular channels of variable width, notably with a linear contraction.

i) *Asymmetrically converging channel* (see Fig. 4.25) :

The supercritical flow, $Fr_1 > 1$, is contracted by a channel wall, inclined under a constant angle of deflection, θ ; the following is observed :

- the (initial) angle of the wave front, β_1, will be calculated, according to eq. 4.56, for the upstream Froude number, Fr_1, and the angle of deflection, θ ; behind the wave front, the Froude number, $Fr_2 < Fr_1$, diminishes and the flow depth, $h_2 > h_1$, increases;

- the wave is reflected by the opposite (straight) wall; the angle of the wave front, $\beta_2 > \beta_1$, is calculated with eq. 4.56, using Fr_2 ;

- behind this wave front, the Froude number, $Fr_3 < Fr_2$, is further diminished and the flow depth, $h_3 > h_2$, further increased;

- this wave is again reflected, but now by the opposite (inclined) wall, etc.;

- if the channel is long enough, a *normal* hydraulic jump will form itself and the flow becomes subcritical.

NON-UNIFORM FLOW 189

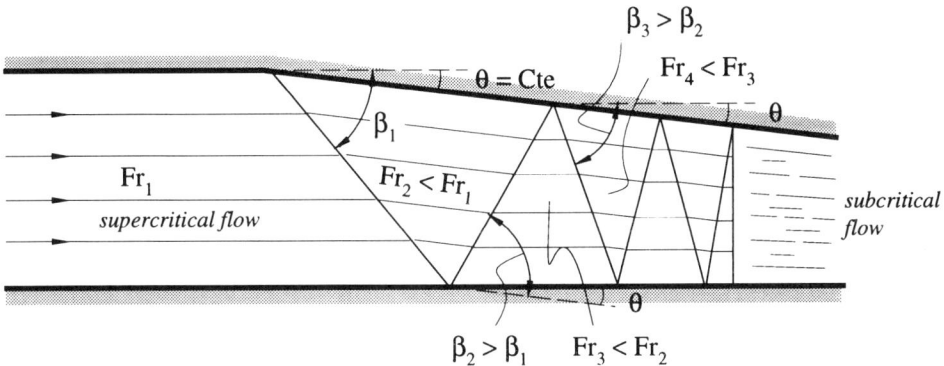

Fig. 4.25   Reflection of a wave front (oblique jump).

ii)   *Symmetrically converging channel* (see Fig. 4.26):

The supercritical flow, $Fr_1 > 1$, which arrives in the converging channel (see *Ippen*, 1949, p. 557) is deviated inwardly by the channel walls; positive waves under an angle of β_1 – to be obtained from Fig. 4.24 – will be generated at the points A and A' at the entrance of the transition:

- these positive waves (perturbations) propagate downstream along the lines ABC' and A'BC; they intersect at a point, B, and the Froude number becomes $Fr_2 < Fr_1$;

- no perturbations will be noticed in the zone AA'B, where the Froude number remains Fr_1;

- in the zone BCC', the Froude number becomes $Fr_3 < Fr_2 < Fr_1$;

- the positive waves, arriving at the points, C and C', are now reflected by the channel walls; a system of (symmetrical) cross waves is formed, moving downstream;

- meanwhile, negative waves are created at the points, D and D', at the end of the transition;

- consequently, a complex interference of the waves created at the points, D and D', and of the waves created at the points, A and A', takes place; the resulting cross waves (often rather disagreeable) may exist for a long distance.

With a certain *good* choice of the angle, $\theta \equiv \theta'$, the positive waves, which touch at the points, C(D) and C'(D'), will be reflected and also compensated by the negative waves, coming from the same points. The flow in the contracted section has a Froude number of Fr_3; no further perturbation (waves) is noticed in the downstream part of the channel.

- The choice of this angle, θ', will obviously depend on the approach Froude number, Fr_1, as well as on the contraction ratio, B_3/B_1 ;

- using the equation of continuity, this ratio is given as :

$$\frac{B_3}{B_1} = \frac{h_1 U_1}{h_3 U_3} = \left(\frac{h_1}{h_3}\right)^{3/2} \left(\frac{Fr_1}{Fr_3}\right) \tag{4.57}$$

with the downstream flow remaining always supercritical, $Fr_3 > 1$;

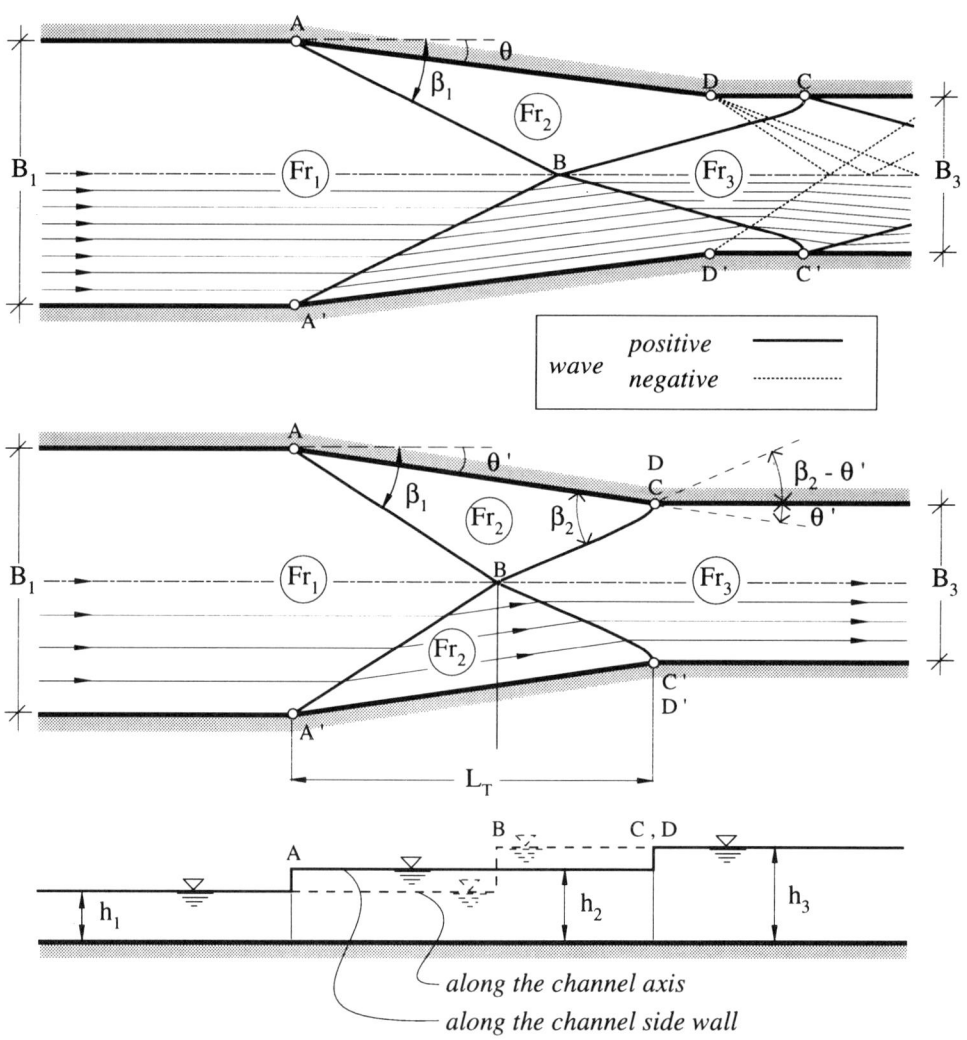

Fig. 4.26 Symmetrically converging channel ; flow is supercritical.

NON-UNIFORM FLOW 191

- an iterative calculation – using eq. 4.57 together with eq. 4.56 – will be necessary to obtain the length of contraction (see *Chow*, 1959, p. 470), which is given by the following geometrical relation :

$$L_T = \frac{(B_1 - B_3)}{2 \, tg\theta}$$

Finally, it should be remarked, that for a *gradual* contraction (and expansion) of the channel, there are simple calculation methods available (see *Ippen*, 1949, p.549 and p.559).

4.6 LATERAL INFLOW

1° To describe the flow in a channel with lateral inflow (or outflow) between two cross sections, the equations of motion must be modified.

2° The *equation of continuity* for steady flow (see eq. 2.7) was given by :

$$\frac{dQ}{dx} = \pm q_\ell(x) \qquad (4.58)$$

where q_ℓ is a supplementary *lateral* discharge per unit length (see Fig. 4.27) which leaves, – , or enters, +. For flow with lateral inflow, one writes :

$$q_l = \frac{dQ}{dx} = U\frac{dA}{dx} + A\frac{dU}{dx} \qquad (4.58a)$$

3° The *equation of motion* for steady flow requires that the sum of the quantity of motion across the surface of a fluid volume be equal to the sum of the applying forces (see *Graf & Altinakar*, 1991, p. 145). By projection into the longitudinal direction of the channel bed with a weak slope (see Fig. 4.27), this equation reads :

$$\Sigma F_x = \Delta F_p + W \sin\alpha - F_F = \rho \, \Delta(QU) \qquad (4.59)$$

The different terms are approximately given (see *Chow*, 1959, sect. 12.2) as :

i) The resulting hydrostatic pressure force is :

$$\Delta F_p \cong (\gamma z_p A) - \gamma(z_p + dh)A \cong -\gamma \, dhA$$

where z_p is the depth of the centroid of the area A, below the water surface.

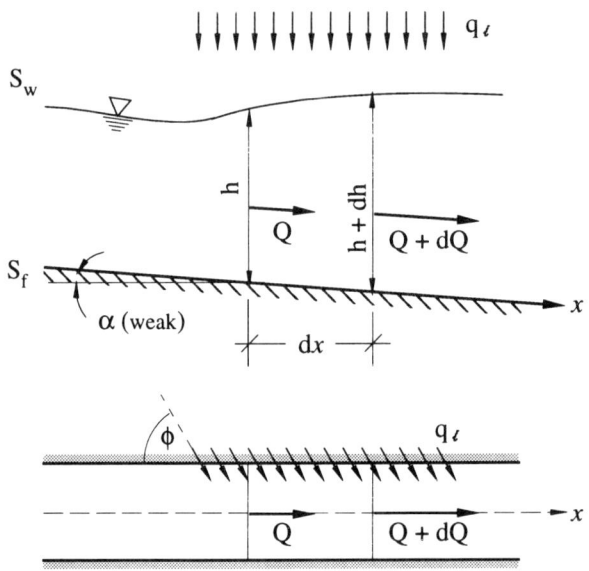

Fig. 4.27 Scheme of a gradually varied inflow with lateral discharge entering, $+ q_\ell$.

ii) The weight of the water between the two sections is :

$$W \sin\alpha \cong \gamma \sin\alpha \, A \, dx \cong \gamma S_f A \, dx$$

where the bed slope is taken as $\sin\alpha \cong S_f$, valid for weak slopes.

iii) The friction force is :

$$F_F = \tau_o P \, dx \cong \gamma S_e A \, dx$$

with the bottom shear stress given by $\tau_o = \gamma R_h S_e$.

iv) The momentum charge is :

$$\rho \Delta (QU) = \rho (Q + dQ)(U + dU) - \rho (QU) - \rho (q_\ell \, dx)(U_\ell \cos\phi)$$

where $q_\ell \, dx = dQ$ (see eq. 4.58a) is the supplementary discharge, *entering* the channel at an angle, ϕ, with a lateral velocity, U_ℓ. For an angle of $\phi = 90°$, and neglecting terms of $dQdU$, one may approximate :

$$\rho \Delta (QU) \cong \rho (dQ \, U + Q \, dU) - 0$$

Consequently, the above equation, eq. 4.59, now yields approximately :

$$-\gamma A \, dh + \gamma A S_f \, dx - \gamma A S_e \, dx = \rho (dQ \, U + Q \, dU) \qquad (4.59a)$$

NON-UNIFORM FLOW

4° From above equation, eq. 4.59a, the *variation in the flow depth* can be obtained :

$$\frac{dh}{dx} = (S_f - S_e) - \frac{U}{g} \left(\frac{1}{A} \frac{dQ}{dx} + \frac{dU}{dx} \right) \tag{4.60}$$

Putting now $U + dU = (Q + dQ)/(A + dA)$, and neglecting some terms, AdA and dQdA, – but maintaining $\phi \neq 90°$ –, one obtains :

$$\frac{dh}{dx} \cong \frac{(S_f - S_e) - q_l \left[2 \dfrac{Q}{gA^2} - \dfrac{1}{gA} (U_\ell \cos \phi) \right]}{1 - \dfrac{(Q/A)^2}{gA/B}} \tag{4.61}$$

This equation of steady, spatially varied flow with the lateral inflow, $q_\ell = dQ/dx$, allows to calculate the water-surface profile with the method of successive approximations (see sect. 4.3.1). Some examples are given in the literature (see *Chow*, 1959, chap. 12, and *French*, 1985, sect. 6.4). Note, that the above equation, eq. 4.61, becomes eq. 4.8, if the lateral inflow, q_ℓ, is zero.

5° The equation of energy can also be used to obtain this equation, eq. 4.61 (see *Chow*, 1959, p. 332).

6° An understanding of spatially varied flow, both with inflow or outflow, becomes important for the design of branching channels, as well as for lateral weirs and spillways (see *Sinniger* et *Hager*, 1989, chaps. 9 and 10).

194   FLUVIAL HYDRAULICS

4.7 EXERCISES

4.7.1 Problems, solved

Ex. 4.A

A trapezoidal channel, having a bottom width of b = 7.0 [m] and side slopes of m = 1.5, conveys a discharge of Q = 28 [m³/s]. The channel has a constant bed slope of S_f = 0.0010; the friction coefficient has already been determined as being n = 0.025 [m$^{-1/3}$s]. The channel is terminated by a sudden drop of the channel bed. Determine what type of water-surface profile is to be expected. Subsequently, calculate and plot this profile upstream from the drop by using:

i)   the method of direct integration,
ii) the *direct* step method,
iii) the *standard* step method.

SOLUTION :

1° Before computing the water-surface profile, the normal depth, h_n, and the critical depth, h_c, of the channel should be determined :

The geometrical characteristics, A, P, R_h, B and D_h, for a trapezoidal channel are given in Table 1.1.

a) *Computation of the normal depth, using the Manning-Strickler formula :*

Expressing the average flow velocity with Manning-Strickler formula, eq. 3.16, the discharge in the channel is :

$$Q = U A = \frac{1}{n} R_h^{2/3} S_f^{1/2} A$$

On a spreadsheet, the normal depth can readily be obtained from the above equation by trial-and-error. The computation sheet is presented below :

Computation of *normal* depth by trial-and-error					
b = 7.0 [m] ; m = 1.5 [-] ; n = 0.025 [m$^{-1/3}$s] ; S_f = 0.001 [-] ; Q = 28.0 [m³/s]					
trial	h_n [m]	A [m²]	P [m]	R_h [m]	$Q_{calc.}$ [m³/s]	Remarks
1	1.500	13.875	12.408	1.118	18.908	no ! 18.908 < 28.000 try : h_n > 1.5 [m]
2	1.900	18.715	13.851	1.351	28.933	no ! 28.933 > 28.000 try : 1.50 < h_n < 1.9 [m]
3	1.866	18.285	13.728	1.332	27.999	yes ! 27.999 ≅ 28.000 h_n = 1.866 [m]

The *normal depth* is therefore : h_n = *1.866* [m]

NON-UNIFORM FLOW

b) *Computation of the critical depth :*

When the flow is critical, one can write (see eq. 2.22) :
$$\frac{U_c}{\sqrt{gD_{h_c}}} = 1$$

where $U_c = Q / A_c$ (see eq. 1.3). By substituting the relationships for U_c and $D_{h_c} = A_c / B$ into eq. 2.22, one obtains : $Q = \sqrt{g\, A_c^3 / B}$.

On a spreadsheet, the critical depth can readily be obtained from the above equation by trial-and-error. This is presented below :

Computation of *critical* depth by trial-and-error						
b = 7.0 [m] ; m = 1.5 [-] ; n = 0.025 [m$^{-1/3}$s] ; S_f = 0.001 [-] ; Q = 28.0 [m³/s]						
trial	h_c [m]	A_c [m²]	B [m]	D_{h_c} [m]	$Q_{calc.}$ [m³/s]	Remarks
1	1.200	10.560	10.600	0.996	33.012	no ! 33.012 > 28.000 try : h_c < 1.2 [m]
2	1.000	8.500	10.000	0.850	24.545	no ! 24.545 < 28.000 try : 1.00 < h_c < 1.2 [m]
3	1.085	9.358	10.254	0.913	27.999	yes ! 27.999 ≅ 28.000 h_c = 1.085 [m]

The *critical depth* is therefore : $h_c = 1.085$ [m]

c) *Determination of the type of the water-surface profile :*

Since $h_n > h_c$, the flow is taking place on a mild slope. The water-surface profile will thus be a *type M curve*.

The same conclusion can also be reached by comparing the bed slope with the critical slope. The critical slope is the one for which $h_n = h_c = 1.085$ [m] (see sect. 4.1.3).

If $h_c = h_n = 1.085$ [m]   ⇒   A = 9.358 [m²]   and $R_h = 0.858$ [m].

These values are introduced in the Manning-Strickler formula, eq. 3.16 :

$$Q = \frac{1}{0.025}\, 0.858^{2/3}\, S_c^{1/2}\, 9.358 = 28.0\ [m^3/s]$$

in order to obtain the critical slope : $S_c = 0.00686$ [-]

Given that : S_f (= 0.001) > 0 and also S_f (= 0.001) < S_c (= 0.00686), it can be concluded that the flow is taking place on a mild slope and the water-surface profile is a *type M curve* (see Fig. 4.2).

2° To determine in which zone the water-surface profile lies (see sect. 4.2.1), a control section has to be identified.

At the upstream extremity of the channel, the flow is subcritical and uniform with a normal depth of $h_n = 1.866$ [m]. The water-surface profile will therefore be controlled by a singularity at the downstream side. As a matter of fact, the channel terminates by a drop (see sect. 4.4.2) where, it can be assumed that the flow depth passes through the critical depth, $h_c = 1.085$ [m]. This section will be the *control section* for the water-surface profile computation.

The water-surface profile falls between the normal and critical depths. By proceeding towards upstream the flow depth increases from the critical depth, $h_c = 1.085$ [m], to the normal depth, $h_n = 1.866$ [m]. *The water-surface profile is a M2 type curve* (see Figs. 4.2 and 4.2a).

3° *Computation of water-surface profile, using the method of direct integration :*

The *method of Chow* (see point 4.3.2; 5°) will now be used since it is free from the simplifying assumptions used in other methods. According to the method of Chow, the abscissa, x_i', (with respect to an arbitrary origin) of a section (i), where the flow depth is equal to h, can be written as follows (see eq. 4.29) :

$$x_i' = \frac{h_n}{S_f} \left\{ \eta_i - \Phi(\eta_i, N) + \left(\frac{h_c}{h_n}\right)^M \frac{J}{N} \left[\Phi(\zeta_i, J) \right] \right\}$$

This equation can be implemented in a spreadsheet program for calculating the water-surface profile in a tabular form.

In order to create a fully automatic computation sheet, the analytical expressions have to be established for the hydraulic exponents, N and M, which are given by :

$$N(h) = \frac{2h}{3A}\left(5B - 2R_h \frac{dP}{dh}\right) \quad , \quad M(h) = \frac{h}{A}\left(3B - \frac{A}{B}\frac{dB}{dh}\right) \quad (4.26)$$

The terms, dP/dh and dB/dh, will be evaluated using the geometrical characteristics of the section. For a trapezoidal section, one has (see Table 1.1) :

$P = b + 2h\sqrt{1 + m^2}$ $\Rightarrow$ $dP/dh = 2\sqrt{1 + m^2}$

$B = b + 2mh$ $\Rightarrow$ $dB/dh = 2m$

Substituting these relationships in eqs. 4.26, and simplifying, the following expressions can be obtained (see *Chow* 1959, p. 131 and p. 66):

$$N(h) = \frac{10}{3}\frac{1 + 2m\,(h/b)}{1 + m\,(h/b)} - \frac{8}{3}\frac{\sqrt{1 + m^2}\,(h/b)}{1 + 2\sqrt{1 + m^2}\,(h/b)}$$

$$M(h) = \frac{3\,[1 + 2m\,(h/b)]^2 - 2m\,(h/b)\,[1 + m\,(h/b)]}{[1 + 2m\,(h/b)]\,[1 + m\,(h/b)]}$$

(4.26a)

NON-UNIFORM FLOW

Method of direct integration : Chow's method										
$b = 7.0$ [m] ; $m = 1.5$ [-] ; $n = 0.025$ [m$^{-1/3}$s] ; $S_f = 0.001$ [-] ; $Q = 28.0$ [m³/s]										
$h_n = 1.866$ [m]				$h_c = 1.085$ [m]				$h_c/h_n = 0.581$ [-]		
1	2	3	4	5	6	7	8	9	10	11
h	h/b	M	N	J	η	ζ	Φ(η,N)	Φ(ζ, J)	x'	x
[m]	[-]	[-]	[-]	[-]	[-]	[-]	[-]	[-]	[m]	[m]
1.085	0.155	3.25	3.48	2.82	0.581	0.512	0.603	0.534	98.243	0.000
1.100	0.157	3.25	3.49	2.82	0.589	0.521	0.613	0.544	97.798	0.445
1.150	0.164	3.26	3.50	2.83	0.616	0.550	0.645	0.580	96.742	1.501
1.200	0.171	3.27	3.51	2.85	0.643	0.581	0.678	0.617	93.192	5.051
1.300	0.186	3.30	3.52	2.87	0.697	0.641	0.748	0.697	81.116	17.127
1.400	0.200	3.32	3.54	2.89	0.750	0.703	0.826	0.787	56.379	41.865
1.500	0.214	3.34	3.56	2.91	0.804	0.765	0.916	0.891	12.509	85.734
1.600	0.229	3.36	3.58	2.93	0.857	0.829	1.026	1.021	-63.247	161.490
1.700	0.243	3.38	3.60	2.95	0.911	0.892	1.177	1.200	-202.640	300.883
1.800	0.257	3.40	3.62	2.97	0.965	0.957	1.457	1.531	-548.543	646.786
1.864	0.266	3.41	3.63	2.98	0.999	0.999	2.454	2.931	-2010.449	*2108.693*

col.	symbol	explanations
1	h	flow depth. The water-surface profile increases from the critical depth, $h_c = 1.085$ [m], to the normal depth, $0.999 h_n = 1.864$ [m]. This difference in flow depth is arbitrarily divided into several intervals.
2	h/b	relative flow depth (with respect to bottom width).
3,4	M , N	hydraulic exponents, calculated according to eqs. 4.26a.
5	J	N / (N-M+1)
6	η	h / h_n , relative flow depth.
7	ζ	$\eta^{N/J}$, dimensionless flow depth.
8	Φ(η,N)	function defined by eq. 4.24a. The values are read from Table 4.1, being functions of η and N .
9	Φ(ζ, J)	function defined by eq. 4.28a. The values are read from Table 4.1, being functions of ζ and J .
10	x'	distance from an *arbitrary* origin, calculated using the expression : $$x_i' = \frac{h_n}{J_f} \left\{ \eta_i - \Phi(\eta_i, N) + \left(\frac{h_c}{h_n}\right)^M \frac{J}{N} \left[\Phi(\zeta_i, J)\right] \right\} \quad (4.29)$$
11	x	distance from the drop ($x = 0$ [m]): $x_i = x'_{(h = 1.085)} - x_i'$.

Step method:
b = 7.0 [m] ; m = 1.5 [-] ; n = 0.025 [m$^{-1/3}$s]
S_f = 1.0E-03 [-] ; Q = 28.0 [m³/s]

1	2	3	4	5	6	7
h	B	P	A	R_h	U	$U^2/2g$
[m]	[m]	[m]	[m²]	[m]	[m/s]	[m]
1.085	10.255	10.912	9.361	0.858	2.991	0.456
1.100	10.300	10.966	9.515	0.868	2.943	0.441
1.150	10.450	11.146	10.034	0.900	2.791	0.397
1.200	10.600	11.327	10.560	0.932	2.652	0.358
1.300	10.900	11.687	11.635	0.996	2.407	0.295
1.400	11.200	12.048	12.740	1.057	2.198	0.246
1.500	11.500	12.408	13.875	1.118	2.018	0.208
1.600	11.800	12.769	15.040	1.178	1.862	0.177
1.700	12.100	13.129	16.235	1.237	1.725	0.152
1.800	12.400	13.490	17.460	1.294	1.604	0.131
1.866	12.598	13.728	18.285	1.332	1.531	0.120

col.	symbol	expression	explanations
1	h		flow depth. The water-surface profile increases from the critical depth, h_c = 1.085 [m], to the normal depth, h_n = 1.866 [m]. This difference in flow depth is arbitrarily divided into several *small* intervals.
2	B	b + 2mh	water-surface width (see Table 1.1) for the flow depth in column 1.
3	P	b + 2h $\sqrt{1+m^2}$	wetted perimeter (see Table 1.1) for the flow depth in column 1.
4	A	h (b + m h)	wetted surface (see Table 1.1) for the flow depth in column 1.
5	R_h	A / P	hydraulic radius for the flow depth in column 1.
6	U	Q / A	average flow velocity.
7	$U^2/2g$		dynamic head.

NON-UNIFORM FLOW

Direct step method (Δh is fixed)						
h_n = 1.866 [m] ; h_c = 1.085 [m]						
convention : $C = 1$, towards upstream (Fr < 1) ; $C = -1$, towards downstream (Fr > 1)						
8	9	10	11	12	13	14
H_s [m]	ΔH_s [m]	S_e [-]	$\overline{S_e}$ [-]	$\overline{S_e} - S_f$ [-]	Δx [m]	x [m]
1.541	—	6.86E-03	—	—	—	0.000
1.541	0.000	6.54E-03	6.70E-03	5.70E-03	0.060	0.060
1.547	0.006	5.60E-03	6.07E-03	5.07E-03	1.093	1.153
1.558	0.011	4.82E-03	5.21E-03	4.21E-03	2.713	3.866
1.595	0.037	3.64E-03	4.23E-03	3.23E-03	11.396	15.262
1.646	0.051	2.80E-03	3.22E-03	2.22E-03	22.962	38.224
1.708	0.061	2.19E-03	2.50E-03	1.50E-03	40.978	79.202
1.777	0.069	1.74E-03	1.97E-03	9.67E-04	71.432	150.634
1.852	0.075	1.40E-03	1.57E-03	5.71E-04	131.246	281.880
1.931	0.079	1.14E-03	1.27E-03	2.70E-04	294.203	576.083
1.986	0.054	1.00E-03	1.07E-03	6.98E-05	779.917	1356.000

col.	symbol	expression	explanations
8	H_s	$h + U^2/2g$	*specific* head, eq. 2.14.
9	ΔH_s	$(H_s)_i - (H_s)_{i-1}$	specific head difference between section (i) and previous section (i-1).
10	S_e	$\dfrac{U^2 n^2}{R_h^{4/3}}$	slope of energy-grade line with respect to horizontal, calculated using the Manning-Strickler formula, eq. 3.16.
11	$\overline{S_e}$	$\dfrac{(S_e)_i + (S_e)_{i-1}}{2}$	average slope of energy-grade line between section (i) and section (i-1).
12	$\overline{S_e} - S_f$		average slope of energy-grade line with respect to channel bed.
13	Δx	$C \dfrac{\Delta H_s}{(\overline{S_e} - S_f)}$	distance between section (i) section (i-1), eq. 4.14. The flow being subcritical, one has $C = +1$.
14	x	$\sum_{1}^{i} \Delta x$	accumulated distance from the control section to section (i).

As can be seen, the hydraulic exponents, N and M, depend only on m and h/b.

Proceeding towards the upstream, the M2 profile increases from the critical depth, $h_c = 1.085$ [m], to the normal depth, $h_n = 1.866$ [m]. The water-surface profile is determined by calculating the distances, x', for different flow depths in this range. Note, that the approach to the normal depth is asymptotic :

for $\eta \to 1$, namely $h \to h_n$, one has $\begin{array}{l} \Phi(\eta_i, N) \to \infty \\ \Phi(\zeta_i, J) \to \infty \end{array}$ (see Table 4.1).

Because of this asymptotic approach, the computations are carried out up to a flow depth of $0.999 h_n = 1.864$ [m].

The water-surface profile computation sheet, prepared using the equation for x_i' and the analytical expressions for N and M, is presented above. The calculation of the distance is direct and does not require a trial-and-error procedure. The explanations of the columns and the calculation procedure are given at the bottom of the computation sheet. Note that the flow depths in column 1 were not chosen by dividing the flow depth in the range $1.085 \leq h$ [m] ≤ 1.864 into equal intervals. The curvature of the water-surface profile being more pronounced near the critical depth, smaller flow-depth intervals are used near the drop at the downstream end of the channel.

According to the results presented in the computation sheet, the gradually varied flow depth reaches 99.9% of the normal depth at a distance $x = 2108.7$ [m] from the control section. The water-surface profile, h(x), is plotted in point 6°.

4° *Computation of water-surface profile, using direct step method* :

The direct step method is particularly suitable for prismatic channels. The water-surface profile is computed by calculating the distance between the pre-selected flow depths in the range $1.085 \leq h$ [m] ≤ 1.866. Since the curvature of the water-surface profile is more pronounced near the critical depth, smaller flow-depth intervals should be used near this region. The calculation of the distance between two sections with known water depth is based on the specific-energy balance given by eq. 4.14.

The computation sheet, prepared on a spreadsheet, is presented above. The explanations of the columns and the calculation procedure are given at the bottom of this computation sheet.

According to the results obtained in the computation sheet, the gradually varied flow reaches the normal depth at a distance of $x = 1356.0$ [m] from the control section. The water-surface profile, h(x), is plotted in point 6°.

NON-UNIFORM FLOW

5° *Computation of water-surface profile, using standard step method :*

The standard step method can be used for water-surface profile calculations, both in prismatic and non prismatic channels. In this method, the flow depth is calculated at pre-selected sections along the channel by considering the conservation of the total energy with respect a reference height (see eq. 4.12a).

The calculation of the flow depth at successive sections is done by trial-and-error. The computation sequence can readily be programmed on a spreadsheet. The computation sheet prepared for the present problem is presented below. The explanations of the columns and the calculation procedure are given at the bottom of this computation sheet.

The computations start at the control section (1st line). The flow depth in this section being known, $h = h_c = 1.085$ [m], the other informations in columns 8 to 15 can be calculated. The next section is selected arbitrarily at a distance of $x = 0.05$ [m] from the control section (1st column, 2nd line). The trial-and-error procedure for this section is given in detail on the computation sheet.

For the first trial, the flow depth is estimated as $h' = 1.110$ [m] (column 7) and the information in columns 8 to 19 are calculated. Subsequently, the total heads in column 13 ($H = 1.5420$ [m]) and column 18 ($H_{calc.} = 1.5414$ [m]) are compared. The difference between these two total heads is displayed in column 19 with a precision of 10^{-4}. The absolute value of the difference between the total heads in columns 13 and 18 is larger than the desired precision, $|\Delta H'| = 0.0006 > 10^{-4}$. The calculations should be repeated with a new estimation for the flow depth. The second estimation, $h'' = 1.090$ [m] (column 7), yields a difference of $|\Delta H''| = 0.0003 > 10^{-4}$. A last trial with $h''' = 1.098$ [m] (columns 7) allows finally to reach an equality between the total heads in columns 13 and 18 with the desired precision, namely $|\Delta H'''| < 10^{-4}$.

The curvature of the water-surface profile near the critical depth being more pronounced, the distance between the selected sections should be smaller.

Unlike the other two methods presented above, the standard step method does not allow to calculate directly the distance at which the gradually varied flow reaches the normal flow depth, $h_n = 1.866$ [m]. The calculations presented in the computation sheet show that the gradually varied flow reaches the normal flow depth almost (asymptotically) at a distance of $x = 1400.0$ [m] from the control section. However, the computation sheet does not say if this distance has been reached at a closer distance or not. The answer to this question can only be given by searching for the smallest distance at which the calculated water depth is equal to the normal depth. The water-surface profile, $h(x)$, calculated using the standard step method is plotted in the figure presented below in point 6°.

Step method :
$S_f = 1.0E-3$ [-] ; $Q = 28.0$ [m³/s] ; $h_n = 1.866$ [m] ; $h_c = 1.085$ [m]

1	2	3	4	5	6	7	8	9	10
x	Δx	z	b	m	n	h	B	P	A
[m]	[m]	[m]	[m]	[-]	[m$^{-1/3}$s]	[m]	[m]	[m]	[m²]
0.000			7.000	1.500	0.025	*1.085*	10.255	10.912	9.361
0.050	0.050	0.000	7.000	1.500	0.025	1.110	10.330	11.002	9.618
						1.090	10.270	10.930	9.412
						1.098	10.294	10.959	9.494
1.000	0.950	0.001	7.000	1.500	0.025	1.146	10.437	11.130	9.987
4.000	3.000	0.004	7.000	1.500	0.025	1.202	10.606	11.334	10.581
20.000	16.000	0.020	7.000	1.500	0.025	1.328	10.984	11.789	11.942
40.000	20.000	0.040	7.000	1.500	0.025	1.406	11.218	12.069	12.807
80.000	40.000	0.080	7.000	1.500	0.025	1.502	11.505	12.414	13.892
180.000	100.000	0.180	7.000	1.500	0.025	1.631	11.892	12.880	15.404
300.000	120.000	0.300	7.000	1.500	0.025	1.710	12.129	13.164	16.353
600.000	300.000	0.600	7.000	1.500	0.025	1.804	12.412	13.505	17.511
1400.000	800.000	1.400	7.000	1.500	0.025	*1.867*	12.600	13.730	18.293

col.	symbol	expression	explanations
1	x		distance between a section and the origin. The choice of sections is arbitrary. It is admitted here that the origin is located at the control section and the x-axis points upstream.
2	Δx	$x_i - x_{i-1}$	distance between section (i) and section (i-1)
3	z	$z_{i-1} + S_f \Delta x$	the elevation of the channel bed with respect to an arbitrary reference plane. Here the reference plane is located on the channel bottom at the control section. For S_f = Cte, z is calculated using the given expression.
4	b		bottom width of channel
5	m		side slopes of channel.
6	n		Manning's friction coefficient for the section.
7	h		*estimation* for flow depth (the trial-and-error procedure is shown only for $x = 0.05$ [m]). The trials are repeated with new estimated values for h until the total heads, H, in columns 13 and 18 are sufficiently close.
8	B	$b + 2mh$	width at the free-water surface (see Table 1.1) for the flow depth in column 7.

NON-UNIFORM FLOW

Standard step method (Δx is fixed)									
convention: $C = 1$, towards upstream (Fr < 1) ; $C = -1$, towards downstream (Fr > 1)									
11	12	13	14	15	16	17	18	19	20
U	$U^2/2g$	H	R_h	S_e	$\overline{S_e}$	h_r	$H_{calc.}$	ΔH	Remarks
[m/s]	[m]	[m]	[m]	[-]	[-]	[m]	[m]	[m]	
2.991	0.456	1.5410	0.858	6.86E-03			1.5410	0.0000	control section
2.911	0.432	1.5420	0.874	6.34E-03	6.60E-03	3.30E-04	1.5414	0.0006	no $H > H_{calc.}$
2.975	0.451	1.5411	0.861	6.75E-03	6.81E-03	3.40E-04	1.5414	-0.0003	no $H < H_{calc.}$
2.949	0.443	1.5413	0.866	6.58E-03	6.72E-03	3.36E-04	1.5414	0.0000	√ $H = H_{calc.}$
2.804	0.401	1.5472	0.897	5.68E-03	6.13E-03	5.82E-03	1.5472	0.0000	√
2.646	0.357	1.5629	0.934	4.80E-03	5.24E-03	1.57E-02	1.5629	0.0000	√
2.345	0.280	1.6283	1.013	3.38E-03	4.09E-03	6.54E-02	1.6283	0.0000	√
2.186	0.244	1.6896	1.061	2.76E-03	3.07E-03	6.14E-02	1.6896	0.0000	√
2.016	0.207	1.7885	1.119	2.19E-03	2.47E-03	9.89E-02	1.7886	0.0000	√
1.818	0.168	1.9791	1.196	1.63E-03	1.91E-03	1.91E-01	1.9791	0.0000	√
1.712	0.149	2.1591	1.242	1.37E-03	1.50E-03	1.80E-01	2.1591	0.0000	√
1.599	0.130	2.5344	1.297	1.13E-03	1.25E-03	3.75E-01	2.5345	0.0000	√
1.531	0.119	3.3861	1.332	9.99E-04	1.06E-03	8.52E-01	3.3861	0.0000	√ $h \cong h_n$

col.	symbol	expression	explanations
9	P	$b + 2h\sqrt{1+m^2}$	wetted perimeter (see Table 1.1) for flow depth in column 7.
10	A	$h(b + mh)$	wetted surface (see Table 1.1) for flow depth in column 7.
11	U	Q/A	average flow velocity.
12	$U^2/2g$		dynamic head.
13	H	$z + h + U^2/2g$	*total* head.
14	R_h	A/P	hydraulic radius.
15	S_e	$\dfrac{U^2 n^2}{R_h^{4/3}}$	slope of energy grade line with respect to horizontal, calculated using the Manning-Strickler formula, eq. 3.16.
16	$\overline{S_e}$	$\dfrac{(S_e)_i + (S_e)_{i-1}}{2}$	average slope of energy-grade line between section (i) and section (i-1).
17	h_r	$\overline{S_e} \Delta x$	head loss between section (i) and section (i-1).
18	$H_{calc.}$	$(H_{i-1})_{calc.} + C(h_r)_i$	total head calculated with eq. 4.12a. The flow being subcritical, one has $C = +1$.
19	ΔH	$(H_i) - (H_i)_{calc.}$	convergence criteria. Trial-and-error, by estimating h-values, stops when $\Delta H \cong 0$.

6° *Conclusion :*

The water-surface profiles calculated, using the three methods, are plotted in the figure below. Although the profile calculated by the method of direct integration (Chow's method) converges more slowly to the normal depth, all the calculated water-surface profiles are practically identical. Note that the water-surface profile approaches the normal depth asymptotically.

By using the direct step method, the distance at which the water-surface profile is influenced by the drop, namely the gradually varied flow zone has been found to be 1356.0 [m]. The calculations using the direct integration method (Chow's method) yield a distance of 2108.7.2 [m], which is considerably larger. The reason for this difference lies in the asymptotic nature of the functions, $\Phi(\eta, N)$ and $\Phi(\zeta, J)$, used in Chow's method.

The length of the gradually varied flow cannot be calculated directly with the method of standard steps. Nevertheless, the calculations show that, at a distance of $x = 1400.0$ [m] from the control section, the flow already reaches the normal depth. This agrees with the result obtained, using direct step method.

There exist different methods for the water-surface profile calculations. Each method has its advantages and its disadvantages. In the practice, one should pay attention to choose the method best suited for the solution of the problem in hand.

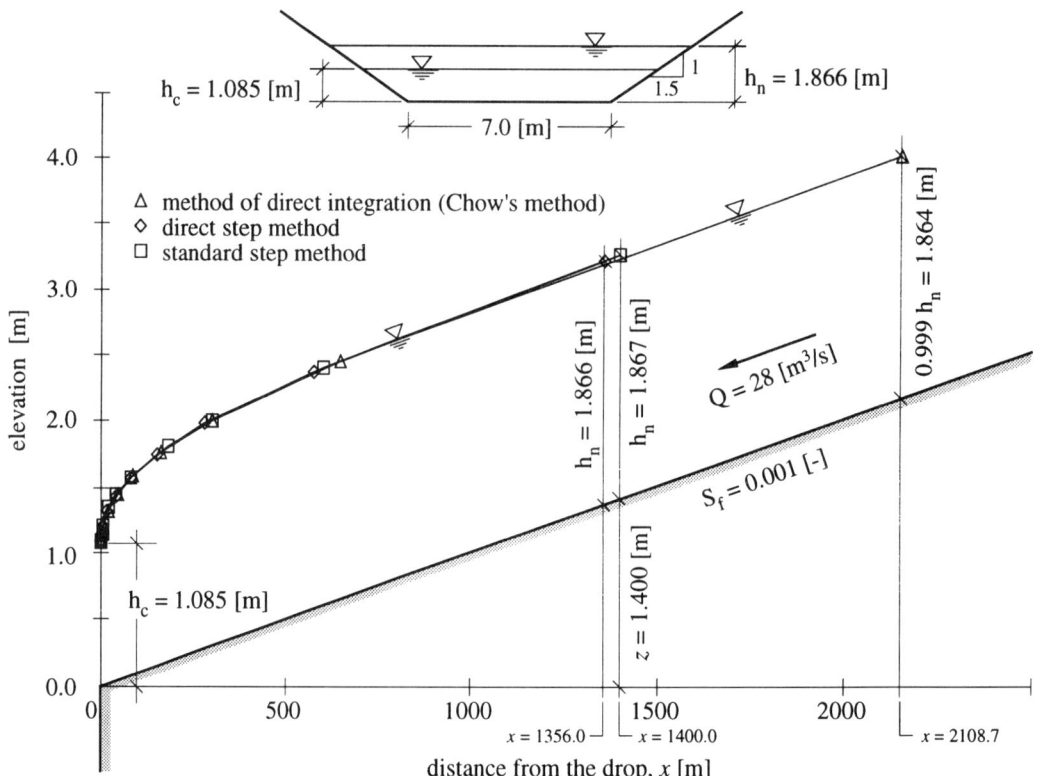

NON-UNIFORM FLOW

Ex. 4.B

In a fish-breeding station, a 105.0 [m] long rectangular channel having a width of b = 5.0 [m] is conveying water from a *lake* to a *reservoir* with a constant surface area of A_R = 5000 [m^2] (see figure below). The lake's water level is maintained constant at H_L = 371.500 [m]. The connecting channel has a bed slope of S_f = 0.003 [-] and a friction coefficient of n = 0.020 [m$^{-1/3}$s]. The local head losses at the entrance and the exit of the channel are small and can be neglected.

a) Determine the relationship between the discharge, Q, flowing in the channel and the water level in the reservoir, H_R. Draw the water-surface profiles corresponding to different water-levels in the reservoir.

b) Compute the time necessary to increase the water level in the reservoir from an elevation of 369.685 [m] to one of 371.500 [m].

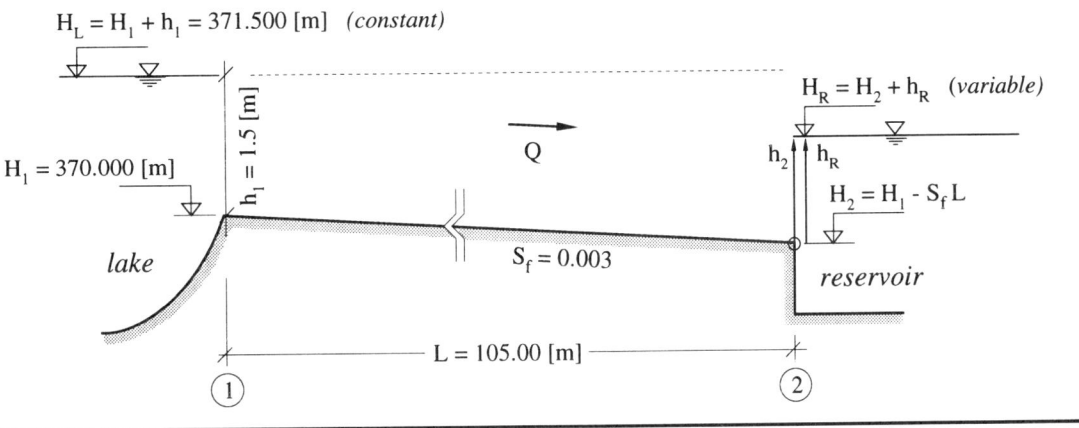

SOLUTION :

Since the water level in the lake remains constant, H_L = 371.500 [m] = Cte, the water depth at the entrance of the channel, section 1, is thus always the same, h_1 = 1.5 [m]. At its downstream end, section 2, the channel discharges into the reservoir. Since the local head loss at the channel exit is neglected, the water depth at section 2, $h_2 = H_R - H_2$, is directly given by the water level in the reservoir, H_R, which is variable. According to the type of the flow regime in the channel, subcritical (Fr < 1) or supercritical (Fr > 1), the water level in the reservoir, H_R, may or may not control the discharge in the channel :

- If the flow is subcritical (Fr < 1), the control section is located at the downstream, at section 2. The flow is then influenced by any variation of the water level above the critical depth at section 2. The discharge flowing in the channel is therefore controlled by the downstream water depth, h_2.

- If the flow is supercritical (Fr > 1), the control section is located at the upstream, at section 1. The discharge flowing in the channel is therefore controlled by the upstream water depth, h_1. This water depth being constant (the water level of the lake, H_L, is constant), the discharge will remain independent of the downstream water depth, h_2.

- No flow will of course take place when the water levels in the reservoir and in the lake are the same, $H_R = H_L = 371.500$ [m].
- If the water level in the reservoir becomes higher than the one in the lake, $H_R > H_L$, the flow takes place from the reservoir towards the lake.

The uniform flow discharge for a normal depth, $h_n = 1.5$ [m], can be readily calculated using Manning's formula (see eq. 3.16). The geometrical characteristics for a rectangular channel are given in Table 1.1.

$$Q_n = U A = \frac{1}{n} R_h^{2/3} S_f^{1/2} A = \frac{1}{n} \left(\frac{b\, h_n}{b + 2 h_n}\right)^{2/3} S_f^{1/2} (b\, h_n)$$

$$Q_n = \frac{1}{0.020} \left(\frac{5 \times 1.5}{5 + 2 \times 1.5}\right)^{2/3} (0.003)^{1/2} (5 \times 1.5) = 19.675 \text{ [m}^3\text{/s]}$$

The flow regime can be checked by calculating the Froude number:

$$\text{Fr} = \frac{U}{\sqrt{g\,h}} = \frac{Q}{A\sqrt{g\,h}} = \frac{Q}{b\,h\sqrt{g\,h}} = \frac{19.675}{5 \times 1.5 \sqrt{9.81 \times 1.5}} = 0.68 < 1$$

The flow is thus *subcritical*. The channel discharge is thus controlled by the variation of downstream water depth, h_2. *One should therefore determine the relationship, $Q = f(h_2)$* namely $Q = f(h_R)$, according to the definition sketch given below.

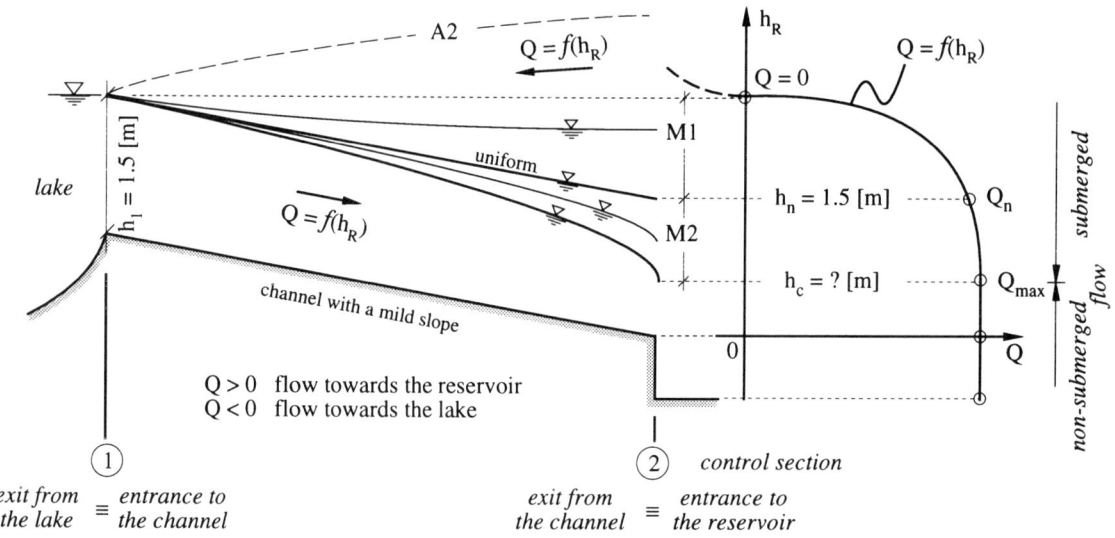

Different types of flow are possible, depending on the water level in the reservoir, H_R, with respect to the water depth at section 2 corresponding to the normal depth, $H_n = H_2 + h_n$, and the critical depth, $H_c = H_2 + h_c$, of the channel. To be able to determine this relationship, $Q = f(h_R)$, all these different possibilities must be studied (see the definition sketch above):

NON-UNIFORM FLOW

1° The *uniform* flow takes place when $h_1 = h_2 = h_n = 1.5$ [m], namely $H_R = H_n$. Using the Manning-Strickler formula (see eq. 3.16), the flow discharge corresponding to the normal depth, $h_n = 1.5$ [m], has already been calculated (see above) as being $Q_n = 19.675$ [m³/s].

2° The discharge in the channel reaches its *maximum* value, $Q = Q_{max}$, when the water depth at section 2 is equal to critical depth, $h_2 = h_c$ (see sect. 2.3.3). The water depth at section 2 cannot be less than the critical depth even if the water level in the reservoir gets lower. In this case, a free overfall takes place at section 2 with a water depth equal to the critical depth. The knowledge about the water level corresponding to this critical depth, $H_c = H_2 + h_c$, is particularly important, since :

for $H_R \leq H_c$ the flow is *non-submerged* ; the discharge stays constant and equal to the maximum discharge,

for $H_R > H_c$ the flow is *submerged*; the discharge varies as a function of H_R (or h_2).

The maximum discharge, Q_{max}, cannot be calculated directly. It is the discharge which corresponds to the water-surface profile that joins the constant water level of the lake, H_L, by starting with the critical depth, h_c, at section 2. A trial-and-error calculation should be carried out by varying the discharge. The computational procedure is as follows :

- assume a discharge, $Q \geq Q_n$,
- compute the critical flow at section 2, corresponding to this discharge,
- starting at section 2 with this critical depth and proceeding towards upstream, calculate the water-surface profile in the channel,
- if the water-surface profile joins the constant water level of the lake, H_L, at section 1 the computation ends. The assumed discharge, Q, is the maximum one, Q_{max}. If the water-surface profile does not join the water level of the lake, H_L, go to the first step to choose another discharge, Q, and repeat the calculations.

This procedure is followed below for solving the present problem :

i) Assume: $Q' = 20.00$ [m³/s]. The critical depth for a rectangular section is given by eq. 2.23, where $q = Q / b$:

$$h_c = \sqrt[3]{\frac{q^2}{g}} = \sqrt[3]{\frac{Q^2}{gb^2}} = \sqrt[3]{\frac{20.00^2}{9.81 \times 5.0^2}} = 1.177 \text{ [m]}$$

The water-surface profile for $Q' = 20.00$ [m³/s] is computed with the *direct step method* (see point 4.3.1,4°) starting with the critical depth, $h_c = 1.177$ [m]. The water surface profile is of *type M2*. The computation sheet is presented below:

b = 5.0 [m] ; n = 0.020 [m$^{-1/3}$s] ; S$_f$ = 3.0E-03 [-] ; Q' = 20.00 [m³/s]										
C = 1, calculation towards upstream (Fr < 1) ; h$_c$ = 1.177 [m]										
h	P	A	R$_h$	U	U²/2g	H$_s$	ΔH$_s$	S$_e$	$\overline{S_e}$	$\overline{S_e}$ - S$_f$	Δx	x
[m]	[m]	[m²]	[m]	[m/s]	[m]	[m]	[m]	10³	10³	10³	[m]	[m]
1.177	7.354	5.886	0.800	3.398	0.589	1.766		6.22				
1.190	7.380	5.950	0.806	3.361	0.576	1.766	0.000	6.02	6.12	3.12	0.067	0.067
....	....	....	....	....	....	....	....	....	....	....	....	....
....	....	....	....	....	....	....	....	....	....	....	....	....
1.421	7.843	7.107	0.906	2.814	0.404	1.825	0.013	3.61	3.73	0.73	18.119	50.765
1.500	8.000	7.500	0.938	2.667	0.362	1.862	0.037	3.10	3.36	0.36	105.036	*155.801*

One obtains : x = 155.80 [m] > L = 105.00 [m]. The computation must therefore be repeated with another discharge.

ii) Assume: Q" = 21.00 [m³/s]. It can be calculated that : h$_c$ = 1.216 [m].

b = 5.0 [m] ; n = 0.020 [m$^{-1/3}$s] ; S$_f$ = 3.0E-03 [-] ; Q" = 21.00 [m³/s]										
C = 1, calculation towards upstream (Fr < 1) ; h$_c$ = 1.216 [m]										
h	P	A	R$_h$	U	U²/2g	H$_s$	ΔH$_s$	S$_e$	$\overline{S_e}$	$\overline{S_e}$ - S$_f$	Δx	x
[m]	[m]	[m²]	[m]	[m/s]	[m]	[m]	[m]	10³	10³	10³	[m]	[m]
1.216	7.432	6.080	0.818	3.454	0.608	1.824		6.24				
1.227	7.455	6.137	0.823	3.422	0.597	1.824	0.000	6.07	6.15	3.15	0.050	0.050
....	....	....	....	....	....	....	....	....	....	....	....	....
....	....	....	....	....	....	....	....	....	....	....	....	....
1.431	7.862	7.154	0.910	2.935	0.439	1.870	0.010	3.91	4.02	1.02	10.206	31.519
1.500	8.000	7.500	0.938	2.800	0.400	1.900	0.030	3.42	3.66	0.66	44.635	*76.154*

One obtains : x = 76.15 [m] < L = 105.00 [m]. The computation must therefore be repeated with yet another discharge.

iii) Assume: Q''' = 20.507 [m³/s]. It can be calculated that : h$_c$ = 1.197 [m].

b = 5.0 [m] ; n = 0.020 [m$^{-1/3}$s] ; S$_f$ = 3.0E-03 [-] ; Q''' = 20.507 [m³/s]										
C = 1, calculation towards upstream (Fr < 1) ; h$_c$ = 1.197 [m]										
h	P	A	R$_h$	U	U²/2g	H$_s$	ΔH$_s$	S$_e$	$\overline{S_e}$	$\overline{S_e}$ - S$_f$	Δx	x
[m]	[m]	[m²]	[m]	[m/s]	[m]	[m]	[m]	10³	10³	10³	[m]	[m]
1.197	7.394	5.985	0.809	3.427	0.598	1.795		6.23				
1.209	7.418	6.045	0.815	3.392	0.587	1.796	0.000	6.05	6.14	3.14	0.058	0.058
1.224	7.448	6.121	0.822	3.350	0.572	1.796	0.001	5.83	5.94	2.94	0.246	0.304
1.244	7.489	6.222	0.831	3.296	0.554	1.798	0.002	5.56	5.70	2.70	0.659	0.963
1.275	7.549	6.374	0.844	3.218	0.528	1.802	0.004	5.19	5.38	2.38	1.806	2.769
1.305	7.610	6.525	0.857	3.143	0.503	1.808	0.006	4.85	5.02	2.02	3.012	5.782
1.335	7.671	6.677	0.870	3.071	0.481	1.816	0.008	4.54	4.70	1.70	4.550	10.332
1.366	7.731	6.828	0.883	3.003	0.460	1.825	0.009	4.26	4.40	1.40	6.577	16.909
1.396	7.792	6.980	0.896	2.938	0.440	1.836	0.011	4.00	4.13	1.13	9.361	26.270
1.426	7.853	7.131	0.908	2.876	0.421	1.848	0.012	3.76	3.88	0.88	13.418	39.687
1.500	8.000	7.500	0.938	2.734	0.381	1.881	0.033	3.26	3.51	0.51	65.313	*105.000*

One obtains : x = L = 105.00 [m]. End of the computations.

The maximum discharge is therefore: Q$_{max}$ = 20.507 [m³/s].

The search for the maximum discharge, using a trial-and-error calculation is illustrated in the figure below :

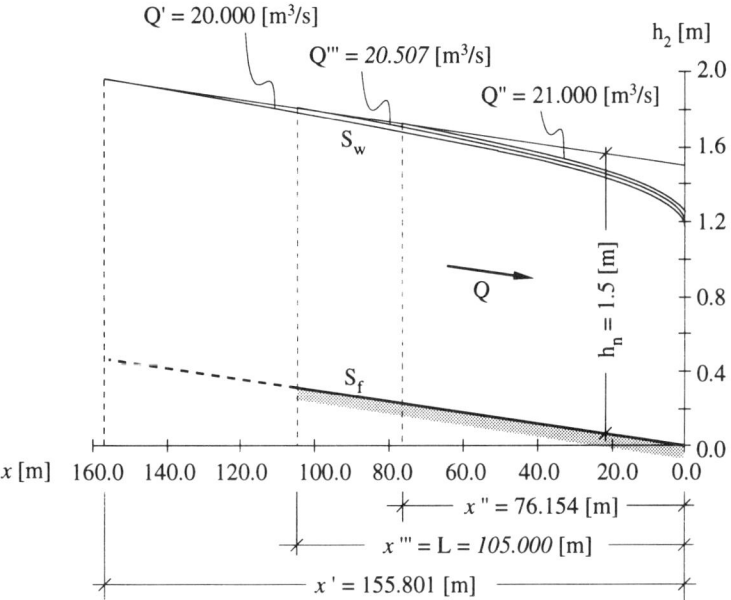

3° If the water level in the reservoir is equal to one of the lake, $H_L = H_R$, namely $h_R = 1.815$ [m], no flow will take place, $Q = 0$. If the water level in the reservoir is higher than the one of the lake, $H_R > H_L$, the flow reverses its direction. This case will not be studied here.

4° If $H_L > H_R > H_n$, namely $1.815 > h_R$ [m] > 1.5 , the flow is submerged and the water-surface profile in the channel is of *type M1*. If the water level in the reservoir varies in this range, it is taken that $h_2 \equiv h_R$ (the local head loss at the exit is neglected). For all flows in this range, the flow depth, h_R , and the corresponding discharge must be determined by trial-and-error. One can either fix the discharge, Q, and calculate the corresponding flow depth, h_R, or fix the flow depth, h_R, and determine the corresponding discharge, Q. The first method is similar to the trail-and-error procedure used previously for determining the maximum discharge. The second method will now be adopted. The computational procedure can be described as follows :
- choose a flow depth in the range : $1.815 > h_R$ [m] > 1.5 ,
- assume a discharge in the range : $0 < Q$ [m³/s] $< Q_n$,
- starting at section 2 with the flow depth, $h_2 = h_R$, and compute the water-surface profile towards upstream,
- if at section 1 the computed water-surface profile joins the constant lake level, H_L, the calculations are terminated. The assumed discharge, Q, is the discharge corresponding to the flow depth, h_2, fixed at the beginning of calculations. If the computed water-surface profile does not join the constant lake level, H_L, go back to the second step to assume a new discharge and repeat the computations.

This procedure is demonstrated for calculating the discharge corresponding to a flow depth of $h_R = 1.55$ [m].

i) Assume : $Q' = 18.000$ [m³/s] .

| \multicolumn{13}{l}{$b = 5.0$ [m] ; $n = 0.020$ [m$^{-1/3}$s] ; $S_f = 3.0E\text{-}03$ [-] ; $h_2 = h_R = 1.55$ [m]} |
| \multicolumn{13}{l}{$C = 1$, calculation towards upstream (Fr < 1) ; $Q' = 18.000$ [m³/s]} |
h	P	A	R_h	U	$U^2/2g$	H_s	ΔH_s	S_e 10^3	$\overline{S_e}$ 10^3	$\overline{S_e} - S_f$ 10^4	Δx	x
[m]	[m]	[m²]	[m]	[m/s]	[m]	[m]	[m]				[m]	[m]
1.550	8.100	7.750	0.957	2.323	0.275	1.825		2.29				
1.545	8.090	7.725	0.955	2.330	0.277	1.822	-0.003	2.31	2.30	-7.01	4.591	4.591
....	....	....	....	....	....	....	....	....	....	....	....	....
....	....	....	....	....	....	....	....	....	....	....	....	....
1.505	8.010	7.525	0.939	2.392	0.292	1.797	-0.003	2.49	2.48	-5.24	5.861	46.465
1.500	8.000	7.500	0.938	2.400	0.294	1.794	-0.003	2.51	2.50	-5.01	6.096	*52.561*

$x = 52.561$ [m] $< L = 105.00$ [m]. Repeat the calculations with another discharge.

ii) Assume : $Q'' = 19.400$ [m³/s] .

| \multicolumn{13}{l}{$b = 5.0$ [m] ; $n = 0.020$ [m$^{-1/3}$s] ; $S_f = 3.0E\text{-}03$ [-] ; $h_2 = h_R = 1.55$ [m]} |
| \multicolumn{13}{l}{$C = 1$, calculation towards upstream (Fr < 1) ; $Q'' = 19.400$ [m³/s]} |
h	P	A	R_h	U	$U^2/2g$	H_s	ΔH_s	S_e 10^3	$\overline{S_e}$ 10^3	$\overline{S_e} - S_f$ 10^4	Δx	x
[m]	[m]	[m²]	[m]	[m/s]	[m]	[m]	[m]				[m]	[m]
1.550	8.100	7.750	0.957	2.503	0.319	1.869		2.66				
1.545	8.090	7.725	0.955	2.511	0.321	1.866	-0.003	2.68	2.67	-3.29	8.896	8.896
....	....	....	....	....	....	....	....	....	....	....	....	....
....	....	....	....	....	....	....	....	....	....	....	....	....
1.505	8.010	7.525	0.939	2.578	0.339	1.844	-0.003	2.89	2.88	-1.24	22.254	122.825
1.500	8.000	7.500	0.938	2.587	0.341	1.841	-0.003	2.92	2.90	-0.97	28.271	*151.096*

$x = 151.096$ [m] $> L = 105.00$ [m]. Repeat the calculations with yet another discharge.

iii) Assume : $Q''' = 19.126$ [m³/s] .

| \multicolumn{13}{l}{$b = 5.0$ [m] ; $n = 0.020$ [m$^{-1/3}$s] ; $S_f = 3.0E\text{-}03$ [-] ; $h_2 = h_R = 1.55$ [m]} |
| \multicolumn{13}{l}{$C = 1$, calculation towards upstream (Fr < 1) ; $Q''' = 19.126$ [m³/s]} |
h	P	A	R_h	U	$U^2/2g$	H_s	ΔH_s	S_e 10^3	$\overline{S_e}$ 10^3	$\overline{S_e} - S_f$ 10^4	Δx	x
[m]	[m]	[m²]	[m]	[m/s]	[m]	[m]	[m]				[m]	[m]
1.550	8.100	7.750	0.957	2.468	0.310	1.860		2.58				
1.545	8.090	7.725	0.955	2.476	0.312	1.857	-0.003	2.61	2.60	-4.04	7.391	7.391
....	....	....	....	....	....	....	....	....	....	....	....	....
....	....	....	....	....	....	....	....	....	....	....	....	....
1.505	8.010	7.525	0.939	2.542	0.329	1.834	-0.003	2.81	2.80	-2.05	13.791	89.291
1.500	8.000	7.500	0.938	2.550	0.331	1.831	-0.003	2.84	2.82	-1.78	15.714	*105.004*

$x = 105.004$ [m] $\cong L = 105.00$ [m]. End of the computations.

The discharge is therefore : $Q = 19.126$ [m³/s].

NON-UNIFORM FLOW   211

5° If $H_n > H_R > H_c$, namely $1.500 > h_R$ [m] > 1.197, the flow is always submerged and the water-surface profile is of *type M2*. If the water level in the reservoir falls in this range, one has $h_2 \equiv h_R$ (the local head loss at section 2 is neglected). For all flows in this range, the flow depth, h_R, and the corresponding discharge should be determined by trial-and-error, using one of the methods described above. The discharge range being very small, only a discharge corresponding to a flow depth of $h_2 = h_R = 1.350$ [m] has been calculated (not shown here), it is $Q = 20.428$ [m³/s].

6° If $H_R < H_c$, namely $h_R < 1.197$ [m], the flow is a free (non-submerged) one. The flow depth at the downstream end of the channel, section 2, will be always equal to the critical depth, $h_2 = h_c = 1.197$ [m], thus independent of the water level in the reservoir, H_R (or h_R). The discharge remains then at its maximum value, $Q_{max} = 20.507$ [m³/s].

7° The discharges, Q, corresponding to 11 different water levels in the reservoir, falling in the range of $H_L \geq H_R \geq H_2$, namely $1.815 \geq h_R$ [m] ≥ 0.0, have been calculated using the trial-and-error calculation described above. The relationship $Q = f(h_R)$ is plotted in the last figure, where the calculated values are listed in a tabular form. The water-surface profiles for different types of flows are also plotted on this same figure.

The curve representing this relationship, $Q = f(h_R)$, shows that the difference between the maximum discharge, $Q_{max} = 20.507$ [m³/s], and the discharge at the uniform flow, $Q_n = 19.675$ [m³/s], is not very large. This situation is typical for very long channels with a mild slope; this observation allows in the practice to assume $Q_{max} \cong Q_n$ in order to simplify the calculations.

8° The *time required*, t_R, to raise the water level in the reservoir from the water level of $H_R = 369.865$ [m] to the one of $H_R = 371.500$ [m] will now be calculated. A part of the reservoir and the supply channel are represented schematically on the figure below.

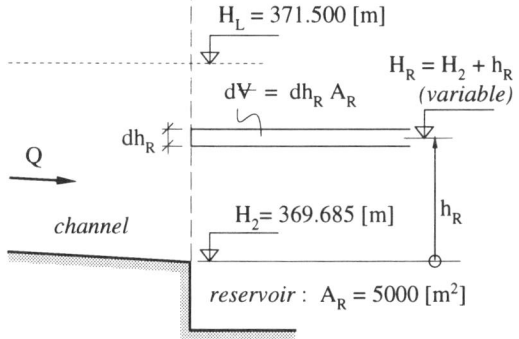

The reservoir has a constant surface area, $A_R = 5000$ [m²]. Referring to the figure above, a slice of water with an infinitesimal thickness, dh_R, between the elevations h_R and $(h_R + dh_R)$, will be considered.

The volume of this slice of water is : $d\mathcal{V} = A_R\, dh_R$

The time required for filling this infinitesimal volume is : $dt = \dfrac{d\mathcal{V}}{Q} = \dfrac{A_R\, dh_R}{Q(h_R)}$

For calculating the total time, t_r, required for raising the water level in the reservoir from $h_R = 0.0$ [m] to $h_R = 1.815$ [m], namely from $H_R = 369.865$ [m] to $H_R = 371.500$ [m], this above relationship must be integrated as follows :

$$t_r = \int_0^t dt = \int_{0.0}^{1.815} \dfrac{A_R\, dh_R}{Q(h_R)}$$

This integration has been carried out numerically in the table below by considering 10 slices of water with varying finite thicknesses :

H_R [m]	h_R [m]	Q [m³/s]	Slice no	$\overline{Q}$ [m³/s]	Δh_R [m]	$\Delta \mathcal{V}$ [m³]	Δt [s]
369.685	0.000	20.507					
			1	20.507	1.197	5985.0	291.9
370.882	1.197	20.507					
			2	20.468	0.153	765.0	37.4
371.035	1.350	20.428					
			3	20.052	0.150	750.0	37.4
371.185	1.500	19.675					
			4	19.401	0.050	250.0	12.9
371.235	1.550	19.126					
			5	18.725	0.050	250.0	13.4
371.285	1.600	18.324					
			6	17.734	0.050	250.0	14.1
371.335	1.650	17.144					
			7	16.250	0.050	250.0	15.4
371.385	1.700	15.356					
			8	13.905	0.050	250.0	18.0
371.435	1.750	12.453					
			9	9.477	0.050	250.0	26.4
371.485	1.800	6.500					
			10	3.250	0.015	75.0	23.1
371.500	1.815	0.000					

$$t_r\ [s] = \sum_1^{10} \Delta t = \sum_1^{10} \dfrac{A_R\, \Delta h_R}{Q(h_R)} = 489.8$$

The time required for raising the water level is therefore : $t_r = 489.8$ [s].

In the following figure the variation of the water depth in the reservoir, h_R, is plotted as a function of time, t.

NON-UNIFORM FLOW

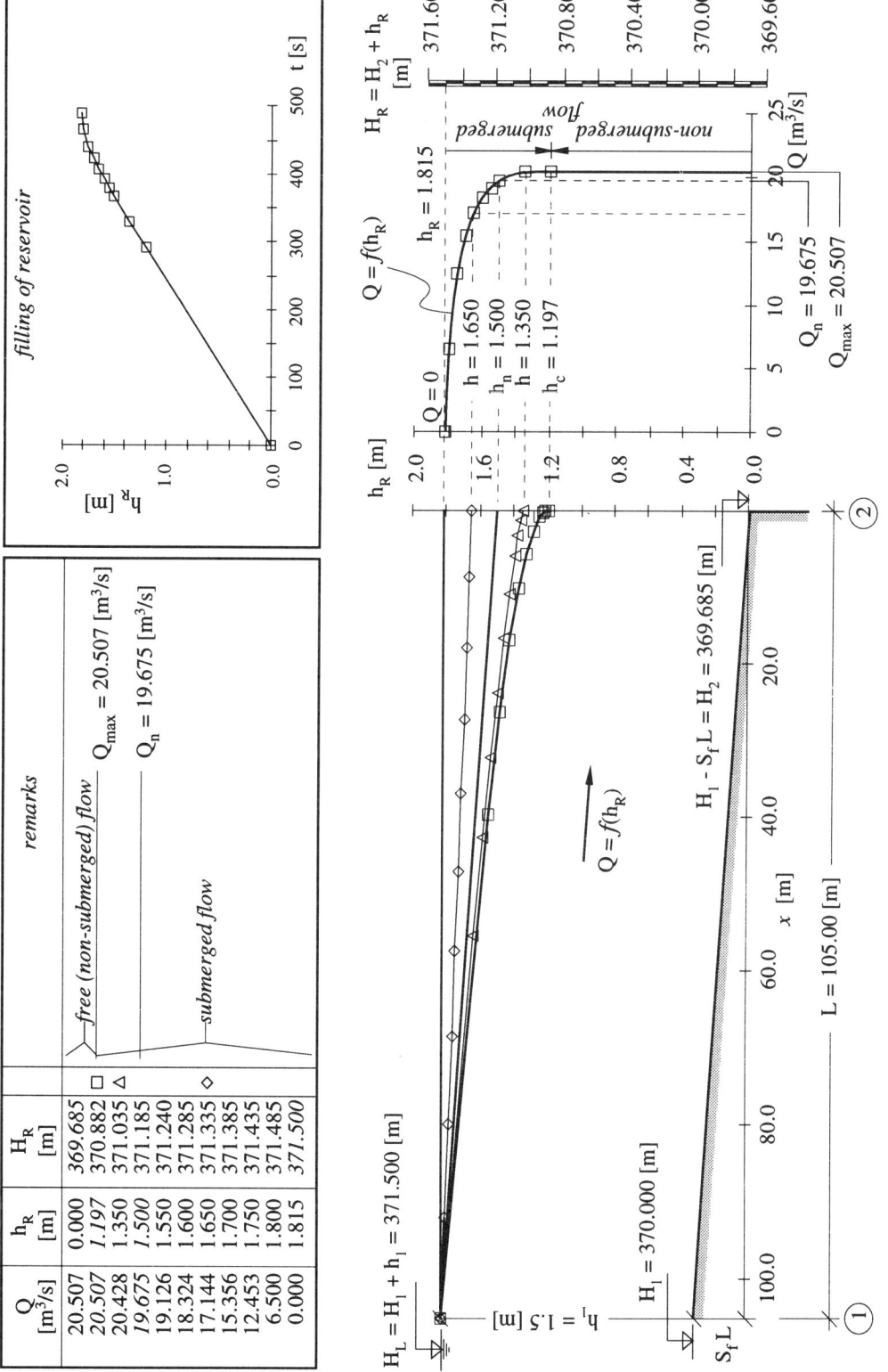

Ex. 4.C

The small river *Wulka*, shown on the map below, conveys a discharge of $Q = 3 \, [m^3/s]$. The geometry of the sections 1 to 6 is given on the figure below, together with the estimated Manning coefficient, n. Compute the water-surface profile in the reach between the sections 1 to 6 for two different water-surface elevations at section 1; namely for 401.200 [m] and 402.400 [m]. Comment the results obtained.

The local (singular) head-loss coefficients, K_{ss}, between the different sections are :

reach	1 - 2	2 - 3	3 - 4	4 - 5	5 - 6
K_{ss} [-]	0.10	0.30	0.30	0.15	0.10

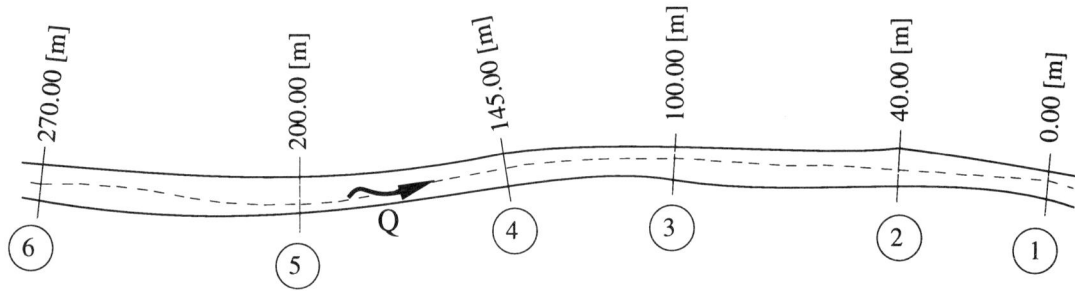

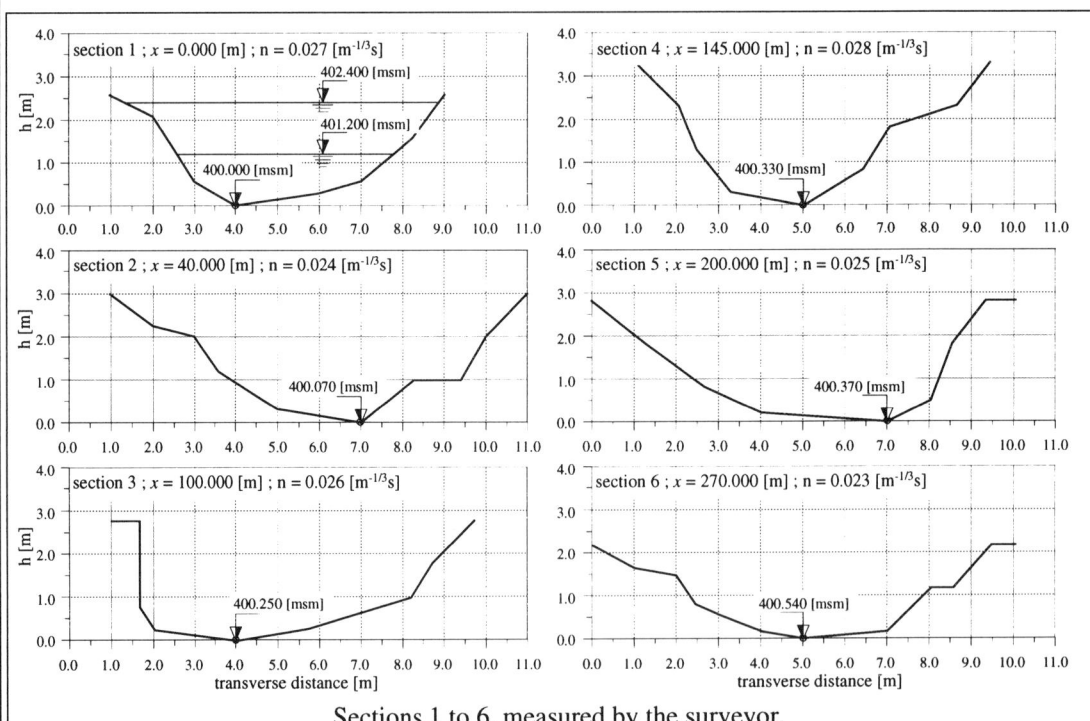

Sections 1 to 6, measured by the surveyor.

NON-UNIFORM FLOW

SOLUTION :

1° The calculations for the water-surface profile will start at section 1, where the water-surface elevation is known. The *standard step* method (see point 4.3.1, 3°) will be used for these calculations. The choice of this method is largely dictated by the fact that the reaches of the river are already fixed by the given six stations, whose geometric characteristics are already available. The standard step method will have to be slightly modified in order to be able to take into account the singular head losses between the stations and to work with the river sections.

2° The river sections are irregular in shape and their *geometric characteristics* — wetted surface, A, wetted perimeter, P, hydraulic radius, R_h, etc. — cannot be given with simple algebraic expressions for introduction in a spreadsheet program. To overcome this difficulty, using the map and the sections provided by the surveyor, it is necessary to prepare tables and/or graphs showing the variation of the useful geometric characteristics, A, P and R_h, as a function of the flow depth, h.

Section 1			
h [m]	P [m]	A [m²]	B [m]
0.000	0.000	0.000	0.000
0.290	2.650	0.363	2.500
0.590	4.180	1.332	3.960
1.560	6.940	6.099	5.870
2.080	8.190	9.339	6.590
2.570	9.920	12.913	8.000

Section 4			
h [m]	P [m]	A [m²]	B [m]
0.000	0.000	0.000	0.000
0.280	2.310	0.305	2.180
0.810	4.000	1.816	3.520
1.130	5.190	3.059	4.250
1.810	6.370	6.109	4.720
2.290	8.530	8.811	6.540
3.290	11.220	16.281	8.400

Section 2			
h [m]	P [m]	A [m²]	B [m]
0.000	0.000	0.000	0.000
0.310	2.510	0.377	2.430
0.970	4.770	2.598	4.300
1.180	6.550	3.669	5.900
2.000	8.510	8.933	6.940
2.250	9.910	10.829	8.230
3.000	12.210	17.665	10.000

Section 5			
h [m]	P [m]	A [m²]	B [m]
0.000	0.000	0.000	0.000
0.220	3.500	0.382	3.470
0.490	4.810	1.471	4.600
0.810	5.940	3.087	5.500
1.810	8.770	9.457	7.240
2.810	11.630	17.722	9.290

Section 3			
h [m]	P [m]	A [m²]	B [m]
0.000	0.000	0.000	0.000
0.260	3.670	0.467	3.590
0.680	6.160	2.430	5.760
0.980	7.060	4.268	6.490
1.780	8.790	9.668	7.010
2.780	11.190	17.163	7.980

Section 6			
h [m]	P [m]	A [m²]	B [m]
0.000	0.000	0.000	0.000
0.150	2.980	0.222	2.960
0.770	5.520	2.724	5.110
1.160	6.520	4.855	5.820
1.460	7.830	6.751	6.820
1.640	9.070	8.082	7.970
2.160	10.910	12.869	10.440

The geometric characteristics of the six section are summarized in the table above. The wetted perimeter, P, and the section width, B, corresponding to different flow

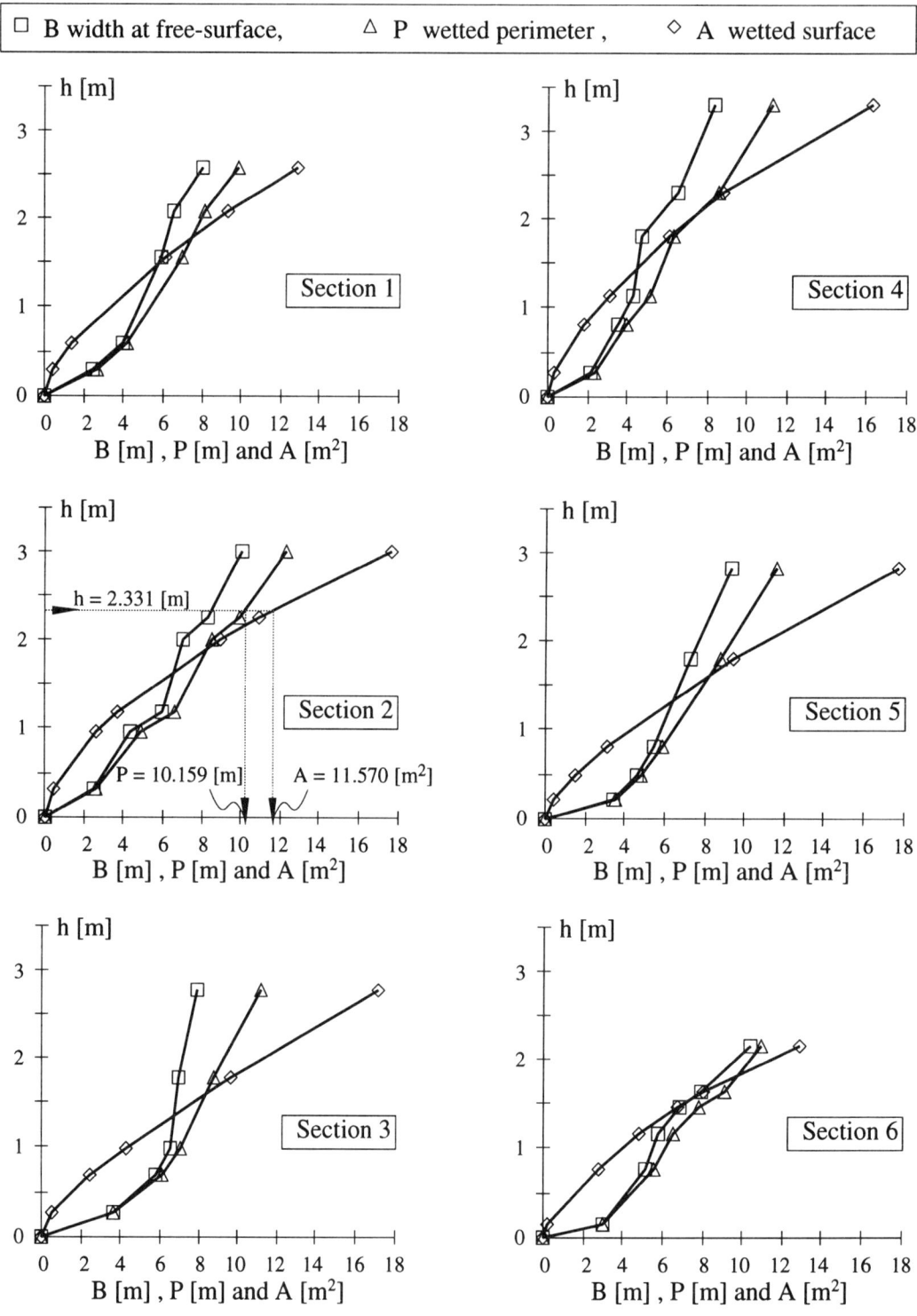

Geometric characteristics, B(h), P(h) and A(h), for the six stations.

NON-UNIFORM FLOW 217

depths, h, were directly measured on the drawings supplied by the surveyor. The wetted surfaces, A, were then determined by summing up the surface of trapezoidal slices between successive flow depths.

The variation of the geometric characteristics, B, P and A, as functions of the flow depth, h, are plotted in the figure above. They will be called *characteristic curves*.

3° The fictitious line joining the lowest points of the transversal sections of a water course is called the *thalweg line*. At a given section, the flow depth, h, is measured with respect to the thalweg elevation, z, of the section. The thalweg elevation for each section is also indicated in the drawings prepared by the surveyor. In preparing the characteristic curves, in general, one should choose the flow depths between which the section characteristics, B, P and A, linearly.

The average bed slope can be approximately evaluated from the thalweg elevations of the section, z(x), and the distance between these two sections :

$$\widetilde{S}_f \approx \frac{\Delta z}{\Delta x} = \frac{400.54 - 400.00}{270.00 - 0.00} = 0.002 \; [-]$$

4° To determine the *type of the water-surface profile*, the normal, h_n, and the critical depths, h_c, should first be determined. For prismatic channels, the section being the same everywhere, this calculation is possible and has a physical meaning. For the natural watercourses having irregular and variable sections, it is not possible to draw the lines of the normal and the critical depth. Nevertheless, by choosing for each section a representative geometry, a very approximate calculation can be made in order to find out the type of the water-surface profile and thus determine the direction for water-surface profile calculations.

The normal depths, h_n, of the sections 1 to 6 can be calculated by solving the equation of Manning-Strickler, eq. 3.16, by trial-and-error as it was shown in exercise Ex. 4.A (see point 1° a).

The critical depths, h_c, of the sections 1 to 6 can be calculated by solving eq. 2.22, valid for critical flow, by trial-and-error as it was shown in Ex. 4.A (see point 1° b).

During the trial-and-error calculations, the geometric characteristics of the sections can be read from the characteristic curves plotted previously. The normal and critical depths calculated for the sections 1 to 6, using the method described here, are presented in the table below. As it can be seen, for all sections, the critical depth, h_c, is smaller than the normal depth, h_n. It is important to note that both starting flow depths, $h_1 = 1.2$ [m] and $h_1 = 2.4$ [m], given in section 1 are larger than the normal depth at this section. It can therefore be concluded that the flow will be *subcritical* over the entire length of the reach between sections 1 and 6.

The type of the water-surface profile – having made some reasonable approximations – is therefore a M1 curve (see Fig. 4.2). The water-surface computations will be done by proceeding from downstream to upstream.

section	1 x [m]	2 z [m]	3 h_n [m]	4 $h_n + z$ [m]	5 h_c [m]	6 $h_c + z$ [m]
1	0.000	400.000	0.871	400.871	0.634	400.634
2	40.000	400.070	0.928	400.998	0.630	400.700
3	100.000	400.250	0.762	401.012	0.513	400.763
4	145.000	400.330	1.052	401.382	0.674	401.004
5	200.000	400.370	0.740	401.110	0.521	400.891
6	270.000	400.540	0.723	401.263	0.481	401.021

<u>col.</u> <u>symbol</u> <u>explanations</u>

1 x distance between a section and an arbitrary origin. It is assumed here that the origin is located at section 1 and the x-axis points towards upstream.

2 z lowest elevation (thalweg) of the section with respect to an arbitrary reference plane. This elevation was given by the surveyor.

3 h_n normal depth for an average bottom slope of $\widetilde{S}_f \approx 0.002$ [-].

4 $h_n + z$ water-surface elevation for normal depth.

5 h_c critical depth.

6 $h_c + z$ water-surface elevation for critical depth.

5° The geometric characteristics of the sections have been prepared as functions of the flow depth and the type of the water-surface profile has been determined. The computation of the water-surface profile can now be carried out using the *standard step* method. The development of this method was already given in point 4.3.1, 3°.

The dynamic equation, eq. 4.12b, is rewritten below for the present problem between the running section (i) and the preceding section (i-1):

$$H_i - H_{i-1} = \overline{S_e}\,(x_i - x_{i-1}) + K_{ss}\frac{Q^2}{2g}\left|\frac{1}{A_i^2} - \frac{1}{A_{i-1}^2}\right|$$

where $H = z + h + U^2/2g$ elevation of the energy-grade line,

$$\overline{S_e} = \frac{(S_e)_i + (S_e)_{i-1}}{2}$$ average slope of the energy-grade line,

$$S_e = \frac{U^2 n^2}{R_h^{4/3}}$$ average slope of the energy-grade line with respect to horizontal.

NON-UNIFORM FLOW   219

The application of the standard step method to the case of a prismatic channel without singular head losses was already presented in solving the problem Ex. 4.A (see point 5°). For the present problem, the trial-and-error calculations for the determination of the water-surface profile are again carried out in a tabular form using a spreadsheet program. The computation of the water-surface profile for a starting water depth of $(h + z) = 402.400$ [m] at section 1 is presented in the following table. The organization of the computation sheet is slightly different from the one used for solving the problem Ex. 4.A. This new computation sheet takes into account the fact that the geometric characteristics of the sections will be interpolated from the plots or the tables prepared previously. It also enables one to take into account the singular head losses between the sections.

The columns 1 to 3 are reserved for the coordinate of the sections, x, the distance between the sections, Δx, and the thalweg elevations, z. The column 7 contains the Manning coefficients, n, and the column 16 the singular head-loss coefficients, K_{ss}. These known values are filled in these columns before beginning the water-surface profile computations.

The computations start at section 1 (line 1). The elevation of the flow depth, $(h + z) = 402.400$ [m], (column 4), thus the flow depth, h, (column 5), are known at section 1. The wetted perimeter, P, (column 6) and the wetted surface, A, (column 8) corresponding to the flow depth, h, (column 5) are read from the plotted geometric characteristics curves (or interpolated from the table of the geometric characteristics). The information contained in columns 9 to 18 are then calculated directly by applying the respective expressions.

For the following sections, the computation of the flow depth is done by trial-and-error. The details of this calculation are only given for the section 2 ($x = 40$ [m]). For section 2, first it is estimated that the water-surface elevation is : $(h + z) = 402.420$ [m] (column 4). The flow depth with respect to the thalweg elevation is then : $h = 402.420 - 400.070 = 2.35$ [m]. The wetted perimeter $P = 10.217$ [m] and the wetted surface $A = 11.741$ [m^2] for this depth are read from the plotted curves of geometric characteristics. Knowing these values the computation of the information in columns 9 to 19 is then straightforward. Next, the elevations of the total energy grade-line contained in columns 11 ($H = 402.423$ [m]) and 18 ($H_{calc.} = 402.405$ [m]) are compared. The difference between these two values is indicated in column 19 with a precision of 10^{-4}. Since the absolute value of the difference between the total energy elevations in columns 11 and 18 is larger than the desired precision, $|\Delta H'| = 0.0186 > 10^{-4}$, the computations must be repeated by selecting a new water-surface elevation, $(h + z)$, at section 2. The second estimation, $(h + z) = 402.370$ [m] (column 4), yields a difference of $|\Delta H''| = |-0.0312| > 10^{-4}$. A third and last trial with $(h + z) = 402.401$ [m] (column 4) allows finally to reach an equality between the total-energy elevations in columns 11 and 18 with the desired precision, namely $|\Delta H'''| < 10^{-4}$.

220   FLUVIAL HYDRAULICS

Method of successive approximations :										
Q = 3.0 [m³/s] ; river sections (see the plots of P *vs* h and A *vs* h)										
1	2	3	4	5	6	7	8	9	10	11
x	Δx	z	h + z	h	P	n	A	U	$U^2/2g$	H
[m]	[m]	[m]	[m]	[m]	[m]	$[m^{-1/3}s]$	[m²]	[m/s]	[m]	[m]
0.000		400.000	*402.400*	2.400	9.320	0.027	11.673	0.257	0.003	402.403
40.000	40.000	400.070	402.420	2.350	10.217	0.024	11.741	0.256	0.003	402.423
			402.370	2.300	10.063		11.285	0.266	0.004	402.374
			402.401	2.331	10.159		11.570	0.259	0.003	402.405
100.000	60.000	400.250	402.404	2.154	9.687	0.026	12.469	0.241	0.003	402.407
145.000	45.000	400.330	402.404	2.074	7.556	0.028	7.593	0.395	0.008	402.412
200.000	55.000	400.370	402.413	2.043	9.436	0.025	11.383	0.264	0.004	402.417
270.000	70.000	400.540	402.415	1.875	9.901	0.023	10.244	0.293	0.004	402.419

col.	symbol	expression	explanations
1	x		distance between a section and an origin. The choice of the origin is arbitrary. It is assumed here that the origin is located at section 1 and the x-axis points upstream.
2	Δx	$x_i - x_{i-1}$	distance between the running section (i) and the preceding section (i-1).
3	z		lowest elevation of the section (thalweg), with respect to an arbitrary reference plane. This elevation is given by the surveyor.
4	h + z		water-surface elevation at section 1 is *known*. For other sections, it has to be estimated. The trial-and-error calculation (here given only for x = 40.0 [m]) is repeated by varying this value until the total-energy line elevations in columns 11 and 18, H and $H_{calc.}$, are sufficiently close.
5	h		flow depth with respect to the lowest elevation of the section (thalweg).
6	P	*plot of* P *vs* h	wetted perimeter of section, read from plot of P *vs* h for flow depth, h, given in column 5.
7	n		coefficient of Manning for the section.
8	A	*plot of* A *vs* h	wetted surface of section, read from plot of A *vs* h for flow depth, h, given in column 5.

NON-UNIFORM FLOW

Standard step method (Δx is fixed)								
convention : $C = 1$, towards upstream (Fr < 1) ; $C = -1$, towards downstream (Fr > 1)								
12	13	14	15	16	17	18	19	20
R_h	S_e	$\overline{S_e}$	h_r	K_{ss}	h_s	$H_{calc.}$	ΔH	Remarks
[m]	[-]	[-]	[m]	[-]	[m]	[m]	[m]	
1.253	3.57E-05					402.403	0.0000	control section
1.149	3.12E-05	3.35E-05	1.34E-03	0.10	3.87E-06	402.405	0.0186	no H > $H_{calc.}$
1.121	3.49E-05	3.53E-05	1.41E-03	0.10	2.36E-05	402.405	-0.0312	no H < $H_{calc.}$
1.139	3.26E-05	3.41E-05	1.36E-03	0.10	6.01E-06	402.405	0.0000	√ H = $H_{calc.}$
1.287	2.79E-05	3.03E-05	1.82E-03	0.30	1.43E-04	402.407	0.0000	√
1.005	1.22E-04	7.48E-05	3.37E-03	0.30	1.50E-03	402.412	0.0000	√
1.206	3.38E-05	7.77E-05	4.27E-03	0.15	6.62E-04	402.416	0.0000	√
1.035	4.34E-05	3.86E-05	2.70E-03	0.10	8.31E-05	402.419	0.0000	√

col.	symbol	expression	explanations
9	U	Q / A	average flow velocity.
10		$U^2/2g$	dynamic head.
11	H	$z + h + U^2/2g$	total head.
12	R_h	A / P	hydraulic radius.
13	S_e	$\dfrac{U^2 n^2}{R_h^{4/3}}$	slope of energy-grade line with respect to horizontal, calculated using the Manning-Strickler formula, eq. 3.16.
14	$\overline{S_e}$	$\dfrac{(S_e)_i + (S_e)_{i-1}}{2}$	average slope of energy-grade line between section (i) and section (i-1).
15	h_r	$\overline{S_e} \Delta x$	head loss between section (i) and section (i-1) due to friction.
16	K_{ss}		local (singular) head-loss coefficient (given) between section (i) and section (i-1) (see eq. 4.12b).
17	h_s	$K_{ss} \dfrac{Q^2}{2g} \left\| \dfrac{1}{A_i^2} - \dfrac{1}{A_{i-1}^2} \right\|$	local (singular) head loss between section (i) and section (i-1).
18	$(H_i)_{calc}$	$(H_{i-1})_{calc.} + C (h_r + h_s)_i$	total head, as calculated with eq. 4.12b. The flow being subcritical, one takes $C = +1$.
19	ΔH	$(H_i) - (H_i)_{calc.}$	convergence criteria. Trial-and-error, by estimating h-values, stops if $\|\Delta H\| \leq 10^{-4}$.

The water-surface profile calculations for the other starting water-surface elevation at section 1, (h + z) = *401.200* [m], are also done following the same trial-and-error procedure. They are presented in the table below without any comments.

6° The water-surface profiles for *both* starting water-surface elevations at section 1 are plotted in the following figure. The profile corresponding to the starting depth *402.400* [m] is almost horizontal. The undulations of the bed being small compared to the flow depth, their influence on the water-surface profile is, in this case, negligible. On the other hand, the flow depths for the starting water-surface elevation *401.200* [m] being smaller, the influence of the undulations of the bed on the water-surface profile is quite evident.

In the present problem, the coefficients of Manning, n, although different from one section to another, are independent of the flow depth. In natural water courses, the coefficient of Manning, n, for a section may also vary with the flow depth. Without much trouble this can also be incorporated into the computational procedure and the same computation sheet can be used for the calculation of the water-surface profile. The only thing to do is to prepare another plot showing the relationship n = f(h) in the same way as it was done for the wetted perimeter, P, and the wetted surface, A. At every calculation step, the value of the coefficient of Manning, n, corresponding to the running flow depth, h, is simply read from this plot.

The use of the standard step method for water-surface calculations presents some other advantages. It was already mentioned (see point 4.2, 7°) that the subcritical flow, Fr < 1, is controlled by a downstream control section and that the computation of the water-surface profile should proceed from downstream to upstream. However, if the average velocities, thus dynamic heads, are small, the computations can even be carried out in the wrong direction without introducing much error in the results (see *Chow*, 1959, pp. 274-278).

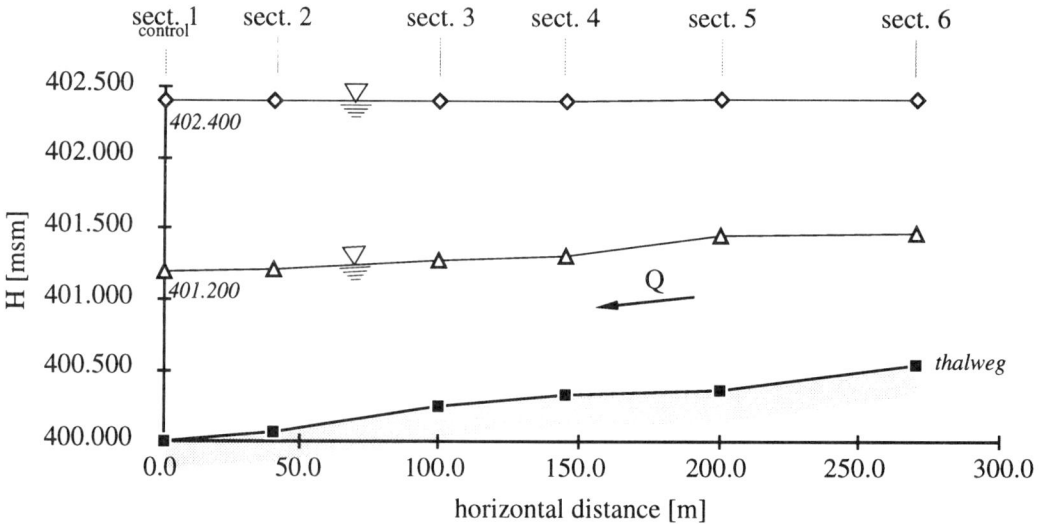

Method of successive approximations :										
$Q = 3.0$ [m³/s] ; river sections (see the plots of P vs h and A vs h)										
1	2	3	4	5	6	7	8	9	10	11
x	Δx	z	$h+z$	h	P	n	A	U	$U^2/2g$	H
[m]	[m]	[m]	[m]	[m]	[m]	[m$^{-1/3}$s]	[m²]	[m/s]	[m]	[m]
0.000		400.000	401.200	1.200	5.916	0.027	4.330	0.693	0.024	401.224
40.000	40.000	400.070	401.217	1.147	6.274	0.024	3.502	0.857	0.037	401.255
100.000	60.000	400.250	401.281	1.031	7.170	0.026	4.612	0.650	0.022	401.303
145.000	45.000	400.330	401.315	0.985	4.649	0.028	2.494	1.203	0.074	401.388
200.000	55.000	400.370	401.459	1.089	6.729	0.025	4.863	0.617	0.019	401.478
270.000	70.000	400.540	401.482	0.942	5.961	0.023	3.664	0.819	0.034	401.516

Standard step method (Δx is fixed)							
convention : $C = 1$, towards upstream (Fr < 1) ; $C = -1$, towards downstream (Fr > 1)							
12	13	14	15	16	17	18	19	20
R_h	S_e	$\overline{S_e}$	h_r	K_{ss}	h_s	$H_{calc.}$	ΔH	Remarks
[m]	[-]	[-]	[m]	[-]	[m]	[m]	[m]	
0.732	5.31E-04					401.224	0.0000	sect. de control
0.558	9.19E-04	7.25E-04	2.90E-02	0.10	1.29E-03	401.255	0.0000	√ $H = H_{calc.}$
0.643	5.15E-04	7.17E-04	4.30E-02	0.30	4.75E-03	401.303	0.0000	√
0.536	2.60E-03	1.56E-03	7.01E-02	0.30	1.57E-02	401.388	0.0000	√
0.723	3.67E-04	1.48E-03	8.17E-02	0.15	8.15E-03	401.478	0.0000	√
0.615	6.78E-04	5.23E-04	3.66E-02	0.10	1.48E-03	401.516	0.0000	√

It is also interesting to note that in using the standard step method, if the computation is done in the correct direction, even starting with a wrong water depth at the first section, the results converge to more and more correct flow depths, as one gets away from the starting section. This particularity of the standard step method is very useful in practice for the computation of water-surface profiles in natural water courses where often it is difficult to determine precisely a control section (see *Chow*, 1959, pp. 274-278).

If the flow depth at the beginning of the reach to be studied is not known, all that is needed is to start the computations at a section sufficiently away (at the upstream or the downstream side according to the direction of the calculation), so that the computed flow depths auto-correct themselves until the computation reaches the reach under study. Whether the starting section was chosen sufficiently away from the reach under study or not can be verified by recalculating the water surface starting from the same section but with a slightly different water-surface elevation. If these two computations with different starting flow depths give the same results, the computed water-surface profile is the correct one (see *Chow*, 1959, pp. 274-278).

Ex. 4.D

At a hydroelectric power plant, a discharge of $Q = 6 \, [m^3/s]$ flows in a very long channel (1) of rectangular section having a width of $b = 4.00 \, [m]$. The bed slope of the channel is $S_{f_1} = 0.010 \, [-]$ and the friction coefficient of the concrete is $n = 0.012 \, [m^{-1/3} \, s]$. At some point, the bed slope changes suddenly to $S_{f_2} = 0.001 \, [-]$. In this second channel (2), a sluice gate is placed at a distance of $L = 140 \, [m]$ downstream of the slope change to control the flow. The width of the gate is equal to the width of the channel, $B_v = 4.00 \, [m]$. The discharge coefficient for the gate is estimated to be $K_v = 0.55 \, [-]$.

Determine the water-surface profile for the two following gate openings :
i) $h_o = 1.20 \, [m]$
ii) $h_o = 0.50 \, [m]$
Should a hydraulic jump develop, its length may be neglected.

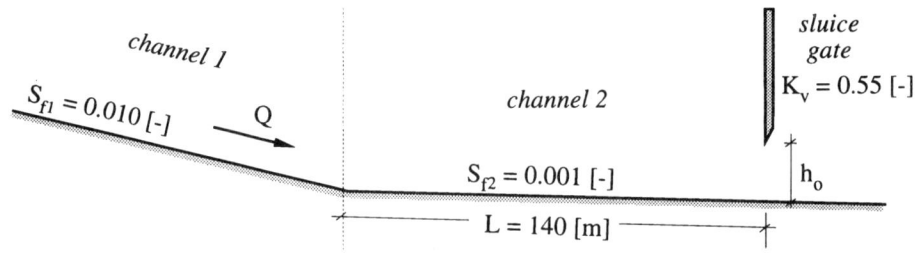

SOLUTION :

1° To determine the type of the water-surface profile, the *critical* depth, h_c, and the *normal* depth, h_n, should first be calculated for the channels 1 and 2.

The critical depth for a rectangular channel is given with eq. 2.23, where $q = Q/b$ is the unit discharge. Both channels have the same width, thus their critical depth is the same and equal to :

$$h_c = \sqrt[3]{\frac{q^2}{g}} = \sqrt[3]{\frac{Q^2}{gb^2}} = \sqrt[3]{\frac{6.0^2}{9.81 \times 4.0^2}} = 0.612 \, [m]$$

The normal depths for the channels 1 and 2 can be calculated by using the equation of Manning-Strickler, eq. 3.16 :

$$Q = U \, A = \frac{1}{n} \, R_h^{2/3} \, S_f^{1/2} \, A$$

The solution of this equation by trial-and-error on a spreadsheet, for finding the normal depth, has already been presented in detail in Ex.4.A (see point 1°). The normal depths computed for a discharge of $Q = 6.0 \, [m^3/s]$ as well as the corresponding flow regimes are summarized in the table below :

NON-UNIFORM FLOW

	b [m]	n [m$^{-1/3}$s]	S_f [-]	h_n [m]	U [m/s]	Fr [-]	Remarks
Channel 1	4.00	0.012	0.010	0.383	3.91	2.02	$h_c > h_{n_1}$: Fr > 1 supercritical flow
Channel 2	4.00	0.012	0.001	0.818	1.83	0.65	$h_{n_2} > h_c$: Fr < 1 subcritical flow

The water-surface profile will be of *type S* in channel 1 and of *type M* in channel 2.

In passing from a supercritical regime in channel 1 to a subcritical regime in channel 2, the water-surface profile will have to cross the line of critical depth, h_c. One should therefore expect the formation of a hydraulic jump (see sect. 4.4.4). The location of the jump should be determined by drawing the water-surface profiles both at the upstream (type S) and downstream (type M) of the point where the slope changes.

2° First, the *influence of the gate* on the flow must be determined.

i) The opening of the gate is $h_o = 1.20$ [m].

For a discharge of $Q = 6.0$ [m³/s], the normal depth in channel 2 was computed as $h_{n_2} = 0.82$ [m]. This flow depth being smaller than the opening of the gate, $h_{n_2} < h_o$, it can be concluded that the gate does *not influence* the flow.

ii) The opening of the gate is $h_o = 0.50$ [m].

For a discharge of $Q = 6.0$ [m³/s], the normal depth in channel 2 was computed as $h_{n_2} = 0.82$ [m]. This flow depth being larger than the opening of the gate, $h_{n_2} > h_o$, it can be concluded that the gate *influences* the flow. The flow depth must increase behind the gate up to a depth sufficient to allow the passage of the required discharge under the gate. Moreover, the opening of the gate being less than the critical depth, $h_o < h_c < h_{n_2}$, a hydraulic jump will form downstream of the gate. This hydraulic jump may be submerged or not. The conjugate depth, h_o^{cj}, corresponding to the opening of the gate, h_o, should be :

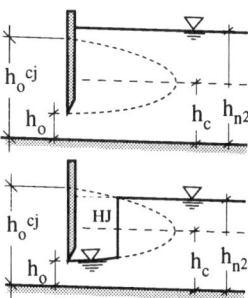

$h_o^{cj} < h_{n_2}$ for a *submerged* jump,

$h_o^{cj} > h_{n_2}$ for an *non-submerged* jump.

For the flow under the gate, one can calculate :

for $\quad h_o = 0.50$ [m] $\quad \Rightarrow \quad U_o = 3.0$ [m/s] $\quad$ and $\quad Fr_o = 1.36$ [-]

The conjugate depth can be obtained using eq. 4.41 :

$$h_o^{cj} = \frac{1}{2} h_o (\sqrt{1 + 8 Fr_o^2} - 1) = \frac{1}{2} 0.50 (\sqrt{1 + 8 \times 1.36^2} - 1) = 0.74 \text{ [m]}$$

$h_o^{cj} = 0.74$ [m] $< h_{n_2} = 0.82$ [m] $\quad \Rightarrow \quad$ the jump is thus *submerged*.

Now, the flow depth upstream of the gate, h_{v_1}, which will allow the passage of a discharge of $Q = 6.0$ [m³/s] through a gate opening of $h_o = 0.50$ [m] can be calculated by taking into account that, due to the submergence of the jump, the flow depth downstream of the gate is $h_{v_2} = h_{n_2} = 0.82$ [m]. The discharge coefficient of the gate being given, $K_v = 0.55$ [-], the discharge under the gate can be calculated using eq. 4.37 :

$$Q = A_o K_v \sqrt{2g (h_{v_1} - h_{v_2})} = B_v h_o K_v \sqrt{2g (h_{v_1} - h_{v_2})}$$

$6.0 = 4.0 \times 0.50 \times 0.55 \sqrt{2g (h_{v_1} - 0.82)} \quad \Rightarrow \quad h_{v_1} = 2.34$ [m]

The flow depth immediately upstream of the gate is therefore : $h_{v_1} = 2.34$ [m].

3° Having determined the *influence of the gate* on the flow and on the flow depths upstream and downstream of the gate, a sufficient number of control sections are now known for calculating the water-surface profiles for both gate openings, h_o, mentioned in the problem statement.

i) **Opening of gate is $h_o = 1.20$ [m].**

The preceding calculations have shown, that the gate *does not* control the flow in channel 2, if its opening is $h_o = 1.20$ [m]. It is reasonable to assume that, if this channel is sufficiently long, a uniform flow with a normal depth of $h_{n_2} = 0.82$ [m] will be established in channel 2.

In channel 1, the uniform flow has a normal depth of $h_{n_1} = 0.38$ [m] and the flow regime is supercritical. The flow will then change from supercritical to subcritical by forming a *hydraulic jump*. The position of the jump which, depending on the hydraulic conditions, may form in either channel 1 or in channel 2 must also be determined for the computation of the water-surface profile.

To predict in which channel the hydraulic jump will be formed, it will be now assumed that in channel 1 the supercritical, uniform flow with a normal depth of

NON-UNIFORM FLOW

$h_{n_1} = 0.38$ [m] continues up to the point of change in slope. The conjugate depth, $h_{n_1}^{cj}$, of the normal flow depth $h_{n_1} = 0.38$ [m] at the point of slope change, is then compared with the normal depth in channel 2, $h_{n_2} = 0.82$ [m] :

if $\quad h_{n_1}^{cj} < h_{n_2} \quad$ the hydraulic jump is formed in channel 1,

if $\quad h_{n_1}^{cj} > h_{n_2} \quad$ the hydraulic jump is formed in channel 2.

The conjugate depth, $h_{n_1}^{cj}$, is now computed using eq. 4.41 :

$$h_{n_1}^{cj} = \frac{1}{2} h_{n_1} (\sqrt{1 + 8 \, Fr_1^2} - 1) = \frac{1}{2} 0.38 (\sqrt{1 + 8 \times 2.02^2} - 1) = 0.92 \text{ [m]}$$

$h_{n_1}^{cj} = 0.92$ [m] $> h_{n_2} = 0.82$ [m] $\quad \Rightarrow \quad$ *the hydraulic jump is formed in channel 2.*

At the point of slope change, the flow enters into the channel 2 with a depth of $h_{n_1} = 0.38$ [m], which is less than the critical depth of $h_c = 0.61$ [m]. Once in channel 2, the flow will first follow a M3-type surface profile (see sect. 4.2.1 and Fig. 4.2a) to join its normal depth, h_{n_2}. To attain this flow depth, the flow has to cross the critical depth, $h_c = 0.61$ [m], which of course will lead to the formation of a hydraulic jump somewhere along the M3-type water-surface profile.

The location of the hydraulic jump can be predicted by following the procedure described below :

- First, the M3-type water-surface profile in channel 2 is computed starting with a flow depth of $h_{n_1} = 0.38$ [m] at the point of the slope change. The computation proceeds towards downstream until reaching the critical depth, $h_c = 0.61$ [m]. This computation was carried out here on a spreadsheet using the standard step method. The computation sheet is presented in the following table. Since the method has already been described in detail when solving the problem Ex. 4.A 5°, only the most important columns are included in the table. In addition, a column containing the Froude numbers corresponding to flow depths has been added (see column 11). The computed M3-type water-surface profile is plotted in the figure at the end of this problem.

- Next, the conjugate depths, h^{cj}, corresponding to flow depths at all the stations are calculated (see column 12). These points are also plotted in the same figure and joined to form a curve that we will be name as *curve of conjugate depths for M3* and will denote it by $M3^{cj}$. Note that the curve of conjugate depths for the uniform supercritical flow in channel 1 is a line parallel to the channel bed. This curve joins the curve $M3^{cj}$ at the point of slope change (see figure).

M3-type water-surface profile

Method of successive approximations : standard step method (Δx is fixed)

$S_f = 1.0\text{E-}3$ [-] ; $Q = 6.0$ [m³/s] ; $b = 4.0$ [m] ; $m = 0$ [-] ; $n = 0.012$ [m$^{-1/3}$s]

$h_n = 0.818$ [m] ; $h_c = 0.612$ [m] ; $C = -1$, proceed towards downstream (Fr > 1)

1	2	3	4	5	6	7	8	9	10	11	12
x	Δx	z	h	U	H	R_h	S_e	h_r	H_{calc}	Fr	h^{cj}
[m]	[m]	[m]	[m]	[m/s]	[m]	[m]	[-]	[m]	[m]	[-]	[m]
0.000		0.140	*0.383*	3.912	1.304	0.322	1.00E-02		1.304	2.02	0.919
7.646	7.646	0.132	0.406	3.692	1.233	0.338	8.35E-03	7.01E-02	1.233	1.85	0.879
14.997	7.350	0.125	0.429	3.495	1.177	0.353	7.04E-03	5.66E-02	1.177	1.70	0.841
21.979	6.983	0.118	0.452	3.319	1.131	0.369	6.00E-03	4.55E-02	1.131	1.58	0.806
28.513	6.534	0.111	0.475	3.159	1.095	0.384	5.15E-03	3.64E-02	1.095	1.46	0.774
34.502	5.989	0.105	0.498	3.014	1.066	0.399	4.46E-03	2.88E-02	1.066	1.36	0.743
39.831	5.329	0.100	0.521	2.881	1.044	0.413	3.89E-03	2.22E-02	1.044	1.27	0.714
44.361	4.530	0.096	0.543	2.760	1.027	0.427	3.41E-03	1.65E-02	1.027	1.20	0.686
47.920	3.558	0.092	0.566	2.649	1.016	0.441	3.01E-03	1.14E-02	1.016	1.12	0.660
50.287	2.367	0.090	0.589	2.546	1.009	0.455	2.67E-03	6.71E-03	1.009	1.06	0.636
51.176	0.889	0.089	0.612	2.451	1.007	0.469	2.38E-03	2.24E-03	1.007	1.00	0.612

col.	symb.	expression	explanations
1	x		distance from origin.
2	Δx	$x_i - x_{i-1}$	distance between section (i) and section (i-1).
3	z	$z_{i-1} \pm S_f \Delta x$	bed elevation with respect to an arbitrarily chosen reference plane. Here, the reference plane is placed at the bed elevation of the place where the sluice gate is located. For S_f = Cte, z is calculated using the expression shown on the left.
4	h		flow depth is calculated by trial-and-error until the total head, H, in columns 6 and 10 are sufficiently close.
5	U	Q/A	average velocity in section.
6	H	$z + h + U^2/2g$	*total* head.
7	R_h	A/P	hydraulic radius.
8	S_e	$\dfrac{U^2 n^2}{R_h^{4/3}}$	slope of the energy-grade line with respect to the horizontal, calculated using the equation of Manning-Strickler, eq. 3.16.
9	h_r	$\overline{S_e} \Delta x$	head loss between section (i) and section (i-1).
10	$H_{calc.}$	$(H_{i-1})_{calc.} + C\,(h_r)_i$	calculated total head.
11	Fr	$U/\sqrt{gh}$	Froude number.
12	h^{cj}	eq. 4.41	conjugate depth.

NON-UNIFORM FLOW   229

- At its downstream end, the hydraulic jump joins the normal depth in channel 2. One draws now a line parallel to the bed at the depth of $h_{n_2} = 0.82\,[m]$. Neglecting the length of the hydraulic jump (see figure 4.19), the jump is formed at the intersection of this line (parallel) with the curve of conjugate depths M3^{cj} (see column 12, numbers in a box).

On the following figure, the resulting water-surface profile is drawn with a thick line. In channel 1, the flow is uniform, h_{n_1}, and supercritical. Downstream of the point of slope change the flow is supercritical and follows an M3-type water-surface profile up to the point where the hydraulic jump takes place. Downstream of the jump, the flow is uniform, h_{n_2}, and subcritical.

ii) **Opening of gate is $h_o = 0.50\,[m]$.**

It has already been shown that for this opening the gate influences the flow. As in the previous case, first the channel in which the hydraulic jump takes place must be determined.

Assuming that the supercritical flow continues all along channel 1, the conjugate depth at the point of slope change was already determined (see the previous case) as being : $h_{n_1}^{cj} = 0.92\,[m]$. On the other hand, in channel 2, the flow depth immediately upstream of the gate was also calculated, being : $h_{v_1} = 2.34\,[m]$. Consequently :

$h_{n_1}^{cj} = 0.92\,[m] < h_{v_1} = 2.34\,[m] \quad \Rightarrow \quad$ *the hydraulic jump is formed in channel 1.*

In channel 2, where $h_{n_2} > h_c$, the flow takes place on a mild slope. The water-surface profile will have a *M-type profile*. To determine the branch (see sect. 4.2.1) the control point has to be identified. The flow depth at the downstream end is known : $h_{v_1} = 2.34\,[m] > h_{n_2}$. The water-surface profile is located between h_{v_1} and h_{n_2}. *The water-surface profile is therefore of the type* M1 (see Figs. 4.2 and 4.2a) and can be computed from downstream to upstream, starting with a flow depth of h_{v_1}.

The computation of the M1-type water-surface profile on a spreadsheet using the standard-step method (see Ex. 4.A, 5°) is summarized in the table below. Only the most important columns are included in this table. The M1-type water-surface profile is plotted on the figure at the end of the problem. The computed profile yields a flow depth of $h_{br} = 2.20\,[m]$ at the point of slope change at a distance of $x = 140.00\,[m]$ (see column 4, last line).

In channel 1 where $h_{n_1} < h_c$, the flow is taking place on a steep slope. The water-surface profile will therefore be a *S-type profile*. To determine the branch (see sect. 4.2.2) one should adopt the following line of reasoning : at the point of slope change the flow depth in channel 1 is obviously the same as in channel 2, namely $h_{br} = 2.20\,[m]$. The water-surface profile will be between h_{br} and h_c. *The water-surface profile is therefore of the type* S1 (see Figs. 4.2 and 4.2a).

M1-type water-surface profile

Method of successive approximations: standard step method (Δx is fixed)

S_f = 1.0E-3 [-] ; Q = 6.0 [m³/s] ; b = 4.0 [m] ; m = 0 [-] ; n = 0.012 [m$^{-1/3}$s]

h_n = 0.818 [m] ; h_c = 0.612 [m] ; C = +1 , proceed towards upstream (Fr < 1)

1	2	3	4	5	6	7	8	9	10	11	12
x	Δx	z	h	U	H	R_h	S_e	h_r	H_{calc}	Fr	h^{cj}
[m]	[m]	[m]	[m]	[m/s]	[m]	[m]	[-]	[m]	[m]	[-]	[m]
0.000		0.000	*2.335*	0.643	2.356	1.077	5.38E-05		2.356	0.13	
13.963	13.963	0.014	2.321	0.646	2.356	1.074	5.47E-05	7.57E-04	2.356	0.14	
27.934	13.971	0.028	2.308	0.650	2.357	1.071	5.55E-05	7.69E-04	2.357	0.14	
41.912	13.979	0.042	2.294	0.654	2.358	1.069	5.64E-05	7.82E-04	2.358	0.14	
55.899	13.987	0.056	2.281	0.658	2.359	1.066	5.72E-05	7.94E-04	2.359	0.14	
69.894	13.995	0.070	2.267	0.662	2.359	1.063	5.81E-05	8.07E-04	2.359	0.14	
83.897	14.003	0.084	2.254	0.666	2.360	1.060	5.90E-05	8.20E-04	2.360	0.14	
97.909	14.012	0.098	2.240	0.670	2.361	1.057	6.00E-05	8.34E-04	2.361	0.14	
111.930	14.021	0.112	2.227	0.674	2.362	1.054	6.09E-05	8.48E-04	2.362	0.14	
125.959	14.030	0.126	2.213	0.678	2.363	1.051	6.19E-05	8.62E-04	2.363	0.15	
140.000	14.039	0.140	*2.200*	0.682	2.364	1.048	6.29E-05	8.76E-04	2.364	0.15	

S1-type water-surface profile

Method of successive approximations: standard step method (Δx is fixed)

S_f = 1.0E-2 [-] ; Q = 6.0 [m³/s] ; b = 4.0 [m] ; m = 0 [-] ; n = 0.012 [m$^{-1/3}$s]

h_n = 0.383 [m] ; h_c = 0.612 [m] ; C = -1 , proceed towards downstream (Fr > 1)

1	2	3	4	5	6	7	8	9	10	11	12
x	Δx	z	h	U	H	R_h	S_e	h_r	H_{calc}	Fr	h^{cj}
[m]	[m]	[m]	[m]	[m/s]	[m]	[m]	[-]	[m]	[m]	[-]	[m]
140.000		0.140	*2.200*	0.682	2.364	1.048	6.29E-05	8.76E-04	2.364	0.15	0.091
155.604	15.605	0.296	2.041	0.735	2.365	1.010	7.67E-05	1.09E-03	2.365	0.16	0.105
171.132	15.529	0.451	1.882	0.797	2.366	0.970	9.53E-05	1.34E-03	2.366	0.19	0.122
186.554	15.422	0.606	1.724	0.870	2.368	0.926	1.21E-04	1.67E-03	2.368	0.21	0.143
201.822	15.268	0.758	1.565	0.959	2.370	0.878	1.57E-04	2.12E-03	2.370	0.24	0.169
216.861	15.039	0.909	1.406	1.067	2.373	0.826	2.12E-04	2.77E-03	2.373	0.29	0.203
231.542	14.681	1.055	1.247	1.203	2.376	0.768	2.96E-04	3.73E-03	2.376	0.34	0.246
245.629	14.087	1.196	1.088	1.378	2.382	0.705	4.36E-04	5.16E-03	2.382	0.42	0.303
258.650	13.021	1.327	0.930	1.613	2.389	0.635	6.87E-04	7.31E-03	2.389	0.53	0.377
269.521	10.871	1.435	0.771	1.946	2.399	0.556	1.19E-03	1.02E-02	2.399	0.71	0.477
275.081	5.560	1.491	*0.612*	2.451	2.409	0.469	2.38E-03	9.91E-03	2.409	1.00	0.612

The computation of the S1-type water-surface profile is presented in the table above. The computation starts at $x = 140.00$ [m] with a flow depth of $h_{br} = 2.20$ [m] and continues up to $x = 275.08$ [m], where the critical depth, h_c, is reached (it is obvious that, depending on the location of the jump, only a portion of this curve represents the true water-surface profile).

To reach its normal depth, h_{n_1}, the flow has to cross the critical depth line, h_c; this will lead to the formation of a hydraulic jump somewhere along the S1 profile. Below two different methods are proposed for locating the hydraulic jump. Both methods give, of course, the same result :

- The undisturbed flow in channel 1 is uniform, $h_{n_1} = 0.38$ [m] and the water-surface profile is parallel to the bed. The conjugate depth for this normal depth was already calculated as being $h_{n_1}^{cj} = 0.92$ [m]. The curve of conjugate depths is also a line parallel to the channel bed. If its length is neglected, the location of the hydraulic jump is point located at the intersection of this curve with the calculated S1-profile (see column 4, numbers in a box).

- The conjugate depths corresponding to the flow depths of the calculated S1-profile are given in column 12 of the computation sheet. Using these values, the curve of conjugate depth is plotted on the figure at the end of the problem. The intersection of this curve with the normal depth, $h_{n_1} = 0.38$ [m] gives the location of the jump (see column 12, numbers in a box).

On the following figure, the resulting water-surface profile is drawn with a thick line. In channel 1 the flow is uniform, h_{n_1}, and supercritical up to the point where the hydraulic jump takes place. Downstream of the jump, up to the point of change in slope, the flow is gradually varied and subcritical with a S1-profile. Between the point of slope change and the sluice gate there is a M1-type gradually varied flow. Immediately downstream of the gate there is a submerged hydraulic jump. Further downstream, the flow becomes uniform, h_{n_2}, and subcritical.

232　　　　　　　　　　　　　　　　　　　　　　　　　　　　　　　　FLUVIAL HYDRAULICS

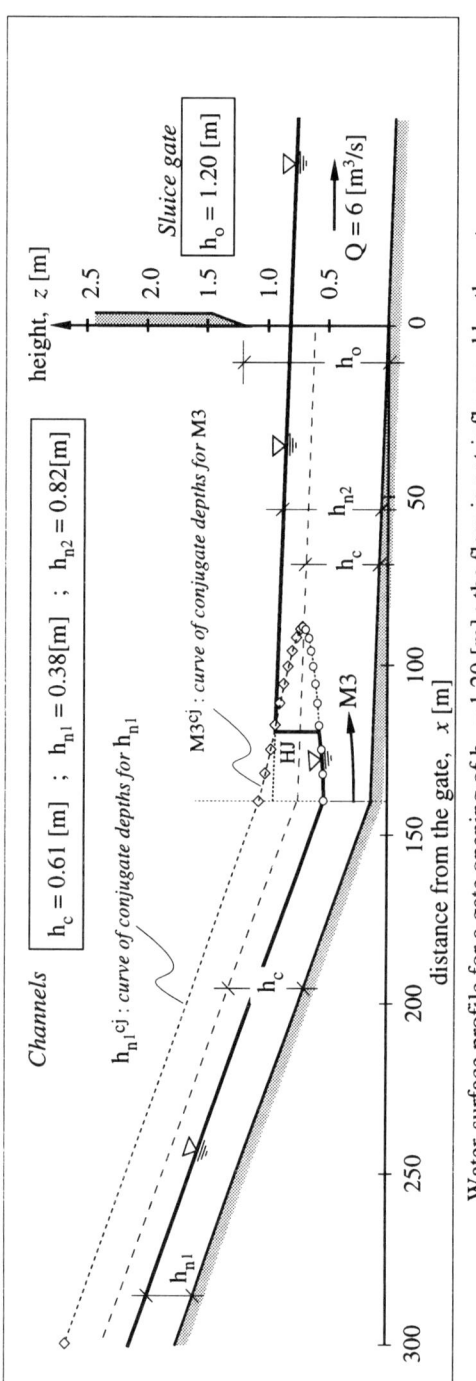

Water-surface profile for a gate opening of $h_o = 1.20$ [m] : the flow is not influenced by the gate.

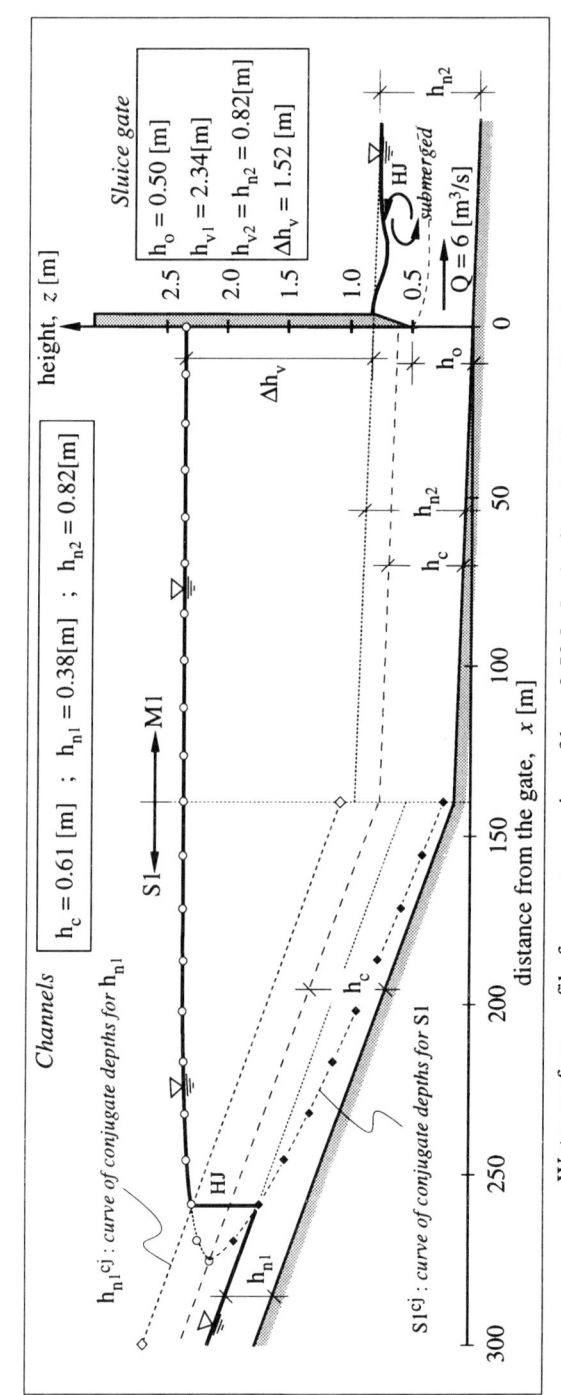

Water-surface profile for a gate opening of $h_o = 0.50$ [m] : the flow is influenced by the gate.

NON-UNIFORM FLOW 233

Ex. 4.E

In an irrigation channel, the discharge can be assumed to be constant and equal to $Q = 0.094$ [m³/s]. This long channel has a rectangular cross section and the bed slope is $S_f = 0.001177$ [-]. The channel bed is made of compacted earth with an average grain diameter of $d_{50} = 0.005$ [m]. The vertical side walls are in rubble masonry and have a coefficient of Strickler of $K_{s_m} = 53$ [m$^{1/3}$s^{-1}]. Neither erosion nor deposition are allowed in this channel.

i) Compute the normal depth and determine the flow regime.

In order to stabilize the bed, at a certain point along the channel, it is proposed to construct a concrete sill having a positive step of height, $\Delta z = + 0.08$ [m].

ii) Determine the flow depth downstream of the positive sill.

iii) What is the maximum height, Δz_{max}, that can be given to this positive sill without influencing the upstream flow conditions ?

iv) What will happen if the positive sill is constructed with a step of height, $1.3\,\Delta z_{max}$?

It is also desired to measure the discharge, Q, by creating a contraction of the channel width at a distance well upstream of this sill.

v) What should be the width of the contracted section ?

SOLUTION :

i) The average grain diameter of the bed material is given as $d_{50} = 0.005$ [m]. For a mono-disperse granular bed material the critical velocities for the deposition, U_D, and erosion, U_{cr}, are given by the Hjulström diagram as a function of the grain diameter (see Fig. 3.12). Although the bed material of the irrigation channel is not mono-disperse, the Hjulström diagram shall still be used for determining the order of magnitude of the velocity to be adopted for the design. The following values are read on the Hjulström diagram :

for $\quad d_{50} = 0.005$ [m] : $\quad U_{cr} \cong 0.60$ [m/s]
$\quad\quad\quad\quad\quad\quad\quad\quad\quad\quad\quad\quad U_D \cong 0.35$ [m/s]

An average velocity of $U \cong 0.50$ [m/s] will be adopted for having neither erosion nor deposition in the channel. In order to compute the uniform flow depth, h_n, the channel width, B, should also be determined. Since there are two unknowns, B and h_n, there should also be *two* equations to solve them.

The first equation can be obtained by writing the equation of Manning-Strickler :

$$U \cong 0.50 = K_s R_h^{2/3} S_f^{1/2} \tag{3.16}$$

The velocity, U, and the bed slope, S_f, are known. The friction coefficient of the bed composed of granular material can be calculated using eq. 3.18 :

$$K_{s_f} = \frac{21.1}{d_{50}^{1/6}} = \frac{21.1}{0.005^{1/6}} = 51 \ [m^{1/3}s^{-1}].$$

The bottom of the channel and the side walls have therefore coefficients of Strickler which are different (51 and 53 $[m^{1/3}s^{-1}]$, respectively), but very close. To simplify the calculations a *global Strickler coefficient* of $K_s = 51 \ [m^{1/3}s^{-1}]$ will be assumed. The hydraulic radius of the channel can now be calculated by substituting these values into the Manning-Strickler equation :

$$R_h = \left(\frac{U}{K_s S_f^{1/2}}\right)^{3/2} = \left(\frac{0.5}{51 \times (0.001177)^{1/2}}\right)^{3/2} = 0.1527 \ [m]$$

According to the definition of hydraulic radius for a rectangular channel, one has :

$$R_h = 0.1527 \ [m] = \frac{B\,h}{B + 2h} = \frac{A}{B + 2h}$$

which gives the first relationship between h and B.

The second equation is given by the expression for the discharge passing through a cross section (see eqs. 1.3 and 2.2), $Q = U\,A$. The discharge in the channel is $Q = 0.094 \ [m^3/s]$. The average velocity, $U = 0.50 \ [m/s]$, was determined from the Hjulström diagram. By recalling also the definition of the wetted surface for a rectangular cross section, one can write :

$$A = B\,h = \frac{Q}{U} = \frac{0.094}{0.50} = 0.188 \ [m^2]$$

The width, B, of the channel and the normal depth, h, can be solved from these two equations :

$$B = \frac{A}{h} \quad \Rightarrow \quad R_h = \frac{A}{\frac{A}{h} + 2h} \quad \Rightarrow \quad 2h^2 - \frac{A}{R_h}h + A = 0$$

This quadratic equation yields two values for h. There are, therefore, two combinations of h and B, satisfying the conditions imposed :

$h' = 0.288 \ [m]$, $B' = 0.65 \ [m] \Rightarrow Fr' = \frac{U}{\sqrt{gh}} = \frac{0.5}{\sqrt{9.81 \times 0.288}} = 0.30 \ [-]$

$h'' = 0.325 \ [m]$, $B'' = 0.58 \ [m] \Rightarrow Fr'' = \frac{U}{\sqrt{gh}} = \frac{0.5}{\sqrt{9.81 \times 0.325}} = 0.28 \ [-]$

NON-UNIFORM FLOW 235

The first combination, which yields a better ratio between the flow depth and the channel width, will be adopted (h'/B' = 0.44 for the first combination, and h"/B" = 0.56 for the second one).

The width of the channel is taken as : B = 0.65 [m]
The *normal* depth for a discharge of Q = 0.094 [m³/s] is : h_n = 0.288 [m]
The flow regime is subcritical, because : Fr = 0.3 [-]

It will also be useful to compute the critical depth, which is the same for the channel with or without the sill. The *critical* depth for a rectangular channel is given by eq. 2.23, where q = Q / B :

$$h_c = \sqrt[3]{\frac{q^2}{g}} = \sqrt[3]{\frac{Q^2}{gB^2}} = \sqrt[3]{\frac{0.094^2}{9.81 \times 0.65^2}} = 0.129 \text{ [m]}$$

ii) The flow depth downstream of the sill (see sect. 4.5.1) can best be calculated graphically from the specific-energy curve, $H_s = f(h)$. The specific energy, or the specific head, is defined as follows (see sect. 2.3, eqs. 2.14):

$$H_s = \frac{U^2}{2g} + h = \frac{Q^2}{2gA^2} + h = \frac{Q^2}{2gB^2 h^2} + h$$

Since the width of the channel, B = 0.65 [m], and also the discharge, Q = 0.094 [m³/s], are constant, the specific energy, H_s, is only a function of the flow depth, h. The *specific-energy curve* can readily be plotted by computing the values of H_s for different h. This computation can easily be done on a spreadsheet program. The computation sheet is presented below :

| Computation of specific-energy curve, $H_s = f(h)$ |||||
| B = 0.65 [m] | | , | Q = 0.094 [m³/s] ||
h [m]	A [m²]	$U^2/2g$ [m]	H_s [m]	Remarks
0.040	0.026	0.666	0.706	
0.100	0.065	0.107	0.207	
0.120	0.078	0.074	0.194	
0.129	0.084	0.064	*0.193*	$h = h_c$
0.140	0.091	0.054	0.194	
0.160	0.104	0.042	0.202	
0.200	0.130	0.027	0.227	
0.250	0.163	0.017	0.267	
0.288	0.187	0.013	*0.301*	$h = h_n$
0.300	0.195	0.012	0.312	
0.350	0.228	0.009	0.359	
0.450	0.293	0.005	0.455	
0.550	0.358	0.004	0.554	

Note, that the characteristic flow depths, h_c and h_n, are included in the column for h. In this way, the specific energies corresponding to these two flow depths are also determined in the same computation :

$h_c = 0.129$ [m] , $H_{sc} = 0.193$ [m] ; $h_n = 0.288$ [m] , $H_{sn} = 0.301$ [m]

The specific-energy curve computed in the preceding computation sheet is plotted in the figure below :

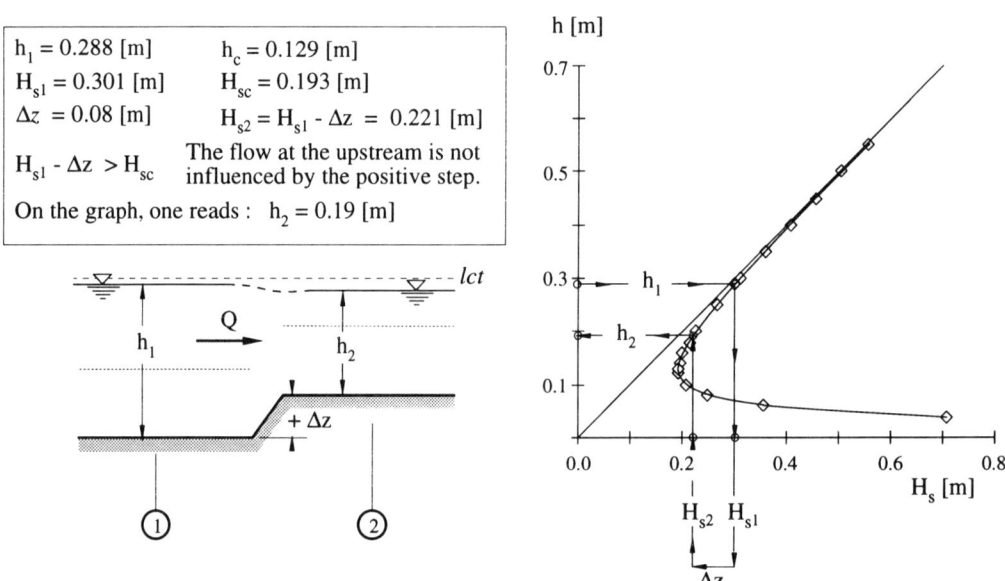

In this figure, the procedure for the graphical solution of the flow depth downstream of the sill, h_2, is indicated by the arrows. One enters the plot with the flow depth upstream of the sill (section 1) : $h_1 = h_n = 0.288$ [m]. The corresponding specific energy is $H_{s_1} = 0.301$ [m]. Assuming that there is no head loss due to the sill, the line of total energy (*lct*) remains constant and is parallel to the channel bed. Given that the channel bed is raised by an amount of Δz (the positive step), the distance between the channel bed and the line of total energy, which is exactly the specific energy itself, is diminished by an amount of Δz.

The specific energy downstream of the positive step (section 2) is thus :

$H_{s_2} = H_{s_1} - \Delta z = 0.301 - 0.08 = 0.221$ [m]

The flow depth corresponding to this specific energy is read on the branch of the specific-energy curve corresponding to subcritical flow. One obtains :

$h_2 = 0.19$ [m].

NON-UNIFORM FLOW   237

iii) The specific energy downstream of the positive step can thus be calculated directly using the expression : $H_{s_2} = H_{s_1} - \Delta z$. In the preceding method of computation, this corresponded to a left-shift on the specific-energy axis by an amount of Δz.

It should be recalled, that (see sect. 2.3.1) for $h = h_c$, the specific-energy curve goes through a minimum, H_{sc}. The maximum permissible height of the positive step, Δz_{max}, which will *not* influence the flow on the upstream, is therefore :

$$H_{s_2} \equiv H_{sc} = H_{s_1} - \Delta z_{max} \Rightarrow \Delta z_{max} = H_{s_1} - H_{sc} = 0.301 - 0.193 = 0.108 \text{ [m]}$$

For a sill with a positive step of height, Δz_{max}, the specific energy at section 2 downstream of the step is equal to the minimum specific energy, $H_{s_2} \equiv H_{sc} = 0.193$ [m]. The flow depth downstream of the step is equal to the critical depth, $h_2 = h_c = 0.129$ [m]. The procedure for the graphical solution is indicated on the figure below.

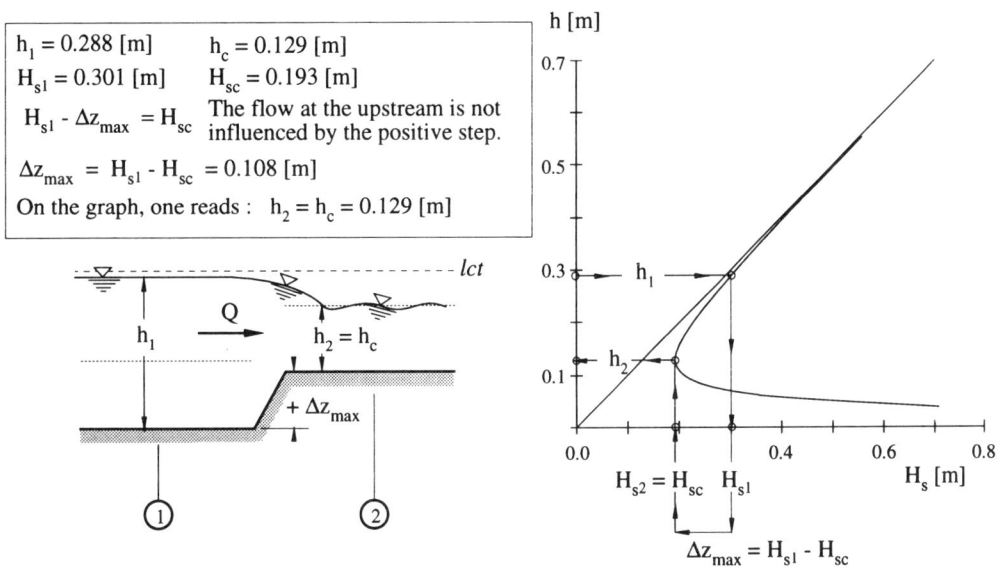

iv) Let there be a positive step with a height of :

$$\Delta z^* = 1.3 \, \Delta z_{max} = 1.3 \times 0.108 = 0.14 \text{ [m]}$$

For the existence of such a flow, the specific energy downstream of the positive step should be :

$$H'_{s_2} = H_{s_1} - \Delta z^* = 0.301 - 0.140 = 0.161 \text{ [m]}$$

However, for the given cross section, the specific energy for a constant discharge of Q = 0.094 [m^3/s] *cannot* go below the minimum energy, H_{sc} = 0.193 [m] (see preceding table), which occurs when $h_2 = h_c$. Consequently, the flow downstream of the positive sill should have a flow depth equal to the critical flow depth, $h_2 \equiv h_c = 0.129$ [m] and a specific energy equal to $H_{s_2} \equiv H_{sc} = 0.193$ [m].

$h_1 = 0.288$ [m] $h_c = 0.129$ [m] $\Delta z_{max} = H_{s1} - H_{sc} = 0.108$ [m]
$H_{s1} = 0.301$ [m] $H_{sc} = 0.193$ [m] $\Delta z^* = 1.3 \Delta z_{max} = 0.14$ [m]
$H_{s1} - \Delta z^* < H_{sc}$ ⇒ The flow at the upstream is influenced by the positive step.
At section 2, the flow depth is critical : $h_2 = h_c = 0.129$ [m], $H_{s2} = H_{sc} = 0.193$ [m]
Specific energy at section 1 : $H_{s1}^m = H_{sc} + \Delta z^* = 0.333$ [m]
The flow depth at section 1 is read from the graph : $h_1^m = 0.32$ [m] $> h_1$
Increase in upstream flow depth : $h_1^m - h_1 = 0.32 - 0.288 = 0.032$ [m]

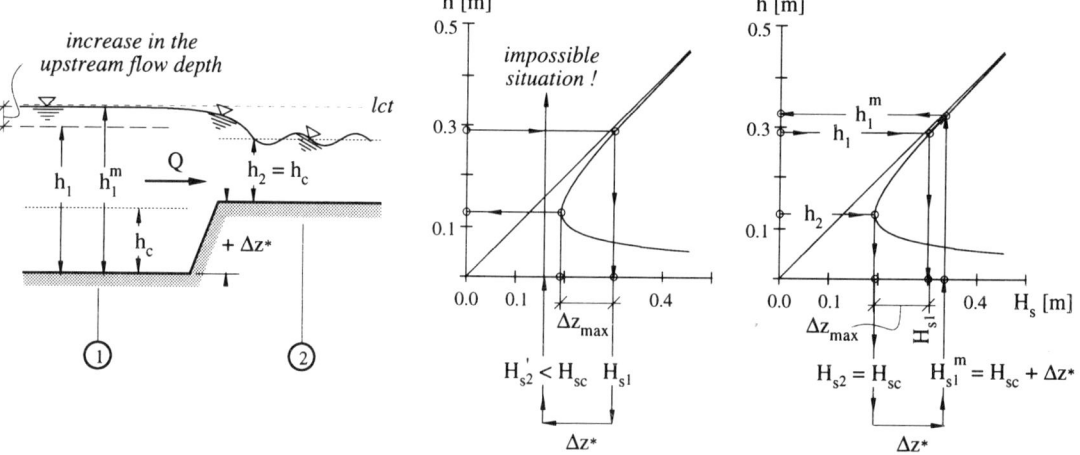

On the above figure, a graphical representation of this situation is given. When one makes a left-shift by an amount $\Delta z^* = 0.14$ [m] starting from $H_{s_1} = 0.301$ [m], The vertical line drawn upwards at $H_{s_2}' = 0.161$ [m] does not intersect the specific-energy curve. This situation is thus impossible!

At the upstream of the positive step, a minimum specific energy of, $H_{s_1}^m$, is needed to pass the required discharge over the step height, $+ \Delta z^*$. This minimum energy is expressed by :

$$H_{s_1}^m = H_{sc} + \Delta z^* = 0.193 + 0.14 = 0.333 \text{ [m]}$$

This increase in the specific energy is obtained by an increase in the flow depth upstream of the step; this results in a transient flow in the channel for a certain time

NON-UNIFORM FLOW 239

period, when a discharge less than Q = 0.094 [m³/s] flows towards section 2. During this time, the surplus discharge is stored for increasing the flow depth, and consequently the specific energy, at the upstream. The transient flow disappears and the flow once again becomes steady when the difference between the specific energies becomes equal to Δz_m^*. At this moment the upstream flow depth has stabilized at $h_1 = 0.32$ [m]; this can be read on the graph.

The graphical solution of the problem is indicated on the figure above. The specific energy at section 2 is equal to the minimum specific energy, $H_{s_2} \equiv H_{sc}$. A right-shift by an amount of Δz^* on the axis of H_{s_m} must be made to find the specific energy at section 1, upstream of the step, $H_{s_1} = 0.333$ [m]. From this point, a vertical line is drawn upwards, parallel to h-axis, up to the point where it intersects the subcritical flow branch of the specific energy curve; this yields $h_1^m = 0.32$ [m].

v) The aim is to create a control section, where the flow passes through critical depth, by narrowing down (contraction) the channel section. What is trying to be achieved here is the construction of a critical-flow measuring-flume (Venturi canal or Parshall flume) (see point 4.4.2, 5°).

Upstream of the contraction, the flow is uniform having a normal depth of $h_1 = 0.288$ [m]. The corresponding specific energy was already computed as $H_{s_1} = 0.301$ [m]. The minimum specific energy, H_{sc_2}, in the contracted section should be equal to H_{s_1}; thus :

$$H_{s_1} \equiv H_{sc_2} = 0.301 \text{ [m]}.$$

The flow depth at this contracted section is equal to the critical depth, h_c. The critical depth in a rectangular section can be calculated with eq. 2.19 :

$$h_{c_2} = \frac{2}{3} H_{sc_2} = \frac{2}{3} \times 0.301 = 0.201 \text{ [m]}$$

According to the definition of the specific energy, eq. 2.14, one writes :

$$H_{sc_2} = \frac{U_2^2}{2g} + h_{c_2} = \frac{Q^2}{2gA_2^2} + h_{c_2} = \frac{Q^2}{2gB_2^2 h_{c_2}^2} + h_{c_2} = 0.301 \text{ [m]}$$

from which the width of the contracted section can be obtained :

$$B_2 = \sqrt{\frac{Q^2}{2g\, h_{c_2}^2 (0.301 - h_{c_2})}}$$

By introducing $h_{c_2} = 0.201$ [m] in this expression, one gets :

$$B_2 = \sqrt{\frac{0.094^2}{2 \times 9.81 \times 0.201^2 \times (0.301 - 0.201)}} = 0.334 \text{ [m]}$$

Since the channel is rectangular, one can also use eq. 2.24 to obtain :

$$B_2^{-1} = \frac{1}{Q} \sqrt{g \, h_{c_2}^3}$$

from which one gets :

$$B_2 = \sqrt{\frac{Q^2}{g h_{c_2}^3}} = \sqrt{\frac{0.094^2}{9.81 \times 0.201^3}} = 0.334 \text{ [m]}$$

the width of the contracted section should therefore be : $B_2 = 0.334$ [m].

The specific-energy curves for the channel and the contracted section are plotted on the figure below.

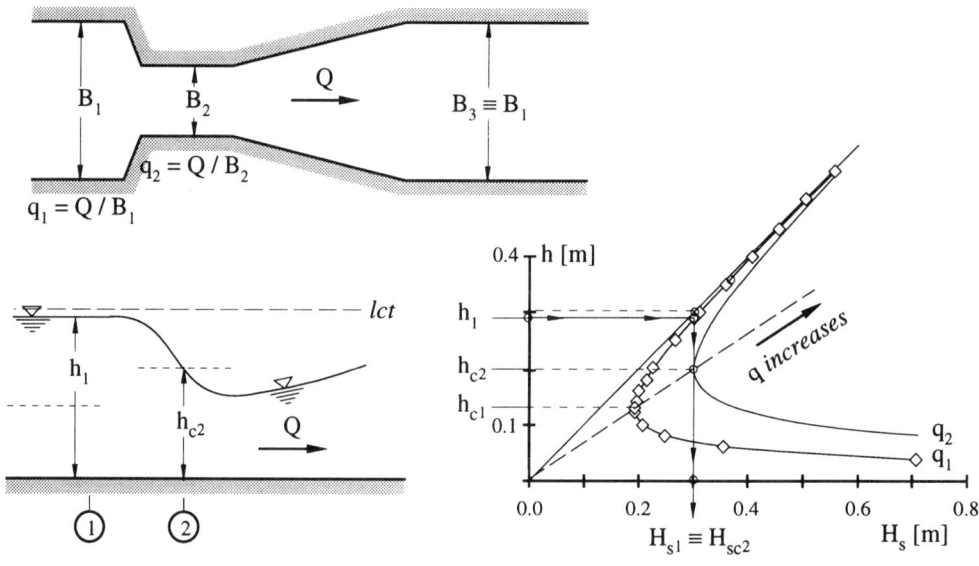

NON-UNIFORM FLOW 241

> Ex. 4.F
>
> A rectangular channel with a weak bed slope and a width of $B_1 = 5.0$ [m] carries a high velocity flow of $Q = 6$ [m³/s] at a flow depth of $h_1 = 0.2$ [m]. The width of this channel should be reduced to $B_3 = 2.5$ [m] by means of a linear symmetrical transition. Assuming that the flow depth downstream of the contraction, h_3, will be twice the one at the upstream, $h_3 = 2 h_1$, design a transition which will avoid any perturbations at the downstream water surface. What will be the length of this transition ?

SOLUTION :

The channel width and the flow depth at a section located at the upstream of the contraction is known :
$$B_1 = 5.0 \text{ [m]} \quad , \quad h_1 = 0.2 \text{ [m]} .$$

For this section, one can readily calculate the following :

- wetted surface : $\quad A_1 = h_1 B_1 = 0.2 \times 5.0 = 1.0 \text{ [m}^2\text{]}$

- average flow velocity : $\quad U_1 = \dfrac{Q}{A_1} = \dfrac{6.0}{1.0} = 6.0 \text{ [m/s]}$

- Froude number : $\quad Fr_1 = \dfrac{U_1}{\sqrt{g\,h_1}} = \dfrac{6.0}{\sqrt{9.81 \times 0.2}} = 4.28 \text{ [-]} > 1$

The flow regime is thus *supercritical*. The reduction of the channel width should be done in a way that no cross waves are generated downstream of the symmetrical transition (see point 4.5.3, 5 °).

The flow depth at section 3, h_3, should be twice the one at section 1, h_1, namely $(h_3 / h_1) = 2$. As it was already mentioned (see point 4.5.3, 5 °, *ii*), when designing a contraction for a supercritical flow care should be taken to avoid a change in the flow regime; the Froude number at the contracted section, Fr_3, should therefore remain superior to 1.

The Froude number at section 3 can be calculated using eq. 4.57 :
$$\frac{B_3}{B_1} = \left(\frac{h_1}{h_3}\right)^{3/2} \left(\frac{Fr_1}{Fr_3}\right)$$

which yields : $Fr_3 = Fr_1 \left(\dfrac{h_3}{h_1}\right)^{-3/2} \left(\dfrac{B_1}{B_3}\right) = 4.28 \; (2)^{-3/2} \; \dfrac{5.0}{2.5} = 3.03$ [-]

It can be seen that at section 3 the flow does not change the regime and remains *supercritical*. The proposed contraction is therefore possible.

The determination of the angle of deflection, θ, of the vertical side walls can now be started. As shown in Fig. 4.26 (see also the figure at the end of this problem), to go from section 1 at the upstream to section 3 at the downstream, the flow has to go through a transition zone.

The determination of the deflection angle, θ, of the side walls can only be done using a trial-and-error procedure :

step 1   Select *arbitrarily* an angle of deflection; for example $\theta = 15°$.

step 2   Determine the angle of the positive waves, β_1, generated at the points A and A' at the beginning of the transition, using the diagram in Fig. 4.24, which is redrawn below.

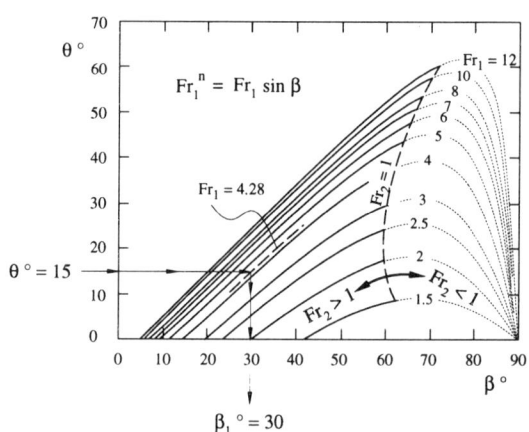

Entering the diagram with $\theta = 15°$ and $Fr_1 = 4.28$, one reads that the angle for the positive wave is $\beta_1 \cong 30°$.

step 3   The ratio of the flow depths of the zones 2 and 1 can be determined using eq. 4.55, or :

$$\frac{h_2}{h_1} = \frac{tg\ \beta_1}{tg\ (\beta_1 - \theta)} = \frac{tg\ 30}{tg\ (30 - 15)} = 2.15$$

Thus the flow depth at section 2 is : $h_2 = 2.15\ h_1 = 2.15 \times 0.2 = 0.43$ [m]

It is important to note that the velocity, U_2, in the zone 2 cannot be calculated using the continuity equation, $U_2 = Q / A_2$. There are two reasons for that. One reason is that the width in the transition zone is variable. The second reason is that the flow in the transition zone is not unidirectional (see the streamlines drawn in Fig. 4.26). Combining eqs. 4.53 and 4.55, one can write :

$$\frac{h_2}{h_1} = \frac{U_1^n}{U_2^n} = \frac{U_1 \sin \beta}{U_2 \sin (\beta - \theta)}$$

from which the following is obtained :

$$U_2 = U_1 \frac{\sin \beta}{\sin (\beta - \theta)} \left(\frac{h_2}{h_1}\right)^{-1} = 6.0 \frac{\sin 30}{\sin (30 - 15)} (2.15)^{-1} = 5.38\ [m/s]$$

which yields : $Fr_2 = \dfrac{U_2}{\sqrt{g\ h_2}} = \dfrac{5.38}{\sqrt{9.81 \times 0.43}} = 2.62$ [-]

NON-UNIFORM FLOW

step 4 Now, the passage from zones 2 to 3 will be studied, using again Fig. 4.24, re-drawn below, in order to determine the angle β_2.

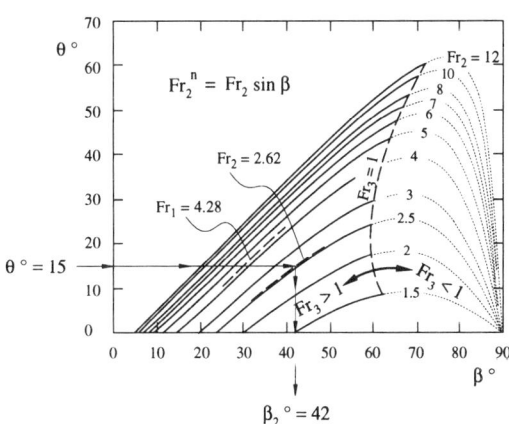

Entering the diagram with $\theta = 15°$ and $Fr_2 = 2.62$, the angle of the positive wave can be read as $\beta_2 \cong 42°$.

step 5 The ratio of the flow depths between the zones 3 and 2 can be determined, using once more eq. 4.55, or :

$$\frac{h_3}{h_2} = \frac{tg\ \beta_2}{tg\ (\beta_2 - \theta)} = \frac{tg\ 42}{tg\ (42 - 15)} = 1.77$$

step 6 The ratio of the flow depths between zones 2 and 1, (h_2 / h_1), and the one between the zones 3 and 2, (h_3 / h_2), being both known, one can now calculate the ratio of the flow depths between zones 3 and 1, or :

$$\frac{h_3}{h_1} = \frac{h_2}{h_1}\frac{h_3}{h_2} = 2.15 \times 1.77 = 3.81 > 2 = \left.\frac{h_3}{h_1}\right)_{imposed}$$

This value is larger than the value imposed, $(h_3 / h_1) = 2$, in the problem statement. The selected angle of deflection for the side walls, $\theta = 15°$, is therefore not appropriate. One must go back to *step 1*, select a smaller angle of deflection and repeat the calculations in a way until the ratio of the flow depths, $(h_3 / h_1) = 2$, imposed by the problem statement is achieved.

The computations with different deflection angles, including the first choice of $\theta = 15°$, are presented below in a tabular form.

The trial-and-error computation yields that the imposed ratio of the flow depths is achieved, if the angle of deflection is $\theta = 6°$. In this case, the positive waves generated at the entrance section 1 are at an angle of $\beta_1 = 19°$ with the parallel side walls of the channel upstream. Thes positive waves intersect each other at point B and continue, arriving at an angle of $\beta_2 = 23°$ at point C.

Computation of a linear channel contraction for supercritical flow										
problem data										
$Q = 6 \, [\text{m}^3/\text{s}]$,		$B_1 = 5.0 \, [\text{m}]$,		$h_1 = 0.2 \, [\text{m}]$,			$B_3 = 2.5 \, [\text{m}]$,			$h_3 / h_1 = 2$
hydraulic characteristics of section 1										
$A_1 = h_1 B_1 = 1.0 \, [\text{m}^2]$				$U_1 = Q / A_1 = 6.0 \, [\text{m/s}]$				$Fr_1 = 4.28 \, [\text{-}]$		
hydraulic characteristics of section 3 (admitting that $h_3 / h_1 = 2$)										
$h_3 = 2 h_1 = 0.40 \, [\text{m}]$,			$A_3 = h_3 B_3 = 1 \, [\text{m}^2]$,				$U_3 = Q / A_3 = 6.0 \, [\text{m/s}]$			
$Fr_3 = Fr_1 \, (h_3 / h_1)^{-3/2} \, (B_1 / B_3) = 3.03 \, [\text{-}]$										
step 1	step 2	step 3					step 4	step 5		step 6
θ	β_1	h_2 / h_1	h_2	U_2	Fr_2	β_2	h_3 / h_2	h_3 / h_1		Remarks
[°]	[°]	[-]	[m]	[m/s]	[-]	[°]	[-]	[-]		
15	30	2.15	0.43	5.38	2.62	42	1.77	3.81	no	$h_3 / h_1 > 2$
10	23	1.84	0.37	5.67	2.98	30	1.59	2.92	no	$h_3 / h_1 > 2$
7	20	1.58	0.32	5.79	3.29	28	1.39	2.18	no	$h_3 / h_1 > 2$
5	17	1.44	0.29	5.87	3.49	22	1.32	1.90	no	$h_3 / h_1 < 2$
6	19	1.49	0.30	5.82	3.40	23	1.39	2.07	yes	$h_3 / h_1 \cong 2$ √

The length of the transition where the channel's side walls are converging is computed (see point 4.5.3, 5°) using eq. 4.58 as follows :

$$L_T = \frac{(B_1 - B_3)}{2 \, \text{tg} \theta} = \frac{(5.0 - 2.5)}{2 \, \text{tg} \, 6} = 11.90 \, [\text{m}]$$

The linear symmetrical contraction obtained with above computations is illustrated below.

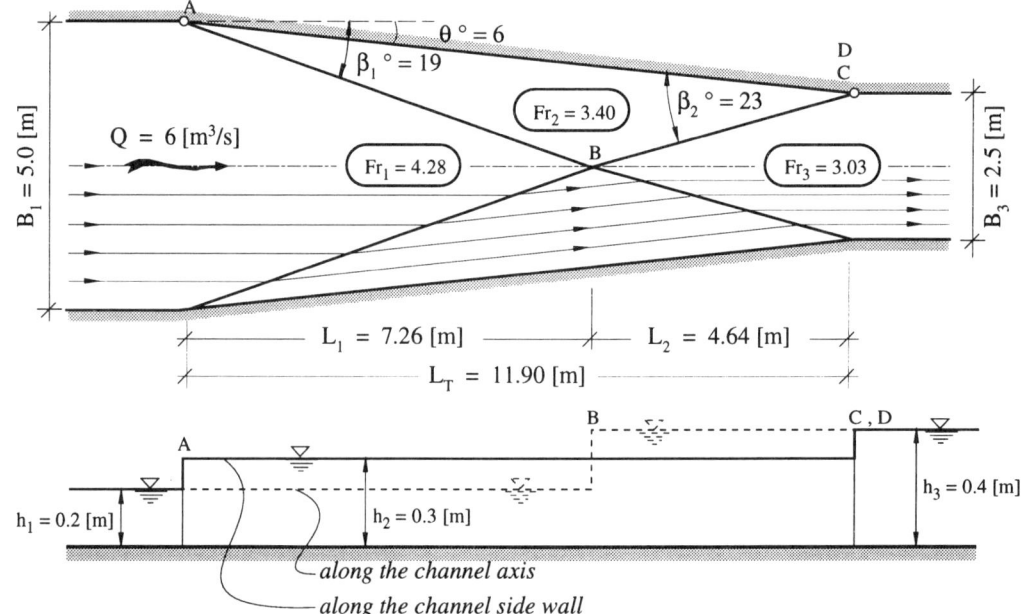

4.7.2 Problems, unsolved

Ex. 4.1
In a rectangular canal, having a width of $B = 2$ [m], a discharge of $Q = 4480$ [l/s] is conveyed. The bed slope is $S_f = 0.002$ [-] and the coefficient of friction was estimated as being $n = 0.012$ [m$^{-1/3}$ s].
Caused by an hydraulic structure, the flow becomes a gradually varied one; at a certain cross section, a flow depth of $h_1 = 1.00$ [m] was measured. Compute (in three steps) at what upstream distance the flow depth will reach $h_2 = 1.10$ [m]. What *water-surface profile* does one expect ?

Ex. 4.2
The following measurements in a rectangular laboratory canal have been performed :
$S_f = 0.0027$ [-] , $K_s = 45$ [m$^{1/3}$/s] , $b = 0.3$ [m] ; $Q = 0.02$ [m^3/s] , $h_b = 0.077$ [m].
Starting with the measured depth of h_b at the end of this canal (control section), compute the water-surface profile towards the upstream till the normal flow depth is reached. Use the method of reaches as well as the method of depths variation.

Ex. 4.3
A canal of a rectangular section has a width of $B = 1.8$ [m]; the discharge is $Q = 1.9$ [m^3/s]. The constant bed slope is $S_f = 0.0004$ [-] and the Manning coefficient is estimated as being $n = 0.013$ [m$^{-1/3}$ s^1]. At a certain section, the flow depth is $h_1 = 1.0$ [m]. Determine at what distance a flow depth of $h_2 = 0.85$ [m] will be attained. Use the method of successive approximations and control these results by using a method of direct integration.

Ex. 4.4
At a particular section in a rectangular channel, being 15.2 [m] wide, a maximum discharge of 110 [m^3/s] can pass. There the bed is at an elevation of 717.50 [m] and the water level is at an elevation of 721.50 [m] (see figure below). The longitudinal bed slope is $S_f = 0.001$ [-] and the Manning coefficient is $n = 0.025$ [m$^{-1/3}$ s]. Upstream at a distance of $L = 1.09$ [km], a bridge is to be constructed; the free board of 0.5 [m] should be respected. Determine the lowest point of the bridge. (For the computation of the water-surface profile, use the method of depths variation, taking $\Delta h = 0.1$ [m] .) Also calculate the slope of the energy-grade line using the Manning coefficient.

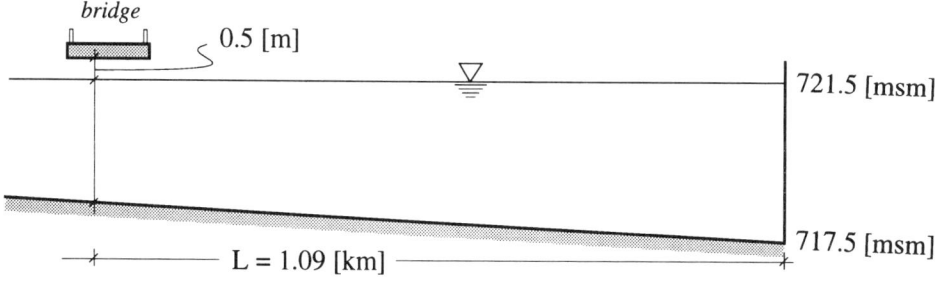

Ex. 4.5
A water discharge of $Q = 25.5$ [m³/s] transits in a rectangular channel whose width is $B = 12.2$ [m] ; the bed slope was measured as being $S_f = 0.00283$ [-] . Flow is non-uniform between two sections, separated by 91.4 [m], whose flow depths were measured as 1.37 [m] upstream and 1.52 [m] downstream. Determine the Manning coefficient for this reach of the channel.

Ex. 4.6
A reservoir is emptied by a bottom sluice gate ($K_v = 0.8$) whose opening is 0.40 [m]. The evacuated water enters into a canal 1, whose bed slope is $S_f = 0.01$ [-] and has a Manning coefficient of $n = 0.02$ [m$^{-1/3}$ s^1]. The unit discharge in the canal is $q = 1.8$ [m²/s]. This canal 1 is followed by another canal 2, where the normal flow depth is $h_n = 1.0$ [m].

i) Determine the normal flow depth in canal 1, assuming that the width of the canal is very large with respect to the flow depth.
ii) Determine the flow regime in both canals.
iii) Compute (in two steps) the distance between the exit flow at the gate and the section, where uniform flow is reached.
iv) Knowing that the both canals are sufficiently long to reach normal flow, sketch the water-surface profiles (without making calculations).

Ex. 4.7
A canal of rectangular section is built to evacuate a unit discharge of $q = 5$ [m²/s]. The bottom slope of this canal shows a sudden change in slope, going from 1% to 0.04 %. The entire canal has a Manning coefficient of $n = 0.015$ [m$^{-1/3}$ s^1].

i) Compute the normal flow depths, occurring upstream and downstream of the change in slope (taking $R_h \cong h$). Determine the flow regimes. Subsequently, compute the water-surface profile between these normal flow depths.
ii) Downstream of the change in slope, a bridge is to be built, whose piers will reduce the width of the section. Compute the maximum contraction possible, so that the flow regime will not be modified.

Ex. 4.8
The river *Merapi* can be approximated by a channel of large width. The normal and critical flow depth have been determined as being $h_n = 1.30$ [m] and $h_c = 0.55$ [m]. The hydraulic roughness is given by the Manning coefficient of $n = 0.025$ [m$^{-1/3}$ s^1]. At a particular section a small dam is built, which increases the water level by 2 [m] ; consequently, the total water height at the obstruction will be $H = 1.30 + 2.00 = 3.30$ [m]. Compute and sketch the water-surface profile using the method of Bresse.

Ex. 4.9
On the river, described in Ex. 4.8, a dam of a height of $H_D = 5.50$ [m] is built. The spillway has an effective width of $L_D = 10$ [m] and a discharge coefficient of $K_D = 0.95$. Compute the water depth at a distance of 50 [m] and 150 [m], upstream from the dam.

NON-UNIFORM FLOW   247

Ex. 4.10
A river, having a bed slope of $S_f = 0.25‰$, will be assimilated by a trapezoidal channel; its bottom width is b = 10 [m], having side slopes of m = 2. For a discharge of Q = 30 [m³/s], the normal flow depth was calculated as being h_n = 2.30 [m]. It is foreseen to install on this river a small dam, creating a flow depth of h = 3.50 [m]. Compute the resulting water-surface profile, using two methods of successive approximations. What will be the flow depth at 25 [m], upstream from the dam ?

Ex. 4.11
A rectangular channel of a width of B = 4.5 [m] having a bed slope and a depth of $S_f = 10^{-4}$ [-] and h_n = 1.65 [m], respectively, conveys a discharge in uniform flow. The roughness coefficient is n = 0.016 [$m^{-1/3}$ s^1]. This canal ends with an abrupt drop. Determine at what distance from the drop the flow depth will be 1.00 [m] ; use the method of direct integration.

Ex. 4.12
A bottom sluice gate, having a coefficient of K_v = 0.75, allows the flow of Q = 10 [m³/s] running into a trapezoidal channel, whose bottom width is b = 5.5 [m], having side slopes of m = 1. This channel has a bottom slope of S_f = 0.015 [-] and a friction coefficient of n = 0.015 [$m^{-1/3}$ s]. The flow depth, immediately after the gate, is h = 0.2 [m]. Compute and sketch the water-surface profile downstream of this sluice gate.

Ex. 4.13
In a rectangular channel made of wood, the width is b = 2.5 [m] and the bed slope is S_f = 0.01 [-]. A discharge of Q = 2.800 [l/s] should be conveyed. At a point, where the flow is retarded, a flow depth of h = 0.75 [m] is measured. Determine (in three steps) at what distance the depths of h = 0.65 [m] and of h = 0.78 [m] will take place.

Ex. 4.14
Taking the data of exercise, Ex. 4.B : it will now be assumed that the water level in the reservoir remains *constant*, and this in such a way that the canal – at the entrance into the reservoir – has a flow depth of h_2 = 1.50 [m]. The lake has a variable water level of ($h_1 \pm 0.5$) [m]. Determine the relation between the discharge, Q, in the canal and the water depth at the exit of the lake, h_1. It is advised to use the method of Bakhmeteff.

Ex. 4.15
A Venturi canal is conceived to measure the discharge, which is Q = 2.25 [m³/s]. What will be the largest width, B_2, of the throat, which produces critical flow ? Energy losses should not be taken in account. In the entrance section, a depth of h_1 = 1.50 [m] and a width of B_1 = 2.50 [m] were measured.

Ex. 4.16
The drainage ditch of a highway has a triangular form with side slopes of m = 4. During a flood, an hydraulic jump is formed, whose respective depths are measured as being

$h_1 = 25$ [cm] and $h_2 = 75$ [cm]. What is the discharge of this flow and what energy was dissipated in this jump ?

Ex. 4.17
A rectangular channel, being B = 6 [m] wide, transports a discharge of $Q = 11$ [m^3/s] at an average velocity of U = 6 [m/s]. The flow in this channel enters another channel having the same width, but its bed slope is horizontal. There is a possibility of the formation of an hydraulic jump; what energy dissipation may one expect ?

Ex. 4.18
A rectangular canal of a width of B = 1.0 [m] with a friction coefficient of $K_s = 100$ [m$^{1/3}$/s] changes its slope. The flow depth downstream and upstream are $h_1 = 0.2$ [m] and $h_2 = 0.4$ [m], respectively. The upstream channel slope is $S_f = 4\%$.

i) Will there be a change in the flow regime ?
ii) For what discharge, will there be a change in the flow regime ?

Ex. 4.19
A canal of rectangular cross section of width B = 4.9 [m] conveys a discharge of $Q = 5.4$ [m^3/s]. An hydraulic jump is formed, whose downstream depth was measured, being $h_2 = 1.3$ [m]. Determine the resulting head loss.

Ex. 4.20
A rectangular channel transports a discharge of $Q = 200$ [m^3/s]. The channel width changes from $B_1 = 60$ [m] to $B_2 = 30$ [m]. The bed slope is $S_f = 0.0016$ [-] and the friction coefficient is C = 30 [m$^{1/2}$/s]. Compute the resulting change in the flow depth, neglecting singular head losses.

Ex. 4.21
Water runs with a velocity of U = 3 [m/s] in a rectangular channel at a flow depth of h = 3.0 [m].

i) A raised step of a height of $\Delta z = 0.3$ [m] is foreseen ; how will it affect the flow depth ?
ii) What maximum step height can be envisioned without perturbing the upstream flow ? The head losses are to be neglected .

Ex. 4.22
A trapezoidal channel, whose bottom width is b = 6.1 [m] and side slopes are m = 0.5, conveys a discharge of $Q = 56,6$ [m^3/s] at a flow depth of 2.4 [m]. This channel becomes gradually a rectangular one, whose width remains at b = 6.1 [m] ; the channel bed is lowered by a step of $\Delta z = 0.6$ [m]. Find the flow depth in the rectangular channel as well as the variation in the water level in the reach of the two sections of the channel. What would be the minimum step, if the flow upstream remains undisturbed ?

NON-UNIFORM FLOW   249

Ex. 4.23
A rectangular channel, whose width is b = 5.0 [m], transports a discharge of
Q = 20.2 [m^3/s]. The corresponding normal flow depth is h = 2.42 [m].
i)   What are the flow regime and specific energy in this channel ?
ii) A step of a height of Δz = 1.0 [m] is put on the bed of the channel. Control if the
     upstream flow will be modified.
iii) If it is modified, compute the flow depth upstream and downstream of the step.

Ex. 4.24
The flow in a rectangular channel, having a width of b = 5.0 [m], a friction coefficient of
n = 0.02 [m$^{-1/3}$ s^1] and a bed slope of S_f = 0.04 [-], has a flow depth of h = 0.5 [m]. An
obstruction put downstream into the channel, causes a rise in the water level of 2.0 [m].
i)   Is an hydraulic jump to be expected ?
ii) What will be the conjugate depths ?
iii) Compute the distance between the obstruction and the hydraulic jump (in four
     steps) neglecting the length of the jump.

Ex. 4.25
A trapezoidal channel, whose bottom width is b = 6.0 [m] and the side slopes are
m = 2, conveys a discharge of Q = 150 [m^3/s] at a flow depth of h = 3.9 [m]. The
construction of a bridge with two piers will modify the flow. If each pier has a width of
s = 1.0 [m], what will be the flow depth underneath the bridge ? Determine also the
width of the piers for the case when the flow is critical underneath the bridge.

Ex. 4.26
Torrential flow in a wide channel is at a Froude number of Fr_1 = 2.5 ; the corresponding
flow depth of h_1 = 0.5 [m] was measured.
A bridge pier of a triangular front section produces standing waves under an angle
of β = 38°.
i)   Determine the flow depth downstream of this pier.
ii) What form (angle θ) should one envision for the triangular section of the pier ?

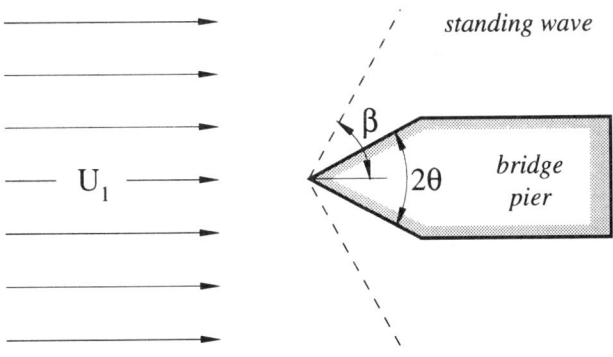

Ex. 4.27
Torrential flow with a Froude number of $Fr_1 = 3.5$ has a flow depth of $h_1 = 0.15$ [m]. This flow arrives at a vertical wall, positioned at an angle of $\theta = 10°$ against the flow. An oblique jump is formed. Determine the angle under which the jump forms as well as the downstream Froude number.

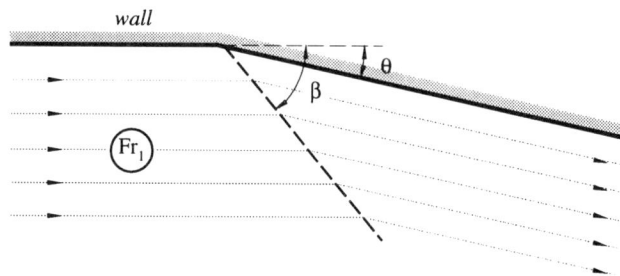

Ex. 4.28
Supercritical flow meets a wall with an angle of deflection of $\theta = 15°$; consequently, an oblique jump forms itself. The upstream and downstream flow depths are measured as being $h_1 = 0.6$ [m] and $h_2 = 1.5$ [m], respectively. What is the unit discharge in the canal ?

Ex. 4.29
Between two rectangular channels of widths $B_1 = 4.0$ [m] and $B_3 = 2.0$ [m], a reach of a symmetrical transition must be projected. In the approach channel, where a large Froude number, $Fr_1 = 5.0$, was calculated, the flow depth is $h_1 = 0.25$ [m]. What will be the length of the transition, for the case that cross waves should not be produced ? The condition is that the Froude number in the narrower channel, Fr_3 , will not go below a value of 1.5 .

5. UNSTEADY FLOW

Flow is unsteady, if the flow depth as well as other hydraulic parameters vary with time. An unsteady flow is usually also non-uniform.

In this chapter, the one-dimensional hydrodynamic equations, known as the equations of Saint-Venant, are developed. Subsequently, the different methods of solution are presented : the method of characteristics, the explicit and implicit method.

The simplified equations of Saint-Venant will be used to treat the kinematic wave as well as the diffusive wave. The flood wave will also be discussed. A sudden variation of the discharge, thus the flow depth, may be the cause of translatory waves.

TABLE OF CONTENTS

5.1 HYDRODYNAMIC EQUATIONS
    5.1.1 Equations of Saint-Venant
    5.1.2 Simplified Equations of Saint-Venant

5.2 METHODS OF SOLUTION
    5.2.1 The Characteristics
    5.2.2 Method of Characteristics
    5.2.3 Explicit Method
    5.2.4 Implicit Method

5.3 KINEMATIC WAVE
    5.3.1 Hydrodynamic Equations
    5.3.2 Celerity of Propagation

5.4 DIFFUSIVE WAVE
    5.4.1 Hydrodynamic Equations
    5.4.2 Attenuation

5.5 FLOOD WAVE

5.6 TRANSLATORY WAVES
    5.6.1 Notions and Examples
    5.6.2 Hydrodynamic Equations
    5.6.3 Celerity of Propagation
    5.6.4 Negative Waves

5.7 EXERCISES
    5.7.1 Problems, solved
    5.7.2 Problems, unsolved

5.1 HYDRODYNAMIC EQUATIONS

5.1.1 Equations of Saint-Venant

1° Considered (see Fig. 5.1) will be a channel with flow having a free-water surface, being one-dimensional, unsteady, non-uniform (gradually varied) and almost rectilinear. The bed slope, S_f, is fix and permanent, but also weak; the discharge of the incompressible fluid is given by $Q = UA$, with $U(x,t)$ as the velocity averaged over the cross section, $A(x,t)$. There is no lateral inflow or outflow.

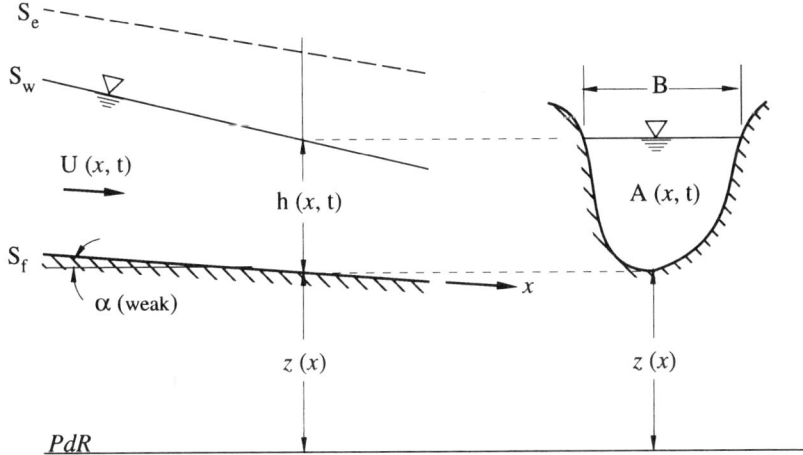

Fig. 5.1 Scheme of an unsteady and non-uniform flow over a slope of a fixed bed, $z(x)$.

2° The *equation of continuity*, developed in sect. 2.1, is given (see Fig. 2.1) by :

$$\frac{\partial Q}{\partial x} + \frac{\partial A}{\partial t} = A\frac{\partial U}{\partial x} + U\frac{\partial A}{\partial x} + B\frac{\partial h}{\partial t} = 0 \tag{5.1}$$

More correct, notably for a non-prismatic section, would be to write the second term as :

$$U\frac{\partial A}{\partial x} = U\left(\frac{\partial A}{\partial h}\frac{\partial h}{\partial x} + \frac{\partial A}{\partial x}\bigg|_{h = Cte}\right)$$

because $A = f[\,h(x,t), x\,]$; with $\partial A/\partial h = B$.

For a rectangular, prismatic channel, above equation, eq. 5.1, is given by :

$$h\frac{\partial U}{\partial x} + U\frac{\partial h}{\partial x} + \frac{\partial h}{\partial t} = 0 \qquad B = Cte \tag{5.2}$$

3° The *dynamic equation* for an unsteady and gradually varied flow, developed in sect. 2.2, is given (see Fig. 2.3) by :

$$\frac{1}{g}\frac{\partial U}{\partial t} + \frac{U}{g}\frac{\partial U}{\partial x} + \frac{\partial h}{\partial x} + \frac{\partial z}{\partial x} = -S_e \qquad (5.3)$$

where $S_f = -(\partial z/\partial x)$ is the bed slope, $h_r = S_e\,dx$ is the head loss and S_e is the energy slope.

As an approximation, it is assumed that the energy slope, S_e, can be expressed by a relation established for uniform, steady flow (see sect. 3.2), of the type of Weisbach-Darcy :

$$S_e = f\,\frac{1}{4R_h}\frac{U^2}{2g} \qquad (3.9)$$

or of the type of Chézy :

$$S_e = \frac{8g}{C^2}\frac{1}{4R_h}\frac{U^2}{2g} \qquad (3.10)$$

The dynamic equation, eq. 5.3, can also be written, after replacing U by Q/A, as :

$$\frac{1}{g}\frac{\partial Q}{\partial t} + \frac{1}{g}\frac{\partial}{\partial x}\left(\frac{Q^2}{A}\right) + A\frac{\partial h}{\partial x} + A\frac{\partial z}{\partial x} = -A\,S_e \qquad (5.3a)$$

4° The different terms in eq. 5.3 may be considered as representing different slopes. In the first and second term, the slope is due to a variation of velocity with time and in space. The third and fourth term give the slope of the water surface, S_w. The fifth term stands for the energy slope, S_e.

Each term in eq. 5.3 has its relative importance depending on the hydraulic situation. Some indicative values for the slope terms are given, expressed in order of magnitude, by *Cunge* et al. (1980, p. 45) for a flood on the Rhone river in France, such as :

$$\frac{1}{g}\frac{\partial U}{\partial t}\quad O(\sim 10^{-5}) \quad ; \quad \frac{U}{g}\frac{\partial U}{\partial x}\quad O(\sim 10^{-5})$$

$$S_f \quad O(\sim 10^{-3}) \quad ; \quad S_e \quad O(\sim 10^{-3})$$

This shows (see also *Henderson*, 1966, p. 364) that for a river with a weak bed slope, S_f, the two terms due to acceleration can readily be neglected. Consequently, one may write eq. 5.3 as being :

$$\frac{\partial h}{\partial x} + \frac{\partial z}{\partial x} = -S_e \quad \text{where} \quad -S_w = \frac{\partial h}{\partial x} + \frac{\partial z}{\partial x}$$

If the variation of the flow depth is weak compared to the bed slope, $\partial h/\partial x < \partial z/\partial x$ – this may be the case in waterways of more or less steep bed slopes – the dynamic equation, eq. 5.3, reduces to:

$$\frac{\partial z}{\partial x} = -S_e \quad \text{where} \quad -S_f = \frac{\partial z}{\partial x}$$

5° By using the form of the relation of Chézy, the dynamic equation, eq. 5.3, can be expressed as:

$$U = C\sqrt{R_h S_e} = C\sqrt{R_h \left(S_f - \frac{\partial h}{\partial x} - \frac{U}{g}\frac{\partial U}{\partial x} - \frac{1}{g}\frac{\partial U}{\partial t}\right)} \quad (5.4)$$

For a steady and uniform flow, one evidently writes:

$$U = C\sqrt{R_h S_f} \quad (3.11)$$

The relation between the velocity, U, or the discharge, Q, and the flow depth, h, or the hydraulic radius, R_h, (see Fig. 5.2a) is given by:

i)   a unique relation for eq. 3.11,

ii)   a non-unique relation with *a loop* for eq. 5.4, where the width of the loop is an indication of the importance of the terms of inertia and of pressure in eq. 5.4.

The relation, $Q = f(h)$, (see Fig. 5.2a), called the *gauging (or rating) curve of the section*, will provide the following information (see *Forchheimer*, 1930, p. 297 and *Flamant*, 1923):

i)   For unsteady flow, the discharge, Q, has two different values for the same flow depth, h, depending on the increase or decrease of the water level.

ii)   When the discharge in a section has its maximum value, $Q = Q_{max}$, thus $\partial Q/\partial t = 0$, the flow depth, h, is still on the increase, since it has as yet not reached its maximum value (see Fig. 5.2b).

iii)   Considering the relation of:

$$\frac{\partial Q}{\partial t} = (UB)\frac{\partial h}{\partial t} + (hB)\frac{\partial U}{\partial t} = 0$$

it is seen, that either the two derivatives annul themselves when $\partial Q/\partial t = 0$, or they take opposite signs.

If the flow depth, h, increases, the term ∂h/∂t is positive and the term ∂U/∂t is negative; consequently the velocity, U, decreases.

iv) Thus at a given section, one observes (see Fig. 5.2b) consequently :

the maximum of the average velocity, U_{max},　　then

the maximum of the discharge, Q_{max},　　then

the maximum of the flow depth, h_{max} ;

all of these curves are preceded by the maximum water-surface slope $S_{w\,max}$.

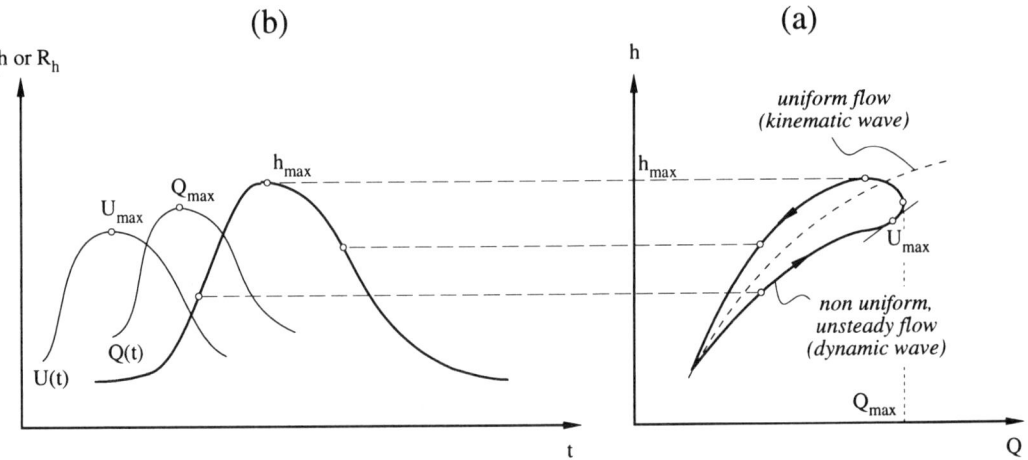

Fig. 5.2 Schematic representation of the relation of Q = *f*(h) and of h = *f*(t) for an unsteady and non-uniform flow.

6°　These equations, eq. 5.1 and eq. 5.3, established first by *Saint-Venant* (1870), render solutions to problems of unsteady flow, if integration is possible.

7°　The equations of Saint-Venant constitute a system of two equations of partial derivatives — of the hyperbolic type — in x and t, introducing two unknown functions, h and U. The initial and boundary conditions must be selected to describe adequately the physical problem.

8°　An exact integration of the equations of Saint-Venant is however very complicated; analytical solutions are rare. Nevertheless, there exist different numerical (and graphical) techniques making solutions (see sect. 5.2) possible.

UNSTEADY FLOW

5.1.2 Simplified Equations of Saint-Venant

1° It happens that a problem, where flow is unsteady and gradually varied, can be formulated in a way that one or more of the terms in the dynamic equation can be neglected.

2° Considering the different terms in the dynamic equation, eq. 5.3, one usually distinguishes the following types of unsteady flow, here called *waves* :

$$\underbrace{\underbrace{\underbrace{\underbrace{\frac{1}{g}\frac{\partial U}{\partial t} + \underbrace{\underbrace{\underbrace{\frac{U}{g}\frac{\partial U}{\partial x} + \underbrace{\frac{\partial h}{\partial x}}_{\text{kinematic wave}\ (5.5)}}_{\text{diffusive wave}\ (5.6)}}_{\text{dynamic, quasi-steady wave}\ (5.7)}}_{\text{dynamic wave}\ (5.3)} = S_f - S_e}_{\text{simple wave}\ (5.8)}} \qquad (5.3)$$

3° The equation of continuity, eq. 5.1, remains of course valid for all types of these waves.

4° Now it might be interesting to point out the difference between a kinematic wave, where many simplifications were done, and a dynamic wave, where no simplifications are necessary. An observer, who remains stationary on the side of the channel, will see between two sections, positioned at a distance of Δx from each other (see Fig. 5.3), the following :

- the bed slope, S_f , and the energy slope, $S_e \equiv S_w$, (see eq. 5.5) will be parallel for a kinematic wave,
- the bed slope, S_f , the piezometric slope, S_w , and the energy slope , S_e , (see eq. 5.3) are no more parallel to each other for a dynamic wave.

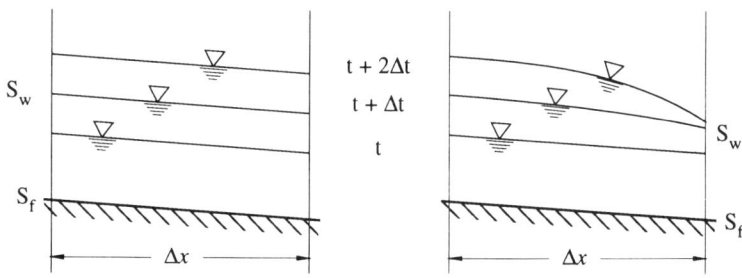

Fig. 5.3 Kinematic and dynamic wave, seen by a stationary observer.

5.2 METHODS OF SOLUTION

1° Numerical methods exist to solve the equations of Saint-Venant. The following ones will herewith be presented :

   i)   the method of characteristics (see sect. 5.2.2)
   ii) the explicit method (see sect. 5.2.3)
   iii) the implicit method (see sect. 5.2.4)

   Usually a *scheme of finite differences* is employed.

2° The numerical methods can well be dealt with the use of the computer.

3° These methods are presented in detail in the books of *Liggett* et *Cunge* (1975), *Cunge* et al. (1980) and *Abbott* et *Basco* (1989). Practical aspects dealing with fluvial hydraulics are treated by *Cunge* et al. (1980).

5.2.1 The Characteristics

1° For the equations of Saint-Venant, eq. 5.1 and eq. 5.3, one obtains rarely analytical solutions, excepting some rather simple cases. Therefore it becomes necessary to use other techniques, such as the method of characteristics.

2° For a rectangular channel, the equations of Saint-Venant read :

$$h\frac{\partial U}{\partial x} + U\frac{\partial h}{\partial x} + \frac{\partial h}{\partial t} = 0 \tag{5.2}$$

$$\frac{1}{g}\frac{\partial U}{\partial t} + \frac{U}{g}\frac{\partial U}{\partial x} + \frac{\partial h}{\partial x} = S_f - S_e \tag{5.3}$$

with :

$$\frac{\partial h}{\partial x} dx + \frac{\partial h}{\partial t} dt = dh \tag{5.9}$$

$$\frac{\partial U}{\partial x} dx + \frac{\partial U}{\partial t} dt = dU \tag{5.10}$$

where eq. 5.9 and eq. 5.10 are the total differentials of the flow depth, dh, and the average velocity, dU, respectively.

The above system of equations, eq. 5.2 to eq. 5.10, proposed by *Massau* (see *Chow*, 1959, p. 587) gives a total of four equations of partial derivatives with four unknowns : $\partial h/\partial x$, $\partial h/\partial t$, $\partial U/\partial x$ and $\partial U/\partial t$.

UNSTEADY FLOW

3° If one expresses the flow depth, h, by the wave celerity, c, (see sect. 2.4.1) one writes:

$$c^2 = gh \tag{2.27}$$

and consequently, upon differentiation:

$$d(c^2) = 2c\,dc = d(gh)$$

The celerity, c, becomes thus a measure of the flow depth, h.

4° The equations of Saint-Venant can (see *Stoker*, 1975, p. 470) thus be written in a system of the variables, U and c, as follows:

$$c\frac{\partial U}{\partial x} + 2U\frac{\partial c}{\partial x} + 2\frac{\partial c}{\partial t} = 0 \tag{5.11}$$

$$\frac{\partial U}{\partial t} + U\frac{\partial U}{\partial x} + 2c\frac{\partial c}{\partial x} = g(S_f - S_e) \tag{5.12}$$

5° Upon taking the sum and the difference of these two equations, one obtains an equivalent system of the form:

$$\left[\frac{\partial}{\partial t} + (U+c)\frac{\partial}{\partial x}\right](U+2c) = g(S_f - S_e)$$

$$\left[\frac{\partial}{\partial t} + (U-c)\frac{\partial}{\partial x}\right](U-2c) = g(S_f - S_e) \tag{5.13}$$

The arguments on the left side represent the total derivatives; consequently one may also write:

$$\frac{d}{dt}(U+2c) = g(S_f - S_e) \quad , \quad \frac{d}{dt}(U-2c) = g(S_f - S_e) \tag{5.14}$$

with $\quad (U+c) = \dfrac{dx}{dt} \quad , \quad (U-c) = \dfrac{dx}{dt} \tag{5.14a}$

The latter relation, eqs. 5.14a, is the absolute celerity, defined as:

$$\frac{dx}{dt} = c_w = (U \pm c) \tag{2.34}$$

which is the velocity with respect to the bed. The double sign indicates, that a propagation in the upstream and downstream direction is possible.

These four equations, eqs. 5.14 and eqs. 5.14a, form a system of equations of total derivatives for the characteristics; they replace the two equations of partial derivatives of Saint-Venant, eq. 5.1 and eq. 5.3.

6° If one (observer) follows the curves defined by $dx/dt = (U \pm c)$, the preceding equations – assuming that the channel is frictionless and is also horizontal, thus $(S_f - S_e) = 0$ – take the simple forms of :

$U + 2c = $ Cte   along a positive characteristic, C^+,   defined by $\dfrac{dx}{dt} = (U + c)$

$U - 2c = $ Cte   along a negative characteristic, C^-,   defined by $\dfrac{dx}{dt} = (U - c)$

The curves of the characteristics, C^+ and C^-, can be represented graphically on a plane, x and t (see Fig. 5.4).

5.2.2 Method of Characteristics

1° The four ordinary differential equations, eqs. 5.14, can also be written (see *Viessman et al.*, 1972, p. 194) as :

$$\left. \begin{array}{l} dU + \sqrt{g/h}\, dh + dt\left[g(S_e - S_f)\right] = 0 \\[4pt] dx = (U + \sqrt{gh})\, dt \end{array} \right\} C^+$$

$$\left. \begin{array}{l} dU - \sqrt{g/h}\, dh + dt\left[g(S_e - S_f)\right] = 0 \\[4pt] dx = (U - \sqrt{gh})\, dt \end{array} \right\} C^-$$

(5.15)

A solution to these equations, eqs. 5.15, can be obtained numerically; it will give a description of the unsteady motion of the flow in a channel, from one position (point) to another one.

2° The curves, which are given by $dx = (U \pm \sqrt{gh}) dt$, are the characteristic curves, being positive, C^+, and negative, C^-, emanating from common points, $L(x_L, t_L)$ and $R(x_R, t_R)$, where (see Fig. 5.4b) the velocities, U_L and U_R, and the flow depths, h_L and h_R, are known.

Thus, by solving the equations for U and h in different points, P, P_1, P_2, etc., which are common to the two characteristic curves throughout the plane, x and t, one obtains a description of the unsteady motion of the flow.

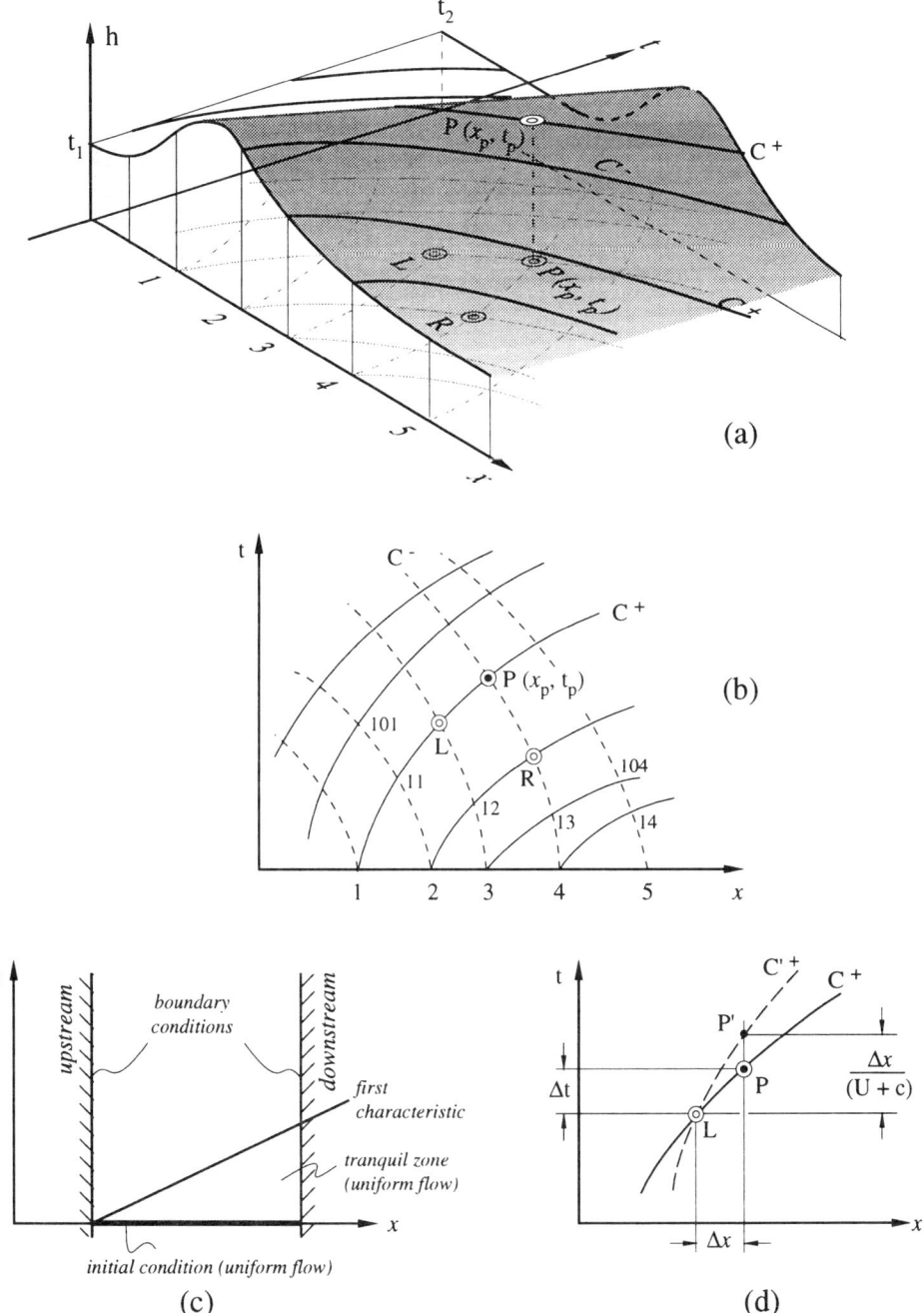

Fig. 5.4 Scheme of the characteristics (*subcritical* flow) :
    a) Profile of an unsteady flow at instants, t_1 and t_2.
    b) Surface around the point, $P(x_p, t_p)$, projected into a plane, x and t, (irregular grid).
    c) Boundary conditions.
    d) Condition of Courant.

3° Upon substitution of first-order finite differences for differentials (see *Viessman et al.*, 1972, p. 195), the above equations, eqs. 5.15, are written as (see Fig. 5.4b):

$$(U_P - U_L) + \sqrt{g/h_L}\,(h_P - h_L) + (t_P - t_L)\left[g(S_{e_L} - S_f)\right] = 0 \quad \Big\}\ C^+ \quad (5.16a)$$

$$(x_P - x_L) = (U_L + \sqrt{gh_L})(t_P - t_L) \quad (5.16b)$$

$$(U_P - U_R) - \sqrt{g/h_R}\,(h_P - h_R) + (t_P - t_R)\left[g(S_{e_R} - S_f)\right] = 0 \quad \Big\}\ C^- \quad (5.16c)$$

$$(x_P - x_R) = (U_R - \sqrt{gh_R})(t_P - t_R) \quad (5.16d)$$

There are four unknowns, namely: x_P, t_P and U_P, h_P.

4° One determines now the time step, t_P, by elimination of x_P between eq. 5.16b and eq. 5.16d:

$$t_P = \frac{(x_L - x_R) + t_R(U_R - \sqrt{gh_R}) - t_L(U_L + \sqrt{gh_L})}{(U_R - U_L) - (\sqrt{gh_L} + \sqrt{gh_R})} \quad (5.17)$$

5° One obtains then the distance, x_P, with eq. 5.16b:

$$x_P = x_L + (U_L + \sqrt{gh_L})(t_P - t_L) \quad (5.18)$$

The position of the point, $P(x_P, t_P)$, is now known (see Fig. 5.4b).

6° One obtains now the flow depth, h_P, at the point, P, by elimination of U_P between eq. 5.16a and eq. 5.16c:

$$h_P = \left\{(U_L - U_R) + \left(\sqrt{gh_L} + \sqrt{gh_R}\right) - (t_P - t_L)\left[g(S_{eL} - S_f)\right] + (t_P - t_R)\left[g(S_{eR} - S_f)\right]\right\}\left[\sqrt{g/h_L} + \sqrt{g/h_R}\right]^{-1} \quad (5.19)$$

7° One determines then the average velocity, U_P, for the corresponding flow depth at this point, P, with eq. 5.16a:

$$U_P = U_L - \sqrt{g/h_L}\,(h_P - h_L) - (t_P - t_L)\left[g(S_{eL} - S_f)\right] \quad (5.20)$$

Thus the velocity and the flow depth, U_P and h_P, at this point, $P(x_p, t_P)$, are also known.

8° This method allows the *explicit* calculation of the velocity, U_P, and the flow depth, h_P, for all points within the plane network, x and t.

For example, with a choice (initial condition) of a series of discrete points (1, 2, 3, etc.) on the time level, t = 0 (see Fig. 5.4b), it is possible to calculate another series of points (11, 12, 13, etc.) at another (advanced) time level. Another series of points (101, 104... also L and R) is subsequently calculated, using the previously known series of points (11, 12, 13, etc.). This type of calculation can readily repeat itself.

9° Before beginning with the calculations, it is evidently necessary to specify the *initial* and *boundary conditions* (see Fig. 5.4c).

The initial condition refers to the initial state of the flow, describing the depth, the average velocity or the discharge of the flow at all points in the channel at time t = 0. (Often is assumed that the initial flow is a uniform one, $h \equiv h_n$.)

The boundary conditions refer to the depth, the velocity or the discharge of the flow at the *upper* and *lower* ends of the reach at all times, t > 0, after the calculations begin. For subcritical flow, two boundary conditions are necessary, one at each end of the reach; for supercritical flow, two boundary conditions are necessary, both at the upper end, since downstream effects cannot propagate upstream. (An upstream boundary condition could be typically the specification of an hydrograph, h = $f(t)$ or U = $f(t)$, but also a relation of the type U = $f(h)$. A downstream boundary condition could be any type of control section, such as the critical depth, h_c, or the junction of two channels.)

The choice of the initial and boundary conditions will evidently define the physical problem under consideration (see *Cunge* et al., 1980, p.29, and *Abbott* et *Basco*, 1989, p. 208).

10° The method of characteristics can however not be used for problems, where discontinuities (for example : a hydraulic jump) in the flow are encountered. In the neighbourhood of such discontinuities the equations of Saint-Venant are no more valid; the curvature of the streamlines is very strong and the pressure distribution is no more hydrostatic (see sect. 2.2). Consequently, special methods must be employed (see *Cunge* et al., 1980, p. 37 and p. 57), when the flow presents itself with a discontinuity.

11° The method of characteristics is a method having an *explicit scheme*. This wants to say : if at a certain time level, $t = t_1$, all values of h and U are known, at another time level, $t = t_1 + \Delta t$, all values of h and U can be determined in a direct and explicit way with the equations of the characteristics, eqs. 5.14.

However, such an explicit scheme only remains numerically stable, if the following stability *condition of Courant* :

$$\Delta t \leq \frac{\Delta x}{(U + c)} \tag{5.21}$$

is satisfied (see Fig. 5.4d). Once Δx has been fixed, this condition can become very restrictive, since it imposes an upper limit on the choice of Δt.

12° Many applications of the method of characteristics can be found in the literature (see *Viessman* et al., 1972, p. 198-207, *Streeter*, 1971, p. 690 and *Abbott*, 1979).

For an in-depth development of this method, the literature (see *Abbott*, 1979, or *Liggett* et *Cunge*, 1975) should be consulted.

5.2.3 Explicit Method

1° The most direct method for the solution of the equations of Saint-Venant, eq. 5.1 and eq. 5.3, is the one which uses an *explicit finite-difference scheme* with fixed time intervals.

2° The schematic definition of a rectangular scheme is given in Fig. 5.5 (see also Fig. 5.4) : this is a network of nodes, including the point, P, projected on a plane, x and t ; Δx are the spatial distances between nodes and Δt are the time intervals (time steps) between nodes.

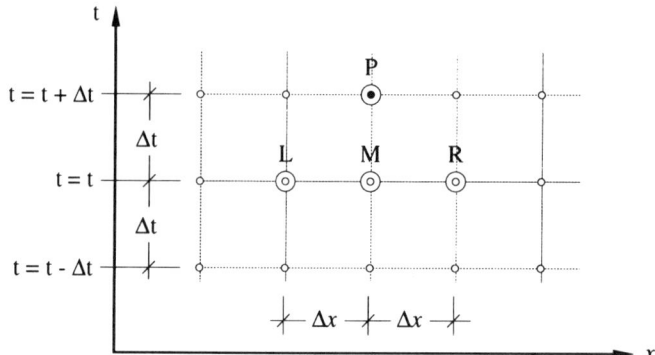

Fig. 5.5 Scheme of a finite-difference network, around the point, P, projected on a plane, x and t (see also Fig. 5.4a).

The known conditions, U_L, h_L and U_R, h_R, at the time level, $t = t$, are used to express explicitly the unknowns conditions, U_P, h_P, after a time step, Δt, namely at the time level, $t = t + \Delta t$.

3° The finite-difference approximations for the derivatives in the equations of Saint-Venant, eq. 5.2 and eq. 5.3, are given (see *Viessman* et al., 1972, p. 192) by :

$$\left.\frac{\partial U}{\partial x}\right|_M = \frac{U_R - U_L}{2\Delta x} \qquad \left.\frac{\partial h}{\partial x}\right|_M = \frac{h_R - h_L}{2\Delta x}$$

$$\left.\frac{\partial U}{\partial t}\right|_P = \frac{U_P - U_M}{\Delta t} \qquad \left.\frac{\partial h}{\partial t}\right|_P = \frac{h_P - h_M}{\Delta t}$$

(5.22)

4° For a rectangular channel, a substitution of the above relations, eqs. 5.22, into the equation of continuity, eq. 5.2, yields :

$$h_P = h_M + \frac{\Delta t}{2\Delta x}\left[U_M(h_L - h_R) + h_M(U_L - U_R)\right] \qquad (5.23)$$

UNSTEADY FLOW

5° Subsequently, a substitution of these relations, eqs. 5.22, into the equation of motion, eq. 5.3, yields:

$$\frac{U_P - U_M}{\Delta t} + U_M \frac{(U_R - U_L)}{2\Delta x} + g \frac{(h_R - h_L)}{2\Delta x} = g(S_f - S_e) \tag{5.24}$$

using the expressions of:

$$S_e = \frac{|U_P|U_P}{R_{h_P}^{4/3}} n^2 \quad \text{and} \quad \frac{n^2}{R_{h_P}^{4/3}} g \Delta t = \Gamma^{-1}$$

with n as the Manning coefficient (see sect. 3.2.3).

Simplifying and rearranging (see *Viessman* et al., 1972, p. 193) the above equation, eq. 5.24, renders now:

$$U_P^2 + \Gamma U_P - \Gamma \beta = 0 \tag{5.24a}$$

where $\beta = \left[U_M + \frac{\Delta t}{2\Delta x} U_M (U_L - U_R) + \frac{g\Delta t}{2\Delta x}(h_L - h_R) + g\Delta t\, S_f \right]$

This equation is quadratic in U_P, resulting in:

$$U_P = \frac{1}{2}\left[-\Gamma + (\Gamma^2 + 4\Gamma\beta)^{1/2}\right] \tag{5.25}$$

6° The explicit finite-difference method allows to find a value for the flow depth, h_P, with a fixed forward time step, Δt, using eq. 5.23, and subsequently a value for the velocity, U_P, at this point, P, using eq. 5.25.

7° The number and the kind of boundary conditions necessary for obtaining solutions have been described previously in point 5.2.2, 9°, (see also *Liggett* et *Cunge*, 1975, p. 135).

8° In order to obtain a stable solution it is always necessary, but often not sufficient, to respect everywhere the condition of:

$$\Delta t \leq \frac{\Delta x}{(|U| + c)} \tag{5.21}$$

where $\Delta t = (t_P - t_R)$ or $\Delta t = (t_P - t_L)$

and $\Delta x = (x_P - x_R)$ or $\Delta x = (x_P - x_L)$.

The relation given with eq. 5.21 is the stability condition of Courant; detailed discussions are available (see *Liggett* et *Cunge*, 1975 and *Cunge* et al., 1980, p. 77).

9° The above presented method is the one proposed by *Stoker* (1957). Other explicit methods are presented by *Liggett* et *Cunge*, 1975, pp. 100-141.

10° Some applications, using the different explicit methods, are given by *Viessman* et *al.* (1972, pp. 197 and 207).

5.2.4 Implicit Method

1° It should be remembered, that a numerical solution to the equations of Saint-Venant, eq. 5.2 and eq. 5.3, can be obtained in two ways:

  i) by putting the system of the equations of partial derivatives of Saint-Venant into a system of equations of ordinary derivatives, using the method of characteristics (see sect. 5.2.2) ;

  ii) by replacing the partial derivatives in the equations of Saint-Venant by quotients of finite differences, using either the *explicit* methods (see sect. 5.2.3) or the *implicit* methods.

2° An explicit scheme, having the advantage of simplicity, remains only stable as long as the condition of Courant, eq. 5.21, is satisfied. The condition of stability is however not a problem, if an implicit scheme is used.

3° There exist different implicit schemes (see *Liggett* et *Cunge*, 1975, pp. 142-172); herewith only the *scheme* of *Preissmann* (1961) will be presented.

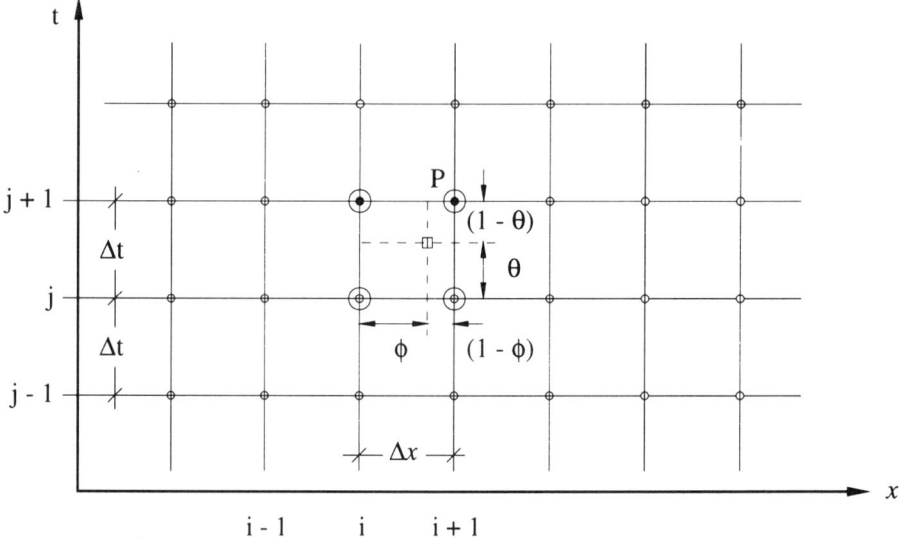

Fig. 5.6 Scheme of a finite difference network around the point, P , projected on a plane, x and t (see also Fig. 5.4a).

UNSTEADY FLOW 267

4° Take a continuous function, f, such as the flow velocity, U, or the flow depth, h. This function and its derivatives (see *Liggett* et *Cunge*, 1975, p. 142 and *Cunge* et *al.*, 1980, p. 65) will be approximated — by replacement of differential expressions with finite-difference expressions a truncation error will subsist — (see Fig. 5.6; see also Fig. 5.4a) as follows :

$$f(x,t) \cong \theta \left[\phi \, f_{i+1}^{j+1} + (1-\phi) \, f_{i}^{j+1} \right] + (1-\theta) \left[\phi \, f_{i+1}^{j} + (1-\phi) \, f_{i}^{j} \right]$$

$$\frac{\partial f}{\partial x} \cong \frac{1}{\Delta x} \left[\theta \, (f_{i+1}^{j+1} - f_{i}^{j+1}) + (1-\theta) \, (f_{i+1}^{j} - f_{i}^{j}) \right] \tag{5.26}$$

$$\frac{\partial f}{\partial t} \cong \frac{1}{\Delta t} \left[\phi \, (f_{i+1}^{j+1} - f_{i+1}^{j}) + (1-\phi) \, (f_{i}^{j+1} - f_{i}^{j}) \right]$$

Here, θ and ϕ are weighting coefficients for temporal and spatial distances ; they take values between 0 and 1.

For $\phi = 0.5$, these expressions, eqs. 5.26, give the classical scheme of *Preissmann*, resulting in :

i)   for $\theta = 0$:   a scheme, being completely explicit (see sect. 5.2.3),

ii)   for $\theta = 1$:   a scheme, being completely implicit,

iii)   for $\theta = 0.5$:   a scheme, being "four point" centre-implicit.

To acertain that the scheme remains numerically precise and stable, a value of $\theta \geq 0.66$ has been recommended (see *Liggett* et *Cunge*, 1975, p. 163 and *Krishnappan*, 1981, p. 7).

5° The equations of Saint-Venant, eq. 5.1 and eq. 5.3a, can also be written (see *Cunge et al.*, 1980, p. 16) in the following way :

$$\frac{\partial h}{\partial t} + \frac{1}{B} \frac{\partial Q}{\partial x} = 0 \tag{5.1a}$$

$$\frac{\partial Q}{\partial t} + \frac{\partial}{\partial x}\left(\frac{Q^2}{A}\right) + gA \frac{\partial h}{\partial x} + gA \frac{Q|Q|}{K^2} = 0 \tag{5.3b}$$

where K is the conveyance (see eq. 3.34) thus $S_f K^2 = Q|Q|$ and $h = (h + z)$ is the water level with respect to a fixed reference.

An application of the *Preissmann scheme* with $\phi = 0.5$, using the above expressions, eqs. 5.26, renders the partial derivatives in the above equations, eq. 5.1a and eq. 5.3b, as follows (see *Cunge et al.*, 1980, p. 94):

$$\frac{\partial h}{\partial t} \approx \frac{h_{i+1}^{j+1} - h_{i+1}^{j}}{2\Delta t} + \frac{h_{i}^{j+1} - h_{i}^{j}}{2\Delta t}$$

$$\frac{\partial Q}{\partial t} \approx \frac{Q_{i+1}^{j+1} - Q_{i+1}^{j}}{2\Delta t} + \frac{Q_{i}^{j+1} - Q_{i}^{j}}{2\Delta t}$$

(5.27a)

$$\frac{\partial h}{\partial x} \approx \theta \frac{h_{i+1}^{j+1} - h_{i}^{j+1}}{\Delta x} + (1-\theta)\frac{h_{i+1}^{j} - h_{i}^{j}}{\Delta x}$$

$$\frac{\partial Q}{\partial x} \approx \theta \frac{Q_{i+1}^{j+1} - Q_{i}^{j+1}}{\Delta x} + (1-\theta)\frac{Q_{i+1}^{j} - Q_{i}^{j}}{\Delta x}$$

(5.27b)

Here the spatial distance is given by $\Delta x = x_{i+1} - x_i$.

The 'coefficients', B and A, in above equations, eq. 5.1a and eq. 5.3b, are also expressed by one of the above expressions, eq. 5.26, $f(x, t) \approx \ldots$; for example, for the width, B, one writes:

$$B \approx \frac{\theta}{2}(B_{i+1}^{j+1} + B_{i}^{j+1}) + \frac{1-\theta}{2}(B_{i+1}^{j} + B_{i}^{j})$$

(5.27c)

6° With the expressions for the derivatives, eqs. 5.27a and eqs. 5.27b, and for the coefficients, eq. 5.27c, the equations of Saint-Venant, eq. 5.1a and eq. 5.3b (see *Cunge et al.*, 1980, p. 95), can be written in the following form:

$$\left[\frac{h_{i+1}^{j+1} - h_{i+1}^{j}}{2\Delta t} + \frac{h_{i}^{j+1} - h_{i}^{j}}{2\Delta t}\right] +$$

$$+ \frac{2}{\Delta x} \frac{\theta(Q_{i+1}^{j+1} - Q_{i}^{j+1}) + (1-\theta)(Q_{i+1}^{j} - Q_{i}^{j})}{\theta(B_{i+1}^{j+1} + B_{i}^{j+1}) + (1-\theta)(B_{i}^{j} + B_{i+1}^{j})} = 0$$

(5.28)

$$\left[\frac{Q_{i+1}^{j+1} - Q_{i+1}^{j}}{2\Delta t} + \frac{Q_{i}^{j+1} - Q_{i}^{j}}{2\Delta t} \right] +$$

$$+ \left\{ \frac{\theta}{\Delta x} \left[\left(\frac{Q^2}{A}\right)_{i+1}^{j+1} - \left(\frac{Q^2}{A}\right)_{i}^{j+1} \right] + \frac{1-\theta}{\Delta x} \left[\left(\frac{Q^2}{A}\right)_{i+1}^{j} - \left(\frac{Q^2}{A}\right)_{i}^{j} \right] \right\} +$$

$$+ g \left[\frac{\theta}{2} (A_{i+1}^{j+1} + A_{i}^{j+1}) + \frac{1-\theta}{2} (A_{i+1}^{j} + A_{i}^{j}) \right] \times$$

$$\times \left\{ \left[\frac{\theta}{\Delta x} (h_{i+1}^{j+1} - h_{i}^{j+1}) + \frac{1-\theta}{\Delta x} (h_{i+1}^{j} - h_{i}^{j}) \right] + \right.$$

$$+ \left[\frac{\theta}{2} \left(Q_{i+1}^{j+1} | Q_{i+1}^{j+1} | + Q_{i}^{j+1} | Q_{i}^{j+1} | \right) + \right.$$

$$+ \left(\frac{1-\theta}{2} \right) \left(Q_{i+1}^{j} | Q_{i+1}^{j} | + Q_{i}^{j} | Q_{i}^{j} | \right) \right] \times$$

$$\left. \times \left[\frac{\theta}{2} \left((K_{i+1}^{j+1})^2 + (K_{i}^{j+1})^2 \right) + \left(\frac{1-\theta}{2} \right) \left((K_{i+1}^{j})^2 + (K_{i}^{j})^2 \right) \right]^{-1} \right\} = 0 \quad (5.29)$$

Thus were obtained two non-linear algebraic equations, eq. 5.28 and eq. 5.29, in terms of h_i^j, Q_i^j and h_i^{j+1}, Q_i^{j+1}.

7° The approximate expressions for a continuous function, f, and its derivatives have been given with the above relations, eqs. 5.26. One can evaluate such a function at a time level (j+1), as follows :

$$f^{j+1} = f^j + \Delta f$$

Since superscripts are no more necessary, one takes now :

$$f^j \equiv f.$$

8° Using this expansion, the equation of continuity, eq. 5.28, for example, can be written as :

$$\left[\frac{\Delta h_{i+1} + \Delta h_i}{2\Delta t} \right] +$$

$$+ \left[\frac{2}{\Delta x} \frac{\theta(\Delta Q_{i+1} - \Delta Q_i) + (Q_{i+1} - Q_i)}{\theta(\Delta B_{i+1} + \Delta B_i) + (B_{i+1} + B_i)} \right] = 0 \tag{5.28a}$$

Subsequently (see *Liggett* et *Cunge*, 1980, p. 144 and p. 149) this equation is linearised — assuming that $\Delta f < f$ — and one obtains :

$$\mathcal{A}_i \Delta h_{i+1} + \mathcal{B}_i \Delta Q_{i+1} = \mathcal{C}_i \Delta h_i + \mathcal{D}_i \Delta Q_i + \mathcal{G}_i \tag{5.28b}$$

Performing the same for the equation of motion, eq. 5.29, one will obtain :

$$\mathcal{A}'_i \Delta h_{i+1} + \mathcal{B}'_i \Delta Q_{i+1} = \mathcal{C}'_i \Delta h_i + \mathcal{D}'_i \Delta Q_i + \mathcal{G}'_i \tag{5.29b}$$

The coefficients, $\mathcal{A}_i$, $\mathcal{A}'_i$, $\mathcal{B}_i$, $\mathcal{B}'_i$, etc., can be evaluated at a time step, t^j, — representing the initial conditions — where the values of h_i^j, Q_i^j and h_{i+1}^j, Q_{i+1}^j are known.

9° The two linear algebraic equations, eq. 5.28b and eq. 5.29b, can be used for a pair of adjacent points, (i, i + 1). This is however not yet sufficient to find the four unknown values, Δh_i, ΔQ_i, and Δh_{i+1}, ΔQ_{i+1}. With addition of two boundary conditions, upstream and/or downstream, the system can be solved.

10° The method of *double sweep* is often and efficiently used to obtain a solution (see *Liggett* et *Cunge*, 1975, p. 149, *Krishnappan*, 1981, p. 12 and *Abbott* et *Basco*, 1989, p. 117), notably for one-dimensional reaches.

5.3 KINEMATIC WAVE

The kinematic wave represents a special and simple case of unsteady flow.

5.3.1 Hydrodynamic Equations

1° The equations of Saint-Venant, eq. 5.1 and eq. 5.3, can be simplified, when considering the kinematic wave (see sect. 5.1.2).

UNSTEADY FLOW

2° The equation of continuity remains valid:

$$\frac{\partial A}{\partial t} + \frac{\partial Q}{\partial x} = B\frac{\partial h}{\partial t} + A\frac{\partial U}{\partial x} + U\frac{\partial A}{\partial x} = 0 \tag{5.1}$$

3° but the equation of motion can be considerably simplified:

$$S_f = S_e \tag{5.5}$$

This implies, that in the complete equation of motion, eq. 5.3, the terms due to the variations in velocity, U, and in depth, h, are negligible (when compared to the term due to the variation of the channel bed, $\partial z/\partial x = -S_f$).

4° This equation, eq. 5.5, can be written, using the relation of Chézy, as:

$$S_f = \frac{U^2}{C^2}\frac{1}{R_h} \quad \text{or} \quad U = C\sqrt{R_h S_f} \tag{3.11}$$

The discharge is thus given by:

$$Q = UA = CA\sqrt{R_h S_f}$$

implying that $Q = f(A)$ or $A = f(Q)$. Thus, it is evident that a single-valued function (see Fig. 5.2a) exists between the discharge, Q, and the wetted surface, A, in a cross section at a given abscissa, $x = x_0$.

The first term, $\partial A/\partial t$, in the above equation, eq. 5.1, can also be expressed as:

$$\frac{\partial A}{\partial t} = \left(\frac{\partial A}{\partial Q}\right)_{x_0} \frac{\partial Q}{\partial t}$$

5° Consequently the equation of continuity, eq. 5.1, can be written (see *Forchheimer*, 1930, p. 298) as follows:

$$\frac{\partial Q}{\partial t} + \left(\frac{\partial Q}{\partial A}\right)_{x_0} \frac{\partial Q}{\partial x} = \frac{dQ}{dt} = 0 \tag{5.30}$$

This is the equation of the *kinematic wave,* where the term:

$$\left(\frac{\partial Q}{\partial A}\right)_{x_0} = c_k \quad \text{or} \quad c_k = -\frac{\partial Q/\partial t}{\partial Q/\partial x} \tag{5.31}$$

is the *celerity of propagation* (or the wave speed) of the kinematic wave for a given discharge ; it will be different for each section of abscissa, $x = x_0$. According to eq. 5.30, the discharge, Q, is convectively displaced with the celerity, c_k.

An observer moving with the crest of the wave — thus with the celerity, c_k, — will notice no variation in discharge (see Fig. 5.7) ; the discharge will appear to be constant :

$$dQ = \frac{\partial Q}{\partial t} dt + \frac{\partial Q}{\partial x} dx = 0$$

This expression should be compared with eq. 5.30.

The equation of continuity, eq. 5.1, can be written as :

$$\frac{\partial A}{\partial t} = -\frac{\partial Q}{\partial x} \tag{5.1a}$$

This shows (see Fig. 2.1) that the temporal variation of the surface area, A, in a particular cross section of a channel is equal to the longitudinal variation of the discharge, Q.

The kinematic wave is a type of wave, whose properties follow essentially from the law of conservation of mass, namely the equation of continuity (see eq. 5.1a and eq. 5.30).

6° The above equation, eq. 5.30, can be transformed — using the single-valued functional relationship between Q and h — into an equation where the flow depth, h, is the dependent variable, namely :

$$\frac{\partial h}{\partial t} + c_k \frac{\partial h}{\partial x} = 0 \qquad \text{with} \qquad c_k = f(h) \tag{5.30a}$$

This is a pure convection (advection) equation ; the dependent variable, h, is convected in the longitudinal direction and in time with a celerity c_k.

7° Since the variation in flow depth, $\partial h/\partial x$, (see eq. 5.3 and eq. 5.5) is not accounted for, the kinematic wave will not be damped (attenuated) — the maximum flow depth remains constant over the distance travelled — but will be deformed, changing its curvature (see Fig. 5.8).

The geometric form of a kinematic wave, namely its wave profile — being of limited interest for the engineer — is given by *Henderson* (1966, p. 371).

8° For a numerical simulation of kinematic waves, the solution methods presented in sect. 5.2 can readily be used :

i) the explicit method, as given by *Jansen* et al. (1979, p. 263) and *Abbott* et *Basco* (1989, p. 58) ;

ii) the implicit method, as given by *Chow* et al. (1988, p. 294), *Jansen* et al. (1979, p. 265) and *Abbott* et *Basco* (1989, p. 86).

For the numerical simulation, one generally assumes (see *Dingman*, 1984, p. 268), that the kinematic wave, eq. 5.30, is a good approximation to the complete dynamic wave, eq. 5.3, if :

$$g L S_f / U^2 > 10$$

Here L represents the length of the channel under investigation and U is the average velocity of the uniform flow.

5.3.2 Celerity of Propagation

1°   Take a wide rectangular channel, with a constant and weak channel slope, S_f. Assume (see Fig. 5.7) that before the arrival of the wave, the flow is uniform, having a flow depth, h_1, and an average velocity, U_1 ; the discharge, Q_1, is initially constant.

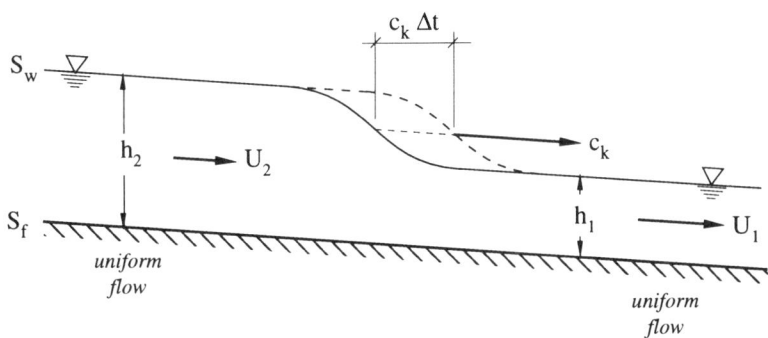

Fig. 5.7 Monocline wave (front of a kinematic wave).

2°   The flow increases suddenly to another discharge, Q_2 ; it stabilises itself at another uniform flow, h_2 and U_2. This step increase in discharge propagates downstream — after a certain time of its formation — and forms a translatory wave of stable (geometric) shape, called the monocline (rising) wave.

3°   During an interval of time, Δt , the wave displaces itself over a distance of $\Delta x = c_k \Delta t$. Applying the concept of the conservation of volume to a control volume moving with the celerity, c_k , yields :

$$c_k (Bh_2) - Q_2 = c_k (Bh_1) - Q_1 \tag{5.32}$$

which defines the celerity of propagation, being :

$$c_k = \frac{Q_2 - Q_1}{A_2 - A_1} = \frac{dQ}{dA} = \frac{1}{B}\frac{dQ}{dh} \tag{5.33}$$

This relation is to be compared with eq. 5.31. The celerity of propagation of a kinematic wave is proportional to the rate of change of discharge, Q, with the surface area, A, or the flow depth, h.

4° The above equation, eq. 5.32, can also be written as :

$$(c_k - U_2) A_2 = (c_k - U_1) A_1 \tag{5.32a}$$

This equation shows that :

i) in absence of an initial flow, $Q_1 = 0$, implying $U_1 = 0$ and $A_1 = 0$, one has :

$$c_k = U_2$$

ii) in presence of an initial flow, $Q_1 > 0$, one has :

$$c_k > U_1 \quad , \quad c_k > U_2 \tag{5.34}$$

The celerity of the kinematic wave, c_k, is thus always larger than the flow velocities, U_1 and U_2, upstream and downstream of the wave front.

5° The celerity of propagation, eq. 5.33, can also be expressed by :

$$c_k = \frac{d(UA)}{dA} = U + A\frac{\partial U}{\partial A} \tag{5.33a}$$

This relation confirms the above conclusions as given by the relations, eq. 5.34, namely $\partial U/\partial A$ is always positive.

According to the above relation, eq. 5.33, the celerity, $c_k(h)$, changes with the flow depth, h ; however, often this change is disregarded (see *Henderson*, 1966, p. 370) and $c_k(h) \equiv c_k$ is taken.

6° If the flow in a wide rectangular channel is turbulent, the average velocity can be obtained, using the relation of Chézy :

$$U = C h^{1/2} S_f^{1/2} \tag{3.11}$$

The above equation, eq. 5.33a, can thus been expressed (see *Forchheimer*, 1930, p. 298) by :

$$c_k = U + h\frac{\partial U}{\partial h} = U + h\frac{C S_f^{1/2}}{2h^{1/2}} = U + \frac{U}{2}$$

UNSTEADY FLOW

which results in :

$$c_k = \frac{3}{2} U \qquad (5.35)$$

For a parabolic or triangular channel, one gets respectively :

$$c_k = \frac{4}{3} U \qquad \text{or} \qquad c_k = \frac{5}{4} U \qquad (5.35a)$$

Should the same reasoning be done, but using the relation of Manning, eq. 3.16, one obtains for a wide rectangular channel :

$$c_k = \frac{5}{3} U \qquad (5.36)$$

The celerity of propagation of a kinematic wave is 3/2 or 5/3 times the flow velocity. This conclusion was arrived at using only two relations, namely the equation of continuity, eq. 5.1, and the momentum equation, eq. 5.5, expressed with a relation of friction.

7° A comparison of the properties of a kinematic wave and the ones of a gravity wave (see sect. 2.4.3) is certainly interesting ; the table below summarises the most important features :

	Gravity wave	*Kinematic wave*
Cause	all perturbations at water surface	local accumulation of water
Celerity	$c_w = U \pm \sqrt{gh}$	$c_k = (5/3) U$
Direction	downstream, upstream	downstream
Dissipation	considerable	not possible

8° If in a wide rectangular channel, a gravity wave propagates downstream, having a celerity of a kinematic wave, one puts :

$$c_k = c_w$$

which upon using the above equations, eq. 5.36 and eq. 2.34, becomes :

$$\frac{5}{3} U = U + \sqrt{gh}$$

resulting in (see eq. 2.27) the expression of :

$$c = \sqrt{gh} = \frac{2}{3} U$$

Using the definition of the Froude number, $U/\sqrt{gh}$, one obtains:

$Fr = \frac{3}{2}$ for $c_w = c_k$

$Fr > \frac{3}{2}$ for $c_w < c_k$

$Fr < \frac{3}{2}$ for $c_w > c_k$

Since in fluvial hydraulics, the Froude number is often $Fr < 3/2$, the kinematic (flood) wave will travel less fast than the gravity wave. (Note, if the relation of Chézy, eq. 5.35, is taken the limiting value of the Froude number would be $Fr = 2$).

5.4 DIFFUSIVE WAVE

The diffusive wave represents another special and simple case of unsteady flow.

5.4.1 Hydrodynamic Equations

1°   The equations de Saint-Venant, eq. 5.1 and eq. 5.3, can be simplified, when considering the diffusive wave (see sect. 5.1.2).

2°   While the equation of continuity remains valid:

$$\frac{\partial A}{\partial t} + \frac{\partial Q}{\partial x} = B\frac{\partial h}{\partial t} + A\frac{\partial U}{\partial x} + U\frac{\partial A}{\partial x} = 0 \tag{5.1}$$

3°   the equation of motion can be simplified:

$$S_f - \frac{\partial h}{\partial x} = S_e \tag{5.6}$$

This implies, that in the complete equation of motion, eq. 5.3, the terms of inertia are negligible (when compared to the term due to the variation of the water surface, $\partial h/\partial x + \partial z/\partial x = - S_w$).

This equation, eq. 5.6, can also be written (see eq. 5.4), using the relation of Chézy, eq. 3.11, as:

UNSTEADY FLOW

$$U = C \sqrt{R_h \left(S_f - \frac{\partial h}{\partial x}\right)} \tag{5.4a}$$

4° The diffusive wave undergoes attenuation (subsidence) due to a (possible) variation of the flow depth, $\partial h/\partial x$; this is not possible for the kinematic wave (see sect. 5.3.1) which cannot subside, but will deform, changing its curvature (see Fig. 5.8).

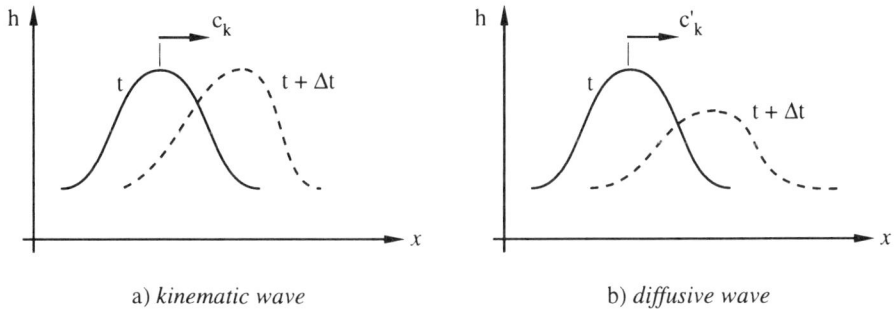

a) *kinematic wave* b) *diffusive wave*

Fig. 5.8 Attenuation of a wave; successif positions.

5° The dynamic equation, eq. 5.6, can also be written as :

$$\frac{\partial h}{\partial x} + \frac{\partial z}{\partial x} + \frac{Q|Q|}{K^2} = 0 \tag{5.37}$$

whereby the conveyance of the channel is taken as :

$$K(h) = \frac{1}{n} R_h^{2/3} A \tag{3.34}$$

which in turn gives $Q = K \sqrt{S_f}$ (see eq. 3.35).

Supposing the channel is rectangular of width $B = $ Cte :

i) differentiating the equation of continuity, eq. 5.1, with respect to x, yields :

$$\frac{\partial^2 h}{\partial x \partial t} + \frac{1}{B} \frac{\partial^2 Q}{\partial x^2} = 0$$

ii) differentiating the equation of motion, eq. 5.37, with respect to t, and assuming $\partial z/\partial t = 0$, yields :

$$\frac{\partial^2 h}{\partial x \partial t} + \frac{2|Q|}{K^2} \frac{\partial Q}{\partial t} - \frac{2|Q|Q}{K^3} \frac{\partial K}{\partial t} = 0$$

with :
$$\frac{\partial K}{\partial t} = \frac{\partial K}{\partial h}\frac{\partial h}{\partial t} = \frac{\partial K}{\partial h}\left(-\frac{1}{B}\frac{\partial Q}{\partial x}\right)$$

iii) eliminating now (see *Cunge et al.*, 1980, p. 45) the term $\partial^2 h/\partial x \partial t$ from the above two equations, renders a single equation :

$$\frac{\partial Q}{\partial t} + \left(\frac{Q}{BK}\frac{dK}{dh}\right)\frac{\partial Q}{\partial x} - \left(\frac{K^2}{2B|Q|}\right)\frac{\partial^2 Q}{\partial x^2} = 0 \qquad (5.38)$$

This is the equation of the *diffusive wave* (in analogy with the diffusion of heat). The partial differential equation is parabolic, having only one dependent variable, $Q(x,t)$. The discharge, Q, is convected with the celerity (see also eq. 5.33) of :

$$c_k' = \left(\frac{Q}{BK}\frac{dK}{dh}\right) = \left(\frac{1}{B}\frac{dQ}{dh}\right) \qquad (5.39)$$

and diffused (or attenuated) with a coefficient of :

$$C_D = \left(\frac{K^2}{2B|Q|}\right) = \frac{Q}{2BS_f} \qquad (5.40)$$

The equation of the diffusive wave, eq. 5.38, can now be written as :

$$\frac{\partial Q}{\partial t} + c_k'\frac{\partial Q}{\partial x} - C_D\frac{\partial^2 Q}{\partial x^2} = 0 \qquad (5.38a)$$

This equation, eq. 5.38, should be compared with the one of the kinematic wave, eq. 5.30. Using the flow depth, h, as the dependent variable (see *Dingman*, 1984, p. 265) the above equation, eq. 5.38a, can be put into the form of :

$$\frac{\partial h}{\partial t} + c_k'\frac{\partial h}{\partial x} - C_D\frac{\partial^2 h}{\partial x^2} = 0 \qquad (5.38b)$$

iv) assuming that the bottom slope remains constant, *Henderson* (1966, p. 383) gives the following relation (see also eq. 5.35 and eq. 5.4a) for the celerity, using the relation of Chézy, or :

$$c_k' = \frac{3}{2}U = \frac{3}{2}C\sqrt{h\left(S_f - \frac{\partial h}{\partial x}\right)} \qquad (5.39a)$$

and for the coefficient of diffusion :

$$C_D = \frac{c_k' h}{3\left[S_f - (\partial h/\partial x)\right]} \approx \frac{c_k' h}{3 S_f} \qquad (5.40a)$$

UNSTEADY FLOW 279

6° For a numerical simulation of diffusive waves, the solution methods presented in sect. 5.2 can readily be used :

   i)   the explicit method, as given by *Abbott* et *Basco* (1989, pp. 100 and 140) ;

   ii) the implicit method, as given by *Abbott* et *Basco* (1989, pp. 114 and 151) and *Jansen* et al. (1979, p. 269).

5.4.2 Attenuation

1° The coefficient of diffusion, C_D, expresses the rate of attenuation (or subsidence) of the wave. This can be shown by assuming that $\partial h/\partial x = 0$ at the *crest* of the wave (see Fig. 5.8). (Upstream of the crest, where $\partial h/\partial x > 0$, the flow depth increases, while downstream, where $\partial h/\partial x < 0$, the flow depth decreases.) One may now use the relation of h and write eq. 5.38b as follows :

$$\frac{\partial h}{\partial t} + 0 = C_D \frac{\partial^2 h}{\partial x^2} \tag{5.41}$$

The crest of the wave is obviously the maximum point in the relation between h and x ; this gives — using calculus — a curvature of $\partial^2 h/\partial x^2 < 0$, at this very point, being a *subtended* arc. Since the right-hand term in eq. 5.41 is now negative, it implies that flow depth, h , of the diffusive wave decreases with time (see *Dingman*, 1984, p. 265). Because the wave travels already downstream with the celerity, c_k', this expresses the downstream rate of decrease (or subsidence). Schematically this is shown in Fig. 5.8b.

2° For a determination of the value of attenuation, different approximate approaches have been advanced (see *Forchheimer*, 1930, p. 299, *Hayashi*, 1953 and *Henderson*, 1966, p. 377). A simple approximation was given by *Jansen* et al. (1979, p. 70), being :

$$h = h_o \exp(-k^2 C_D t) \sin\left[k(x - c_k' t)\right] \tag{5.42}$$

where h_o is the initial flow depth. The wave is assimilated by a sinusoidal function, exponentially decreasing and progressing downstream with the celerity, c_k'.

3° The diffusion term in the above equation, eq. 5.38, may be neglected if $\partial^2 Q/\partial x^2$ or/and $Q/2BS_f$ (see eq. 5.40) are small. The last condition implies that :

$$Q \ll 2BS_f \tag{5.43}$$

or, when using the Manning relation, eq. 3.16, for a wide rectangular channel :

$$h^{5/3} \ll 2n\, S_f^{1/2} \tag{5.43a}$$

This relation shows, that for bottom slopes, S_f, being :

i) sufficiently *steep*, the diffusion term can be eliminated ; the diffusive wave is thus well approximated by a kinematic wave (see eq. 5.30, as compared to eq. 5.38a) ;

ii) *gentle*, the diffusion term must remain ; the diffusive wave is maintained (see eq. 5.38a).

4° In steady state, the discharge, Q, can be calculated according to eq. 5.4a, by :

$$Q = C A \sqrt{R_h \left(S_f - \frac{\partial h}{\partial x} \right)} \qquad (5.44)$$

which gives a not-single-valued relation, with a loop (see Fig. 5.2a). The size of this loop is due to the diffusion coefficient, C_D, eq. 5.40, in the equation of the diffusive wave, eq. 5.38.

5.5 FLOOD WAVE

1° Flood waves in watercourses are phenomena of greatest importance for the hydraulician. A flood is a type of unsteady flow, whose theoretical study must be deduced from the complete equations of Saint-Venant ; methods of solution have been presented in sect. 5.2.

2° The displacement — rising and falling — of a flood is generally very *slow* ; thus certain terms in equation of motion, eq. 5.3, can be neglected. Consequently the equations of Saint-Venant can be simplified (see sect. 5.1.2). Usually one makes a distinction (see *Henderson*, 1966, pp. 364 - 374) between :

i) floods in channels on *steep slopes*, Fr < 1;

ii) floods in channels on *weak slopes*, Fr << 1.

A flood wave is thus considered as being a slowly variable flow.

3° Floods in channels on *steep* slopes can be approximated by *kinematic waves* (see sect. 5.3). The in-situ observations by Seddon (1900) as well as theoretical considerations by Kleitz (1858) have this confirmed ; it is known as the principle of *Kleitz-Seddon* (see *Chow*, 1950, p. 529).

One supposes that a flood rises gradually to a (unique) maximum, and subsequently descends till the initial uniform flow is reached again. Before the rise and after the fall of the flood, the flow is in permanent regime and has the same discharge in all cross sections. This wave, being a kinematic one (see sect. 5.3), is characterised by :

UNSTEADY FLOW

i) a relation of $Q = f(h)$ which is single-valued (see Fig. 5.2a) ;

ii) the celerity of propagation, given by :

$$c_k = \frac{3}{2} U = \frac{3}{2} C \sqrt{h\, S_f} \tag{5.35}$$

for a wide rectangular section, using the relation of Chézy ;

iii) a geometric form, being invariable and without attenuation.

For such type of a flood (see Fig. 5.9), the major part — the main body of the flood — is well described by a kinematic wave, eq. 5.32. In front and behind the main body of the flood, there travel gravity waves, eq. 2.34. According to *Henderson* (1966, p. 369), if Fr < 3/2 (or Γr < 2), these gravity waves are usually of little importance, since they subside rapidly (see also point 5.3.2, 8°).

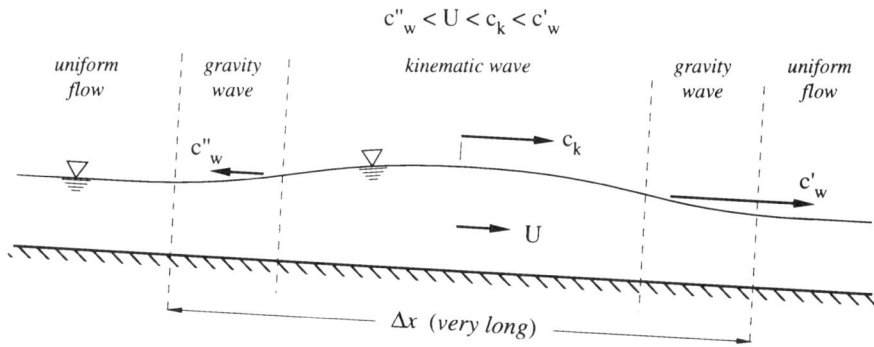

Fig. 5.9 Flood wave (Fr < 3/2).

4° Floods in channels on *weak* slopes can be approximated by *diffusive waves*, being characterised (see sect. 5.4) by :

i) a relation of $Q = f(h)$ which is not single-valued (see Fig. 5.2a) ;

ii) the celerity of propagation, given by :

$$c'_k = \frac{3}{2} C \sqrt{h \left(S_f - \frac{\partial h}{\partial x} \right)} \tag{5.39a}$$

where C is the Chézy coefficient ;

iii) a geometric form, being variable, thus an attenuation is present.

Various different aspects of such waves are presented by *Henderson* (1966, pp. 374-383).

5.6 TRANSLATORY WAVES

5.6.1 Notions and Examples

1° In *rapidly* varied unsteady flow a sudden change in the flow depth is encountered; a translatory wave, also referred as *surge*, is formed.

The curvature of the streamlines may not be neglected and the pressure distribution is no more hydrostatic. Consequently, the equations of Saint-Venant, eq. 5.1 and eq. 5.3, are no longer valid. Note, that up till now in this chapter, gradually varied unsteady flow was considered.

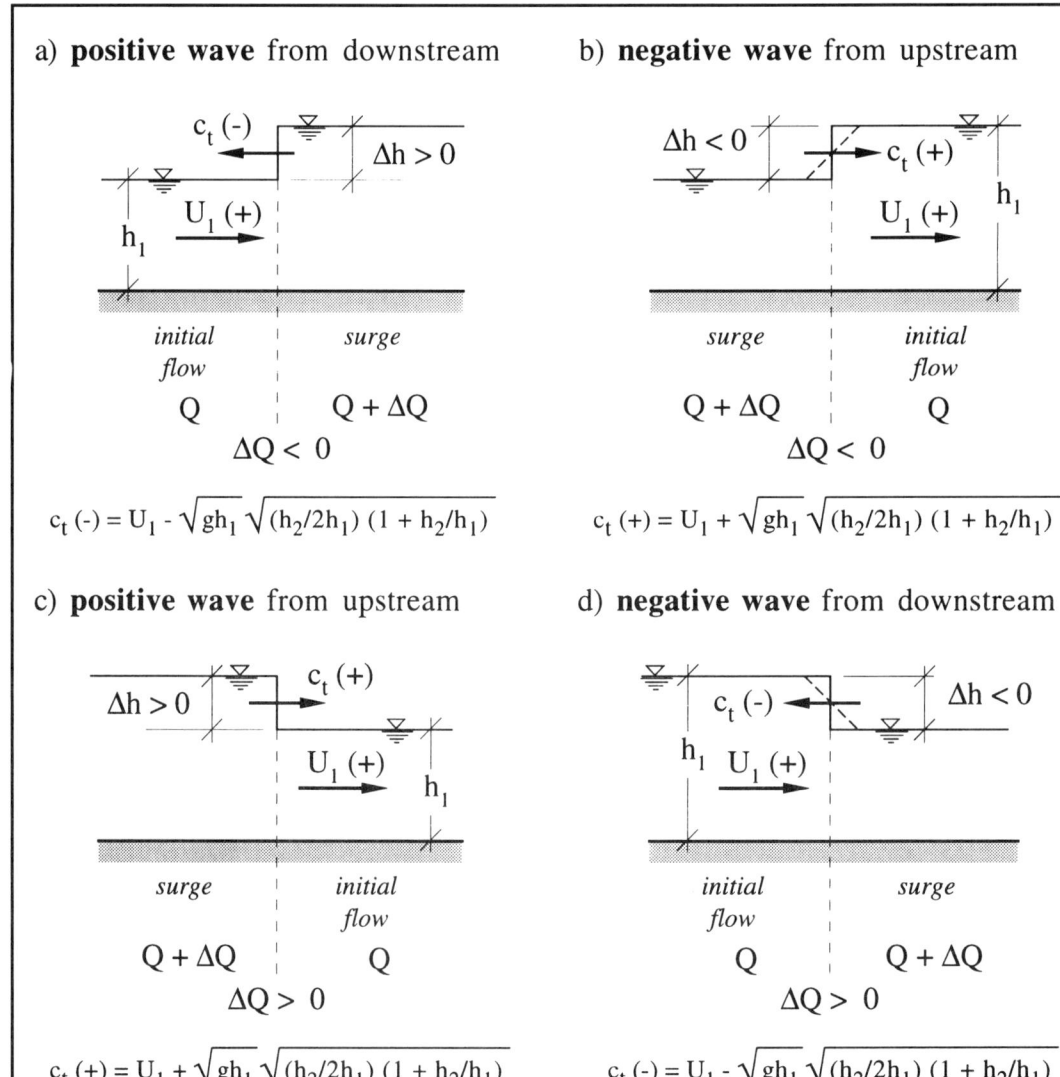

Fig. 5.10 Translatory waves (surges) due to a sudden variation in discharge, ΔQ.

UNSTEADY FLOW 283

2° Such a sudden variation on the water surface is usually caused by a sudden change in the discharge, ΔQ.

3° Due to this sudden variation (intumescences or perturbation) a *translatory wave*, also referred as surge or bore, is formed, consisting of a discontinuity, Δh, called the *front* (head) of the wave (see Figs. 5.10 and 5.12). After this discontinuity, whose length is often negligible, the *body* of the wave remains reasonably parallel to the water surface of the initial uniform flow, Q. Consequently, within the body of the wave the discharge is equal to Q + ΔQ.

4° There are (see *Favre*, 1935, pp. 33-44) four different types of translatory waves (see Fig. 5.10).

   i) The perturbation be caused by a variation in discharge, ΔQ, at an upstream section ; the front of the wave moves downstream :

   - for an increase in discharge, ΔQ > 0, the perturbation will be above the initial water level ; this is a *positive wave* coming *from upstream* (see Fig. 5.10c);
   - for a decrease in discharge, ΔQ < 0 , the perturbation will be below the initial water level ; this is a *negative wave* coming *from upstream* (see Fig. 5.10b).

   ii) The perturbation be now caused by a variation in discharge, ΔQ , at a downstream section ; the front of the wave moves upstream :

   - for an increase in discharge, ΔQ > 0 , the perturbation will be below the initial water level ; this is a *negative wave* coming *from downstream* (see Fig. 5.10d);
   - for a decrease in discharge, ΔQ < 0, the perturbation will be above the initial water level ; this is a *positive wave* coming *from downstream* (see Fig. 5.10a).

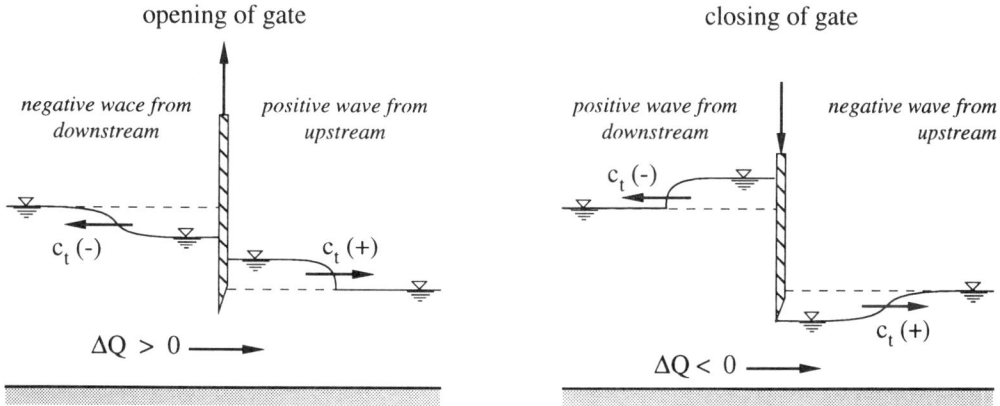

Fig. 5.11 Example of the four different types of translatory waves. (operation of a gate, situated in a channel)

5° The different types of translatory waves (see Fig. 5.10) are often encountered in hydraulic engineering. Some examples are herewith given :

i) During the control of discharge by a gate, one encounters (see Fig. 5.11) :
- by suddenly opening the gate, an increase in discharge; a negative wave from downstream forms itself in the upstream channel and a positive wave from upstream forms itself in the downstream channel.
- by suddenly closing the gate, a decrease in discharge; a positive wave from downstream forms itself in the upstream channel and a negative wave from upstream forms itself in the downstream channel.

ii) The situation is analogous, during the operation of a lock in a navigation channel :
- during the emptying of the lock (opening the downstream gate), when the water level in the lock's chamber is larger than the one in the downstream channel, a positive wave from upstream moves downstream (see Fig. 5.10c) ;
- during the filling of the lock (opening the upstream gate) a positive wave from the upstream approach channel will enter the lock's chamber (see Fig. 5.10c).

iii) During the operation of a hydro-electric (low pressure) power plant, sudden increases or decreases in power output will cause variations in the discharge and consequently considerable perturbations of the uniform flow in the upstream and downstream channels.

iv) Each large perturbation of the uniform flow, such as the arrival of a flood wave as well as earth slides or earthquakes causing obstruction of the flow, will be responsible of different types of translatory waves.

v) Still another phenomenon, where a translatory wave of large amplitude is encountered, is the tidal bore, caused by the oceanic tidal flow entering the mouth of a river estuary.

6° Above, it was assumed that the front of the wave represents a discontinuity at the water surface. In reality, this may present itself under different forms (see Fig. 5.10) :

i) if the wave is positive, $h_1 + \Delta h > h_1$, the front can be formed by one single wave or by a series of wavelets, being steep, breaking or not ; the front, being steep and rather concave, remains stable ;

ii) if the wave is negative, $h_1 + \Delta h < h_1$, the front has the form of a curve ; the front, being rather convex, becomes unstable (see sect. 5.6.4).

This can be explained as follows (see *Chow*, 1959, p. 556) : a wave can be considered as being a sum of wavelets, placed one on top of the other. Each wavelet displaces itself with its celerity of $c = \sqrt{gh}$ (see eq. 2.27) ; the wavelets on top have a greater celerity than the ones below. Consequently one observes :

UNSTEADY FLOW

i) for a positive wave, the wavelets on top will absorb (overtake) the ones below ; a single large front is formed, being rather steep (see Fig. 5.12);

ii) for a negative wave, the wavelets on top moving faster than the ones below ; the front becomes sloping and eventually flattens out (see Fig. 5.14).

5.6.2 Hydrodynamic Equations

1° Considered will be a *positive wave* from downstream (see Fig. 5.12a or Fig. 5.10a), which produces itself on an (initial) uniform flow having a velocity, U_1, and is due to a sudden variation in discharge, ΔQ. The front of the wave is rather steep. The flow velocity, U_2, following the front of the wave is accompanied by an increase in the flow depth, Δh. Further, c_t is the velocity of propagation of the front, called the *celerity of the translatory wave*.

2° Imagine now an observer (see Fig. 5.12b) moving with this celerity, c_t ; for this observer, this phenomenon appears to be stationary.

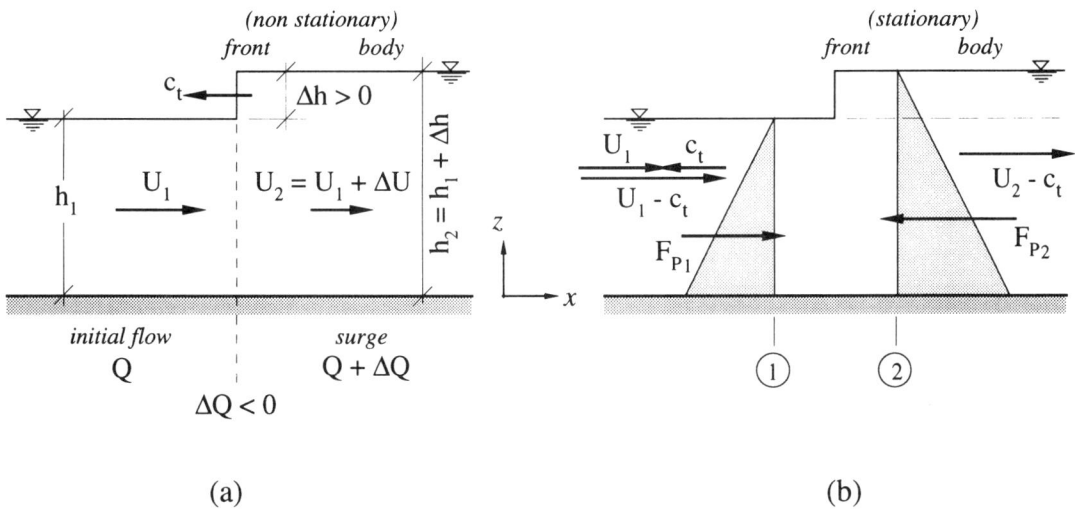

Fig. 5.12 Profile along a perturbation or surge :
positive wave from downstream ;
a) observer stationary ; b) observer not stationary, moving with a celerity, c_t.

3° The *continuity equation* (see Fig. 5.12b) yields :

$$(U_1 - c_t) A_1 = (U_2 - c_t) A_2 = (U_1 + \Delta U - c_t)(A_1 + \Delta A) \tag{5.45}$$

and for a rectangular channel, with B = Cte :

$$(U_1 - c_t) h_1 = (U_2 - c_t) h_2 \tag{5.45a}$$

The above equation, eq. 5.45, can also be written as :

$$(A_1 + \Delta A)(U_1 + \Delta U) - U_1 A_1 = c_t \Delta A$$

where the first member represents an expression of the variation of discharge, ΔQ. A simple, but general expression for the celerity can now be obtained :

$$c_t = \frac{Q_2 - Q_1}{(h_2 - h_1) B} = \frac{\Delta Q}{\Delta A} = \frac{\Delta q}{\Delta h} = \frac{d(Uh)}{dh} \tag{5.46}$$

4° The *momentum equation* (see Fig. 5.12b) yields :

$$F_P + F_G + F_F = \rho A_2 (U_2 - c_t)(U_2 - c_t) - \rho A_1 (U_1 - c_t)(U_1 - c_t) \tag{5.47}$$

Assuming, that the forces of friction, F_F, and of gravity, F_G, cancel and using the above relation, eq. 5.45, one obtains :

$$F_{P_1} - F_{P_2} = \rho A_1 (U_1 - c_t)(U_2 - c_t - U_1 + c_t) \tag{5.47a}$$

and written for a rectangular channel :

$$\rho g \frac{h_1^2}{2} - \rho g \frac{h_2^2}{2} = \rho h_1 (U_1 - c_t)(U_2 - U_1) \tag{5.48}$$

5° From the above equations, eq. 5.48 and eq. 5.45a, one gets :

$$(U_1 - c_t) = + \sqrt{gh_1} \sqrt{\frac{h_2}{2h_1}\left(1 + \frac{h_2}{h_1}\right)} \tag{5.49}$$

resulting in an expression for the celerity of the translatory wave — here it is a positive wave from downstream — or :

$$c_t = U_1 - \sqrt{gh_1} \sqrt{\frac{h_2}{2h_1}\left(1 + \frac{h_2}{h_1}\right)} \tag{5.50}$$

6° For a channel of rectangular cross section these relations, eqs. 5.50 and 5.46, allow computation of c_t and $\Delta h = (h_2 - h_1)$, if U_1, h_1 and Δq are known in advance.

7° For the other types of translatory waves, the same development can be done (see *Favre*, 1935, p. 43) ; the results, notably for the celerity of these waves, $c_t (\pm)$, are summarised in Fig. 5.10. The following should be noted :

UNSTEADY FLOW 287

i) for positive and negative waves from upstream (see Fig. 5.10c and b), eq. 5.50 keeps the same form, and one takes a positive sign (+) before the root ;

for positive and negative waves from downstream (see Fig. 5.10a and d), eq. 5.50 keeps the same form, and one takes a negative sign (-) before the root ;

ii) for negative waves (see Fig. 5.10b and d), the established relations, eqs. 5.46 and 5.50, keep the same form, but the arguments, Δh and ΔS, will carry negative signs, (-) ;

iii) in eq. 5.46, ΔQ and Δq will carry a positive sign, (+) , for positive waves from upstream and for negative waves from downstream and will carry a negative sign, (-) , for negative waves from upstream and for positive waves from downstream (see Fig. 5.10).

8° Throughout the above development, it was assumed that the perturbation superposes itself on a uniform (initial) flow. For the case when the initial flow is non-uniform (gradually varied), approximate solutions are available (see *Favre*, 1935, p. 45 and *Jaeger* , 1954, p. 369).

9° Numerical simulations of translatory waves have been advanced (see *Cunge* et al., 1980, pp. 37, 121 and 233).

5.6.3 Celerity of Propagation

1° For a positive and negative translatory wave, propagating in a rectangular channel, the celerity was given by :

$$c_t (\pm) = U_1 \pm \sqrt{gh_1} \sqrt{\frac{h_2}{2h_1}(1 + \frac{h_2}{h_1})} \qquad (5.50a)$$

The upper sign, (+) , corresponds to a wave from upstream, the lower sign, (-) , to a wave from downstream (see Fig. 5.10).

2° Using the expression, $\Delta h = (h_2 - h_1)$, where Δh is height of the wave, and neglecting the quadratic terms, the above relation, eq. 5.50a, can be approximated (see *Favre,* 1935, p. 38 and p.41) by :

$$c_t \cong U_1 \pm \sqrt{gh_1} \sqrt{1 + \frac{3}{2}\frac{\Delta h}{h_1}} \qquad (5.52)$$

but also by :

$$c_t \cong U_1 \pm \sqrt{gh_1}\,(1 + \frac{3}{4}\frac{\Delta h}{h_1})$$

This relation, eq. 5.52, has already been given (see eq. 2.28) for a long wave — whose amplitude is not small — in a stagnant water, $U_1 = 0$.

3° For a wave of small amplitude, where one takes :

$$\frac{h_2 - h_1}{h_1} \ll 1 \quad \text{or} \quad h_2 \cong h_1 = h$$

the above relation, eq. 5.50a, yields :

$$c_t = U \pm \sqrt{gh} \tag{5.51}$$

This equation should be compared with the equation (see eq. 2.34) for the absolute celerity, c_w, of a wave in a channel having an average velocity, U, and with :

$$c = \pm \sqrt{gh} \tag{2.27}$$

which is the celerity of a long wave (or a gravity wave of small depth) of small amplitude in stagnant water, when U = 0 (see sect. 2.4).

4° Consider the positive wave from downstream, whose celerity was given with eq. 5.50. If one now takes the celerity to be zero, $c_t = 0$, the wave remains fixed in space ; thus :

$$U_1 = \sqrt{gh_1} \sqrt{\frac{h_2}{2h_1}\left(1 + \frac{h_2}{h_1}\right)} \tag{5.53}$$

Such a relation was derived (see eq. 4.40a) to describe the hydraulic jump, a phenomenon which takes place in steady and rapidly varied flow (see sect. 4.4.4).

5° The above calculations might seem a bit schematic, since one assumed the front as being vertical and neglected the curvature of the streamlines. A more exact computation (see *Flamant*, 1923, p. 433 and *Kozeny*, 1953, p. 256) gives the following expression for the celerity :

$$c_t = U_1 \pm \sqrt{gh_1}\left(1 + \frac{3}{4}\frac{\Delta h}{h_1} + \frac{h_1^2}{6\Delta h}\frac{d^2\Delta h}{dx^2}\right) \tag{5.52a}$$

6° The above developed relations are all limited to channels of rectangular cross sections. For a trapezoidal or parabolic cross section, similar relations have been elaborated by *Chow* (1959, p. 555). The above equation, eq. 5.50a, is thus written as :

$$c_t = U_1 \pm \sqrt{g}\sqrt{(A_2 h_{G_2} - A_1 h_{G_1}) / A_1(1 - A_1/A_2)} \tag{5.50b}$$

where h_{G_1} and h_{G_2} are the centroidal depths of the areas, A_1 and A_2.

7° The influence of the bed resistance, F_F, and of the bed slope, F_G, on the formation of translatory waves was considered by *Henderson* (1966, pp. 297-304).

UNSTEADY FLOW

8° The *meeting* of two positive translatory waves, being opposite in direction, represents an interesting practical case (see *Jaeger*, 1954, p. 378 and *Martin* 1980, p.431) :

i) Assumed is a subcritical and steady flow, U_1 and h_1.

ii) Considered will be that one wave, U' and h', and another one, U'' and h'', move into the downstream and upstream, respectively (see Fig. 5.13a). If these waves are simple waves of small height (see eq. 5.54, later), one may write :

$$U' - 2\sqrt{gh'} = U_1 - 2\sqrt{gh_1}$$

$$U'' + 2\sqrt{gh''} = U_1 + 2\sqrt{gh_1}$$

iii) These waves meet each other (see Fig. 5.13b).

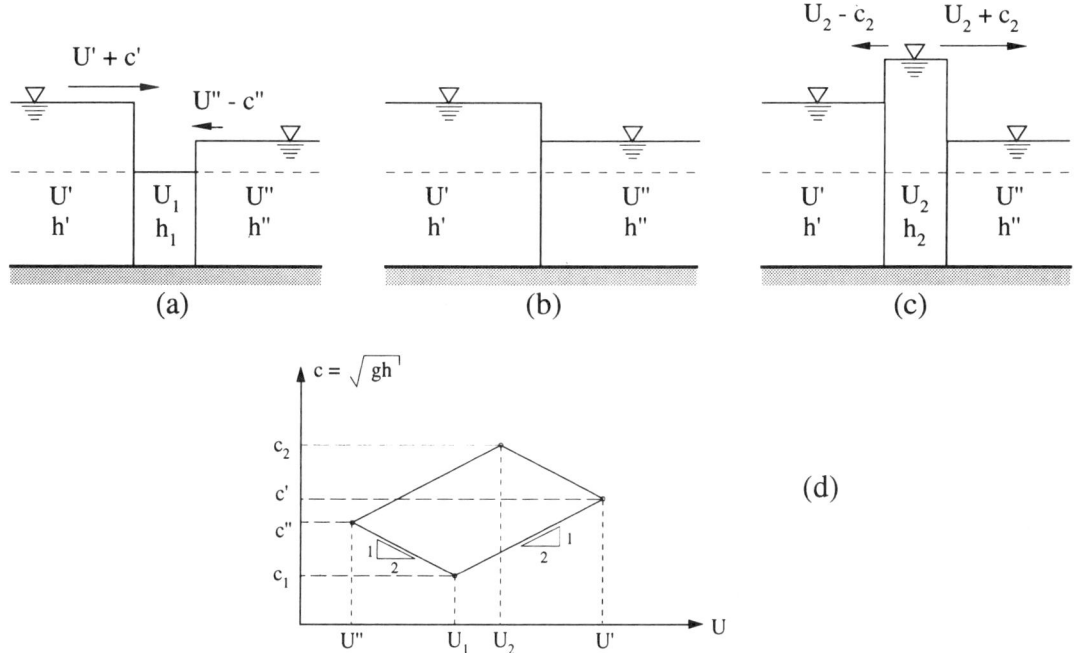

Fig. 5.13 Meeting of two translatory waves.

iv) Consequently (see Fig. 5.13c) a new wave regime, U_2 and h_2, is established, and these waves propagate into both directions ; one may write :

$$U_2 - 2\sqrt{gh_2} = U'' - 2\sqrt{gh''}$$

$$U_2 + 2\sqrt{gh_2} = U' + 2\sqrt{gh'}$$

v) All these equations are linear and are usually represented in a graphical form (see Fig. 5.13d) on a plane of, U and $\sqrt{gh}$; they are lines with an inclination of ± 1/2. For the resulting wave, U_2 and h_2, which propagates upstream and downstream, the velocity is given by :

$$U_2 = \frac{1}{2}\left[(U' + U'') + (2\sqrt{gh'} - 2\sqrt{gh''})\right]$$

while the flow depth, thus the celerity, is :

$$\sqrt{gh_2} = \frac{1}{4}\left[(U' - U'') + (2\sqrt{gh'} + 2\sqrt{gh''})\right]$$

vi) Meeting of two waves and the influence of a step in the bed is exposed in *Jaeger* (1954, p. 380) and *Martin* (1980, p.433). The reflection and transmission of long perturbations were treated by *Favre* (1935, p. 85).

5.6.4 Negative Waves (see Fig. 5.14)

1° A negative wave, steep at the beginning, will deform itself much faster than is the case for a positive wave. A true front does not establish itself and the notion of a wave height, Δh, makes less (physical) sense than for the positive wave.

2° The negative wave can however be considered as being an unsteady, gradually varied flow. The equation of Saint-Venant, eq. 5.1 and eq. 5.3, shall be used and assumed will be that the friction forces are practically compensated by the gravity forces ; for a rectangular channel, this yields :

$$\frac{\partial h}{\partial t} + \frac{\partial (Uh)}{\partial x} = 0 \tag{5.2}$$

$$\frac{\partial U}{\partial t} + U\frac{\partial U}{\partial x} + g\frac{\partial h}{\partial x} = 0 \tag{5.8}$$

This represents now a *simple wave* (see sect. 5.1.2).

3° For *small* waves, when Δh and U are only dependent on the flow depth, h, one may write (see *Jaeger*, 1954, p. 364) :

$$\frac{\partial (Uh)}{\partial x} = \frac{\partial (Uh)}{\partial h}\frac{\partial h}{\partial x} \quad ; \quad \frac{\partial U}{\partial x} = \frac{\partial U}{\partial h}\frac{\partial h}{\partial x}$$

as well as :

$$\frac{\partial U}{\partial t} = \frac{\partial U}{\partial h}\frac{\partial h}{\partial t}$$

UNSTEADY FLOW

Consequently, the expression of ∂h/∂t in the above equations, eq. 5.2 and eq. 5.8, can be eliminated, which gives :

$$g - h\left(\frac{\partial U}{\partial h}\right)^2 = 0$$

and, by taking the root of the two members :

$$\frac{\partial U}{\partial h} = \pm \sqrt{\frac{g}{h}}$$

which, upon integration, renders :

$$U(h) = \pm 2\sqrt{gh} + Cte \qquad (5.54)$$

The upper sign, (+), corresponds to an elementary negative wave — but it could also be positive — which propagates in direction of the initial flow (see Fig. 5.10b and Fig. 5.14) ; the lower sign, (-), corresponds to a wave propagating into the opposite direction (see Fig. 5.10d).

For the evaluation of the constant of integration (see *Favre*, 1935, p. 29), one takes $U = U_1$ and $h = h_1$ in this part of the channel, which has as yet not been attained by the wave (see Fig. 5.14) ; thus one gets :

$$U(h) = U_1 \pm 2\sqrt{gh} \mp 2\sqrt{gh_1} \qquad (5.55)$$

Using eq. 5.51, the celerity of the small wave is given by :

$$c_t(h) = U \pm \sqrt{gh} = U_1 \pm 3\sqrt{gh} \mp 2\sqrt{gh_1} \qquad (5.55a)$$

This celerity is the velocity of propagation of a wave of small amplitude.

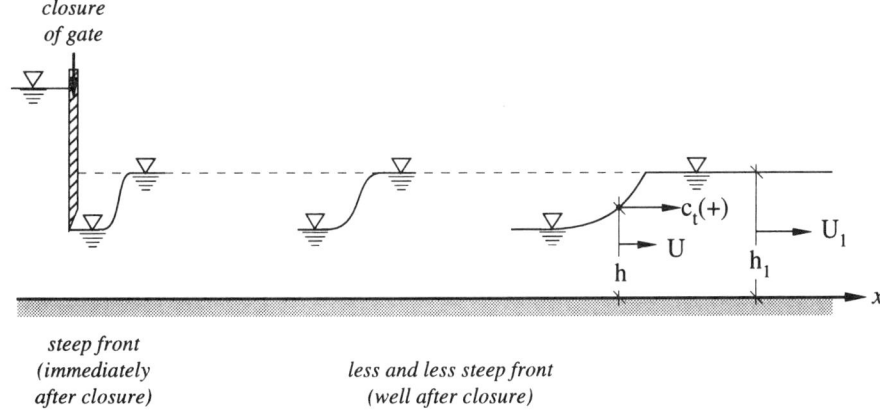

Fig. 5.14 Negative wave from upstream.

4°   If one considers a negative wave from upstream (see Fig. 5.14), this equation, eq. 5.55a, is written as :

$$c_t(h) = U + \sqrt{gh} = U_1 + 3\sqrt{gh} - 2\sqrt{gh_1} \tag{5.56}$$

By putting the reference — creation of wave — at time, t = 0, the position of the wave is given by $x = c_t t$; thus one gets :

$$x(h) = (U_1 + 3\sqrt{gh} - 2\sqrt{gh_1})\, t \tag{5.57}$$

being an expression for the curve of the water surface of the wave. Between eq. 5.56 and eq. 5.57, one finds a relation for the velocity, in terms of x and t, or :

$$U(h) = \frac{U_1}{3} + \frac{2}{3}\frac{x}{t} - \frac{2}{3}\sqrt{gh_1} \tag{5.58}$$

5°   If one considers now a negative wave from downstream (see Fig. 5.15), this relation, eq. 5.55a, is written as :

$$c_t(h) = U_1 - 3\sqrt{gh} + 2\sqrt{gh_1} \tag{5.59}$$

6°   The hydraulic problem of a *dam break* (see Fig. 5.15) can thus be addressed, considering it to be a negative wave from downstream propagating on a frictionless, horizontal and rectangular channel (see *Streeter*, 1971, p. 681).

   i)   Before the failure of the gate (or dam), the water depth upstream of the dam is $h = h_1$ and there is no water on the other side, $h = 0$.

   ii) After a sudden failure of the gate, an expression for the velocity at any section is obtained using the above equation, eq. 5.55, and assuming that $U_1 = 0$, or :

$$U(h) = 0 - 2\sqrt{gh} + 2\sqrt{gh_1} \tag{5.60}$$

   iii) The celerity is given (see eq. 5.59) by :

$$c_t(h) = 0 - 3\sqrt{gh} + 2\sqrt{gh_1} \tag{5.59a}$$

   which renders (see Fig. 5.15) :

   for $h = 0$: $c_t = +2\sqrt{gh_1}$

   for $h = h_1$: $c_t = -\sqrt{gh_1}$

   iv) The shape of the water surface is given (see eq. 5.57) by :

$$x(h) = c_t t = (0 - 3\sqrt{gh} + 2\sqrt{gh_1})\, t \tag{5.57a}$$

UNSTEADY FLOW

Subsequently, one obtains for the section at the gate, $x = 0$, for all time instants, t, the flow depth as being $h_{xo} = \frac{4}{9} h_1$. This flow depth remains thus independent of time. The water-surface profile for all time instants pivots around this point of $h_{xo} = \frac{4}{9} h_1$.

v) The velocity, U, in this section, $x = 0$, is obtained from eq. 5.60, or:

$$U_{xo} = -2\sqrt{g}\left(\frac{2}{3}\sqrt{h_1} - \sqrt{h_1}\right) = +\frac{2}{3}\sqrt{gh_1}$$

showing that it is also independent of time.

vi) The unit discharge, q, can readily be computed:

$$q = hU = \frac{8}{27} h_1 \sqrt{gh_1}$$

being constant with time.

vii) In reality, the shape of the water surface, eq. 5.57a, should be modified, since due to friction on the bed a positive wave from upstream will establish itself (see Fig. 5.10c) in the downstream reach of the channel.

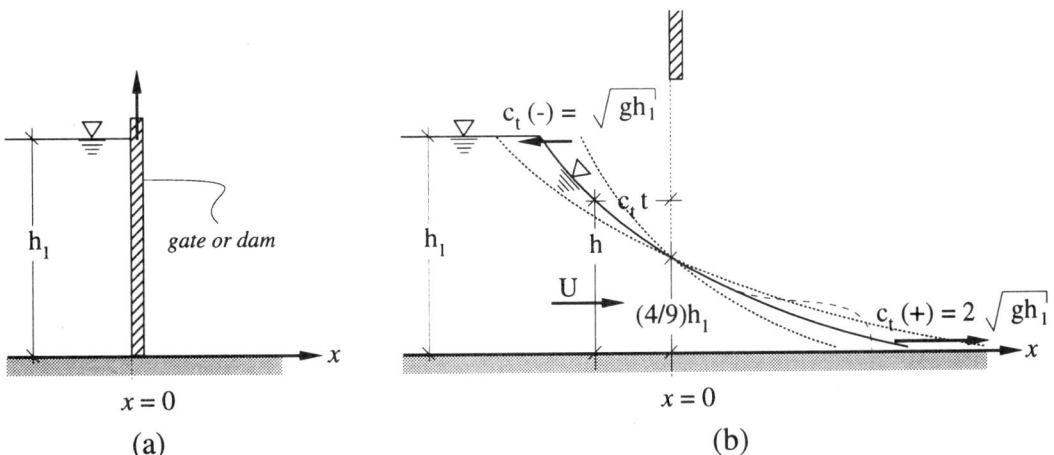

Fig. 5.15 Negative wave from downstream; problem of a dam break : a) before ; b) after.

7° Solutions to the problem of a simple wave, eq. 5.2 and eq. 5.8, thus for the problem of dam break, can be obtained using the method of characteristics (see sect. 5.2.2) presented by *Stoker* (1957), *Henderson* (1966, p. 304) and *Martin* (1989, p. 590).

5.7 EXERCICES

5.7.1 Problems, solved

Ex. 5.A

At a hydroelectric power plant, a long channel with a rectangular cross section and a weak bed slope enters into a reservoir. The flow in this channel has a normal depth of 1.5 [m] and an average velocity of 1.2 [m/s]. It shall be assumed that the slope of the channel bed and the one of the energy-grade line are negligible, $S_f = S_e \cong 0$. Initially, at the channel mouth, the water level is the same, both in the channel and in the reservoir. During a partial emptying of the reservoir the water level is lowered at a rate of 0.25 [m/h], creating a gradual and weak perturbation on the initial steady (and uniform) flow in the channel.

a) Derive the characteristic equations by assuming that the perturbation can be treated as a *simple wave* (see eq. 5.8).
b) Determine the time required for recording a decrease of 0.5 [m] in the water level at a station located 1 [km] upstream of the channel mouth.

SOLUTION :

a) Derivation of equations :

Assuming that the channel is horizontal and without friction, $S_f = S_e \cong 0$, the characteristic equations, eqs. 5.14 and 5.14a, can be written in the following simple form (see point 5.2.1, 6°) :

$U + 2c = $ Cte   along a positive characteristic, C^+,   defined by $\dfrac{dx}{dt} = (U + c)$

$U - 2c = $ Cte   along a negative characteristic, C^-,   defined by $\dfrac{dx}{dt} = (U - c)$

In the plane, x and t, (see *Henderson*, 1966, p. 289) consider the points, A, B, D and E, on the characteristic lines, C^+ and C^- (see Fig. Ex.5.A.1).

i) Along C^- one has : $U_A - 2c_A = U_D - 2c_D$
$U_B - 2c_B = U_E - 2c_E$

ii) If $\overline{DE}$ is a *straight* line, there will be along C^+ :

$U(t) + c(t) = $ Cte

with a slope of : $\dfrac{dt}{dx} = \dfrac{1}{U(t) + c(t)}$

$U(t) + 2c(t) = $ Cte

U and c should be thus constant.

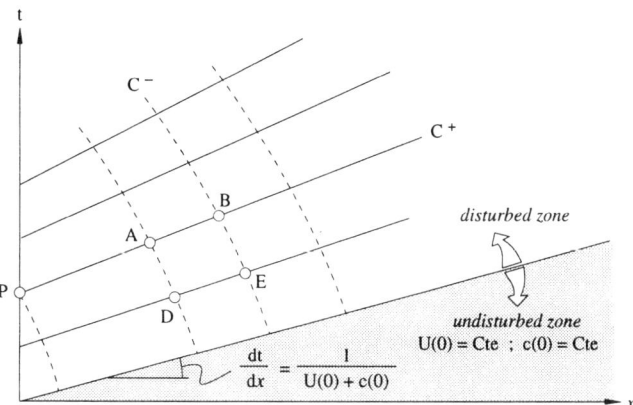

Fig. Ex.5.Λ.1 Characteristics for a simple wave.

Consequently, the following can be written :

$U_D = U_E$ and $c_D = c_E$

which allows now one to write :

$U_A - 2c_A = U_B - 2c_B$

iii) Since $\overline{AB}$ is also on C^+, one has :

$U_A + 2c_A = U_B + 2c_B$

which is only possible if :

$U_A = U_B$ and $c_A = c_B$

thus, $\overline{AB}$ is also a *straight* line.

iv) For a point, $P(x = 0, t = t)$, one writes for :

C^+ : $\dfrac{dx}{dt} = U_p(t) + c_p(t)$

C^- : $U_p(t) - 2c_p(t) = U(0) - 2c(0)$

v) As a result, the slope of C^+ can be written as :

$$\dfrac{dx}{dt} = 3c_p(t) - 2c(0) + U(0) \tag{5.α}$$

or, by expressing $c_p(t)$ in terms of $U_p(t)$:

$$\dfrac{dx}{dt} = \dfrac{3}{2} U_p(t) + c(0) - \dfrac{1}{2} U(0) \tag{5.β}$$

This relation, eq. 5.α , gives the celerity of propagation of an elementary negative wave from downstream (see eq. 5.56).

296   FLUVIAL HYDRAULICS

- vi) With these last two relations the values of U_p and c_p can be obtained at every point, $P(x = 0, t = t)$.

- vii) An initial (weak) perturbation can be created either by varying the initial flow depth, h or c, or by varying the initial velocity, U.

b) Computation of the consequences of partial emptying of the reservoir:

- i) The perturbation is created here by a variation of the water level in the reservoir; it will be assumed that this perturbation results in a wave of small amplitude (see sect. 2.4.3).

- ii) Computation of the *initial* parameters (before partial emptying):

  Average velocity in the channel: $U(0) = -1.2$ [m/s]

  Celerity of the wave: $c(0) = \sqrt{g\,h(0)} = \sqrt{9.81 \times 1.5} = 3.84$ [m/s]

  The absolute celerity of this wave, c_{w_0}, which is also the slope of the characteristic, C^+, emanating from the channel mouth at time $t = 0$ [h], is given by:

  $$c_{w_0} = \frac{dx}{dt} = U(0) + c(0) = -1.2 + 3.84 = 2.64 \text{ [m/s]}$$

- iii) At the channel mouth, the rate of lowering the water level in the reservoir is known as being:

  $$\frac{\partial h}{\partial t} = -0.25 \text{ [m/h]}$$

  At this rate, a time period of $t_v = 2.0$ [h] is needed to lower the water level in the reservoir by 0.5 [m], and consequently the flow depth in the channel mouth.

- iv) Computation of parameters at *time* $t_v = 2$ [h] (at the end of partial emptying):

  Flow depth in the channel, at the channel mouth:

  $h(2) = 1.5 - 0.5 = 1.0$ [m]

  Celerity of the waves at this depth:

  $c(2) = \sqrt{g\,h(2)} = \sqrt{9.81 \times 1.0} = 3.13$ [m/s]

  The absolute celerity of this wave, c_{w_2}, which is also the slope of the characteristic, C^+, emanating from the channel mouth at time $t_v = 2$ [h], is given by:

  $$c_{w_2} = \frac{dx}{dt} = U(2) + c(2)$$

UNSTEADY FLOW 297

The average flow velocity, U(2), at the junction being an unknown, the absolute celerity can not be calculated using above expression. On the other hand, one of the relation, eq. 5α, derived in the preceding section gives the slope of the C^+ characteristic, dx/dt, at time t as a function of the celerity c(t), as well as the initial parameters, U(0) and c(0). By using this relation, one obtains :

$$c_{w_2} = \frac{dx}{dt} = 3\,c(2) - 2\,c(0) + U(0) = 3 \times 3.13 - 2 \times 3.84 - 1.2 = 0.51\,[m/s]$$

The average velocity of the flow is then :

$$U(2) = c_{w_2} - c(2) = 0.51 - 3.13 = -2.62\,[m/s]$$

The Froude number, Fr = U(2) / c(2) = 2.62 / 3.13 = 0.86, is such that the flow at the channel mouth remains subcritical and submerged (see Ex. 4.B p. 206 and p. 213). The water-surface profile is therefore a M2-type profile.

v) The time necessary for the wave to travel a distance of $x = 1.0$ [km] since the instant when the water level at the channel mouth has been lowered to h(2) = 1.0 [m], is given by :

$$t' = \frac{1000}{0.51} = 1960.8\,[s] = 0.54\,[h]$$

vi) The time necessary for the wave to travel a distance of $x = 1.0$ [km] since the beginning of the partial emptying, can now be obtained :

T' = 2.0 + 0.54 = 2.54 [h]

At this instant, the front of the wave traveling along the first characteristic line is at a distance of :

$$L' = T'\,c_{w_0} = 2.54 \times 3600 \times 2.64 / 1000 = 24.14\,[km]$$

vii) The graphical representation of the solution in the plane, x and t, is given with Fig. Ex.5.A.2.

The graph on the left shows the variation of the flow depth, h, at the channel mouth during the partial emptying. Initially, at time t = 0 [h], the flow depth is at h(0) = 1.5 [m]. During two hours the flow depth is lowered at a rate of dh/dt = –0.25 [m/h]. It is finally stabilized at h(2) = 1.0 [m].

The development of the flow in the channel is studied in x-t plane. At time t = 0 [h], when the flow depth at the channel mouth starts decreasing, the first characteristic line emanates with a slope of $c_{w_0} = dx/dt = 2.64$ [m/s] in order to carry the information to the upstream. The undisturbed zone is located between the first characteristic line and the x-axis. No information concerning the perturbation is transmitted to this zone.

Two hours after the beginning of the partial emptying, t_v, the flow depth at the channel mouth has been lowered to $h(2) = 1.0$ [m]. The information concerning this final flow depth is carried to the upstream with an absolute celerity of $c_{w_2} = dx/dt = 0.51$ [m/s] along the characteristic line emanating at this point. This characteristic line arrives at a distance of $x = 1$ [km] in a time period of $t' = 0.54$ [h]. A time period of $T' = 2.0 + 0.54 = 2.54$ [h] has elapsed since the beginning of the partial emptying during which the initial wave, traveling with an absolute celerity of $c_{w_0} = dx/dt = 2.64$ [m/s] along the first characteristic line, has arrived at a distance of $L' = 24.14$ [km]. For all distances larger than L', the uniform flow in the channel at a velocity of $U_o = -1.2$ [m/s] has not yet been influenced by this initial wave.

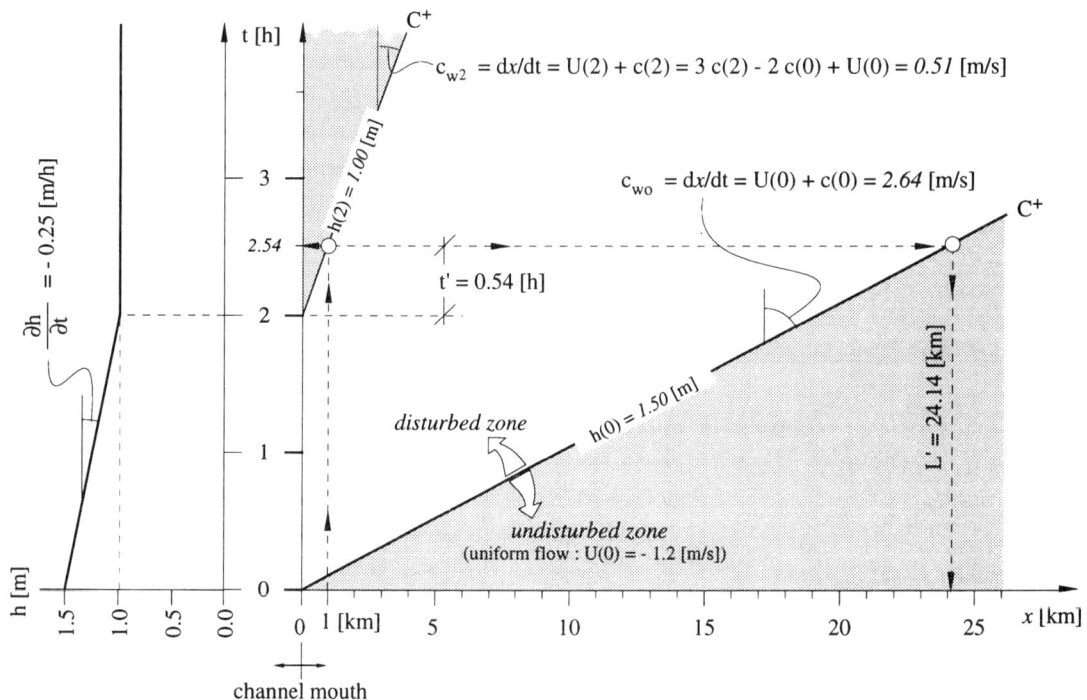

Fig. Ex.5.A.2 Graphical solution of the simple wave problem using the method of characteristics.

viii) An axonometric view of the variation of the flow depth, $h(x,t)$, over the entire space-time plane, x and t, is presented in Fig. Ex.5.A.3 (note that the continuous linear decrease of the flow depth is simulated by small rapid variations of $\Delta h = 0.05$ [m]). The observations made for Fig. Ex.5.A.2 apply here also. Moreover, one can also see the initial water-surface profile, $h(x, t = 0$ [h]), as well as the profile, $h(x, t = 2.54$ [h]), when the information on the flow depth $h(2) = 1.0$ [m] arrives at a distance of $x = 1$ [km].

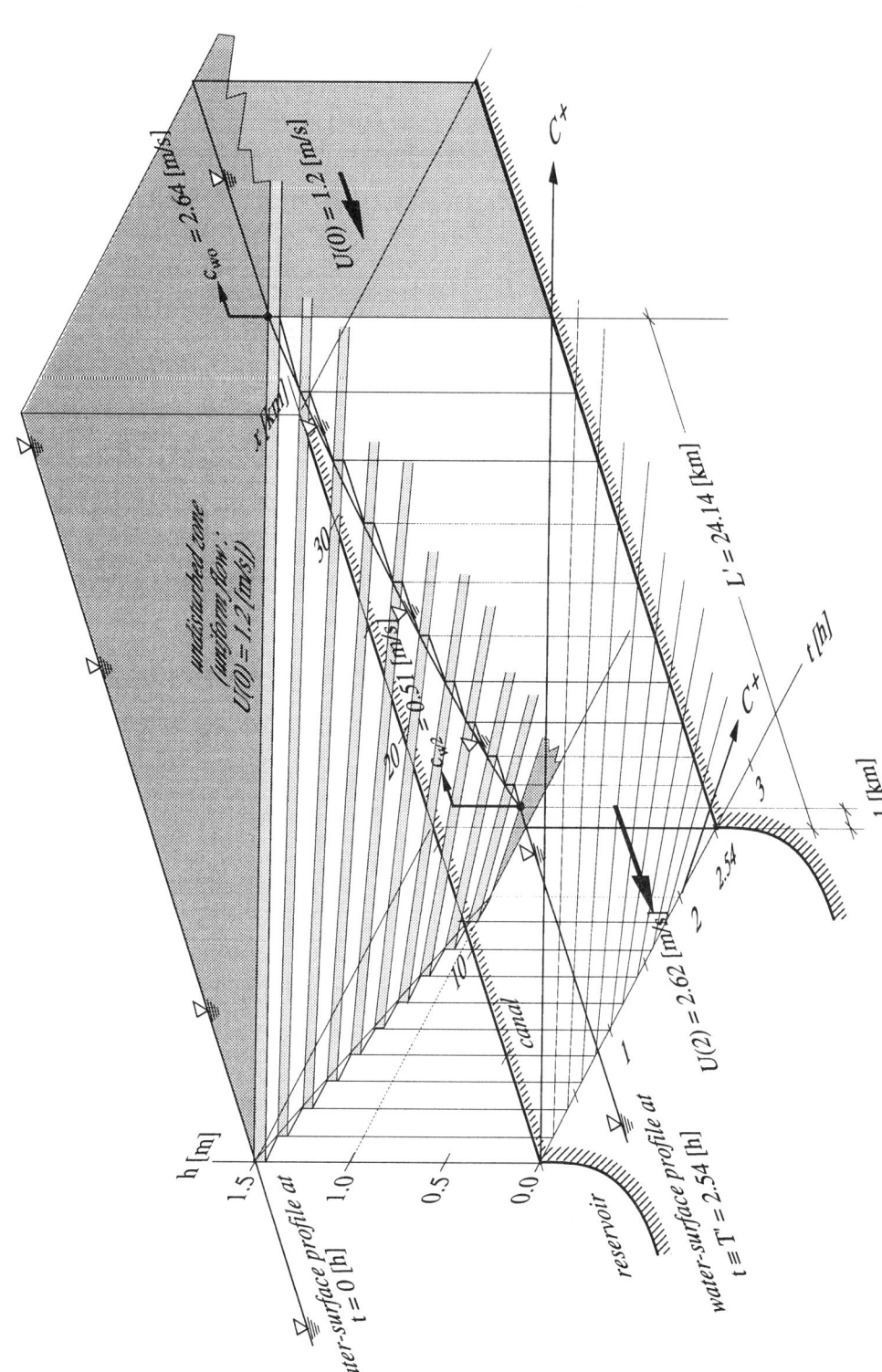

Fig.Ex.5.A.3 Axonometric view of the variation of the flow depth, h(x,t), over the entire space-time plane, x-t.

Ex. 5.B

A rectangular prismatic channel, made of poor quality concrete, has a width of B = 5 [m] and conveys a uniform flow at a depth of h_n = 1.20 [m]. During a flood event, the discharge in the channel increases linearly to Q = 50 [m³/s] in a time period of t' = 20 [min] and decreases linearly back to its initial value during a time period of t" = 60 [min]. The channel has a bed slope of S_f = 0.001 [-] and the friction coefficient is estimated to be n = 0.020 [m$^{-1/3}$s]. The length of the channel reach to be studied is 3 [km]. Using the *explicit* finite difference method, determine :

i)   the hydrograph at the two stations, L_1 = 1.5 [km] and L_2 = 3 [km];

ii) the instants when the maximum discharge, Q_{max}, occurs at stations L_1 and L_2, respectively;

iii) the attenuation of the maximum flow depth at the stations L_1 and L_2;

iv) the time necessary for the flow to become uniform again at the station L_2.

SOLUTION :

a) *General comments on use of the computer and the programming :*

The finite difference methods as well as the method of characteristics involve a large amount of calculations and are ideally suited for work on the computer. In this exercise, a program in FORTRAN IV language is presented to solve the present problem, and other similar problems, using the explicit finite difference method introduced in sect. 5.2.3. The program is written in standard FORTRAN and can be run on most of the personal computers.

Although a basic knowledge of computers and of programming in languages like FORTRAN, BASIC or PASCAL can be certainly helpful in understanding the solution of the exercise, it is not essential. Extreme care has been given to make the general programming techniques understandable to everybody, even to those who lack any experience in programming.

The program code is written in a pedagogical style and does not have the pretention of being optimized. Numerous comments inserted in the code explain the flow of the program almost step by step. As far as possible, the names of the variables are chosen to recall the notation used in sect. 5.2. An exhaustive list of variables together with the type of variables and their explanations, are provided at the beginning of the main program and of the related subprograms.

The user enters the data of the problem into the computer in an interactive manner (see Fig. Ex.5.B.5) by answering the questions asked by the program. The numerous comments provided in the dialog recall the subject matter treated in the book and guides the users in making their choices. The possible errors during the data entry are checked by the program. In case of an error, the question is repeated until the user answers it correctly. The errors concerning the stability of the solution (violation of Courant's stability condition) are also detected and flagged by an error message giving also a clue to the type of corrective action to be taken.

UNSTEADY FLOW

b) *Definition of solution domain:*

For the numerical solution of a problem using a finite difference method, be it implicit or explicit, one has to define first a space-time, x-t, and discretize it. The space-time can be visualized as an orthogonal coordinate system, x-t, on a horizontal plane (see Fig. Ex.5.B.1 and Fig.5.4). The channel axis having a length of LT = 3 [km], coincides with the x-axis. It is divided into several reaches of length Δx (variable DX). The number of nodes obtained in this way is NN.

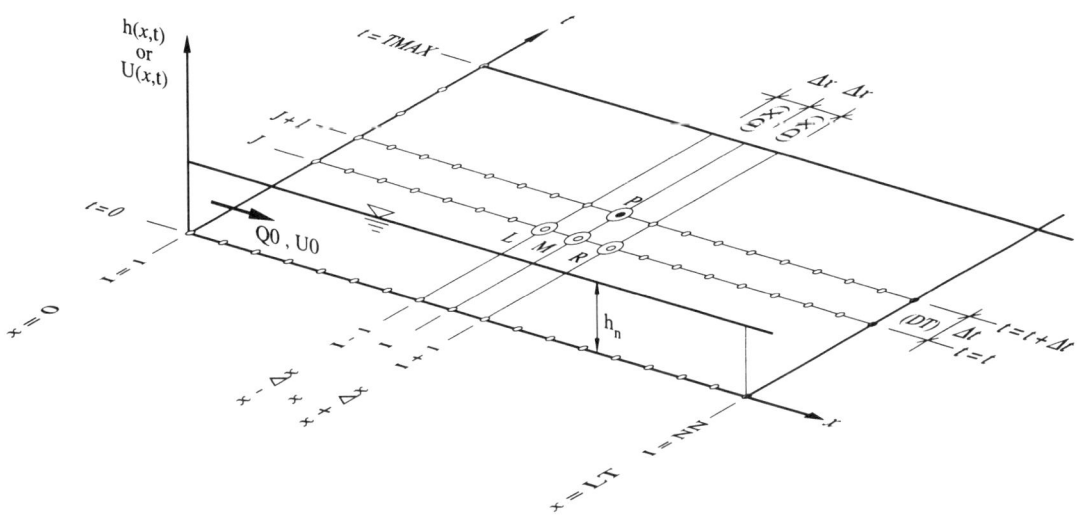

Fig. Ex.5.B.1 Solution domain and its discretization.

Perpendicular to the space axis, x, the time axis, t, (variable T) begins at time t = 0 at the origin and extends to the infinity by time steps of Δt (variable DT). For the present problem, it will be assumed that the solution stops at a time t = TMAX, to be fixed by the user.

Solving the problem means to know the flow depth, h(x,t) (variable H), and the flow velocity, U(x,t) (variable U), at every point in the space-time. These variables can be represented by a third axis perpendicular to the plane of space-time, x-t; this yields a tridimensional space, providing some useful information :
- a section parallel to the plane h(t)–x [or U(t)–x] , and cutting the time axis at time t, gives an instantaneous picture of the variation of that variable along the channel length (if, for example, the third axis represents the flow depth, h(t), one obtains the *water-surface profile* in the channel at time t) ;
- a section parallel to the plane h(x)–t [or U(x)–t] , cutting the x-axis at x, gives the variation of that variable in time at the station x (if, for example, the third axis represents the flow depth, h(x), one obtains the *variation of the flow depth* at the station x as a function of time t) .

c) *Discretization of equations for the explicit method :*

c.1) **Initial conditions**

The explicit method has been developed in detail in sect. 5.2.3. According to this method, the flow depth and the velocity at a given point in the channel are calculated using eqs. 5.23 and 5.25, respectively, from the values known at the preceding time. To initiate the calculations, the initial values of the flow depth, h_o (variable H0), and the average velocity, U_o (variable U0), must be known at every node at time t = 0, namely just before the beginning of the flood event (initial conditions).

The data of the problem contain the necessary information for writing the initial conditions. Initially the flow in the channel is uniform and has a depth of h_n (variable HN) = 1.2 [m]. At time t = 0, the flow depth at every node along the x-axis is equal to this value. The geometry of the cross section, the bed slope and the friction coefficient of the channel being known, the corresponding discharge and the velocity can be calculated using a formula for computing uniform flow (see sect. 3.2). In the present case, the Manning-Strickler equation will be used :

$$U = \frac{1}{n} R_h^{2/3} S_f^{1/2} \tag{3.16}$$

For a rectangular channel, one has (see Table 1.1) :

A (variable A) = $b \, h_n$ = 5 × 1.2 = 6.0 [m²]   and

R_h (variable RH) = $\dfrac{b \, h_n}{b + 2h_n}$ = $\dfrac{5 \times 1.2}{5 + 2 \times 1.2}$ = 0.81 [m]

At every node, at time t = 0, the velocity, U_o (variable U0), the discharge, Q_o (variable Q0), and the Froude number, Fr, are therefore :

U_o (= U0) = $\dfrac{1}{0.02}$ $(0.81)^{2/3}$ $(0.001)^{1/2}$ = 1.374 [m/s]

Q_o (= Q0) = $A \times U_o$ = 6.0 × 1.374 = 8.243 [m³/s]

$Fr = \dfrac{U_o}{\sqrt{gh_n}} = \dfrac{1.374}{\sqrt{(9.81)(1.2)}}$ = 0.40 [-]   ;   flow is subcritical.

c.2) **Storage of various information in the program**

The program, based on the explicit method, will be presented later. In the program, the solution domain in the space-time, x-t, is represented by a two dimensional matrix. The values of the flow depth and the velocity are stored in two arrays, called H(J,I) and U(J,I) respectively. The subscript J (row number) stands for the time whereas the subscript I (column number) represents the node number, namely the position along the x-axis.

According to the rules of FORTRAN, the maximum number of rows and columns must be specified at the beginning of the program in order to reserve a sufficient storage place in the computer memory. The number of columns is dictated by the maximum number of divisions the user needs to have along x-axis. In the present program this number was arbitrarily fixed as MAXDIV = 100, which means that the maximum number of nodes will be NNMAX = MAXDIV + 1 = 101 (see Fig. Ex.5.B.5, PARAMETER statements at the beginning of the program). To increase the maximum number of divisions, and consequently the maximum number of nodes, it is necessary to change the value of the parameter MAXDIV and recompile the program.

The situation for the number of rows is somewhat different. In problems involving a large number of nodes and large number of time steps, one would need huge arrays in order to be able to represent all the space time, x-t, by an array. This would require a large amount of computer memory. It is possible to bypass this problem by considering that at a given time step the computation algorithm requires to know the values only for the preceding time. It is therefore sufficient to declare the arrays (H, U and Q) of two lines in length only. At the beginning of calculations, the initial values of H, U and Q are stored in the first row (values at time J) by using the initial conditions. At every time step, the calculations are made to obtain the values to be stored in the second row (values at time J+1) and the results are immediately written on the output file. Next, the values in the first row are replaced by those in the second row in order to be used for computing the values of the variables at the next time step, and so on.

c.3) **Discretisation of equations** (see Fig. Ex.5.B.2)
The eqs. 5.23 and 5.25 with the subscripts L, M and R should be rewritten in a slightly different form which is more suitable for programming in FORTRAN IV. The discretized equations of the explicit method, rewritten for the purposes of programming, are given herewith :

flow depth (see eq. 5.23) :

$$h_i^{j+1} = h_i^j + \frac{\Delta t}{2\Delta x} \left[U_i^j (h_{i-1}^j - h_{i+1}^j) + h_i^j (U_{i-1}^j - U_{i+1}^j) \right]$$

velocity (see eq. 5.25) :

$$U_i^{j+1} = \frac{1}{2} \left[-\Gamma + (\Gamma^2 + 4\Gamma\beta)^{1/2} \right]$$

with : $\quad \beta = \left[U_i^j + \frac{\Delta t}{2\Delta x} U_i^j (U_{i-1}^j - U_{i+1}^j) + \frac{g\Delta t}{2\Delta x} (h_{i-1}^j - h_{i+1}^j) + g\Delta t \, S_f \right]$

$$\Gamma = \frac{\left(R_{h\,i}^{j+1} \right)^{4/3}}{n^2 \, g \, \Delta t}$$

The equations given above can be used directly in the program to calculate the flow depth and the velocity at the nodes from I = 2 to I = NN–1. They cannot be used, however, for the nodes I = 1 and I = NN, since the nodes I – 1, for the first, and I + 1, for the second, do not exist (see Fig. Ex.5.B.2). The computations for these nodes should use the boundary conditions given in the problem statement.

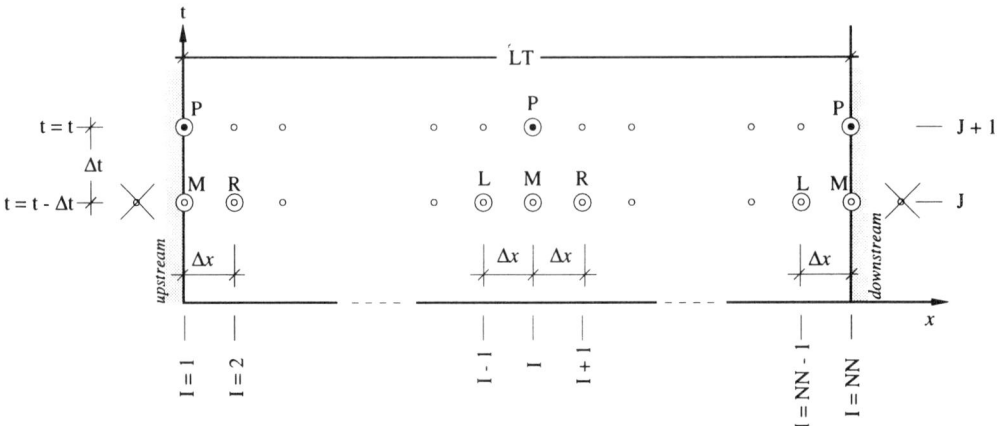

Fig. Ex.5.B.2 Numerical scheme and boundary conditions at the upstream and downstream ends.

c.4) **Upstream boundary conditions** (see Fig. Ex.5.B.2)
At the upstream end of the channel, I = 1 according to the notation adopted here, the flood hydrograph, namely the discharge as a function of time, is known and is shown in Fig. Ex.5.B.3.

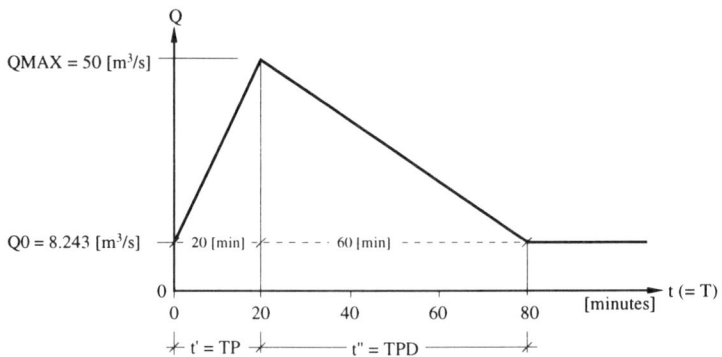

Fig. Ex.5.B.3 Triangular hydrograph imposed at the upstream end of the channel.

At the node I = 1, the discharge is known at every instant, t. Assuming that (see eqs. 5.22):

$$\left.\frac{\partial h}{\partial x}\right|_M = \frac{h_R - h_M}{\Delta x}$$

UNSTEADY FLOW

the eq. 5.23 for the flow depth is modified such that it contains only the terms with the subscripts I and I+1 (namely, M and R) :

$$h_i^{j+1} = h_i^j + \frac{\Delta t}{\Delta x} \left[U_i^j (h_i^j - h_{i+1}^j) + h_i^j (U_i^j - U_{i+1}^j) \right]$$

The flow depth at time J+1 being already calculated, the velocity can be obtained from :

$$U_i^{j+1} = \frac{Q_i^{j+1}}{A_i^{j+1}}$$

where the wetted surface, A_i^{j+1}, corresponding to the flow depth, h_i^{j+1}, is computed according to the shape of the cross section, being rectangular in the present case (see Table 1.1).

The discharge at time T (running time step, J+1), Q_i^{j+1}, is given by the shape of the hydrograph imposed at the entrance. In the present case the hydrograph has a triangular shape (see Fig. Ex.5.B.3). Using linear interpolation, one writes :

if $\quad\quad\quad T \leq \quad TP \quad : \quad Q_i^{j+1} = Q0 + \dfrac{QMAX - Q0}{TP} T$

if $\quad TP < T \leq TP+TPD \quad : \quad Q_i^{j+1} = QMAX - \dfrac{QMAX - Q0}{TPD} [T - TP]$

if $\quad\quad\quad T > TP+TPD \quad : \quad Q_i^{j+1} = Q0$

c.5) **Downstream boundary conditions** (see Fig. Ex.5.B.2)
At the downstream end of the channel, I = NN, the situation is the inverse of the one at the upstream, I = 1. Assuming that (see eqs. 5.22) :

$$\left. \frac{\partial h}{\partial x} \right|_M = \frac{h_M - h_L}{\Delta x}$$

the eq. 5.23 for the flow depth is modified such that it contains only the terms with the subscripts I and I–1 (namely, M and L) :

$$h_i^{j+1} = h_i^j + \frac{\Delta t}{\Delta x} \left[U_i^j (h_{i-1}^j - h_i^j) + h_i^j (U_{i-1}^j - U_i^j) \right]$$

The discharge at the node I = NN, at the time step J+1, is not known. However, the velocity can be calculated by assuming the validity of the Manning-Strickler formula (see eq. 3.16) :

$$U_i^{j+1} = \frac{1}{n}\left(R_{h\,i}^{j+1}\right)^{2/3}(S_f)^{1/2}$$

and writing the discharge as : $\quad Q_i^{j+1} = U_i^{j+1} A_i^{j+1}$

where the wetted surface, A_i^{j+1}, corresponding to the flow depth, h_i^{j+1}, is computed according to the shape of the cross section, which is rectangular in the present case (see Table 1.1). With this boundary condition, one is implicitly assuming that the stage discharge curve, $Q = f(h_n)$, for the uniform flow is valid at that station. At the end of the exercise, the error introduced by this assumption and its consequences on the results obtained will be discussed.

d) *Structure of program written in FORTRAN IV :*

The program in FORTRAN IV, based on the explicit scheme described above, is presented in the following pages. The program has a *modular* structure. It is composed of of a *main program* and several *subprograms*, each one accomplishing a specific pre-defined task. The flowchart, presented in Fig. Ex.5.B.4, shows the relations between different program units. We advise the reader to follow the logic of the program directly from the code given in Figs Ex.5.B.8. An exhaustive list of variables, together with the type of variables and their explanations, is provided at the beginning of the main program and the related subprograms (see Fig. Ex.5.B.8). The specific tasks carried out by each subprogram are described below in detail (see Fig. Ex.5.B.4).

- The main program EXPLIC controls the flow of the entire program by calling five subroutine subprograms :

  - The main program first calls the subroutine subprogram DREAD to read the program data by questioning the user. The subprogram DREAD computes also various useful parameters as well as the values of the discharge, Q0, and the velocity, U0, at time t = 0 according to the formula of Manning, eq. 3.16.

  - The subroutine subprogram INIT, as its name indicates, assigns the initial values to all the nodes at time t = 0. This short subroutine receives the value of the variable HN given by the user and the values of Q0 and U0 computed by the subroutine subprogram DREAD and assigns them to all nodes by filling the first lines of the three arrays H, Q and U.

  - The subroutine subprogram CALCUL controls the flow of the unsteady flow calculations according to the explicit scheme. This subroutine subprogram advances the time, T, by constant increments, DT, and stops the computations when T = TMAX. It calls six other subprograms for carrying out various tasks :

    - The subroutine subprogram TITLES is called only once at the beginning of the program for writing the titles and the given problem data on the output file.

    - The subroutine subprogram RWRITE, as its name indicates, writes the results on the output file at the end of every time step.

UNSTEADY FLOW

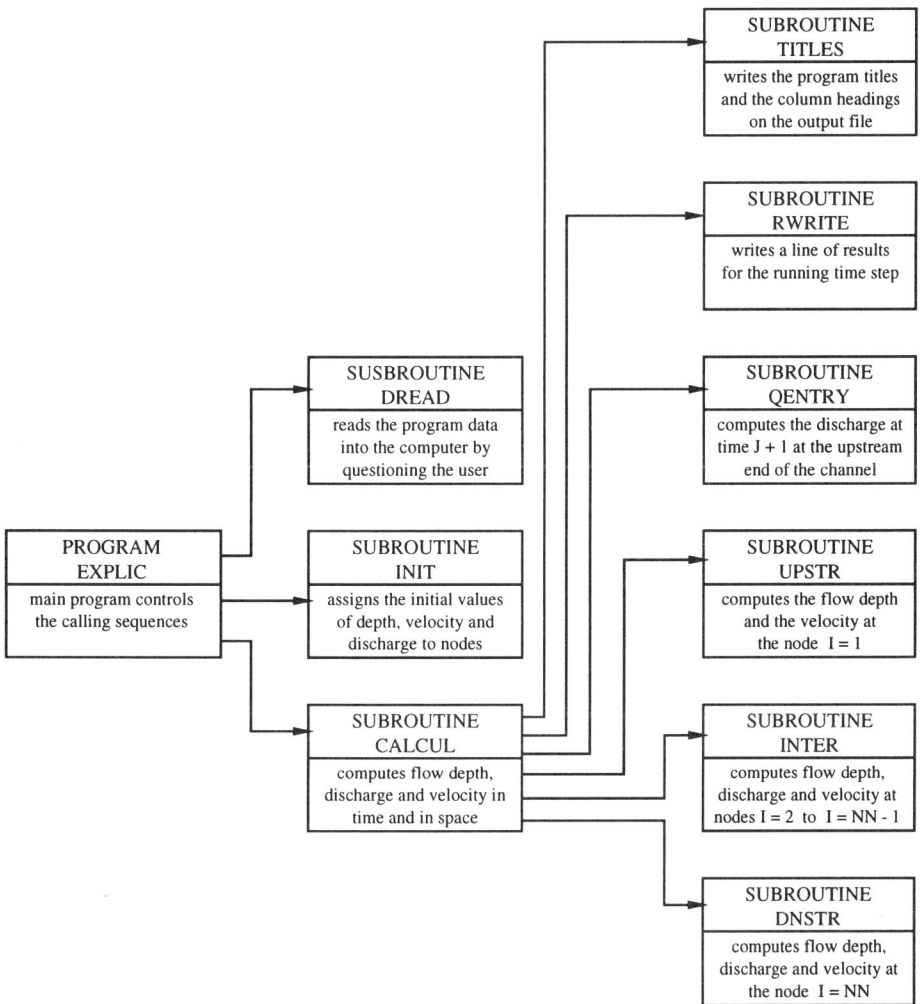

Fig. Ex.5.B.4 Flowchart of the program EXPLIC.

- The subroutine subprogram QENTRY computes the discharge entering at the upstream end according to the triangular hydrograph specified by the user.

- The subroutine subprogram UPSTR takes the value of the discharge computed by the subroutine subprogram QENTRY and computes the flow depth and the velocity at the node $I = 1$.

- The subroutine subprogram INTER computes the flow depth, the discharge and the velocity at the intermediate nodes $I = 2$ to $I = NN - 1$, where NN is the total number of nodes.

- The last subroutine subprogram DNSTR computes the flow depth, the discharge and the velocity at the node $I = NN$.

At the end of each time step, the subroutine subprogram CALCUL replaces the values of the variables for the preceding time step by those newly computed.

e) *Use of program :*

The source code of the program EXPLIC presented in the Fig. Ex.5.B.8 is first *compiled* and then *linked* to obtain an executable code. A FORTRAN compiler is of course necessary to do these operations on your computer. The reader should consult the manuals of the computer to learn the procedure to follow.

The dialog between the user and the computer for solving the present problem is presented in Fig. Ex.5.B.5. The values typed in by the user are highlighted by a white background. They are followed by a sign representing the RETURN (CR or ENTER on some computers) key on the keyboard. The text of the dialog is considered to be sufficiently self-explanatory and will not be commented any further. Nevertheless, there are two points that still need to be clarified :

The first point concerns the time step specified for the computations. As seen, the value of DT = 0.1 [s] entered in the program is much smaller than 6.242 [s] required by the stability condition of Courant, calculated using the initial conditions. The reader should run the program with different combinations of DX and DT to see that, in all cases, the value of DT necessary to have a stable solution is considerably smaller than the one indicated by the stability condition of Courant, obtained using the initial values. The necessity to use very small time steps in order to have a stable solution is a disadvantage inherent to the use of the explicit scheme.

It should be noted that the time step given by the stability condition of Courant is not constant but varies from one time step to the other and also from one station to the other according to the newly computed values of the velocity, U, and the celerity of the waves, c (the latter depends on the calculated value of the flow depth, h). The stability condition of Courant is constantly verified in the subroutine subprogram INTER for every node. In case of violation of the stability condition at a node (negative flow depth), an error message is displayed and the program stops.

The second point concerns the results of the calculations to be written on the output file. It is clear that writing the flow depth, the discharge and the velocity for all nodes at every time step is not very practical. Hence, the user is invited to select three node numbers for which the results should be printed as a function of the time. In the present case the nodes 1 , 51 and 101 (representing the upstream end, the middle and downstream end of the channel) were selected. The program asks the user to specify the frequency with which the results will be written on the output file. In the present case a printing frequency of 1200 time steps is adopted (NPSTEP). One time step being equal to 0.1 [s], the results will be printed at every 2 minutes.

f) *Anaylsis of computational results and comments :*

The output file of the program EXPLIC is presented in Fig. Ex.5.B.6. The maximum values of the discharge, the flow depth and the velocity for the three stations are marked by an asterisc (*). A graphical representation of the results is given in Fig. Ex.5.B.7.

The Fig. Ex.5.B.7a is an axonometric representation of the flow depth, h(x,t), over the entire space-time, x-t , of the solution domain streching from x = 0 [m] to x = 3000 [m] and from t = 0 [s], namely the beginning of the flood at x = 0 [m], until t = 12500 [s], the value specified by the user.

```
                          PROGRAM EXPLIC
UNSTEADY FLOW COMPUTATION IN A PRISMATIC CHANNEL USING EXPLICIT METHOD

                       SYSTEM OF UNITS : SI

READING OF DATA RELATED TO THE CHANNEL :
The channel may have a trapezoidal, rectangular (m = 0) or
triangular (b = 0) cross section.


Bottom width of channel ?                b (m) = 5.0      [RETURN]
Channel side slope ?                     m (-) = 0.0      [RETURN]
Channel bed slope ?                      Sf (-) = 0.001   [RETURN]
Manning coefficient ?           n (m^-1/3 s) = 0.02       [RETURN]
Uniform flow depth ?                    hn (m) = 1.2      [RETURN]
Total length of channel ?               LT (M) = 3000.0   [RETURN]

Uniform flow discharge can be computed using the formula of
Manning-Strickler (eq. 3.16 or eq. 3.33)
Uniform flow discharge is,            Qo (m3/s) =     8.249


READING OF DATA RELATED TO THE UPSTREAM BOUNDARY CONDITION :

It is assumed that the flood hydrograph is triangular defined by :

Rising time ?            t' (s) = 1200.0    [RETURN]
Peak discharge ?     QMAX (m3/s) = 50.0     [RETURN]
Falling time ?          t'' (s) = 3600.0    [RETURN]

ATTENTION ! ...
For  t > (t' + t'') =   4800.000 (s), it is assumed that Q = Qo



READING OF DATA RELATED TO THE DISCRETISATION OF SPACE, DX :
Total length of channel is                       LT (m) = 3000.00
Maximum number of reaches allowed by the program,    MAXDIV = 100
Minimum value of DX is therefore,         LT / MAXDIV =   30.00(m)

(Note ! If you need to divide the channel into more reaches,
 change the value of the parameter MAXDIV in the main program)

Do you agree with the proposed value ?   Answer Y/CR or N =  [RETURN]

READING OF DATA RELATED TO THE TIME STEP, DT :

Uniform flow discharge is,                 Qo (m3/s) =     8.249
Average velocity of flow is,            Uo (m/s) = Q / S =  1.375
Celerity of gravity waves is,         Co (m/s) = (g*hn)^0.5 = 3.431

Stability condition of Courant requires that :  DT < (DX / (ABS(Uo) + Co)
Maximum value of DT is therefore: DTMAX= DX / (ABS(Uo) + Co) :   6.242 s


Value of time step,    DT (s) ? = 0.1    [RETURN]

Duration of computation,   TMAX (s) ? = 12500    [RETURN]

There are  101 stations along the channel length.
Where would you like to print the results ?
Write the numbers of 3 stations in free format   = 1 51 101  [RETURN]

Frequency for writing the results (number of steps) ? = 1200  [RETURN]

Name of output file ?       = RESULT.DAT   [RETURN]


NORMAL END OF PROGRAM
PRESS RETURN TO END PROGRAM
FORTRAN STOP
```

Fig. Ex.5.B.5 Dialog for introducing the problem's data into the computer.

Fig. Ex.5.B.6 Output file of the program EXPLIC.

```
7320.09    8.249    1.206    1.368    8.428    1.233    1.367    9.031    1.278    1.414
7440.01    8.249    1.206    1.369    8.400    1.228    1.368    8.934    1.268    1.409
7560.02    8.249    1.205    1.370    8.373    1.223    1.369    8.830    1.258    1.404
7680.04    8.249    1.205    1.369    8.355    1.220    1.369    8.741    1.249    1.400
7800.06    8.249    1.203    1.371    8.336    1.216    1.371    8.663    1.241    1.396
7920.08    8.249    1.204    1.371    8.321    1.214    1.371    8.601    1.235    1.393
8040.09    8.249    1.203    1.371    8.312    1.212    1.372    8.545    1.230    1.390
8160.01    8.249    1.203    1.371    8.301    1.210    1.372    8.496    1.225    1.387
8280.10    8.249    1.203    1.372    8.291    1.208    1.372    8.457    1.221    1.385
8400.03    8.249    1.203    1.372    8.284    1.207    1.373    8.423    1.217    1.384
8520.05    8.249    1.202    1.372    8.278    1.206    1.373    8.392    1.214    1.382
8640.08    8.249    1.202    1.373    8.269    1.205    1.373    8.371    1.212    1.381
8760.01    8.249    1.202    1.373    8.269    1.204    1.373    8.353    1.210    1.380
8880.04    8.249    1.202    1.373    8.268    1.203    1.374    8.338    1.209    1.379
9000.07    8.249    1.202    1.373    8.263    1.203    1.374    8.322    1.207    1.379
9120.00    8.249    1.201    1.374    8.264    1.203    1.374    8.310    1.206    1.378
9240.03    8.249    1.202    1.373    8.258    1.202    1.374    8.301    1.205    1.377
9360.06    8.249    1.202    1.372    8.255    1.202    1.374    8.292    1.204    1.377
9480.09    8.249    1.201    1.374    8.263    1.202    1.375    8.285    1.204    1.377
9600.02    8.249    1.202    1.372    8.263    1.202    1.375    8.281    1.203    1.377
9720.05    8.249    1.202    1.373    8.259    1.201    1.375    8.274    1.203    1.376
9840.08    8.249    1.201    1.373    8.253    1.201    1.374    8.270    1.202    1.376
9960.01    8.249    1.201    1.374    8.254    1.201    1.375    8.266    1.202    1.376
9480.09    8.249    1.201    1.374    8.263    1.202    1.375    8.285    1.204    1.377
9600.02    8.249    1.202    1.372    8.263    1.202    1.375    8.281    1.203    1.377
9720.05    8.249    1.202    1.373    8.259    1.201    1.375    8.274    1.203    1.376
9840.08    8.249    1.201    1.373    8.253    1.201    1.374    8.270    1.202    1.376
9960.01    8.249    1.201    1.374    8.254    1.201    1.375    8.266    1.202    1.376
10080.04   8.249    1.202    1.372    8.253    1.201    1.374    8.263    1.201    1.376
10200.07   8.249    1.201    1.374    8.250    1.201    1.374    8.262    1.201    1.375
10320.10   8.249    1.202    1.373    8.248    1.200    1.375    8.259    1.201    1.375
10440.03   8.249    1.201    1.373    8.247    1.200    1.374    8.259    1.201    1.375
10560.05   8.249    1.201    1.373    8.256    1.201    1.375    8.257    1.201    1.375
10680.08   8.249    1.201    1.373    8.257    1.201    1.375    8.256    1.201    1.375
10800.01   8.249    1.201    1.374    8.250    1.200    1.375    8.251    1.200    1.375
10920.04   8.249    1.201    1.374    8.253    1.200    1.375    8.251    1.200    1.375
11040.07   8.249    1.200    1.375    8.248    1.200    1.375    8.250    1.200    1.375
11160.00   8.249    1.202    1.373    8.247    1.200    1.375    8.249    1.200    1.375
11280.03   8.249    1.201    1.374    8.246    1.200    1.374    8.252    1.200    1.375
11400.06   8.249    1.201    1.373    8.250    1.200    1.375    8.248    1.200    1.375
11520.09   8.249    1.201    1.374    8.248    1.200    1.375    8.250    1.200    1.375
11640.02   8.249    1.200    1.375    8.250    1.200    1.375    8.249    1.200    1.375
11760.05   8.249    1.201    1.374    8.249    1.200    1.375    8.250    1.200    1.375
11880.08   8.249    1.201    1.374    8.251    1.201    1.375    8.249    1.200    1.375
12000.01   8.249    1.201    1.373    8.254    1.200    1.375    8.247    1.200    1.375
12120.04   8.249    1.201    1.373    8.251    1.201    1.375    8.250    1.200    1.375
12240.07   8.249    1.201    1.373    8.247    1.200    1.375    8.249    1.200    1.375
12360.10   8.249    1.201    1.373    8.249    1.200    1.375    8.250    1.200    1.375
12480.03   8.249    1.201    1.373    8.250    1.200    1.375    8.249    1.200    1.375
```

back to uniform flow conditions

NORMAL END OF PROGRAM.

Fig. Ex.5.B.6 Output file of the program EXPLIC (continuation and end).

By studying the contents of the output file, Fig. Ex.5.B.6, and the graphs, Fig. Ex.5.B.7a-f, the questions in the problem statement can be answered:

i) The stations, L_1 = 1.5 [km] and L_2 = 3 [km], correspond to nodes 51 and 101. The hydrographs, Q(t), for these two stations, as well as for the station 1 at the upstream end of the channel, are shown in Fig. Ex.5.B.7c. The variation of the velocity, U(t), and the flow depth, h(t), with time at these stations are given in Figs. Ex.5.B.7d and b, respectively.

ii) At the *node 1*, the maximum discharge, Q_{max} = 50 [m³/s], and the maximum velocity, U_{max} = 2.66 [m/s], occur at the same time, t = 1200 [s], after the beginning of the flood, whereas the maximum flow depth, h_{max} = 3.90 [m], appears 360 [s] later, at a time t = 1560 [s].

312　　FLUVIAL HYDRAULICS

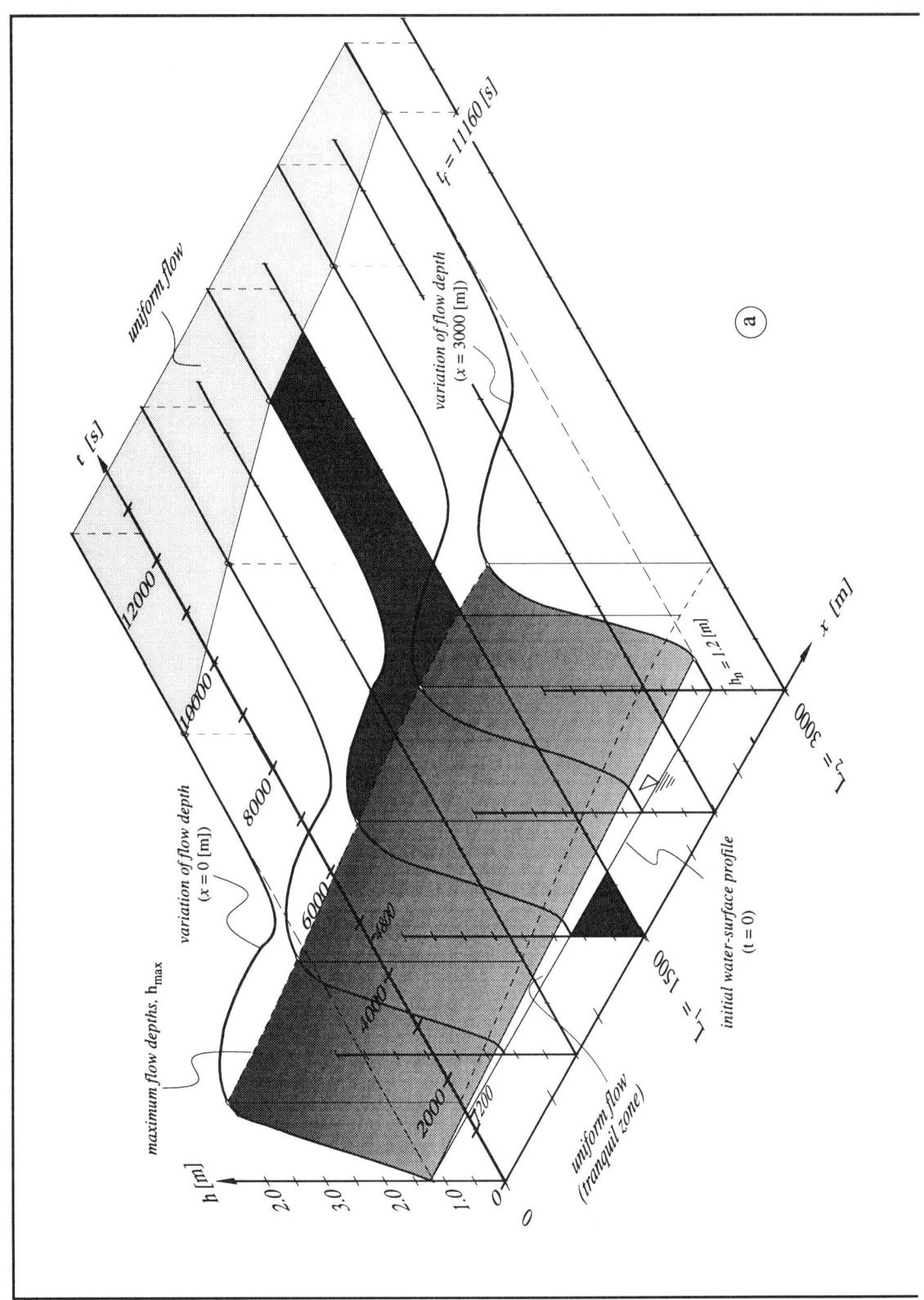

UNSTEADY FLOW

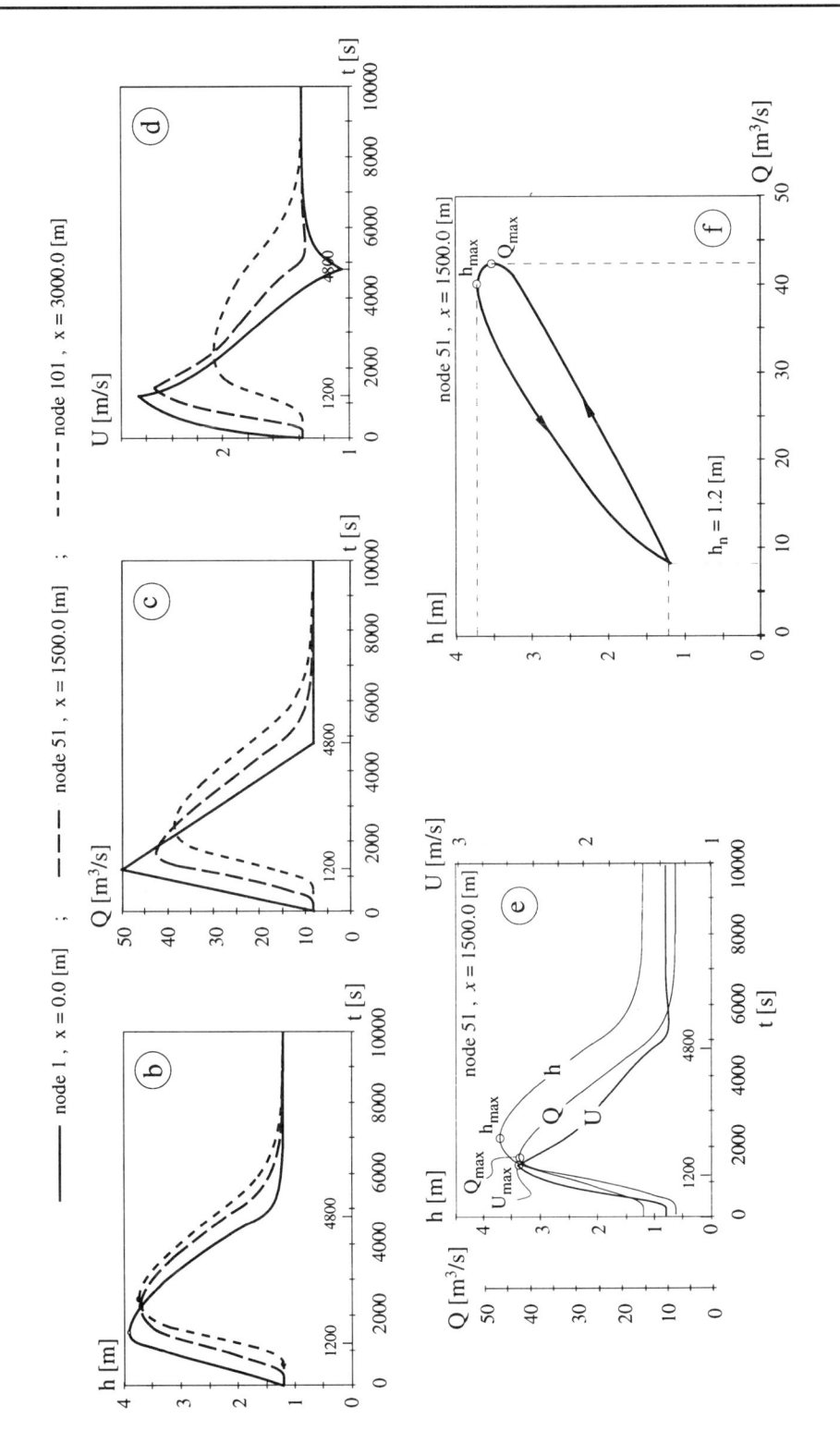

Fig.Ex.5.B.7 Graphical representation of the following relations : $h = f(x,t)$; $h(x) = f(t)$; $Q(x) = f(t)$ and $U(x) = f(t)$ as well as $Q(x) = f(h)$ for $x = L_1$.

At the *node 51*, station L_1 = 1.5 [km], one observes first the maximum velocity, U_{max} = 2.54 [m/s], at time t = 1440 [s]. It is followed by the maximum discharge, Q_{max} = 42.47 [m³/s], arriving at time t = 1680 [s] and the maximum flow depth, h_{max} = 3.71 [m], at time t = 2280 [s]. The variation of Q, U and h at this station is plotted in Fig. Ex.5.B.7e (see also Fig. 5.2). The Fig. Ex.5.B.7f shows the hysteresis loop of the stage-discharge curve (see also Fig. 5.2).

At the *node 101*, station L_2 = 3.0 [km], the maximum values of the discharge, Q_{max} = 38.64 [m³/s], and the velocity, U_{max} = 2.07 [m/s], and the depth, h_{max} = 3.73 [m], occur simultaneously at time t = 2400 [s] (see Fig. Ex. 5.B.6 and the *remarks* at the end of exercise).

NOTE : In Fig. Ex.5.B.7d, it is interesting to note that at the node 1, at time t = 4800 [s], the velocity goes down to U = 1.065 [m/s], which is considerably smaller than the initial velocity of U0 = 1.374 [m/s], corresponding to the uniform flow case. This is due to the fact that when the flood situation is terminated at the node 1, the discharge decreases back to Q = Q0 = 8.25 [m³/s], however the flood conditions still prevail at the downstream nodes. The flow is therefore slowed down by adopting a flow depth of h = 1.55 [m], which is larger than the one corresponding to the uniform flow. In fact, one should wait until t = 8520 [s] to see that the flow returns back to uniform flow values at the node 1 (see Fig. Ex.5.B.6).

iii) Attenuation of the maximum value of the flow depth, the discharge, and the velocity can be clearly seen in Fig. Ex.5.B.7b, c et d, respectively. The exact values can be read from the output file presented in Fig. Ex.5.B.6 ; they are :

node	x [m]	h_{max} [m]	Q_{max} [m³/s]	U_{max} [m/s]
1	0	3.904	50.00	2.66
51	1500	3.705	42.47	2.54
101	3000	3.734	38.64	2.07

In Fig. Ex.5.B.7b, one observes an important attenuation, 3.90 – 3.71 = 0.19 [m], of the flow depth from the node 1 to the node 51, which clearly indicates the diffusive character of the flood wave in this reach (see also Fig. 5.8). There is practically no difference between the maximum values of the flow depth at the nodes 51 and 101, since the dynamic wave has been transformed into a kinematic wave. The maximum flow depth at the node 101 appears to be even 0.03 [m] larger than the one at the node 51.

iv) In order to say that the flow became again uniform at a given station, it is necessary that the discharge, Q, the flow depth, h, and the velocity, U, all return back to their initial values before the arrival of the flood, namely Q0 = 8.25 [m³/s], H0 = 1.20 [m] and U0 = 1.38 [m/s], respectively. For the station L_2 = 3000 [m], these conditions are reasonably satisfied at the time t_f = 11160 [s] (see Fig. Ex.5.B.7a-b and the numbers in a rectangular frame in Fig. Ex.5.B.6).

UNSTEADY FLOW

REMARKS :

For the downstream boundary condition (see point c.5), it was assumed that at the station $L_2 = 3$ [km], the formula of Manning for uniform flow is valid. That particular choice for the boundary condition implicitely forces simulatenous occurance of maximum velocity, U_{max}, maximum discharge, Q_{max}, and maximum flow depth, h_{max}, at the time, $t = 2400$ [s], (see Fig. Ex. 5.B.7a and Fig. Ex. 5.B.6) at this station.

It seems however, that a more realistic solution can be obtained by pushing the downstream boundary, namely the point where Manning's formula is assumed to be valid, to a distance considerably farther downstream.

In order to show the influence of the downstream boundary condition, a new run was made by specifying the downstream of the channel to be at a distance of $L = 6$ [km]. The variation of $Q(t)$, $h(t)$ and $U(t)$ for the station at $L_2 = 3$ [km], are shown in Fig. Ex. 5.B.9 for comparaison (see also Fig. Ex. 5.B.7).

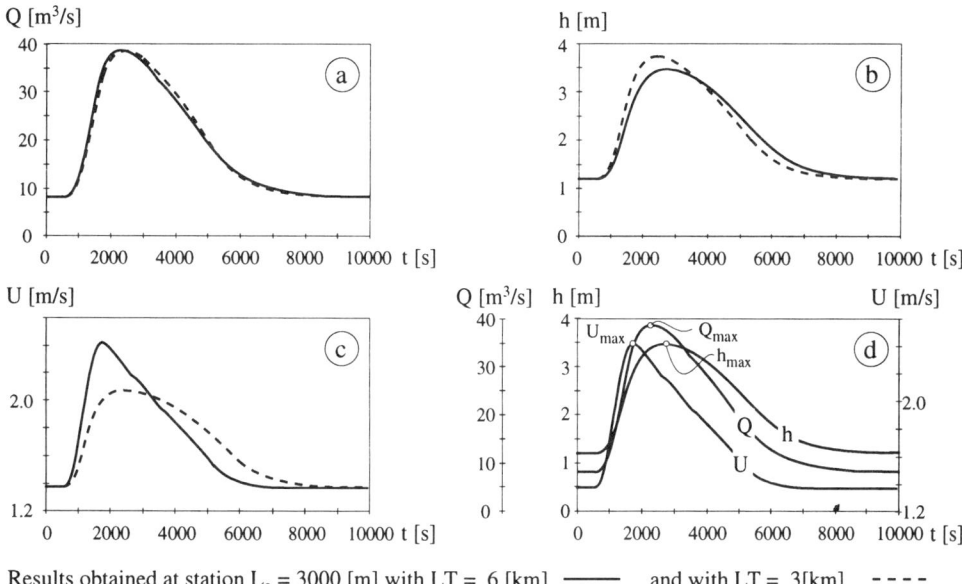

Results obtained at station $L_2 = 3000$ [m] with LT = 6 [km] ——— and with LT = 3[km] - - - -

Fig. Ex.5.B.9 Comparison of the relations $Q(t)$, $h(t)$ and $U(t)$ at the station $L_2 = 3$ [km] obtained for different channel lengths, LT.

The table below shows a comparison of the results obtained during these two different runs for the station $L_2 = 3$ [km]. It is seen that the numerical values are slightly different. The maximum values of velocity, discharge and flow depth occur now in the order (see Fig. 5.2) to be expected for a dynamic wave (see Fig. Ex. 5.B.9d).

node	x [m]	h_{max} [m]		Q_{max} [m³/s]		U_{max} [m/s]	
		LT = 3 [km]	LT = 6 [km]	LT = 3 [km]	LT = 6 [km]	LT = 3 [km]	LT = 6 [km]
51	1500	3.705	3.651	42.47	42.564	2.54	2.534
101	3000	3.734	3.465	38.64	38.691	2.07	2.415

```
      PROGRAM EXPLIC
C******************************************************************
C
C     MAIN PROGRAM FOR THE SOLUTION OF UNSTEADY FLOW IN A PRISMATIC CHANNEL OF  *
C     TRAPEZOIDAL, RECTANGULAR, OR TRIANGULAR CROSS SECTION, USING A FINITE     *
C     DIFFERENCE EXPLICIT SCHEME.                                               *
C
C******************************************************************
C
C     LIST OF VARIABLES DEFINED GLOBALLY FOR THE MAIN PROGRAM AND THE
C     SUBPROGRAMS. VARIABLES DEFINED LOCALLY IN SUBPROGRAMS ARE LISTED AT
C     THE BEGINNING OF EACH SUBPROGRAM.
C
C     TYPE  NAME       DIMEN.      EXPLANATIONS
C     ====  ====       ======      ============
C     R*4   B                    = WIDTH AT THE CHANNEL BOTTOM
C     R*4   BETA                 = SEE EQ. 5.24a
C     R*4   C0                   = CELERITY OF GRAVITY WAVES AT TIME T = 0
C     R*4   DT                   = TIME STEP
C     R*4   DTMAX                = MAXIMUM VALUE OF DT ACCORDING TO STABILITY CONDITION
C                                  OF COURANT
C     R*4   DX                   = STEP LENGTH ALONG THE CHANNEL AXIS (REACH LENGTH)
C     CHAR  PICH                 = NAME OF THE OUTPUT FILE
C     R*4   G                    = GRAVITATIONAL ACCELERATION (G = 9.81 m/s2)
C     R*4   GAMA                 = SEE EQ. 5.24
C     R*4   H         (2,101)    = FLOW DEPTH AT THE NODES FOR THE LAST TWO TIME STEPS
C     R*4   HN                   = UNIFORM FLOW DEPTH
C     I*4   I                    = COUNTER FOR NODE NUMBERS
C     I*4   IL                   = COUNTER FOR THE LINE NUMBERS OF THE TITLE
C     R*4   SF                   = BED SLOPE
C     I*4   K                    = DO-LOOP VARIABLE
C     R*4   LT                   = TOTAL CHANNEL LENGTH
C     R*4   M                    = SIDE SLOPE OF THE CHANNEL
C     I*4   MAXDIV               = MAXIMUM NUMBER OF REACHES ALLOWED BY THE PROGRAM (100)
C     R*4   N                    = MANNING COEFFICIENT
C     I*4   NDIV                 = NUMBER OF REACHES
C     I*4   NN                   = NUMBER OF NODES
C     I*4   NNMAX                = MAXIMUM NUMBER OF NODES ALLOWED BY THE PROGRAM (100+1)
C     I*4   NOST      (3)        = NUMBERS OF THE NODES WHERE THE RESULTS ARE PRINTED
C     I*4   NP                   = TIME STEP NUMBER AT TIME T
C     I*4   NPSTEP               = FREQUENCY (NUMBER OF TIME STEPS) OF PRINTING
C     I*4   NUNIT                = UNIT NUMBER FOR THE OUTPUT FILE
C     R*4   P                    = WETTED PERIMETER
C     R*4   Q         (2,101)    = DISCHARGES AT THE NODES FOR THE LAST TWO TIME STEPS
C     R*4   QE                   = DISCHARGE ENTERING THE CHANNEL AT TIME T
C     R*4   Q0                   = INITIAL DISCHARGE (UNIFORM FLOW DISCHARGE)
C     R*4   QMAX                 = PEAK FLOOD DISCHARGE
C     CHAR  REP                  = ALPHANUMERICAL ANSWER
C     R*4   RH                   = HYDRAULIC RADIUS
C     R*4   A                    = WETTED SURFACE
C     R*4   T                    = DURATION OF COMPUTATIONS
C     CHAR  TITLE     (24)       = TITLE TO BE PRINTED AT THE BEGINNING OF OUTPUT FILE
C     R*4   TMAX                 = TIME LIMIT FOR THE COMPUTATIONS
C     R*4   TP                   = RISE TIME FOR THE HYDROGRAPH
C     R*4   TPD                  = FALL TIME FOR THE HYDROGRAPH
C     R*4   TTH                  = TOTAL DURATION OF FLOOD (= TP + TPD)
C     R*4   U         (2,101)    = VELOCITIES AT THE NODES FOR THE LAST TWO TIME STEPS
C     R*4   U0                   = INITIAL VELOCITY (UNIFORM FLOW VELOCITY)
C******************************************************************
C
C     PARAMETERS
      PARAMETER ( NUNIT = 5 )
      PARAMETER ( MAXDIV = 100 , NNMAX = MAXDIV+1 )
      PARAMETER ( G = 9.81 )
C
C     DECLARATION OF VARIABLES
      CHARACTER*50 PICH
      REAL         M , SF , N , LT
      DIMENSION    H(2,NNMAX) , U(2,NNMAX) , Q(2,NNMAX) , NOST(3)
C
C     CALL SUBROUTINE "DREAD" TO READ INTERACTIVELY THE DATA OF THE PROBLEM
C
      CALL DREAD( MAXDIV , NNMAX , G ,
     1            B , M , SF , N , HN , LT , Q0 , U0 , C0 ,
     2            TP , QMAX , TPD ,
     3            NDIV , NN , DX , DT , TMAX ,
     4            NOST , NPSTEP , NUNIT , PICH )
C
C     CALL SUBROUTINE "INIT" TO COMPUTE THE INITIAL CONDITIONS (AT TIME T = 0)
C     AT ALL THE NODES
C
      CALL INIT ( NNMAX , NN , H , U , Q , HN , U0 , Q0 )
C
C     CALL SUBROUTINE "CALCUL" FOR COMPUTING THE UNSTEADY FLOW
C
      CALL CALCUL ( NNMAX , NN , H , U , Q , HN , LT ,
     1              B , M , SF , N , Q0 , U0 , C0 , G ,
     2              TP , QMAX , TPD , DX , DT , TMAX ,
     3              NOST , NPSTEP , NUNIT )
C
C     END OF THE PROGRAM
C
      STOP
      END
```

Fig.Ex.5.B.8 Program EXPLIC (continued).

UNSTEADY FLOW

```fortran
      SUBROUTINE DREAD( MAXDIV , NNMAX , G ,
     1                 B , M , SF , N , HN , LT , Q0 , U0 , C0 ,
     2                 TP , QMAX , TPD , DT , TMAX ,
     3                 NDIV , NN , DX , DT , TMAX ,
     4                 NOST, NPSTEP , NUNIT , FICH )
C*****************************************************************
C
C     SUBPROGRAM FOR INTERACTIVE READING OF THE PROBLEM DATA
C
C*****************************************************************
C
C     DECLARATION OF VARIABLES
      CHARACTER*1  REP
      CHARACTER*50 FICH
      REAL         M, SF , N , LT
      DIMENSION    NOST(3)
C
    5 FORMAT(A)
      OPEN( UNIT = 7 , FILE = 'DIALOG.DAT' , STATUS = 'NEW')
C
C     DISPLAY THE PROGRAM TITLE
C
      WRITE(*,20)
      WRITE(7,20)
      WRITE(*,20)
      WRITE(*,10)
      WRITE(7,10)
   10 FORMAT(//
     1' UNSTEADY FLOW COMPUTATION IN A PRISMATIC CHANNEL USING E'
     2'XPLICIT METHOD.'
     3             //                SYSTEM OF UNITS : SI'//)
C
C     READING THE DATA ON THE CHANNEL CHARACTERISTICS
C
      WRITE(*,20)
      WRITE(7,20)
   20 FORMAT(' READING OF DATA RELATED TO THE CHANNEL :'/
     1'  The channel may have a trapezoidal, rectangular (m = 0) o'
     2'r'/
     3' triangular (b = 0) cross section.'//)
      WRITE(*,40)
      WRITE(7,40)
   40 FORMAT(' Bottom width of channel ?                 b (m) = ',$)
      READ(*,*,ERR=30)B
      WRITE(*,60)
      WRITE(7,60)
   60 FORMAT(' Channel side slope ?                      m (-) = ',$)
      READ(*,*,ERR=50)M
      WRITE(*,80)
      WRITE(7,80)
   80 FORMAT(' Channel bed slope ?                      Sf (-) = ',$)
      READ(*,*,ERR=70)SF
      WRITE(*,100)
      WRITE(7,100)
  100 FORMAT(' Manning coefficient ?            n (m^-1/3 s) = ',$)
      READ(*,*,ERR=90)N
      WRITE(*,120)
      WRITE(7,120)
  120 FORMAT(' Uniform flow depth ?                     hn (m) = ',$)
      READ(*,*,ERR=110)HN
      WRITE(*,140)
      WRITE(7,140)
  140 FORMAT(' Total length of channel ?                LT (M) = ',$)
      READ(*,*,ERR=130)LT
C
C     COMPUTATION OF UNIFORM FLOW DISCHARGE (INITIAL CONDITION)
C
      A  = HN * (B + M * HN)
      P  = B + 2 * HN * SQRT(1 + M**2)
      RH = A / P
      U0 = RH**(2./3.) * SQRT(SF) / N
      Q0 = U0 * A
      WRITE(*,150) Q0
      WRITE(7,150) Q0
  150 FORMAT(// ' Uniform flow discharge can be computed using the formu'
     2'la of:'/
     2        '     . Manning-Strickle: (eq. 3.16 or eq. 3.33)'/
     3        '     . Uniform flow discharge is,       Qo (m3/s) = ',
     4           F8.3)
C
C     READING THE DATA ON THE BOUNDARY CONDITIONS
C
      WRITE(*,200)
      WRITE(7,200)
  200 FORMAT(///' READING OF DATA RELATED TO THE UPSTREAM BOUNDARY COND'
     1'ITION : '//
     2'         . It is assumed that the flood hydrograph is triangular'
     3'defined by :'//)
      WRITE(*,220)
      WRITE(7,220)
  220 FORMAT(' Rising time ?                         t'' (s) = ',$)
      READ(*,*,ERR=210)TP
      WRITE(*,240)
      WRITE(7,240)
  240 FORMAT(' Peak discharge ?              QMAX (m3/s) = ',$)
      READ(*,*,ERR=230)QMAX
      WRITE(*,260)
      WRITE(7,260)
  260 FORMAT(' Falling time ?                        t'''' (s) = ',$)
      READ(*,*,ERR=250)TPD
      TTH = TP + TPD
      WRITE(*,270) TTH
      WRITE(7,270) TTH
  270 FORMAT(' ATTENTION !...'/' For t > (t'' + t'''') = ',F10.3,
     1'    (s), it is assumed that Q = Qo'///)
C
C     READING THE STEP LENGTHS IN SPACE AND IN TIME
C
      NDIV = MAXDIV
      NN   = NNMAX
      DX   = LT / MAXDIV
      WRITE(*,310) LT , MAXDIV , LX
      WRITE(7,310) LT , MAXDIV , LX
  310 FORMAT(// ' READING OF DATA RELATED TO THE DISCRETISATION OF SPACE,'
     1'DX :'/
     2         ' Total length of channel is                          LT'
     3' (m) = ',F7.2/
     4         ' Maximum number of reaches allowed by the program.   MA'
     5'XDIV = ',I3/
     6         ' Minimum value of DX is therefore,               LT / MA'
     7'XDIV = ',F7.2,' (m)'//)
```

Fig.Ex.5.B.8 Program EXPLIC (continued).

```fortran
      8                ' (Note ! If you need to divide the channel into more reach
     9es,'/'           change the value of the parameter MAXDIV in the main progr
     Aam'/
     B                 ' // Do you agree with the proposed value ?   Answer Y/CR or N
     C = ',$)
            READ(*,5)REP
            IF(REP.EQ.'N'.OR.REP.EQ.'n')THEN
              WRITE(*,330)
              WRITE(7,330)
 320          READ(*,*,ERR=320)NDIV
              IF(NDIV.LE.0.OR.NDIV.GT.MAXDIV) GO TO 300
              DX = LT / NDIV
              NN = NDIV + 1
              WRITE(*,340) DX , NN
              WRITE(7,340) DX , NN
 330     FORMAT(' How many reaches will there be, NDIV ? = ',$)
 340     FORMAT(/' Value of DX is therefore,            LT',
     1           ' / NDIV = ',F7.2,' (m)'/
     2           ' Number of nodes is,                  NN =',
     3           ' NDIV + 1 = ',I3)
            ENDIF
C
            C0   = SQRT(G * HN)
            DTMAX = DX / (U0 + C0)
            DT   = DTMAX
 400        WRITE(*,410) Q0 , U0 , C0 , DTMAX
            WRITE(7,410) Q0 , U0 , C0 , DTMAX
 410     FORMAT(//' READING OF DATA RELATED TO THE TIME STEP, DT :'//
     1           ' Uniform flow discharge is,              Q0 (m3
     2/s) = ',F8.3/
     3           ' Average velocity of flow is,            U0 (m/s) = Q
     4/ S = ',F8.3/
     5           ' Celerity of gravity waves is,           C0 (m/s) = (g*hn)^
     60.5 = ',F8.3/
     7           ' Stability condition of Courant requires that : DT < (DX
     8/ (ABS(U0) + C0)/
     9           ' Maximum value of DT is therefore: DTMAX= DX / (ABS(U0) +
     AC0) : ',F8.3,' s'//)
            WRITE(*,430)
            WRITE(7,430)
 420        READ(*,*,ERR=420)DT
 430     FORMAT(' Value of time step,           DT (s) ? = ',$)
            IF(DT.LE.0.0.OR.DT.GT.DTMAX) GO TO 400
C
C WHEN SHOULD THE COMPUTATIONS BE TERMINATED ?
            WRITE(*,450)
            WRITE(7,450)
 440        READ(*,*,ERR=440)TMAX
 450     FORMAT(//' Duration of computation,         TMAX (s) ? = ',$)
            IF(TMAX.LT.DT) GO TO 440
C
C HOW TO PRINT THE RESULTS ?
 500        WRITE(*,510)NN
            WRITE(7,510)NN
 510     FORMAT(//' There are ',I3,' stations along the channel length.'/
     1           ' Where would you like to print the results ?'/
     2           ' Write the numbers of 3 stations in free format = ',$)
            READ(*,*,ERR=500) ( NOST(K) , K=1,3)
            DO 520 K=1,3
            IF(NOST(K).LT.1 .OR. NOST(K).GT.NN)GO TO 500
 520        CONTINUE
C
C FREQUENCY OF PRINTING THE RESULTS IN THE OUTPUT FILE
 550        WRITE(*,560)
            WRITE(7,560)
 560     FORMAT(//' Frequency for writing the results (number of steps) ? =
     1 ',$)
            READ(*,*,ERR=550)NPSTEP
            IF(NPSTEP.LT.1 .OR. NPSTEP.GT.(TMAX/DT))GO TO 550
C
C OUTPUT FILE
 600        WRITE(*,610)
            WRITE(7,610)
 610     FORMAT(//' Name of output file ?              = ',$)
            READ(*,5) FICH
            OPEN( UNIT = NUNIT , FILE = FICH , STATUS = 'NEW' , ERR = 600 )
C
            RETURN
            END
C**********************************************************************
            SUBROUTINE INIT ( NNMAX , NN , H , U , Q , HN , U0 , Q0 )
C**********************************************************************
C
C THIS SUBROUTINE ASSIGNS THE INITIAL (T = 0) VALUES OF DEPTH, VELOCITY, AND
C DISCHARGE TO ALL NODES.
C
C DECLARATION OF VARIABLES
            DIMENSION H(2,NNMAX) , U(2,NNMAX) , Q(2,NNMAX)
C
C ASSIGN THE UNIFORM FLOW VALUES TO ALL NODES
            DO 10 I = 1,NN
              H(1,I) = HN
              U(1,I) = U0
              Q(1,I) = Q0
 10         CONTINUE
C
            RETURN
            END
C**********************************************************************
            SUBROUTINE CALCUL ( NNMAX , NN , H , U , Q , HN , LT ,
     1                          B , M , SF , N , Q0 , U0 , C0 , G ,
     2                          TP , QMAX , TPD , DX , DT , TMAX
     3                          NOST , NPSTEP , NUNIT )
C**********************************************************************
C
C THIS SUBPROGRAM USES THE EXPLICIT FINITE DIFFERENCE METHOD (SEE CHAP. 5.2.3)
C TO SOLVE THE UNSTEADY FLOW IN AN OPEN CHANNEL. IT COMPUTES DEPTH, VELOCITY
C AND THE DISCHARGE AT ALL THE NODES OF A DISCRETIZED SOLUTION DOMAIN BY SMALL
C TIME INCREMENTS.
C
C DECLARATION OF VARIABLES
            REAL     M , N , SF
            DIMENSION H(2,NNMAX) , U(2,NNMAX) , Q(2,NNMAX) , NOST(3)
C
C INITIALIZE THE TIME
            T = 0
```

Fig.Ex.5.B.8 Program EXPLIC (continued).

```
C     WRITE THE TITLES IN OUTPUT FILE
      CALL TITLES ( B  , N  , M  , HN , SF , LT , Q0 , U0 , C0 ,
     1              TP , QMAX , TPD , NDIV , DX , NN , DT ,
     2              TMAX , NPSTEP , NOST , NUNIT )
C
C     WRITE THE RESULTS FOR THE TIME T
 100  CALL RWRITE ( NNMAX , NN , H , U , Q , T , DT ,
     1              NOST , NPSTEP , NUNIT )
C
C     INCREMENT THE TIME BY DT IF THE END OF COMPUTATIONS IS NOT YET REACHED
      IF(T+DT.GT.TMAX) GO TO 900
      T = T + DT
C
C     COMPUTE THE DICHARGE ENTERING INTO THE CHANNEL AT THE UPSTREAM END, AT
C     THE TIME T
      CALL QENTRE ( TP , QMAX , TPD , T , Q0 , QE )
C
C     COMPUTE THE VALUES OF DEPTH, VELOCITY AND DISCHARGE FOR THE NODE AT
C     THE UPSTREAM END OF THE CHANNEL, AT THE RUNNING TIME T
      CALL UPSTR ( NNMAX , NN , H , U , Q ,
     1              B , M , QE , DX , DT )
C
C     COMPUTE THE VALUES OF DEPTH, VELOCITY AND DISCHARGE FOR THE INTERME-
C     DIATE NODES, AT THE RUNNING TIME T
      CALL INTER ( NNMAX , NN , H , U , Q ,
     1              B , M , SF , N , G , DX , DT , NUNIT )
C
C     COMPUTE THE VALUES OF DEPTH, VELOCITY AND DISCHARGE FOR THE NODE AT
C     THE DOWNSTREAM END OF THE CHANNEL, AT THE RUNNING TIME T
      CALL DNSTR ( NNMAX , NN , H , U , Q ,
     1              B , M , SF , N , DX , DT )
C
C     REPLACE THE VALUES COMPUTED AT THE PRECEDING TIME STEP BY THE NEW
C     VALUES COMPUTED AT THE RUNNING TIME STEP
      DO 200 I = 1, NN
      H(1,I) = H(2,I)
      U(1,I) = U(2,I)
      Q(1,I) = Q(2,I)
 200  CONTINUE
C
C     GO TO INCREMENT THE TIME AND REPEAT THE CALCULATIONS
      GO TO 100
C
C     IF THE END OF COMPUTATIONS IS NOT YET REACHED
 900  WRITE(*,910)
      WRITE(NUNIT,910)
 910  FORMAT(//' NORMAL END OF PROGRAM')
      WRITE(*,*) ' PRESS RETURN TO END PROGRAM'
      READ(*,*)
      RETURN
      END
C
      SUBROUTINE TITLES ( B , N , M , HN , SF , LT , Q0 , U0 , C0 ,
     1                    TP , QMAX , TPD , NDIV , DX , NN , DT ,
     2                    TMAX , NPSTEP , NOST , NUNIT )
C
C********************************************************************
C     THIS SUBROUTINE SUBPROGRAM WRITES THE TITLES AND ECHO-PRINTS THE INPUT DATA
C     ONTO THE OUTPUT FILE.
C********************************************************************
C
C     DECLARATION OF VARIABLES
      CHARACTER*131 TITLE(24)
      REAL          M , SF , N , LT
      DIMENSION NOST(3)
C
      DATA TITLE(1)/'                            RESULTS OF THE UNSTEADY FLOW COMP
     1UTATION IN A PRISMATIC CHANNEL USING THE EXPLICIT METHOD'/
      DATA TITLE(2)/'==================================================================
     1================================================================'/
      DATA TITLE(3),TITLE(4)/' ',' '/
      DATA TITLE(5)/'                                                   DATA CONCERNING THE CHA
     1NNEL AND THE UNIFORM FLOW (INITIAL CONDITION)'/
      DATA TITLE(6)/'                                                   ----------------
     1--------------------------------------------'/
      DATA TITLE(7)/' BOTTOM WIDTH OF CHANNEL,              B (m) = ',
     1'                 MANNING COEFFICIENT,             n (m^-1/3 s) = '/
      DATA TITLE(8)/' CHANNEL SIDE SLOPE,                   m (-) = ',
     1'                 UNIFORM FLOW DEPTH,               hn (m) = '/
      DATA TITLE(9)/' CHANNEL BED SLOPE,                    sf (-) = ',
     1'                 TOTAL LENGTH OF CHANNEL,          LT (m) = '/
      DATA TITLE(10)/' ',' '/
      DATA TITLE(11)/'                INITIAL DISCHARGE,    Q0 (M3/S) = ',
     1'                INITIAL VELOCITY, U0 (m/s) =               CELERITY OF
     1GRAVITY WAVES C0 (m/s) = '/
      DATA TITLE(12)/' ',' '/
      DATA TITLE(13)/'                                                   t'' (s) = '/
      DATA TITLE(14)/'                                                   
     1RISING TIME,                                       t'' (s) = '/
      DATA TITLE(15)/' DEFINITION OF TRIANGULAR HYDROGRAPH :
     1PEAK DISCHARGE, QMAX (m3/s) = '/
      DATA TITLE(16)/'                                                   
     2FALLING TIME,                                      t'' (s) = '/
      DATA TITLE(17)/' ',' '/
      DATA TITLE(18)/' NUMBER OF REACHES, NDIV =           REACH LENGTH
     1,                                           NUMBER OF NODES, NN=NDIV+1
     2 = '/
      DATA TITLE(19)/' TIME STEP,                       DT (s) =           DURATION OF
     1COMPUTATIONS, TMAX (s) = '/
      DATA TITLE(20)/'                                                       FREQUENCY OF WRITING (step)
     2 = '/
      DATA TITLE(21)/' ',' '/
      DATA TITLE(22)/'       STATION 1                     STATION 2                     STATION 3 '/
      DATA TITLE(23)/'   NODE NO =            X =            NODE NO =            X =            NODE NO =            X =
     1                    TIME
     2 '/
      DATA TITLE(24)/'     T        Q        H        U          Q        H        U          Q        H        U
     1     (s)     (m3/s)     (m)     (m/s)      (m3/s)     (m)    (m/s)      (m3/s)    (m)    (m/s)
     2(m/s) '/
C
      WRITE(TITLE(7)(45:51),'(F7.3)')B
      WRITE(TITLE(7)(109:114),'(F5.4)')N
      WRITE(TITLE(8)(45:51),'(F7.3)')M
      WRITE(TITLE(8)(109:114),'(F5.3)')HN
      WRITE(TITLE(9)(45:52),'(F8.5)')SF
      WRITE(TITLE(9)(109:116),'(F3.3)')LT
      WRITE(TITLE(11)(49:56),'(F8.3)')Q0
      WRITE(TITLE(11)(103:110),'(78.3)')U0
      WRITE(TITLE(12)(76:83),'(F8.3)')C0
```

Fig.Ex.5.B.8 Program EXPLIC (continued).

```
      WRITE(TITLE(14)(85:93),'(F9.2)')TP
      WRITE(TITLE(15)(85:92),'(F8.3)')QMAX
      WRITE(TITLE(16)(85:93),'(F9.2)')TPD
      WRITE(TITLE(18)(29:31),'(I3)')NN-1
      WRITE(TITLE(18)(76:83),'(F8.3)')DX
      WRITE(TITLE(18)(120:122),'(I3)')NN
      WRITE(TITLE(19)(29:35),'(F7.3)')DT
      WRITE(TITLE(19)(76:86),'(F11.3)')TMAX
      WRITE(TITLE(19)(120:124),'(I5)')NPSTEP
C
      WRITE(TITLE(22)(30:32),'(I3)')NOST(1)
      WRITE(TITLE(22)(43:50),'(F8.3)')DX*(NOST(1)-1)
      WRITE(TITLE(22)(67:69),'(I3)')NOST(2)
      WRITE(TITLE(22)(80:87),'(F8.3)')DX*(NOST(2)-1)
      WRITE(TITLE(22)(104:106),'(I3)')NOST(3)
      WRITE(TITLE(22)(117:124),'(F8.3)')DX*(NOST(3)-1)
C
      DO 200 IL = 1 , 24
      WRITE(NUNIT,100)TITLE(IL)
100   FORMAT(A)
200   CONTINUE
      RETURN
      END
C
      SUBROUTINE RWRITE ( NNMAX , NN , H , U , Q , T , DT ,
     1                   NOST , NPSTEP , NUNIT )
C********************************************************************
C THIS SUBROUTINE WRITES THE VALUES OF H , U AND Q FOR THE RUNNING TIME T ON  *
C THE OUTPUT FILE.                                                             *
C********************************************************************
C DECLARATION OF VARIABLES
      DIMENSION H(2,NNMAX) , U(2,NNMAX) , Q(2,NNMAX) , NOST(3)
C
      NP = T / DT
      IF( MOD(NP,NPSTEP).NE.0 ) GO TO 100
      WRITE(NUNIT,10) T , Q(1,NOST(1)) , H(1,NOST(1)) , U(1,NOST(1)) ,
     1                    Q(1,NOST(2)) , H(1,NOST(2)) , U(1,NOST(2)) ,
     2                    Q(1,NOST(3)) , H(1,NOST(3)) , U(1,NOST(3))
10    FORMAT(3X,F10.2,3(5X,F10.3,1X,F10.3,1X,F10.3))
100   RETURN
      END
C
      SUBROUTINE QENTRE ( TP , QMAX , TPD , T , Q0 , QE )
C********************************************************************
C THIS SUBPROGRAM CALCULATES THE DISCHARGE ENTERING THE CHANNEL FROM THE       *
C UPSTREAM END AT THE RUNNING TIME T, BY ASSUMING A TRIANGULAR HYDROGRAPH.     *
C********************************************************************
C
      IF( T.LE.TP)THEN
C
C RISING PART OF THE HYDROGRAPH
      QE = Q0 + T * (QMAX - Q0) / TP
      RETURN
      ENDIF
C
      IF(T.LT.TP+TPD)THEN
C
C FALLING PART OF THE HYDROGRAPH
      QE = QMAX - (T - TP) * (QMAX - Q0) / TPD
      RETURN
      ENDIF
C
C T >= (TP + TPD), TRIANGULAR HYDROGRAPH IS TERMINATED.
C ENTERING DISCHARGE IS ASSUMED TO BE EQUAL TO THE UNIFORM FLOW DISCHARGE.
      QE = Q0
C
      RETURN
      END
C
      SUBROUTINE UPSTR( NNMAX , NN , H , U , Q ,
     1                 B , M , QE , DX , DT )
C********************************************************************
C THIS SUBPROGRAM CALCULATES DEPTH, VELOCITY AND DICHARGE AT THE NODE 1,       *
C LOCATED AT THE UPSTREAM END OF THE CHANNEL, FOR THE RUNNING TIME T.          *
C THE NEWLY CALCULATED VALUES ARE STORED IN THE SECOND LINE OF CORRESPONDING   *
C ARRAYS WHEREAS THE VALUES FOR THE PRECEDING TIME ARE IN THE FIRST LINE.      *
C********************************************************************
C DECLARATION OF VARIABLES
      REAL     M
      DIMENSION H(2,NNMAX) , U(2,NNMAX) , Q(2,NNMAX)
C
C COMPUTE FIRST THE WATER DEPTH AT THE NODE "1" AT THE RUNNING TIME
      H(2,1) = H(1,1) + (DT/DX) *
     1        ( U(1,1)*(H(1,1)-H(1,2)) + H(1,1)*(U(1,1)-U(1,2)) )
C
C THEN COMPUTE THE WETTED SURFACE FOR THIS NEW WATER DEPTH
      A = H(2,1) * (B + M * H(2,1))
C
C THE NEW DISCHARGE AT THE NODE "1" IS EQUAL TO THE ENTERING DISCHARGE
C INTERPOLATED FROM THE TRIANGULAR FLOOD HYDROGRAPH
      Q(2,1) = QE
C
C NEW VELOCITY AT THE NODE "1" AT THE RUNNING TIME
      U(2,1) = Q(2,1) / A
C
      RETURN
      END
C
      SUBROUTINE INTER( NNMAX , NN , H , U , Q ,
     1                 B , M , SF , N , G , DX , DT , NUNIT )
C********************************************************************
C THIS SUBPROGRAM CALCULATES DEPTH, VELOCITY AND DISCHARGE AT ALL THE INTER-   *
C MEDIATE NODES (NODES BETWEEN THE BOUNDARY NODES) FOR THE RUNNING TIME T.     *
C THE NEWLY CALCULATED VALUES ARE STORED IN THE SECOND LINE OF CORRESPONDING   *
C ARRAYS WHEREAS THE VALUES FOR THE PRECEDING TIME ARE IN THE FIRST LINE.      *
C********************************************************************
C DECLARATION OF VARIABLES
      REAL     M , N , SF
      DIMENSION H(2,NNMAX) , U(2,NNMAX) , Q(2,NNMAX)
C
      DO 300 I = 2 , NN-1
```

Fig.Ex.5.B.8 Program EXPLIC (continued).

UNSTEADY FLOW

```
C
C COMPUTE FIRST THE DEPTH AT THE NODE "NN" AT THE RUNNING TIME
      H(2,I) = H(1,I) + 0.5 * (DT/DX) *
     1          ( U(1,I)*(H(1,I-1)-H(1,I+1)) +
     2            H(1,I)*(U(1,I-1)-U(1,I+1)) )
C
C VERIFY THE CONVERGENCE
      IF(H(2,I).LE.0)THEN
        WRITE(*,100)
100     FORMAT(' ERROR ! THE COMPUTED DEPTH IS NEGATIVE !'/
     1         ' TRY TO RERUN THE PROGRAM USING A SMALLER TIME STEP'/
     2         ' ABNORMAL END OF PROGRAM !'//)
        WRITE(NUNIT,100)
        WRITE(*,*)' PRESS RETURN TO END PROGRAM'
        READ(*,*)
        STOP
      ENDIF
C
C COMPUTE ALSO THE VELOCITY AT THE RUNNING TIME
      BETA = U(1,I) +
     1    0.5 * (DT/DX) * U(1,I)*(U(1,I-1)-U(1,I+1)) +
     2    0.5 * G * (DT/DX) * (H(1,I-1)-H(1,I+1)) +
     3    G * DT * SF
      RH = (B + M * H(2,I)) * H(2,I) / (B + 2 * H(2,I) * SQRT(1 + M**2))
      GAMA = RH**(4./3.) / (N**2 * G * DT)
      U(2,I) = 0.5 * (SQRT(GAMA**2 + 4.0*GAMA*BETA) - GAMA)
C
C VERIFY THE STABILITY CONDITION OF COURANT
      IF(DX/(ABS(U(2,I))+SQRT(H(2,I)*G)).LT.4*DT)THEN
        WRITE(*,200)
200     FORMAT(' ERROR ! THE STABILITY CONDITION OF COURANT IS'/
     1         ' NOT RESPECTED. TRY TO RERUN THE PROGRAM WITH A'/
     2         ' SMALLER TIME STEP.'//
     3         ' ABNORMAL END OF PROGRAM !'//)
        WRITE(NUNIT,200)
        WRITE(*,*)' PRESS RETURN TO END PROGRAM'
        READ(*,*)
        STOP
      ENDIF
C
C COMPUTE THE WETTED SURFACE
      A = H(2,I) * (B + M * H(2,I))
C
C COMPUTE THE DISCHARGE AT THE NODE "I" AT THE RUNNING TIME
      Q(2,I) = U(2,I) * A
C
300   CONTINUE
C
      RETURN
      END

      SUBROUTINE DNSTR( NNMAX , NN , H , U , Q ,
     1                  B , M , SF , N , DX , DT )
C***************************************************************
C
C THIS SUBPROGRAM CALCULATES DEPTH, VELOCITY AND DISCHARGE AT THE NODE "NN",
C LOCATED AT THE DOWNSTREAM END OF THE CHANNEL, FOR THE RUNNING TIME T.
C THE NEWLY CALCULATED VALUES ARE STORED IN THE SECOND LINE OF CORRESPONDING
C ARRAYS WHEREAS THE VALUES FOR THE PRECEDING TIME ARE IN THE FIRST LINE.
C
C***************************************************************
C DECLARATION OF VARIABLES
      REAL    M , N , SF
      DIMENSION H(2,NNMAX) , U(2,NNMAX) , Q(2,NNMAX)
C
C COMPUTE FIRST THE DEPTH AT THE NODE "NN" AT THE RUNNING TIME T
      H(2,NN) = H(1,NN) + (DT/DX) *
     1          ( U(1,NN)*(H(1,NN-1)-H(1,NN)) +
     2            H(1,NN)*(U(1,NN-1)-U(1,NN)) )
C
C NEXT COMPUTE THE WETTED SURFACE AND THE WETTED PERIMETER
      A = H(2,NN) * (B + M * H(2,NN))
      P = B + 2 * H(2,NN) * SQRT(1 + M**2)
      RH = A / P
C
C COMPUTE NOW THE VELOCITY AT THE NODE "NN" AT THE RUNNING TIME T, USING
C THE FORMULA OF MANNING
      U(2,NN) = RH**(2./3.) * SQRT(SF) / N
C
C COMPUTE THE DISCHARGE AT THE NODE "NN" AT THE RUNNING TIME T
      Q(2,NN) = U(2,NN) * A
C
      RETURN
      END
```

Fig.Ex.5.B.8 Program EXPLIC (end).

Ex. 5.C

The same flood wave, as investigated in Ex. 5.B, shall be studied.

A rectangular prismatic channel, made of poor quality concrete, has a width of B = 5 [m] and conveys a uniform flow at a depth of h_n = 1.20 [m]. During a flood event, the discharge in the channel increases linearly to Q = 50 [m³/s] in a time period of t' = 20 [min] and decreases linearly back to its initial value during a time period of t" = 60 [min]. The channel has a bed slope of S_f = 0.001 [-] and the friction coefficient is estimated to be n = 0.020 [m$^{-1/3}$s]. The length of the channel reach to be studied is 3 [km].

i) Assume that this flood wave can be simulated by a kinematic wave (see sect. 5.3). Determine the hydrograph at station L_2 = 3 [km];

ii) Compare this hydrograph with the one obtained in Ex. 5.B, when the same flood wave was simulated as a dynamic wave.

SOLUTION :

i) Simulation of the flood wave by a kinematic wave :

It has been shown (see sect. 5.3), that the kinematic wave represents a special and simplified form of the Saint-Venant equations, eq. 5.1 and eq. 5.3, where the inertia terms and the terms due to the change in depth, hence in pressure, are neglected.

The dynamics of the kinematic wave is described by eq. 5.30. This equation can be solved by numerical finite difference methods (explicit or implicit). For channels with a simple geometry, a solution based on the method of characteristics can be used and has the advantage of allowing a better insight and comprehension of the physical principles of the phenomenon. Thus it will be used here to solve the present problem.

a) Celerity of kinematic wave

A kinematic wave is essentially described by the continuity equation. It is created by a variation of the discharge, Q, and its celerity, c_k, is given by (see sect. 5.3.2) :

$$c_k = \frac{dQ}{dA} = \frac{1}{B}\frac{dQ}{dh} \qquad (5.33)$$

According to eq. 5.30, the discharge, Q, is convected with the celerity c_k. For an observer moving with this celerity, the discharge remains constant and equal to Q. It can therefore be said that a line having a slope of c_k in the plane of space-time, x-t, constitutes a characteristics describing the propagation of this discharge, Q.

By using the formula of Manning, eq. 3.16, the celerity of kinematic waves in a wide, rectangular channel was obtained as ($R_h \cong h$) :

$$c_k = \frac{5}{3}U \qquad (5.36)$$

UNSTEADY FLOW

In the present case, the channel cannot be considered as being a wide one ($R_h \neq h$). The expression for the celerity of kinematic wave must therefore be derived from eq. 5.33 by using the full expression for the hydraulic radius.

According to formula of Manning, eq. 3.16, the discharge is given by :

$$Q = U A = \frac{1}{n} R_h^{2/3} S_f^{1/2} A$$

where U is average velocity, A is wetted surface, n is friction coefficient of Manning, R_h represents hydraulic radius and S_f is bed slope. For a rectangular channel, one has (see Table 1.1) :

$$A = h B \qquad \text{and} \qquad R_h = \frac{h B}{(B + 2h)}$$

By introducing these relations into the formula of Manning, one obtains :

$$Q = \frac{h B}{n} \frac{(h B)^{2/3}}{(B + 2h)^{2/3}} S_f^{1/2} = \frac{S_f^{1/2} B^{5/3}}{n} \frac{h^{5/3}}{(B + 2h)^{2/3}}$$

The expression for the celerity of kinematic wave, eq. 5.33, can therefore be used, being written as :

$$c_k = \frac{1}{B} \frac{dQ}{dh} = \frac{1}{B} \frac{d}{dh}\left(\frac{S_f^{1/2} B^{5/3}}{n} \frac{h^{5/3}}{(B + 2h)^{2/3}}\right) = \frac{S_f^{1/2} B^{2/3}}{n} \frac{d}{dh}\left(\frac{h^{5/3}}{(B + 2h)^{2/3}}\right)$$

By taking the derivative of this equation with respect to h, and simplifying, the general expression for the celerity of kinematic waves in any rectangular channel is obtained :

$$c_k = \frac{S_f^{1/2} B^{2/3}}{n} \left(\frac{h^{2/3} (5B + 6h)}{3 (B + 2h)^{5/3}}\right) \tag{5.33α}$$

b) Determination of the hydrograph at the station $L_2 = 3$ [km]

The initial discharge in the channel, $Q_o = 8.24$ [m³/s] , was computed in exercise Ex.5.B by using the formula of Manning, eq. 3.16. The triangular shaped hydrograph at the station $x = 0$ [m] was then constructed using information on the time of rise and time of fall given in problem statement (see Fig. Ex.5.C.2 and also Fig. Ex.5.B.3).

The initial hydrograph is now divided into several *individual* discharges and each of these individual discharges is propagated to the downstream with its own celerity.

Let us first consider the initial discharge, $Q_o = 8.24$ [m³/s]. The uniform flow depth for this discharge is known, $h_n = 1.2$ [m]. Its celerity can be calculated using the expression previously established for any rectangular channel, eq. 5.33α :

Computation of propagation of a flood wave (simulated as a kinematic wave)					
B = 5.0 [m]	n = 0.020 [m$^{-1/3}$s]		S_f = 0.001 [-]		L_2 = 3000 [m]
1	2	3	4	5	6
Q [m³/s]	t_d [s] departure	h_n [m]	c_k [m/s]	Δt [s] travel time	t_a [s] arrival
8.24	*0*	*1.20*	*1.99*	*1505*	*1505*
17.00	252	2.00	2.33	1290	1542
20.00	338	2.26	2.39	1254	1592
25.00	482	2.67	2.48	1210	1692
30.00	625	3.07	2.54	1180	1805
35.00	769	3.46	2.59	1157	1926
40.00	913	3.84	2.63	1139	2052
45.00	1056	4.22	2.67	1126	2182
50.00	*1200*	*4.59*	*2.69*	*1114*	*2314*
45.00	1631	4.22	2.67	1126	2757
40.00	2062	3.84	2.63	1139	3202
35.00	2493	3.46	2.59	1157	3650
30.00	2924	3.07	2.54	1180	4104
25.00	3355	2.67	2.48	1210	4566
20.00	3786	2.26	2.39	1254	5040
15.00	4217	1.83	2.27	1321	5538
8.24	*4800*	*1.20*	*1.99*	*1505*	*6305*

col.	symbol	expression	explanations
1	Q		discharge. The hydrograph at station $x = 0$ [m] is idealized as a succession of *individual* discharges, each of which occurs at a particular instant. The correct representation of the hydrograph shape and the degree of precision desired in obtaining the hydrograph at the point of arrival are the only criteria guiding the choice of the number of individual discharges.
2	t_d		departure time for individual discharges.
3	h_n		uniform flow depth corresponding to the individual discharge. The computation of h_n is done by trial-and-error using the formula of Manning, eq. 3.16.
4	c_k	eq. 5.33α	celerity with which an individual discharge is propagated.
5	Δt	L_2 / c_k	travel time for an individual discharge.
6	t_a	$t_d + \Delta t$	arrival time of an individual discharge at $L_2 = 3000$ [m].

Fig. Ex.5.C.1 Computations on a spreadsheet.

$$c_k = \left(\frac{S_f^{1/2} B^{2/3}}{n}\right)\left(\frac{h^{2/3} (5B + 6h)}{3 (B + 2h)^{5/3}}\right) = \left(\frac{(0.001)^{1/2} 5^{2/3}}{0.02}\right) \left(\frac{1.2^{2/3} (5 \times 5 + 6 \times 1.2)}{3 (5 + 2 \times 1.2)^{5/3}}\right)$$

$c_k = 1.99$ [m/s]

This initial discharge, Q_o, will be propagated to the station $L_2 = 3000$ [m] with a celerity of $c_k = 1.99$ [m/s]. In the plane of space-time, x-t, the propagation of the wave can be represented by a line having a slope of c_k (see Fig. Ex.5.C.2). The time necessary for this discharge to arrive at the station L_2, which will be called the *travel time*, Δt, is simply computed from :

$$\Delta t = \frac{L_2}{c_k} = \frac{3000}{1.99} \cong 1505 \text{ [s]}$$

The initial discharge, Q_o, starts from station at $x = 0$ [m] at time $t_d = 0$ [s]. The travel time, Δt, is added to the departure time, t_d, to obtain the arrival time, t_a, of this discharge at the station L_2 :

$$t_a = t_d + \Delta t = 0 + 1505 = 1505 \text{ [s]}$$

The computations for the following individual discharges are made in the same manner. The uniform flow depth for each discharge must be determined by a trial-and-error computation. The computational procedure can be implemented in a spreadsheet; this will also facilitate the trial-and-error computation of the uniform flow depths for the individual discharges.

The computation sheet prepared on a personal computer spreadsheet program is presented in Fig. 5.C.1. The explanations for the columns and for the computational procedure are given at the bottom of the table.

The Fig. Ex.5.C.2 shows the evolution of the discharge in the plane of space-time, x-t, both as a two-dimensional plot and a tridimensional axonometric view.

ii) Comparison of hydrographs; kinematic wave and dynamic wave :

The hydrograph at the station $L_2 = 3000$ [m], calculated by kinematic wave approximation is plotted in Fig. Ex.5.C.3, together with the flood hydrograph at the channel entrance, $x = 0$ [m].

The hydrograph obtained in exercise Ex.5.B for the station $L_2 = 3000$ [m], by assuming the flood wave to propagate as a dynamic wave, is superposed on this same figure. The comparison of these hydrographs leads to following remarks:

- The computation of the flood wave as a kinematic wave leads to an over estimation of the tranquil zone for the station L_2. According to the kinematic wave assumption the flood wave arrives at the station L_2 at time $t = 1505$ [s] whereas according to the dynamic wave the flood at this station starts already at time $t = 720$ [s]. This is an error which may have important consequences, especially in developing flood emergency schemes.

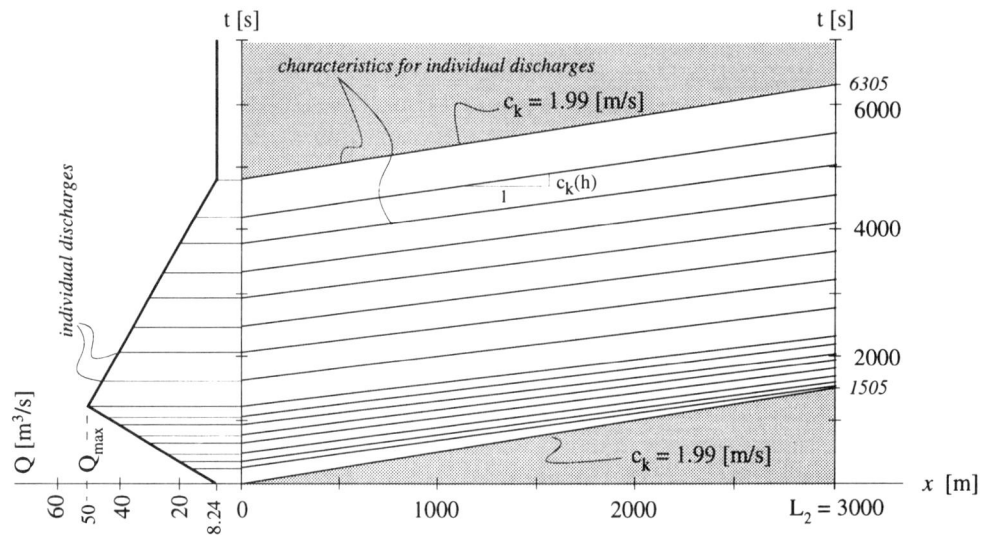
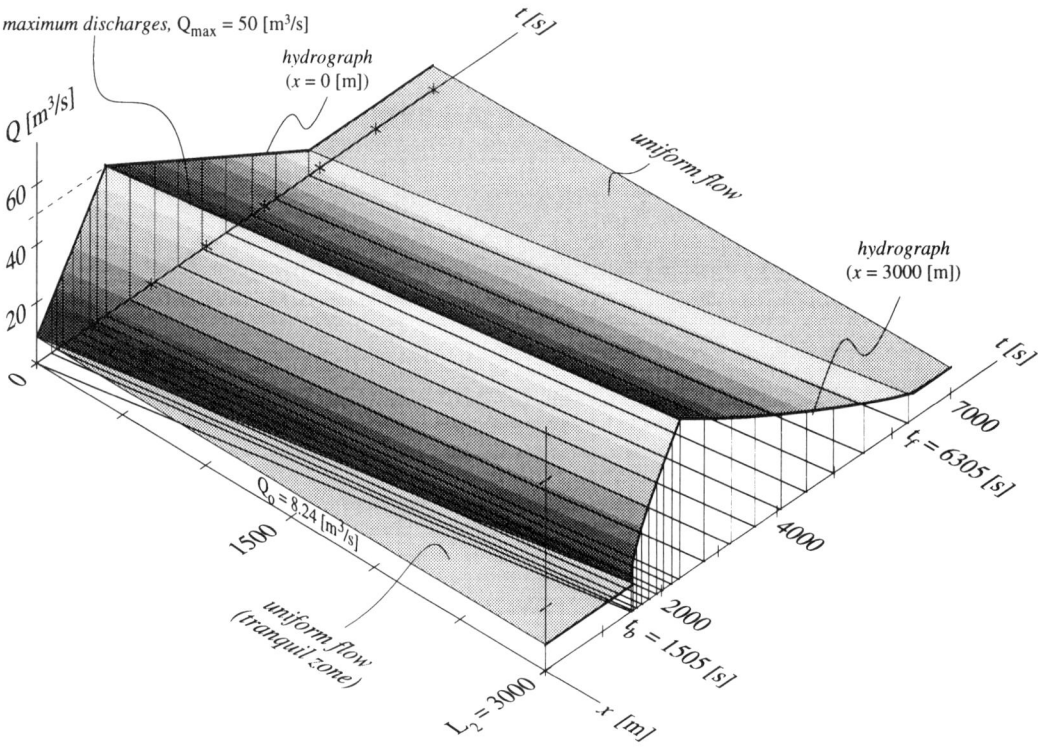

Fig. Ex.5.C.2
Simulation of the propagation of the flood wave in the space time, x-t, by a kinematic wave and its axonometric representation.

- According to solution with kinematic wave assumption, the flood hydrograph at the station L_2 terminates at $t_f = 6305$ [s]. At this moment, the computation with dynamic wave assumption indicates a discharge of $Q \cong 11.3$ [m³/s] (see Fig. Ex.5.B.6), which is larger than the initial uniform flow discharge of $Q_o = 8.24$ [m³/s]. The hydrograph simulated by a dynamic wave returns back to the initial discharge asymptotically at time $t_f = 11160$ [s].

- The hydrograph computed by a kinematic wave does not show any attenuation of the peak discharge, Q_{max}. The rising branch of the hydrograph becomes steeper whereas the falling branch becomes flatter and spreads. The width of the hydrograph at its base remains however the same as the initial hydrograph at the entrance (6305 − 1505 = 4800 [s]).

The hydrograph computed by assuming a dynamic wave on the other hand spreads considerably and its base become much larger. However, more important is the the fact that there is an attenuation of the peak discharge : $\Delta Q_{max} = 50.00 - 38.64 = 11.36$ [m³/s].

- According to the solution with kinematic wave assumption the maximum discharge, Q_{max}, arrives at the station L_2 at time $t = 2314$ [s], whereas the dynamic wave reaches the same station at time $t = 2400$ [s]. Thus, the error made in estimating the arrival time of the peak discharge, Q_{max}, is less than the one made in estimating the beginning of the flood.

The results obtained by considering the flood as a dynamic wave are certainly closer to the real situation than those obtained by kinematic wave assumption. Nevertheless, as can be seen in comparing the hydrographs in Fig. Ex.5.C.3, the hydrograph obtained with kinematic wave assumption gives a reasonable approximation of the reality with a considerable economy in the computational efforts.

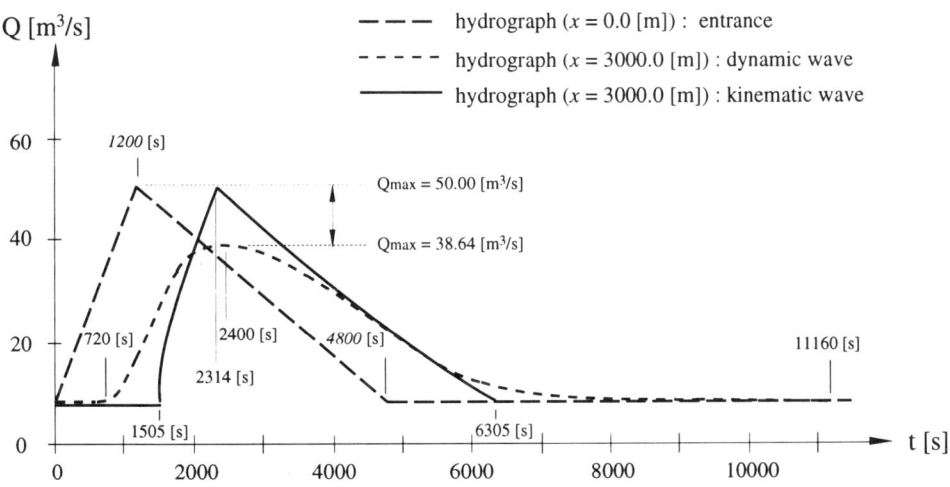

Fig. Ex.5.C.3
Comparison of hydrographs : kinematic wave and dynamic wave.

Ex. 5.D

The headrace channel in a hydroelectric power plant has a rectangular cross section with a width of B = 10.0 [m]. This channel has a bed slope of $S_f = 0.002$ [-] and the coefficient of Manning is n = 0.02 [$m^{-1/3}$s]. During the normal turbine operation the flow is uniform, having a discharge of Q = 40 [m^3/s]. Due to a sudden load rejection, the gate in the headrace channel is partially, but rapidly closed, thus decreasing the discharge to $Q_f = 0.5$ [m^3/s].

i) Determine the hydraulics of the waves traveling downstream and upstream from the gate;

ii) Determine the type of front for these waves;

iii) How long will it take for the wave to arrive at a station located at a distance of 0.5 [km] upstream from the gate.

SOLUTION :

The generation of waves resulting from the closure of a gate are presented in Fig. 5.11; this figure is re-drawn below :

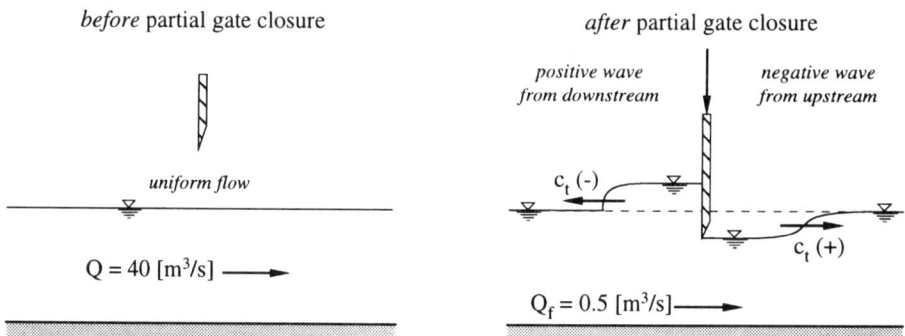

Fig. Ex. 5.D.1

Computation of the uniform flow using the equation of Manning, eq. 3.16 :

The geometrical characteristics of a rectangular cross section are (see Table 1.1) :

wetted surface : $A = B h_n$; hydraulic radius : $R_h = \dfrac{B h_n}{B + 2 h_n}$

Known are : B = 10.0 [m]   ; S_f = 0.002 [-]
              n = 0.02 [$m^{-1/3}$s] ;   Q = U A = 40 [m^3/s]

The normal depth can be calculated using the equation of Manning-Strickler :

$$Q = U A = \frac{1}{n} R_h^{2/3} S_f^{1/2} A = \frac{1}{n} \left(\frac{B h_n}{B + 2h_n}\right)^{2/3} S_f^{1/2} B h_n = \frac{S_f^{1/2} B^{5/3}}{n} \frac{h_n^{5/3}}{(B + 2h_n)^{2/3}}$$

UNSTEADY FLOW

Introducing the known values; one gets :

$$40 = \frac{(0.002)^{1/2} (10.0)^{5/3}}{0.02} \frac{h_n^{5/3}}{(10.0 + 2h_n)^{2/3}} = 103.79 \frac{h_n^{5/3}}{(10.0 + 2h_n)^{2/3}}$$

This equation can be solved by trial-and-error :

h [m]	Q [m³/s]
1.55	38.77
1.65	42.60
1.58 (= h_n)	40.00

The *normal depth* is therefore: $h_n = 1.58$ [m]

Wetted surface : $A = B \, h_n = 10.0 \times 1.58 = 15.8 \, [m^2]$

Average velocity : $U = Q/A = 40.0/15.8 = 2.53 \, [m/s]$

Celerity of the wave : $c = \sqrt{gh_n} = \sqrt{9.81 \times 1.58} = \pm 3.94 \, [m/s]$

Froude number : $Fr = U/c = \dfrac{2.53}{3.94} = 0.64 \, [-] < 1 \quad \Rightarrow \quad$ subcritical flow

i) *Determination of the downstream and upstream waves*

a) Computation of *positive wave from downstream* :

The continuity equation, eq. 5.46, yields :

$$c_t = \frac{\Delta q}{\Delta h} = \frac{(Q_f - Q_1)/B}{(h_2 - h_1)} = \frac{(0.5 - 40.0)/10.0}{(h_2 - h_1)} = \frac{-3.95}{(h_2 - h_1)}$$

with $\Delta h = (h_2 - h_1)$ where $h_1 = h_n = 1.58$ [m].

Here the unknowns are: c_t and h_2.

Considering a positive wave from downstream, the momentum equation, eq. 5.50, can be written as :

$$c_t = U_1 - \sqrt{gh_1} \sqrt{\frac{h_2}{2h_1}\left(1 + \frac{h_2}{h_1}\right)}$$

By taking $h_1 = h_n$ and $U_1 = U$, this equation becomes:

$$c_t = 2.53 - 3.94 \sqrt{\frac{h_2}{2 \times 1.58}\left(1 + \frac{h_2}{1.58}\right)}$$

Here the unknowns are also : c_t and h_2.

The computation is done by trial-and-error using different values of h_2, until the same value is obtained for c_t using the two relationships given above. The trial-and-error computation is illustrated below :

Trial-and-error computation of *positive wave from downstream*					
$Q = 40$ [m³/s] $\quad Q_f = 0.5$ [m³/s] $\quad h_1 = h_n = 1.58$ [m]					
$U_1 = U = +2.53$ [m/s] $\quad \sqrt{gh_n} = \sqrt{gh_1} = \pm 3.94$ [m/s]					
				trials	
		units	1	2	3
flow depth :	h_2 (estimated)	[m]	2.650	2.780	2.714
wave height :	$\Delta h = h_2 - h_1$	[m]	+1.070	+1.200	+1.134
$c_t = \dfrac{-3.95}{(h_2 - h_1)}$		[m/s]	−3.692	−3.292	−3.483
$c_t = 2.53 - 3.94 \sqrt{\dfrac{h_2}{3.16}\left(1 + \dfrac{h_2}{1.58}\right)}$		[m/s]	−3.367	−3.603	−3.483

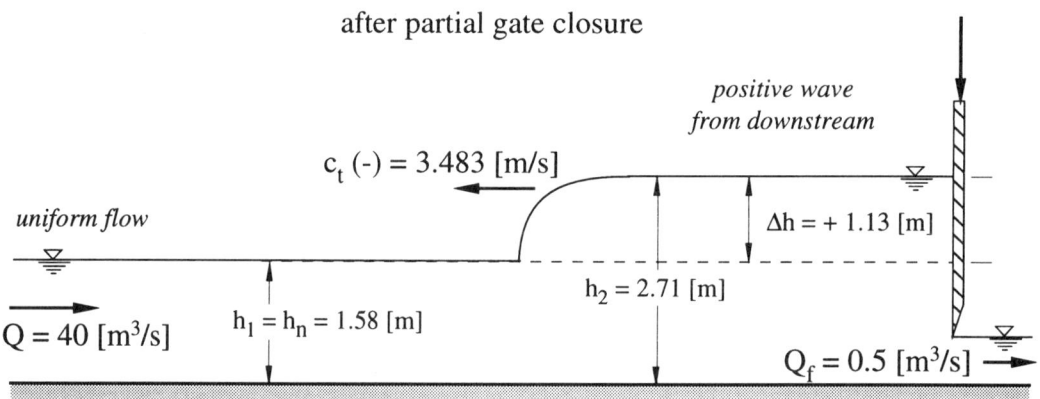

Fig. Ex.5.D.2 (see also Fig. Ex.5.D.1)

A *positive wave* of height, $\Delta h = +1.134$ [m], propagates from the gate towards the upstream with a celerity of $c_t = 3.483$ [m/s] (see Fig. Ex.5.D.2). The wave profile presents a wave front which is steep and reasonably stable.

b) **Computation of *negative wave from upstream, immediately after its creation* :**

The continuity equation, eq. 5.46, yields :

$$c_t^* = \frac{\Delta q}{\Delta h} = \frac{(Q_f - Q_1)/B}{(h_2 - h_1)} = \frac{(0.5 - 40.00)/10.0}{(h_2 - h_1)} = \frac{-3.95}{(h_2 - h_1)}$$

Here the unknowns are : c_t^* and h_2.

UNSTEADY FLOW

Considering a negative wave from upstream, the momentum equation, eq. 5.50a, can be written as :

$$c_t^* = U_1 + \sqrt{gh_1} \sqrt{\frac{h_2}{2h_1}\left(1 + \frac{h_2}{h_1}\right)}$$

Here the unknowns are also : c_t^* and h_2.

Again a trial-and-error computation is to be done, which is illustrated below :

Trial-and-error computation of *negative wave from upstream*					
$Q = 40$ [m³/s] $\quad Q_f = 0.5$ [m³/s] $\quad h_1 = h_n = 1.58$ [m]					
$U_1 = U = +2.53$ [m/s] $\quad \sqrt{gh_n} = \sqrt{gh_1} = \pm 3.94$ [m/s]					
				trials	
		units	1	2	3
flow depth :	h_2 (estimated)	[m]	0.700	0.85	0.776
wave height :	$\Delta h = h_2 - h_1$	[m]	–0.880	–0.730	–0.804
$c_t^* = \dfrac{-3.95}{(h_2 - h_1)}$		[m/s]	+4.489	+5.411	*+4.915*
$c_t^* = 2.53 + 3.94 \sqrt{\dfrac{h_2}{3.16}\left(1 + \dfrac{h_2}{1.58}\right)}$		[m/s]	+4.758	+5.064	*+4.915*

A *negative wave* of height, $\Delta h = -0.804$ [m], propagates from the gate towards the downstream (turbine) with a celerity of $c^*_t = 4.915$ [m/s] (see Fig. Ex.5.D.3). Note that the wave front is unstable and it tends to flatten out.

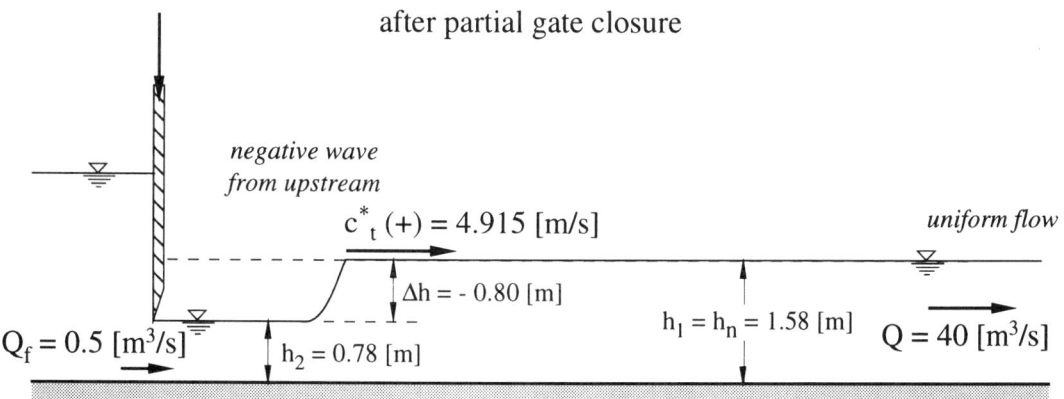

Fig. Ex.5.D.3 (see also Fig. Ex.5.D.1)

332   FLUVIAL HYDRAULICS

ii)   *Determination of the form of the "front" (see sect. 5.6.4)*

a) The *positive wave from downstream*, which is formed at the upstream side of the gate after partial closure of the gate, presents a steep front. The upper part (top) of the front propagates faster than the lower part (toe), it catches up with it, and breaks. A balance is reached between the two levels and the wave front looks like a moving hydraulic jump (see Fig. Ex.5.D.2).

b) Immediately after partial closure of the gate, the *negative wave from the upstream* has also a steep front. In the present case however, the top of the wave, which propagates faster than the toe, tends to move away from the latter (see Fig. 5.14). The profile of this negative wave is therefore not stable, but varies with time as the wave travels away from the gate.

For a negative wave from upstream, the celerity as a function of the height is given by eq. 5.56 :

$$c_t(h) = U + \sqrt{gh} = U_1 + 3\sqrt{gh} - 2\sqrt{gh_1}$$

computation of the profile of the negative wave from upstream				
$U_1 = +2.53$ [m/s]		$h_1 = h_n = 1.58$ [m]		
flow depth	celerity	distance from gate, x [m],		
h	$c_t(h)$	for 3 different instants, t, after partial closure		
[m]	[m/s]	$t = 10$ [s]	$t = 60$ [s]	$t = 120$ [s]
$h_1 = 1.58$	$c_{t_1} = 6.47$	64.7	388.0	776.0
1.52	6.24	62.4	374.4	748.9
1.46	6.01	60.1	360.6	721.2
1.40	*5.77*	*57.7*	*346.4*	*692.9*
1.34	5.53	55.3	332.0	664.0
1.28	5.29	52.9	317.2	634.4
1.22	5.03	50.4	302.1	604.2
1.16	4.78	47.8	286.6	573.1
1.10	4.51	45.1	270.7	541.3
1.04	4.24	42.4	254.3	508.6
0.98	3.96	39.6	237.5	475.0
0.92	3.67	36.7	220.0	440.2
0.86	3.37	33.7	202.2	404.4
0.80	3.06	30.6	183.6	367.2
$h_2 = 0.78$	$c_{t_2} = 2.95$	29.6	177.3	354.6

The celerities for the top, c_{t_1}, and toe, c_{t_2}, of the wave can be calculated using the above expression :

for $h = h_1 = 1.58$ [m/s]
        $c_{t_1} = 2.53 + \sqrt{9.81 \times 1.58} = 6.47$ [m/s]

UNSTEADY FLOW

for $h = h_2 = 0.78$ [m/s]

$c_{t_2} = 2.53 + 3\sqrt{9.81 \times 0.78} - 2\sqrt{9.81 \times 1.58} = 2.95$ [m/s]

The celerities for other flow depths are calculated in the same way. One can then determine the position, x, of each depth, h, at time, t, after partial closure of the gate by writing (see eq. 5.57) :

$$x = c_t t = (U_1 + 3\sqrt{gh} - 2\sqrt{gh_1})t$$

The above table shows the computation of the profile of the negative wave from upstream at instants t = 10, 60 and 120 [s] after partial closure of the gate.

For these three instants, being selected arbitrarily, the profiles of the negative wave from upstream are plotted in Fig. Ex. 5.D.4.

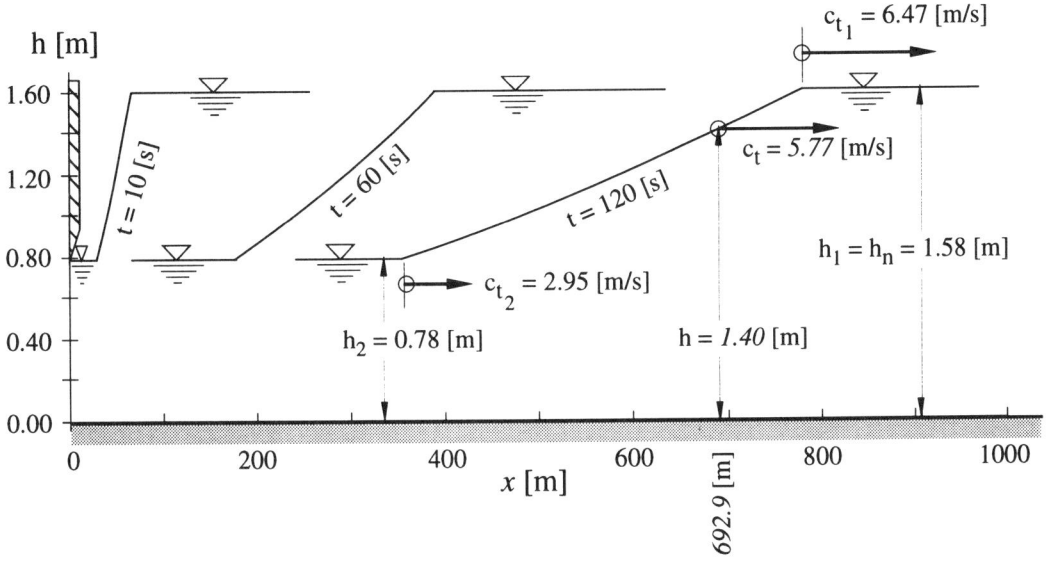

Fig. Ex. 5.D.4

iii) *Determination of the time of arrival of the positive wave from downstream to a station, located at 0.5 [km] upstream of the gate :*

celerity for the positive wave from downstream : $c_t = -3.483$ [m/s]
distance to travel : $x = 500$ [m]

$$t = \frac{x}{c_t} = \frac{500}{3.483} = 143.6 \text{ [s]}$$

The arrival time is therefore : $t = 143.6$ [s].

Ex. 5.E

In a long, rectangular irrigation channel, stagnant water at a depth of $h_1 = 3.2$ [m] is stored behind a sluice gate. The channel has a smooth bed of negligible slope. Due to a maneuvering error the gate is suddenly and completely opened.

Determine the water-surface profiles at different instants, t = 60, 120, 180, 240 and 300 [s], after the accident, and this for the following two cases :

i) the channel downstream of the gate is dry ;

ii) the channel downstream of the gate is filled with stagnant water at a depth of $h_o = 0.32$ [m].

SOLUTION :

i) The flow situation resulting from the sudden and complete opening of the gate, due to a maneuvering error, is drawn below (see also Fig. 5.15) :

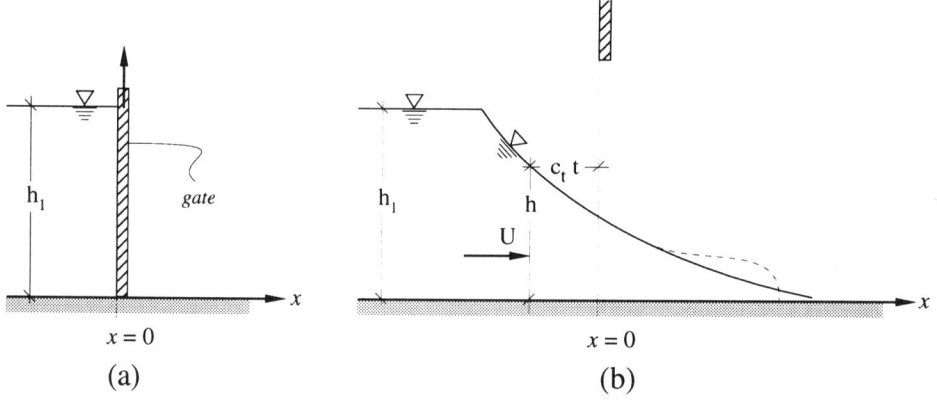

Fig. Ex.5.E.1 Negative wave from downstream, resulting from the sudden and complete opening of a gate : a) before and b) after.

This problem can be assimilated to the problem of *dam break* (see point 5.6.4 6°); it can be solved as a negative wave from downstream traveling in a horizontal channel without friction.

Before the sudden opening of the gate, the water depth at the upstream is h_1 and there is no water in the channel downstream of the gate.

a) The absolute celerity of the wave after the sudden and complete opening of the gate is given by :

$$c_t(h) = U_1 - 3\sqrt{gh} + 2\sqrt{gh_1} \qquad (5.59)$$

Initially the water is at rest, thus $U_1 = 0$. The celerities of the downstream and upstream ends of the negative wave from downstream are therefore (see Fig. Ex.5.E.3) :

UNSTEADY FLOW

at the downstream end :

$h = 0 \Rightarrow c_t = +2\sqrt{gh_1} = 2\sqrt{9.81 \times 3.2} = +11.21$ [m/s]

at the upstream end :

$h = h_1 \Rightarrow c_t = -\sqrt{gh_1} = -\sqrt{9.81 \times 3.2} = -5.60$ [m/s]

b) The water-surface profile is given by :

$$x(h) = c_t t = (U_1 - 3\sqrt{gh} + 2\sqrt{gh_1})t \qquad (5.57a)$$

where c_t is the wave celerity at any given depth, h, and $U_1 = 0$. Since the friction on the bed is neglected, the surface of the negative wave will be tangent to the channel bed at its downstream end.

The table in Fig. Ex.5.E.2 shows the computation of the water-surface profile at different instants, t = 60, 120, 180, 240 and 300 [s], after the sudden opening of the gate. A certain number of flow depths, arbitrarily chosen within the limits of $0.0 \le h$ [m] ≤ 3.2, are listed in the first column. The head loss due to friction being neglected, the wave celerity, $c_t(h)$, for any depth is constant. The wave celerities for the selected flow depths are therefore calculated using eq. 5.59 and are listed in the second column of the table. The explanations on the next two columns will be given later. The remaining five columns give the position, $x = c_t t$, (see eq. 5.57a) of the chosen flow depths at 5 different instants, t, after the gate opening. Each column represents therefore a water-surface profile of the negative wave at a given instant.

The water-surface profiles of the negative wave from downstream, h(x), at the 5 instants mentioned in the problem statement are plotted in Fig. Ex. 5.E.3, together with the characteristics on the plane, x-t.

c) It is interesting to note that the absolute celerity, c_t, of the wave for the flow depth $h_{xo} = 1.42$ [m] remains null at all times. For the larger flow depths, $h > h_{xo}$, the wave celerity is negative, $c_t < 0$, and the wave travels upstream, whereas for the smaller flow depths, $h < h_{xo}$, the wave celerity is positive, $c_t > 0$, and the wave travels downstream. The water-surface profile of the negative wave swivels around this point whose depth, h_{xo}, is constant. On the plane, x-t (see Fig. Ex.5.E.3), the characteristic for this depth, h_{xo}, coincides thus with the time-axis. To determine this depth, one writes (see eq. 5.59) :

$$c_t(h_{xo}) = \frac{dx}{dt} = \frac{x}{t} = U - \sqrt{gh_{xo}} = U_1 - 3\sqrt{gh_{xo}} + 2\sqrt{gh_1} = 0$$

and obtains : $h_{xo} = \frac{4}{9}h_1 = \frac{4}{9}3.2 = 1.42$ [m]

computation of the *negative wave from downstream*									
$U_1 = +0.0$ [m/s]				$h_1 = 3.2$ [m]				
flow depth	absolute celerity	relative celerity	velocity	distance from the gate, x [m] $= c_t\,t$,				
h	$c_t(h)$	$\sqrt{gh}$	$U(h)$	for 5 different instants, t, after sudden gate opening				
[m]	[m/s]	[m/s]	[m/s]	t = 60 [s]	t = 120 [s]	t = 180 [s]	t = 240 [s]	t = 300 [s]
$h_1 = 3.20$	−5.60	5.60	0.00	−336.2	−672.3	−1008.5	−1344.7	−1680.9
3.00	−5.07	5.42	0.36	−304.1	−608.3	−912.4	−1216.6	−1520.7
2.80	−4.52	5.24	0.72	−271.0	−542.1	−813.1	−1084.1	−1355.2
2.60	−3.95	5.05	1.11	−236.7	−473.4	−710.2	−946.9	−1183.6
2.40	−3.35	4.85	1.50	−201.1	−402.1	−603.2	−804.2	−1005.3
2.20	−2.73	4.65	1.91	−163.9	−327.7	−491.6	−655.5	−819.4
2.00	−2.08	4.43	2.35	−125.0	−249.9	−374.9	−499.8	−624.8
1.80	−1.40	4.20	2.80	−84.0	−168.1	−252.1	−336.2	−420.2
1.60	−0.68	3.96	3.28	−40.8	−81.6	−122.4	−163.1	−203.9
$h_{xo} = 1.42$	0.00	3.74	3.74	0.0	0.0	0.0	0.0	0.0
1.40	0.09	3.71	3.79	5.3	10.5	15.8	21.1	26.4
$h_2 = 1.27$	0.63	3.53	4.15	37.6	75.1	112.7	150.2	187.8
1.20	0.91	3.43	4.34	54.8	109.5	164.3	219.0	273.8
1.00	1.81	3.13	4.94	108.6	217.1	325.7	434.3	542.8
0.80	2.80	2.80	5.60	168.1	336.2	504.3	672.3	840.4
0.60	3.93	2.43	6.35	235.6	471.3	706.9	942.6	1178.2
0.40	5.26	1.98	7.24	315.8	631.6	947.3	1263.1	1578.9
0.20	7.00	1.40	8.40	420.2	840.4	1260.6	1680.9	2101.1
0.00	11.21	0.00	11.21	672.3	1344.7	2017.0	2689.4	3361.7

Fig. Ex.5.E.2 Summary of the computation of the water-surface profile of the negative wave from downstream at five different instants, t, after the sudden gate opening.

The velocity at this section, $x = 0$, is also constant and thus independent of the time. It can be obtained using eq. 5.60 :

$$U_{xo} = -2\sqrt{gh_{xo}} + 2\sqrt{gh_1} = -2\sqrt{g\,\tfrac{4}{9}h_1} + 2\sqrt{gh_1} = +\tfrac{2}{3}\sqrt{gh_1}$$

$$U_{xo} = +\tfrac{2}{3}\sqrt{9.81 \times 3.2} = 3.74 \text{ [m/s]}$$

Since the flow depth and velocity at this section, $x = 0$, are both constant, the discharge, flowing towards downstream, is also constant. The discharge per unit width of the channel is therefore :

$$q = U\,h = (\tfrac{2}{3}\sqrt{gh_1})(\tfrac{4}{9}h_1) = \tfrac{8}{27}h_1\sqrt{gh_1}$$

$$q = \tfrac{8}{27}\,3.2\,\sqrt{9.81 \times 3.2} = 5.31 \text{ [m}^2\text{/s]}$$

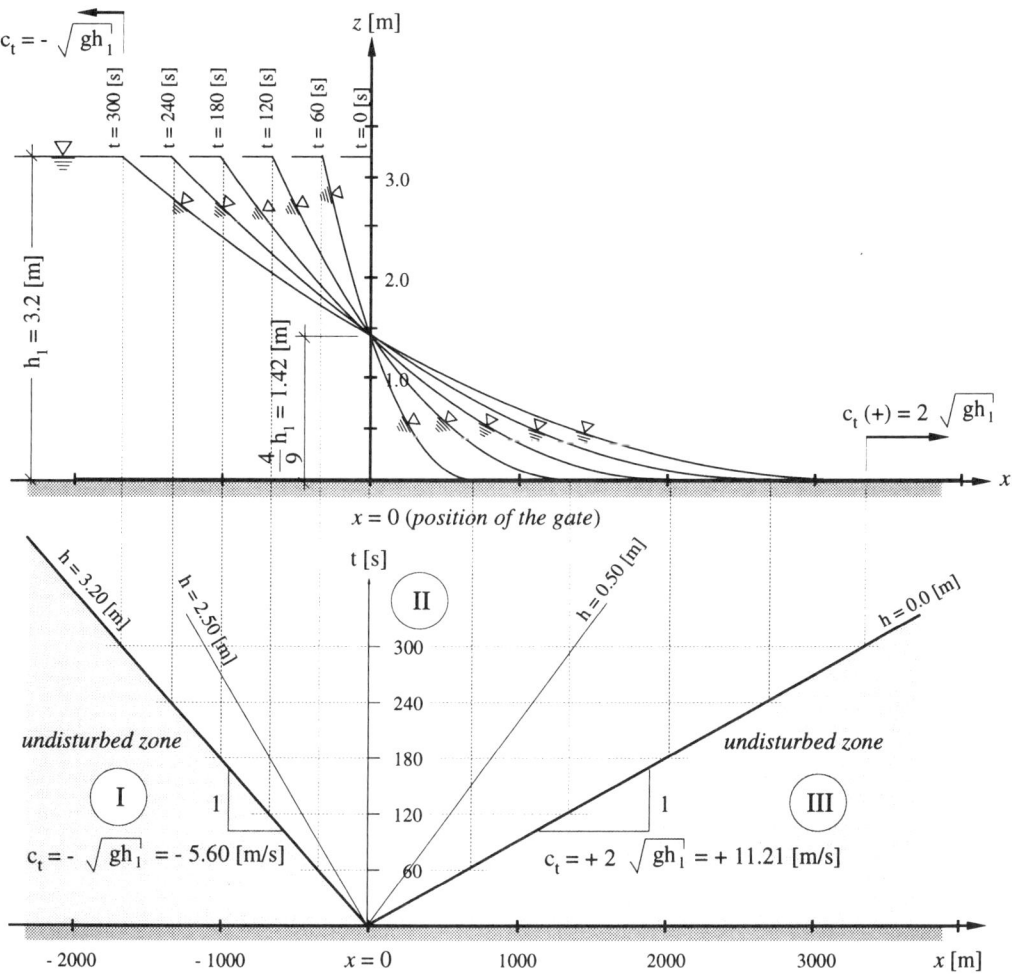

Fig. Ex.5.E.3 The water-surface profile of the negative wave from downstream at different instants after the sudden opening of the gate and the corresponding plane, x-t.

d) Fig. Ex.5.E.3 shows also the characteristics in the plane, x-t, for the negative wave from downstream, resulting from the sudden opening of the gate. The origin is located at the initial position of the gate, $x = 0$ [m]. For each flow depth, $0.00 < h[m] < 3.20$, a characteristics initiates at this origin with a slope whose inverse is equal to the celerity of the wave for the corresponding depth. It is to be noted that the characteristics for the depth h_{xo} coincides with the time-axis.

Three zones can be distinguished. The zones I and III are the undistrubed zones at the upstream and the downstream, where the negative wave has no influence. The zone II, between the characteristics of the upstream and downstream ends of the negative wave, is the zone where the negative wave develops itself. A tridimensional representation of the propagation of the negative wave is drawn in Fig. Ex. 5.E.4.

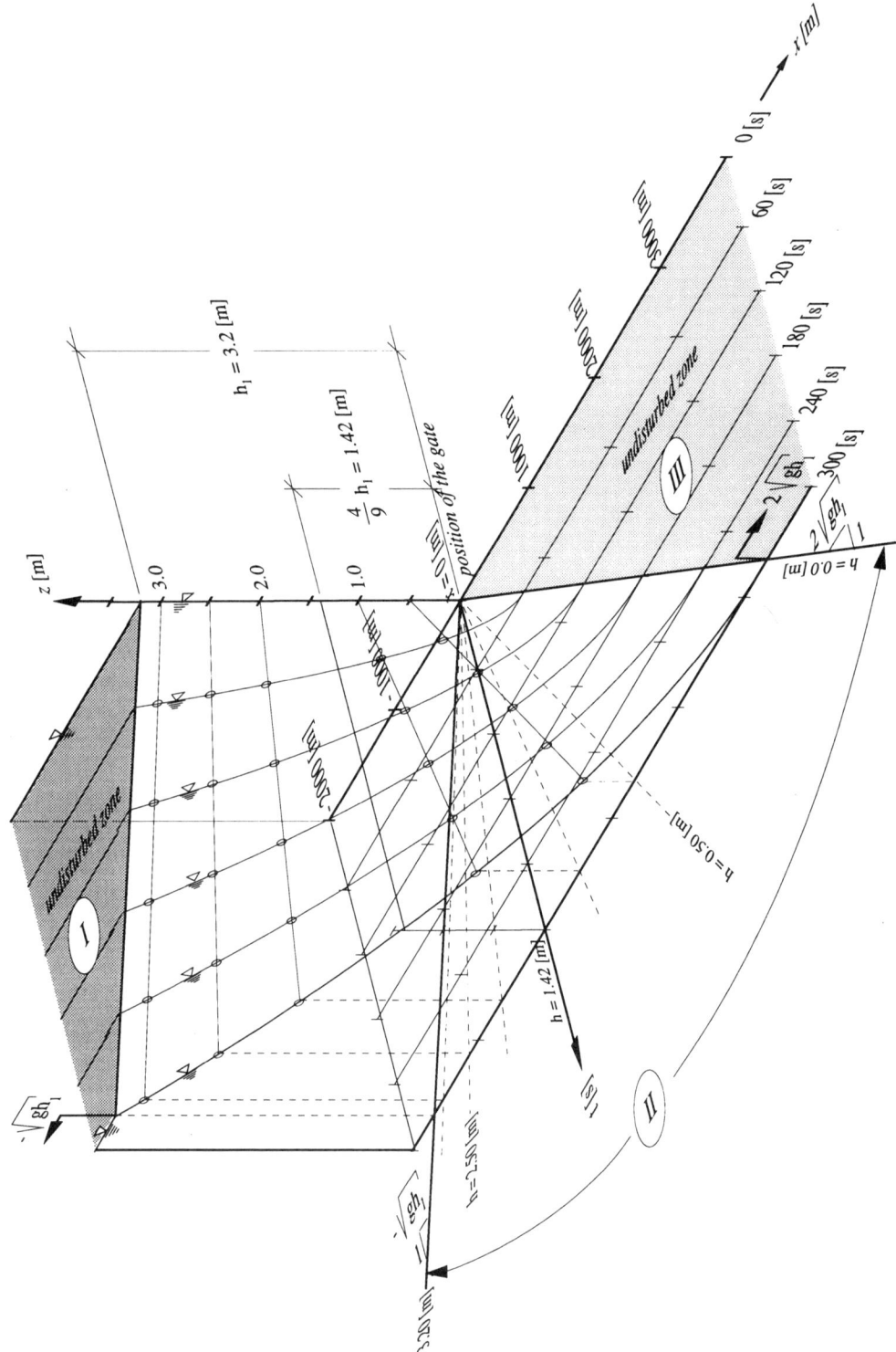

Fig. Ex.5.E.4 Tridimensional representation of the negative wave from downstream in the space, x-t-z.

e) Consider again the table in Fig. Ex.5.E.2, in order to analyze in more detail the information listed in the first four columns. The explanations of the first two columns, h and c_t, were given previously. The wave celerities, $c = \sqrt{gh}$ (see eq. 2.27), corresponding to the different flow depths are listed in the third column of the table. The fourth column contains the flow velocities due to the negative wave at different depths, calculated with :

$$U(h) = U_1 - 2\sqrt{gh} + 2\sqrt{gh_1} = 0 - 2\sqrt{gh} + 2\sqrt{gh_1} \qquad (5.60)$$

The absolute celerity of a negative wave having a small amplitude is then given by :

$$c_t(h) = U - \sqrt{gh} \qquad (5.55a)$$

It will be thus interesting to plot the variation of the average velocity, $U(h)$, and the celerities, $c(h)$ and $c_t(h)$, as a function of the flow depth, h, in order to better comprehend the different aspects of a negative wave from downstream. Such a plot is given in Fig. Ex.5.E.5, where the eq. 5.55a is represented graphically. By subtracting point by point the curve $c(h)$ from the curve $U(h)$, the distribution of the absolute celerity, $c_t(h)$, for the negative wave from downstream is obtained. A few interesting remarks are in order :

- At the depth $h = h_1 = 3.20$ [m], the average velocity is null, $U_1 = 0$ [m/s]. The wave celerity for this depth is $c = 5.60$ [m/s]; this gives an absolute wave celerity of $c_t = -5.60$ [m/s].

- On the downstream channel, being dry and thus at $h = 0.00$ [m], the velocity reaches its maximum value, $U = 11.21$ [m/s]. The wave celerity for this height being null, $c = 0.00$ [m/s], the downstream tip of the negative wave has an absolute celerity of $c_t = +11.21$ [m/s].

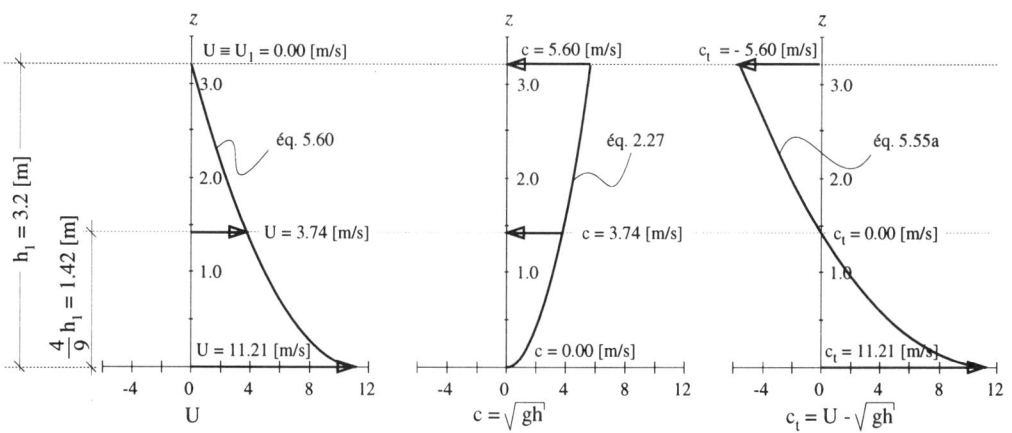

Fig. Ex.5.E.5
Distribution of the absolute celerity, $c_t(h)$, for the negative wave from upstream as a function of depth, h, and its decomposition into the average velocity, $U(h)$, and the relative wave celerity, $c(h) = \sqrt{gh}$.

- At the depth of $h_{xo} = \frac{4}{9} h_1 = 1.42$ [m] an interesting phenomena is observed. At this depth, the velocity, $U_{xo} = 3.74$ [m/s] — flowing towards downstream — is equal to the relative wave celerity, $c = \sqrt{gh_{xo}} = 3.74$ [m/s] — traveling towards upstream —. The resulting absolute celerity is zero, $c_t = +0.00$ [m/s]. This explains also why this point remains in the place of the initial position of the gate.

ii) Consider now the case when the downstream channel contains stagnant water, $U_o = 0$, at a water depth of $h_o = 0.32$ [m] (see *Henderson*, 1966, p. 309 and *Stoker*, 1957, p. 333-341).

a) First of all, the water-surface profile should be determined. To start with, it shall be assumed that the profile of the negative wave, computed in the preceding case, is also valid down to the water surface of the stagnant water (see Fig. Ex.5.E.6a). This profile, however, leads to a physically impossible situation and it is not acceptable. At point C, located at the intersection of the negative wave and the stagnant water surface at the downstream, a contradiction appears. Immediately upstream of point C —because of the negative wave —, the water should have a velocity equal to the wave celerity at a depth of, h_o. Immediately downstream of point C, on the other hand, the velocity should be zero because the water in the downstream channel is at rest.

b) The only possibility for linking the negative wave profile with the surface of the water at rest at the downstream is to create *a discontinuity in the form of a positive wave from upstream*, having a depth of h_2. This situation is presented in Fig. Ex.5.E.6b. At point C, there is a discontinuity in the flow depth. Upstream of point C, the flow depth is h_2 and the water has a velocity of $U_2 \neq 0$, whereas downstream of point C, the water depth is h_o and the velocity of the water is $U_o = 0$. This discontinuity in depth makes it also possible to have a discontinuity in the velocity.

In order to compute the complete waters-surface profile, one should first determine the depth of the positive wave from upstream, h_2.

At the upstream of point B, the profile of the negative wave from downstream is always valid. One can thus write :

$$U(h) = U_1 - 2\sqrt{gh} + 2\sqrt{gh_1} \tag{5.55}$$

Initially, water is at rest behind the gate, $U_1 = 0$. At point C, the flow depth is h_2. The velocity at this depth is now :

$$U(h_2) = U_2 = 0 - 2\sqrt{gh_2} + 2\sqrt{gh_1} \tag{5.55α}$$

This equation establishes a *link* between the negative wave from downstream (profile $\overline{AB}$) and the positive wave from upstream (profile $\overline{BC}$).

UNSTEADY FLOW 341

Let us now consider a positive wave from upstream of depth h_2, traveling on a layer of stagnant water of a depth h_o. The upstream end of this positive wave is at point B.

The continuity equation for the positive wave from upstream is given by eq. 5.45. This can be written as follows:

$$(U_o - c_{t_C}) bh_o = (U_2 - c_{t_C}) bh_2$$

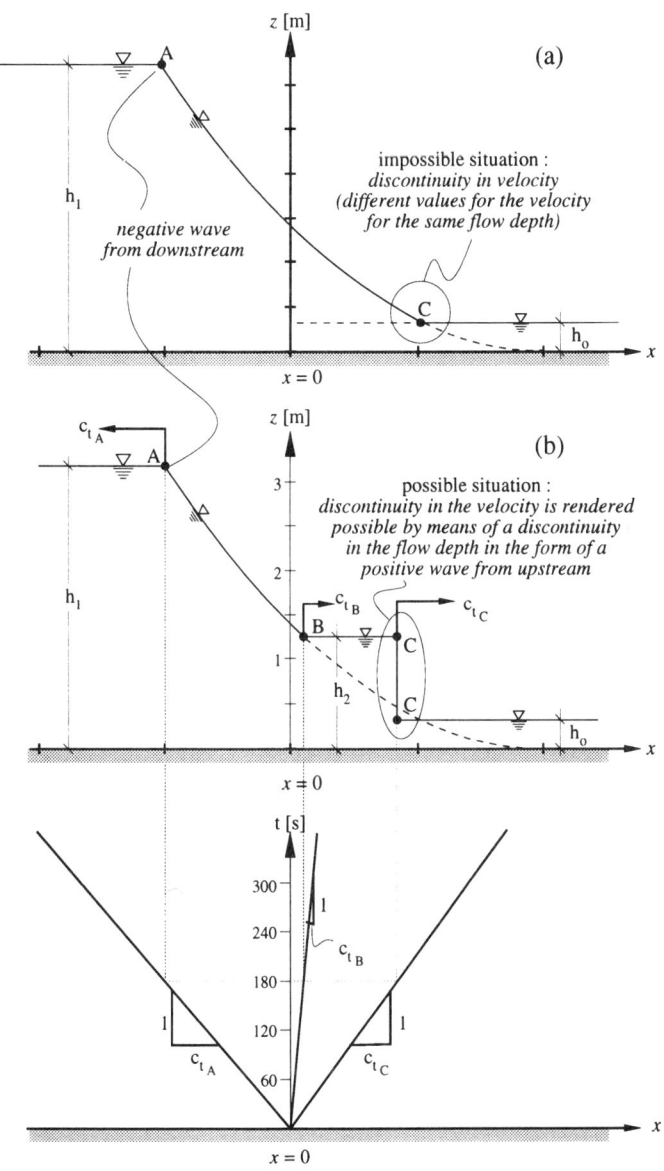

Fig. Ex.5.E.6 The water-surface profile of the negative wave from downstream when there is stagnant water in the downstream channel.

The width, b, of the channel is constant and the water in the downstream channel is at rest, $U_o = 0$; this allows the following simplification of the above expression :

$$c_{t_C} (h_2 - h_o) = U_2 h_2$$

By substituting the relationship for U_2, eq. 5.55α, an expression for the celerity, c_{t_C}, of the positive wave from upstream is obtained, containing only the depths, h_1, h_2 and h_o :

$$c_{t_C} = \frac{(-2\sqrt{gh_2} + 2\sqrt{gh_1}) h_2}{(h_2 - h_o)}$$

A second expression for the celerity, c_{t_C}, of the positive wave from upstream can be obtained by writing the momentum equation (see eq. 5.50a and Fig. 5.10c) with $U_1 = 0$:

$$c_{t_C} = 0 + \sqrt{gh_o} \sqrt{\frac{h_2}{2h_o}(1 + \frac{h_2}{h_o})}$$

The flow depths, h_1 and h_o, are known. Using the above two relationships for the celerity, c_{t_C}, the unknown depth, h_2, can be obtained by trial-and-error. These computations are presented in the table below :

computation of the *positive wave from upstream*			
h_1 = 3.20 [m/s] h_o = 0.32 [m/s] h_o / h_1 = 0.1 [-]			
trial no :	1	2	3
h_2 [m] estimated	1.20	1.30	1.27
U_2 [m/s] $= -2\sqrt{gh_2} + 2\sqrt{gh_1}$	4.34	4.06	4.15
c_{t_C} [m/s] $= \frac{(-2\sqrt{gh_2} + 2\sqrt{gh_1}) h_2}{(h_2 - h_o)}$	5.92	5.39	5.55
c_{t_C} [m/s] $= +\sqrt{gh_o}\sqrt{\frac{h_2}{2h_o}(1 + \frac{h_2}{h_o})}$	5.29	5.68	5.55
Δc_{t_C}	0.64	−0.29	0.00
	no	no	yes

The flow depth for the positive wave from upstream is therefore $h_2 = 1.27$ [m]. The celerity of the positive wave is $c_{t_C} = 5.55$ [m/s].

UNSTEADY FLOW   343

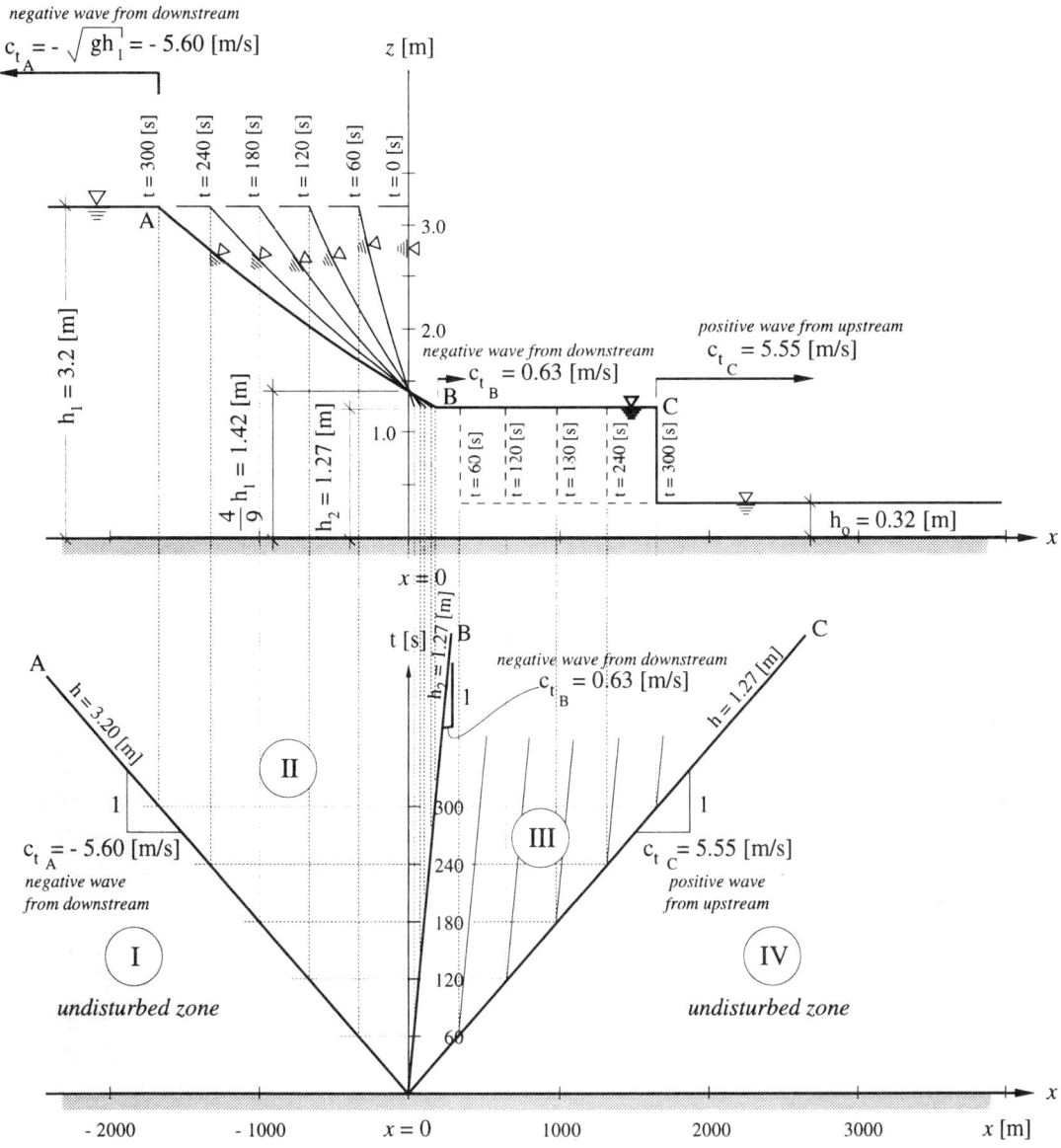

Fig. Ex.5.E.7 The water-surface profile of the negative wave from downstream when there is stagnant water in the downstream channel, at different instants after the opening of the gate. Also shown is the corresponding plane, x-t.

c) When there is water in the channel downstream, the water-surface profile resulting from a complete and sudden opening of a gate is composed of *a negative wave from downstream* and *a positive wave from upstream*. This entire profile is plotted in Fig. Ex.5.E.7 for the five different instants mentioned in the problem statement together with the corresponding plane, x-t. The reader is invited to compare this figure with Fig. Ex.5.E.3 (the case with the dry downstream channel, $h_o = 0$).

Four zones can be distinguished on the plane, x-t. The zones I and IV are the undisturbed zones where the waves generated by the gate opening, negative wave from downstream and positive wave from upstream, have no influence.

The negative wave from downstream develops in the zone II, delimited by the characteristics of the points A and B between which the profile of the negative wave from downstream is always valid. The profile of the negative wave computed for the case with a dry downstream channel was already given in the table of Fig. Ex.5.E.2. In this table, only the portion of the profile between the depths $h_2 \leq h \leq h_1$ ($1.27 \leq h$ [m] ≤ 3.20), is presently valid. The celerities for the depths $h_1 = 3.20$ [m] (point A) and $h_2 = 1.27$ [m] (point B) are taken from this table. The characteristics of point A, which separates zone II from the undisturbed zone at the upstream (zone I), propagates the depth $h_A = h_1 = 3.2$[m] in the space, x-t. The table in Fig. Ex.5.E.2 gives $c_{t_A} = -5.60$ [m/s]. The characteristics for the point B separates the zone II from the zone III where the positive wave from upstream travels and propagates the flow depth $h_B = h_2 = 1.27$ [m] in the space, x-t. The table of Fig. Ex.5.E.2 gives $c_{t_B} = 0.63$ [m/s]. These two characteristics are plotted in Fig. Ex.5.E.7.

The zone III, where the positive wave from upstream having a depth of h_2 develops, is limited at the upstream by the characteristics for point B. At the downstream, it is the characteristics for point C, corresponding to the celerity of the positive wave, which separates it from the undisturbed zone IV. This celerity, which is equal to $c_{t_C} = 5.55$ [m/s], has been computed above by trial-and-error. The corresponding line is also plotted in Fig. Ex.5.E.7.

For a better comprehension, a tridimensional representation of the space x-t, and the height z is drawn in Fig. Ex. 5.E.8. The reader is invited to compare this figure with Fig. Ex.5.E.4.

d) The origin of the space, x-t, is always located at the initial position of the gate, $x = 0.0$ [m] (see Fig. Ex.5.E.7). In the zone II, for every flow depth in the range $1.27 \leq h$ [m] ≤ 3.20, a characteristics initiates at this origin with a slope whose inverse is equal to the celerity of the wave for the corresponding depth (see table of Fig. Ex.5.E.2).

Since the friction is neglected, the flow depth in the zone III remains constant and equal to $h_2 = 1.27$ [m]. In the space, x-t, this zone is defined by characteristics being parallel to the characteristics of point B ($c_{t_B} = 0.63$ [m/s]). These lines are delimited by the characteristics of point C. The celerities of the upstream end (point B) and downstream end (point C) in the zone III are different. Although both points travel in the same direction — the celerity of the point C being faster than that of the point B — the distance between them increases with time.

It should be noted that the time-axis, t, is always located in the zone of the negative wave. It represents the characteristics for a certain flow depth.

UNSTEADY FLOW 345

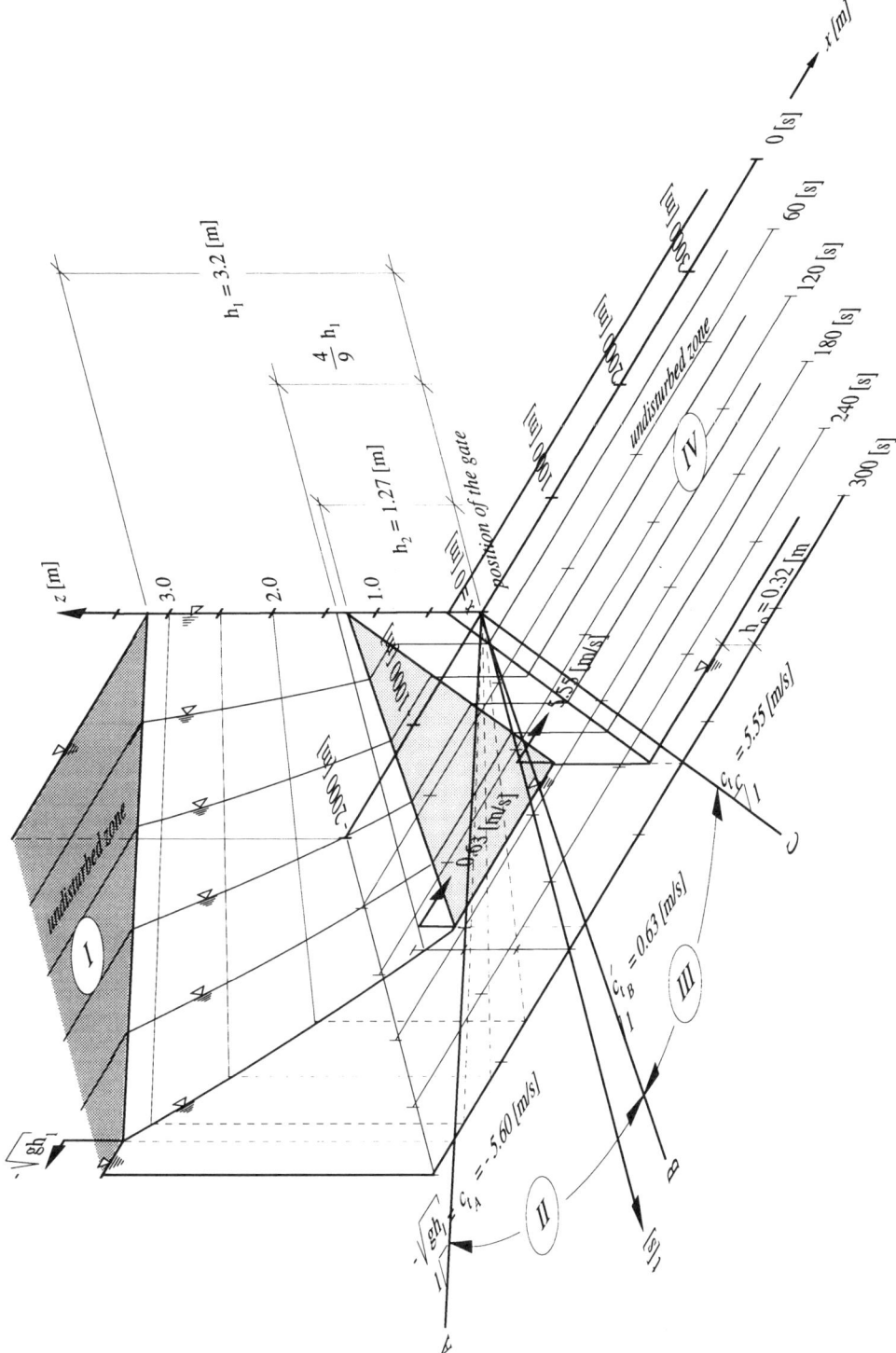

Fig. Ex.5.E.8 Tridimensional representation of the wave profile in the space, x-t-z, when there is stagnant water in the downstream channel.

This depth was calculated in the first part of problem as being $h_{xo} = \frac{4}{9}h_1 = 1.42$ [m]. The average velocity in this section is always $U_{xo} = +\frac{2}{3}\sqrt{gh_1} = 3.74$ [m/s]. The unit discharge is therefore not influenced by the presence of stagnant water in the downstream channel and remains equal to $q = \frac{8}{27} h_1 \sqrt{gh_1} = 5.31$ [m²/s].

e) It can therefore be seen that the presence of stagnant water in the downstream channel, $h_o \neq 0$, has modified the water-surface profile, $h(x)$ (compare Fig. Ex.5.E.3 with Fig. Ex.5.E.7), but it has not influenced the unit discharge, q. Is this related to a particular value of the downstream water depth, h_o? Is there a limiting value above which the depth in the downstream channel influences the unit discharge ? The answers to these questions are worth to be studied. Let us consider :

- $h_2 < h_{xo}$:
  In this case, for a downstream water depth of $h_o = 0.32$ [m], the flow depth for the positive wave from upstream was found to be $h_2 = 1.27$ [m]; this is less than the flow depth, $h_{xo} = 1.42$ [m], around which the negative wave profile swivels. Since the point of swivel is always the same, it is the negative wave from downstream, which controls the unit discharge, $q = \frac{8}{27} h_1 \sqrt{gh_1}$.

- $h_2 = h_{xo}$:
  Let us consider now a downstream flow depth of $h_o = h_o^*$, such that the flow depth of the positive wave from upstream is $h_2 = h_{xo}$. The characteristics for point B coincides then with the time-axis, t. The point B will stay at the initial position of the gate, while the front of the positive wave from upstream (point C) will travel towards downstream. A limiting situation has been reached, but the unit discharge remains the same : $q = \frac{8}{27} h_1 \sqrt{gh_1}$.

- $h_2 > h_{xo}$:
  For larger water depths, $h_o > h_o^*$, there will be $h_2 > h_{xo}$. For these flow depths, the celerity being negative (see Fig. Ex.5.E.2), the point B will travel towards upstream. The characteristics for point B, separating the zones II and III, will be on the upstream side of the time-axis t. The flow depth, h_2, will be larger, but the velocity U_2 will be lesser (see Fig. Ex.5.E.5). The unit discharge, $q = h_2 U_2$, will diminish.

These three situations are represented in Fig. Ex. 5.E.9.

f) For a given water depth h_1, the downstream stagnant water depth, h_o, which corresponds to the critical situation, can be calculated using the two expressions derived previously for the absolute celerity, c_{t_C}, of the positive wave from upstream. By eliminating this celerity, c_{t_C}, between these two expressions, developed previously, one can write :

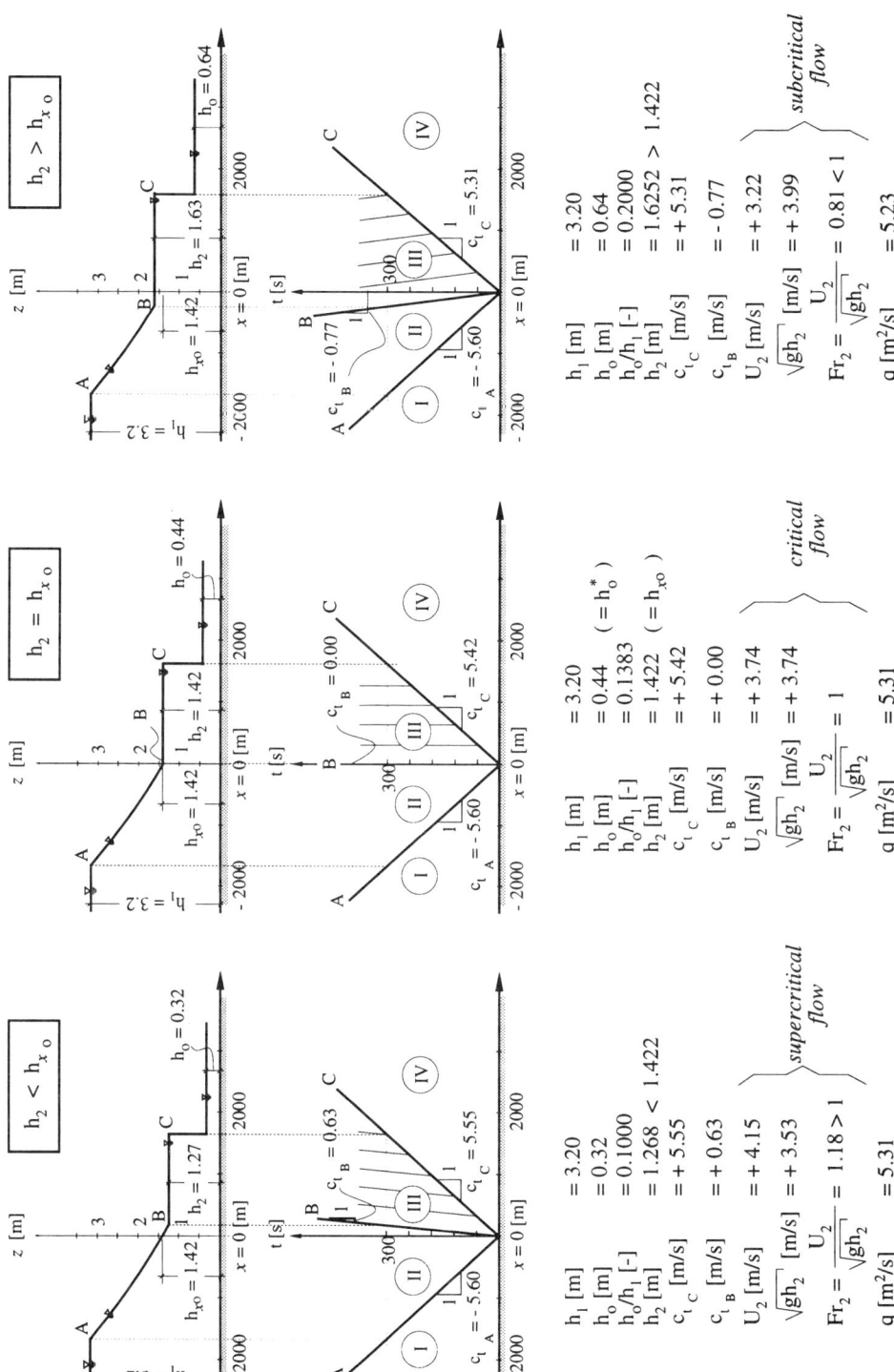

Fig. Ex.5.E.9 Water-surface profile resulting from a sudden gate opening and the corresponding plane, x-t, for three different depths of the stagnant water, h_o, in the downstream channel.

$$U_2 h_2 = \sqrt{gh_o} \sqrt{\frac{h_2}{2h_o}(1 + \frac{h_2}{h_o})} (h_2 - h_o)$$

In the limiting situation, one has (see the first part of the problem):

$$h_2 \equiv h_{xo} = \frac{4}{9} h_1 \quad \text{and} \quad U_2 \equiv U_{xo} = \frac{2}{3}\sqrt{gh_1}.$$

By substituting these relations in the above expression and rearranging the terms, the following will be obtained:

$$f(h_1/h_o) = \frac{8}{27}\left(\frac{h_1}{h_o}\right)^{3/2} - \left(\frac{4}{9}\frac{h_1}{h_o} - 1\right)\sqrt{\frac{4}{18}\frac{h_1}{h_o}(1 + \frac{4}{9}\frac{h_1}{h_o})} = 0$$

The value of the relative depth, (h_1/h_o), which satisfies this equation can be obtained by trial-and-error. The computations are presented in the table below:

	(h_1/h_o)	(h_o/h_1)	$f(h_1/h_o)$	
trial 1	7.1429	0.14000	0.05852	$f(h_1/h_o) \neq 0$, no
trial 2	7.6923	0.13000	–0.32638	$f(h_1/h_o) \neq 0$, no
trial 3	7.2322	0.13827	0.00000	$f(h_1/h_o) = 0$, yes

The limiting situation, where $h_2 = h_{xo} = \frac{4}{9} h_1$, is thus reached if the downstream stagnant water depth is $h_o = 0.13827\, h_1$, thus $h_o = 0.44$ [m].

g) The water-surface profiles resulting from the sudden opening of the gate and the corresponding representations of the plane, x-t, are presented in Fig. Ex.5.E.9 for three different relative depths, $h_o/h_1 = 0.10000$, 0.13827 and 0.20000. As can be seen, the absolute celerity of the positive wave, c_{t_C}, decreases when the relative depth, h_o/h_1, increases.

According to *Stoker* (1957, p. 338), the three types of flow, shown in Fig. Ex.5.E.9, can also be classified as functions of the Froude number, $Fr_2 = h_2/\sqrt{gh_2}$, in the zone III:

- $\dfrac{h_o}{h_1} < 0.13827 \quad \Rightarrow \quad Fr_2 > 1 \quad \Rightarrow \quad$ supercritical flow

- $\dfrac{h_o}{h_1} = 0.13827 \quad \Rightarrow \quad Fr_2 = 1 \quad \Rightarrow \quad$ critical flow

- $\dfrac{h_o}{h_1} > 0.13827 \quad \Rightarrow \quad Fr_2 < 1 \quad \Rightarrow \quad$ subcritical flow.

UNSTEADY FLOW

5.7.2 Problems, unsolved

Ex. 5.1
A river joins a lake. Flow in this river is fluvial, having a velocity of U = 0.6 [m/s] ; the flow depth is h = 2.2 [m]. Initially, the water level at the juncture, both in the river and in the lake are identical. Now the level of the lake is lowered by Δh = 0.3 [m] in a time period of ΔT = 0.5 [h]. Determine the time necessary till this lowering becomes noticeable in the river at 2 [km] from the juncture. It may be assumed that the slope of the bed and of the energy-grade line are negligible.

Ex. 5.2
Study the canal described in Ex. 5.B.
i) Determine the hydrographs for the stations L_3 = 2 [km] and L_4 = 4 [km].
ii) Determine the attenuation of the flow depth at the stations L_1, L_2, L_3 and L_4.

Ex. 5.3
The bed slope of the canal, described in Ex. 5.B, is changed to S_f = 0.0005 [-].
i) Determine the hydrographs at the station L_2.
ii) Develop also the rating curve.

Ex. 5.4
An alternate study for the canal, described in Ex. 5.B, is to be done : the cross section be trapezoidal with b = 5 [m] and m = 1.
i) Determine the hydrographs at the two stations, L_1 and L_2.
ii) Develop the rating curve for the station L_1.

Ex. 5.5
Study the flow in the canal, described in Ex. 5.B, by using the method of *characteristics* (see sect. 5.2.2).

Ex. 5.6
Study the flow in the canal, described in Ex. 5.B, by using the *implicit* method (see sect. 5.2.4).

Ex. 5.7
A river discharges into the ocean. This river, being rather wide, flows with a velocity of U = 0.5 [m/s] at a flow depth of h_n = 1.5 [m]. An ocean tide moves upstream the river with a celerity of c_t = 5 [m/s]. Determine the height, Δh, and the velocity of the flow after the passing of the tide.

Ex. 5.8
The rectangular section of an irrigation canal has a width of B = 2.0 [m] and a canal depth of H = 2.0 [m]. Uniform flow with a discharge of Q = 3.0 [m^3/s] establishes itself at a normal depth of h_n = 1.0 [m]. In case of an instantaneous closure of the gate, installed downstream in the canal, a wave will form. Will the canal depth, H, be sufficient ?

350 FLUVIAL HYDRAULICS

Ex. 5.9

In a smooth rectangular canal, the steady uniform flow, having an average velocity of $U_1 = 1.1$ [m/s], is suddenly stopped at the downstream by the closure of a sluice gate. Subsequently to this closure, the wave propagating towards upstream had a celerity of $c_t = 7.2$ [m/s] ; the water depth was h = 6.5 [m]. What had been the normal depth of the uniform flow in this canal ?

Ex. 5.10

In a rectangular canal, having a width of B = 3.0 [m], the uniform flow with its normal depth of $h_n = 0.8$ [m] has a discharge of Q = 10 [m³/s]. Instantaneously this discharge is reduced by 40%. Determine the celerity and the height of the thus-created wave.

Ex. 5.11

A river flows at a velocity of U = 1.2 [m/s] at a normal flow depth of $h_n = 30$ [m]. A tide moves against the river flow with a celerity of c = 7.0 [m/s]. What will be the wave height caused by this tide ?

Ex. 5.12

In a rectangular (smooth and horizontal) canal, water runs with an average velocity of U = 1.1 [m/s] at the corresponding flow depth of h = 1.2 [m]. Suddenly, the discharge of the flow at the upstream end is doubled. Investigate the consequences of the change in discharge.

6. TRANSPORT OF SEDIMENTS

The study of flow in a watercourse is a particularly difficult task, since the channel bed is usually of a form which varies in space and time. The movement of the sediments, which make up the bed, represents a rather complex phenomenon.

In this chapter, the hydrodynamic equations of the flow over a mobile bed are developed ; some solutions are given. The different modes of the transport of (non-cohesive) sediments as bed load and as suspended load are presented. The formulae for the calculations of the transport of the total load will be exposed, as well as their domain of application.

TABLE OF CONTENTS

6.1 GENERALITIES
    6.1.1 Notions
    6.1.2 Flow of a Mixture
    6.1.3 Modes of Transport
    6.1.4 Types of Problems

6.2 HYDRODYNAMIC EQUATIONS
    6.2.1 Equations of Saint-Venant - Exner
    6.2.2 Propagation of Perturbations
    6.2.3 Analytical Solutions
    6.2.4 Degradation and Aggradation
    6.2.5 Numerical Solutions

6.3 BED - LOAD TRANSPORT
    6.3.1 Notions
    6.3.2 Theoretical Considerations
    6.3.3 Bed-Load Relations
    6.3.4 Granulometry, Armouring

6.4 SUSPENDED - LOAD TRANSPORT
    6.4.1 Notions
    6.4.2 Theoretical Considerations
    6.4.3 Suspended-Load Relation

6.5 TOTAL - LOAD TRANSPORT
    6.5.1 Notions
    6.5.2 Total-Load Relations
    6.5.3 Applications of Relations
    6.5.4 Wash Load

6.6 EXERCISES
    6.6.1 Problems, solved
    6.6.2 Problems, unsolved

6.1 GENERALITIES

6.1.1 Notions

1° The flow of water over a mobile bed has the ability to entrain the sediments (solid particles) ; a water-sediment mixture will consequently displace itself in the water-course. The movement of the sediments — *erosion, transport, deposition* — will modify the flow, but also the channel bed, thus its elevation, its slope and its roughness. The interaction between the water and the sediments makes the problem a coupled one.

2° When the bed is a *mobile* one, the fluvial hydraulics must concern itself with both the flow of the liquid phase, namely the mixture, and the movement of the solid phase, namely the sediments in the mixture.

3° A characterisation of the liquid and the solid phase of a water-sediment mixture is a difficult task.

 The *liquid* phase is rather well described by :
 i) its density, ρ ,
 ii) its viscosity, μ ,
 iii) the average velocity of the flow, U , and
 iv) the friction velocity, u_*.

 The *solid* phase is more difficult to characterise ; considered should be :
 i) the size of the solid particles, given by its granulometric curve, which includes different types of diameters such as d_{50} , d_{90} , d_{35} , etc.,
 ii) the form of these particles,
 iii) the density of the particles, ρ_s ,
 iv) together, these parameters can be defined by the settling velocity of the particles, v_{ss} , and
 v) possibly, the cohesion between the particles.

 All these parameters could vary along the watercourse. Furthermore, they will depend on the way the bed samples (in the nature) are taken and are analysed.

4° The dimensions of the sediments are relatively small compared to the ones of the flow ; thus the *turbulence* will play an essential role in all flows of a water-sediment mixture.

5° The transport of these sediments plays an important, if not the most important role in all problems of fluvial hydraulics. This phenomenon is very complex and consequently a theoretical study can only be performed in simple or simplified cases. The formulae, developed for the quantitative determination of the transport of sediments, are based on experimental results, being often limited, and thus should be used with much caution. Such formulae are of great value for the hydraulic engineer, but must be applied within hydraulic conditions under which they have been established.

6.1.2 Flow of a Mixture

1° For gravitational flow of a water-sediment mixture, one may distinguish three types of movement (see Table 6.1). :

   i)   The mixture may be considered Newtonian, if the volumic concentration of the particles is very small, $C_s \ll 1\%$. The difference between the density of the mixture and of the water, $\Delta\rho = (\rho_m - \rho) = (\rho_s - \rho) C_s$ (see eq. 7.2), remains also small, $\Delta\rho \ll 16$ [kg/m^3].

   The *transport of sediments* (see chap. 6), as bed load and as suspended load, falls into this category. It is this type of transport of solid particles, which is most often encountered in watercourses.

   ii) The mixture behaves quasi-Newtonian, if the volumic concentration remains small, $C_s < 8\%$. The difference between the density of the mixture and of the water becomes important, $\Delta\rho < 130$ [kg/m^3].

   The transport of sediments as *concentrated suspension* (see *Graf*, 1971, p. 182-186) notably close to the bed, as well as the *turbidity currents* (see chap. 7) fall into this category.

Table 6.1 Classification of flows of a mixture.

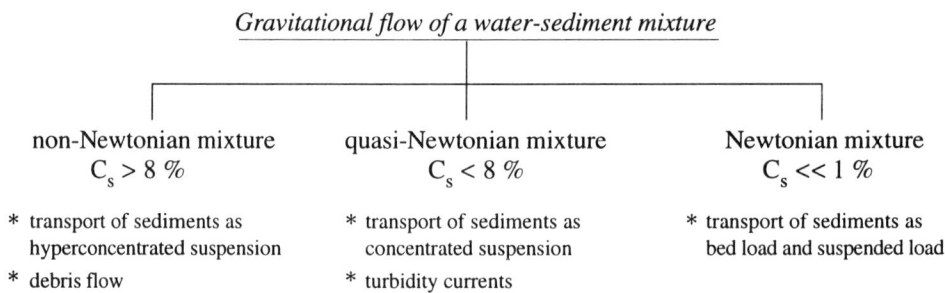

   iii) The mixture behaves non-Newtonian, if the volumic concentration becomes of importance, $C_s > 8\%$. The difference between the density of the mixture and of the water is also very large, $\Delta\rho > 130$ [kg/m^3].

   The flow of a non-Newtonian fluid modifies all concepts of Newtonian hydraulics, such as the resistance to the flow, as well as the distribution of velocity and of concentration ; the settling velocity is also influenced and the solid particles stay longer in suspension.

   The transport of sediments as *hyperconcentrated suspension* (see *Wan* et *Wang*, 1994), the *debris flow* (see *Takahashi*, 1991), as well as *hyperconcentrated turbidity currents* (see *Wan* et *Wang*, 1994) fall into this category.

TRANSPORT OF SEDIMENTS

- The transport of sediments as an hyperconcentrated suspension is encountered in watercourses of small slopes. Usually enormous quantities of sediments — being of small sizes — enter the channel due to surface erosion caused by extensive rainfalls in the catchment basin. These solid particles stay usually for long time periods in suspension, as wash load.

- Torrential flows of debris may establish themselves at rather steep slopes, $\alpha > 15°$. All kinds of particles, from the finest (having cohesion) to the largest (blocks of 1 $[m^3]$), participate in the movement, which is rather rare in occurrence and of short duration, and is usually caused by severe rainfalls.

2° It should be stressed, that the above schematic classification (see Table 6.1) is a simplification of the reality, where limits can often not readily be defined and where the different cases can coexist.

6.1.3 Modes of Transport (see Fig. 6.1)

1° The (total) transport of sediments by flow of water is the entire solid transport (of the particles) which passes through a cross section of a watercourse.

Traditionally (but a bit artificially) the transport of sediments is classified in different modes of transport (see Table 6.2) which correspond to distinctly different physical mechanisms.

Table 6.2 Modes of transport of sediments (T.S.).

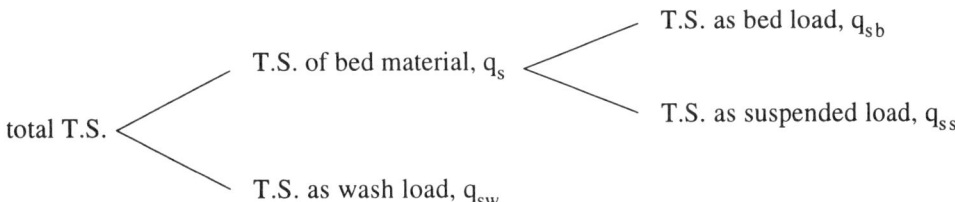

2° In a watercourse the sediments, namely the solid phase, are transported :

i) as *bed load*, q_{sb}, — volumic solid discharge per unit width $[m^3/sm]$ — when the particles stay in close contact with the bed ; the particles displace themselves by gliding, rolling or (shortly) jumping ; this type of transport concerns the relatively larger particles ;

ii) as *suspended load*, q_{ss}, when the particles stay occasionally in contact with the bed ; the particles displace themselves by making more or less large jumps and remain often surrounded by water ; this type of transport concerns the relatively smaller particles ;

iii) as bed load + suspended load, being the (total) *bed-material load*, $q_s = q_{sb} + q_{ss}$, when the particles stay more or less in continuous contact with the bed.

iv) as *wash load*, q_{sw}, when the particles are almost never in contact with the bed ; the particles are washed through the cross section by the flow ; this type of transport concerns the relatively finest particles.

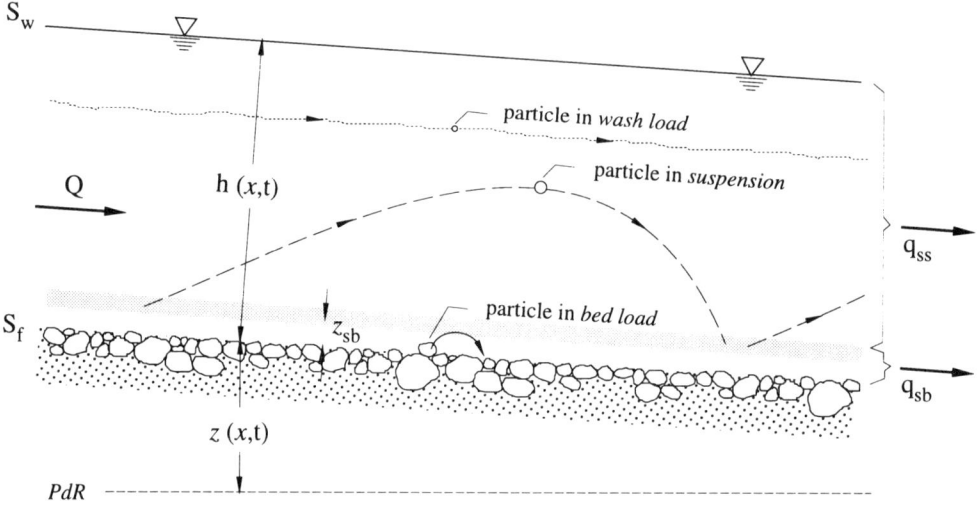

Fig. 6.1 Scheme of the modes of transport.

3° The transport of sediments, namely the erosion of the bed (see sect. 3.4.2), commences upon attainment of a certain critical value, which can be parametrised, for example, by the critical shear stress, τ_{ocr}.

4° It will be useful, but it is also rather imprecise, to give limiting values for the separation of the different modes of transport. Given here are purely indicative values, which use the ratio of the shear velocity of the flow, u_*, and the settling velocity of the particles, v_{ss} (see *Graf*, 1971) :

- $\dfrac{u_*}{v_{ss}} > 0.10$ beginning of bed-load transport,

- $\dfrac{u_*}{v_{ss}} > 0.40$ beginning of suspended load transport.

5° To determine quantitatively the transport of sediments, there are three possibilities available, namely :
  - using existing formulae (see sects. 6.3, 6.4 and 6.5),
  - obtaining field measurements with adequate instruments (see *Graf*, 1971, chap. 13),
  - performing physical models (see *Graf*, 1971, chap. 14).

6.1.4 Types of Problems

1° Many of the hydraulic problems, which require a knowledge of the transport of sediments, can readily be put into one of the following categories :

- determination of a sedimentological rating curve, $q_s = f(q)$, for a given cross section of the channel (see Fig. 6.2);
- determination of the stability of the bed in a given cross section (see sect. 3.4.4);
- determination of the stability of the channel slope (aggradation and degradation) in a given reach of the channel (see sect. 6.2.4).

2° The different modes of transport of sediments, quantified in form of solid discharge, q_{sb}, q_{ss} and q_s, should be related to the liquid discharge, q. This will give the relation of the "sedimentological" rating curve (see Fig. 6.2) for a given cross section of the channel. This curve together with the "liquid" rating curve (see Fig. 3.8) give a rather complete hydraulic description for a given cross section of a channel having a mobile bed.

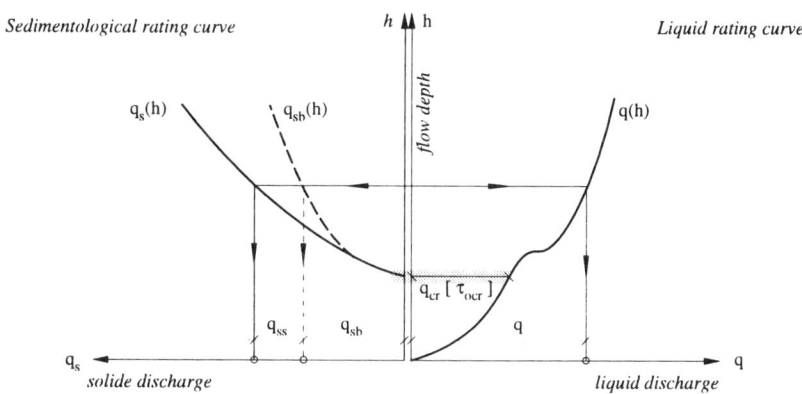

Fig. 6.2 Rating curves for the liquid discharge and the solid discharge.

3° The formulae, which are used to calculate the solid discharge, q_s, allow to know the *capacity* of the transport of sediments for a given flow. Under such conditions, the transport of sediments is said to be in equilibrium.

However it could happen, that the supply of solid discharge is not equal to the capacity of the transport. The transport of sediments is then not in equilibrium :

- if the capacity is larger than the supply, erosion and transport occurs,
- if the supply is larger than the capacity, deposition and transport occurs,
- if the supply is equal to the capacity, transport without erosion or deposition occurs,
- if the bed is armoured, the capacity may not be satisfied (see sect. 6.3.4).

One sees here the complexity of the problem, where along a watercourse the different scenarios can coexist or overlap.

6.2 HYDRODYNAMIC EQUATIONS

Presented will be the hydrodynamic equations, and some solutions, for flow in an open channel over a mobile bed, when entrainment of sediments is possible.

6.2.1 Equations of Saint-Venant - Exner

1° The equations of Saint-Venant (see sect. 5.11) for unsteady and non-uniform flow over a *fixed bed* in a prismatic open channel with a small bed slope (see Fig. 5.1), have been given before (see eq. 5.2 and eq. 5.3) ; for flow over a *mobile bed* they can be written as :

$$\frac{\partial h}{\partial t} + h\frac{\partial U}{\partial x} + U\frac{\partial h}{\partial x} = 0 \qquad B = Cte \qquad (6.1)$$

$$\frac{\partial U}{\partial t} + U\frac{\partial U}{\partial x} + g\frac{\partial h}{\partial x} + g\frac{\partial z}{\partial x} = - g\, S_e \qquad (6.2)$$

The energy slope, S_e, shall be expressed with a relationship established for uniform flow, by using a friction coefficient, f, for a mobile bed (see sect. 3.2.6), or :

$$S_e = f(f, U, h) \qquad (6.3)$$

where h is the flow depth, U is the average velocity of the flow and $z(x,t)$ gives the elevation of the channel bed.

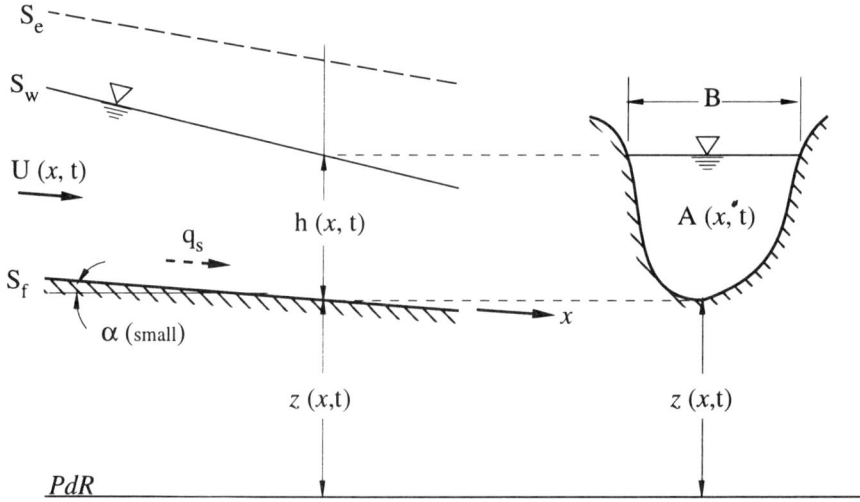

Fig. 6.3 Scheme of unsteady and non-uniform flow over a mobile bed, $z(x,t)$.

TRANSPORT OF SEDIMENTS 359

2° For flow over a *mobile bed* (see Fig. 6.3), the elevation (level) of the channel bed, $z(x,t)$, may vary. According to the relation of *Exner* (see *Exner*, 1925, and *Graf*, 1971, p. 288), such a variation can be expressed by :

$$\frac{\partial z}{\partial t} = - a_E \frac{\partial U}{\partial x} \qquad (6.4)$$

where a_E is an erosion coefficient. This relationship, eq. 6.4, can be written (see *Graf*, 1971, p. 152 and *Krishnappan*, 1981, p. 91) in form of a continuity equation for the solid phase, namely :

$$\frac{\partial z}{\partial t} + (\frac{1}{1-p}) \left[\frac{\partial}{\partial t} (\tilde{C}_s h) + \frac{\partial}{\partial x} (C_s U h) \right] \cong \frac{\partial z}{\partial t} + (\frac{1}{1-p}) \frac{\partial q_s}{\partial x} = 0 \qquad (6.4a)$$

where p is the porosity of the sediments of the bed, being defined as the ratio of the volume of empty space (occupied by water) and of the total volume. $q_s = C_s U h$ is the volumic solid discharge per unit width and C_s ($\tilde{C}_s$) is the volumic concentration (in the cross section) of the solid phase, being defined by the ratio of the volume of the sediments and of the volume of the mixture. In general one admits that the solid discharge, q_s , is a function — still to be determined (see sect. 6.3 to sect. 6.5) — of the liquid discharge, $q = U h$, or :

$$q_s = f(U, h ; \text{sediment}) \qquad (6.5)$$

3° The three basic (differential) equations, eqs. 6.1, 6.2 and 6.4a, contain three unknowns, $U(x, t)$, $h(x, t)$ and $z(x,t)$, with their independent variables, x and t. U and h are the average velocity and the flow depth of the water-sediment mixture (the liquid phase), or of the water only, if the concentration of the sediments, C_s, is negligible.

The two other unknowns, S_e (see eq. 6.3) and q_s (see eq. 6.5), have to be expressed with semi-empirical relationships.

4° The five relations, eqs. 6.1, 6.2 and 6.4a together with eq. 6.3 and eq. 6.5, are the *equations of Saint-Venant - Exner*.

The three relations, eqs. 6.1, 6.2 and 6.3, describe the flow of the liquid phase over a mobile bed ; the two other relations, eqs. 6.4a and 6.5, describe the transport (erosion and deposition) of the solid phase.

5° The liquid and the solid phase are *implicitly* coupled by the semi-empirical relations, eqs. 6.3 and 6.5. After the solution for the liquid phase, eqs. 6.1 and 6.2, a solution for the solid phase, eq. 6.4a, can be obtained, giving the variation of the bed elevation, $z(x,t)$.

The equations of Saint-Venant - Exner can be *explicitly* coupled, if the equation of continuity for the liquid phase, eq. 6.1, is expressed (see *Krishnappan*, 1981, p. 93) as follows :

$$\frac{\partial h}{\partial t} + \frac{\partial z}{\partial t} + \frac{\partial}{\partial x}(Uh) = 0 \tag{6.1a}$$

A direct coupling is thus achieved (see *Correia et al.*, 1992), since the term, $\partial z/\partial t$ — which is however often rather small — exists now in both eq. 6.1a and eq. 6.4a. One looks now for a solution by solving simultaneously the equations for the liquid and for the solid phase.

6° To obtain solutions to the equations of Saint-Venant - Exner, use can be made of :

   i)   analytical methods (see sect. 6.2.3) for simple problems, and
   ii) numerical methods (see sect. 6.2.5) for complex problems.

6.2.2 Propagation of Perturbations

1° The propagation of a perturbation, being a wave of small amplitude on the mobile bed, can now be investigated by using the equations of Saint-Venant - Exner, eq. 6.1 to eq. 6.5.

2° For a rectangular channel, these equations — see also the system of equations, eq. 5.2 to eq. 5.10 — are written as six equations of partial derivatives :

$$\frac{\partial h}{\partial t} + h\frac{\partial U}{\partial x} + U\frac{\partial h}{\partial x} = 0 \tag{6.1}$$

$$\frac{1}{g}\frac{\partial U}{\partial t} + \frac{U}{g}\frac{\partial U}{\partial x} + \frac{\partial h}{\partial x} + \frac{\partial z}{\partial x} = -S_e \tag{6.2}$$

$$(1-p)\frac{\partial z}{\partial t} + \frac{\partial q_s}{\partial U}\frac{\partial U}{\partial x} = 0 \tag{6.4b}$$

with :
$$\frac{\partial h}{\partial x}dx + \frac{\partial h}{\partial t}dt = dh$$

$$\frac{\partial U}{\partial x}dx + \frac{\partial U}{\partial t}dt = dU$$

$$\frac{\partial z}{\partial x}dx + \frac{\partial z}{\partial t}dt = dz$$

TRANSPORT OF SEDIMENTS

The three last equations are expressions for the total derivatives of the three dependent variables, h(x,t), U(x,t) and z(x,t).

In writing eq. 6.4b, it was assumed that the solid discharge is only a function of the flow velocity, $q_s = f(U)$, thus :

$$\frac{\partial q_s}{\partial x} = \frac{\partial q_s}{\partial U} \frac{\partial U}{\partial x}$$

3° Upon mathematical manipulations of these six equations — the determinant of the matrix of the coefficients must become zero — the following cubic relationship (see *de Vries*, 1965 and 1973, p. 2) is obtained :

$$-c_w^3 + 2U c_w^2 + (gh - U^2 + g\frac{\partial q_s}{\partial U}) c_w - gU\frac{\partial q_s}{\partial U} = 0 \qquad (6.6)$$

where the absolute celerity (the characteristic) is defined by :

$$c_w = \frac{dx}{dt} \qquad (2.34)$$

4° This equation, eq. 6.6, has evidently three real roots, thus three characteristics :

i) Two roots, c_{w_1} and c_{w_2}, are an expression of the celerity of the perturbation (wave) on the water surface ; the third root, c_{w_3}, gives the celerity of a perturbation (undulation) on the mobile bed.

ii) c_{w_3} is positive for subcritical flow, $U < \sqrt{gh}$; the form (undulation) of the bed, usually called dunes (see sect. 3.2.5), displaces itself in the same direction as the flow (see *de Vries*, 1973, p. 3).

c_{w_3} is negative for supercritical flow, $U > \sqrt{gh}$; the form (undulation) of the bed, usually called antidunes (see sect. 3.2.5), displaces itself in the opposite direction of the flow.

iii) For a fixed bed without solid discharge, $q_s = 0$, one gets :

$$c_{w_3} = 0$$
$$c_{w_1} = U + \sqrt{gh} \qquad \text{and} \qquad c_{w_2} = U - \sqrt{gh} \qquad (5.14a)$$

This solution has already been presented in sect. 5.2.1 and sect. 2.4.3 (see Fig. 2.9).

5° It seems reasonable (see *de Vries*, 1973, p. 4, and *Jansen et al.*, 1979, p. 96) to assume that for $Fr = U/\sqrt{gh} \neq 1$ the celerities, c_{w_1} and c_{w_2}, of the waves on the surface are much larger than the celerity, c_{w_3}, of the undulations on the bed, or :

$$c_{w_1} \text{ and } c_{w_2} \gg c_{w_3}$$

When studying the perturbations on the bed having a weak celerity, c_{w_3}, it is now possible to consider the flow of the liquid phase as quasi-steady ; thus :

$$\partial U/\partial t = 0 \quad \text{and} \quad \partial h/\partial t = 0$$

Consequently, combining eq. 6.1 with eq. 6.2, one can write a single differential equation, which is the equation of the free-surface flow (see eq. 4.7), or :

$$\frac{\partial U}{\partial x}(U - \frac{h}{U}g) + g\frac{\partial z}{\partial x} = -gS_e \qquad (6.7)$$

By eliminating $\partial U/\partial x$ between eq. 6.7 and eq. 6.4b, one obtains :

$$\frac{\partial z}{\partial t} + c_{w_3}\frac{\partial z}{\partial x} = -c_{w_3}S_e = \mathbf{F}(U) \qquad (6.8)$$

where $\mathbf{F}(U)$ is a friction (roughness) term — being responsible for the decay of the perturbation on the bed — and

$$c_{w_3} = \frac{g}{(1-p)}\frac{(\partial q_s/\partial U)}{(gh/U - U)} = \frac{1}{(1-p)}\frac{U(\partial q_s/\partial U)}{h(1-Fr^2)} \qquad (6.9)$$

where $Fr^2 = U^2/gh$. For subcritical flow, when $Fr^2 \ll 1$, the following approximation is possible :

$$c_{w_3} \cong \frac{1}{(1-p)}\frac{U}{h}\frac{\partial q_s}{\partial U} \qquad (6.9a)$$

If the solid discharge is expressed by a power law of the form :

$$q_s = a_s U^{b_s} \quad \text{and} \quad \frac{dq_s}{dU} = b_s(\frac{q_s}{U}) \qquad (6.5a)$$

one may write :

$$c_{w_3} = \frac{1}{(1-p)}b_s\frac{q_s}{h} \qquad (6.9b)$$

It is to be noted (see eq. 6.9a), that the celerity of propagation of the undulations, c_{w_3}, on the bed is usually rather small compared to the average velocity, U, of the flow itself.

TRANSPORT OF SEDIMENTS 363

6.2.3 Analytical Solutions

1° To obtain analytical solutions to the equations of Saint-Venant - Exner, which are non-linear and hyperbolic, is a very difficult and often impossible task. However simplifications are nevertheless possible, if one assumes that for flow at small Froude numbers, Fr < 0.6, a *quasi-steadiness* is maintained. This hypothesis of a steadiness of flow can be justified : in general a variation of liquid discharge, $\partial(Uh)/\partial t$, is a short-term phenomenon, while a variation of the bed elevation, $\partial z/\partial t$, is a long-term phenomenon, which produces itself when the variation of the discharge has already disappeared ; thus the flow may be considered reasonably constant, q = Uh = Cte. Under such conditions solutions are of great interest, notably if one studies the variation of the bed, $z(x,t)$, as a long-term phenomenon.

2° Using the hypothesis of quasi-steadiness of the flow, a system of two differential equations can be written as :

$$\frac{\partial U}{\partial x}(U - g\frac{h}{U}) + g\frac{\partial z}{\partial x} = -g S_e \tag{6.7}$$

$$(1-p)\frac{\partial z}{\partial t} + \frac{\partial q_s}{\partial U}\frac{\partial U}{\partial x} = 0 \tag{6.4b}$$

These two equations are non-linear ones ; only numerical solutions are possible. For certain special cases, analytical solutions (after linearisation) can be of help to understand the problem, notably the relative importance of the different parameters.

3° If one further assumes (see *Vreugdenhil* et *de Vries*, 1973, p. 8) that the quasi-steady flow is also quasi-uniform, $\partial U/\partial x = 0$, the above equation, eq. 6.7, becomes :

$$0 + g\frac{\partial z}{\partial x} = -g S_e = -g\frac{U^2}{C^2 h} = -g\frac{U^3}{C^2 q} \tag{6.10}$$

where C is the coefficient of Chézy and q = Uh is the unit discharge.

By eliminating $\partial U/\partial x$ after differentiation of eq. 6.10 with respect to x, one obtains for the above equation, eq. 6.4b :

$$\frac{\partial z}{\partial t} - K(t)\frac{\partial^2 z}{\partial x^2} = 0 \tag{6.11}$$

where the coefficient (of diffusion), K(t), being a function of time, is given by :

$$K = \frac{1}{3}\frac{\partial q_s}{\partial U}\frac{1}{(1-p)}\frac{C^2 h}{U} \tag{6.12}$$

This model, eq. 6.11, is a *parabolic* one and is limited to large values of x and of t, namely for $x > 3h/S_e$ (see *de Vries*, 1973, p. 9). The expression for the coefficient, eq. 6.12, can also be written (see *de Vries*, 1973, p. 6) in the following way:

$$K = \frac{1}{3} \frac{\partial q_s}{\partial U} \frac{1}{(1-p)} \frac{U}{S_{e_o}} \left(\frac{U_o}{U}\right)^2 \quad (6.12a)$$

and upon linearisation (possible for $U \simeq U_o$), one obtains:

$$K \equiv K_o \approx \frac{1}{3} \frac{\partial q_s}{\partial U} \frac{1}{(1-p)} \frac{U_o}{S_{e_o}} \quad (6.12b)$$

where the index, o, refers to the uniform (initial) condition. Using the power-law expression for the solid discharge, namely:

$$q_s = a_s U^{b_s} \quad (6.5a)$$

one obtains:

$$K \cong \frac{1}{3} b_s q_s \frac{1}{(1-p)} \frac{1}{S_{e_o}} \quad (6.12c)$$

The parabolic model — obtained by using the different important assumptions — is of interest, since it allows to obtain analytical solutions in certain well-defined cases.

Depending on the applied mathematical techniques and on the hypothesis used, some solutions — which often are rather similar — have been communicated by *Vreugdenhil* et *de Vries* (1973, p. 9), *Ashida* et *Michiue* (1971), *Jaramillo* et *Jain* (1984), *Ribberink* et *Sande* (1985) and *Gill* (1987).

4° A *hyperbolic* model for quasi-steady flow, but being non-uniform, has been proposed (see *Vreugdenhil* et *de Vries*, 1973, p. 5 and *Exner*, 1925):

$$\frac{\partial z}{\partial t} - K \frac{\partial^2 z}{\partial x^2} - \frac{K}{c_{w_3}} \frac{\partial^2 z}{\partial x \, \partial t} = 0 \quad (6.13)$$

where K and c_{w_3} are respectively given by eq. 6.12 and eq. 6.9. Solving this equation, one fixes K and c_{w_3} at their initial values, K_o and c_{w_30}. However, since an analytical solution to eq. 6.13 is rarely possible (see *de Vries*, 1985, *Ribberink* et *Sande*, 1985 and *Lenau* et *Hjelmfeld*, 1992), this model turns out to be not all that useful.

TRANSPORT OF SEDIMENTS

5° A model of a *simple wave* (see *Vreugdenhil* et *de Vries*, 1973, and *Exner*, 1925) is obtained by reduction of eq. 6.8, or :

$$\frac{\partial z}{\partial t} + c_{w_3} \frac{\partial z}{\partial x} = 0 \qquad (6.14)$$

where c_{w_3} is given by eq. 6.9. Since the friction term, $\mathbf{F}(U)$, is now neglected, the application of this model remains limited (see *Ribberink* et *Sande*, 1985) to small values of x and of t, namely for $x \ll 3h/S_{e_0}$.

6.2.4 Degradation and Aggradation

1° A *degradation* (or *aggradation*) in a reach of a watercourse is encountered if the entering solid discharge is smaller (or larger) than the capacity of the transport of sediments. The sediments of the bed will be eroded (or deposited) and as a consequence the elevation of the channel bed decreases (or increases).

Degradation (erosion) and aggradation (deposition) are long-term processes of the evolution of the channel bed, $z(x,t)$.

The flow, being steady and uniform at the beginning, will also be steady and uniform at the end of the process ; in between the flow becomes non-uniform and quasi-steady. If one assumes, that during this transition the flow can be considered as being quasi-uniform, $\partial U/\partial x = 0$, one may make use of the *parabolic model*.

2° The equation for this parabolic model was given as :

$$\frac{\partial z}{\partial t} - K \frac{\partial^2 z}{\partial x^2} = 0 \qquad (6.11)$$

where K is a function of time ; the other variables (see eq. 6.12) must be kept constant to facilitate a resolution of eq. 6.11.

Note that this model is limited to large values of x and of t, namely for $x > 3h/S_e$, and for Froude numbers of Fr < 0.6.

3° Analytical solutions to the parabolic model, eq. 6.11, can be obtained for cases, where the initial and boundary conditions are well specified. Such solutions will not only clarify a physical problem, but can often be considered as a first tentative for an understanding of the problem. Caution is however necessary since this model was established using different assumptions.

4° The analytical solutions will certainly help to explain the long-term evolution of the bed of the channel, when the variation of the liquid discharge can be readily neglected. The following examples can be cited (see Fig. 6.4) :

Degradation :
- the supply of solid discharge is reduced (interrupted) at the upstream ;
- the liquid discharge is increased ;
- a lowering of a fixed point on the channel bed at the downstream.

Aggradation :
- the supply of solid discharge is increased at the upstream ;
- the liquid discharge is decreased ;
- a mounting of a fixed point on the channel bed at the downstream.

Some applications of the parabolic model will be presented in the following pages.

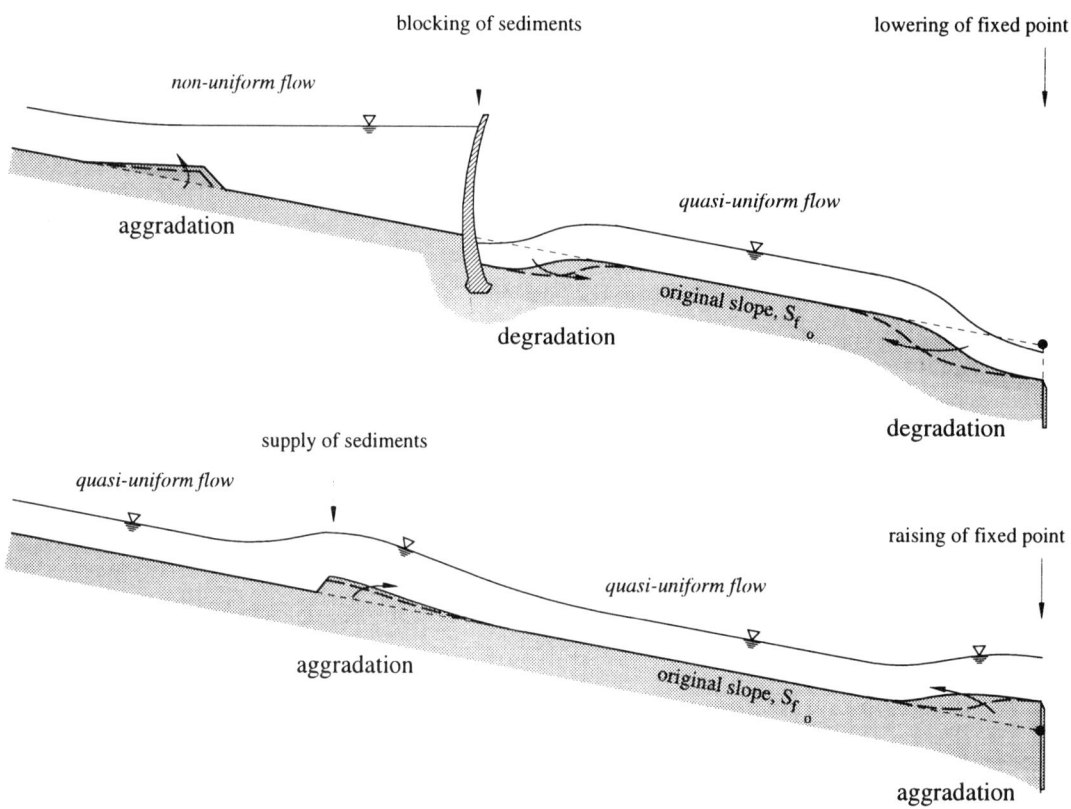

Fig. 6.4 Scheme of a degradation or an aggradation.

TRANSPORT OF SEDIMENTS

5° *Degrading channel* (see Fig. 6.5) :

i) Consider a channel with a mobile bed, having a uniform flow of constant unit discharge, q , at an initial time, t = 0 , and a flow depth of h = h°. This discharge enters into a reservoir, whose water level is lowered by Δh_w, causing a local lowering of the fixed point of the bed of Δh. Consequently, a degradation of the bed is initiated and a long time after, t = ∞ , one will notice throughout the reach of the channel a lowering of the bed and of the water surface ; the flow depth will then be again the initial depth, $h° \equiv h^\infty$. During the period of degradation, t = t , the discharge, q , as well as the flow depth, h , remain quasi-constant.

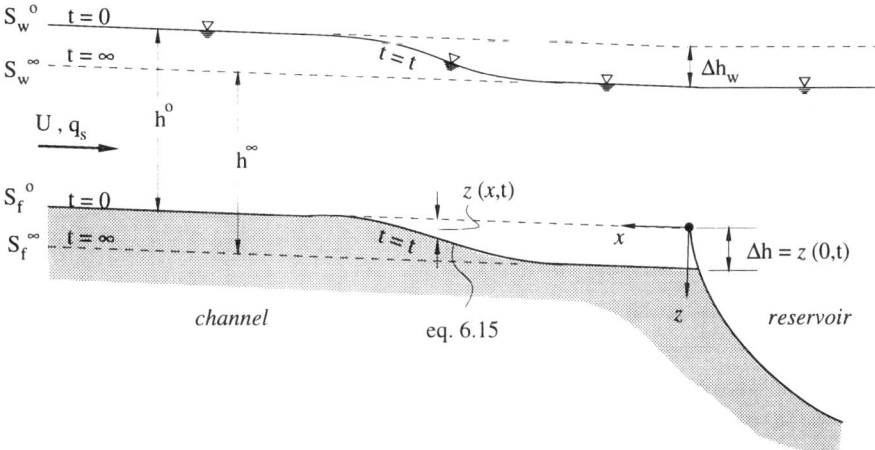

Fig. 6.5 Degradation due to lowering, Δh, of the fixed point on the bed.

ii) The flow is considered as being steady and quasi-uniform ; the use of the parabolic model, eq. 6.11, seems justified. Since the discharge stays constant, the coefficient K (see eq. 6.12) remains also constant.

iii) The problem has to be mathematically specified. The *x*-axis is put into the initial bed pointing upstream ; z stands now for the variation of the bed elevation with respect to the initial bed slope, $S_f°$.

The initial and boundary conditions are :

$z(x,0) = 0$; $\lim_{x \to \infty} z(x,t) = 0$

$z(0,t) = \Delta h$

iv) The solution to eq. 6.11 — making use of Laplace transformations (see *Vreugdenhil* et *de Vries*, 1973, p. 9 and p. 11) — is :

$$z(x,t) = \Delta h \, erfc \left(\frac{x}{2\sqrt{Kt}} \right) \qquad (6.15)$$

where the complementary error function is given as :

$$erfc\,(Y) = \frac{2}{\sqrt{\pi}} \int_Y^\infty e^{-\xi^2}\, d\xi \qquad (6.16)$$

which can be found in mathematical tables (see Ex.6.A).

v) One may now ask (see *de Vries*, 1973), after what time period, $t_{50\%}$, at a certain section, $x_{50\%}$, the bed elevation will be lowered by 50 % with respect to the final bed elevation, namely $z/\Delta h = 1/2$. Using eq. 6.15, one writes :

$$\frac{z(x,t)}{\Delta h} = \frac{1}{2} = erfc\left(\frac{x_{50\%}}{2\sqrt{Kt_{50\%}}}\right) = erfc\,(Y)$$

and, upon consulting the tables of the error function, one finds $Y \cong 0.48$, thus obtaining :

$$x_{50\%} = 0.48\,(2\sqrt{Kt_{50\%}}) \quad \text{where} \quad t_{50\%} \approx x^2_{50\%}/(0.96^2\,K) \qquad (6.17)$$

vi) It has been shown (see *Vreugdenhil et de Vries*, 1973, p. 11) that the parabolic model — which approaches itself to the hyperbolic one, being more correct but also more complex — is rather valuable, if the value of $(x^2/(2Kt)) < 1/2$ is small, or $x > 3h/S_{e_0}$, namely if the time, t, and the distance, x, are large and/or if the bed slope, S_{e_0}, is relatively large.

6° *Aggrading Channel* (see Fig. 6.6)

i) Consider a channel with a mobile bed, having a uniform flow. A particular cross section, till now in equilibrium, is overloaded, namely the supply of solid discharge, Δq_s, is increased (caused by an earth slide or by a mining operation). Aggradation on the channel bed will take place. Subsequently, after a lapse of time, Δt, the elevation of the bed as well as the water surface will increase by Δh. During the period of aggradation, $t = t$, the discharge, q, will remain essentially constant.

ii) The flow is considered as being steady and quasi-uniform ; the use of the parabolic model, eq. 6.11, seems justified. Since the discharge remains constant, the coefficient K (see eq. 6.12) stays also constant.

iii) The problem has to be mathematically specified. The x-axis is put into the initial bed, being positive towards the downstream ; z stands now for the variation of the bed elevation with respect to the initial bed slope, $S_f{}^\circ$.

The initial and boundary conditions are :

$$z\,(x,0) = 0 \quad ; \quad \lim_{x \to \infty} z\,(x,t) = 0$$

$$z\,(0,t) = \Delta h(t)$$

TRANSPORT OF SEDIMENTS

iv) The solution to eq. 6.11 is :

$$z(x,t) = \Delta h(t)\, erfc \left(\frac{x}{2\sqrt{Kt}}\right) \qquad (6.15a)$$

Evidently this solution is of the same type as the one in the preceding problem ; however now $\Delta h(t)$ is a function of time, to be determined. The coefficient $K \equiv K_o$, eq. 6.12, must be evaluated for the initial situation, thus not taking into account the overloading, Δq_s .

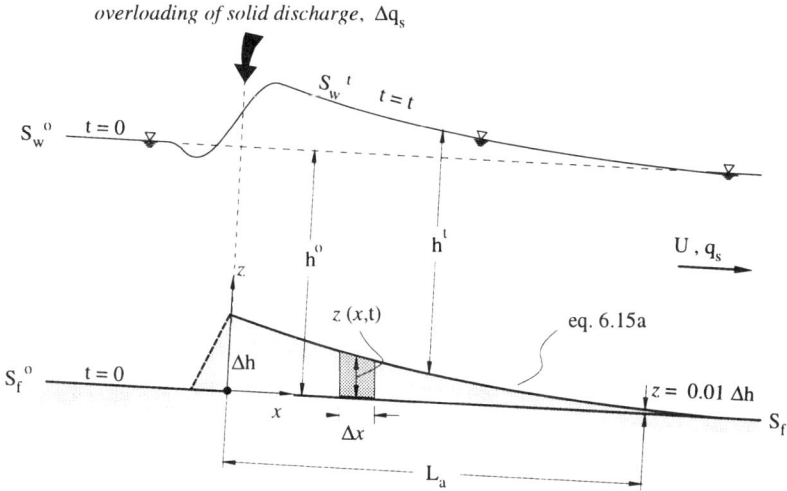

Fig. 6.6 Aggradation by overloading the supply of solid discharge, Δq_s .

v) It is necessary to define the length of the zone of aggradation, L_a , taken as being the one corresponding to a deposition of $z/\Delta h = 0.01$ ($Y \approx 1.80$); according to eq. 6.15a, one writes :

$$L_a = x_{1\%} \cong 3.65 \sqrt{Kt_{1\%}} \qquad (6.19)$$

The volume of the supply of solid discharge, Δq_s , during a certain time, Δt , is given by $\Delta q_s \cdot \Delta t$; this quantity is distributed over the bed of the channel (see Fig. 6.6) as follows :

$$\Delta q_s \cdot \Delta t = (1-p) \int_0^{L_a} z\, dx \qquad (6.18)$$

Subsequently, eq. 6.15a can be used to calculate (see *Soni* et al., 1980, p. 122) the thickness of the layer, Δh , due to the aggradation :

$$\Delta h(t) = \frac{\Delta q_s \cdot \Delta t}{1.13\,(1-p)\,\sqrt{K\Delta t}} \tag{6.20}$$

where it becomes evident that Δh increases with time, t.

vi) Agreement of experimental work (in the laboratory) with eq. 6.15a has been communicated (see *Soni* et al., 1980); but the coefficient of aggradation, K, had to be slightly adjusted. Also, it was remarked, that the parabolic model, eq. 6.11, remains valid over the entire region; thus not limited to $x > 3h/S_f$.

7° Computation of a degradation and of an aggradation, such as was discussed in the present section, is only possible if the conditions assumed for a parabolic model are well fulfilled, namely :

i) quasi-steadiness of the flow (long-term variation of the bed) ;

ii) quasi-uniformity of the flow at Fr < 0.6 ;

iii) validity for $x > 3h/S_e$.

If these conditions (hypothesis) are not fulfilled, it is obviously necessary to solve the equations of Saint-Venant - Exner using numerical methods.

6.2.5 Numerical Solutions

1° Analytical solutions of the equations of Saint-Venant - Exner are only possible when the hypothesis of quasi-steadiness of the flow is justified. Furthermore, it is often necessary to assume also quasi-uniformity of the flow. However these assumptions are *no* more possible, if the temporal variation of the discharge, $\partial(Uh)/\partial t$, and the one of the elevation of the bed, $\partial z/\partial t$, are of the same order of magnitude, namely relatively rapid.

2° If the flow is unsteady and non-uniform (see chap. 5) or steady and non-uniform (see chap. 4), no analytical solutions, which are reasonably simple, are available.

The system of the equations of Saint-Venant - Exner, eq. 6.1b, eq. 6.2 and eq. 6.4a, together with eq. 6.3 and eq. 6.5, can be resolved — without making too severe assumptions — by numerical methods ; this may be well achieved with the use of computers.

3° The numerical methods are essentially the same which are used to solve the equations of Saint-Venant, namely for flow over a *fixed* bed (see sect. 5.2). They become however rather complicated, if they are applied for the modelisation of flow over *mobile* bed.

The *implicit* methods (see sect. 5.2.4) using finite differences are the ones which are at the present frequently used to solve the equations of Saint-Venant - Exner.

4° Here we shall only give reference to a selection of the existing literature, which employ numerical methods for the solutions of the equations of Saint-Venant - Exner for steady and unsteady flow over mobile bed : *Chen* et al. (1975) and *Cunge* et al. (1980, p. 271); *Yucel* et *Graf* (1971), *Krishnappan* (1981), *Holly* et *Rahuel* (1990) and *Correia* et al. (1992).

6.3 BED-LOAD TRANSPORT

6.3.1 Notions

1° Transport as bed-load is the mode of transport of sediments (see Fig. 6.1) where the solid particles glide, roll or (briefly) jump, but stay very close to the bed, $0 < z < z_{sb}$, which they may leave only temporarily. The displacement of the particles is intermittent ; the random concept of the turbulence plays an important role.

2° There exist a number of formulae, which can be used for the prediction of the bed-load transport (see *Graf*, 1971, chap. 7, *Yalin*, 1972, chap. 5, and *Raudkivi*, 1976, chap. 7).

3° Many of theses formulae are of empirical nature, but often have incorporated dimensionless numbers. This allows to make experiments in the laboratory, where the hydraulic conditions can be well controlled ; subsequently it is possible to use such formulae for field conditions.

6.3.2 Theoretical Considerations

1° Considered will be that the bed of a channel (see Fig. 6.1) is plane but mobile, composed of solid particles of uniform size and being non-cohesive. These particles displace themselves under the action of the flow, which be uniform and steady.

For such simplified conditions — bed forms (see sect. 3.2.5) may form, the granulometric distribution may be non-uniform (see sect. 6.3.4) and cohesion may exist —, one tries to obtain functional relations, such as the ones given by eq. 3.40 and eq. 6.29. The form of such functions, being often rather complex, will be established by experiments, which more or less will take care of the reality of the problem.

2° The forces, which enter (see *Graf*, 1971, chap. 6) into the description of the uniform and steady motion of a single particle, isolated and without cohesion, are :

- the hydrodynamic force : $\quad F_H \propto f(\dfrac{u_* d}{\nu}) \rho d^2 u_*^2$

- the submerged weight of the particle : $\quad W_p \propto g (\rho_s - \rho) d^3$

where u_* is the friction velocity, considered as being proportional to the velocity of the particle.

3° The components of this two-phase flow are :

- the *fluid*, by its density, ρ, and its viscosity, ν ;

- The *solid material*, by its density, ρ_s, and a characteristic diameter, d ;

- the *flow*, by its flow depth, h or R_h, the slope, S_f, and the gravity, g ; thus by the friction velocity, $u_* = \sqrt{\rho g R_h S_f}$, which characterises the turbulence (see sect. 2.6.4).

In all, there are thus 7 parameters.

4° A dimensionless analysis, using the Π-theorem (see *Yalin*, 1972, p. 61), shows that the arguments which quantify the two-phase flow, such as the bed-load transport, can now be expressed by 4 dimensionless quantities, namely :

- a Reynolds number of the particle :

$$Re_* = \dfrac{u_* d}{\nu} \qquad (6.21)$$

- a dimensionless shear stress (see eq. 3.38) :

$$\tau_* = \dfrac{\rho u_*^2}{(\gamma_s - \gamma)d} = \dfrac{\tau_o}{(\gamma_s - \gamma)d} = \dfrac{\gamma R_h S_f}{(\gamma_s - \gamma)d} \qquad (6.22)$$

or a densimetric Froude number of the particle :

$$Fr_{*D} = \dfrac{u_*}{\sqrt{(s_s - 1) gd}} = \dfrac{\sqrt{\tau_o}}{\sqrt{(\gamma_s - \gamma)d}} = \sqrt{\tau_*} \qquad (6.23)$$

- a relative depth :

$$\dfrac{h}{d} \quad \text{ou} \quad \dfrac{R_h}{d} \qquad (6.24)$$

- a relative density :

$$s_s = \dfrac{\rho_s}{\rho} \qquad (6.25)$$

TRANSPORT OF SEDIMENTS

- In addition, a dimensionless particle diameter can be obtained by combining eq. 6.21 with eq. 6.22 (see point 3.4.2.7°), or :

$$d_* = d\left((s_s-1)\frac{g}{v^2}\right)^{1/3} \qquad (6.26)$$

5° Combining eq. 6.22 and eq. 6.21, a relation was proposed by *Shields* (see sect. 3.4.2 and Fig. 3.13), such as :

$$\tau_* = f(\text{Re}_*) \quad \text{or} \quad \tau_* = f(d_*) \qquad (3.40)$$

for the study of the commencement of erosion, expressed by the dimensionless shear stress, τ_{*cr}. Furthermore, a relation of the form :

$$\tau_{*cr} = f(\text{Re}_*) \qquad (6.27)$$

gives a delimitation of the zone of "motion" from the zone of "no motion" of the particles ; this was developed experimentally from laboratory date, showing a rather large spread. The function of Shields, eq. 6.27, is generally agreed upon as being valuable and useful, notably for the hydraulic engineers, if the granulometry is uniform or almost so.

6° The transport of sediments can be expressed as a function of these 4 dimensionless quantities, namely :

$$\Phi = f(d_*, \tau_*, R_h/d, \rho_s/\rho)$$

Utilising the Π-theorem, one obtains (see *Yalin*, 1972, p. 67) an expression for a dimensionless *intensity of the solid discharge* as the bed load, or :

$$\Phi \equiv q_{sb*} = \frac{q_{sb}}{\sqrt{(s_s-1)gd^3}} \qquad (6.28)$$

with q_{sb} [m²/s] as the volumic solid discharge per unit width.

Expressions, which are similar to the one of eq. 6.28, can be written (see *Yalin*, 1972, p. 65) as :

$$\Phi' = \frac{q_{sb}}{u_* d} \quad \text{or} \quad \Phi'' = \frac{q_{sb}}{Ud} \qquad (6.28a)$$

Since some terms, R_h/d and ρ_s/ρ, are included in the term of τ_*, and taking $\tau_* = f(\text{Re}_*)$, one can formulate now a rather simple relationship :

$$\Phi = f(\tau_*) \quad \text{or} \quad \frac{q_{sb}}{\sqrt{(s_s-1)gd^3}} = f\left(\frac{\tau_o}{(\gamma_s-\gamma)d}\right) \qquad (6.29)$$

which is often written as :

$$\Phi = f(\Psi) \qquad (6.29a)$$

where $\tau_* \equiv \Psi^{-1}$ and Ψ is called the dimensionless *intensity of shear stress*, applied upon the solid particles.

This expression, eq. 6.29, links the solid transport, q_{sb} (see eq. 6.28), to the shear stress, τ_* (see eq. 6.22). Thus an increase in τ_* — passing by τ_{*cr}, where erosion begins — is responsible for an increase in q_{sb}.

The form of this function, eq. 6.29, must still be established ; it is given by the formulae of bed-load transport, which are established by experiments performed in the laboratory and in the field.

7° One often assumes that this relation, eq. 6.29, can be expressed in form of a power law, or :

$$\Phi = \alpha(\tau_*)^\beta \qquad (6.30)$$

Making use of the ratio, which defines the coefficient of friction :

$$\frac{U}{\sqrt{\tau_o/\rho}} = \sqrt{\frac{8}{f}} \qquad (2.55)$$

one can formulate the following proportionalities :

$$U^2 \propto \tau_o \propto \tau_*$$

Thus it is possible to express the above equatioin, eq. 6.29, by an approximate relation (see *de Vries*, 1973) in the form of :

$$q_{sb} = a_s U^{b_s} \qquad (6.5a)$$

where a_s, α and $b_s = 2\beta$, β are the coefficients which depend essentially on the granulometry. This simple, but often useful relation, shows that the average velocity, U, of the flow is the predominant parameter for the determination of the solid discharge, q_{sb}.

6.3.3 Bed-load Relations

1° At the present, the formulae for a determination of the solid discharge as bed load give only reasonably satisfying results within a domain of the parameters for which the chosen formula has been established. Consequently, the application and use of such formulae has to be done with great care.

Here will be given a selection (in chronological order) of some of the many available formulae ; their most characteristic hydraulic aspects will be pointed out.

2° From the different empirical formulae, proposed by *Schoklitsch* in 1934 and 1950 (see *Graf*, 1971, p. 133), the last one is presented, namely :

$$q_{sb} = \frac{2.5}{s_s} S_e^{3/2} (q - q_{cr}) \tag{6.31}$$

The critical liquid discharge, q_{cr}, characterises the commencement of erosion — usually expressed by τ_{cr} — ; it is given with the use of the Manning-Strickler formula, eq. 3.16 and eq. 3.18, such as :

$$q_{cr} = 0.26 \, (s_s - 1)^{5/3} \frac{d^{3/2}}{S_e^{7/6}} \tag{6.31a}$$

valid for $d \geq 0.006$ [m] (see *Bathurst* et al., 1987); for a non-uniform granulometric mixture one takes $d = d_{40}$, as the equivalent diameter.

This relation, eq. 6.31, is applicable for larger grain sizes, $d \geq 6$ [mm], being rather uniform and for bed slopes being moderate to strong (see Table 6.3).

3° From the different empirical formulae — using the condition of similitude of Froude — which *Meyer-Peter* et al. have developed in 1934 and 1948 (see *Graf*, 1971, p. 136 or *Yalin*, 1972, p. 112), the last one is presented, namely :

$$0.25 \, \rho^{1/3} \frac{(g_{sb}')^{2/3}}{(\gamma_s - \gamma)d} = \frac{\gamma R_{hb} \xi_M S_e}{(\gamma_s - \gamma)d} - 0.047 \tag{6.32}$$

where $g_{sb}' = g_{sb} (\gamma_s - \gamma)/\gamma_s$ is the solid discharge in weight under water and $g_{sb}/\gamma_s = q_{sb}$; R_{hb} is the hydraulic radius of the bed. For a non-uniform granulometry, the mean diameter, $d = d_{50}$, is taken as the equivalent diameter.

This relation can be written in the dimensionless form (see eq. 6.29) such as :

$$\Phi = 8 \, (\xi_M \tau_* - \tau_{*cr})^{3/2} \tag{6.32a}$$

where τ_{*cr} is the dimensionless critical shear stress (see eq. 6.27 and Fig. 3.13). ξ_M is a roughness parameter, given by :

$$\xi_M = \left(\frac{K_S}{K_S'}\right)^{3/2}$$

where K_S' is the roughness of the granulates, to be evaluated with the formula of Strickler, eq. 3.18, and K_S is the total roughness of the bed, evaluated with the formula of Manning-Strickler, eq. 3.16, or :

$$K_S = \frac{U}{R_{hb}^{2/3} S_e^{1/2}} \quad \text{and} \quad K_S' = \frac{26}{d_{90}^{1/6}}$$

In the absence of bed forms it is recommended to take $\xi_M = 1$; but $1 > \xi_M > 0.35$, if bed forms are present.

This relation, eq. 6.32, is applicable for rather large grain sizes, d > 2 [mm], being uniform as well as non-uniform, and for bed slopes, being moderate to strong (see Table 6.3).

4° Using extensively the concepts of hydrodynamics, *Einstein* has developed in 1942 and 1950 (see *Graf*, 1971, pp. 139-150) a probabilistic model for the transport of sediments as bed load.

i) *Determination of the probability of erosion*

The probability, p_e, of erosion of a particle at any time instant depends (see *Einstein*, 1950, p. 35) on the hydrodynamic force, F_H, here the lift force, and the submerged weight of the particle, W_P :

$$p_e = f\left[\frac{W_P}{F_H}\right] = f\left[\frac{k_2 g(\rho_s-\rho) d^3}{1/2 \, k_1 C_L(1+\eta) \rho \, d^2 \, (5.75^2 u_*'^2 \beta_x^2)}\right] =$$

$$= f\left[\frac{1}{(1+\eta)} (B\beta_x^{-2}\Psi')\right] \quad (6.33)$$

k_1 and k_2 are form factors of the particle ; $C_L = 0.178$ is a lift coefficient and η is a random variable of lift, where $\eta_o = 0.5$ is the most probable value. According to eq. 6.29a and eq. 6.22, the intensity of shear stress, Ψ', is defined by :

$$\Psi' = \frac{(\gamma_s-\gamma)\,d}{\rho u_*'^2} = \frac{(\gamma_s-\gamma)}{\gamma}\frac{d}{R_{hb}'S_f} = \frac{1}{\tau_*'} \qquad (6.34)$$

where R_{hb}' is the hydraulic radius of the bed due to the granulate (see sect. 3.2.5); for a non-uniform granulometry, $d = d_{35}$ is taken. $B = 2k_2/(5.75^2 C_L k_1)$ is a numerical constant and β_x is a relation which takes in account the logarithmic velocity distribution as well as the roughness, $k_s = d_{65}$. The following expression (see *Einstein*, 1950, p. 36), where $d = X$ is a characteristic diameter of the granulometry, was given :

$$\beta_x = \log(10.6\,X/\Delta) \qquad (6.35a)$$

with $\qquad X \begin{cases} = 0.77\Delta & \text{if} \quad \Delta/\delta > 1.80 \\ = 1.39\delta & \text{if} \quad \Delta/\delta < 1.80 \end{cases}$

where $\quad \Delta = f(k_s/\delta) \qquad$ according to Fig. 6.7a

$$\delta = 11.5\,\nu/u'_*$$

The above functional relation, eq. 6.33, is valid for a uniform granulometry, but can be generalised for a non-uniform one, in the following way :

$$p_e = f\left[\zeta_H\,\zeta_P\,\frac{1}{(1+\eta)}\,(B\beta_x^{-2}\,\Psi')\right] \qquad (6.35)$$

ζ_H is a hiding coefficient — the smaller particles hide behind the larger ones — ; it was obtained experimentally (see Fig. 6.7b). ζ_P is a lift-force correction coefficient, also obtained experimentally (see Fig. 6.7c). This expression, eq. 6.35, can also be written (see *Einstein*, 1950, p. 37) as :

$$p_e = f(B_*\Psi_*)$$

where Ψ_* is the intensity of shear stress after Einstein :

$$\Psi_* = \zeta_H\,\zeta_P\,(\beta^2/\beta_x^2)\,\Psi' \qquad (6.36)$$

and B_* is a constant to be determined experimentally (see eq. 6.42b) :

$$B_* = \frac{B}{\beta^2 \eta_o} \qquad \text{with} \qquad \beta = \log(10.6) \qquad (6.36a)$$

Note, that for a uniform granulometry, one takes (see *Einstein*, 1950, p. 36) $\zeta_H = 1$, $\zeta_P = 1$ and $(\beta^2/\beta_x^2) = 1$; thus one writes :

$$\Psi_* = \Psi' \qquad (6.36b)$$

In order to express the probability of motion, *Einstein* (1950, p. 37) postulated the following function, being rather similar to the normal function, or :

$$p_e = 1 - \frac{1}{\sqrt{\pi}} \int_{-B_* \Psi_* - 1/\eta_o}^{+B_* \Psi_* - 1/\eta_o} e^{-\xi^2} d\xi \qquad (6.37)$$

where ξ is a variable of integration.

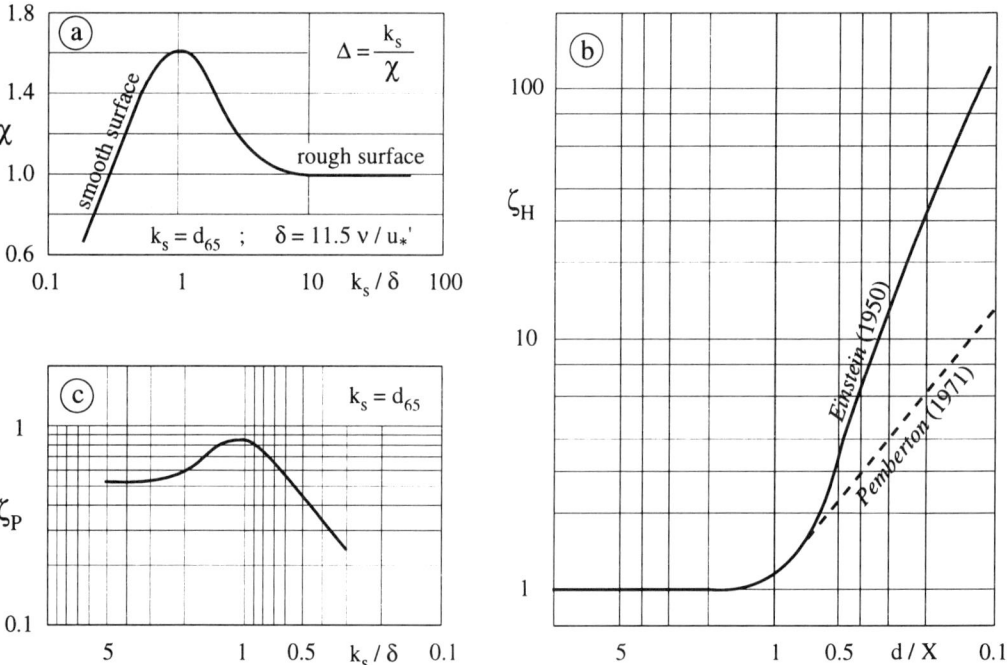

Fig. 6.7 Correction coefficients : (a) of velocity distribution, (b) of hiding and (c) of lift force (see *Graf*, 1971, p. 146).

ii) *Equation of bed load*

The number of particles, which are *deposited* per unity of time and of bed surface, $A_L d \cdot 1$, is given by :

$$N_D = \frac{g_{sb} i_{sb}}{(A_L d)(\gamma_s k_2 d^3)} \tag{6.38}$$

where $(\gamma_s k_2 d^3)$ is the weight of a particle and A_L is a constant. i_{sb} is a fraction (see Fig. 6.9) of the granulometric curve of the unit solid discharge, g_{sb}, in weight.

The number of particles, which are *eroded* per unity of time and of bed surface, is given by :

$$N_E = \frac{i_b}{k_1 d^2} (p_e/t_e) \tag{6.39}$$

where $(i_b/k_1 d^2)$ is the number of particles in a unit bed surface ; i_b is a fraction (see Fig. 6.9) of the granulometric curve of the bed material. p_e is the probability of erosion of a particle ; the exchange time, t_e, necessary for the replacement of a particle of the bed by another particle, is expressed by :

$$t_e \propto \frac{d}{v_{ss}} = k_3 \sqrt{\frac{\rho d}{g(\rho_s - \rho)}}$$

where v_{ss} is the settling velocity of the particle.

The *equation of bed load* after *Einstein* (1950) postulates that the rate of erosion, eq. 6.39, is equal to the rate of deposition, eq. 6.38 ; thus one takes $N_D = N_E$ and consequently :

$$\frac{g_{sb} i_{sb}}{(A_L d)(\gamma_s k_2 d^3)} = \frac{i_b p_e}{k_1 k_3 d^2} \sqrt{\frac{g(\rho_s - \rho)}{\rho d}} \tag{6.40}$$

Furthermore, one admits that a solid particle displaces itself by making jumps of a length of $A_L d$ (see eq. 6.38), which are linked to the exchange probability (see *Einstein*, 1950, p. 34) in the following way :

$$A_L d = \lambda d \left(\frac{1}{1-p_e}\right)$$

where λ is a constant of the jump of the particles. Introducing this expression into the above relation, eq. 6.40, yields :

$$\left(\frac{p_e}{1-p_e}\right) = A_* \left(\frac{i_{sb}}{i_b}\right) \Phi = A_* \Phi_* \tag{6.41}$$

Φ is the intensity of transport after Einstein, given by :

$$\Phi = \frac{q_{sb}}{\sqrt{(s_s-1) g d^3}} \qquad (6.28)$$

where $q_{sb} = g_{sb}/\gamma_s$ is the volumic solid discharge per unit width and A_* is an empirical constant to be determined experimentally (see eq. 6.42b). For a uniform granulometry, one takes (see *Einstein*, 1950, p. 36) simply :

$$\Phi_* = \Phi$$

The above relations, eq. 6.37 and eq.6.41 put together, give now the final form of the *equation of bed load of Einstein* (1950):

$$p_e = 1 - \frac{1}{\sqrt{\pi}} \int_{-B_* \Psi_* -1/\eta_o}^{+B_* \Psi_* -1/\eta_o} e^{-\xi^2} d\xi = \frac{A_* \Phi_*}{1+A_* \Phi_*} \qquad (6.42)$$

namely a functional relation (see eq. 6.29a), such as :

$$\Phi_* = f(\Psi_*) \qquad (6.42a)$$

The (universal) constants have now to be determined experimentally both for uniform and non-uniform granulometries (see *Einstein*, 1950, pp. 37 and 43) ; they are given (see *Graf*, 1971, p. 149) as being :

$$A_* = 43.6 \quad ; \quad B_* = 0.143 \quad ; \quad \eta_o = 0.5 \qquad (6.42b)$$

This relation, eq. 6.42, is plotted in Fig. 6.8 — using the tables of the error-function — together with the data of *Meyer-Peter* et al. and *Gilbert*. The graphical representation facilitates the use of the above relation, eq. 6.42.

Since a non-uniform granulometry can be broken down into its fractions, i_{sb}/i_b, this relation is rather flexible. For a quasi-uniform granulometry, an equivalent diameter of $d = d_{35}$ can be taken.

It is interesting to remark, that the relation of *Einstein*, eq. 6.42, and the one of *Meyer-Peter* et al., eq. 6.32, give rather similar results (see *Graf*, 1971, p. 150), and this notably for $\Phi < 10$. Note also, that in the relation of *Einstein* the notion of a critical value (for erosion), $\Psi_{cr} = \tau_{*cr}^{-1}$, has nowhere been used explicitly. Nevertheless, one may ask now, what numerical value for Ψ one would get, if the value of Φ becomes very small ; for example :

$$\Phi \cong 0.0004 \quad \Rightarrow \quad \Psi \cong 25 \quad \Rightarrow \quad \tau_{*cr} \cong 0.04$$

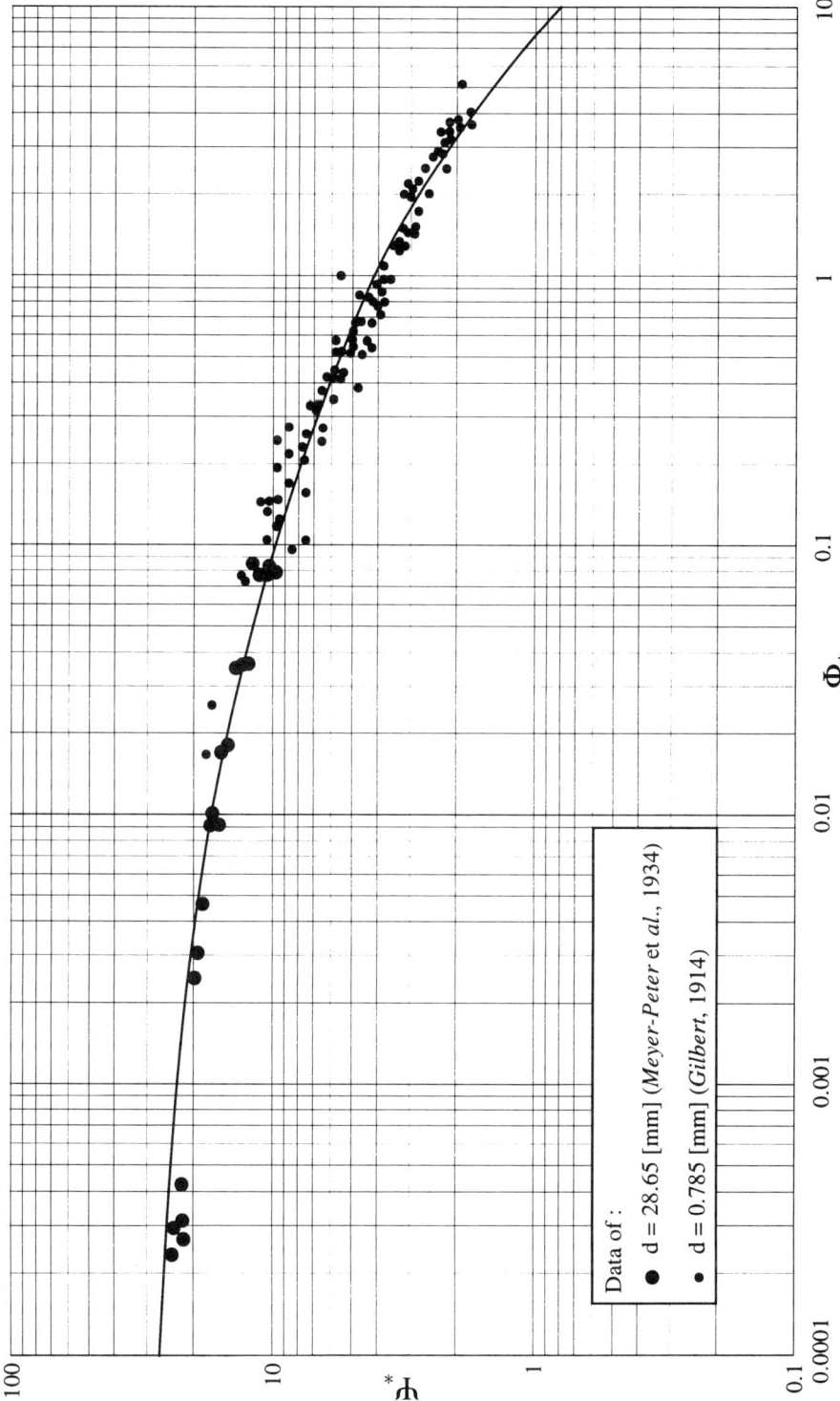

Fig. 6.8 Equation of bed load, $\Phi_* = f(\Psi_*)$, of Einstein (see *Graf*, 1971, p. 148).

Here one sees that there exists a rather good agreement with the critical value, τ_{*cr}, taken from the diagram of Shields (see Fig. 3.13).

The equation of Einstein, eq. 6.42 and Fig. 6.8, is well suitable for uniform and non-uniform granulates over a large range of diameters, d > 0.7 [mm], and of bed slopes (see Table 6.3). It is world-wide used with great success.

6.3.4 Granulometry, Armouring

1° The non-cohesive sediments (solid particles), which make up the bed of a watercourse, are in general of different sizes, being given by the granulometric curve of the bed material (see Fig. 6.9a).

This curve, which is in general half-logarithmic, can be divided into fractions (percentages), i_{b_i}, whose sum is :

$$1(100\%) = \Sigma i_{b_i} = i_{b_1} + i_{b_2} + ... + (5\% + 5\%)$$

Usually this curve is partitioned into 4 or 5 (unequal) fractions, after having eliminated small fractions, namely ≈ 5 %, of the finest particles — they are part of the wash load — and of the coarsest particles.

For each fraction the average diameter, d_i, is determined and the corresponding solid discharge, $i_{sb_i}q_{sb_i}$, is calculated, using one of the formulae for bed-load transport. For the entire granulometric mixture, the solid discharge is now obtained as :

$$q_{sb} = \Sigma i_{sb_i} q_{sb_i} = i_{sb_1} q_{sb_1} + i_{sb_2} q_{sb_2} + ...$$

2° The granulometric curve of the bed material is in general different from the one of the material moving as bed load or as suspended load (see Fig. 6.9b). Consequently, for an average diameter of the granulate, d_i, the given fraction of the granulometric curve of the bed material, i_{b_i}, will be different from the corresponding fraction of the granulometric curve of the solid discharge, i_{sb_i}.

This subtlety was elaborated by *Einstein* (1950, p. 32) by introducing the ratio of i_{sb}/i_b and the hiding factor, ζ_H, into the equation of bed-load transport, eq. 6.42.

For a very intensive sediment transport, all sizes (fractions) of particles will readily participate ; consequently $i_{b_i} \equiv i_{sb_i}$, since the curves L and C become identical (see Fig. 6.9b).

3° For cohesive material, the determination of the solid discharge represents a very difficult task ; literature specialised on this topic should be consulted (see *Graf*, 1971, chap. 12 and *Raudkivi*, 1976, chap. 9).

4° The granulometric curve of the bed material is obtained by taking samples from the bed of the channel. Recommended is (see *Einstein*, 1950, p. 48) to take many samples at different sections of the watercourse under study, and obtain an average granulometric curve. Each sample should be taken at (up to) a, to-be expected, maximal erosion depth, namely at a depth of 0.70 [m] below the bed surface.

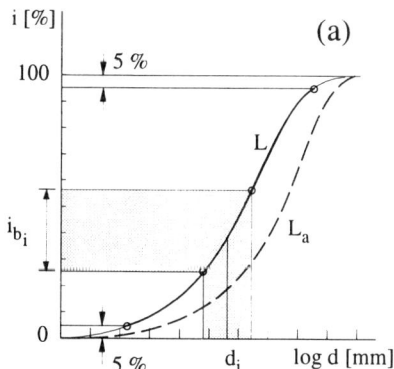
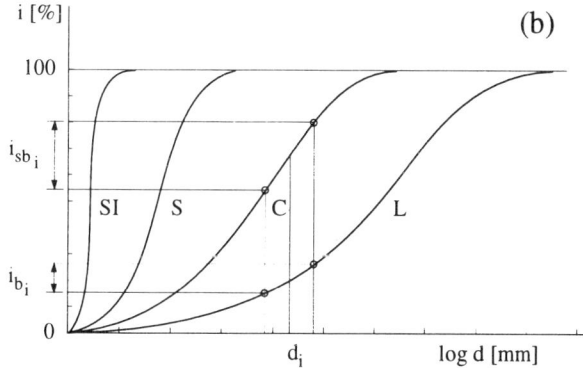

Fig. 6.9 Scheme of granulometric curves for the bed material, L, and the armoured bed material, L_a, the bed-load material, C, and the suspended (wash) load material, S (SI).

5° On a channel bed of non-uniform granulometry, the smaller particles are more easily eroded than the larger ones; a grain-size sorting takes place. An accumulation of the remaining larger particles results in an *armouring* of the bed, which subsequently protects the underlying "original" granulate (see *Graf*, 1971, p. 102).

It can thus happen, that erosion does not take place, if the bed becomes (naturally) *armoured* with the larger particles which remain at the bed surface after an important erosion process during a previous flood event. In such a case the flow cannot take its (full) capacity of sediments transport (see point 6.1.4.3°), and this until another exceptional flood will destroy the armour layer and the original granulate reappears to form once more another new armour layer.

The formulae developed for the capacity of transport are thus only valid for such watercourses, which pass through their own alluvium, namely in a bed being made up of material, which was also transported and can again be transported.

From above it becomes evident that granulometric samples taken in situ — if armouring takes place — have to be interpreted with great caution.

The development of an armour layer is an asymptotic process. When the friction velocity, u_*, increases, the smaller particles are eroded and the larger ones stay in place. The corresponding friction velocity is used to define the critical friction velocity for armouring, $u_{*a,cr}$. The (*maximum* possible) armoured bed will now be formed by the largest particles, d_{90} or larger, which are found in the granulometry of the original bed. For high discharges, when $u_* > u_{*a,cr}$, the armoured bed becomes unstable and will be destroyed. A bed, being composed of the sizes of the granulometric curve of the original composition, arrives at the surface and an extremely active erosion will take place.

There exists only limited conclusive information about the ratio of the original granulometry, d, and the granulometry of the armour layer, d_a. For some Swiss rivers, having large bed slopes, $S_f > 0.03$ [-], and large grain sizes, $d_{50} > 6$ [mm], an indicative relationship of :

$$d_{50_a}/d_{50} \approx 1.4 \quad ; \quad d_{50_a}/d_{90} \leq 0.6$$

was developed by *Correia* et *Graf* (1988).

An empirical relationship for a prediction of the stability of the armour layer was given by *Raudkivi* (1990, p. 113) as :

$$\tau_{*a,cr} = \frac{(u_{*a,cr})^2}{(s_s-1)g\, d_{50_{a,max}}} = \tau_{*cr}\left[0.4\left(\frac{d_{50}}{d_{50_{a,max}}}\right)^{1/2} + 0.6\right]^2$$

$$\text{with} \quad d_{50_{a,max}} \leq 0.55\, d_{100}$$

where $d_{50_{a,max}}$ is the median diameter of the (maximal possible) armour and τ_{*cr} is the dimensionless critical shear stress, taken as $\tau_{*cr} \approx 0.05$ (see Fig. 3.13). Evidently the armouring process is controlled by the largest fraction, d_{90} or larger, of the granulometric curve of the channel bed. No armouring takes place, if the granulometry is uniform.

6.4 SUSPENDED-LOAD TRANSPORT

6.4.1 Notions

1° Transport of sediments in suspension is the mode of transport where the solid particles displace themselves by making large jumps, but remain (occasionally) in contact with the bed load and also with the bed. The zone of suspension is delimited by : $z_{sb} < z < h$ (see Fig. 6.1).

2° Transport as suspended load could be considered as an advanced stage of transport as bed load ; however the analytical methods do not allow a description of these two modes of transport with the same (or single) relationship.

6.4.2 Theoretical Considerations

1° The transport of sediments in suspension can be explained with the concept of diffusion-convection, which gives the vertical distribution of the (local) concentration, $c_s(z)$, of the suspended particles.

TRANSPORT OF SEDIMENTS 385

2° For steady uniform flow, the vertical distribution of the concentration of the suspended particles, $c_s(z)$, in the fluid, can be obtained by using the equation of one-dimensional diffusion-convection (see sect. 8.4 or *Graf*, 1971, p. 166):

$$0 = v_{ss} \frac{\partial c_s}{\partial z} + \frac{\partial}{\partial z}(\varepsilon_s \frac{\partial c_s}{\partial z}) \qquad (6.43)$$

where $c_s(z)$ is the local volumic concentration, ε_s is the diffusivity of the suspended particles, whose units are $[L^2/T]$, and v_{ss} is the settling velocity of the particles.

This equation, eq. 6.43, relates the vertical exchange of solid particles due to the turbulence (upwards) with the gravitational motion (downwards), expressed with the settling velocity, v_{ss}; it is valid only for weak concentrations, namely for $(1 - c_s) \cong 1$ or $c_s < 0.1\ [\%]$.

3° Integration of the above equation, eq. 6.43, yields:

$$v_{ss} c_s + \varepsilon_s \frac{dc_s}{dz} = \text{Cte} = 0 \qquad (6.44)$$

where the constant of integration is taken to be Cte = 0, implying that $c_s = 0$ at the water surface for $\varepsilon_s = 0$.

The above equation expresses that, at all levels, $z_{sb} < z < h$, there is a (vertical) equilibrium between the movement in the direction of gravity and the one due to the concentration gradient in the direction against gravity. In other words, the rate of sedimentation of particles per unit volume is equal to the rate of turbulent diffusion per unit volume.

For not so weak concentrations (see *Graf*, 1971, p. 185) the above equation, eq. 6.44, should be written as:

$$v_{ss} c_s (1 - c_s) + \varepsilon_s \frac{dc_s}{dz} = 0 \qquad (6.44a)$$

4° The following remarks, concerning the diffusivity, should be made:

A relation between the diffusivity of suspended particles in the fluid, ε_s, and the turbulent diffusivity of a (soluble) substance in the fluid, ε_t, is in general admitted (see *Graf*, 1971, pp. 167 and 177):

$$\varepsilon_s = \beta_s \varepsilon_t \qquad (6.45)$$

where β_s is a factor of proportionality. For fine particles, which follow readily the fluid motion, one takes $\beta_s = 1$; for larger particles, one takes $\beta_s \leq 1$. Some researchers (see *Graf*, 1971, p. 178 and *Raudkivi*, 1990, p. 172) advanced arguments to show that $\beta_s \geq 1$.

For weak concentrations it is usually assumed that :

$$\varepsilon_s \approx \varepsilon_t \tag{6.46}$$

thus one takes $\beta_s = 1$.

Furthermore, one may also postulate (see sect. 8.1.3) that :

- diffusion (per unity of surface) of matter, namely of a substance in the fluid, is given by :

$$\rho(\varepsilon_m + \varepsilon_t)\frac{\partial c}{\partial z} \approx \rho\varepsilon_t\frac{\partial c}{\partial z} = q_m$$

- diffusion (per unity of surface) of momentum is given (see eq. 2.49) by :

$$\rho(\nu + \nu_t)\frac{\partial u}{\partial z} \approx \rho\nu_t\frac{\partial u}{\partial z} = \tau_{zx}$$

Here is assumed that the turbulent diffusivity, ε_t, and the turbulent viscosity, ν_t, are far more important than the molecular diffusivity, ε_m, and the viscosity, ν, respectively (see *Graf*, 1971, p. 166).

According to the analogy of Reynolds (see *Taylor*, 1954, p. 451), the transfer of matter (as well as the one of heat) and the transfer of momentum by the turbulence are analogous ; this is strictly correct close to a solid surface (the bed). Consequently, one may also take :

$$\varepsilon_t \cong \nu_t \cong \varepsilon_s \tag{6.47}$$

5° For the case, where the diffusivity is independent of the level, ε_s = Cte, the above equation, eq. 6.44, can be integrated and yields :

$$\frac{c_s}{c_{sa}} = \exp\left[-\frac{v_{ss}}{\varepsilon_s}(z-a)\right] \tag{6.48}$$

where c_{sa} is the concentration at a reference level, a. This relation has been experimentally verified (see *Graf*, 1971, p. 167).

TRANSPORT OF SEDIMENTS

6° In the open-channel flow, the turbulence and thus the diffusivity are vertically distributed, $\varepsilon_s(z)$, (see sect. 2.6). The distribution of the diffusivity, $\varepsilon_s \approx v_t$, is given (see *Graf*, 1971, p. 173) by :

$$\varepsilon_s = \kappa u_*' \frac{z}{h} (h - z) \tag{6.49}$$

This parabolic relation, established for unidirectional flow, has been obtained by assuming :

- the vertical distribution of the tangential turbulent shear stress :

$$\tau_{zx} = \tau_0 \left(\frac{h-z}{h}\right) \tag{2.47b}$$

- the vertical distribution of the velocity (see sect. 2.5.2) :

$$\frac{du}{dz} = \frac{\sqrt{\tau_0/\rho}}{\kappa} \frac{1}{z}$$

where $\kappa = 0.4$ is the Karman constant, which is independent of the concentration (see *Colemann*, 1981) ;

- the expression of the Reynolds stress :

$$\tau_{zx} = \rho v_t \frac{du}{dz} \tag{2.49}$$

- the analogy of Reynolds :

$$\varepsilon_s \cong \varepsilon_t \cong v_t \tag{6.47}$$

This distribution of the diffusivity, eq. 6.49, has been experimentally verified (see *Raudkivi*, 1990, p. 170).

Substitution of eq. 6.49 into eq. 6.44, and separation of the variables, yields :

$$\frac{dc_s}{c_s} = -\frac{v_{ss}}{\kappa u_*'} \left(\frac{h}{h-z}\right) \frac{dz}{z} \tag{6.50}$$

where one defines the *Rouse* exponent as :

$$\zeta = \frac{v_{ss}}{\kappa u_*'} \qquad \left(\text{or } \zeta' = \frac{v_{ss}}{\beta_s \kappa u_*'}\right) \tag{6.50a}$$

This expression, eq. 6.50, can now be integrated by parts, within the limits of $a < z < h$ (see *Rouse*, 1938, p. 341, and *Graf*, 1971, p. 173) and renders:

$$\frac{c_s}{c_{sa}} = \left(\frac{h-z}{z} \cdot \frac{a}{h-a} \right)^{\zeta} \quad (6.51)$$

where c_{sa} is the concentration at a reference level, a. This equation, eq. 6.51, gives the distribution of the relative concentration, c_s/c_{sa}, for one single particle size, v_{ss} and ζ. Note that in the definition of the Rouse exponent, ζ, the friction velocity, u_*', due to the granulate must be used.

In the Rouse exponent, eq. 6.50a, one should take the settling velocity, v_{ss}, of the particle in clear and quiescent water, thus being not influenced by turbulence or by concentration. For *natural* particles of quartz, $s_s = 2.65$ [-], falling in quiescent water at $T = 20$ [C°], the settling velocity can be determined using the Fig. 6.10 (see *Graf*, 1971, p. 45).

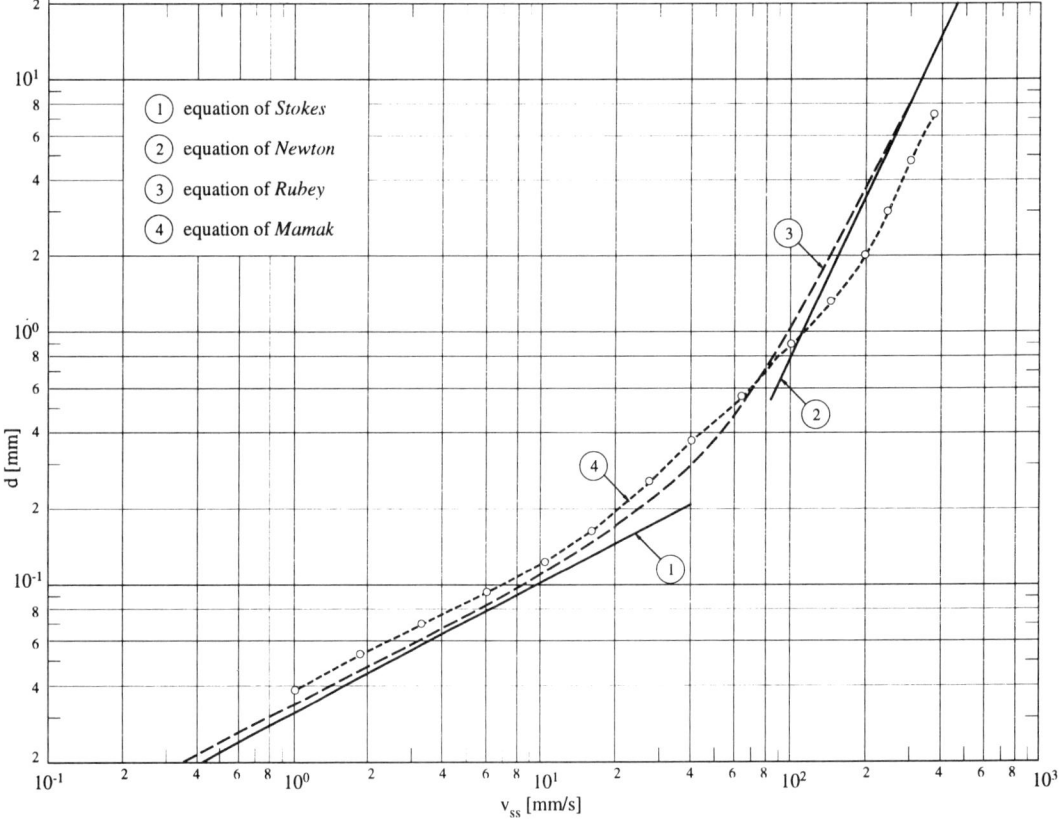

Fig. 6.10 Settling velocity, v_{ss}, as function of particle diameter, d.

The equation, giving the distribution of the relative concentration, eq. 6.51, for different values of the Rouse exponent, ζ, is shown in Fig. 6.11. The following is to be observed :

- For small ζ-values, the relative concentration is large and tends to become uniform over the entire flow depth, h.
- For large ζ-values, the relative concentration is small at the water surface and is large close to the bed.
- The size of the particles, expressed with the settling velocity, v_{ss}, is directly responsible for these distributions.
- Close to the bed, $z \cong 0$, the concentration goes towards infinity, $c_s = \infty$, thus to an impossible value. Thus one delimits this level usually by $a \equiv z_{sb} \cong 0.05h$ or by $z_{sb} = 2d$, below which there exists the bed load (see Fig. 6.1).
- The reference concentration, c_{sa}, is usually taken at a level of $a \equiv z_{sb}$; it will be calculated later (see eq. 6.57) with one of the bed-load formulae, q_{sb}.

Numerous are the investigations, both in laboratory and in situ, which give evidence of the validity of the above equation, eq. 6.51 (see *Graf*, 1971, p. 175).

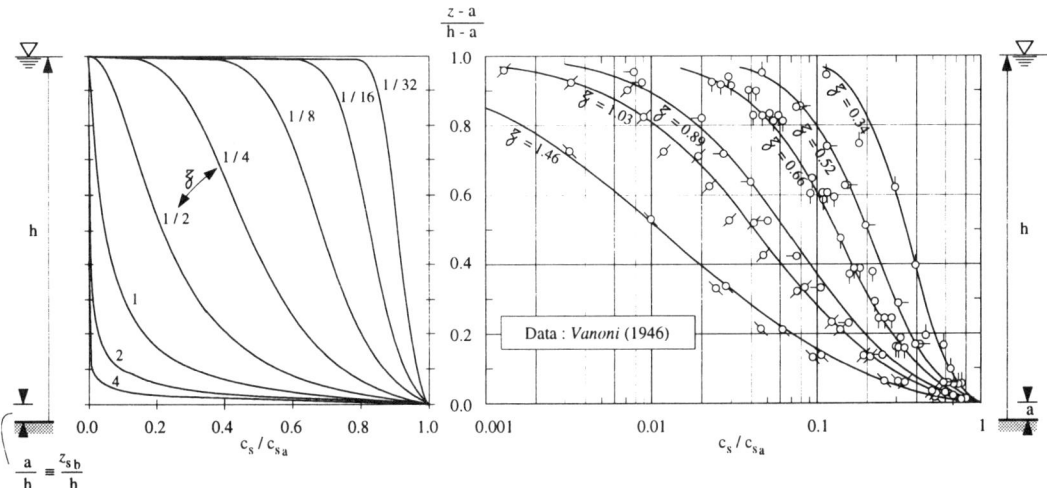

Fig. 6.11 Vertical distribution of the relative concentration, c_s/c_{sa}, in a suspension.

6.4.3 Suspended-Load Relation

1° The volumic solid discharge in suspension per unit width, in a region delimited by $z_{sb} < z < h$, is obtained by :

$$q_{ss} = \int_{z_{sb}}^{h} c_s \, u \, dz \qquad (6.52)$$

where $c_s(z)$ is the local concentration, eq. 6.51, and $u(z)$ is the local velocity. This relation is valid for a single particle size, d or v_{ss}.

There exist different methods for the calculation of the suspended-load transport (see *Graf*, 1971, p. 189), but only the one of *Einstein* (1950) will be presented, being presently the most popular one.

2° The distribution of the velocity shall be given by a logarithmic relation (see *Einstein*, 1950, p. 17), of the form :

$$u(z) = u_*' \, 5.75 \log \left(30.2 \frac{z}{\Delta}\right) \tag{6.53}$$

where Δ is a correction term, given in Fig. 6.7a, and u_*' is the friction velocity due to the granulate.

3° Upon substitution of eq. 6.51 and of eq. 6.53 into the above equation, eq. 6.52, one obtains :

$$q_{ss} = \int_{z_{sb}}^{h} c_{sa} \left(\frac{h-z}{z} \cdot \frac{a}{h-a}\right)^{z} u_*' \, 5.75 \log \left(30.2 \frac{z}{\Delta}\right) dz \tag{6.54}$$

Replacing $a \equiv z_{sb}$ by a dimensionless expression, $z_{sb}/h = A_E$, and using h as the unity of z (see *Einstein*, 1950, p. 18), yields :

$$q_{ss} = \int_{z_{sb}}^{h} c_s \, u \, dz = \int_{A_E}^{1} c_s \, u \, h \, dz \tag{6.52a}$$

After some mathematical manipulations, one gets :

$$q_{ss} = c_{sa} \, u_*' \, 5.75 \, h \left(\frac{A_E}{1-A_E}\right)^{z} \cdot$$

$$\cdot \left\{ \log \left(30.2 \frac{h}{\Delta}\right) \int_{A_E}^{1} \left(\frac{1-z}{z}\right)^{z} dz + 0.434 \int_{A_E}^{1} \left(\frac{1-z}{z}\right)^{z} \ln z \, dz \right\} \tag{6.55}$$

The values of the following integrals :

$$\mathcal{J}_1 = 0.216 \, \frac{A_E^{z-1}}{(1-A_E)^{z}} \int_{A_E}^{1} \left(\frac{1-z}{z}\right)^{z} dz$$

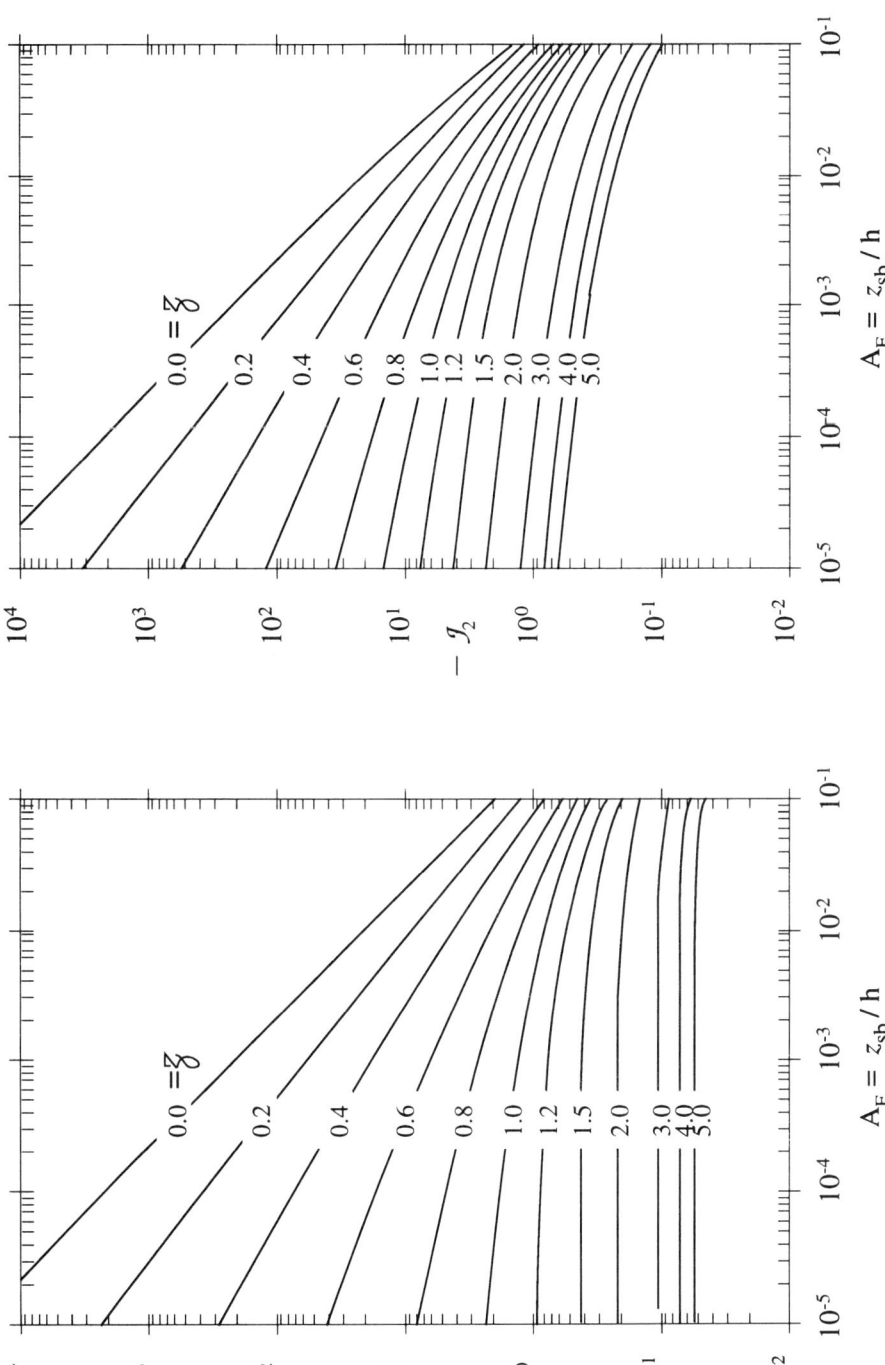

Fig. 6.12 The integrals, $\mathcal{I}_1(A_E, \mathcal{Z})$ and $\mathcal{I}_2(A_E, \mathcal{Z})$, used in the method of *Einstein* (1950).

$$J_2 = 0.216 \frac{A_E^{\mathfrak{z}-1}}{(1-A_E)^{\mathfrak{z}}} \int_{A_E}^{1} \left(\frac{1-z}{z}\right)^{\mathfrak{z}} \ln z \, dz$$

are numerically evaluated (see *Einstein*, 1950, p. 19-24) and this for different values of A_E and $\mathfrak{z}$; they are given in Fig. 6.12.

Finally, the above equation, eq. 6.55, can be put into the following form :

$$q_{ss} = 11.6 \, c_{sa} \, u_*' \, z_{sb} \left[2.303 \log\left(30.2\frac{h}{\Delta}\right) J_1 + J_2\right] \quad (6.56)$$

where q_{ss} is the volumic solid discharge per unit width of the suspended load.

4° The *reference concentration*, c_{sa}, shall be taken there, where the concentration distribution, eq. 6.51, lacks any physical sense, namely very close to the bed. It will thus be positioned within the layer where the bed load moves (see Fig. 6.1).

One usually assumes (see *Graf*, 1972, p. 191), that the thickness of this layer, called *bed layer*, is twice the grain diameter, $z_{sb} \approx 2d$; for a granulometric mixture, the bed layer takes different values for each granulometric fraction.

It is now of the foremost interest, to establish a relation between the bed-load and suspended-load transport ; the reference concentration, c_{sa}, will make this link.

The formula of Einstein for bed-load transport, eq. 6.42, for one single granulometric fraction, $q_{sb} \, i_{sb}$, shall be used for the determination of the (average) concentration in this bed layer ; one writes :

$$c_{sa} = \frac{q_{sb} \, i_{sb}}{u_b \, z_{sb}} \quad (6.57)$$

Exploiting experiments (see *Einstein*, 1950, p. 40) which rendered the velocity of bed load as being $u_b = 11.6 \, u_*'$, one obtains an expression for the reference concentration, such as :

$$c_{sa} = \frac{q_{sb} \, i_{sb}}{11.6 \, u_*' \, z_{sb}} \quad (6.57a)$$

5° Consequently, the solid discharge as suspended load per unit width — using this expression, eq. 6.57a — is given by :

$$q_{ss} \, i_{ss} = q_{sb} \, i_{sb} \left[2.303 \log\left(30.2\frac{h}{\Delta}\right) J_1 + J_2\right] \quad (6.58)$$

TRANSPORT OF SEDIMENTS

$q_{ss}i_{ss}$ being the volumic solid discharge per unit width of the suspended load for one single granulometric fraction.

This relation, eq. 6.58, establishes the link between bed-load and suspended load transport for all particle sizes, which are found in the granulometric fraction of the bed-load.

6.5   TOTAL-LOAD TRANSPORT

6.5.1   Notions

1° Total-load transport of sediments — or better called *total bed-material load transport* — is made up of transport as bed load (see sect. 6.3) and of transport as suspended load :

$$q_s = q_{sb} + q_{ss} \, (+ q_{sw}) \tag{6.59}$$

Added should (possibly) be the transport as wash load, q_{sw}.

2° Different formulae (see *Graf*, 1971, chap. 9 or *White* et al., 1973) exist, which can be used for the prediction of the bed-material load in a watercourse.

The formulae for determination of the total load — just as the ones of the bed load — give only reasonable results in the domain of their established parameters. Thus an application of any formula must be done with great care.

Here will be given a selection (in chronological order) of some existing formulae.

6.5.2   Total-Load Relations

1° The *indirect* methods determine the bed-material load by addition of the calculated bed load and the calculated suspended load. Thus these methods take into account that the hydromechanics of each mode of transport is not the same. However, a clear distinction between the two modes is not easily possible.

The *direct* methods determine the bed-material load directly, without making a distinction between the two modes of transport.

2° *Einstein* (1950, p. 40) proposed a formula for bed-load transport, eq. 6.42, and one for suspended-load transport, eq. 6.58 ; by combining up these two relations, it is possible to get a formula for the bed-material load transport :

$$q_s \, i_s = q_{sb} \, i_{sb} + q_{ss} \, i_{ss} = q_{sb} \, i_{sb} \left[1 + 2.303 \, \log \, (30.2 \, \frac{h}{\Delta}) \, \mathcal{J}_1 + \mathcal{J}_2 \right] \tag{6.60}$$

This relation gives the sediment-transport capacity, but does not, of course, include the wash-load transport.

This formula, eq. 6.60, can be used if the hydraulic and sedimentological parameters are known in advance. If in addition a measurement of the suspended load is also available, there exists a modified version (see *Graf*, 1971, p. 207) of the above relation.

In many ways, the indirect method of *Einstein*, eq. 6.60, is hydraulically rather complete, but its application might seem laborious. Notably, the non-uniformity of the granulometry is accounted for by using the ratio of i_s/i_{sb}. Furthermore, one considers also the influence of the water temperature (see *Graf*, 1971, p. 238), of the velocity distribution, eq. 6.35, and of the concentration distribution, eq. 6.51, using the exponent of Rouse, eq. 6.50a.

3° A relation for the direct prediction of the bed-material transport, valid for open-channel flow [but also for flow in pipes], was developed by *Graf et Acaroglu* (1968).

A *parameter of shear intensity* was elaborated as a criteria of solid transport (see *Graf*, 1971, pp. 218 et 443), such as :

$$\Psi_A = \frac{(s_s - 1)d}{S_e R_h} \qquad (6.61)$$

which is the inverse of the dimensionless shear stress, given by eq. 6.22.

Applying the concept of power (work) of a flow system, a *parameter of transport* was proposed (see *Graf*, 1971, pp. 219 and 446), such as :

$$\Phi_A = \frac{C_s U R_h}{\sqrt{(s_s - 1)gd^3}} = \frac{(q_s/q)UR_h}{\sqrt{(s_s - 1)gd^3}} \qquad (6.62)$$

which is similar to the dimensionless intensity of solid discharge, given by eq. 6.28.

Note, that the hydraulic radius, R_h, is here taken as the total one ; for a narrow channel, the hydraulic radius of the bed, R_{hb}, should be taken. C_s is the volumic concentration in the section and $d = d_{50}$ is the equivalent diameter.

It could be shown, that a functional relation between these parameters, Ψ_A and Φ_A, (see eq. 6.29a) is possible :

$$\Phi_A = f(\Psi_A)$$

whose form was experimentally determined. Using close to 800 experiments from the laboratory and close to 80 experiments in the field (see Table 6.3), all for free-surface flow [and close to 300 experiments for pipe-line flow], the following relationship (see *Graf*, 1971, pp. 220 and 448) was established :

$$\Phi_A = 10.39 \, (\Psi_A)^{-2.52} \qquad (6.63)$$

This relationship is found valid for $10^{-2} < \Phi_A < 10^3$ or for $\Psi_A \leq 14.6$. An extension of this work by *Graf* et *Acaroglu* (1968) has been done by *Graf* et *Suszka* (1987) ; it provided the following relationship :

$$\Phi_A = 10.4 \, K \, (\Psi_A)^{-1.5} \tag{6.63a}$$

with
$$\begin{aligned} K &= \Psi_A^{-1} & \text{if} & \quad \Psi_A \leq 14.6 \\ K &= (1 - 0.045 \, \Psi_A)^{2.5} & \text{if} & \quad 22.2 > \Psi_A > 14.6 \\ K &= 0 & \text{if} & \quad \Psi_A > 22.2 \end{aligned}$$

The trend, for very weak solid transport, $10^{-5} < \Phi_A < 10^{-2}$, with $\Psi_A > 14.6$, is also evident in other experiments (see *Pazis* et *Graf*, 1977).

If one takes in the above relation, eq. 6.61, the energy slope, S_e, defined by eq. 6.2, the functional relation between Ψ_A and Φ_A, eq. 6.63, can also be used for the calculation of the sediment transport during unsteady flow (see *Graf* et *Song*, 1995).

The relations, eq. 6.63 and eq. 6.63a, are also valid when taking an equivalent diameter, $d \cong d_{50}$, if the granulometry is a non-uniform one.

4° For a direct determination of the total-load transport, q_s, *Ackers* et *White* (1973) proposed the use of some sedimentological parameters ; employed were hydraulic considerations and dimensional analysis.

A *parameter of mobility* of sediments was defined as :

$$F_{gr} = \frac{u_*^{n_w}}{\sqrt{(s_s - 1)gd}} \left[\frac{U}{\sqrt{32} \, \log(10 h_m/d)} \right]^{(1-n_w)} \tag{6.64}$$

which becomes $F_{gr} = \sqrt{\tau_*}$ (see eq. 6.23) for very fine particles, where $n_w = 1$.

A *parameter of transport* of sediments was postulated as :

$$G_{gr} = C_w \left(\frac{F_{gr}}{A_w} - 1 \right)^{m_w} \tag{6.65}$$

The total-load transport is calculated according to :

$$C_s = \frac{q_s}{q} = G_{gr} \frac{d}{h_m} \left(\frac{U}{u_*} \right)^{n_w} \tag{6.66}$$

where C_s is the volumic average concentration in a section and $h_m = A/B$ is the average flow depth.

The coefficients in the above relations were determined by regression analysis, using close to 1000 experiments in the laboratory and close to 250 experiments in the field, with sediments having a uniform and a non-uniform granulometry, $0.04 < d_{50}$ [mm] < 4.0 and for flow at Fr < 0.8 (see Table 6.3). The resulting values of these coefficients are the following :

coefficient	$d_* > 60$ $d > 2.5$ [mm]	$1.0 < d_* \leq 60$	$d_* < 1$ $d < 0.04$ [mm]
n_w	0.0	$(1.0 - 0.56 \log d_*)$	1.0
m_w	1.50	$(9.66/d_*) + 1.34$	
A_w	0.17	$(0.23/\sqrt{d_*}) + 0.14$	
C_w	0.025	$\log C_w = 2.86 \log d_* - (\log d_*)^2 - 3.53$	

Above, the dimensionless particle diameter, d_*, is used, defined as :

$$d_* = d \left[(s_s - 1) \frac{g}{v^2} \right]^{1/3} \tag{6.26}$$

For a non-uniform granulometry, one takes $d = d_{35}$ as the equivalent diameter.

6.5.3 Applications of Relations

1° Different formulae for the determination of the solid transport have been presented. However, none of these relations can pretend to translate the intrinsic complexity of the transport of sediments.

2° Most of these formulae should not be used beyond the conditions within which they were established. Table 6.3 contains a summary of the range of the parameters, d and S_f, investigated for the establishment of each formula by their author(s) ; other author(s) may have extended this range. Also listed are the recommendation by the author(s) for the choice of the equivalent diameter, d_x, if the granulometry is quasi or non-uniform.

3° The formulae for the transport of sediments are often established, using laboratory data and less often using field data.

A verification of these formulae in watercourses is a very delicate task, since it is difficult to measure correctly the solid discharge in the field. Furthermore, it is often a rather subjective evaluation, since the zones of the modes of transport cannot easily be separated.

4° Numerous studies have been reported, comparing measurements in watercourses with the different existing formulae.

For a better appreciation of the validity of the above presented formulae, it will now be of interest to compare the computed results with the direct measurements of the solid discharge in the field.

Table 6.3 Parameters used for establishing the different formulae.

Formula	d [mm]	S_f [-]	d_x [mm], equivalent diameter for a non-uniform granulate
Schoklitsch (eq. 6.31)	0.3 – 7.0 (44.0)	0.003 – 0.1	d_{40}
Meyer-Peter et al. (eq. 6.32)	3.1 – 28.6	0.0004 – 0.020	$d_m(d_{50})$
Einstein (eq. 6.42)	0.8 – 28.6	–	d_{35}
Graf et *Acaroglu* (eq. 6.63)	0.3 – 1.7 (23.5)	–	d_{50}
Ackers et *White* (eq. 6.66)	0.04 – 4.0	Fr < 0.8	d_{35}

Many (nineteen) of the existing formulae for the calculation of the total transport have been studied by *White* et al. (1973) and compared with experimental results. They evaluated almost 1000 laboratory experiments with uniform and non-uniform sediments of 0.04 < d_{50} [mm] < 4.9, at flow depth of h < 0.4 [m], and almost 270 experiments in watercourses with sediments of 0.1 < d [mm] < 68.0 and a width/depth ratio of 9 < B/h < 160.

Each formula was applied to all the data of the solid-discharge measurements. Subsequently was established a ratio of the values calculated, C_{calc}, and the values observed, C_{obs}, where $C \equiv C_s$ is the total-load transport, expressed in concentration.

Some results of this investigation are given in Fig. 6.13, where one may see the success of a prediction (in percentage) for different ranges of the ratio, C_{calc}/C_{obs}. For the formulae, which are presented in this book — considering only the range of 1/2 < C_{calc}/C_{obs} < 2 — it can be seen that the percentage is for the formula of

Einstein (1950), eq. 6.60 : 44 % of success

Graf et *Acaroglu* (1968), eq. 6.63 : 40 % of success

Ackers et *White* (1973), eq. 6.66 : 64 % of success

This implies that with the formula of *Ackers* et *White*, 64 % of the experimental data can be predicted in the above-mentioned range. This is usually considered as a good (or a not-so-bad) result ; more than half of the studied (nineteen) formulae give results which are less good, namely < 40 %. Also noticed is that with the formula of *Einstein* there is a slight under-estimation of the solid discharge ; while the one of *Graf* et *Acaroglu* gives a slight over-estimation.

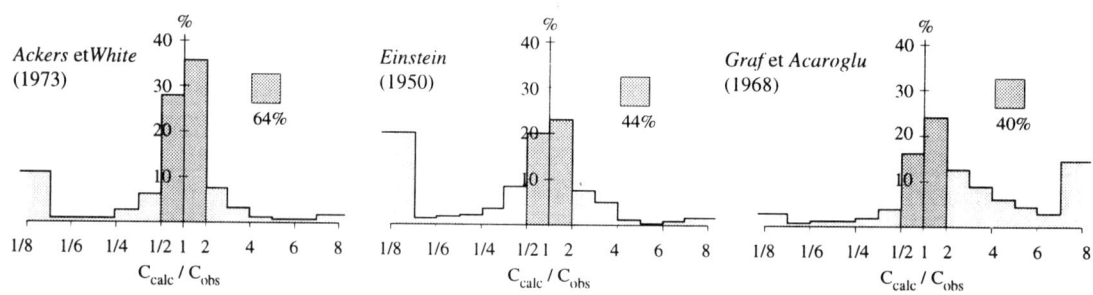

Fig. 6.13 Comparison, with respect to C_{calc}/C_{obs}, of the success of prediction for the presented formulae.

The comparative study of *White* et al. (1973) is reasonably objective, but certainly not conclusive. Other studies exist (see *Raudkivi*, 1976, p. 227) which show clearly that an objective validation is nearly impossible.

5° Amongst the different existing formulae for the determination of the total-load transport, but equally for the ones of the bed-load and suspended-load transport, each one will give an answer, but none will be very precise nor very true.

Finally, it must be said, that the results obtained with these formulae give only valuable guide-lines for the engineer. For practical purposes, it is advised to consult more than one formula ; the obtained result may however render different values (see *Graf*, 1971, p. 156).

6.5.4 Wash Load

1° The wash load, q_{sw}, contains all these particles which are never in contact with the bed and displace themselves by being carried (washed) through the channel by the flow (see Fig. 6.1).

This mode of the transport of sediments (see Table 6.2) is limited to the very finest particles which are rare in the granulometry of the bed material. The distribution of these particles is rather uniform over the entire flow depth (see Fig. 6.11).

Einstein (1950, p. 7) has proposed that the granulometry of the wash load is the fraction of granulometry of the bed which is smaller than 10 %. It was also proposed (see *Raudkivi*, 1976, p. 220) that the wash load is composed of the fine particles having a diameter of d < 0.06 [mm].

2° Since there exists no physical relationship to the flow, it has been difficult to advance an analytical method for the determination of the wash load.

The wash load depends more on the hydrological, geomorphological and meteorological conditions within the drainage basin (see *Graf*, 1971, p. 232), namely on the overland surface erosion and less on the erosion in the stream bed.

3° Thus it is to be remarked, that at the present no methods exist for the prediction of the wash load.

In order to obtain a quantitative information on the wash load, measurements in the field must be performed. One measures thus the total suspended load, $q_{ss} + q_{sw}$. Subsequently is calculated the suspended load, q_{ss}, (see sect. 6.4) and consequently the suspended wash load, q_{sw}, can be obtained.

4° In some watercourses, the transport as wash load can be much more important than the bed-material load, $q_{sw} > q_s$. Obviously, this makes the problem of sediment transport hopelessly complicated.

If the total suspended-load transport, $q_{ss} + q_{sw}$, becomes very large, one may well imagine that this influences on the flow behaviour ; such a mixture of water-sediments is probably not anymore a Newtonian one (see Table 6.1). The flow of such a non-Newtonian mixture will modify the hydraulics, thus the distribution of the velocity and of the concentration, but also the flow resistance as well as the bed forms.

An early version of section 6.2 was published as :
  Graf W.H. (1994) : *Les équations de Saint-Venant-Exner*.
  Österr. Ing. und Arch. Zeitschrift, Jgg. 139, N°9, Wien, A

6.6 EXERCISES

6.6.1 Problems, solved

Ex. 6.A

A rectangular channel has a width of B = 5 [m]. At some point, the bed of the channel changes from a fixed bed to a mobile bed with a uniform sediment of d_{50} = 1 [mm] and s_s = 2.6 [-]. The discharge of Q = 15 [m³/s] remains constant and the water depth is h = 2.2 [m]. In the fixed-bed reach of the channel there is no sediment transport. This flow initiates however erosion in the mobile-bed reach of the channel, where the porosity of the bed material is p = 0.3 [-].

A *degradation* of the channel starts at the junction between the fixed bed and the mobile bed. Determine the time it will take to lower the bed level down to $z = 0.4\Delta h$ at a station located at $L = 6R_h/S_f$ downstream from the junction; subsequently draw the bed profile for this particular moment. Furthermore, show the temporal variation of the degradation at this station. Calculate also the resulting bed profile if the mobile bed is limited to a length of x_f = 90 [km].

SOLUTION :

i) The steady flow will be considered to be quasi-uniform during the phase of degradation (see Fig. Ex. 6.A.1); therefore the *parabolic model* can be used :

$$\frac{\partial z}{\partial t} - K \frac{\partial^2 z}{\partial x^2} = 0 \qquad (6.11)$$

where x is positive towards the downstream and follows the initial bed profile; z represents the bed-level variation with respect to the initial bed, S_f^o. Note that the use of the parabolic model is limited to : Fr < 0.6 and $x > 3R_h/S_e$.

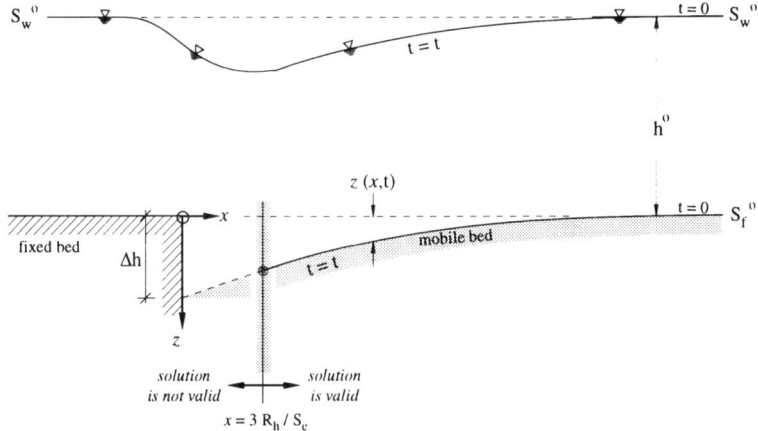

Fig. Ex.6.A.1 Scheme of the degradation.

TRANSPORT OF SEDIMENTS

The initial and boundary conditions are given as :

$$z(x,0) = 0 \quad ; \quad \lim_{x \to \infty} z(x,t) = 0$$

$$z(0,t) = \Delta h(t)$$

The solution to eq. 6.11 is given by :

$$z(x,t) = \Delta h \; erfc \left(\frac{x}{2\sqrt{Kt}} \right) \tag{6.15}$$

ii) Calculation of the quasi-uniform *flow* in the mobile-bed channel.

The slope of the energy line, S_e, is calculated using the Manning-Strickler formula:

$$U = \frac{Q}{Bh} = K_s \, R_h^{2/3} \, S_e^{1/2} \tag{3.16}$$

with
$$K_s = 21.1/d_{50}^{1/6} = 66.7 \; [m^{1/3}/s] \tag{3.18}$$
$$h = 2.2 \; [m] \quad , \quad B = 5.0 \; [m] \quad , \quad R_h = 1.17 \; [m]$$
$$Q = 15.0 \; [m^3/s] \quad , \quad q = Q/B = 3 \; [m^2/s]$$
$$U = q/h = 1.36 \; [m/s]$$

The slope of the energy line : $S_e = 0.00034 \; [-]$

The Froud number is : $Fr = \dfrac{U}{\sqrt{gh}} = 0.29 \; [-]$

It should be emphasized that the Froude number has to be small, Fr < 0.6, being one of the conditions (see sect. 6.2.3) for the validity of the parabolic model, eq. 6.11.

iii) Calculation of the *solid discharge* in the mobile-bed channel.

The solid discharge, $q_s = C_s Uh$, is calculated using the *Graf et al.* (1968) formula :

$$\frac{C_s \, U R_h}{\sqrt{[(\rho_s - \rho)/\rho] \, g \, d_{50}^3}} = 10.39 \left\{ \frac{[(\rho_s - \rho)/\rho] \, d_{50}}{S_f \, R_h} \right\}^{-2.52} \tag{6.63}$$

with
$$(\rho_s - \rho)/\rho = 1.6 \; [-]$$
$$d_{50} = 1 \; [mm]$$
$$S_f \equiv S_e = 0.00034 \; [-]$$

$$C_s \, U R_h = 3.9 \cdot 10^{-5} \; [m^2/s]$$

The solid discharge is : $q_s = C_s \, U h \dfrac{R_h}{R_h} = 3.9 \cdot 10^{-5} \dfrac{2.2}{1.17} = 7.3 \cdot 10^{-5} \; [m^2/s]$

iv) The *coefficient*, K, in the parabolic model, eq. 6.11, is approximately given by :

$$K_o \equiv K \approx \frac{1}{3} b_s q_s \frac{1}{(1-p)} \frac{1}{S_e^o} \tag{6.12c}$$

with $\quad S_e^o = 0.00034\ [-]$
$\quad\quad (1-p) = 0.7\ [-]$
$\quad\quad b_s = 2\ (2.52) \cong 5 \quad$ (where $\beta = 2.52$ is the exponent in eq. 6.63, according to eq. 6.5a and eq. 6.30)

The coefficient is : $\quad K = 0.511\ [m^2/s]$

Table for the complementary error function
(see *Handbook of Mathematical Functions*, 1964, National Bureau of Standards, pp. 310-311, formula 7.1.28)

Y	erfc(Y)	Y	erfc(Y)	Y	erfc(Y)
0.00	1.00000	1.20	0.08969	2.30	0.00114
0.10	0.88754	1.30	0.06599	2.40	0.00069
0.20	0.77730	1.40	0.04772	2.50	0.00041
0.30	0.67137	1.50	0.03390	2.60	0.00024
0.40	0.57161	1.60	0.02365	2.70	0.00013
0.50	0.47950	1.70	0.01621	2.80	0.00008
0.60	0.39614	1.80	0.01091	2.90	0.00004
0.70	0.32220	1.90	0.00721	3.00	0.00002
0.80	0.25790	2.00	0.00468	3.10	0.00001
0.90	0.20309	2.10	0.00298	3.20	0.00001
1.00	0.15730	2.20	0.00186	3.30	0.00000
1.10	0.11979				

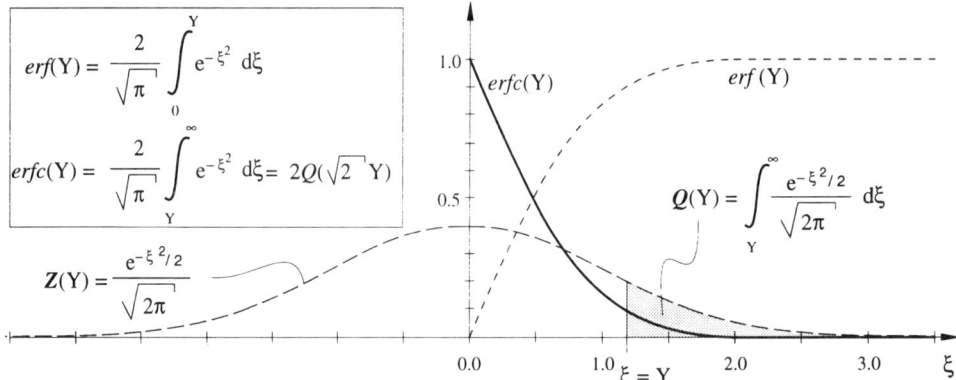

$$erf(Y) = \frac{2}{\sqrt{\pi}} \int_0^Y e^{-\xi^2} d\xi$$

$$erfc(Y) = \frac{2}{\sqrt{\pi}} \int_Y^\infty e^{-\xi^2} d\xi = 2Q(\sqrt{2}\ Y)$$

$$Z(Y) = \frac{e^{-\xi^2/2}}{\sqrt{2\pi}}$$

$$Q(Y) = \int_Y^\infty \frac{e^{-\xi^2/2}}{\sqrt{2\pi}} d\xi$$

The complementary error function can be calculated approximately using the following expression :
$erfc(Y) = 1 / (1 + a_1 Y + a_2 Y^2 + a_3 Y^3 + a_4 Y^4 + a_5 Y^5 + a_6 Y^6)^{16} + \varepsilon(Y) \quad$ where $\quad |\varepsilon(Y)| \leq 3 \cdot 10^{-7}$
$\quad a_1 = 0.0705230784 \quad ; \quad a_2 = 0.0422820123 \quad ; \quad a_3 = 0.0092705272$
$\quad a_4 = 0.0001520143 \quad ; \quad a_5 = 0.0002765672 \quad ; \quad a_6 = 0.0000430638$

TRANSPORT OF SEDIMENTS

v) In the present problem, it is asked to determine the time it takes to lower the bed level down to $z = 0.4\Delta h$, thus :

$$\frac{z(x,t)}{\Delta h} = \frac{0.4\Delta h}{\Delta h} = 0.4$$

The eq. 6.15 is now written as :

$$0.4 = erfc\left(\frac{x}{2\sqrt{Kt}}\right) = erfc(Y)$$

Using the table of the complementary error function yields :

$$Y \cong 0.6 = \left(\frac{x}{2\sqrt{Kt}}\right) \quad\Rightarrow\quad t \cong \frac{x^2}{4\,Y^2\,K} \cong \frac{x^2}{1.44\,K}$$

At the station $x \equiv L = 6R_h/S_e = 20.73$ [km], the lowering of the bed down to a level of $z = 0.4\Delta h$ occurs at the time :

$$t = \frac{(20.73 \cdot 10^3)^2}{(1.44)\,(0.511)} = 5.84 \cdot 10^8 \text{ [s]} = 1.62 \cdot 10^5 \text{ [h]} \cong 18.52 \text{ [years]}$$

To draw the bed profile for the entire channel at this particular moment, $t = 5.84 \cdot 10^8$ [s], the calculations are repeated for different values for the distance x (see following table).

Calculation of the bed profile				
$R_h = 1.17$ [m] ;		$S_f = 0.00034$ [-] ;		$K = 0.511$ [m²/s]
	$\Delta h = \mathbf{3.11}$ [m] ;		$t = 5.84 \cdot 10^8$ [s]	
x [m]	$x\,(S_e / R_h)$ [-]	$Y = x/(2\sqrt{Kt})$ [-]	$z/\Delta h = erfc(Y)$ [-]	z [m]
10500	3.04	0.30	0.66735	2.073
11000	3.18	0.32	0.65253	2.027
13000	3.76	0.38	0.59465	1.847
15000	4.34	0.43	0.53923	1.675
20000	5.79	0.58	0.41299	1.283
20730	6.00	0.60	0.39615	1.231
30000	8.68	0.87	0.21946	0.682
40000	11.58	1.16	0.10157	0.316
50000	14.47	1.45	0.04070	0.126
60000	17.37	1.74	0.01405	0.044
70000	20.26	2.03	0.00417	0.013
80000	23.15	2.32	0.00106	0.003
90000	26.05	2.60	0.00023	0.001
100000	28.94	2.89	0.00004	0.000

The depth of degradation of the channel bed due to a solid discharge of $q_s = 7.3 \cdot 10^{-5}$ [m²/s], during a time period of $t = 5.84 \cdot 10^8$ [s] is given by :

$$\Delta h = \frac{q_s \cdot \Delta t}{1.13 \, (1-p)\sqrt{K \, \Delta t}} = \frac{(7.3 \cdot 10^{-5})\sqrt{5.84 \cdot 10^8}}{(1.13)(0.7)\sqrt{0.511}} = \mathbf{3.11} \text{ [m]} \qquad (6.20)$$

and $z = 0.4 \Delta h = 1.23$ [m].

The bed profile, $z(x)$, for $t = 5.84 \cdot 10^8$ [s] = 18.52 [years], is plotted in Fig. Ex. 6.A.2. This solution is valid only if $x > 3R_h/S_e$. For distances of $x < 3R_h/S_e$, the solution is only an indicative one.

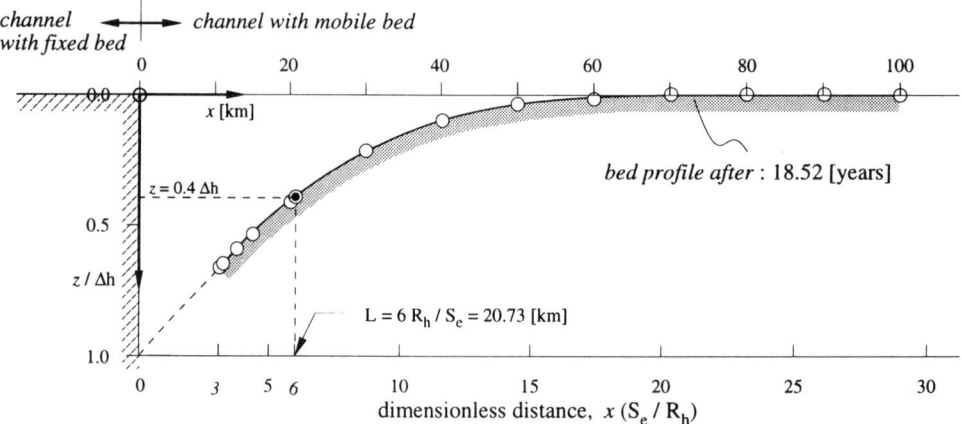

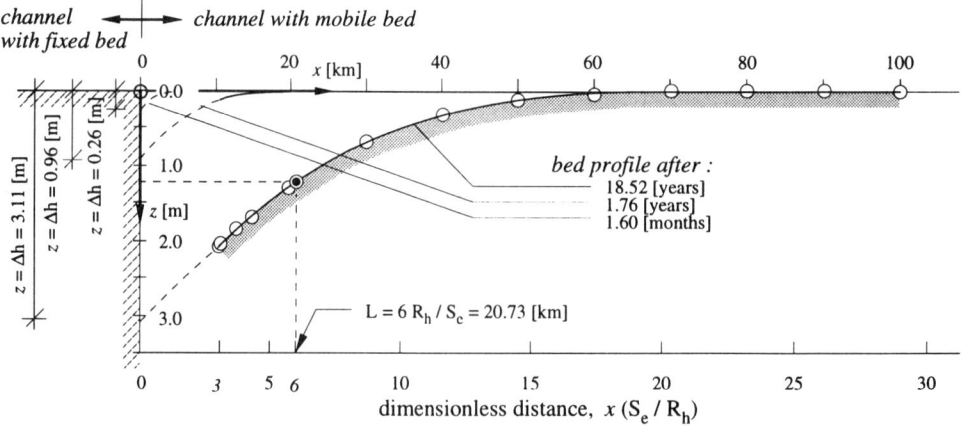

Fig. Ex.6.A.2 Bed profile after *18.52* [years] of degradation.

For sake of comparison, the bed profiles, $z(x)$, for $t = 1.76$ [year] and for $t = 1.6$ [month] are also plotted (without giving the calculations) in Fig. Ex. 6.A.2.

TRANSPORT OF SEDIMENTS

vi) The *temporal evolution of the degradation* at the station located at $x \equiv L = 6R_h/S_e = 20.73$ [km] is given by:

$$z(t) = \Delta h \, erfc \left(\frac{x}{2\sqrt{K \, \Delta t}} \right) = \Delta h \, erfc \left(\frac{20730}{2\sqrt{0.511 \, \Delta t}} \right) \tag{6.15}$$

where, $\Delta h(t)$ can be evaluated by:

$$\Delta h = \frac{q_s \cdot \Delta t}{1.13 \, (1-p)\sqrt{K \, \Delta t}} \tag{6.20}$$

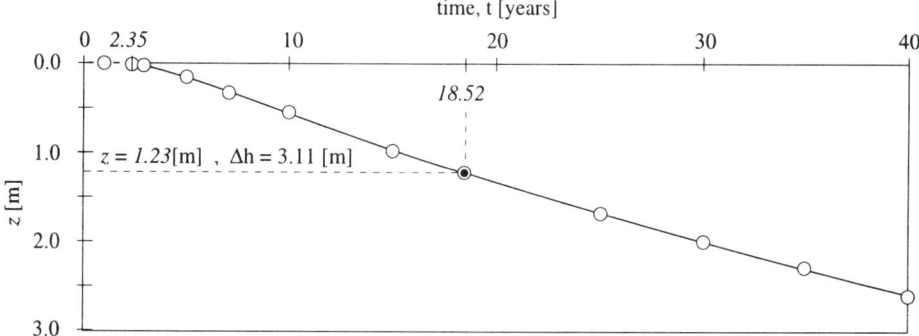

Fig. Ex.6.A.3 Evolution of the degradation at the station $x \equiv L = 6R_h/S_e = 20.73$ [km].

Calculation of the evolution of the degradation					
$R_h = 1.17$ [m] ; $S_f = 0.00034$ [-] ; $K = 0.511$ [m^2/s]					
$x \equiv L = 6R_h/S_{f_o} = 20730$ [m]					
t [years]	t [s]	$Y = x/(2\sqrt{Kt})$ [-]	$z/\Delta h = erfc(Y)$ [-]	Δh [m]	z [m]
1	3.15E+07	2.58	0.00026	0.72	0.0002
3	9.46E+07	1.49	0.03502	1.25	0.0438
5	1.58E+08	1.15	0.10248	1.61	0.1654
7	2.21E+08	0.98	0.16756	1.91	0.3201
10	3.15E+08	0.82	0.24823	2.28	0.5667
15	4.73E+08	0.67	0.34579	2.80	0.9669
18.52	*5.84E+08*	*0.60*	*0.39618*	*3.11*	*1.2309*
25	7.88E+08	0.52	0.46522	3.61	1.6794
30	9.46E+08	0.47	0.50500	3.95	1.9970
35	1.10E+09	0.44	0.53711	4.27	2.2941
40	1.26E+09	0.41	0.56371	4.57	2.5740
45	1.42E+09	0.38	0.58622	4.84	2.8391
50	1.58E+09	0.37	0.60559	5.11	3.0916

The evolution of the bed degradation can now be calculated by assuming different values for $\Delta t \equiv t$. By using the approximate formula for the complementary error function (see before), the calculation can easily be programmed on a spreadsheet. The table above summarizes these calculations; Fig. Ex. 6.A.3 shows the evolution of the erosion, $z(t)$, at the station, $x \equiv L$.

This solution is however only valid (see *Ribberink* et *Sande*, 1984, p. 30) for :

$$t > \frac{40}{30} \frac{R_h^2}{S_f} \frac{1}{q_s} = \frac{40}{30} \frac{1.17^2}{0.00034} \frac{1}{7.3 \cdot 10^{-5}} = 7.42 \cdot 10^7 \text{ [s]} \cong 2.35 \text{ [years]}$$

vii) Calculation of the final bed profile if the channel reach with the mobile bed is *limited* to a length of $x_f = 90$ [km].

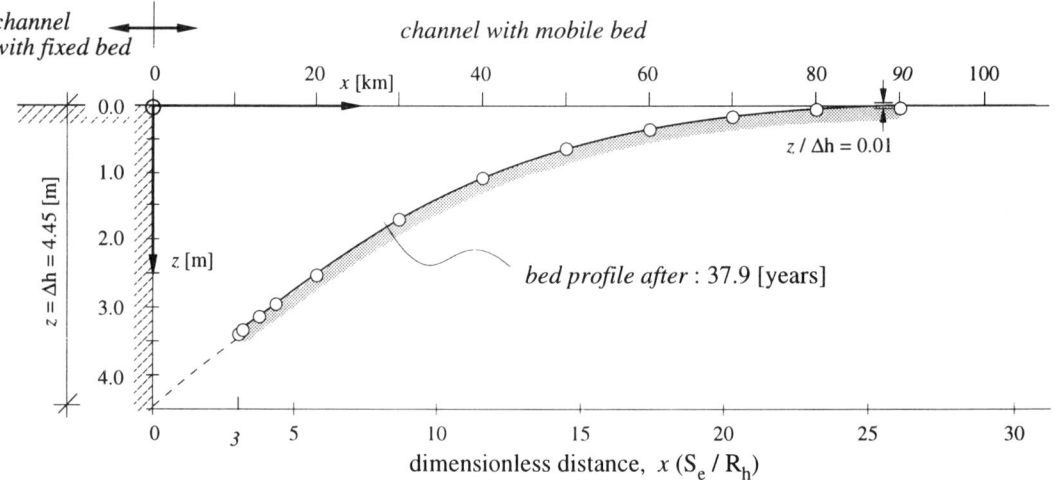

Fig. Ex.6.A.4 The channel-bed profile after *37.9* [years] of degradation.

By assuming a very small amount of erosion, such as $z = 0.01 \Delta h$, at the station $x_f = 90$ [km], one can write :

$$\frac{z(x,t)}{\Delta h} = 0.01 = \text{erfc}\left(\frac{x_f}{2\sqrt{Kt}}\right) = \text{erfc}(Y)$$

Using the table of the complementary error function yields :

$$Y = 1.82 = \left(\frac{x}{2\sqrt{Kt}}\right) \qquad \Rightarrow \qquad t = \frac{x^2}{4 Y^2 K} \cong \frac{x^2}{13.25 K}$$

TRANSPORT OF SEDIMENTS

and with K = 0.511 [m²/s], one calculates :

$$t = \frac{(90 \cdot 10^3)^2}{(13.25)(0.511)} = 1.2 \cdot 10^9 \text{ [s]} = 3.3 \cdot 10^5 \text{ [h]} \cong 37.93 \text{ [years]}$$

To obtain the bed profile for the entire channel at this moment, $t = 1.2 \cdot 10^9$ [s], the calculations for the degradation are repeated using different values for x (see the following table). The final bed profile, calculated in this way, is plotted in Fig. Ex. 6.A.4.

This solution is valid only if $x > 3R_h / S_e$.

The depth of the bed degradation due to a solid discharge, $q_s = 7.3 \cdot 10^{-5}$ [m²/s], during a time period of $t = 1.2 \cdot 10^9$ [s], is given by the eq. 6.20 :

$$\Delta h = \frac{q_s \cdot \Delta t}{1.13 (1-p)\sqrt{K \, \Delta t}} = \frac{(7.3 \cdot 10^{-5}) \sqrt{1.2 \cdot 10^9}}{(1.13)(0.7) \sqrt{0.511}} = \mathbf{4.45} \text{ [m]}$$

Calculation of the final bed profile				
$R_h = 1.17$ [m] ; $S_f = 0.00034$ [-] ; $K = 0.511$ [m²/s]				
$\Delta h = \mathbf{4.45}$ [m] ; $t = 1.2 \cdot 10^9$ [s]				
x [m]	$x \, (S_e / R_h)$ [-]	$Y = x /(2\sqrt{Kt})$ [-]	$z/\Delta h = \mathit{erfc}(Y)$ [-]	z [m]
10500	3.04	0.21	0.76396	3.397
11000	3.18	0.22	0.75307	3.349
13000	3.76	0.26	0.71005	3.157
15000	4.34	0.30	0.66793	2.970
20000	5.79	0.40	0.56734	2.523
30000	8.68	0.61	0.39091	1.738
40000	11.58	0.81	0.25264	1.123
50000	14.47	1.01	0.15273	0.679
60000	17.37	1.21	0.08617	0.383
70000	20.26	1.42	0.04529	0.201
80000	23.15	1.62	0.02214	0.098
90000	26.05	1.82	0.01006	0.045

408   FLUVIAL HYDRAULICS

Ex. 6.B

A river on a bed slope of $S_f = 0.0005$ [-] conveys a unit discharge of $q = 1.5$ [m^2/s]. The river bed is made of granular material of uniform size of $d_{50} = 0.00032$ [m] with a specific gravity of $s_s = 2.6$ [-]; the porosity of the bed material is $p = 0.4$ [-]. There exists a weak transport of sediments.

At a certain station on this river, the solid discharge is locally increased by $\Delta q_s = 0.0001$ [m^2/s] for a time period of $\Delta t = 50$ [h]. Determine the *aggradation* of the bed to be expected.

SOLUTION :

i) The flow is steady and is considered to be quasi-uniform during the period of aggradation (see Fig. Ex. 6.B.1); thus the *parabolic model* can be used :

$$\frac{\partial z}{\partial t} - K \frac{\partial^2 z}{\partial x^2} = 0 \qquad (6.11)$$

where x is positive towards the downstream and follows the initial bed profile; z represents the bed-level variation with respect to the initial bed, S_{f_o}. Note that the use of the parabolic model is limited to : Fr < 0.6 and $x > 3R_h/S_e$.

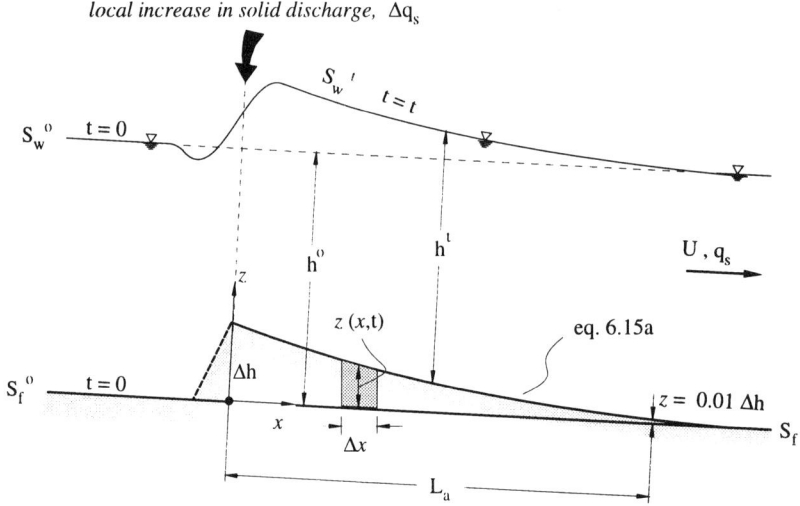

Fig. Ex.6.B.1 Sketch of the aggradation.

The initial and boundary conditions are given as :

$z(x,0) = 0 \qquad ; \qquad \lim_{x \to \infty} z(x,t) = 0$

$z(0,t) = \Delta h(t)$

TRANSPORT OF SEDIMENTS

The solution to eq. 6.11 is given by :

$$z(x,t) = \Delta h \; erfc\left(\frac{x}{2\sqrt{Kt}}\right) \tag{6.15}$$

ii) Calculation of the quasi-uniform *flow* in the river having a mobile bed.

The normal depth is calculated using the Manning-Strickler formula :

$$U = \frac{q}{h} = K_s \, h^{2/3} \, S_f^{1/2} \tag{3.16}$$

with $\quad K_s = 21.1/d_{50}^{1/6} = 80.7 \; [m^{1/3}/s] \tag{3.18}$
$\quad\quad\quad q = 1.5 \; [m^2/s]$
$\quad\quad\quad S_f = 0.0005 \; [-]$

The flow depth is $\quad : \; h = 0.895 \; [m]$
The average velocity is $\quad : \; U = 1.676 \; [m/s]$
The Froude number is $\quad : \; Fr = \dfrac{U}{\sqrt{gh}} = 0.566$

It should be remembered that the Froude number has to be small, namely Fr < 0.6.

iii) Calculation of the *solid discharge* in the river having a mobile bed.

The solid discharge, $q_s = C_s \, Uh$, is calculated using the relationship given by *Graf et al.* (1968) :

$$\frac{C_s \, UR_h}{\sqrt{[(\rho_s-\rho)/\rho]g \, d_{50}^3}} = 10.39 \left\{ \frac{[(\rho_s-\rho)/\rho] \, d_{50}}{S_f \, R_h} \right\}^{-2.52} \tag{6.63}$$

with $\quad (\rho_s-\rho)/\rho = 1.6 \; [-]$
$\quad\quad\quad d_{50} = 0.32 \; [mm]$
$\quad\quad\quad R_h \cong h = 0.895 \; [m]$

The solid discharge is $\quad : \; q_s = 1.678 \cdot 10^{-4} \; [m^2/s]$

iv) The *coefficient*, K, in the parabolic model, eq. 6.11, is approximately given by :

$$K_o \equiv K \approx \frac{1}{3} \, b_s q_s \, \frac{1}{(1-p)} \, \frac{1}{S_e^o} \tag{6.12c}$$

with $\quad S_f^o \equiv S_e^o = 0.0005 \; [-]$
$\quad\quad\quad (1-p) = 0.6 \; [-]$
$\quad\quad\quad b_s = 2\,(2.52) \cong 5 \quad$ (where $\beta = 2.52$ is the exponent in eq. 6.63, according to eq. 6.5a and eq. 6.30)

The coefficient is $\quad : \; K = 0.932 \; [m^2/s]$

v) The thickness of the aggradation of the bed (see Fig. Ex. 6.B.1) due to a local increase in solid discharge, $\Delta q_s = 0.0001$ [m²/s], during a time period of $\Delta t = 50$ [h] $= 1.8 \cdot 10^5$ [s], is given by eq. 6.20, or:

$$\Delta h(t) = \frac{\Delta q_s \cdot \Delta t}{1.13 \, (1-p) \sqrt{K \, \Delta t}} = \frac{(0.0001) \sqrt{1.8 \cdot 10^5}}{(1.13) \, (0.6) \, \sqrt{0.932}} = \mathbf{0.065 \; [m]}$$

The length of the zone of aggradation, L_a, can be calculated with eq. 6.15 by assuming, for example, a precision of $z/\Delta h = 0.01$:

$$\frac{z(x,t)}{\Delta h} = \frac{0.01 \Delta h}{\Delta h} = 0.01 = erfc \left(\frac{x}{2\sqrt{K \, \Delta t}} \right) = erfc \, (Y)$$

Using the table of the complementary error function (see Ex. 6.A), yields:

$$Y = 1.821 = \left(\frac{x}{2\sqrt{K \, \Delta t}} \right)$$

The length of the zone of aggradation (see eq. 6.19) can now be calculated as follows:

$$L_a \equiv x_{1\%} = 2Y \sqrt{K \, \Delta t} = (2) \, (1.821) \, \sqrt{(0.932) \, (1.8 \cdot 10^5)} = \mathit{1492.3 \; [m]}$$

vi) To plot the bed profile after a time period of $\Delta t = 50$ [h] $= 1.8 \cdot 10^5$ [s], calculations are made using eq. 6.15 for different distances, x. (see the following table).

The resulting bed profile, $z(x)$, is plotted in Fig. Ex. 6.B.2.

The calculations, summarized in the following table, are valid only if $x > 3h/S_e$. In the present case, it can be shown that:

$$x = 3h/S_e = (3) \, (0.895) \, / \, (5 \cdot 10^{-4}) = 5370 \; [m] \gg L_a = 1492.3 \; [m]$$

However, experimental data (see *Soni et al.*, 1980), have shown that the calculated value is only indicative, but nevertheless acceptable.

TRANSPORT OF SEDIMENTS

Calculation of the bed profile due to aggradation					
$R_h = h = 0.895$ [m] ; $S_f = 0.0005$ [-] ; $K = 0.932$ [m^2/s]					
$\Delta h = \mathbf{0.065}$ [m] ; $\Delta t = 1.8 \cdot 10^5$ [s]					
x [m]	$x\,(S_e/R_h)$ [-]	$Y = x/(2\sqrt{Kt})$ [-]	$z/\Delta h = erfc(Y)$ [-]	z [m]
---	---	---	---	---
10.0	0.01	0.01	0.98623	0.064
50.0	0.03	0.06	0.93123	0.060
100.0	0.06	0.12	0.86296	0.056
300.0	0.17	0.37	0.60459	0.039
500.0	0.28	0.61	0.38813	0.025
700.0	0.39	0.85	0.22696	0.015
900.0	0.50	1.10	0.12032	0.008
1000.0	0.56	1.22	0.08434	0.005
1100.0	0.61	1.34	0.05761	0.004
1300.0	0.73	1.59	0.02484	0.002
1492.3	*0.83*	*1.82*	*0.01000*	*0.001*
1500.0	0.84	1.83	0.00962	0.001
1600.0	0.89	1.95	0.00575	0.000

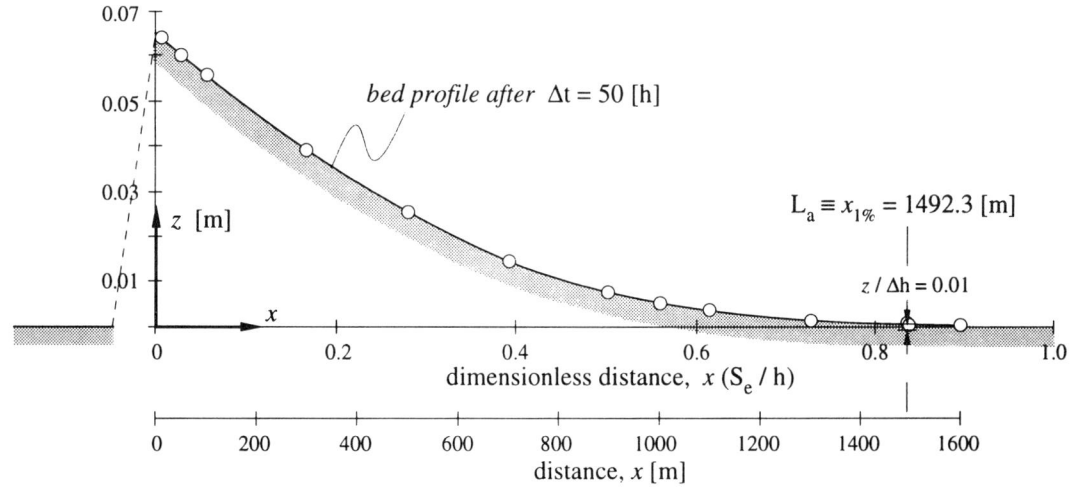

Fig. Ex.6.B.2 Bed profile after 50 [hours] of aggradation.

Ex. 6.C

The unit discharge of a river is kept constant at q = 2.5 [m²/s]. The bed slope is $S_f = 5.4 \cdot 10^{-4}$ [-]. The river bed is composed of quasi-uniform sediments (s_s = 2.65 [-]) with an average grain size of d_{50} = 6 [mm] and a porosity of p = 0.3 [-]. The Manning coefficient of the bed was determined as being n = 0.032 [$m^{-1/3}$s].

This river enters into a reservoir, created by a dam which keeps the water at a height of H = 23.5 [m] at the immediate vicinity of the dam.

Determine the deposition pattern of bed-load material, which is carried by the river into the reservoir, after 20 [years] and 100 [years], respectively.

SOLUTION :

a)   *General comments on the method of solution* :

Fig. Ex.6.C.1 shows the longitudinal profile of this river-reservoir system. The dam creates a backwater curve extending to a certain upstream distance. This curve can be calculated by using one of the methods presented in chap. 4. The backwater calculation enables one to know the hydraulic parameters (average velocity, water depth, slope of energy grade line, etc.) for the entire length of the system.

Let there be two stations, (i) and (i+1), separated by a distance of Δx. If the characteristics of the sediments at the bed are known, one can calculate the bed-load discharge for these two stations, $q_{sb}(i)$ and $q_{sb}(i+1)$, by using one of the bed-load formulae presented in sect. 6.3.3. It will then be seen that the bed-load transport at the upstream station, (i+1) , is larger than the one at the downstream station, (i). In fact the closer a station is to the dam, the larger is the water depth, resulting in a smaller average velocity and as a consequence in a decrease in the bed-load transport capacity. The difference of the transport capacities between the two consecutive stations causes a deposition (or erosion) of the sediments, which in turn modifies the bed level. This modification of the bed level causes a change in the water-surface profile and therefore modifies all hydraulic parameters. This cycle repeats itself.

To calculate the deposition of the sediments, namely the *delta* formation, one has to simulate the process described above. Such a simulation involves a large number of calculations and therefore is particularly well suited to treatment on a computer.

In this exercise, a computer program in FORTRAN IV language has been written to carry out this simulation. The program is written in standard FORTRAN and can be run on most of the personal computers. Although a basic knowledge of computers and of programming in languages like FORTRAN, BASIC or PASCAL will certainly be helpful in understanding this exercise, it is not essential. Special care is taken to make the general programming techniques understandable to everybody, even to those who do not have any experience in programming.

TRANSPORT OF SEDIMENTS 413

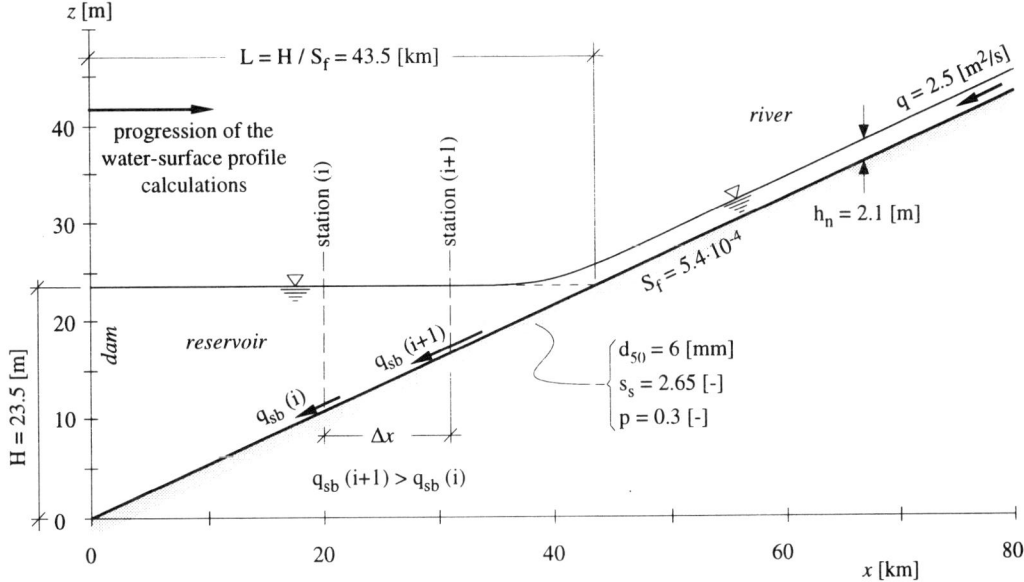

Fig. Ex.6.C.1 Modeling of the river-reservoir system.

b)   *Definition of solution domain and boundary conditions* :

In reality, the cross sections of a natural river have complex forms. Nevertheless, by considering that the length of the river-reservoir system is much larger than the water depth, a simplified one-dimensional approach, where the hydrodynamic equations of the water flow and of the bed-load transport are expressed for a unit width, will be adopted.

The modeled river-reservoir system is presented in Fig. Ex.6.C.1. The origin of the coordinate system coincides with the dam location. The dam constitutes a control section and gives the boundary condition necessary for the water-surface profile calculations. Since the flow regime is subcritical, the calculations start at the dam, where the flow depth is known, and proceed towards the upstream.

The length of the reach to be modeled upstream of the dam can be decided by considering the boundary condition for the sediment transport at the upstream end. In fact it is necessary to extend the calculations up to a point where the river atteins its normal depth, h_n. It is even better to include in the calculations a certain length of the river with the normal depth. This insures a sufficiently long river reach at the upstream end, where the bed-load transport is in equilibrium, namely where the bed-load transport capacity at two consecutive sections will be the same.

For the calculation of the unit discharge, q, one can take $R_h = h$; the Manning-Strickler equation, eq. 3.16, can be written as :

$$q = \frac{Q}{B} = \frac{h}{n} R_h^{2/3} S_f^{1/2} = \frac{h^{5/3}}{n} S_f^{1/2}$$

The normal depth can now be calculated with the following expression:

$$h_n = \left(\frac{q\,n}{S_f^{1/2}}\right)^{3/5} = \left(\frac{(2.5)\,(0.032)}{\sqrt{S_f}}\right)^{3/5} = 2.1 \,[m]$$

The Froude number for the uniform flow is:

$$Fr = \frac{U}{\sqrt{gh_n}} = \frac{q}{h_n\sqrt{gh_n}} = \frac{q}{\sqrt{gh_n^3}} = \frac{2.5}{\sqrt{(9.81)\,(2.1)^3}} = 0.26\,[-]$$

The flow is therefore subcritical.

It is not necessary to compute an initial water-surface profile to guess the point where the river reaches its normal depth. By considering the known values of the water depth at the dam, H = 23.5 [m], and of the bed slope, $S_f = 5.4 \cdot 10^{-4}$ [-], the approximate length of the reservoir can be calculated : L = H / $S_f \cong$ 43.5 [km]. To be able to guarantee a sufficiently long river reach at the upstream, where the flow remains uniform throughout the whole simulation period, namely 100 [years], a computational reach length of, for example, TL = 120 [km] shall be adopted. This total length of the system (TL) is divided in ND reaches having a length of DX; this yields NS = ND +1 stations. Starting from the dam location, the stations are numbered from the downstream to the upstream end.

c)   *Structure of the program DELTA*:

A *decoupled* algorithm has been used in writing the program DELTA. The adjective "decoupled" means that the calculations for the liquid and solid phases are carried out separately and successively (see Fig. Ex.6.C.2).

The calculations start at time t = 0, when the bed-level elevations are known. The water-surface profile is calculated without considering the sediment transport. Once the water-surface profile is calculated, the hydraulic parameters are known at all the stations. The bed-load transport rate is now calculated for all the stations. The balance of the sediments entering and leaving is subsequently calculated for all reaches to find the volume of deposition (or erosion). These volumes are then translated into a deposition height. Finally the bed levels are modified by using these deposition heights. This concludes the computational cycle for the time t = 0. The time is then advanced by Δt, and a new water-surface calculation is carried out with the new bed profile; and so on. It should be noticed that during the calculations for one phase, the characteristics of the other phase are kept constant.

The program DELTA is written in a didactic style and does not have the pretention of being optimized. The complete program code is presented in Fig. Ex.6.C.11. Numerous comments inserted in the code explain the flow of the program almost step by step. As far as possible, the names of the variables are chosen to recall the notation used in the text. An exhaustive list of variables together with the types of variables and explanations, are provided at the beginning of the main program and the related subprograms.

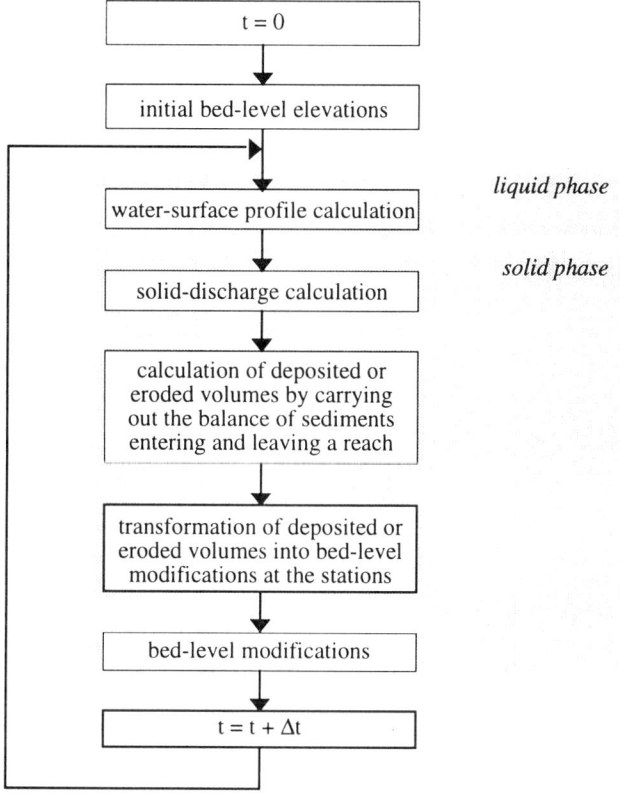

Fig. Ex.6.C.2 Decoupled simulation algorithm.

The program has a modular structure. It is composed of a main program DELTA and nine subprograms, each accomplishing a specific pre-defined task : DREAD; TITLES; RK4; DERIVE; SCHOKL; MEYPET; EINS42; FORMUL and DWRITE. The flowchart, given in Fig. Ex.6.C.3, shows not only the relations between different program units but also the calculation loops inside the principal program. However, it is important to note that the flowchart is somewhat simplified; it does not show all the details of the code. The specific tasks carried out by the main program and each subprogram are described below in detail (see Figs. Ex.6.C.3 and 11).

d) *Working principles of the program DELTA* :

The main program DELTA controls the flow of the entire program. The working principles of the program and the algorithms are described below step by step. The reader is advised to follow these explanations in parallel with the flowchart, given in Fig. Ex.6.C.3, and the program code presented in Fig. Ex.6.C.11.

- The main program first calls the subroutine subprogram DREAD to read the program data by questioning the user. The interactive dialog between the program and the user is presented in Fig. Ex.6.C.5. This dialog will be explained later in detail. The program data are read into the computer in 6 groups :

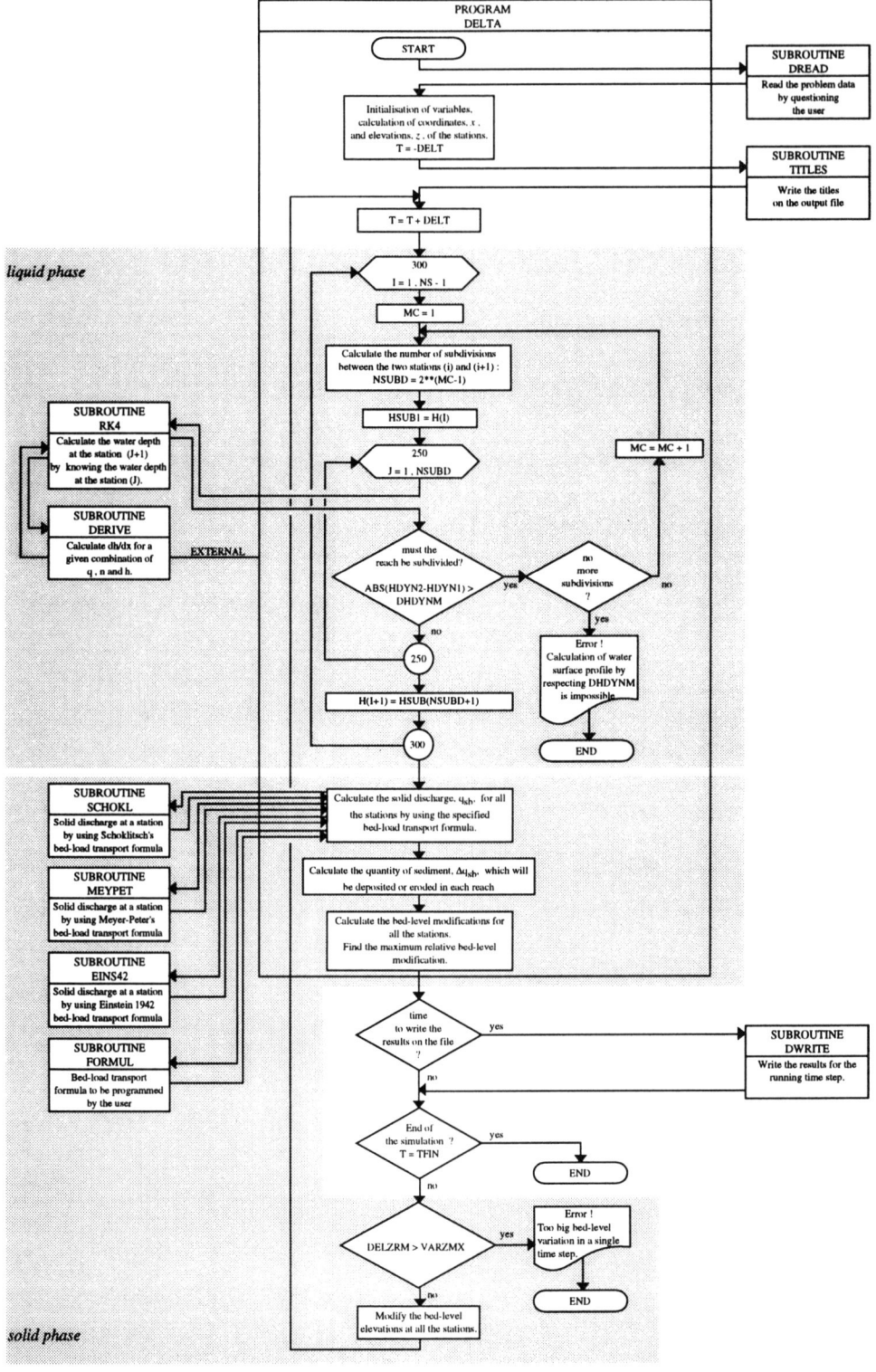

Fig. Ex.6.C.3 Flowchart of the program DELTA.

- *Physical characteristics* (initial bed slope, average sediment diameter, Manning-Strickler coefficient, densities of the water and of the sediments, discharge per unit width),
- *Choice of the bed-load transport formula* (number of the formula to be used),
- *Data concerning the modification of the bed profile* (maximum relative bed-level change tolerated, porosity of the sediments, the ratio of the upstream/downstream heights of the sediment deposition or erosion),
- *Information concerning the computational domain* (coordinates of the first and the last station, space-step length, maximum tolerated dynamic-head variation, maximum number of subdivisions which can be automatically created),
- *Boundary conditions* (water depth at the downstream end),
- *Simulation time and the printing of results* (time step, duration of the simulation, frequency of the printing of the results and the name of the output file).

- According to the data supplied by the user, the program calculates the coordinates of the stations and the initial bed level at these stations. It also initializes certain variables, such as the calculation-steps counter, eroded or deposited cumulative volumes, etc.

- The subprogram TITLES is called to echo-print of the program data on the output file.

- The time is advanced one time step. The calculation loop starts in fact at this point. It can be seen that before entering the calculation loop the time is initialized as T = – DELT; in this way the time for the first calculation step is correctly obtained as being T = 0.

- The calculation of the water-surface profile, using the running bed profile, is carried out using the 4th-order Runge-Kutta method. The differential equation for the free-surface flow in a rectangular channel is given by:

$$\frac{d}{dx}\left(\frac{Q^2}{2g(Bh)^2}\right) + \frac{dh}{dx} - S_f = -S_e \qquad (4.5)$$

For a very wide river, B >> h, with a constant unit discharge of q = Q/B, this equation becomes:

$$\frac{q^2}{2g}\frac{d}{dx}\left(\frac{1}{h^2}\right) + \frac{dh}{dx} - S_f = -S_e$$

The slope of the energy-grade line can be expressed with the Manning-Strickler formula for uniform flow, eq. 3.16. Recalling that for a very wide river one can take $R_h = h$, the following is written:

$$S_e = \frac{q^2 n^2}{h^{10/3}} \qquad (6.\alpha)$$

Substituting this expression into the differential equation for the free-surface flow, one obtains (see eq. 4.8a) :

$$\frac{dh}{dx} = -\frac{S_f - \frac{q^2 n^2}{h^{10/3}}}{1 - \frac{q^2}{gh^3}}$$

It is to be noted that a negative sign is added before the term on the right-hand side of the equation to take into account that the calculation progresses from downstream to upstream. This differential equation is solved using the 4th-order Runge-Kutta method (presented in detail in Ex.7.A.b) by calling the subroutine subprogram RK4. The differential equation is programmed in the subroutine subprogram DERIVE, whose name is passed to the subprogram RK4 in the argument list. For this reason, according to the rules of FORTRAN language, the subprogram DERIVE is declared as EXTERNAL in the main program.

- The calculations for the liquid phase are finished for this time step. The program now proceeds with the calculations for the solid phase. During these calculations it is assumed that the water surface does not vary. The bed-load transport at the stations is calculated by calling the subprogram corresponding to the method specified by the user :

    - The subprogram SCHOKL calculates the bed-load discharge with the method of *Schoklitsch* (1950), whose formula is given by eq. 6.31 :

$$q_{sb} = \frac{2.5}{s_s} S_e^{3/2} (q - q_{cr}) \quad \Rightarrow \quad QSU = \frac{2.5 * SEFF^{3/2} * (QU - QCRIT)}{SS}$$

    The critical liquid discharge is calculated using eq. 6.31a :

$$q_{cr} = 0.26 (s_s - 1)^{5/3} d_{40}^{3/2} S_e^{-7/6} \quad \Rightarrow \quad QCRIT = 0.26 * (SS - 1)^{5/3} * D50^{3/2} * SEFF^{-7/6}$$

    Since the grain-size distribution is uniform, one may take $d_{40} = d_{50}$. If at a station $q_{cr} > q$, the program will assume $q_{sb} = 0$. The slope of the energy-grade line, S_e (SEFF), is calculated by the main program using the Manning-Strickler formula (see the explanations for the water-surface profile calculation) and sent to the subprogram in the argument list.

    - The subprogram MEYPET calculates the bed-load discharge with the method of *Meyer-Peter* et al. (1948), whose formula is given by eq. 6.32 :

$$q_{sb} = \frac{1}{g(\rho_s - \rho)} \left(\frac{g \rho R_{hb} \xi_M S_e - 0.047 g (\rho_s - \rho) d_{50}}{0.25 \rho^{1/3}} \right)^{3/2}$$

    In the program this equation is written as:

$$QSU = \frac{1}{G*(ROS-ROE)} \left(\frac{G*ROE*RH*FCOR*SEFF - 0.047*G*(ROS-ROE)*D50}{0.25*ROE^{1/3}} \right)^{3/2}$$

Since the calculations are done for a unit width, the program uses $R_{hb} = h$ (RH = H). It is assumed that $q_{sb} = 0$, if $(g \rho R_{hb} \xi_M S_e) < (0.047 g (\rho_s - \rho) d_{50})$. The user may or may not use the roughness parameter, ξ_M (FCOR). This parameter is calculated in the subprogram DREAD, during data input, according to the expression :

$$\xi_M = \left(\frac{n'}{n}\right)^{3/2} \quad \Rightarrow \quad FCOR = \left(\frac{CN50}{CN}\right)^{3/2}$$

where n' (CN50) is the grain roughness calculated using the formula of Strickler, eq. 3.18 :

$$n' = \frac{d_{50}^{1/6}}{21.1} \quad \Rightarrow \quad CN50 = \frac{D50^{1/6}}{21.1}$$

and n (CN) represents the total bed roughness, which is introduced by the user during the data input. If the user chooses not to make any corrections, the program takes FCOR = 1.

- The subprogram EINS42 calculates the bed-load discharge with the method of *Einstein* (1942), described by *Graf* (1971, p. 145) :

$$q_{sb} = \frac{\sqrt{(s_s - 1) g d_{50}^3}}{0.465} \exp\left(\frac{-0.391 (s_s - 1) d_{50}}{R_{hb}' S_e}\right)$$

In the program this equation is written as:

$$QSU = \frac{\sqrt{(SS - 1)*G*D50^3}}{0.465} \exp\left(\frac{-0.391*(SS - 1)*D50}{RH*SEFF}\right)$$

Since the calculations are done for a unit width, the program uses RH = H ($R_{hb}' \equiv R_{hb} \equiv h$).

- The subprogram FORMUL does not do any calculations, but simply displays a message, inviting the user to program a bed-load transport formula of his choice in this subprogram.

- Let us now go back to the main program DELTA. The quantity of the sediments to be deposited (or eroded) in a reach, Δq_{sb}, depends on the difference between the bed-load transport capacities at the upstream, $q_{sb}(i+1)$, and at the downstream, $q_{sb}(i)$, stations :

$$\Delta q_{sb}(i) = q_{sb}(i+1) - q_{sb}(i) \quad \Rightarrow \quad DELQS(I) = QSU(I+1) - QSU(I)$$

In the program the reaches between two consecutive stations are numbered from downstream to upstream. The number of a reach is therefore the same as the number of the station at its downstream end (i.e.: the **reach** (i) is limited by the **station** (i) at the downstream and by the **station** (i+1) at the upstream). This fact is used in the program as a *programming trick*. The same variable I is used to denote both a reach and the station. In this way, the solid discharges, QSU(I+1) and QSU(I), refer to the **stations** (I+1) and (I), respectively, whereas DELQS(I) is the difference in transport capacities for the **reach** (I). While reading the program this double role of the variable I must be kept in mind.

- The procedure used by the program to translate the transport-capacity differences between the upstream and the downstream ends of a reach into a deposition height — this is Exner's relationship — is presented in Fig. Ex.6.C.4. The virtual volume of the sediments to be deposited at the **reach** (i), by taking into account the volume increase due to the porosity, p, is the following :

Deposition volume for the **reach** (i) = $\Delta q_{sb}(i) \, \Delta t \, \dfrac{1}{(1-p)}$

The program admits that this volume creates a trapezoidal deposit whose upstream and downstream heights are $\delta z_{am}(i)$ and $\delta z_{av}(i)$, respectively. For the **reach** (i), one can therefore write :

$$\Delta q_{sb}(i) \, \Delta t \, \frac{1}{(1-p)} = \frac{\delta z_{am}(i) + \delta z_{av}(i)}{2} \, \Delta x \tag{6.4a}$$

During data input the user specifies the ratio between the upstream and the downstream heights of this trapezoidal deposit, $\lambda = \delta z_{am}(i)/\delta z_{av}(i)$ (in the program $\lambda \equiv$ HAMHAV). By using this information one obtains :

$$\delta z_{av}(i) = \frac{2}{(1+\lambda)} \left(\frac{\Delta q_{sb}(i) \, \Delta t}{\Delta x \, (1-p)} \right) = \text{CRAV} \left(\frac{\Delta q_{sb}(i) \, \Delta t}{\Delta x \, (1-p)} \right)$$

$$\delta z_{am}(i) = \frac{2\lambda}{(1+\lambda)} \left(\frac{\Delta q_{sb}(i) \, \Delta t}{\Delta x \, (1-p)} \right) = \text{CRAM} \left(\frac{\Delta q_{sb}(i) \, \Delta t}{\Delta x \, (1-p)} \right)$$

For an internal **station** (i), the upstream (i–1) and the downstream (i) **reaches** give two different deposition heights : $\delta z_{am}(i-1)$ and $\delta z_{av}(i)$, respectively. The final value of the deposition height at such a **station** (i) corresponds therefore to the average of these two values :

$$\Delta z(i) = \frac{1}{2} \left[\frac{2\lambda}{(1+\lambda)} \left(\frac{\Delta q_{sb}(i-1) \, \Delta t}{\Delta x \, (1-p)} \right) + \frac{2}{(1+\lambda)} \left(\frac{\Delta q_{sb}(i) \, \Delta t}{\Delta x \, (1-p)} \right) \right]$$

At the end of each time step the program searches for the maximum relative variation of the bed level, $\Delta z(i)/h_i$ (= DELZRM), and compares it with the maximum tolerated value (VARZMX), as specified by the user during data input. If DELZRM > VARZMX, the program displays an error message and stops.

TRANSPORT OF SEDIMENTS

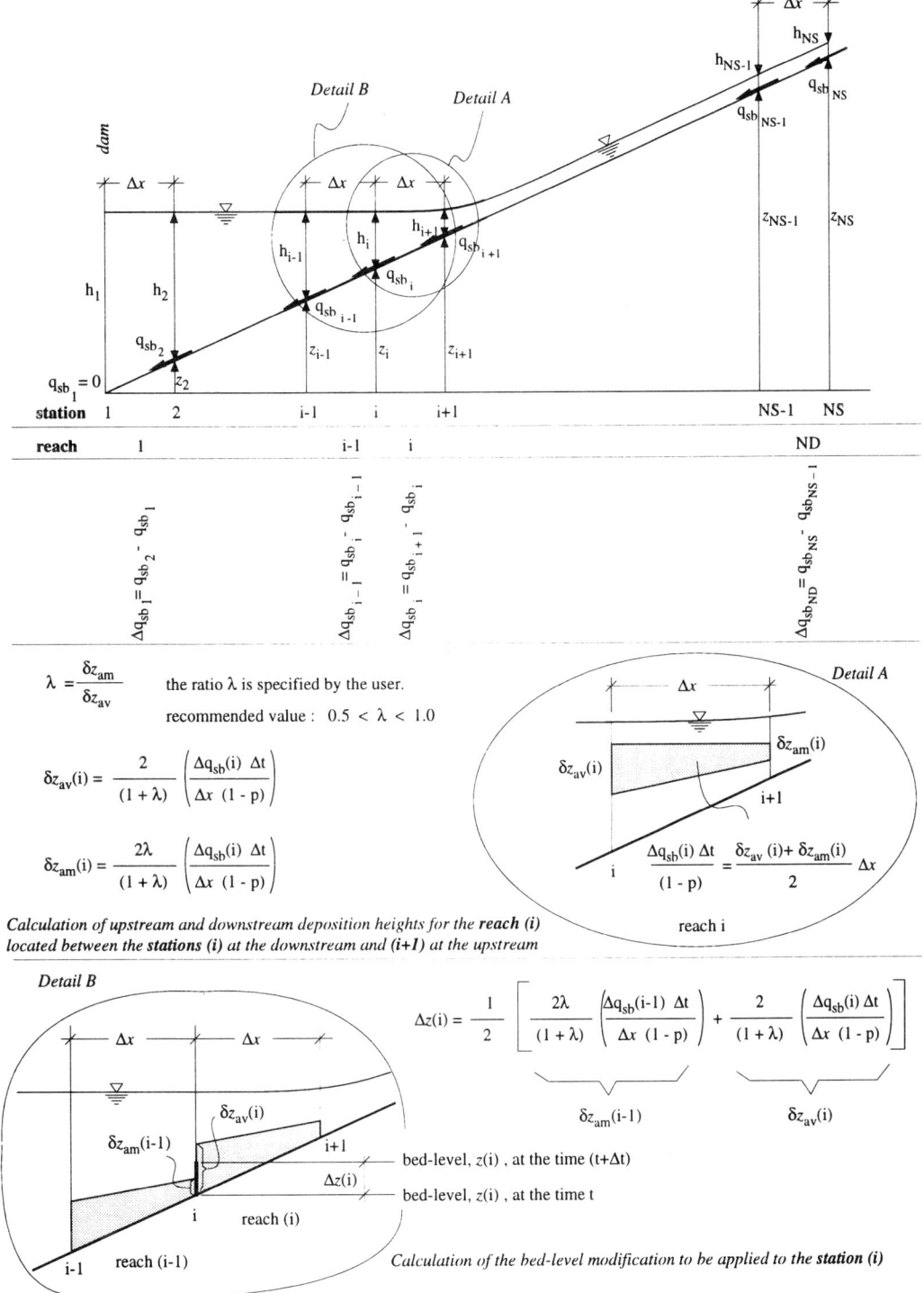

Fig. Ex.6.C.4 Procedure for calculating the volume of the sediments to be deposited at each reach and the modification of the bed level.

- At the end of a time step — if the printing time has come — the main program calls the subprogram DWRITE for printing the results for the current time step on the users specified output file as well as on a second file called GRAPH.DAT.
- If the simulation period specified by the user is not yet reached, the program modifies the bed-level heights at the stations :

$$z_i^{t+\Delta t} = z_i^{t} + \Delta z_i$$

and then goes to the beginning of the calculation loop to start a new time step with a water-surface profile calculation, using this new bed profile.

The program DELTA is a simplified way of solving numerically the equations of Saint-Venant - Exner (see sect. 6.2.1). The equations of Saint-Venant, eq. 6.1 and eq. 6.2, are written for a steady flow in the form of the equation for the free-surface flow, eq. 4.5; the slope of the energy-grade line, eq. 6.3, is expressed here by the Manning-Strickler formula, eq. 6.α; the relationship of Exner, eq. 6.4a, gives the volume of the deposit; the solid discharge, eq. 6.5, is expressed using the formula of *Schoklitsch* (1950), eq. 6.31, and the formula of *Meyer-Peter* et al. (1948), eq. 6.32.

e) *Use of the program DELTA for solving the problem* :

The source code of the program DELTA presented in the Fig. Ex.6.C.11 is first *compiled* and *linked* to obtain an executable code. A FORTRAN compiler is of course necessary to do these operations on your computer. The reader should consult the manuals of his computer to learn the exact procedure to follow.

The user feeds the program data into the computer interactively by answering the questions asked by the program. Numerous comments has been introduced into the conversation to remind the user of certain important points treated in chap. 6 and to guide the user in making his choices. The program also checks some of the likely errors in the data introduction and warns the user. In case of an error, the question is repeated until the user answers correctly.

The dialog between the user and the computer for solving the present problem is presented in Fig. Ex.6.C.5. The values typed in by the user are highlighted by a white background. They are followed by a sign representing the RETURN (CR or ENTER on some computers) key on the keyboard.

The user begins the work by introducing the initial bed slope (SF) and the average grain diameter (D50) of the sediments. Using the average grain diameter the program calculates the Manning-Strickler coefficient due to grain roughness and displays it on the screen, CN50 = 0.0202 (m$^{-1/3}$s). The program then asks the user to enter the total Manning-Strickler coefficient (CN). It is the value of CN that is used later in all calculations (specifying a value of CN = CN50 will mean that the grain roughness is the only cause of the regular head loss). If the user has good reason to think that the head loss is larger then the one given by CN50 — since bed forms (such as dunes) or other irregularities are present — the estimated total value should be entered. For the present problem the value given in the problem will be entered, namely CN = 0.032 (m$^{-1/3}$s).

Supplying the density of the sediments (ROS) and of the water (ROE) as well as the discharge per unit width (QU) is sufficiently clear.

The program proposes now 4 formulae for solid discharge as bed-load transport and asks the user to select one of them. To select the formula of Meyer-Peter et al., one enters "2". Up to this point the dialog with the computer is the same for all cases. Now comes a short conversation with the computer which depends on the selected method. Some methods do not need supplementary data. The program only displays the method used. In case of the selection of the formula of Meyer-Peter et al., if CN > CN50, the program asks if the user wishes to use the roughness parameter, FCOR.

From here on the text of the dialog with the computer is again the same, regardless of the choice of the bed-load transport formula. The conversation continues with the introduction of the data concerning the modification of the bed profile. First, the desired value of the maximum relative variation of the bed level during a single time step, VARZMX = 0.1, must be entered. The program sees to it that at none of the stations, during a single time step, the bed-level modification, DZ, is higher than 10% of the water depth. The user enters also the porosity of the deposited sediments, p (POROS).

The procedure used by the program to transform the volume of the sediments deposited in reaches into upstream and downstream bed-level modification heights was described above in detail. For the ratio of the upstream/downstream heights a value of λ = HAMHAV = 0.75 is entered; this means that the deposition at the downstream end of the reach is higher than the one at the upstream end. It is important to note that this ratio must be between 0.5 and 1.0. In case of high bed-load transport rates, a uniform distribution of the sediments over the reach (HAMHAV = 1.0) may cause instabilities in the calculation of the bed levels at the stations.

The data input continues with the information on the calculation domain. The first station is located on the dam; by entering X1 = 0 [m], the origin of the coordinate system is placed at the dam. The calculations will be carried out up to an upstream distance of XF = 120000 [m].

The choice of the step length in the longitudinal direction, DX, is important. A very small step length necessitates a very small time step; this increases of course the calculation time. For the calculation of the deposition one can choose relatively large time steps. In the present case a step length of DX = 600 [m] is used. A too big step length may cause errors in the calculation of the water-surface profile (especially around the region where the river meets the reservoir) and may cause the program to stop the execution. To overcome this difficulty without loosing the advantages of working with long steps the program uses a clever programming trick.

The water-surface profile is calculated using the 4th-order Runge-Kutta method. The calculations start at the station located at the dam where the water depth is known. By sending this value to the subprogram RK4, the main program obtains in return the water depth at a station immediately upstream. To be able to guarantee a sufficient precision in the water-surface profile calculations, the program checks if the difference of the dynamic heads, $U^2/2g$, between these two successive stations is less than a value specified by the user, in the present case DHDYNM = 0.01 [m], or not. If in a reach this value is not respected, the program divides this reach into $2^1 = 2$ *sub-reaches*, and redoes the calculation in two steps. If the criteria is still not respected the program tries this time $2^2 = 4$ sub-reaches and so on. It is the user who specifies up to which power of 2 the

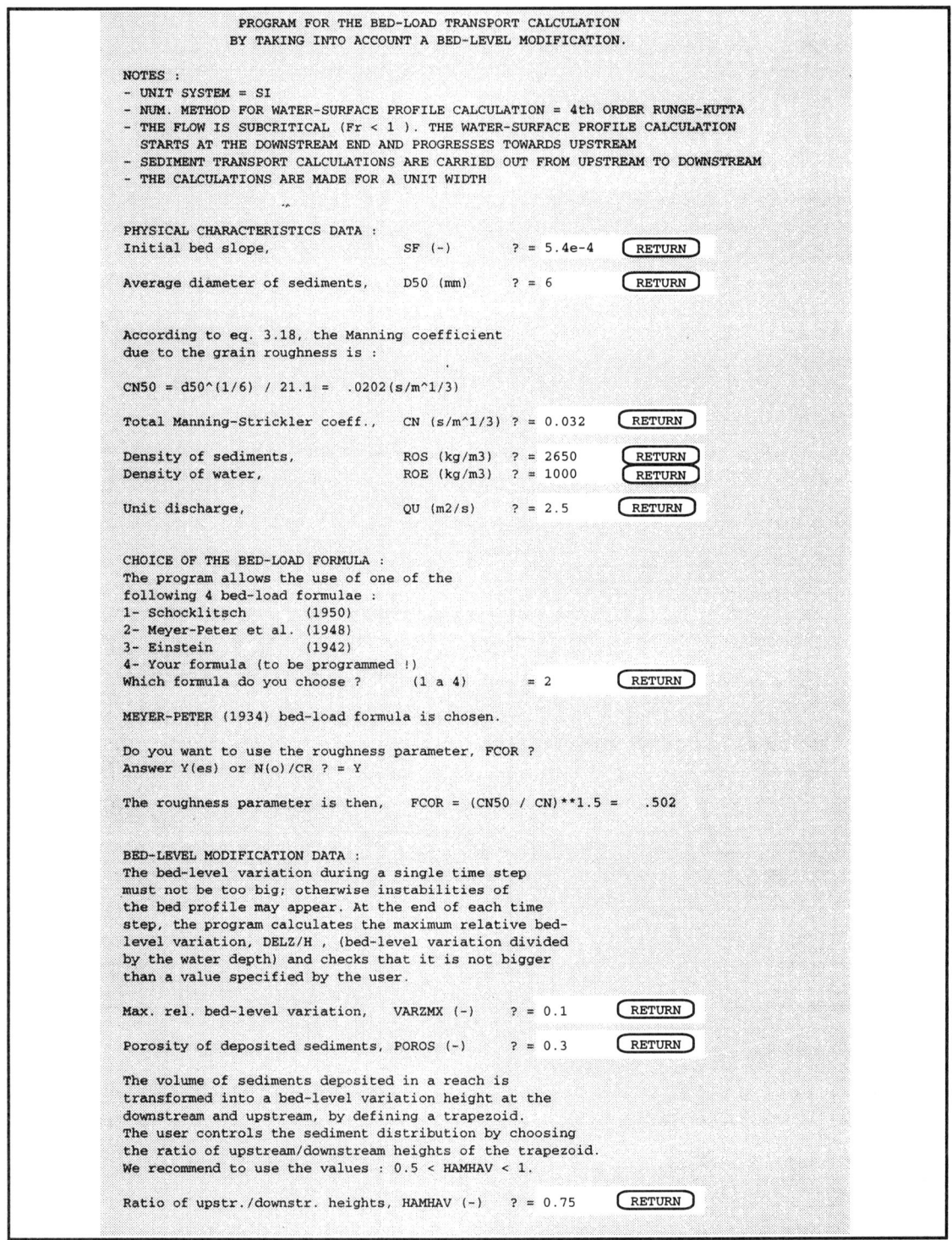

Fig. Ex.6.C.5 Interactive dialog with the computer for the data input (continued).

```
INFORMATION ON COMPUTATIONAL DOMAIN :
x-coordinate of first station,      X1 (m)      ? = 0           [RETURN]
x-coordinate of last station,       XF (m)      ? = 120000      [RETURN]

Total reach length is therefore,    TL (m)        = 120000.00

Now you have to specify the step length in x-direction.
If the step length is too long to guarantee a correct
prediction of the water-surface profile, the program
will automatically add some intermediate stations. The
results, however, are only printed at the stations
specified by the user; others remain invisible. The step
length for the space must therefore be specified to
guarantee a correct representation of physical processes
involved in the bed-load transport. In case of doubt, you
can repeat the simulation with different step lengths and
compare the results.

Step length in x-direction,         DX (m)      ? = 600         [RETURN]

Number of reaches,                  ND            =   200
Number of stations,                 NS            =   201

Max. tolerated var. of dyn. head,   DHDYNM (m)  ? = 0.01        [RETURN]

In case this value is exceeded the reach will be subdivided
in order to refine the calculations.
The number of divisions is specified in powers of 2.
The maximum value is      MCMAX =   7
Which corresponds to  2.0^MCMAX = 128 subdivisions.

Maximum number of subdiv. in powers of 2, NMC   ? = 7           [RETURN]

INFORMATION ON BOUNDARY CONDITIONS :
Note that the sediment transport at downstream end
is automatically taken as zero,     QSU(1)        = 0.0
Water depth at the downstream end,  H1 (m)      ? = 23.5        [RETURN]

PARAMETERS RELATED TO TIME AND PRINTING OF RESULTS :
Time step,                          DELT (days) ? = 10          [RETURN]

Duration of simulation,             TFIN (days) ? = 36500       [RETURN]

Results will be printed every NPP step          ? = 730         [RETURN]

Name of output file (max. 40 char.)             ? = MEYPET.OUT  [RETURN]

TIME STEP     =              1.
TIME (days)   =          .000

TIME STEP     =              2.
TIME (days)   =       10.000
....
....
TIME STEP     =           3651.
TIME (days)   =    36500.000

    2 subdivisions between stations    57 and    58

    4 subdivisions between stations    57 and    58

    8 subdivisions between stations    57 and    58

   16 subdivisions between stations    57 and    58

NORMAL END OF PROGRAM

PRESS ON RETURN KEY TO EXIT                                     [RETURN]
```

Fig. Ex.6.C.5 Interactive dialog with the computer for the data input (end).

program should continue to subdivide the reach. The maximum number of subdivisions provided in the program is $2^7 = 128$, but this can be modified. In the present case the maximum number of subdivisions is chosen as NMC ≤ 7. While running the program, in case of subdivision of a reach, a message is displayed on the screen indicating the number of the reach and the number subdivisions applied (as a power of 2). In this way the user can follow the calculations.

The program needs two boundary conditions to solve the problem. The boundary condition for the calculation of sediment deposition is implicit. The program automatically considers a zero solid discharge at the dam section (station). For the water-surface calculations the user enters the water depth at the dam, H1 = 23.5 [m].

The last group of data concerns the time and the printing of the results. The choice of the time step depends of course on the step length in the longitudinal direction. In the present case a time step of DT = 10 [days] is chosen. The calculations are done for a simulation period of hundred years (TFIN = 36500 [days]). The results are written in the output file MEYPET.OUT every 730 steps; this corresponds to a period of 20 [years].

The program creates two other files in the current directory (the directory in which the program is run) :

DIALOG.DAT : contains the text of the dialog with the computer to enter the data.
GRAPH.DAT : contains the station numbers, the bed level and the water-surface elevation separated by commas. This file can be easily read by any commercial spreadsheet program to draw the longitudinal profile.

Before running the program on a micro-computer, the user should make sure that the current directory does not contain files with these two names; in such a case the program will refuse to start. If you do not want to erase these files you must either rename them or move them to another directory. You may also try to run the program in another directory.

f)   *Results of the calculations with the program DELTA* :

The formation and the advance of the delta in the river-reservoir system were simulated for a period of 100 [years] with the program DELTA, using the methods of *Meyer-Peter* et al. (1948) and of *Schoklitsch* (1950).

The output file MEYPET.OUT containing the results of the simulation of the delta using the method of *Meyer-Peter* et al. (1948) is partially, namely for T = 0, 20 and 100 [years], presented in the Fig. Ex.6.C.6. Due to lack of space the results for T = 40, 60 and 80 [years] are omitted. For the same reason the output file for the simulation using the method of *Schoklitsch* (1950) is not presented either. The results for these two simulations are presented in a graphical form in Figs. Ex.6.C.7 and 8, respectively.

At the beginning of the output file one finds the problem title and a summary of all the data fed in by the user as well as some useful parameters calculated by the program; the results at different time steps follow. Since a time step of 10 [days] was used, and it was decided that the results are to be printed at every 730 steps (see Fig. Ex.6.C.5); the results are printed with a time interval of 20 [years], starting with the initial time, T = 0 [years].

TRANSPORT OF SEDIMENTS

The results for each time step are preceded by a header, indicating the time step-number as well as the time itself in seconds, hours, days and years. The total volumes of the deposited (DQSDET) and eroded (DQSERT) sediments refer to the cumulative sums of the positive and negative values of the variable DELQS, since the beginning of the simulation (T = 0). The absolute (DELZMX) and relative (DELZRM) maximum bed-level variations refer to the bed-level variations (DELZ) for the runnung time step; being the one printed at the top right. The explanations for the different columns of the output file are given below :

STATION NO		:	Station number. There are 201 numbered stations, starting with station 1 (the dam at the downstream end) up to the station 201 (at the upstream end).
X	(m)	:	Distance between the station and the downstream end (station no 1).
ZF	(m)	:	Bed-level elevation with respect to the first station. The bed-level elevations, ZF, at time T = 0 are also stored as initial bed-level elevations in the variable ZFI.
ZF − ZFI	(m)	:	Difference between the running bed-level elevation, ZF (at the time indicated on the top left) and the initial bed-level elevation, ZFI (at time T = 0). It is this column which shows the *height of the deposition* (or the erosion). On Fig. Ex.6.C.6, the region of the delta formation is highlighted by a thin frame.
H	(m)	:	Running actual water depth.
ZF + H	(m)	:	Water-surface elevation with respect to the bed-level elevation at the downstream end (at the dam).
U	(m/s)	:	Average velocity.
Fr	(-)	:	Froude number.
QSU	(m³/s/m)	:	Solid discharge at a station.
DELQS	(m³/s/m)	:	Difference of solid discharge between the extremities of a reach. There are only 200 reaches for NS = 201 stations. This is the reason for having DELQS = 0 on the line 201; however, this value is not used in the calculations. To calculate the deposition height at the upstream end of the computational domain, DELZ(NS), the program assumes implicitly that DELQS(NS) = DELQS(NS−1).
DELZ	(m)	:	Modifications to be applied to the bed, before starting the calculations for the next time step.

The different stages of the formation and the progression of the delta and its influence on the water-surface profile can readily be observed in Figs. Ex.6.C.7 and 8. These figures have been prepared by plotting the information in columns ZF, bed-level elevation, and (ZF+H), water-surface elevation, as a function of the distance given in column X. By studying the contents of the output file, presented in Fig. Ex.6.C.6, and the profiles plotted in Figs. Ex.6.C.7 and 8, one can make several interesting observations.

First the column QSU for the time **T = 0 [years]** in Fig. Ex.6.C.6 will be considered. From the dam up to the station 69 ($x = 40.8$ [km]), there is no solid discharge. The velocity in this reach is equal to or smaller than 1 [m/s]. From the station 71 ($x = 42.0$ [km]) onward the solid discharge increases towards the upstream to attain a constant value of QSU $\cong 0.16 \cdot 10^{-4}$ [m^3/s/m], which is the solid discharge for the river cross-section. The column DELQS shows that the deposition of the sediments occurs between the station 69 ($x = 40.8$ [km]) and station 95 ($x = 56.4$ [km]). Downstream of this reach there is no sediment transport whereas at the upstream constitutes the river reach where the bed-load transport is in equilibrium. The formation of the delta starts therefore at the point where the river enters the reservoir. The column DELZ gives the modifications to be applied to the bed elevation between the station 69 and the station 95, for the next time step.

At time **T = 20 [years]**, as the column ZF–ZFI shows, the downstream end of the delta is at the station 62 ($x = 36.6$ [km]). The maximum height of the delta is $h_d = 1.23$ [m] at the station 64 ($x = 37.8$ [km]). By integrating the data in this column, using the trapezoidal rule, the total volume of deposited sediments is found as being 14396 [m^3/m]. This volume takes into account the porosity. In the header of the results for T = 20 [years], the cumulative volumes of deposition and erosion are given as : DQSDET = 12650.7 [m^3/m] and DQSERT = –2559.5 [m^3/m]. The net volume of the deposited sediments is now calculated as being (DQSDET + DQSERT) = 10091.2 [m^3/m]. By multiplying this volume with the coefficient of swelling CFOI = 1 / (1–p) = 1.4286, one gets the swelled volume of deposited sediment of 14416 [m^3/m]; this value is very close to 14396 [m^3/m] calculated before. In Fig. Ex.6.C.6, the extent of the delta is highlighted by enclosing a portion of the column ZF–ZFI in a rectangle. The bed-level modifications near the upstream end are in fact very small (see Fig. Ex.6.C.7).

```
              PROGRAM FOR THE BED-LOAD TRANSPORT CALCULATION
              BY TAKING INTO ACCOUNT THE BED-LEVEL MODIFICATION.

   NOTES :
   - UNIT SYSTEM = SI
   - NUM. METHOD FOR WATER-SURFACE PROFILE CALCULATION = 4th ORDER RUNGE-KUTTA
   - THE FLOW IS SUBCRITICAL (Fr < 1 ). THE WATER-SURFACE PROFILE CALCULATION
     STARTS AT THE DOWNSTREAM END AND PROGRESSES TOWARDS THE UPSTREAM END
   - SEDIMENT TRANSPORT CALCULATIONS ARE CARRIED OUT FROM UPSTREAM TO DOWNSTREAM
   - CALCULATIONS ARE MADE FOR A UNIT WIDTH

   PHYSICAL CHARACTERISTICS DATA :                    CHOICE OF BED-LOAD FORMULA:
   Initial bed slope,           SF (-)    = .0005400  Bed-load transport formula by Meyer-Peter et al. (1934) is used.

   Average diameter of sediments,  D50 (mm)    =    6.00   Roughness coefficient,  FCOR = (CN50 / CN)^(3/2) =     .502
   Manning coeff. for sed. grains, CN50 (s/m^1/3) =  .0202
   Manning-Strickler coefficient,  CN  (s/m1/3) =    .0320
   Density of sediments,           ROS (kg/m3) = 2650.00
   Density of water,               ROE (kg/m3) = 1000.00
   Unit discharge,                 QU (m2/s)   =    2.50

   INFORMATION RELATED TO WATER-SURFACE CALCULATIONS :  INFORMATION RELATED TO CALCULATION OF DEPOSITION VOLUME :
   Max. tolerated var. of dyn. head, DHDYNM (m) = .010   Max. rel. bed-level variation, VARZMX (-)    =    .10
   Maximum number of subdiv. in powers of 2, NMC = 7     Coefficient of swelling,       CFOI (-)      = 1.4286
                                                         Ratio of upst./dwnst. heights, HAMHAV (-)    =    .750

   INFORMATION ON CALCULATION DOMAIN :                 BOUNDARY CONDITIONS :
   x-coordinate of first station,   X1 (m)  =      .00   Bed-load transport at the downstream end is zero.
   x-coordinate of last station,    XF (m)  = 120000.00  Water depth at downstream end,  H1 (m)    =   23.500
   Total reach length is therefore, TL (m)  = 120000.00
   Step length in x-direction,      DX (m)  =   600.00   PARAMETERS RELATED TO TIME :
   Number of reaches,               ND (m)  =      200   Time step,                 DELT (days)  =     10.000
   Number of stations,              NS      =      201   Duration of simulation,    TFIN (jours) =  36500.000

   PRINTING OF RESULTS :
   Printing of results every NPP step              =    730
   Name of output file (max. 40 char.)             = MEYPET.OUT
```

Fig. Ex.6.C.6 Output file of the program DELTA, resulting from a simulation using the method of *Meyer-Peter* et al. (1948) (continued): file header.

TRANSPORT OF SEDIMENTS

```
============================================================<()>==============================================================
Time step    =      1.              Total volume of deposited sediments,  DQSDET (m3/m) :   0.138047D+02
Time (s)     =  0.0000000D+00       Total volume of eroded sediments,     DQSERT (m3/m) :   0.000000D+00
Time (hours) =       .000
Time (days)  =       .000           Max. absolute bed-level variation,  DELZMX (m) = 0.759922D-02 at the station, IMAXZ =  71
Time (years) =       .000           Max. relative bed-level variation,  DELZRM (-) = 0.343111D-02 at the station, IMAXR =  71
```

STATION NO	X (m)	ZF (m)	ZF - ZFI (m)	H (m)	ZF + H (m)	U (m/s)	Fr (-)	QSU (m3/s/m)	DELQS (m3/s/m)	DELZ (m)
1	.00	0.000000D+00	0.000000D+00	23.500	0.235000D+02	.000	.000	0.000000D+00	0.000000D+00	0.000000D+00
3	1200.00	0.648000D+00	0.000000D+00	22.852	0.235002D+02	.109	.007	0.000000D+00	0.000000D+00	0.000000D+00
5	2400.00	0.129600D+01	0.000000D+00	22.204	0.235004D+02	.113	.008	0.000000D+00	0.000000D+00	0.000000D+00
7	3600.00	0.194400D+01	0.000000D+00	21.557	0.235006D+02	.116	.008	0.000000D+00	0.000000D+00	0.000000D+00
9	4800.00	0.259200D+01	0.000000D+00	20.909	0.235009D+02	.120	.008	0.000000D+00	0.000000D+00	0.000000D+00
11	6000.00	0.324000D+01	0.000000D+00	20.261	0.235011D+02	.123	.009	0.000000D+00	0.000000D+00	0.000000D+00
13	7200.00	0.388800D+01	0.000000D+00	19.613	0.235014D+02	.127	.009	0.000000D+00	0.000000D+00	0.000000D+00
15	8400.00	0.453600D+01	0.000000D+00	18.966	0.235018D+02	.132	.010	0.000000D+00	0.000000D+00	0.000000D+00
17	9600.00	0.518400D+01	0.000000D+00	18.318	0.235022D+02	.136	.010	0.000000D+00	0.000000D+00	0.000000D+00
19	10800.00	0.583200D+01	0.000000D+00	17.671	0.235026D+02	.141	.011	0.000000D+00	0.000000D+00	0.000000D+00
21	12000.00	0.648000D+01	0.000000D+00	17.023	0.235031D+02	.147	.011	0.000000D+00	0.000000D+00	0.000000D+00
23	13200.00	0.712800D+01	0.000000D+00	16.376	0.235036D+02	.153	.012	0.000000D+00	0.000000D+00	0.000000D+00
25	14400.00	0.777600D+01	0.000000D+00	15.728	0.235043D+02	.159	.013	0.000000D+00	0.000000D+00	0.000000D+00
27	15600.00	0.842400D+01	0.000000D+00	15.081	0.235050D+02	.166	.014	0.000000D+00	0.000000D+00	0.000000D+00
29	16800.00	0.907200D+01	0.000000D+00	14.434	0.235059D+02	.173	.015	0.000000D+00	0.000000D+00	0.000000D+00
31	18000.00	0.972000D+01	0.000000D+00	13.787	0.235068D+02	.181	.016	0.000000D+00	0.000000D+00	0.000000D+00
33	19200.00	0.103680D+02	0.000000D+00	13.140	0.235080D+02	.190	.017	0.000000D+00	0.000000D+00	0.000000D+00
35	20400.00	0.110160D+02	0.000000D+00	12.493	0.235094D+02	.200	.018	0.000000D+00	0.000000D+00	0.000000D+00
37	21600.00	0.116640D+02	0.000000D+00	11.847	0.235110D+02	.211	.020	0.000000D+00	0.000000D+00	0.000000D+00
39	22800.00	0.123120D+02	0.000000D+00	11.201	0.235130D+02	.223	.021	0.000000D+00	0.000000D+00	0.000000D+00
41	24000.00	0.129600D+02	0.000000D+00	10.555	0.235153D+02	.237	.023	0.000000D+00	0.000000D+00	0.000000D+00
43	25200.00	0.136080D+02	0.000000D+00	9.910	0.235183D+02	.252	.026	0.000000D+00	0.000000D+00	0.000000D+00
45	26400.00	0.142560D+02	0.000000D+00	9.266	0.235219D+02	.270	.028	0.000000D+00	0.000000D+00	0.000000D+00
47	27600.00	0.149040D+02	0.000000D+00	8.623	0.235265D+02	.290	.032	0.000000D+00	0.000000D+00	0.000000D+00
49	28800.00	0.155520D+02	0.000000D+00	7.980	0.235325D+02	.313	.035	0.000000D+00	0.000000D+00	0.000000D+00
51	30000.00	0.162000D+02	0.000000D+00	7.340	0.235407D+02	.341	.040	0.000000D+00	0.000000D+00	0.000000D+00
53	31200.00	0.168480D+02	0.000000D+00	6.703	0.235507D+02	.373	.046	0.000000D+00	0.000000D+00	0.000000D+00
55	32400.00	0.174960D+02	0.000000D+00	6.069	0.235651D+02	.412	.053	0.000000D+00	0.000000D+00	0.000000D+00
57	33600.00	0.181440D+02	0.000000D+00	5.442	0.235857D+02	.459	.063	0.000000D+00	0.000000D+00	0.000000D+00
59	34800.00	0.187920D+02	0.000000D+00	4.824	0.236160D+02	.518	.075	0.000000D+00	0.000000D+00	0.000000D+00
61	36000.00	0.194400D+02	0.000000D+00	4.223	0.236626D+02	.592	.092	0.000000D+00	0.000000D+00	0.000000D+00
63	37200.00	0.200880D+02	0.000000D+00	3.650	0.237376D+02	.685	.114	0.000000D+00	0.000000D+00	0.000000D+00
65	38400.00	0.207360D+02	0.000000D+00	3.127	0.238630D+02	.799	.144	0.000000D+00	0.000000D+00	0.000000D+00
67	39600.00	0.213840D+02	0.000000D+00	2.691	0.240754D+02	.929	.181	0.000000D+00	0.000000D+00	0.000000D+00
69	40800.00	0.220320D+02	0.000000D+00	2.384	0.244164D+02	1.048	.217	0.000000D+00	0.411567D-07	0.483801D-04
71	42000.00	0.226800D+02	0.000000D+00	2.215	0.248948D+02	1.229	.242	0.338982D-05	0.395311D-05	0.759922D-02
73	43200.00	0.233280D+02	0.000000D+00	2.141	0.254692D+02	1.168	.255	0.104630D-05	0.213865D-05	0.526467D-02
75	44400.00	0.239760D+02	0.000000D+00	2.114	0.260898D+02	1.183	.260	0.139624D-04	0.830167D-06	0.217562D-02
77	45600.00	0.246240D+02	0.000000D+00	2.104	0.267284D+02	1.188	.261	0.152869D-04	0.290155D-06	0.776868D-03
79	46800.00	0.252720D+02	0.000000D+00	2.101	0.273732D+02	1.190	.262	0.157460D-04	0.979595D-07	0.264109D-03
81	48000.00	0.259200D+02	0.000000D+00	2.100	0.280201D+02	1.190	.262	0.159005D-04	0.326765D-07	0.883289D-04
83	49200.00	0.265680D+02	0.000000D+00	2.100	0.286678D+02	1.191	.262	0.159521D-04	0.108409D-07	0.293921D-04
85	50400.00	0.272160D+02	0.000000D+00	2.100	0.293157D+02	1.191	.262	0.159692D-04	0.359736D-08	0.973265D-05
87	51600.00	0.278640D+02	0.000000D+00	2.100	0.299636D+02	1.191	.262	0.159749D-04	0.117617D-08	0.323533D-05
89	52800.00	0.285120D+02	0.000000D+00	2.100	0.306116D+02	1.191	.262	0.159768D-04	0.397835D-09	0.107764D-05
91	54000.00	0.291600D+02	0.000000D+00	2.100	0.312596D+02	1.191	.262	0.159776D-04	0.147028D-09	0.393955D-06
93	55200.00	0.298080D+02	0.000000D+00	2.100	0.319076D+02	1.191	.262	0.159776D-04	0.345948D-10	0.864164D-07
95	56400.00	0.304560D+02	0.000000D+00	2.100	0.325556D+02	1.191	.262	0.159777D-04	0.000000D+00	0.381249D-07
97	57600.00	0.311040D+02	0.000000D+00	2.100	0.332036D+02	1.191	.262	0.159777D-04	0.000000D+00	0.264109D-03
99	58800.00	0.317520D+02	0.000000D+00	2.100	0.338516D+02	1.191	.262	0.159777D-04	0.000000D+00	0.000000D+00
101	60000.00	0.324000D+02	0.000000D+00	2.100	0.344996D+02	1.191	.262	0.159777D-04	0.000000D+00	0.000000D+00
103	61200.00	0.330480D+02	0.000000D+00	2.100	0.351476D+02	1.191	.262	0.159777D-04	0.000000D+00	0.000000D+00
105	62400.00	0.336960D+02	0.000000D+00	2.100	0.357956D+02	1.191	.262	0.159777D-04	0.000000D+00	0.000000D+00
107	63600.00	0.343440D+02	0.000000D+00	2.100	0.364436D+02	1.191	.262	0.159777D-04	0.000000D+00	0.000000D+00
109	64800.00	0.349920D+02	0.000000D+00	2.100	0.370916D+02	1.191	.262	0.159777D-04	0.000000D+00	0.000000D+00
111	66000.00	0.356400D+02	0.000000D+00	2.100	0.377396D+02	1.191	.262	0.159777D-04	0.000000D+00	0.000000D+00
113	67200.00	0.362880D+02	0.000000D+00	2.100	0.383876D+02	1.191	.262	0.159777D-04	0.000000D+00	0.000000D+00
115	68400.00	0.369360D+02	0.000000D+00	2.100	0.390356D+02	1.191	.262	0.159777D-04	0.000000D+00	0.000000D+00
117	69600.00	0.375840D+02	0.000000D+00	2.100	0.396836D+02	1.191	.262	0.159777D-04	0.000000D+00	0.000000D+00
119	70800.00	0.382320D+02	0.000000D+00	2.100	0.403316D+02	1.191	.262	0.159777D-04	0.000000D+00	0.000000D+00
121	72000.00	0.388800D+02	0.000000D+00	2.100	0.409796D+02	1.191	.262	0.159777D-04	0.000000D+00	0.000000D+00
123	73200.00	0.395280D+02	0.000000D+00	2.100	0.416276D+02	1.191	.262	0.159777D-04	0.000000D+00	0.000000D+00
125	74400.00	0.401760D+02	0.000000D+00	2.100	0.422756D+02	1.191	.262	0.159777D-04	0.000000D+00	0.000000D+00
127	75600.00	0.408240D+02	0.000000D+00	2.100	0.429236D+02	1.191	.262	0.159777D-04	0.000000D+00	0.000000D+00
129	76800.00	0.414720D+02	0.000000D+00	2.100	0.435716D+02	1.191	.262	0.159777D-04	0.000000D+00	0.000000D+00
131	78000.00	0.421200D+02	0.000000D+00	2.100	0.442196D+02	1.191	.262	0.159777D-04	0.000000D+00	0.000000D+00
133	79200.00	0.427680D+02	0.000000D+00	2.100	0.448676D+02	1.191	.262	0.159777D-04	0.000000D+00	0.000000D+00
135	80400.00	0.434160D+02	0.000000D+00	2.100	0.455156D+02	1.191	.262	0.159777D-04	0.000000D+00	0.000000D+00
137	81600.00	0.440640D+02	0.000000D+00	2.100	0.461636D+02	1.191	.262	0.159777D-04	0.000000D+00	0.000000D+00
139	82800.00	0.447120D+02	0.000000D+00	2.100	0.468116D+02	1.191	.262	0.159777D-04	0.000000D+00	0.000000D+00
141	84000.00	0.453600D+02	0.000000D+00	2.100	0.474596D+02	1.191	.262	0.159777D-04	0.000000D+00	0.000000D+00
143	85200.00	0.460080D+02	0.000000D+00	2.100	0.481076D+02	1.191	.262	0.159777D-04	0.000000D+00	0.000000D+00
145	86400.00	0.466560D+02	0.000000D+00	2.100	0.487556D+02	1.191	.262	0.159777D-04	0.000000D+00	0.000000D+00
147	87600.00	0.473040D+02	0.000000D+00	2.100	0.494036D+02	1.191	.262	0.159777D-04	0.000000D+00	0.000000D+00
149	88800.00	0.479520D+02	0.000000D+00	2.100	0.500516D+02	1.191	.262	0.159777D-04	0.000000D+00	0.000000D+00
151	90000.00	0.486000D+02	0.000000D+00	2.100	0.506996D+02	1.191	.262	0.159777D-04	0.000000D+00	0.000000D+00
153	91200.00	0.492480D+02	0.000000D+00	2.100	0.513476D+02	1.191	.262	0.159777D-04	0.000000D+00	0.000000D+00
155	92400.00	0.498960D+02	0.000000D+00	2.100	0.519956D+02	1.191	.262	0.159777D-04	0.000000D+00	0.000000D+00
157	93600.00	0.505440D+02	0.000000D+00	2.100	0.526436D+02	1.191	.262	0.159777D-04	0.000000D+00	0.000000D+00
159	94800.00	0.511920D+02	0.000000D+00	2.100	0.532916D+02	1.191	.262	0.159777D-04	0.000000D+00	0.000000D+00
161	96000.00	0.518400D+02	0.000000D+00	2.100	0.539396D+02	1.191	.262	0.159777D-04	0.000000D+00	0.000000D+00
163	97200.00	0.524880D+02	0.000000D+00	2.100	0.545876D+02	1.191	.262	0.159777D-04	0.000000D+00	0.000000D+00
165	98400.00	0.531360D+02	0.000000D+00	2.100	0.552356D+02	1.191	.262	0.159777D-04	0.000000D+00	0.000000D+00
167	99600.00	0.537840D+02	0.000000D+00	2.100	0.558836D+02	1.191	.262	0.159777D-04	0.000000D+00	0.000000D+00
169	100800.00	0.544320D+02	0.000000D+00	2.100	0.565316D+02	1.191	.262	0.159777D-04	0.000000D+00	0.000000D+00
171	102000.00	0.550800D+02	0.000000D+00	2.100	0.571796D+02	1.191	.262	0.159777D-04	0.000000D+00	0.000000D+00
173	103200.00	0.557280D+02	0.000000D+00	2.100	0.578276D+02	1.191	.262	0.159777D-04	0.000000D+00	0.000000D+00
175	104400.00	0.563760D+02	0.000000D+00	2.100	0.584756D+02	1.191	.262	0.159777D-04	0.000000D+00	0.000000D+00
177	105600.00	0.570240D+02	0.000000D+00	2.100	0.591236D+02	1.191	.262	0.159777D-04	0.000000D+00	0.000000D+00
179	106800.00	0.576720D+02	0.000000D+00	2.100	0.597716D+02	1.191	.262	0.159777D-04	0.000000D+00	0.000000D+00
181	108000.00	0.583200D+02	0.000000D+00	2.100	0.604196D+02	1.191	.262	0.159777D-04	0.000000D+00	0.000000D+00
183	109200.00	0.589680D+02	0.000000D+00	2.100	0.610676D+02	1.191	.262	0.159777D-04	0.000000D+00	0.000000D+00
185	110400.00	0.596160D+02	0.000000D+00	2.100	0.617156D+02	1.191	.262	0.159777D-04	0.000000D+00	0.000000D+00
187	111600.00	0.602640D+02	0.000000D+00	2.100	0.623636D+02	1.191	.262	0.159777D-04	0.000000D+00	0.000000D+00
189	112800.00	0.609120D+02	0.000000D+00	2.100	0.630116D+02	1.191	.262	0.159777D-04	0.000000D+00	0.000000D+00
191	114000.00	0.615600D+02	0.000000D+00	2.100	0.636596D+02	1.191	.262	0.159777D-04	0.000000D+00	0.000000D+00
193	115200.00	0.622080D+02	0.000000D+00	2.100	0.643076D+02	1.191	.262	0.159777D-04	0.000000D+00	0.000000D+00
195	116400.00	0.628560D+02	0.000000D+00	2.100	0.649556D+02	1.191	.262	0.159777D-04	0.000000D+00	0.000000D+00
197	117600.00	0.635040D+02	0.000000D+00	2.100	0.656036D+02	1.191	.262	0.159777D-04	0.000000D+00	0.000000D+00
199	118800.00	0.641520D+02	0.000000D+00	2.100	0.662516D+02	1.191	.262	0.159777D-04	0.000000D+00	0.000000D+00
201	120000.00	0.648000D+02	0.000000D+00	2.100	0.668996D+02	1.191	.262	0.159777D-04	0.000000D+00	0.000000D+00

Fig. Ex.6.C.6 Output file of the program DELTA, resulting from a simulation using the method of *Meyer-Peter* et al. (1948) (continued) : **T = 0 [years]**.

```
===============================================<0>==============================================
Time step      =        731.           Total volume of deposited sediments, DQSDET (m3/m) :    0.126507D+05
Time (s)       =  0.6307200D+09        Total volume of eroded sediments,    DQSERT (m3/m) :   -.255946D+04
Time (hours)   =   175200.000
Time (days)    =     7300.000          Max. absolute bed-level variation, DELZMX (m) = 0.532230D-02 at the station, IMAXZ =   63
Time (years)   =       20.000          Max. relative bed-level variation, DELZRM (-) = 0.157857D-02 at the station, IMAXR =   63

STATION      X           ZF           ZF - ZFI         H          ZF + H        U        Fr       QSU          DELQS         DELZ
   NO       (m)         (m)             (m)           (m)          (m)        (m/s)     (-)      (m3/s/m)     (m3/s/m)        (m)
    1         .00   0.000000D+00   0.000000D+00    23.500    0.235000D+02     .000     .000   0.000000D+00  0.000000D+00  0.000000D+00
    3     1200.00   0.648000D+00   0.000000D+00    22.852    0.235002D+02     .109     .007   0.000000D+00  0.000000D+00  0.000000D+00
    5     2400.00   0.129600D+01   0.000000D+00    22.204    0.235004D+02     .113     .008   0.000000D+00  0.000000D+00  0.000000D+00
    7     3600.00   0.194400D+01   0.000000D+00    21.557    0.235006D+02     .116     .008   0.000000D+00  0.000000D+00  0.000000D+00
    9     4800.00   0.259200D+01   0.000000D+00    20.909    0.235009D+02     .120     .008   0.000000D+00  0.000000D+00  0.000000D+00
   11     6000.00   0.324000D+01   0.000000D+00    20.261    0.235011D+02     .123     .009   0.000000D+00  0.000000D+00  0.000000D+00
   13     7200.00   0.388800D+01   0.000000D+00    19.613    0.235014D+02     .127     .009   0.000000D+00  0.000000D+00  0.000000D+00
   15     8400.00   0.453600D+01   0.000000D+00    18.966    0.235018D+02     .132     .010   0.000000D+00  0.000000D+00  0.000000D+00
   17     9600.00   0.518400D+01   0.000000D+00    18.318    0.235022D+02     .136     .010   0.000000D+00  0.000000D+00  0.000000D+00
   19    10800.00   0.583200D+01   0.000000D+00    17.671    0.235026D+02     .141     .011   0.000000D+00  0.000000D+00  0.000000D+00
   21    12000.00   0.648000D+01   0.000000D+00    17.023    0.235031D+02     .147     .011   0.000000D+00  0.000000D+00  0.000000D+00
   23    13200.00   0.712800D+01   0.000000D+00    16.376    0.235036D+02     .153     .012   0.000000D+00  0.000000D+00  0.000000D+00
   25    14400.00   0.777600D+01   0.000000D+00    15.728    0.235043D+02     .159     .013   0.000000D+00  0.000000D+00  0.000000D+00
   27    15600.00   0.842400D+01   0.000000D+00    15.081    0.235050D+02     .166     .014   0.000000D+00  0.000000D+00  0.000000D+00
   29    16800.00   0.907200D+01   0.000000D+00    14.434    0.235059D+02     .173     .015   0.000000D+00  0.000000D+00  0.000000D+00
   31    18000.00   0.972000D+01   0.000000D+00    13.787    0.235068D+02     .181     .016   0.000000D+00  0.000000D+00  0.000000D+00
   33    19200.00   0.103680D+02   0.000000D+00    13.140    0.235080D+02     .190     .017   0.000000D+00  0.000000D+00  0.000000D+00
   35    20400.00   0.110160D+02   0.000000D+00    12.493    0.235094D+02     .200     .018   0.000000D+00  0.000000D+00  0.000000D+00
   37    21600.00   0.116640D+02   0.000000D+00    11.847    0.235110D+02     .211     .020   0.000000D+00  0.000000D+00  0.000000D+00
   39    22800.00   0.123120D+02   0.000000D+00    11.201    0.235130D+02     .223     .021   0.000000D+00  0.000000D+00  0.000000D+00
   41    24000.00   0.129600D+02   0.000000D+00    10.555    0.235153D+02     .237     .023   0.000000D+00  0.000000D+00  0.000000D+00
   43    25200.00   0.136080D+02   0.000000D+00     9.910    0.235183D+02     .252     .026   0.000000D+00  0.000000D+00  0.000000D+00
   45    26400.00   0.142560D+02   0.000000D+00     9.266    0.235219D+02     .270     .028   0.000000D+00  0.000000D+00  0.000000D+00
   47    27600.00   0.149040D+02   0.000000D+00     8.623    0.235265D+02     .290     .032   0.000000D+00  0.000000D+00  0.000000D+00
   49    28800.00   0.155520D+02   0.000000D+00     7.980    0.235325D+02     .313     .035   0.000000D+00  0.000000D+00  0.000000D+00
   51    30000.00   0.162000D+02   0.000000D+00     7.340    0.235402D+02     .341     .040   0.000000D+00  0.000000D+00  0.000000D+00
   53    31200.00   0.168480D+02   0.000000D+00     6.703    0.235507D+02     .373     .046   0.000000D+00  0.000000D+00  0.000000D+00
   55    32400.00   0.174960D+02   0.000000D+00     6.069    0.235651D+02     .412     .053   0.000000D+00  0.000000D+00  0.000000D+00
   57    33600.00   0.181440D+02   0.000000D+00     5.442    0.235857D+02     .459     .063   0.000000D+00  0.000000D+00  0.000000D+00
   59    34800.00   0.187920D+02   0.000000D+00     4.824    0.236160D+02     .518     .075   0.000000D+00  0.000000D+00  0.000000D+00
   61    36000.00   0.194400D+02   0.000000D+00     4.223    0.236626D+02     .592     .092   0.000000D+00  0.000000D+00  0.000000D+00
   63    37200.00   0.203685D+02   0.280500D+00     3.372    0.237401D+02     .741     .129   0.000000D+00  0.452765D-05  0.532230D-02
   65    38400.00   0.218738D+02   0.113785D+01     2.242    0.241159D+02    1.115     .238   0.164146D-05  0.262090D-05  0.536337D-03
   67    39600.00   0.224332D+02   0.104824D+01     2.223    0.246549D+02    1.125     .241   0.283465D-05  0.159323D-05  0.614148D-03
   69    40800.00   0.229922D+02   0.960225D+00     2.209    0.252017D+02    1.132     .243   0.379080D-05  0.999156D-06  0.612838D-03
   71    42000.00   0.235568D+02   0.876814D+00     2.199    0.257550D+02    1.137     .245   0.466231D-05  0.588672D-06  0.579454D-03
   73    43200.00   0.241263D+02   0.798292D+00     2.189    0.263156D+02    1.142     .246   0.545554D-05  0.353738D-06  0.596170D-03
   75    44400.00   0.247007D+02   0.724663D+00     2.182    0.268823D+02    1.146     .248   0.616173D-05  0.270472D-06  0.628673D-03
   77    45600.00   0.252802D+02   0.656174D+00     2.175    0.274550D+02    1.150     .249   0.681102D-05  0.261065D-06  0.640256D-03
   79    46800.00   0.258649D+02   0.592902D+00     2.169    0.280335D+02    1.153     .250   0.743129D-05  0.268734D-06  0.632588D-03
   81    48000.00   0.264547D+02   0.534683D+00     2.163    0.286175D+02    1.156     .251   0.803226D-05  0.272885D-06  0.613688D-03
   83    49200.00   0.270492D+02   0.481242D+00     2.157    0.292066D+02    1.159     .252   0.861361D-05  0.270631D-06  0.590081D-03
   85    50400.00   0.276483D+02   0.432286D+00     2.152    0.298006D+02    1.162     .253   0.917262D-05  0.263546D-06  0.564049D-03
   87    51600.00   0.282515D+02   0.387522D+00     2.148    0.303992D+02    1.164     .254   0.970702D-05  0.253648D-06  0.536845D-03
   89    52800.00   0.288587D+02   0.346676D+00     2.143    0.310019D+02    1.166     .254   0.102153D-04  0.241796D-06  0.508836D-03
   91    54000.00   0.294695D+02   0.309481D+00     2.139    0.316087D+02    1.169     .255   0.106967D-04  0.229128D-06  0.480538D-03
   93    55200.00   0.300837D+02   0.275684D+00     2.135    0.322192D+02    1.171     .256   0.111507D-04  0.215993D-06  0.452203D-03
   95    56400.00   0.307010D+02   0.245043D+00     2.132    0.328331D+02    1.173     .256   0.115775D-04  0.202793D-06  0.424176D-03
   97    57600.00   0.313213D+02   0.217325D+00     2.129    0.334502D+02    1.174     .257   0.119772D-04  0.189604D-06  0.396488D-03
   99    58800.00   0.319443D+02   0.192311D+00     2.126    0.340703D+02    1.176     .257   0.123503D-04  0.176642D-06  0.369450D-03
  101    60000.00   0.325698D+02   0.169791D+00     2.123    0.346931D+02    1.177     .258   0.126975D-04  0.164016D-06  0.343133D-03
  103    61200.00   0.331976D+02   0.149564D+00     2.121    0.353184D+02    1.179     .258   0.130194D-04  0.151758D-06  0.317644D-03
  105    62400.00   0.338274D+02   0.131442D+00     2.119    0.359460D+02    1.180     .259   0.133170D-04  0.139929D-06  0.293080D-03
  107    63600.00   0.344592D+02   0.115247D+00     2.117    0.365758D+02    1.181     .259   0.135912D-04  0.128597D-06  0.269529D-03
  109    64800.00   0.350928D+02   0.100809D+00     2.115    0.372075D+02    1.182     .260   0.138430D-04  0.117814D-06  0.247075D-03
  111    66000.00   0.357280D+02   0.879714D-01     2.113    0.378410D+02    1.183     .260   0.140734D-04  0.107581D-06  0.225786D-03
  113    67200.00   0.363646D+02   0.765857D-01     2.111    0.384761D+02    1.184     .260   0.142837D-04  0.978825D-07  0.205566D-03
  115    68400.00   0.370025D+02   0.665141D-01     2.110    0.391126D+02    1.185     .260   0.144749D-04  0.888144D-07  0.186682D-03
  117    69600.00   0.376416D+02   0.576280D-01     2.109    0.397505D+02    1.185     .261   0.146482D-04  0.803193D-07  0.168877D-03
  119    70800.00   0.382818D+02   0.498086D-01     2.108    0.403896D+02    1.186     .261   0.148047D-04  0.723604D-07  0.152277D-03
  121    72000.00   0.389229D+02   0.429459D-01     2.107    0.410297D+02    1.187     .261   0.149455D-04  0.649903D-07  0.136893D-03
  123    73200.00   0.395649D+02   0.369388D-01     2.106    0.416708D+02    1.187     .261   0.150722D-04  0.582420D-07  0.122703D-03
  125    74400.00   0.402077D+02   0.316948D-01     2.105    0.423128D+02    1.188     .261   0.151855D-04  0.519425D-07  0.109564D-03
  127    75600.00   0.408511D+02   0.271289D-01     2.104    0.429555D+02    1.188     .261   0.152864D-04  0.462304D-07  0.975608D-04
  129    76800.00   0.414952D+02   0.231642D-01     2.104    0.435989D+02    1.188     .262   0.153762D-04  0.409959D-07  0.865511D-04
  131    78000.00   0.421397D+02   0.197307D-01     2.103    0.442429D+02    1.189     .262   0.154557D-04  0.362211D-07  0.765742D-04
  133    79200.00   0.427848D+02   0.167649D-01     2.103    0.448875D+02    1.189     .262   0.155260D-04  0.319380D-07  0.675309D-04
  135    80400.00   0.434302D+02   0.142103D-01     2.102    0.455325D+02    1.189     .262   0.155878D-04  0.280578D-07  0.593819D-04
  137    81600.00   0.440760D+02   0.120155D-01     2.102    0.461779D+02    1.189     .262   0.156422D-04  0.245588D-07  0.520079D-04
  139    82800.00   0.447221D+02   0.101351D-01     2.102    0.468237D+02    1.190     .262   0.156897D-04  0.214616D-07  0.454742D-04
  141    84000.00   0.453685D+02   0.852811D-02     2.101    0.474698D+02    1.190     .262   0.157312D-04  0.186572D-07  0.395796D-04
  143    85200.00   0.460152D+02   0.715849D-02     2.101    0.481162D+02    1.190     .262   0.157672D-04  0.161906D-07  0.343488D-04
  145    86400.00   0.466620D+02   0.599424D-02     2.101    0.487628D+02    1.190     .262   0.157984D-04  0.139949D-07  0.297427D-04
  147    87600.00   0.473090D+02   0.500711D-02     2.101    0.494096D+02    1.190     .262   0.158255D-04  0.120798D-07  0.256906D-04
  149    88800.00   0.479562D+02   0.417269D-02     2.100    0.500567D+02    1.190     .262   0.158488D-04  0.103861D-07  0.221088D-04
  151    90000.00   0.486035D+02   0.346879D-02     2.100    0.507038D+02    1.190     .262   0.158688D-04  0.889736D-08  0.189406D-04
  153    91200.00   0.492509D+02   0.287683D-02     2.100    0.513511D+02    1.190     .262   0.158859D-04  0.763143D-08  0.162457D-04
  155    92400.00   0.498984D+02   0.238013D-02     2.100    0.519985D+02    1.190     .262   0.159006D-04  0.647653D-08  0.138095D-04
  157    93600.00   0.505460D+02   0.196454D-02     2.100    0.526460D+02    1.190     .262   0.159131D-04  0.551939D-08  0.117724D-04
  159    94800.00   0.511936D+02   0.161760D-02     2.100    0.532936D+02    1.190     .262   0.159237D-04  0.468255D-08  0.997497D-05
  161    96000.00   0.518413D+02   0.132877D-02     2.100    0.539412D+02    1.191     .262   0.159327D-04  0.394891D-08  0.845073D-05
  163    97200.00   0.524891D+02   0.108901D-02     2.100    0.545889D+02    1.191     .262   0.159402D-04  0.335320D-08  0.713397D-05
  165    98400.00   0.531369D+02   0.890298D-03     2.100    0.552367D+02    1.191     .262   0.159466D-04  0.280909D-08  0.597667D-05
  167    99600.00   0.537847D+02   0.726173D-03     2.100    0.558845D+02    1.191     .262   0.159520D-04  0.234260D-08  0.500948D-05
  169   100800.00   0.544326D+02   0.591081D-03     2.100    0.565323D+02    1.191     .262   0.159565D-04  0.193650D-08  0.418935D-05
  171   102000.00   0.550805D+02   0.479783D-03     2.100    0.571802D+02    1.191     .262   0.159604D-04  0.164269D-08  0.350114D-05
  173   103200.00   0.557284D+02   0.388683D-03     2.100    0.578281D+02    1.191     .262   0.159634D-04  0.136611D-08  0.293221D-05
  175   104400.00   0.563763D+02   0.314093D-03     2.100    0.584760D+02    1.191     .262   0.159660D-04  0.115001D-08  0.244194D-05
  177   105600.00   0.570243D+02   0.253283D-03     2.100    0.591239D+02    1.191     .262   0.159681D-04  0.951178D-09  0.198719D-05
  179   106800.00   0.576722D+02   0.203941D-03     2.100    0.597719D+02    1.191     .262   0.159699D-04  0.769643D-09  0.166706D-05
  181   108000.00   0.583202D+02   0.163763D-03     2.100    0.604198D+02    1.191     .262   0.159714D-04  0.631278D-09  0.136742D-05
  183   109200.00   0.589681D+02   0.131295D-03     2.100    0.610678D+02    1.191     .262   0.159726D-04  0.527519D-09  0.113685D-05
  185   110400.00   0.596161D+02   0.105340D-03     2.100    0.617157D+02    1.191     .262   0.159736D-04  0.406457D-09  0.904755D-06
  187   111600.00   0.602641D+02   0.844112D-04     2.100    0.623637D+02    1.191     .262   0.159744D-04  0.328630D-09  0.729408D-06
  189   112800.00   0.609121D+02   0.680714D-04     2.100    0.630117D+02    1.191     .262   0.159751D-04  0.294041D-09  0.650631D-06
  191   114000.00   0.615601D+02   0.552977D-04     2.100    0.636597D+02    1.191     .262   0.159756D-04  0.242154D-09  0.498145D-06
  193   115200.00   0.622080D+02   0.454620D-04     2.100    0.643077D+02    1.191     .262   0.159761D-04  0.198914D-09  0.454943D-06
  195   116400.00   0.628560D+02   0.382541D-04     2.100    0.649556D+02    1.191     .262   0.159765D-04  0.216213D-09  0.406656D-06
  197   117600.00   0.635040D+02   0.336183D-04     2.100    0.656036D+02    1.191     .262   0.159769D-04  0.181620D-09  0.365994D-06
  199   118800.00   0.641520D+02   0.310310D-04     2.100    0.662516D+02    1.191     .262   0.159772D-04  0.147027D-09  0.294830D-06
  201   120000.00   0.648000D+02   0.303169D-04     2.100    0.668996D+02    1.191     .262   0.159775D-04  0.000000D+00  0.266873D-06
```

Fig. Ex.6.C.6 Output file of the program DELTA, resulting from a simulation using the method of *Meyer-Peter* et al. (1948) (continued): **T = 20 [years]**.

TRANSPORT OF SEDIMENTS

```
===========================================================<0>=================================================================
Time step      =    3651.              Total volume of deposited sediments, DQSDET (m3/m) :    0.642136D+05
Time (s)       = 0.315360D+10          Total volume of eroded sediments,    DQSERT (m3/m) :   -.138816D+05
Time (hours)   = 876000.000
Time (days)    = 36500.000             Max. absolute bed-level variation, DELZMX (m) = 0.321819D-02 at the station, IMAXZ =  57
Time (years)   = 100.000               Max. relative bed-level variation, DELZRM (-) = 0.599934D-03 at the station, IMAXR =  57


 STATION       X            ZF         ZF - ZFI         H         ZF + H          U        Fr         QSU            DELQS           DELZ
  NO          (m)          (m)            (m)          (m)         (m)          (m/s)      (-)       (m3/s/m)       (m3/s/m)         (m)
    1            .00   0.000000D+00   0.000000D+00    23.500   0.235000D+02     .000      .000    0.000000D+00    0.000000D+00    0.000000D+00
    3        1200.00   0.648000D+00   0.000000D+00    22.852   0.235002D+02     .109      .007    0.000000D+00    0.000000D+00    0.000000D+00
    5        2400.00   0.129600D+01   0.000000D+00    22.204   0.235004D+02     .113      .008    0.000000D+00    0.000000D+00    0.000000D+00
    7        3600.00   0.194400D+01   0.000000D+00    21.557   0.235006D+02     .116      .008    0.000000D+00    0.000000D+00    0.000000D+00
    9        4800.00   0.259200D+01   0.000000D+00    20.909   0.235009D+02     .120      .008    0.000000D+00    0.000000D+00    0.000000D+00
   11        6000.00   0.324000D+01   0.000000D+00    20.261   0.235011D+02     .123      .009    0.000000D+00    0.000000D+00    0.000000D+00
   13        7200.00   0.388800D+01   0.000000D+00    19.613   0.235014D+02     .127      .009    0.000000D+00    0.000000D+00    0.000000D+00
   15        8400.00   0.453600D+01   0.000000D+00    18.966   0.235018D+02     .132      .010    0.000000D+00    0.000000D+00    0.000000D+00
   17        9600.00   0.518400D+01   0.000000D+00    18.318   0.235022D+02     .136      .010    0.000000D+00    0.000000D+00    0.000000D+00
   19       10800.00   0.583200D+01   0.000000D+00    17.671   0.235026D+02     .141      .011    0.000000D+00    0.000000D+00    0.000000D+00
   21       12000.00   0.648000D+01   0.000000D+00    17.023   0.235031D+02     .147      .011    0.000000D+00    0.000000D+00    0.000000D+00
   23       13200.00   0.712800D+01   0.000000D+00    16.376   0.235036D+02     .153      .012    0.000000D+00    0.000000D+00    0.000000D+00
   25       14400.00   0.777600D+01   0.000000D+00    15.728   0.235043D+02     .159      .013    0.000000D+00    0.000000D+00    0.000000D+00
   27       15600.00   0.842400D+01   0.000000D+00    15.081   0.235050D+02     .166      .014    0.000000D+00    0.000000D+00    0.000000D+00
   29       16800.00   0.907200D+01   0.000000D+00    14.434   0.235059D+02     .173      .015    0.000000D+00    0.000000D+00    0.000000D+00
   31       18000.00   0.972000D+01   0.000000D+00    13.787   0.235068D+02     .181      .016    0.000000D+00    0.000000D+00    0.000000D+00
   33       19200.00   0.103680D+02   0.000000D+00    13.140   0.235080D+02     .190      .017    0.000000D+00    0.000000D+00    0.000000D+00
   35       20400.00   0.110160D+02   0.000000D+00    12.493   0.235094D+02     .200      .018    0.000000D+00    0.000000D+00    0.000000D+00
   37       21600.00   0.116640D+02   0.000000D+00    11.847   0.235110D+02     .211      .020    0.000000D+00    0.000000D+00    0.000000D+00
   39       22800.00   0.123120D+02   0.000000D+00    11.201   0.235130D+02     .223      .021    0.000000D+00    0.000000D+00    0.000000D+00
   41       24000.00   0.129600D+02   0.000000D+00    10.555   0.235153D+02     .237      .023    0.000000D+00    0.000000D+00    0.000000D+00
   43       25200.00   0.136080D+02   0.000000D+00     9.910   0.235183D+02     .252      .026    0.000000D+00    0.000000D+00    0.000000D+00
   45       26400.00   0.142560D+02   0.000000D+00     9.266   0.235219D+02     .270      .028    0.000000D+00    0.000000D+00    0.000000D+00
   47       27600.00   0.149040D+02   0.000000D+00     8.623   0.235265D+02     .290      .032    0.000000D+00    0.000000D+00    0.000000D+00
   49       28800.00   0.155520D+02   0.000000D+00     7.980   0.235325D+02     .313      .035    0.000000D+00    0.000000D+00    0.000000D+00
   51       30000.00   0.162000D+02   0.000000D+00     7.340   0.235402D+02     .341      .040    0.000000D+00    0.000000D+00    0.000000D+00
   53       31200.00   0.168480D+02   0.000000D+00     6.703   0.235507D+02     .373      .046    0.000000D+00    0.000000D+00    0.000000D+00
   55       32400.00   0.174960D+02   0.000000D+00     6.069   0.235651D+02     .412      .053    0.000000D+00    0.000000D+00    0.000000D+00
   57       33600.00   0.182214D+02   0.774307D-01     5.364   0.235857D+02     .466      .064    0.000000D+00    0.273770D-05    0.321819D-02
   59       34800.00   0.216124D+02   0.282042D+01     2.254   0.238663D+02    1.109      .236    0.103614D-05    0.119980D-05   -.897741D-04
   61       36000.00   0.221535D+02   0.271346D+01     2.236   0.243891D+02    1.118      .239    0.200523D-05    0.110452D-06   -.735593D-04
   63       37200.00   0.226905D+02   0.260248D+01     2.226   0.249169D+02    1.123      .240    0.258387D-05    0.183810D-06    0.231963D-03
   65       38400.00   0.232284D+02   0.249240D+01     2.221   0.254497D+02    1.125      .241    0.292950D-05   -.670058D-07    0.361552D-03
   67       39600.00   0.237704D+02   0.238638D+01     2.217   0.259871D+02    1.128      .242    0.325280D-05    0.345870D-07    0.384765D-03
   69       40800.00   0.243167D+02   0.228472D+01     2.212   0.265288D+02    1.130      .243    0.359281D-05    0.968953D-07    0.383166D-03
   71       42000.00   0.248671D+02   0.218713D+01     2.207   0.270746D+02    1.133      .243    0.394216D-05    0.130858D-06    0.376400D-03
   73       43200.00   0.254213D+02   0.209331D+01     2.203   0.276244D+02    1.135      .244    0.429269D-05    0.148340D-06    0.368044D-03
   75       44400.00   0.259790D+02   0.200303D+01     2.199   0.281778D+02    1.137      .245    0.463971D-05    0.156522D-06    0.359161D-03
   77       45600.00   0.265401D+02   0.191611D+01     2.195   0.287349D+02    1.139      .245    0.498097D-05    0.159517D-06    0.350385D-03
   79       46800.00   0.271044D+02   0.183239D+01     2.191   0.292953D+02    1.141      .246    0.531550D-05    0.159682D-06    0.342007D-03
   81       48000.00   0.276717D+02   0.175174D+01     2.187   0.298590D+02    1.143      .247    0.564305D-05    0.158305D-06    0.334093D-03
   83       49200.00   0.282420D+02   0.167403D+01     2.184   0.304258D+02    1.145      .247    0.596374D-05    0.156190D-06    0.326756D-03
   85       50400.00   0.288152D+02   0.159916D+01     2.180   0.309955D+02    1.147      .248    0.627783D-05    0.153714D-06    0.319900D-03
   87       51600.00   0.293910D+02   0.152701D+01     2.177   0.315682D+02    1.148      .248    0.658562D-05    0.151101D-06    0.313463D-03
   89       52800.00   0.299696D+02   0.145757D+01     2.174   0.321436D+02    1.150      .249    0.688745D-05    0.148429D-06    0.307363D-03
   91       54000.00   0.305507D+02   0.139067D+01     2.171   0.327217D+02    1.152      .250    0.718351D-05    0.145755D-06    0.301494D-03
   93       55200.00   0.311343D+02   0.132628D+01     2.168   0.333025D+02    1.153      .250    0.747404D-05    0.143120D-06    0.295877D-03
   95       56400.00   0.317203D+02   0.126431D+01     2.165   0.338857D+02    1.155      .250    0.775915D-05    0.140522D-06    0.290368D-03
   97       57600.00   0.323087D+02   0.120470D+01     2.163   0.344715D+02    1.156      .251    0.803896D-05    0.137899D-06    0.284910D-03
   99       58800.00   0.328994D+02   0.114738D+01     2.160   0.350596D+02    1.157      .251    0.831354D-05    0.135295D-06    0.279538D-03
  101       60000.00   0.334923D+02   0.109228D+01     2.158   0.356500D+02    1.159      .252    0.858287D-05    0.132662D-06    0.274118D-03
  103       61200.00   0.340874D+02   0.103935D+01     2.155   0.362426D+02    1.160      .252    0.884693D-05    0.130077D-06    0.268753D-03
  105       62400.00   0.346845D+02   0.988525D+00     2.153   0.368375D+02    1.161      .253    0.910575D-05    0.127386D-06    0.263241D-03
  107       63600.00   0.352837D+02   0.939740D+00     2.151   0.374344D+02    1.162      .253    0.935924D-05    0.124728D-06    0.257800D-03
  109       64800.00   0.358849D+02   0.892940D+00     2.149   0.380334D+02    1.164      .253    0.960736D-05    0.122048D-06    0.252257D-03
  111       66000.00   0.364881D+02   0.848067D+00     2.146   0.386345D+02    1.165      .254    0.985010D-05    0.119331D-06    0.246679D-03
  113       67200.00   0.370931D+02   0.805064D+00     2.144   0.392374D+02    1.166      .254    0.100874D-04    0.116598D-06    0.241061D-03
  115       68400.00   0.376999D+02   0.763876D+00     2.142   0.398423D+02    1.167      .255    0.103192D-04    0.113846D-06    0.235326D-03
  117       69600.00   0.383084D+02   0.724449D+00     2.140   0.404489D+02    1.168      .255    0.105454D-04    0.111037D-06    0.229638D-03
  119       70800.00   0.389187D+02   0.686728D+00     2.139   0.410574D+02    1.169      .255    0.107661D-04    0.108258D-06    0.223923D-03
  121       72000.00   0.395307D+02   0.650661D+00     2.137   0.416675D+02    1.170      .256    0.109812D-04    0.105450D-06    0.218158D-03
  123       73200.00   0.401442D+02   0.616196D+00     2.135   0.422794D+02    1.171      .256    0.111907D-04    0.102655D-06    0.212433D-03
  125       74400.00   0.407593D+02   0.583280D+00     2.134   0.428928D+02    1.172      .256    0.113946D-04    0.998784D-07    0.206661D-03
  127       75600.00   0.413759D+02   0.551866D+00     2.132   0.435078D+02    1.173      .256    0.115930D-04    0.970471D-07    0.200898D-03
  129       76800.00   0.419939D+02   0.521902D+00     2.130   0.441243D+02    1.173      .257    0.117857D-04    0.943131D-07    0.195244D-03
  131       78000.00   0.426133D+02   0.493342D+00     2.129   0.447422D+02    1.174      .257    0.119730D-04    0.915479D-07    0.189545D-03
  133       79200.00   0.432341D+02   0.466137D+00     2.127   0.453616D+02    1.175      .257    0.121547D-04    0.888419D-07    0.183987D-03
  135       80400.00   0.438562D+02   0.440242D+00     2.126   0.459823D+02    1.176      .257    0.123310D-04    0.861432D-07    0.178372D-03
  137       81600.00   0.444796D+02   0.415613D+00     2.125   0.466044D+02    1.177      .258    0.125020D-04    0.835023D-07    0.172933D-03
  139       82800.00   0.451042D+02   0.392204D+00     2.124   0.472277D+02    1.177      .258    0.126677D-04    0.809016D-07    0.167563D-03
  141       84000.00   0.457300D+02   0.369974D+00     2.122   0.478523D+02    1.178      .258    0.128282D-04    0.783171D-07    0.162242D-03
  143       85200.00   0.463569D+02   0.348880D+00     2.121   0.484780D+02    1.179      .258    0.129836D-04    0.757768D-07    0.157032D-03
  145       86400.00   0.469849D+02   0.328883D+00     2.120   0.491049D+02    1.179      .259    0.131339D-04    0.733322D-07    0.151932D-03
  147       87600.00   0.476139D+02   0.309944D+00     2.119   0.497328D+02    1.180      .259    0.132794D-04    0.709132D-07    0.146970D-03
  149       88800.00   0.482440D+02   0.292025D+00     2.118   0.503619D+02    1.180      .259    0.134201D-04    0.685710D-07    0.142092D-03
  151       90000.00   0.488751D+02   0.275088D+00     2.117   0.509919D+02    1.181      .259    0.135561D-04    0.662673D-07    0.137387D-03
  153       91200.00   0.495071D+02   0.259099D+00     2.116   0.516229D+02    1.182      .259    0.136875D-04    0.641275D-07    0.132836D-03
  155       92400.00   0.501400D+02   0.244025D+00     2.115   0.522549D+02    1.182      .260    0.138147D-04    0.619239D-07    0.128357D-03
  157       93600.00   0.507738D+02   0.229832D+00     2.114   0.528878D+02    1.183      .260    0.139375D-04    0.598797D-07    0.124123D-03
  159       94800.00   0.514085D+02   0.216488D+00     2.113   0.535216D+02    1.183      .260    0.140563D-04    0.578756D-07    0.119944D-03
  161       96000.00   0.520440D+02   0.203965D+00     2.112   0.541563D+02    1.184      .260    0.141712D-04    0.559773D-07    0.116018D-03
  163       97200.00   0.526802D+02   0.192232D+00     2.112   0.547917D+02    1.184      .260    0.142822D-04    0.541709D-07    0.112191D-03
  165       98400.00   0.533173D+02   0.181265D+00     2.111   0.554280D+02    1.184      .260    0.143896D-04    0.524163D-07    0.108603D-03
  167       99600.00   0.539550D+02   0.171036D+00     2.110   0.560650D+02    1.185      .260    0.144936D-04    0.507399D-07    0.105072D-03
  169      100800.00   0.545935D+02   0.161520D+00     2.109   0.567023D+02    1.185      .261    0.145943D-04    0.491768D-07    0.101854D-03
  171      102000.00   0.552327D+02   0.152696D+00     2.109   0.573413D+02    1.186      .261    0.146919D-04    0.476611D-07    0.987262D-04
  173      103200.00   0.558725D+02   0.144541D+00     2.108   0.579804D+02    1.186      .261    0.147865D-04    0.463035D-07    0.958561D-04
  175      104400.00   0.565130D+02   0.137036D+00     2.107   0.586203D+02    1.187      .261    0.148785D-04    0.450042D-07    0.931426D-04
  177      105600.00   0.571542D+02   0.130161D+00     2.107   0.592608D+02    1.187      .261    0.149679D-04    0.437895D-07    0.906028D-04
  179      106800.00   0.577959D+02   0.123899D+00     2.106   0.599019D+02    1.187      .261    0.150549D-04    0.426775D-07    0.882884D-04
  181      108000.00   0.584382D+02   0.118236D+00     2.106   0.605436D+02    1.187      .261    0.151397D-04    0.416181D-07    0.861195D-04
  183      109200.00   0.590812D+02   0.113155D+00     2.105   0.611860D+02    1.188      .261    0.152226D-04    0.407738D-07    0.842428D-04
  185      110400.00   0.597246D+02   0.108645D+00     2.104   0.618289D+02    1.188      .261    0.153036D-04    0.399330D-07    0.824853D-04
  187      111600.00   0.603687D+02   0.104693D+00     2.104   0.624724D+02    1.188      .262    0.153831D-04    0.391985D-07    0.809613D-04
  189      112800.00   0.610133D+02   0.101290D+00     2.103   0.631164D+02    1.189      .262    0.154612D-04    0.385883D-07    0.796549D-04
  191      114000.00   0.616584D+02   0.984275D-01     2.103   0.637610D+02    1.189      .262    0.155381D-04    0.380776D-07    0.785149D-04
  193      115200.00   0.623041D+02   0.960977D-01     2.102   0.644062D+02    1.189      .262    0.156140D-04    0.376412D-07    0.775806D-04
  195      116400.00   0.629503D+02   0.942952D-01     2.102   0.650519D+02    1.190      .262    0.156891D-04    0.373573D-07    0.769216D-04
  197      117600.00   0.635970D+02   0.930161D-01     2.101   0.656981D+02    1.190      .262    0.157637D-04    0.370544D-07    0.763675D-04
  199      118800.00   0.642443D+02   0.922567D-01     2.101   0.663448D+02    1.190      .262    0.158377D-04    0.369828D-07    0.760840D-04
  201      120000.00   0.648920D+02   0.920160D-01     2.100   0.669921D+02    1.190      .262    0.159116D-04    0.000000D+00    0.759960D-04
```

Fig. Ex.6.C.6 Output file of the program DELTA, resulting from a simulation using the method of *Meyer-Peter* et al. (1948) (end): **T = 100 [years]**.

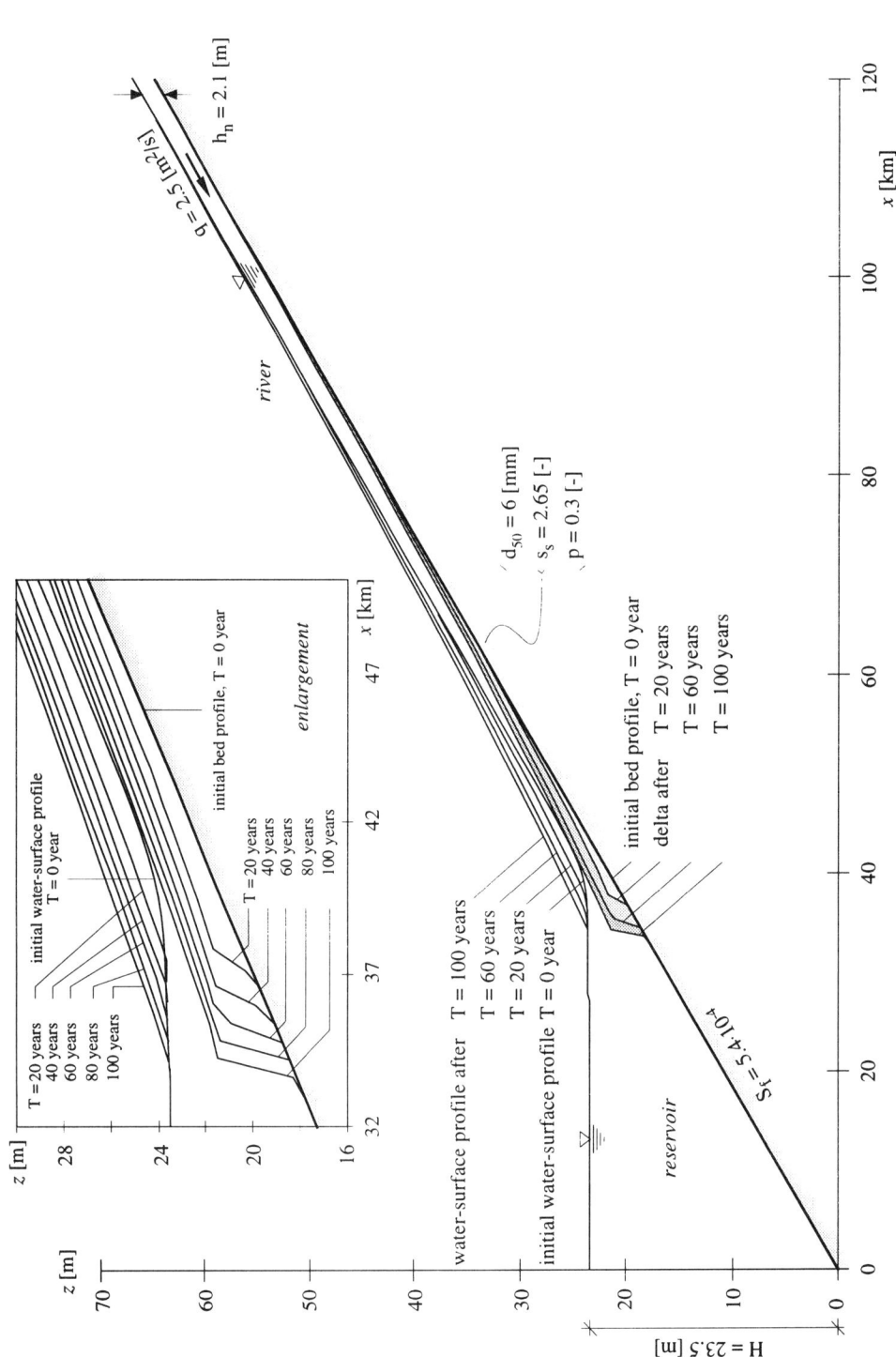

Fig. Ex.6.C.7 Time history of formation and advancement of the delta in a river-reservoir system, simulated for a period of 100 years with the program DELTA by using the bed-load formula of *Meyer-Peter et al.* (1948).

TRANSPORT OF SEDIMENTS

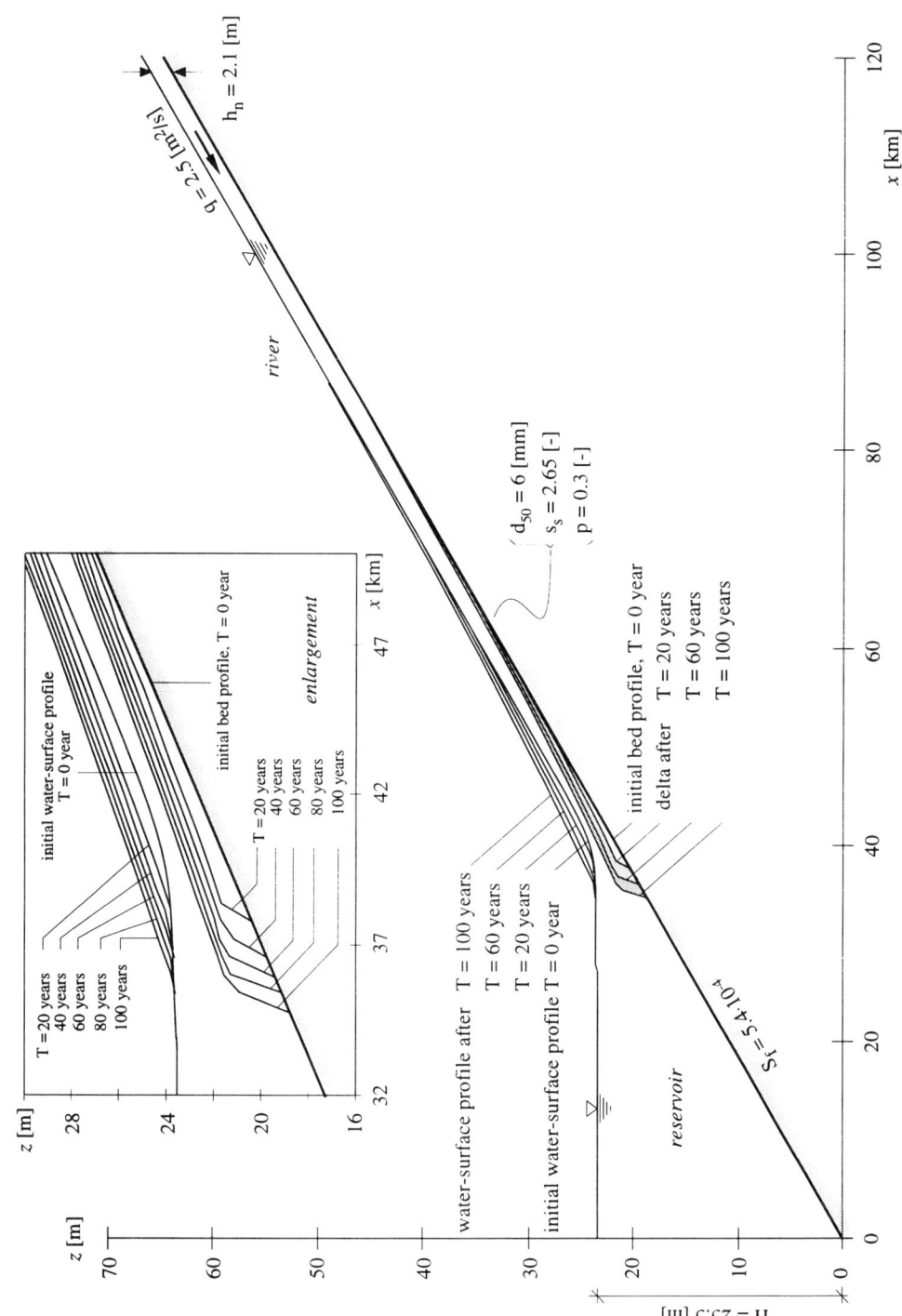

Fig. Ex.6.C.8 Time history of formation and advancement of the delta in a river-reservoir system, simulated for a period of 100 years with the program DELTA by using the bed-load formula of *Schoklitsch* (1950).

At time **T = 100 [years]**, the downstream end of the delta has reached the station 56 ($x = 33.3$ [km]) (see column ZF–ZFI). The maximum height of the delta is 2.91 [m] (station 58, $x = 34.2$ [km]). The total volume of deposited sediments is 71887 [m^3] per unit width.

On Figs. Ex.6.C.7 and 8, it is interesting to note that the influence of the delta can be felt up to a quite long distance upstream.

The table below summarizes the most important characteristics of the deltas, calculated using the two methods.

Results of the calculations, using the bed-load formula of *Meyer-Peter* et al. (1948)						
	Nose of delta			Maximum height of delta		Vol. of
time [years]	station no	x [m]	h_d (m)	station no	x [m]	delta [m^3/m]
20	62	36600	1.23	64	37800	14396
40	60	35400	1.78	62	36600	28789
60	59	34800	2.17	61	36000	43173
80	57	33600	2.56	59	34800	57539
100	56	33000	2.91	58	34200	71887

Results of the calculations, using the bed-load formula of *Schoklitsch* (1950)						
	Nose of delta			Maximum height of delta		Vol. of
time [years]	station no	x [m]	h_d (m)	station no	x [m]	delta [m^3/m]
20	63	37200	0.91	65	38400	7395
40	62	36600	1.31	64	37800	14790
60	61	36000	1.62	63	37200	22185
80	60	35400	1.90	62	36600	29580
100	59	34800	2.14	61	36000	36973

The above table shows that the bed-load formula of *Meyer-Peter* et al. (1948) predicts a delta which is *larger* than the one predicted by the bed-load formula of *Schoklitsch* (1950). After a period of 100 [years], although their height is almost the same, the delta obtained by using the method of *Meyer-Peter* et al. (1948) has a volume 1.94 times the volume of the delta obtained by using the method of *Schoklitsch* (1950). A graphical comparison of the results obtained using these two methods is presented in Fig. Ex.6.C.9. The difference between the two methods comes from the difference between the predicted bed-load transport rates. For the same hydraulic conditions, the formula of *Meyer-Peter* et al. predicts a solid discharge which is larger than the one predicted by the formula of *Schoklitsch* (the same observation is also valid for Ex.6.D). In Fig. Ex.6.C.6., the bed-load formula of *Meyer-Peter* et al. predicts a solid discharge of $q_{sb} = 0.16 \cdot 10^{-4}$ [m^3/s/m] at time T = 0 for the initial river cross-section. The output file for the simulation using the method of *Schoklitsch* is not given here; the user can easily run the program himself and will obtain a solid discharge of $q_{sb} = 0.82 \cdot 10^{-5}$ [m^3/s/m]. The ratio between the two solid discharges is 1.95; this is very close to the ratio between the different values for the volume of the deltas.

TRANSPORT OF SEDIMENTS

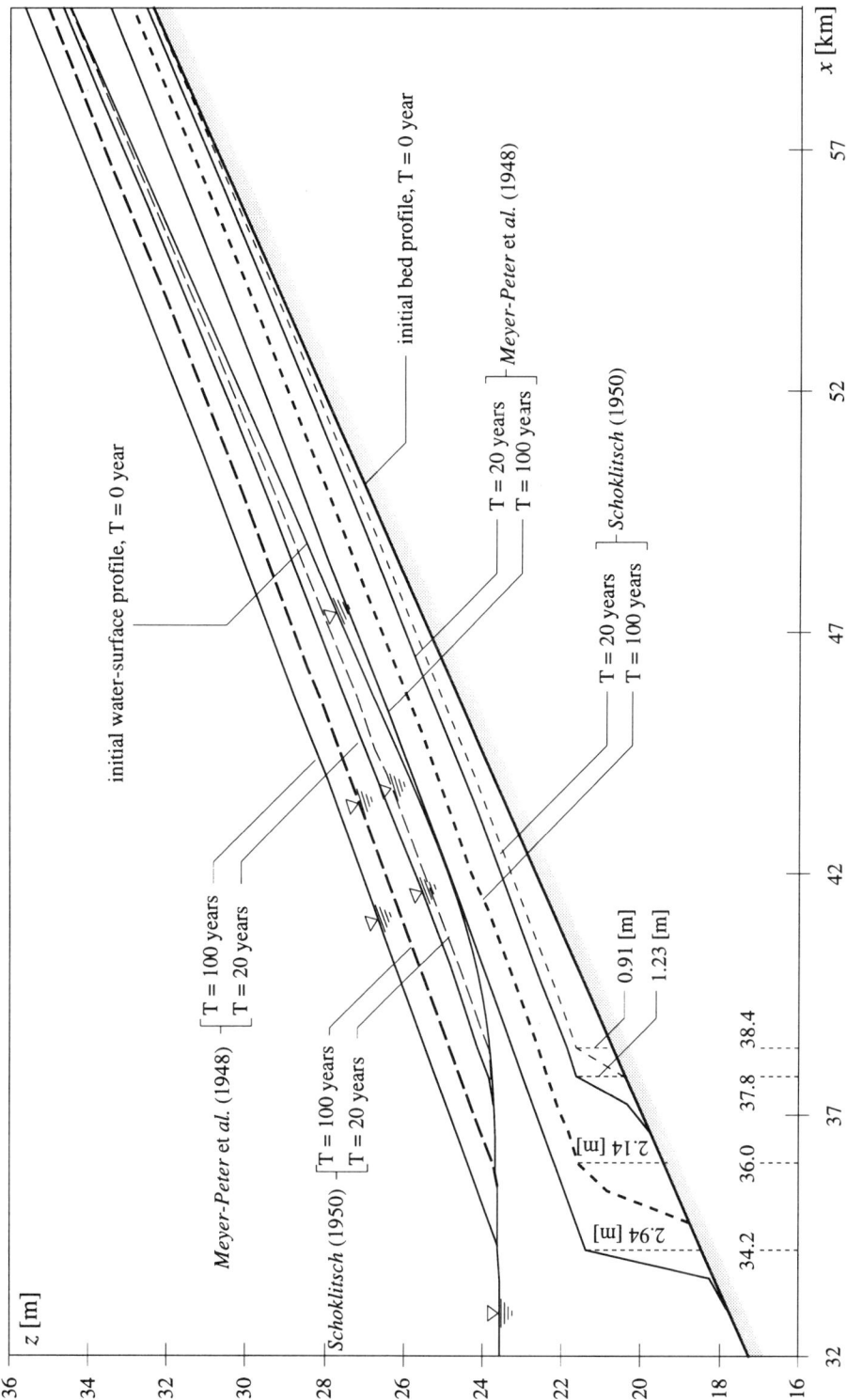

Fig. Ex.6.C.9 Evolution of the formation and advancement of the delta in the river-reservoir system; comparison of results, obtained using the bed-load formula of *Meyer-Peter* et al. (1948) and of *Schoklitsch* (1950).

Differences of this order of magnitude between the results obtained using different bed-load formulae are not rare. In a real study of the sediment transport in a river or in a river-reservoir system, it is necessary to do the calculations using different methods and to compare the results against in-situ measurements, in order to determine the formula which is best suited to the particular case. Sometimes it may even be necessary to calibrate the sediment-transport formula and/or the simulation program to adapt it to the particular problem in hand.

It is left to the user to run the program using the method of *Einstein* (1942) (see point 6.3.3, 4°) and to compare the results with those presented above. It is also important to recall that the option number 4 in choosing the bed-load formula is provided to be programmed by the reader. The user should select a bed-load formula of his choice and program it into this subroutine.

g)   *Remarks* :

Before finishing this exercise, it is in place to call attention of the reader to a few critical points :

- A computer cannot represent a number internally with an absolute precision. Most computers on the market use a set of 32 bits (= 4 bytes), called *word*, for storing a real floating-point number of single precision (R*4). In such a computer a real number is stored according to a standard format as a combination of an integer number (23 bits), called *mantissa*, and an *exponent* of 2 (8 bits). The last remaining bit is reserved for the *sign*. The relative precision that can be obtained with such a storage method is approximately $3 \cdot 10^{-8}$. In many cases this precision may appear to be sufficient. However, the arithmetic operations with real floating-point numbers reserve a few surprises. When a very small number is added on a very large number or when the difference between two numbers being nearly equal is calculated, the round-off errors may become very important. In the present program, very often the difference between two nearly equal numbers are calculated (for example for the calculation of DELQS). We also add very small values contained in the variable DELZ on the variable ZF containing much larger values. To avoid the errors caused by a round off, some of the variables of the program are declared as "DOUBLE PRECISION" (see Fig. Ex.6.C.11). The computer uses then two words (= 8 octets = 64 bits) for representing a real floating-point number (R*8). In this way a relative precision in the order of 10^{-15} is obtained.

- Since the program uses a decoupled algorithm, one cannot talk about a stability condition of the Courant type, for example (see sect. 5.2.3). The choice of the space- and time-step lengths remains nevertheless important from the point of view of above mentioned round-off errors. The step length in the longitudinal direction must be small enough to allow a correct representation of the form of the delta. A too short space step and/or a too small time step may lead to deposition volumes and/or bed-level modification values close to the precision of the internal representation of the floating-point numbers in the computer. The reader is advised to try the program with different space and time step combinations.

- One of the most important variables controlling the delta formation is the HAMHAV-variable, which defines the way the deposited sediments will be distributed over a reach. It is recalled that a value of HAMHAV = 1, which means a uniform distribution of the sediments depositing in the reach, generates instabilities at the upstream and brings the program to a halt either because of an impossibility to calculate the water-surface profile or because of the detection of a deposition height being larger than the one specified by the user as the maximum allowable value. The reader is encouraged to try the program with different values of HAMHAV ≤ 1.

- A sufficiently long river reach, where the flow remains uniform during the whole simulation period, must be provided at the upstream end of the river-reservoir system to insure a correct functioning of the program DELTA. The upstream limit of the delta should not, therefore, touch the upstream end of the computational reach. To illustrate the influence of the computational domain on the results of the delta formation and progression, simulations were done by modeling three different lengths of the river-reservoir system, namely : TL = 60 [km], 80 [km] and 160 [km]. The form of the data for these three simulations are presented in Fig. Ex.6.C.10 (to be also compared with Fig. Ex.6.C.9). The results are summarized in the table below.

Delta after 100 [years] Simulations using the bed-load formula of *Meyer-Peter* et al. (1948) with different computational domain lengths, TL.									
Modeled length	Space step	Nose of delta	Maximum height of delta		Apparent volume of delta after 100 [years] from the downstream end (dam) up to a distance of			
TL	DX	x	h_d	x				
[km]	[m]	[m]	[m]	[m]	60 [km]	80 [km]	120 [km]	160 [km]
60	400	32800	3.30	33600	69617	—	—	—
80	400	33600	2.85	34400	47576	57524	—	—
120	600	33000	2.91	34200	49205	63830	71887	—
160	400	34000	2.87	34400	49131	63789	71302	71977

The solution with TL = 60 [km] predicts a longer and higher delta (see also Fig. Ex.6.C.10). The other three simulations yield similar values. The difference between the simulation with TL = 120 [km] and the simulations with TL = 80 and 160 [km] is probably due to the difference in the space-step length.

Each simulation is done with a different computational domain length. In the first three simulations, the delta touches the most upstream station; it is therefore not simulated in its total length. Thus it is necessary to calculate the volume of the delta at intermediate intervals, as shown in the table above, for a comparison of deposition volumes obtained from different runs. The simulation with TL = 60 [km] overestimates considerably the volume of the delta up to that distance. It is also interesting to note that the last two simulations predict almost identical delta formations, despite the fact that only the solution with TL = 160 [km] has still a river reach with a uniform flow after 100 years of simulation. The solution with TL = 80 [km] gives a good prediction of the maximum height of the delta; the volume of deposition is, however, smaller.

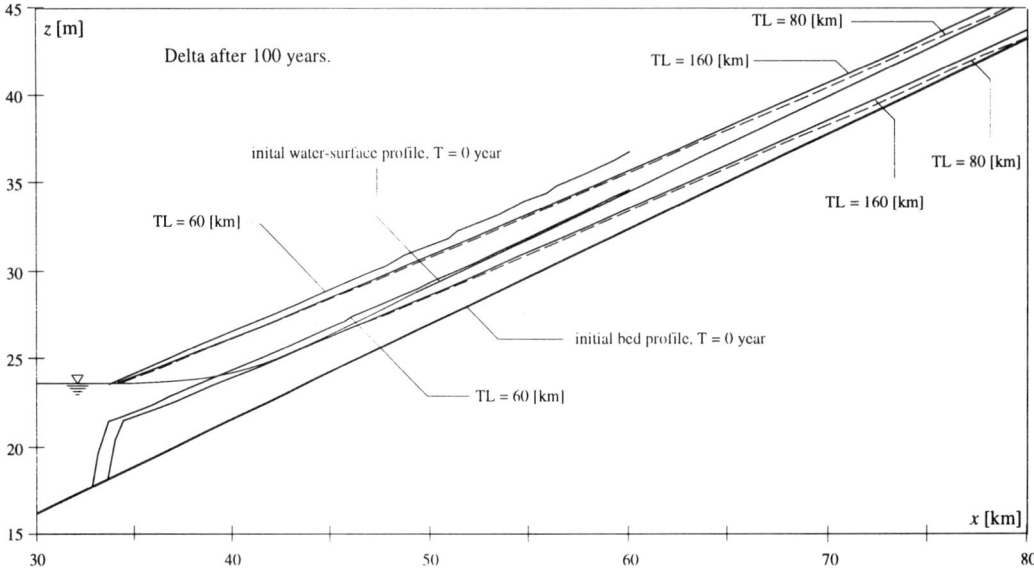

Fig.Ex.6.C.10 Comparison of deltas, simulated with differents lengths of the river-reservoir system by using the bed-load transport formula of *Meyer-Peter et al.*

```
C        PROGRAM DELTA
C        ****************
C
C   MAIN PROGRAM FOR CALCULATING THE BED-LOAD TRANSPORT IN A
C   RIVER-RESERVOIR SYSTEM BY TAKING INTO ACCOUNT THE MODIFICATION
C   OF THE BED LEVEL.
C
C   THE PROGRAM CONSIDERS A SUBCRITICAL FLOW WITH A CONTROL SECTION
C   LOCATED AT THE DOWNSTREAM.
C
C   LIST OF VARIABLES DEFINED GLOBALLY FOR THE MAIN PROGRAM AND THE
C   SUBPROGRAMS. VARIABLES DEFINED LOCALLY IN SUBPROGRAMS ARE LISTED AT
C   THE BEGINNING OF EACH SUBPROGRAM.
C
C   TYPE  NAME          DIMEN.      EXPLANATIONS
C   ====  ====          ======      ============
C   R*4   CFOI                   =  COEFFICIENT OF VOLUME INCREASE DUE TO POROSITY
C   R*4   CN                     =  TOTAL MANNING COEFFICIENT
C   R*4   CN50                   =  MANNING COEFFICIENT DUE TO GRAIN ROUGHNESS
C   R*4   CPTT                   =  TIME-STEP COUNTER
C   R*4   CRAM                   =  WEIGHTING COEFFICIENT FOR UPSTREAM STATION
C                                    USED IN CALCULATION OF DEPOSITION
C   R*4   CRAV                   =  WEIGHTING COEFFICIENT FOR DOWNSTREAM STATION
C                                    USED IN CALCULATION OF DEPOSITION
C   R*4   D50                    =  AVERAGE SEDIMENT DIAMETER
C   R*8   DELQS    (NSMAX-1)     =  SEDIMENTS DEPOSITED OR ERODED IN A REACH
C                                    BOUNDED BY TWO PRINCIPAL SECTIONS
C   R*8   DELT                   =  TIME STEP
C   R*8   DELZ     (NSMAX)       =  ARRAY CONTAINING THE BED LEVEL CHANGES AT ALL
C                                    STATIONS
C   R*8   DELZMX                 =  MAXIMUM BED-LEVEL CHANGE
C   R*8   DELZRM                 =  DIMENSIONLESS MAXIMUM BED-LEVEL CHANGE
C   R*8   DHDYNM                 =  MAXIMUM DIFFERENCE IN DYNAMIC HEAD (U2/2g)
C                                    BETWEEN TWO CONSECUTIVE STATIONS
C   R*8   DQSDET                 =  TOTAL VOLUME DEPOSITED SINCE T = 0
C   R*8   DQSERT                 =  TOTAL VOLUME ERODED SINCE T = 0
C   R*8   DX                     =  DISTANCE BETWEEN TWO PRINCIPAL STATIONS
C   R*8   DXSUB                  =  DISTANCE BETWEEN INTERPOLATED STATIONS
C   R*4   FCOR                   =  ROUGHNESS PARAMETER IN BED-LOAD FORMULA OF
C                                    MEYER-PETER (1948) (K/K'=n'/n)
C*40  FICHS                      =  NAME OF OUTPUT FILE
C   R*4   FRNAM                  =  FROUDE NUMBER AT UPSTREAM STATION OF A REACH
C   R*4   FRNAV                  =  FROUDE NUMBER AT DOWNSTREAM STATION OF A REACH
C   R*4   G                      =  GRAVITATIONAL ACCELERATION
C   R*4   H        (NSMAX)       =  WATER DEPTH AT PRINCIPAL STATIONS
C   R*4   HAMHAV                 =  RATIO OF UPSTREAM/DOWNSTREAM HEIGHTS OF TRAPE-
C                                    ZOID FORMED BY SEDIMENTS DEPOSITED IN A REACH
C   R*4   HDYN1                  =  DYNAMIC HEAD (U^2/2G) AT STATION 1
C   R*4   HDYN2                  =  DYNAMIC HEAD (U^2/2G) AT STATION 2
C   R*4   HSUB    (2**MCMAX+1)   =  WATER DEPTH AT INTERPOLATED STATIONS
C   I*4   I                      =  DO-LOOP COUNTER VARIABLE
C   I*4   IAM                    =  UPSTREAM STATION NUMBER FOR A GIVEN REACH
C   I*4   IAV                    =  DOWNSTREAM STATION NUMBER FOR A GIVEN REACH
C   I*4   II                     =  DO-LOOP COUNTER VARIABLE
C   I*4   IMAXR                  =  NUMBER OF SUBDIVISIONS WHERE WE HAVE "DELZRM"
C   I*4   IMAXZ                  =  NUMBER OF THE STATION WHERE WE HAVE "DELZMX"
C   J*4   J                      =  DO-LOOP COUNTER VARIABLE
C   I*4   MC                     =  NUMBER OF SUBDIVISIONS (IN POWERS OF 2)
C   I*4   MCMAX                  =  MAXIMUM NUMBER OF SUBDIVISIONS (IN POWERS OF 2)
C   I*4   ND                     =  NUMBER OF PRINCIPAL REACHES
C   I*4   NFTS                   =  NUMBER OF BED-LOAD FORMULA TO BE USED
C   I*4   NMC                    =  NUMBER OF SUBDIVISIONS (IN POWERS OF 2)
C                                    SPECIFIED BY USER
C   I*4   NOUT                   =  UNIT NUMBER OF OUTPUT FILE
```

Fig. Ex.6.C.11 Program DELTA.

```
C I*4    NPP               = PRINTING FREQUENCY
C I*4    NPT               = PRINTING TIME COUNTER
C I*4    NS                = NUMBER OF PRINCIPAL STATIONS
C I*4    NSMAX             = NUMBER OF MAXIMUM STATIONS ALLOWED BY THE PROGRAM
C I*4    NSUBD             = NUMBER OF SUBDIVISIONS AT A GIVEN REACH
C R*4    PI                = PI NUMBER
C R*8    QCRIT             = CRITICAL DISCHARGE IN FORMULA OF SCHOKLITSCH
C R*8    QSU     (NSMAX)   = SEDIMENTS TRANSPORTED THROUGH A STATION
C R*4    QU                = UNIT DISCHARGE (WATER)
C C*1    QUIT              = READ FOR TERMINATING THE PROGRAM
C R*4    ROE               = DENSITY OF WATER
C R*4    ROS               = DENSITY OF SEDIMENTS
C R*4    SEFF              = SLOPE OF ENERGY GRADE LINE AT A STATION
C R*4    SF                = INITIAL BED SLOPE
C R*4    SFTR              = LOCAL BED SLOPE AT A GIVEN REACH
C R*4    SS                = SPECIFIC DENSITY OF THE SEDIMENTS
C R*8    T                 = TIME
C R*4    TFIN              = FINAL TIME
C R*4    TJ                = TIME (NUMBER OF DAYS)
C R*4    TL                = TOTAL LENGTH OF RIVER REACH STUDIED
C R*4    VARZMX            = MAXIMUM RELATIVE BED-LEVEL CHANGE ALLOWED BY
C                            THE PROGRAM IN A SINGLE TIME STEP
C R*4    X       (NSMAX)   = X COORDINATES OF PRINCIPAL STATIONS
C R*4    X1                = X COORDINATE OF MOST DOWNSTREAM STATION
C R*4    XF                = X COORDINATE OF MOST UPSTREAM STATION
C R*4    XSUB  (2**MCMAX+1)= X-COORDINATES OF INTERPOLATED STATIONS
C R*8    ZF      (NSMAX)   = BED-LEVEL ELEVATIONS AT ANY TIME
C R*8    ZFI     (NSMAX)   = INITIAL BED-LEVEL ELEVATIONS
C
C**********************************************************
C
C PARAMETERS
      PARAMETER ( NOUT = 10 , NSMAX = 1000 , MCMAX = 7 )
C
C LABELLED COMMON BLOCKS (SHARED VARIABLES)
      COMMON /DONNE1/ SFTR , QU , CN
C
C THE MAIN PROGRAM "DELTA" SENDS THE NAME OF THE SUBROUTINE SUBPROGRAM
C "DERIVE" TO THE SUBROUTINE SUBPROGRAM "RK4" IN THE LIST OF ARGUMENTS.
C ACCORDING TO THE FORTRAN PROGRAMMING RULES, THE SUBROUTINE SUBPROGRAM
C "DERIVE" SHOULD BE DECLARED AS "EXTERNAL".
      EXTERNAL DERIVE
C
C DECLARATION OF VARIABLES
      CHARACTER*1  QUIT
      CHARACTER*40 FICHS
      DOUBLE PRECISION ZF (NSMAX) , ZFI (NSMAX)
      DOUBLE PRECISION QSU(NSMAX) , DELZI(NSMAX) , DELQS(NSMAX-1)
      DOUBLE PRECISION DQSDET , DQSERT , DELZMX , DELZRM
      DOUBLE PRECISION DELT , TFIN , T
      DIMENSION    X(NSMAX) , H(NSMAX)
      DIMENSION    XSUB(2**MCMAX+1) , HSUB(2**MCMAX+1)
C
      OPEN( UNIT = 8 , FILE = 'GRAPH.DAT' , STATUS = 'NEW' )
C
C READ THE PROBLEM DATA
      CALL DREAD( NOUT , NSMAX , MCMAX ,
     1            QU , SF , CN , D50 , ROS , ROE , SS , CN50 ,
     2            X(1) , XF , TL , DX , ND , NS , NMC ,
     3            H(1) , DHDYNM , VARZMX ,
     4            DELT , TFIN ,
     5            NFTS , CFOI , FCOR , HAMHAV , CRAM , CRAV ,
     6            FICHS , NPP )
C
C INITIALISATIONS
      PI     = 3.1415927
      G      = 9.81
      T      = -DELT
      CPTT   = 0.0
      NPT    = -1
      DQSDET = 0.0
      DQSERT = 0.0
C
C CALCULATION OF X COORDINATES AND INITIAL BED LEVELS OF COMPUTATIONAL
C SECTIONS
      X(1)  = 0.0
      ZF(1) = 0.0
      DO 10 I = 2 , NS
      X(I)  = X(I-1) + DX
      ZF(I) = ZF(I-1) + SF * DX
      ZFI(I)= ZF(I)
   10 CONTINUE
C
C WRITE THE TITLES ON OUTPUT FILE
      CALL TITLES (NOUT , SF , D50 , CN50 , CN , ROS , ROE , QU ,
     1             DHDYNM , NMC , VARZMX , CFOI , HAMHAV , FCOR ,
     2             NFTS , X1 , XF , TL , DX , ND , NS , H(1) ,
     3             DELT , TFIN , NPP , FICHS)
C
C CALCULATION LOOP
  100 T    = T + DELT
      CPTT = CPTT + 1.0
      NPT  = NPT + 1
      TJ   = T / 86400
      WRITE(*,110) CPTT , TJ
  110 FORMAT(/' TIME STEP   = ',F15.0/
     1        ' TIME (days) = ',F10.3)
C
C==================================================================
C            BEGINNING OF LIQUID PHASE CALCULATIONS
C==================================================================
C
C BACKWATER CALCULATION
      DO 300 I = 1 , NS-1
      SFTR = (ZF(I+1) - ZF(I)) / DX
      MC   = 0
  200 NSUBD = 2**MC
      DXSUB = (X(I+1) - X(I)) / NSUBD
      HSUB(1) = H(I)
      XSUB(1) = X(I)
C
      DO 250 J = 1 , NSUBD
      XSUB(J+1) = X(I) + DXSUB * NSUBD
      CALL RK4( G , HSUB(J) , DXSUB , HSUB(J+1) , DERIVE )
      HDYN1 = QU**2 / (2*G*HSUB(J)**2)
      HDYN2 = QU**2 / (2*G*HSUB(J+1)**2)
      IF (ABS(HDYN2-HDYN1).GT.DHDYNM)THEN
C
C INTERMEDIATE SECTIONS MUST BE GENERATED BY INTERPOLATION
      MC = MC+1
      WRITE(*,223) 2**MC , I , I+1
  223 FORMAT(/1X,I2,' subdivisions between sections ',I5,' and ',I5)
      IF(MC.GT.NMC)THEN
      FRNAM = QU / (H(I) * SQRT(G * H(I)))
```

Fig.Ex.6.C.11 Program DELTA (continued).

```
               FRNAV = QU / (H(I+1) * SQRT(G * H(I+1)))
               WRITE(*,225)  T , CPTT , I , I+1 , DHDYNM , NMC ,
      1                      FRNAM , FRNAV
               WRITE(NOUT,225) T , CPTT , I , I+1 , DHDYNM , NMC ,
      1                        FRNAM , FRNAV
  225      FORMAT(//
      1 ' ERROR !   Computation of the backwater curve by respecting'/
      2 '           the condition on Del(U^2/2g) is impossible.'/
      3 '           Time (s)    = ',D15.7/
      4 '           Step no     = ',F15.0/
      5 '           Reach between sections           = ',I5,' and',I5/
      6 '           Del(U^2/2g) tolerated (m)        = ',F7.5/
      7 '           Number of calculated subdivisions = ',I2/
      8 '           Either increase the tolerated Del(U^2/2g) value,'/
      9 '           or ask for more subdivisions'/
      A '           Verify also the Froude Number (Fr < 1) :'/
      B '           Fr upstream section   (-) : ',F5.2/
      C '           Fr downstream section (-) : ',F5.2//
      D '  ABNORMAL END ! '//)
               CALL DWRITE( NOUT , NS , T , X , ZF , ZFI , H , QU , G ,
      1                    QSU , DELQS , DQSDET , DQSERT , CPTT ,
      2                    DELZ , DELZMX , IMAXZ , DELZRM , IMAXR )
               WRITE(*,1000)
               READ(*,1001)QUIT
               STOP
           ENDIF
           GO TO 200
  250      CONTINUE
           H(I+1) = HSUB(NSUBD+1)
  300      CONTINUE
C=======================================================================
C=====           END OF LIQUID PHASE CALCULATIONS                  =====
C=======================================================================
C=======================================================================
C=====           BEGINNING OF SOLID PHASE CALCULATIONS             =====
C=======================================================================
C
C COMPUTATION OF SOLID DISCHARGE AT PRINCIPAL SECTIONS STARTING WITH
C MOST UPSTREAM SECTION AND GOING DOWNSTREAM.
           I = NS + 1 - II
           SEFF = (QU*CN)**2/H(I)**(10./3.)
           SFTR = (ZF(I) - ZF(I-1)) / DX
           IF(NFTS.EQ.1)THEN
C USE THE FORMULA OF SCHOKLITSCH (1950)
               CALL SCHOKL( QU , D50 , SS , SEFF , QCRIT , QSU(I) )
           ELSE IF(NFTS.EQ.2)THEN
C USE THE FORMULA OF MEYER-PETER et al. (1948)
               CALL MEYPET( G , SEFF , ROE , ROS , D50 , FCOR , H(I) , QSU(I) )
           ELSE IF(NFTS.EQ.3)THEN
C USE THE FORMULA OF EINSTEIN (1942)
               CALL EINS42( SS , G , D50 , H(I) , SEFF , QSU(I) )
           ELSE IF(NFTS.EQ.3)THEN
C USE THE FORMULA PROGRAMMED BY THE USER
               CALL FORMUL(NOUT)
           ELSE
               WRITE(*,450)NFTS
               WRITE(NOUT,450)NFTS
  450          FORMAT(' ERROR !  TRANSPORT FORMULA NFTS = ',I2,' DOES NOT'/
      1        '          EXIST. VERIFY THE INPUT DATA'//
      2        '  ABNORMAL END'//)
               WRITE(*,1000)
               READ(*,1001)QUIT
               STOP
           ENDIF
  400      CONTINUE
C
C SOLID DISCHARGE AT DOWNSTREAM END IS ZERO
           QSU(1) = 0.0
C
C COMPUTE THE QUANTITY OF SEDIMENT TO BE DEPOSITED OR ERODED IN
C EACH REACH LOCATED BETWEEN TWO MAIN SECTIONS. COMPUTE ALSO THE
C TOTAL VOLUMES OF DEPOSITED AND ERODED SEDIMENTS.
           DO 500 II = 1 , NS-1
               IAV = NS + 1 - II
               IAM = IAV - 1
               DELQS(IAV) = QSU(IAM) - QSU(IAV)
               IF(DELQS(IAV).GE.0.0)THEN
                   DQSDET = DQSDET + DELQS(IAV) * DELT
               ELSE
                   DQSERT = DQSERT + DELQS(IAV) * DELT
               ENDIF
  500      CONTINUE
C
C COMPUTE BED-LEVEL VARIATION AT EACH SECTION
           DELZ(1) = DELQS(1) * DELT * CFOI * CRAV / DX
           DELZMX = DELZ(1)
           DELZRM = DELZ(1) / H(1)
           IMAXZ = 1
           IMAXR = 1
           DO 600 I = 2 , NS-1
               DELZ(I) = (DELQS(I) * CRAV + DELQS(I-1) * CRAM) * DELT / (DX * 2)
               DELZ(I) = DELZ(I) * CFOI
C
C SEARCH FOR MAXIMUM BED-LEVEL VARIATION
               IF(DELZMX.LT.DELZ(I))THEN
                   DELZMX = DELZ(I)
                   IMAXZ = I
               ENDIF
C
C SEARCH FOR RELATIVE MAXIMUM BED-LEVEL VARIATION
               IF(DELZRM.LT.DELZ(I)/H(I))THEN
                   DELZRM = DELZ(I) / H(I)
                   IMAXR = I
               ENDIF
  600      CONTINUE
           DELZ(NS) = DELQS(NS-1) * DELT * CFOI / DX
           IF(DELZMX.LT.DELZ(NS))THEN
               DELZMX = DELZ(NS)
               IMAXZ = NS
           ENDIF
           IF(DELZRM.LT.DELZ(NS)/H(NS))THEN
               DELZRM = DELZ(NS) / H(NS)
               IMAXR = NS
           ENDIF
C
C WRITE THE COMPUTATIONAL RESULTS FOR TIME T
           IF(CPTT.EQ.1.0.OR.NPT.EQ.NPP.OR.T.GE.TFIN)THEN
               CALL DWRITE( NOUT , NS , T , X , ZF , ZFI , H , QU , G ,
```

Fig.Ex.6.C.11 Program DELTA (continued).

```fortran
C LIST OF VARIABLES LOCALLY DEFINED FOR SUBPROGRAM SUBROUTINE "DREAD"
C
C TYPE NAME    DIMEN.    EXPLANATIONS
C ==== ====    ======    ============
C C*1  ANSWER            = ANSWER (Yes or No) TO A QUESTION
C R*4  H1                = WATER DEPTH AT THE MOST DOWNSTREAM STATION
C                          (CONTROL SECTION)
C R*4  POROS             = POROSITY OF DEPOSITED SEDIMENTS
C
C SEE THE BEGINNING OF THE MAIN PROGRAM FOR A LIST OF SIMPLE VARIABLES,
C VECTORS AND MATRICES DEFINED GLOBALLY.
C**********************************************************************
C DECLARATION OF VARIABLES
       CHARACTER*1    ANSWER
       CHARACTER*40   FICHS
       DOUBLE PRECISION DELT , TPIN
C
       OPEN( UNIT = 7 , FILE = 'DIALOG.DAT' , STATUS = 'NEW' )
C
C WRITE THE PROGRAM TITLE ON THE SCREEN
       WRITE(*, 10)
       WRITE(7, 10)
10     FORMAT(//, '          PROGRAM FOR BED-LOAD SEDIMENT TRANSPORT CALC
      1ULATION',/,
      2       '                 BY TAKING INTO ACCOUNT BED-LEVEL MODIFICAT
      3IONS.',//, '  NOTES :',//,'- UNIT SYSTEM = SI',/,
      4       '       - NUM. METHOD FOR WATER SURFACE PROFILE CALCULATION = 4
      5                             6th ORDER RUNGE-KUTTA',/,
      6       '       - THE FLOW BEING SUBCRITICAL (Fr < 1 ), THE WATER-SURFA
      7CE PROFILE CALCULATION',/,
      8       '         STARTS AT THE DOWNSTREAM END AND PROGRESSES TOWARDS U
      9PSTREAM',/,
      B       '       - SEDIMENT TRANSPORT CALCULATIONS ARE CARRIED OUT FROM
      C        UPSTREAM TO DOWNSTREAM',/,
      D       '       - THE CALCULATIONS ARE MADE FOR A UNIT WIDTH',///)
C
C READ THE PHYSICAL CHARACTERISTICS DATA
       WRITE(*,90)
       WRITE(7,90)
90     FORMAT(' PHYSICAL CHARACTERISTICS DATA :')
       WRITE(*,110)
110    FORMAT(' Initial bed slope,               SF (-)            ? = ',$)
       READ(*,*,ERR=100)SF
       WRITE(7,110)
       WRITE(7,*)SF
C
       WRITE(*,130)
120    FORMAT(/' Average diameter of sediments,   D50 (mm)          ? = ',$)
130    READ(*,*,ERR=120)D50
       WRITE(7,130)
       WRITE(7,*)D50
       D50 = D50 * 0.001
C
       CN50 = D50**(1./6.) / 21.1
140    WRITE(*,150)CN50
150    FORMAT(//' According to eq. 3.18, the Manning coefficient'/
      1        '  due to the grain roughness is :'//
      2        '  CN50 = d50^(1/6) / 21.1 = ',F6.4,' (s/m^1/3)'///
      3        '  Total Manning-Strickler coeff.,  CN (s/m^1/3)    ? = ',$)

      1                QSU  , DELZ  , DELZMX , IMAXZ , DELZRM , IMAXR
      2                DELZ , DELZMX , IMAXZ , DELZRM , IMAXR )
       IF(NPT.EQ.NPP)NPT = 0
       ENDIF
C
C VERIFY THE END OF COMPUTATIONS
       IF(T.GE.TPIN)GO TO 999
C
C VERIFY IF THE BED-LEVEL MODIFICATIONS ARE WITHIN THE LIMITS FIXED
C BY THE USER
       IF(DELZRM.GT.VARZMX)THEN
          WRITE(*,650)T , DELZRM , DELZMX , IMAXR
          WRITE(NOUT,650) T , DELZRM , DELZMX , IMAXR
650       FORMAT(' ERROR ! Bed-level variation at a single step is too'/
      1          '                  large. Use a smaller time step',/
      2          '                   T      (s) = ',D15.7/
      3          '                   DELZRM (m) = ',D12.6/
      4          '                   DELZRM (-) = ',F10.6/
      5          '                   IMAX       = ',I6//
      6          '                             ABNORMAL END !',//)
          CALL DWRITE( NOUT , NS , T , X , ZF , ZFI , H , QU , G ,
      1                QSU , DELQS , DQSDET , CPTT ,
      2                DELZ , DELZMX , IMAXZ , DELZRM , IMAXR )
          WRITE(*,1000)
          READ(*,1001)QUIT
          STOP
       ENDIF
C
C MODIFY THE BED-LEVEL ELEVATIONS AT ALL STATIONS
       DO 700 I = 1,NS
          ZF(I) = ZF(I) + DELZ(I)
700    CONTINUE
C=====================================================================
C                  END OF SOLID PHASE CALCULATIONS
C=====================================================================
C
C GO TO COMPUTE A NEW TIME STEP
       GO TO 100
C
C END OF COMPUTATIONS
999    WRITE(*,1000)
1000   FORMAT(/' NORMAL END OF PROGRAM.'/
      1        '  PRESS RETURN TO END THE PROGRAM.'/)
       READ(*,1001)QUIT
1001   FORMAT(A)
       STOP
       END

       SUBROUTINE DREAD( NOUT , NSMAX , MCMAX ,
      1                  QU , SF , CN , D50 , ROS , ROE , SS , CN50 ,
      2                  X1 , XF , TL , DX , ND , NS , NMC ,
      3                  H1 , DHDYNM , VARZMX ,
      4                  DELT , TPIN ,
      5                  NFTS , CFOI , PCOR , HAMHAV , CRAM , CRAV ,
      6                  FICHS , NPP )
C*********************************************************************
C SUBROUTINE SUBPROGRAM FOR INTERACTIVE READING OF THE PROBLEM DATA
C*********************************************************************
```

Fig.Ex.6.C.11 Program DELTA (continued).

```
            READ(*,*,ERR=140)CN
            IF(CN.LT.CN50)THEN
              WRITE(*,155)
              WRITE(7,155)
  155         FORMAT(/' ERROR ! CN should be greater than CN50'/)
              GO TO 140
            ENDIF
            WRITE(7,150)
            WRITE(7,*)CN50
  C
  160       WRITE(*,170)
  170       FORMAT(/' Density of sediments,          ROS (kg/m3) ? = ',$)
            READ(*,*,ERR=160)ROS
            WRITE(7,170)
            WRITE(7,*)ROS
  C
  180       WRITE(*,190)
  190       FORMAT(/' Density of water,              ROE (kg/m3) ? = ',$)
            READ(*,*,ERR=180)ROE
            WRITE(7,190)
            WRITE(7,*)ROE
  C
  C  COMPUTE THE SPECIFIC DENSITY OF SEDIMENTS
            SS = ROS / ROE
  C
  200       WRITE(*,210)
  210       FORMAT(/' Unit discharge,                QU (m2/s)   ? = ',$)
            READ(*,*,ERR=200)QU
            WRITE(7,210)
            WRITE(7,*)QU
  C
  C  SELECT THE BED-LOAD TRANSPORT FORMULA TO BE USED
  300       WRITE(*,310)
  310       FORMAT(///' CHOICE OF BED-LOAD TRANSPORT FORMULA :'/
       1          '   The program allows the use of one of the'/
       2          '   following 4 bed-load formulae :'/
       3          '     1- Schoklitsch      (1950)'/
       4          '     2- Meyer-Peter et al. (1948)'/
       5          '     3- Einstein          (1942)'/
       6          '     4- Your formula (to be programmed )'/
       7          '   Which formula do you choose ?      (1 a 4) = ',$)
            READ(*,*,ERR=300)NFTS
            WRITE(7,310)
            WRITE(7,*)NFTS
            IF(NFTS.EQ.1)THEN
  C  USE THE BED-LOAD TRANSPORT FORMULA OF SCHOKLITSCH (1950)
  320         WRITE(*,330)
              WRITE(7,330)
  330         FORMAT(/' SCHOKLITSCH (1950) bed-load formula is chosen.'/)
            ELSE IF(NFTS.EQ.2)THEN
  C  USE THE BED-LOAD TRANSPORT FORMULA OF MEYER-PETER (1948)
  340         WRITE(*,340)
              WRITE(7,340)
  340         FORMAT(/' MEYER-PETER (1948) bed-load formula is chosen.')
              IF(CN.GT.CN50)THEN
                WRITE(*,342)
  342           FORMAT(/' Do you want to use the roughness parameter, PCOR ?'/
       3              '     Answer Y(es) or N(o)/CR ? = ',$)
                READ(*,345)ANSWER
  345           FORMAT(A)
                WRITE(7,342)
                WRITE(7,345)ANSWER
                IF(ANSWER.EQ.'Y'.OR.ANSWER.EQ.'y')THEN
                  PCOR = (CN50 / CN)**1.5
                ELSE
                  PCOR = 1.0
                ENDIF
              ELSE
                PCOR = 1.0
              ENDIF
              WRITE(7,346) PCOR
  346         FORMAT(/' The roughness parameter is then,    PCOR = (CN50 / CN)*
       1          *1.5 = ',F6.3/)
              WRITE(7,346) PCOR
            ELSE IF(NFTS.EQ.3)THEN
  C  USE THE BED-LOAD TRANSPORT FORMULA OF EINSTEIN (1942)
              WRITE(*,350)
              WRITE(7,350)
  350         FORMAT(/' EINSTEIN (1942) bed-load formula is chosen.'/)
            ELSE IF(NFTS.EQ.4)THEN
  C  USE THE BED-LOAD TRANSPORT FORMULA PROGRAMMED BY THE USER
              WRITE(*,360)
              WRITE(7,360)
  360         FORMAT(/' Bed-load formula programmed by the user is chosen.'/)
            ELSE
              WRITE(*,380)
              WRITE(7,380)
  380         FORMAT(' ERROR !.... NON EXISTING CHOICE !'//)
              GO TO 300
            ENDIF
  C
  C
  385       WRITE(*,390)
  390       FORMAT(/' BED-LEVEL MODIFICATION DATA :'/
       1          '   The bed-level variation during a single time step'/
       2          '   may not be too big; otherwise, instabilities of'/
       3          '   the bed profile may appear. At the end of each time'/
       4          '   step, the program calculates the maximum relative bed-'/
       5          '   level variation, DELZ/H , (bed-level variation divided'/
       6          '   by the water depth) and checks that it is not bigger'/
       7          '   than a value specified by the user.'//
       8          '   Max. rel. bed-level variation,       VARZMX (-)   ? = ',$)
            READ(*,*,ERR=385)VARZMX
            WRITE(7,390)
            WRITE(7,*)VARZMX
  C
  391       WRITE(*,392)
  392       FORMAT(/' Porosity of deposited sediments,     POROS  (-)   ? = ',$)
            READ(*,*,ERR=391)POROS
            WRITE(7,392)
            WRITE(7,*)POROS
            CPOI = 1.0 / (1.0 - POROS)
  C
  395       WRITE(*,396)
  396       FORMAT(/' The volume of sediments deposited in a reach is'/
       1          '   transformed into a bed-level variation height at the'/
       2          '   downstream and upstream, by defining a trapezoid.'/
       3          '   The user controls the sediment distribution by choosing'/
       4          '   the ratio of upstream/downstream heights of the trapezoid'/
       5          '   We recommend to use the values : 0.5 < HAMHAV < 1.'//
       6          '   Ratio of upstr./downstr. heights, HAMHAV (-)   ? = ',$)
            READ(*,*,ERR=395)HAMHAV
            WRITE(7,396)
```

Fig.Ex.6.C.11 Program DELTA (continued).

```fortran
      WRITE(7,*)HAMHAV
C
C COMPUTATION OF THE COEFFICIENTS FOR CALCULATING CONTRIBUTIONS OF THE
C UPSTREAM AND DOWNSTREAM DEPOSITS
      CRAM = 2.0 * HAMHAV / (1 + HAMHAV)
      CRAV = 2.0 / (1 + HAMHAV)
C
C READ THE INFORMATION ON COMPUTATIONAL DOMAIN
      WRITE(*,400)
      WRITE(7,400)
  400 FORMAT(//' INFORMATION ON COMPUTATIONAL DOMAIN :')
C
      WRITE(*,410)
  410 FORMAT(' x-coordinate of first station,       X1 (m)      ? = ',$)
      READ(*,*,ERR=405)X1
      WRITE(7,410)
      WRITE(7,*)X1
C
      WRITE(*,430)
  430 FORMAT(' x-coordinate of last station,        XF (m)      ? = ',$)
      READ(*,*,ERR=420)XF
      WRITE(7,430)
      WRITE(7,*)XF
      TL = XF - X1
      WRITE(*,450)TL
  450 FORMAT(/' Total reach length is therefore,    TL (m)         = ',
     1F10.2//
     2        ' Now you have to specify the step length in x-direction.'/
     3        ' If the step length is too long to guarantee a correct'/
     4        ' prediction of the water-surface profile, the program'/
     5        ' will automatically add some intermediate stations. The'/
     6        ' results, however, are only printed at the stations'/
     7        ' specified by the user; others remain invisible. The step'/
     8        ' length for the space must therefore be specified to'/
     9        ' guarantee a correct representation of physical processes'/
     A        ' involved in the bed-load transport. In case of doubt, you'/
     B        ' can repeat the simulation with different step lengths and'/
     C        ' compare the results.'/
     D        ' Step length in x-direction,          DX (m)      ? = ',$)
      READ(*,*,ERR=440)DX
      WRITE(7,450)TL
      WRITE(7,*)DX
C
      ND = (XF -X1) / DX
      NS = ND + 1
C
      IF(NS.GT.NSMAX)THEN
         WRITE(*,460) NS , NSMAX
         WRITE(7,460) NS , NSMAX
  460    FORMAT(//' ERROR ! The maximum number of principal sections'/
     1            '         is defined as : NSMAX = ',I5/
     2            '         If you want to work with more principal'/
     3            '         sections, you should change the parameter'/
     4            '         NSMAX and recompile the program.')
         GO TO 440
      ENDIF
C
      WRITE(*,463) ND , NS
      WRITE(7,463) ND , NS
  463 FORMAT(/' Number of reaches,                   ND          = ',I5/
     2        ' Number of stations,                  NS          = ',I5/)
C
C NUMBER OF ALLOWED SUBDIVISIONS FOR REFINING WATER SURFACE PROFILE
C
C CALCULATIONS
  465 WRITE(*,470)
  470 FORMAT(' Max. tolerated var. of dyn. head,    DHDYNM (m)  ? = ',$)
      READ(*,*,ERR=465)DHDYNM
      WRITE(7,470)
      WRITE(7,*)DHDYNM
  475 WRITE(*,480) MCMAX , 2**MCMAX
      WRITE(7,480) MCMAX , 2**MCMAX
  480 FORMAT(/' In case this value is exceeded, the reach will be subdiv
     1ided'/
     2        ' in order to refine the calculations.'/
     3        ' The number of d-visions is specified in powers of 2.'/
     4        ' The maximum value is         MCMAX = ',I3/
     5        ' which corresponds to  2^MCMAX = ',I3,' subdivisions.')
      WRITE(*,485)
  485 FORMAT(/' Maximum number of subdiv. in powers of 2, NMC  ? = ',$)
      READ(*,*,ERR=475)NMC
      IF(NMC.GT.MCMAX)GO TO 475
      WRITE(7,485)
      WRITE(7,*)NMC
C
      WRITE(*,490)
      WRITE(7,490)
  490 FORMAT(//' INFORMATION ON BOUNDARY CONDITIONS :')
  500 WRITE(*,510)
  510 FORMAT(/' Note that the sediment transport at downstream end'/
     2        ' is automatically taken as zero,     QSU(1)      = 0.0'//
     2        ' Water depth at downstream end,      H1 (m)      ? = ',$)
      READ(*,*,ERR=500)H1
      WRITE(7,510)
      WRITE(7,*)H1
C
C READ THE TIME STEP AND THE COMPUTATION TIME
      WRITE(*,590)
      WRITE(7,590)
  590 FORMAT(//' PARAMETERS RELATED TO TIME AND PRINTING OF RESULTS :')
      WRITE(*,610)
  600 WRITE(*,610)
  610 FORMAT(' Time step,                           DELT (days)  ? = ',$)
      READ(*,*,ERR=600)DELT
      WRITE(7,610)
      WRITE(7,*)DELT
      DELT = DELT * 24.0 * 60.0 * 60.0
C
  620 WRITE(*,630)
  630 FORMAT(/' Duration of the simulation,         TPIN (days) ? = ',$)
      READ(*,*,ERR=620)TPIN
      WRITE(7,630)
      WRITE(7,*)TPIN
      TPIN = TPIN * 24.0 * 60.0 * 60.0
C
C FREQUENCY OF PRINTING THE RESULTS
  850 WRITE(*,860)
  860 FORMAT(/' Results will be printed every NPP step         ? = ',$)
      READ(*,*,ERR=850)NPP
      WRITE(7,860)
      WRITE(7,*)NPP
C
C READ THE NAME OF OUTPUT FILE
  900 WRITE(*,910)
  910 FORMAT(/' Name of output file (max. 40 char.)           ? = ',$)
      READ(*,345)FICHS
      OPEN(UNIT = NOUT , FILE = FICHS , STATUS = 'NEW' , ERR = 900 )
      WRITE(7,910)
```

Fig.Ex.6.C.11 Program DELTA (continued).

```
      WRITE(7,345)FICHS
C
      RETURN
      END
C
      SUBROUTINE RK4( G , Y , DX , YP , DERIVE )
C**********************************************************************
C                                                                     *
C   CALCULATE THE VALUES OF DEPENDENT VARIABLES AT STATION X+DX       *
C                                                                     *
C**********************************************************************
C
C   LIST OF VARIABLES LOCALLY DEFINED FOR SUBROUTINE SUBPROGRAM "RK4"
C
C   TYPE NAME    DIMEN.    EXPLANATIONS
C   ==== ====    ======    ============
C   R*4  DX2              = DX*0.5
C   R*4  DX6              = DX/6
C   R*4  K1               = K1 IN THE 4TH ORDER RUNGE-KUTTA METHOD
C   R*4  K2               = K2 IN THE 4TH ORDER RUNGE-KUTTA METHOD
C   R*4  K3               = K3 IN THE 4TH ORDER RUNGE-KUTTA METHOD
C   R*4  K4               = K4 IN THE 4TH ORDER RUNGE-KUTTA METHOD
C   R*4  Y                = WATER DEPTH KNOWN AT THE DOWNSTREAM STATION
C                           OF A REACH
C   R*4  YP               = WATER DEPTH CALCULATED AT THE UPSTREAM STATION
C                           OF A REACH
C   R*4  YT               = TEMPORARY VALUE OF Y AT (X+DX/2) AND/OR AT
C                           (X+DX) USED FOR CALCULATING K2, K3 AND K4
C
C   SEE THE BEGINNING OF THE MAIN PROGRAM FOR A LIST OF SIMPLE VARIABLES,
C   VECTORS AND MATRICES DEFINED GLOBALLY
C
C   ATTENTION : THE NAME OF THE SUBROUTINE SUBPROGRAM "DERIVE" IS SENT
C   BY THE PRINCIPAL PROGRAM IN THE LIST OF ARGUMENTS
C
C**********************************************************************
C
C   DECLARATION OF VARIABLES
      REAL K1 , K2 , K3 , K4
C
      DX2=DX*0.5
      DX6=DX/6.0
C
C   Y IS THE WATER DEPTH KNOWN AT THE DOWNSTREAM STATION OF REACH
C   COMPUTE K1
      CALL DERIVE( G , Y , K1 )
C
C   COMPUTE K2
      YT = Y + DX2 * K1
      CALL DERIVE( G , YT , K2 )
C
C   COMPUTE K3
      YT = Y + DX2 * K2
      CALL DERIVE( G , YT , K3 )
C
C   COMPUTE K4
      YT = Y + DX * K3
      CALL DERIVE( G , YT , K4 )
C
C   COMPUTE THE WATER DEPTH AT THE UPSTREAM STATION OF REACH
      YP = Y + DX6 * ( K1 + 2*K2 + 2*K3 + K4)
C
      RETURN
      END
C
      SUBROUTINE DERIVE( G , Y , DYDX )
C**********************************************************************
C                                                                     *
C   SUBROUTINE SUBPROGRAM FOR CALCULATING THE DERIVATIVES OF NEQ      *
C   EQUATIONS                                                         *
C                                                                     *
C**********************************************************************
C
C   LIST OF VARIABLES LOCALLY DEFINED FOR SUBROUTINE SUBPROGRAM "DERIVE"
C
C   TYPE NAME    DIMEN.    EXPLANATIONS
C   ==== ====    ======    ============
C   R*4  DYDX             = THE DERIVATIVE dh/dx CALCULATED ACCORDING TO
C                           DIFFERENTIAL EQUATION OF GRADUALLY VARIED FLOW
C                           (SEE THE TEXT AND EQ. 4.5)
C   R*4  Y                = WATER DEPTH AT THE STATION WHERE dh/dx IS BEING
C                           CALCULATED (REPLACES ALTERNATELY Y AND YT)
C
C   SEE THE BEGINNING OF THE MAIN PROGRAM FOR A LIST OF SIMPLE VARIABLES,
C   VECTORS AND MATRICES DEFINED GLOBALLY
C
C**********************************************************************
C
C   LABELED COMMON BLOCKS (SHARED VARIABLES)
      COMMON /DONNE/ SFTR , QU , CN
C
C   COMPUTATION OF dh/dx
      DYDX = SFTR - (QU*CN)**2 / Y**(10./3.)
      DYDX = -DYDX / (1 - QU**2 / (G * Y**3))
C
      RETURN
      END
C
      SUBROUTINE TITLES( NOUT , SF , D50 , CN50 , CN , ROS , ROE , QU ,
     1                   DHDYNM , NMC , VARZMX , CFOI , HAMHAV , FCOR ,
     2                   NPTS , X1 , XF , TL , DX , ND , NS , H1 ,
     3                   DELT , TFIN , NPP , FICHS)
C**********************************************************************
C                                                                     *
C   SUBROUTINE SUBPROGRAM FOR WRITING THE TITLES ON OUTPUT FILE       *
C                                                                     *
C**********************************************************************
C
C   LIST OF VARIABLES LOCALLY DEFINED FOR SUBROUTINE SUBPROGRAM "TITLES"
C
C   TYPE NAME    DIMEN.    EXPLANATIONS
C   ==== ====    ======    ============
C   R*4  TITLE   (20)     = LINES OF TITLE
C
C   SEE THE BEGINNING OF THE MAIN PROGRAM FOR A LIST OF SIMPLE VARIABLES,
C   VECTORS AND MATRICES DEFINED GLOBALLY
C
C**********************************************************************
C
C   DECLARATION OF VARIABLES
      CHARACTER*40    FICHS
```

Fig.Ex.6.C.11 Program DELTA (continued).

TRANSPORT OF SEDIMENTS

```fortran
      CHARACTER*131 TITLE(25)
      DOUBLE PRECISION DELT , TFIN
C DATA
C       DATA TITLE(1)/' PHYSICAL CHARACTERISTICS DATA :
     1                  CHOICE OF BED-LOAD FORMULA :'/
     1=' ',
        DATA TITLE(2)/'     Initial bed slope,                        SF   (-)
     1=' ',
        DATA TITLE(3)/'     Average diameter of sediments,            D50  (mm)
     1=' ',
        DATA TITLE(4)/'     Manning coeff. for sed. grains,           CN50 (s/m^1/3)
     1=' ',
        DATA TITLE(5)/'     Manning-Strickler coefficient,            CN   (s/m1/3)
     1=' ',
        DATA TITLE(6)/'     Density of sediments,                     ROS  (kg/m3)
     1=' ',
        DATA TITLE(7)/'     Density of water,                         ROE  (kg/m3)
     1=' ',
        DATA TITLE(8)/'     Unit discharge,                           QU   (m2/s)
     1=' ',
        DATA TITLE(9)/' '/
        DATA TITLE(10)/' INFORMATION RELATED TO WATER-SURFACE CALCULATIONS
     1    :       INFORMATION RELATED TO THE CALCULATION OF DEPOSIT
     2ION VOLUME :    '/
        DATA TITLE(11)/'     Max. tolerated var. of dyn. head,        DHDYNM
     1=' ,
     2                  '     Max. rel bed-level variation,          VARZMX (-)
     2=' ',
        DATA TITLE(12)/'     Maximum number of subdiv. in powers of 2,  NMC
     1=' ,
     2                  '     Coefficient of swelling,                CFOI (-)
     2=' ',
        DATA TITLE(13)/'
     1                  '     Ratio of upst./dwnst. heights, HAMHAV  (-)
     2=' ',
        DATA TITLE(14)/' '/
        DATA TITLE(15)/' INFORMATION ON CALCULATION DOMAIN :
     1                  BOUNDARY CONDITIONS : '/
        DATA TITLE(16)/'     x-coordinate of first station,           X1   (m)
     1=' ,
     2                  '     The bed-load transport at the downstream end is z
     2ero.                                                              '
        DATA TITLE(17)/'     x-coordinate of last station,            XF   (m)
     1=' ,
     2                  '     Water depth at the downstream end,      H1   (m)
     2=' ',
        DATA TITLE(18)/'     Total reach length is therefore,         TL   (m)
     1=' ',
        DATA TITLE(19)/'     Step length in x-direction,              DX   (m)
     1=' ',
        DATA TITLE(20)/' PARAMETERS RELATED TO TIME : '/
     1                  '     Time step,                         DELT      (days)
     1=' ,
     2                  '     Number of reaches,                      ND   (m)
     2=' ',
        DATA TITLE(21)/'     Number of stations,                      NS
     1=' ,
     2                  '     Duration of simulation,            TFIN      (jours)
     2=' ',
        DATA TITLE(22)/' '/
        DATA TITLE(23)/' PRINTING OF RESULTS : '/
        DATA TITLE(24)/'     Printing of the results every NPP step
     1=' ',
        DATA TITLE(25)/'     Name of the output file (max. 40 char.)
     1=' ',
C
C WRITE THE PROGRAM TITLE ON THE OUTPUT FILE
      WRITE(NOUT,10)
 10   FORMAT(//'          PROGRAM FOR BED-LOAD SEDIMENT TRANSPORT CALC
     1ULATION'/
     2      '          ',
     3                                        BY TAKING INTO ACCOUNT BED-LEVEL MODIFICAT
     3IONS.'//
     4      '          NOTES : '/'    - UNIT SYSTEM = SI'/
     5      '                    - NUM. METHOD FOR WATER-SURFACE PROFILE CALCULATION = 4
     6th ORDER RUNGE-KUTTA'/
     7      '                    - THE FLOW IS SUBCRITICAL (Fr < 1). THE WATER SURFACE
     8PROFILE CALCULATION '/
     9      '                      STARTS AT THE DOWNSTREAM END AND PROGRESSES TOWARDS T
     AHE UPSTREAM END'/
     B      '                    - SEDIMENT TRANSPORT CALCULATIONS ARE CARRIED OUT FROM
     CUPSTREAM TO DOWNSTREAM'/
     D      '                    - CALCULATIONS ARE MADE FOR A UNIT WIDTH'//)
C
C FILL IN THE TITLE ARRAY WITH THE INFORMATION
      WRITE(TITLE(2)(53:61),'(F9.7)') SP
      WRITE(TITLE(3)(53:56),'(F4.2)') D50*1000.0
      WRITE(TITLE(4)(53:58),'(F5.4)') CN50
      WRITE(TITLE(5)(53:58),'(F5.4)') CN
      WRITE(TITLE(6)(53:59),'(F7.2)') ROS
      WRITE(TITLE(7)(53:59),'(F7.2)') ROE
      WRITE(TITLE(8)(53:60),'(F3.2)') QU
C
      WRITE(TITLE(11)(53:57),'(F5.3)') DHDYNM
      WRITE(TITLE(11)(119:122),'(F4.2)') VARZMX
      WRITE(TITLE(12)(53:53),'(I1)') NMC
      WRITE(TITLE(12)(119:124),'(F6.4)') CFOI
      WRITE(TITLE(13)(119:123),'(F5.3)') HAMHAV
C
      WRITE(TITLE(16)(53:61),'(F9.2)') X1
      WRITE(TITLE(17)(53:61),'(F9.2)') XF
      WRITE(TITLE(17)(119:125),'(F7.3)') H1
      WRITE(TITLE(18)(53:61),'(F9.2)') TL
      WRITE(TITLE(19)(53:61),'(F9.2)') DX
      WRITE(TITLE(20)(53:57),'(I5)') ND
      TEMP = DELT / 86400.0
      WRITE(TITLE(20)(119:128),'(F10.3)') TEMP
      WRITE(TITLE(21)(53:57),'(I5)') NS
      TEMP = TFIN / 86400.0
      WRITE(TITLE(21)(119:128),'(F10.3)') TEMP
C
      WRITE(TITLE(24)(53:57),'(I5)') NPP
C
      TITLE(25)(53:92) = PICHS
C
      IF(NFTS.EQ.1)THEN
C
C  SCHOKLITSCH (1950) FORMULA IS USED
         TITLE(2)(68:131) = 'Bed-load transport formula by Schoklitsch (195
     10) is used.
        ELSE IF(NFTS.EQ.2)THEN
C
C  MEYER-PETER (1948) FORMULA IS USED
         TITLE(2)(68:131) = 'Bed-load transport formula by Meyer-Peter et a
     11. (1948) is used.
         TITLE(3)(68:118) = 'Roughness parameter,        FCOR = (CN50 / CN)^(3/
     12) =
         WRITE(TITLE(3)(119:125),'(F7.3)') FCOR
        ELSE IF(NFTS.EQ.3)THEN
C
C  EINSTEIN (1942) FORMULA IS USED
         TITLE(2)(68:131) = 'Bed-load transport formula by Einstein (1942)
     1is used.
        ELSE IF(NFTS.EQ.4)THEN
```

Fig.Ex.6.C.11 Program DELTA (continued).

```
C     USER PROGRAMMED FORMULA IS USED
        TITLE(2)(68:131) = 'User''s programmed bed-load transport formula
     1 is used.'
      ENDIF
C
      DO 110 I = 1 , 25
      WRITE(NOUT,100)TITLE(I)
  100 FORMAT(A)
  110 CONTINUE
C
      RETURN
      END

      SUBROUTINE DWRITE( NOUT , NS , T , X , ZF , ZFI , H , QU , G ,
     1                  QSU , DELQS , DQSDET , DQSERT , CPTT ,
     2                  DELZ , DELZMX , IMAXZ , DELZRM , IMAXR )
C*********************************************************************
C     SUBROUTINE SUBPROGRAM FOR WRITING THE RESULTS AT A GIVEN TIME.   *
C*********************************************************************
C     LIST OF VARIABLES LOCALLY DEFINED FOR SUBROUTINE SUBPROGRAM "DWRITE"
C     ================================================================
C     TYPE  NAME     DIMEN.    EXPLANATIONS
C     ====  ====     ======    ============
C     C*131 HEADER   (20)    = LINES OF THE TITLE
C     R*4   FRN              = FROUDE NUMBER AT A STATION
C     R*4   TH               = TIME IN HOURS
C     R*4   U                = AVERAGE VELOCITY AT A STATION
C
C     SEE THE BEGINNING OF THE MAIN PROGRAM FOR A LIST OF SIMPLE VARIABLES,
C     VECTORS AND MATRICES DEFINED GLOBALLY.
C*********************************************************************
C     DECLARATION OF VARIABLES
      CHARACTER*131    HEADER(9)
      DOUBLE PRECISION T , ZF (NS) , ZFI (NS)
      DOUBLE PRECISION QSU(NS) , DELQS(NS) , DQSDET , DQSERT
      DOUBLE PRECISION DELZ(NS) , DELZMX , DELZRM
      DIMENSION        X(NS) , H(NS)
C
C     INITIALIZE HEADER
      HEADER(1) = ' Time step           =                 Total vol
     lume of deposited sediments, DQSDET (m3/m) :    '
      HEADER(2) = ' Time (s)            =                 Total vol
     lume of eroded sediments,   DQSERT (m3/m) :    '
      HEADER(3) = ' Time (hours)        =                                '
      HEADER(4) = ' Time (days)         =                 Max. absolute
     be at the station, IMA
     2XZ =     '
      HEADER(5) = ' Time (years)        =                 Max. relative
     be at the station, IMA
     2XR =     '
      HEADER(6) = '                                                      '
      HEADER(7) = '                                                      '
      HEADER(8) = '  STATION      X         U         Fr          ZF
     1         ZF + H          ZF - ZPI         QSU          DELQS
     2DELZ                                                             '
      HEADER(9) = '    NO         (m)      (m/s)      (-)          (m)
     1         (m)             (m)             (m3/s/m)     (m3/s/m)
     2  (m)                                                            '
C
C     FILL IN THE HEADER WITH THE INFORMATION
      WRITE(HEADER(1)(19:28),'(F10.0)')CPTT
      WRITE(HEADER(2)(19:33),'(D15.7)')T
      TH = T / 3600.0
      WRITE(HEADER(3)(19:28),'(F10.3)')TH
      TJ = T / 86400.0
      WRITE(HEADER(4)(19:28),'(F10.3)')TJ
      TA = T / (86400.0 * 365)
      WRITE(HEADER(5)(19:28),'(F10.3)')TA
C
      WRITE(HEADER(1)(101:112),'(D12.6)')DQSDET
      WRITE(HEADER(2)(101:112),'(D12.6)')DQSERT
C
      WRITE(HEADER(4)(89:100),'(D12.6)')DELZMX
      WRITE(HEADER(4)(124:127),'(I4)')IMAXZ
      WRITE(HEADER(5)(89:100),'(D12.6)')DELZRM
      WRITE(HEADER(5)(124:127),'(I4)')IMAXR
C
C     WRITE THE RESULTS FOR THE RUNNING TIME
      WRITE(NOUT,10)
   10 FORMAT(//1X,'==================================================
     1===================================<0>============================
     2===========')
      DO 30 I = 1 , 9
      WRITE(NOUT,20)HEADER(I)
   20 FORMAT(A)
   30 CONTINUE
      DO 50 I = 1 , NS
      U = QU / H(I)
      IF(I.EQ.1) U = 0.0
      FRN = U / SQRT(G * H(I))
      WRITE(NOUT,40) I , X(I) , ZF(I) , ZF(I)-ZFI(I) , H(I) ,
     1                ZF(I)+H(I) , U , FRN , QSU(I) , DELQS(I) , DELZ(I)
   40 FORMAT(3X , I4 , 3X , F9.2 , 2X , D12.6 , 2X , D12.6 , 2X , F7.3 ,
     1       2X , D12.6 , 2X , F6.3 , 2X , F6.3 , 2X , D12.6 , 2X , 2X ,
     2       D12.6 , 2X , D12.6 )
   50 CONTINUE
C
C     WRITE THE RESULTS IN THE GRAPHICS FILE
      WRITE(8,100)TA
  100 FORMAT('TIME (years) = ',F10.3)
      DO 120 I = 1 , NS
      WRITE(8,110) I , X(I) , ZF(I) , ZF(I)+H(I)
  110 FORMAT(I4 , ' , ' , F9.2 , ' , ' , D12.6 , ' , ' , D12.6 )
  120 CONTINUE
      WRITE(8,130)
  130 FORMAT(/)
C
      RETURN
      END

C*********************************************************************
C     SUBROUTINE SCHOKLI( QU , D50 , SS , SEFF , QCRIT , QSU )
C*********************************************************************
C     SUBROUTINE SUBPROGRAM FOR CALCULATING THE BED-LOAD TRANSPORT AT A *
C     SECTION USING THE FORMULA OF SCHOKLITSCH (1950)                  *
C*********************************************************************
```

Fig.Ex.6.C.11 Program DELTA (continued).

TRANSPORT OF SEDIMENTS

```
C  LIST OF VARIABLES LOCALLY DEFINED FOR SUBROUTINE SUBPROGRAM "SCHOKL"
C  ================================================================
C  TYPE NAME.   DIMEN.    EXPLANATIONS
C  ====  ====   ======    ============
C  R*4   QCRIT            = CRITICAL DISCHARGE
C
C  SEE THE BEGINNING OF THE MAIN PROGRAM FOR A LIST OF SIMPLE VARIABLES,
C  VECTORS AND MATRICES DEFINED GLOBALLY.
C**********************************************************************
C
C  DECLARATION OF VARIABLES
      DOUBLE PRECISION QSU
C
C  COMPUTE THE CRITICAL DISCHARGE
      QCRIT = 0.26 * (SS - 1)**(5./3.) * D50**1.5 / SEFF**(7./6.)
C
C  COMPUTE THE BED-LOAD DISCHARGE
      IF(QU-QCRIT.LE.0)THEN
        QSU = 0.0
      ELSE
        QSU = 2.5 * SEFF**1.5 * (QU - QCRIT) / SS
      ENDIF
C
      RETURN
      END

      SUBROUTINE MEYPET(G , SEFF , ROE , ROS , D50 , FCOR , RH , QSU)
C**********************************************************************
C  SUBROUTINE SUBPROGRAM FOR CALCULATING THE BED-LOAD TRANSPORT AT A
C  SECTION USING THE FORMULA OF MEYER-PETER et al. (1948) (see eq 6.32)
C
C  LIST OF VARIABLES LOCALLY DEFINED FOR SUBROUTINE SUBPROGRAM "MEYPET"
C  ================================================================
C  TYPE NAME.   DIMEN.    EXPLANATIONS
C  ====  ====   ======    ============
C  R*4   RH               = HYDRAULIC RADIUS (= H)
C  R*8   TERM1            = NOMINATOR IN MEYER-PETER'S FORMULA
C  R*8   TERM2            = DENOMINATOR IN MEYER-PETER'S FORMULA
C
C  SEE THE BEGINNING OF THE MAIN PROGRAM FOR A LIST OF SIMPLE VARIABLES.
C  VECTORS AND MATRICES DEFINED GLOBALLY.
C**********************************************************************
C
C  DECLARATION OF VARIABLES
      DOUBLE PRECISION TERM1 , TERM2 , QSU
C
C  COMPUTE THE BED-LOAD DISCHARGE
      TERM1 = (G * ROE * RH * FCOR * SEFF) / D50
      TERM1 = TERM1 - (0.047 * G * (ROS - ROE))
      IF(TERM1.LE.0.0)THEN
        QSU = 0.0
      ENDIF
      TERM2 = D50 / (0.25 * ROE**(1./3.))
      QSU = (TERM2 * TERM1)**(3./2.) / (G * (ROS - ROE))
C
      RETURN
      END

      SUBROUTINE EINS42(SS , G , D50 , RH , SEFF , QSU)
C**********************************************************************
C  SUBROUTINE SUBPROGRAM FOR CALCULATING THE BED-LOAD TRANSPORT AT A
C  SECTION USING THE FORMULA OF EINSTEIN (1942)
C
C  LIST OF VARIABLES LOCALLY DEFINED FOR SUBROUTINE SUBPROGRAM "EINS42"
C  ================================================================
C  TYPE NAME.   DIMEN.    EXPLANATIONS
C  ====  ====   ======    ============
C  R*4   RH               = HYDRAULIC RADIUS (= H)
C  R*8   TERM1            = TERM MULTIPLYING THE EXPONENTIAL TERM
C  R*8   TERM2            = EXPONENTIAL TERM
C
C  SEE THE BEGINNING OF THE MAIN PROGRAM FOR A LIST OF SIMPLE VARIABLES.
C  VECTORS AND MATRICES DEFINED GLOBALLY.
C**********************************************************************
C
C  DECLARATION OF VARIABLES
      DOUBLE PRECISION TERM1 , TERM2 , QSU
C
C  COMPUTE THE BED-LOAD DISCHARGE
      TERM1 = SQRT((SS - 1.0) * G * D50**3) / 0.465
      TERM2 = 0.391 * (SS - 1.0) * D50 / (RH * SEFF)
      QSU = TERM1 * EXP(-TERM2)
C
      RETURN
      END

      SUBROUTINE FORMUL(NOUT)
C**********************************************************************
C  SUBROUTINE SUBPROGRAM FOR CALCULATING THE BED-LOAD TRANSPORT AT A
C  SECTION USING THE FORMULA PROGRAMMED BY THE USER
C**********************************************************************
C
      WRITE(NOUT,*)' User should program here a formula of his/her choic
     le.'
      WRITE(7,*)' User should program here a formula of his/her choice.'
C
      RETURN
      END
```

Fig.Ex.6.C.11 Program DELTA (end).

Ex. 6.D

An artificial channel has been constructed to divert a certain discharge from a river. This channel has an approximately rectangular cross section with a width of B = 46.5 [m] and a bed slope of $S_f = 6.5 \cdot 10^{-4}$ [-]. The uniform flow is established when the flow depth is h_n = 5.6 [m]. The velocity-profile measurements carried out in this channel allowed to obtain the average velocity of U = 1.8 [m/s] for a friction coefficient of n' = 0.0212 [$m^{-1/3}s$]. The granulometry of the bed material has not been analyzed.

Estimate the bed-load transport in this channel. Subsequently, express the solid discharge as a concentration. Is suspended-load transport to be expected ?

SOLUTION :

i) First, preliminary calculations concerning the hydraulics of the channel and the sedimentology of the bed material should be carried out.

Unit discharge : $q = Uh = 1.8\,(5.6) = 10.08$ [m^3/sm]

Discharge : $Q = qB = 10.08\,(46.5) = 468.72$ [m^3/s]

Hydraulic radius : $R_h = \dfrac{Bh}{B + 2h} = \dfrac{(46.5)\,(5.6)}{46.5 + 2\,(5.6)} = 4.51$ [m]

Hydraulic radius of the channel bed (the channel banks are assumed to be smooth) :

$$R_{hb} \cong h_n = 5.6 \text{ [m]}$$

The granulometry can be estimated using the calculated friction coefficient, n', which, being obtained from a measured velocity profile, corresponds to the friction coefficient due to grain roughness. By using the Strickler formula :

$$\frac{1}{n'} = K_s' = \frac{1}{0.0212} = \frac{26}{d_{90}^{1/6}} \quad \text{and} \quad K_s' = \frac{21.1}{d_{50}^{1/6}} \qquad (3.18)$$

one obtains :

$$d_{90} = 0.0280 \text{ [m]} \quad ; \quad d_{50} = 0.0080 \text{ [m]}$$

By assuming a granulometric distribution to be logarithmic, one finds :

$d_{35} = 0.0055$ [m] ; $d_{40} = 0.0062$ [m] ; $d_{65} = 0.0117$ [m]

The total friction coefficient, n , which is due to the combined effect of grain roughness and bed forms, can now be obtained using the Manning-Strickler formula :

TRANSPORT OF SEDIMENTS 449

$$K_s = \frac{1}{n} = \frac{U}{R_{hb}^{2/3} S_f^{1/2}} = \frac{1.8}{(5.6)^{2/3}(0.00065)^{1/2}} = 22.41 \ [m^{1/3}/s] \quad (3.16)$$

whereas :

$$K_s' = \frac{1}{n'} = \frac{1}{0.0212} = 47.17 \ [m^{1/3}/s].$$

The roughness parameter is therefore :

$$\xi_M = \left(\frac{K_s}{K_s'}\right)^{3/2} = \left(\frac{22.41}{47.17}\right)^{3/2} = 0.327$$

The settling velocity for d_{50} is (see Fig. 6.10) : $\quad v_{ss} \cong 0.4 \ [m/s]$

ii) Three different bed-load equations will be used to estimate the solid discharge; namely the bed-load equation of *Schoklitsch*, eq. 6.31, of *Meyer-Peter* et al., eq. 6.32, and of *Einstein*, eq. 6.42.

a) The bed-load equation of *Schoklitsch* is given by :

$$q_{sb} = \frac{2.5}{s_s} S_e^{3/2} (q - q_{cr}) \quad (6.31)$$

The critical liquid discharge is calculated using :

$$q_{cr} = 0.26 \ (s_s - 1)^{5/3} \ d^{3/2} \ S_e^{-7/6} \quad (6.31a)$$

where $d = d_{40} = 0.0062$ [m] for a non-uniform granulometry and assuming that the specific density of the bed material is $s_s = 2.65$ [-] :

$$q_{cr} = 0.26 \ (1.65)^{5/3} \ (0.0062)^{3/2} \ (0.00065)^{-7/6} = 1.53 \ [m^2/s]$$

The volumic solid discharge for a unit width is then :

$$q_{sb} = \frac{2.5}{2.65} \ (0.00065)^{3/2} \ (10.08 - 1.53) = \mathit{1.33 \cdot 10^{-4}} \ [m^2/s].$$

b) The bed-load equation of *Meyer-Peter* et al. is given by :

$$\frac{\gamma R_{hb} \xi_M S_e}{(\gamma_s - \gamma)d} - 0.047 = 0.25 \ \rho^{1/3} \frac{g'_{sb}^{2/3}}{(\gamma_s - \gamma)d} \quad (6.32)$$

where $d = d_{50} = 0.008$ [m] for a non-uniform granulometry .

The solid discharge by submerged weight for a unit width can then be calculated as :

$$g'^{2/3}_{sb} = \left(\frac{9.81\ (1650)\ (0.008)}{0.25\ (1000)^{1/3}}\right) \left(\frac{5.6\ (0.327)\ (0.00065)}{(2.65-1.0)\ 0.008} - 0.047\right)$$

$$g'_{sb} = [(51.91)\ (0.090 - 0.047)]^{3/2} = (2.23)^{3/2} = 3.35\ [\text{N/ms}]$$

The volumic solid discharge for a unit width is :

$$q_{sb} = \frac{g_{sb}}{\gamma_s} = \frac{g'_{sb}}{(\gamma_s - \gamma)} = \frac{3.35}{9.81\ (1650)} = 2.07 \cdot 10^{-4}\ [\text{m}^2/\text{s}]$$

c) The bed-load equation of *Einstein* is given by :

$$\Phi_* = f(\Psi_*) \tag{6.42a}$$

Considering the non-uniform granulometry with $d = d_{35} = 0.0055$ [m], one shall take :

$$\Phi = f(\Psi') \qquad \text{or} \qquad \frac{q_{sb}}{\sqrt{(s_s - 1)gd^3_{35}}} = f\left(\frac{(\gamma_s - \gamma)\ d_{35}}{\tau_o'}\right)$$

First the shear-stress intensity parameter should be calculated :

$$\Psi' = \frac{(\gamma_s - \gamma)d_{35}}{\rho u_*'^2} = (s_s - 1)\frac{d_{35}}{R_{hb}'S_f} \tag{6.34}$$

The friction velocity due to the grain roughness will be calculated using the logarithmic velocity distribution :

$$\frac{U}{u_*'} = 5.75\ \log\left(\frac{h}{k_s}\right) + 6.25 = 21.66 \tag{3.13b}$$

with $k_s = d_{65} = 0.0117$ [m] and $h = 5.6$ [m]. It is found that :

$$u_*' = \frac{U}{21.66} = \frac{1.8}{21.66} = 0.083\ [\text{m/s}].$$

The value of Ψ' can then be calculated as follows :

$$\Psi' = \frac{g(\rho_s - \rho)d_{35}}{\rho\ u_*'^2} = \frac{9.81(2.65 - 1)0.0055}{(0.083)^2} = 12.92\ [-]$$

TRANSPORT OF SEDIMENTS

The hydraulic radius due to grain groughness is:

$$u_*' = \sqrt{g\, R_{hb}'\, S_f} \quad \Rightarrow \quad R_{hb}' = \frac{u_*'^2}{g\, S_f} = 1.08\ [m]$$

whereas the hydraulic radius due to the bed forms is:

$$R_{hb}'' = R_{hb} - R_{hb}' = 5.6 - 1.08 = 4.52\ [m]$$

This shows the importance of bed forms in this cross section of the channel. (If these values, $\Psi' = 12.92$ and $U/u_*'' = 10.60$, are compared with the relationship of Einstein-Barbarossa, represented in Fig. 3.6, a slight difference will be observed).

One can now either evaluate the function given by eq. 6.42, or read it directly on Fig. 6.8, to find:

for $\Psi' = 12.92 \quad \Rightarrow \quad \Phi \cong 0.033$

The volumic solid discharge per unit width can then be calculated as:

$$q_{sb} = \Phi \sqrt{(s_s-1)\, g d_{35}^3} = 0.033\, [(1.65)(9.81)(0.0055^3)]^{1/2} \quad (6.28)$$

$$= 5.41 \cdot 10^{-5}\ [m^2/s]$$

iii) The transport of sediments as bed load, obtained using these three bed-load equations, is presented in the table below, both by volume and by mass.

The difference between the values obtained using the different formulae is considerable but this is not surprising; one should in fact never expect to find exactly the same values using different formulae.

Formula	Solid discharge as bed load			
	$q_{sb}\ [m^2/s]$	$g_{sb}\ [kg/ms]$	$Q_{sb}\ [m^3/s]$	$G_{sb}\ [kg/s]$
Schoklitsch eq. 6.31	$1.33 \cdot 10^{-4}$	0.35	$6.18 \cdot 10^{-3}$	16.39
Meyer-Peter et al. eq. 6.32	$2.07 \cdot 10^{-4}$	0.55	$9.62 \cdot 10^{-3}$	25.50
Einstein eq. 6.42	$0.54 \cdot 10^{-4}$	0.14	$2.51 \cdot 10^{-3}$	6.65

iv) The liquid discharge of

$$Q = 468.72 \ [m^3/s] \quad \text{or} \quad G = 468.72 \cdot 10^3 \ [kg/s]$$

is responsible for the solid discharge — taking the average of the values listed in the above table — of :

$$Q_{sb} = 6.1 \cdot 10^{-3} \ [m^3/s] \quad \text{or} \quad G_{sb} = 16.2 \ [kg/s]$$

The average sediment *concentration*, C_s, can be expressed in different manners. Here are the possible definitions :

concentration by volume : $C_s = \dfrac{\text{volume of sediments}}{\text{total volume}} \quad \dfrac{[m^3/s]}{[m^3/s]}$

concentration by mass : $C_s' = \dfrac{\text{mass of sediments}}{\text{total mass}} \quad \dfrac{[kg/s]}{[kg/s]}$

concentration by unit mass : $C_s'' = \dfrac{\text{mass of sediments}}{\text{total volume}} \quad \dfrac{[kg/s]}{[m^3/s]}$

These definitions are related to one another in the following way :

$$C_s'' = \rho_s C_s \quad ; \quad C_s' = \frac{\rho_s C_s}{\rho + (\rho_s - \rho) C_s} = \frac{\rho_s}{\rho_m} C_s$$

where ρ_m is the average density of the water-sediment mixture, defined by :

$$\rho_m = C_s \rho_s + (1 - C_s) \rho = \rho + (\rho_s - \rho) C_s$$

With above definitions, the concentrations can be obtained.

The density of the solid particles is : $\rho_s = 2650 \ [kg/m^3]$ or $2.65 \ [g/cm^3]$.

The concentrations can now be calculated :

$$C_s = \frac{6.1 \cdot 10^{-3}}{468.72} = 0.000013 \ [-]$$

$$C_s' = \frac{16.2}{468.72 \cdot 10^3} = 0.000035 \ [-]$$

$$C_s'' = \frac{16.2}{468.72} = 0.0345 \ \left[\frac{kg}{m^3}\right] \text{ or } \left[\frac{g}{l}\right] \text{ or } \frac{1}{1000} \left[\frac{g}{cm^3}\right]$$

TRANSPORT OF SEDIMENTS

The average density of the mixture is :

$$\rho_m = 1000.00 + 0.02 = 1000.02 \ [kg/m^3]$$

v) It is already shown that there is a strong bed-load transport in this channel. One can now ask, if there will be also suspended-load transport.

According to an indicative criteria, given in sect. 6.1.3, the suspended-load transport starts when :

$$\frac{u_*}{v_{ss}} > 0.40$$

For the present problem one obtains :

$$\frac{u_*}{v_{ss}} = \frac{0.189}{0.400} = 0.47$$

Therefore, a weak transport of sediments as suspended load is to be expected.

This expectation can still be controlled by determining the Rouse exponent, eq. 6.50a :

$$\mathcal{Z} = \frac{v_{ss}}{\kappa u_*'} = \frac{0.4}{0.4 \ (0.083)} = 12.05$$

On Fig. 6.11, it can be seen that, for this $\mathcal{Z}$-value, the relative concentration distribution, consequently the suspended-load transport, will be indeed weak.

Ex. 6.E

A mountain river with a bottom slope of $S_f = 0.0062$ [-] has an approximately rectangular cross section, being $B = 23.5$ [m] wide. Analysis of the sediment samples taken from well below the armour layer show that $d_{50} = 60$ [mm] and $d_{90} = 200$ [mm] and the density of sediments is $s_s = 2.65$ [-].

Determine the diameter, d_{50_a}, of the maximum possible armouring. At which flow depth does the armour layer become unstable ?

SOLUTION :

i) The grain diameter of the armour layer is calculated (see point 6.3.4.5°) using the relationship :

$$d_{50_a} \cong 0.6 \, d_{90}$$

$$d_{50_a} \cong 0.6 \, (200) = 120 \text{ [mm]}$$

ii) The stability of the armour layer (see point 6.3.4.5°) can be estimated using the expression :

$$\tau_{*a,cr} = \frac{u_{*a,cr}^2}{(s_s-1) \, g \, d_{50_a}} = \tau_{*cr} \left[0.4 \left(\frac{d_{50}}{d_{50_a}} \right)^{1/2} + 0.6 \right]^2$$

$$\tau_{*a,cr} = \tau_{*cr} \left[0.4 \left(\frac{60}{120} \right)^{1/2} + 0.6 \right]^2 = 0.05 [0.88]^2 = 0.04 \text{ [-]}$$

from which one obtains :

$$u_{*a,cr} = \sqrt{0.04 \, [(s_s-1) \, g \, d_{50_a}]} = \sqrt{0.04 \, [1.94]} = 0.28 \text{ [m/s]}.$$

According to the definition of the friction velocity :

$$u_* = \sqrt{g \, R_h \, S_f} \quad \Rightarrow \quad R_h = u_*^2 / g \, S_f$$

$$R_h = \frac{0.28^2}{9.81 \, (0.0062)} = 1.25 \text{ [m]}$$

Now the limiting depth for the stability of the armour layer can be calculated as :

$$R_h = \frac{h \, B}{2h + B} \quad \text{or} \quad 1.25 = \frac{h \, (23.5)}{2h + 23.5} \quad \Rightarrow \quad h = 1.40 \text{ [m]}$$

If the flow depth becomes larger than $h = 1.40$ [m], the armour layer is no longer stable and an important erosion of the bed material may be expected.

TRANSPORT OF SEDIMENTS 455

> **Ex. 6.F**
>
> The river *Happy* — whose *stage—water-discharge curve* was established in Ex. 3.B — has a variable discharge in the range of $10 < Q \ [m^3/s] < 1000$. The width of the river bed is $b = 90$ [m] and its non-erodible banks have a slope of 1:1. The topographical survey of the river showed that the bed slope is $S_f = 0.0005$ [-]. The sediment forming the bed has a specific density of $s_s = 2.652$ [-] and the grain-size analysis yielded: $d_{50} = 0.32$ [mm], $d_{35} = 0.29$ [mm] and $d_{90} = 0.48$ [mm]. The water temperature in the river is $T = 14$ [°C].
>
> Determine the *stage—sediment-discharge curve*, $Q_s = f(h)$, for this river.

SOLUTION :

First the hydraulic calculations should be done to determine the *stage—water*-discharge curve, $Q = f(h)$, for the river. Once this is done, one can carry out the sediment-transport calculations to determine the *stage—solid*-discharge curve, $Q_s = f(h)$.

i)   *Hydraulic calculations* :
     The hydraulic calculations were presented and commented in Ex. 3.B. The calculation table, where each line represents the calculation of a discharge, Q, and other useful hydraulic parameters, R_h', R_h'', R_h, etc., is partially reprinted in Table Ex.6.F.1, together with the explanations for the columns. On every line, the calculations start by *assuming* a hydraulic radius due to grain roughness, R_h'. The corresponding discharge is obtained by following the procedure described in Ex. 3.B. The values for R_h' are selected such that the calculations cover the entire range of the desired water discharges in the river, $10 < Q \ [m^3/s] < 1000$.

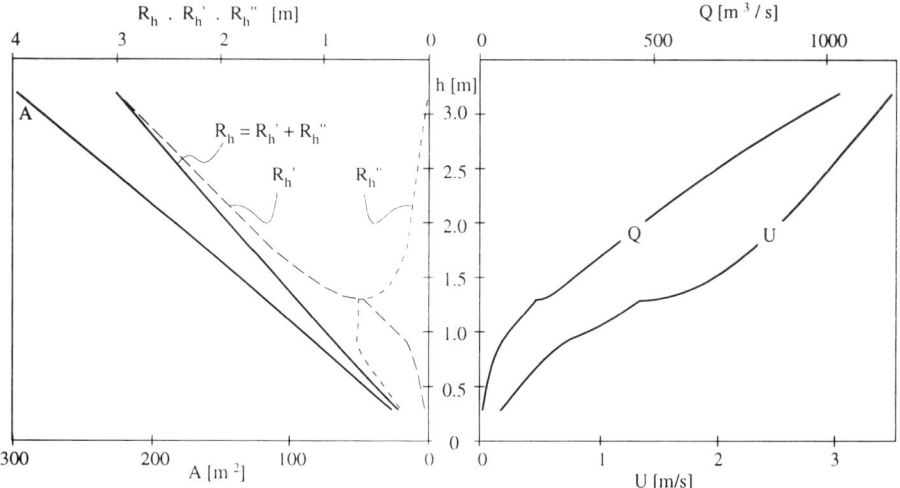

Fig. Ex. 6.F.1 *Stage—liquide*-discharge curve.

The *stage—water*-discharge curve, $Q = f(h)$, and the variation of other parameters, U, A, R_h', R_h'' et R_h, as a function of the flow depth, h, are presented in Fig. Ex. 6.F.1.

Table Ex.6.F.1

Computation sheet for determining the *stage*—**water**-discharge curve, using the method of *Einstein-Barbarossa* (1952)

$b = 90$ [m] $T = 14$ [°C] $\rho_s = 2650$ [kg/m³]
$m = 1$ $\rho = 999.1$ [kg/m³] $d_{35} = 0.00029$ [m]
$S_f = 0.0005$ [-] $\nu = 1.186 \times 10^{-6}$ [m²/s] $k_s = d_{50} = 0.00032$ [m]

1	2	3	4	5	6	7	8	9	10	11
R_h'	u_*'	U	Ψ'	U/u_*''	u_*''	R_h''	R_h	u_*	h	Q
[m]	[m/s]	[m/s]	[-]	[-]	[m/s]	[m]	[m]	[m/s]	[m]	[m³/s]
0.02	0.01	0.16	47.92	4.5	0.04	0.26	0.28	0.04	0.29	4.2
0.05	0.02	0.29	19.17	6.6	0.04	0.39	0.44	0.05	0.44	11.7
0.10	0.02	0.45	9.58	8.7	0.05	0.55	0.65	0.06	0.65	26.5
0.15	0.03	0.58	6.39	10.4	0.06	0.63	0.78	0.06	0.79	41.4
0.20	0.03	0.69	4.79	11.9	0.06	0.68	0.88	0.07	0.89	55.8
0.40	0.04	1.05	2.40	18.3	0.06	0.67	1.07	0.07	1.09	103.7
0.60	0.05	1.33	1.60	23.4	0.06	0.66	1.26	0.08	1.29	157.4
0.80	0.06	1.58	1.20	32.4	0.05	0.49	1.29	0.08	1.32	191.3
1.00	0.07	1.81	0.96	42.6	0.04	0.37	1.37	0.08	1.41	232.4
1.25	0.08	2.06	0.77	56.2	0.04	0.27	1.52	0.09	1.57	297.3
1.50	0.09	2.30	0.64	73.1	0.03	0.20	1.70	0.09	1.76	371.6
2.00	0.10	2.72	0.48	107.2	0.03	0.13	2.13	0.10	2.23	559.3
2.50	0.11	3.11	0.38	163.0	0.02	0.07	2.57	0.11	2.71	780.9
3.00	0.12	3.46	0.32	6844.0	0.00	0.00	3.00	0.12	3.19	1026.7

col.	symbol	explanations	expression
1	R_h'	hydraulic radius due to grain roughness (*assumed* value)	
2	u_*'	friction velocity due to grain roughness, eq. 3.24,	$\sqrt{g\,R_h'\,S_f}$
3	U	average velocity in the cross section	$u_*'\sqrt{8/f'}$
		with (see eq. 3.13b) : $\sqrt{8/f'} = 5.6 \log(R_h'/k_s) + 6.25$	
4	Ψ'	parameter of Einstein-Barbarossa, eq. 3.31,	$\dfrac{(s_s - 1)\,d_{35}}{R_h'\,S_f}$
5	U/u_*''	ratio of velocities corresponding to Ψ' (see eq. 3.31 and Fig. 3.6)	
6	u_*''	friction velocity due to bed forms	$U/(U/u_*'')$
7	R_h''	hydraulic radius due to bed forms	$(u_*'')^2/(g\,S_f)$
8	R_h	total hydraulic radius, eq. 3.24,	$R_h' + R_h''$
9	u_*	total friction velocity, eq. 3.7,	$\sqrt{g\,R_h\,S_f}$
10	h	flow depth (see Tableau 1.1)	
11	Q	water discharge, eq. 3.2a,	$Uh(b + mh)$

TRANSPORT OF SEDIMENTS 457

ii) *Sediment-transport calculations* :
The sediment-transport calculations will be made for the representative grain diameter, using three different total-load relations, namely : (1) *Einstein*, (2) *Graf* and *Acaroglu* and also (3) *Ackers* et *White*.

1° The formula of *Einstein* (1950), which allows the calculation of the total load transported by the flow, is given by :

$$q_s = q_{sb} + q_{ss} = q_{sb} \left[1 + 2.303 \log(30.2 \, h / \Delta) \, \mathcal{I}_1 + \mathcal{I}_2 \right] \qquad (6.60)$$

Since the grain-size distribution is *quasi uniform*, the calculations can be done using an equivalent grain diameter (see Table 6.3) of :

$$d = d_{35} = 0.00029 \, [m]$$

The intensity of transport is :

$$\Phi_* \equiv \Phi = \frac{q_{sb}}{\sqrt{(s_s-1)g d_{35}^3}} \qquad (6.28)$$

where $q_{sb} = g_{sb}/\gamma_s$ is the volumic solid discharge for a unit width and g_{sb} the solid discharge by weight; both transported as bed load.

The intensity of shear is :

$$\Psi_* \equiv \Psi' = (s_s-1) \, \frac{d_{35}}{R_{hb}' S_f} \qquad (6.34)$$

where $R_{hb}' \equiv R_h'$ is the hydraulic radius of the bed due to the granulats.

With the functional relationship of :

$$\Phi_* = f(\Psi_*) \qquad (6.42a)$$

given by eq. 6.42 and presented in Fig. 6.8, one can obtain the solid discharge, q_{sb}, transported as bed load.

Next, the integrals, $\mathcal{I}_1$ and $\mathcal{I}_2$, which appear in the suspended-load formula, eq. 6.56 (see also eq. 6.60), should be determined to calculate the solid discharge, q_{ss}, transported as suspended load.

The total load, $q_s = q_{sb} + q_{ss}$, transported by the river can then be calculated using eq. 6.60.

The calculations can be programmed on a microcomputer using a spreadsheet program. The table of calculations prepared in this way is presented in Table Ex. 6.F.2. Each line of this table gives the calculations of the solid discharge (by volume, mass and weight), as well as other useful parameters calculated for each flow depth, h. The detailed explanations on the contents of the columns are given below the table of calculations.

Table Ex.6.F.2

Computation sheet for determining the *stage*—**solid**-discharge curve, using the method of *Einstein* (1950)											
$b = 90$ [m] $S_f = 0.0005$ [-]						$\rho = 999.1$ [kg/m³] $\nu = 1.186 \times 10^{-6}$ [m²/s]					
1	2	3	4	5	6	7	8	9	10	11	12
h	R_h'	u_*'	δ	k_s/δ	χ	Δ	P_e	Ψ'	Φ	q_{sb}	Q_{sb}
[m]	[m]	[m/s]	[m]	[-]	[-]	[m]	[-]	[-]	[-]	[m³/s/m]	[m³/s]
0.29	0.02	0.01	1.39E-03	0.256	0.91	3.92E-04	10.01	47.92	0.00	0.00E+00	*0.00E+00*
0.44	0.05	0.02	8.81E-04	0.405	1.25	2.86E-04	10.76	19.17	0.00	7.96E-08	*7.16E-06*
0.65	0.10	0.02	6.23E-04	0.573	1.48	2.42E-04	11.32	9.58	0.10	2.00E-06	*1.80E-04*
0.79	0.15	0.03	5.09E-04	0.702	1.56	2.29E-04	11.56	6.39	0.35	6.91E-06	*6.22E-04*
0.89	0.20	0.03	4.41E-04	0.811	1.60	2.24E-04	11.71	4.79	0.71	1.41E-05	*1.27E-03*
1.09	0.40	0.04	3.12E-04	1.147	1.60	2.23E-04	11.91	2.40	2.51	5.00E-05	*4.50E-03*
1.29	0.60	0.05	2.54E-04	1.404	1.57	2.28E-04	12.06	1.60	4.32	8.59E-05	*7.73E-03*
1.32	0.80	0.06	2.20E-04	1.621	1.49	2.40E-04	12.03	1.20	6.06	1.21E-04	*1.08E-02*
1.41	1.00	0.07	1.97E-04	1.813	1.42	2.51E-04	12.05	0.96	7.77	1.54E-04	*1.39E-02*
1.57	1.25	0.08	1.76E-04	2.027	1.37	2.61E-04	12.12	0.77	9.86	1.96E-04	*1.76E-02*
1.76	1.50	0.09	1.61E-04	2.220	1.31	2.72E-04	12.19	0.64	11.94	2.37E-04	*2.14E-02*
2.23	2.00	0.10	1.39E-04	2.564	1.24	2.87E-04	12.37	0.48	16.06	3.19E-04	*2.87E-02*
2.71	2.50	0.11	1.25E-04	2.866	1.19	2.99E-04	12.53	0.38	20.18	4.01E-04	*3.61E-02*
3.19	3.00	0.12	1.14E-04	3.140	1.16	3.07E-04	12.66	0.32	24.28	4.83E-04	*4.34E-02*

col.	symbol	explanations	expression
1	h	flow depth (see Table Ex.6.F.1)	
2	R_h'	hydraulic radius due to grain roughness (see Table Ex.6.F.1)	
3	u_*'	friction velocity due to grain roughness (see Table Ex.6.F.1)	
4	δ	thickness of viscous sublayer	$\delta = 11.5\, \nu/u_*'$
5	k_s / δ	relative roughness (see Fig. 6.7a)	$k_s / \delta = d_{65} / \delta$
6	χ	correction term for logarithmic velocity distribution (see Fig. 6.7a)	
7	Δ	apparent roughness diameter (see Fig. 6.7a)	$\Delta = d_{65} / \chi$
8	P_e	transport parameter (see eq. 6.58)	$P_e = 2.303 \log(30.2 h / \Delta)$
9	Ψ'	intensity of shear, eq. 6.34,	$\Psi' = (s_s - 1) \dfrac{d_{35}}{R_h' S_f}$
10	Φ	intensity of transport, eq. 6.42,	$\Phi = f(\Psi')$
11	q_{sb}	solid discharge, as bed load, by volume and by unit width, eq. 6.28,	$q_{sb} = \Phi \sqrt{(s_s - 1) g\, d_{35}^3}$
12	Q_{sb}	solid discharge, as bed load, by volume	$Q_{sb} = q_{sb}\, b$

TRANSPORT OF SEDIMENTS

Table Ex.6.F.2 (suite)

Computation sheet for determining the *stage*—**solid**-discharge curve, using the method of *Einstein* (1950)			
$d_{35} = 0.00029$ [m]	$v_{ss}(d_{35}) = 0.0365$ [m/s]		
$d_{65} = 0.00036$ [m]	$\rho_s = 2650$ [kg/m³]		$\kappa = 0.4$ [-]

13	14	15	16	17	18	19	20	21
A_E	z	$\mathcal{J}_1$	$\mathcal{J}_2$	q_{ss}	Q_{ss}	Q_s	G_s	G_s
[-]	[-]	[-]	[-]	[m³/s/m]	[m³/s]	[m³/s]	[kg/s]	[N/s]
2.03E–03	9.222	2.62E–02	–1.59E–01	0.00E+00	0.00E+00	*0.00E+00*	0	0
1.30E–03	5.832	4.46E–02	–2.87E–01	1.54E–08	1.38E–06	*8.54E–06*	0	0
8.87E–04	4.124	6.90E–02	–4.63E–01	6.36E–07	5.73E–05	*2.37E–04*	1	6
7.32E–04	3.367	9.11E–02	–6.19E–01	3.00E–06	2.70E–04	*8.92E–04*	2	23
6.48E–04	2.916	1.13E–01	–7.67E–01	7.77E–06	6.99E–04	*1.97E–03*	5	51
5.32E–04	2.062	2.02E–01	–1.34E+00	5.35E–05	4.81E–03	*9.31E–03*	25	242
4.49E–04	1.684	3.10E–01	–1.96E+00	1.52E–04	1.37E–02	*2.14E–02*	57	557
4.39E–04	1.458	4.43E–01	–2.62E+00	3.27E–04	2.94E–02	*4.03E–02*	107	1047
4.12E–04	1.304	6.10E–01	–3.37E+00	6.14E–04	5.52E–02	*6.91E–02*	183	1798
3.69E–04	1.166	8.74E–01	–4.47E+00	1.20E–03	1.08E–01	*1.26E–01*	333	3268
3.29E–04	1.065	1.21E+00	–5.73E+00	2.13E–03	1.92E–01	*2.13E–01*	566	5548
2.61E–04	0.922	2.14E+00	–8.90E+00	5.62E–03	5.05E–01	*5.34E–01*	1416	13888
2.14E–04	0.825	3.47E+00	–1.29E+01	1.23E–02	1.10E+00	*1.14E+00*	3019	29619
1.82E–04	0.753	5.25E+00	–1.77E+01	2.35E–02	2.12E+00	*2.16E+00*	5727	56183

col.	symbol	explanations	expression
13	A_E	dimensionless height, eq. 6.52a,	$A_E = \dfrac{z_{sb}}{h} = \dfrac{2d_{35}}{h}$
14	z	Rouse exponent, eq. 6.50a, v_{ss}: settling velocity (see Fig. 6.10)	$z = \dfrac{v_{ss}}{\kappa\, u_*'}$
15	$\mathcal{J}_1$	Einstein's first integral (see Fig. 6.12)	
16	$\mathcal{J}_2$	Einstein's second integral (see Fig. 6.12)	
17	q_{ss}	solid discharge, as suspended load, by volume and by unit width, eq. 6.58,	$q_{ss} = q_{sb}(P_e I_1 + I_2)$
18	Q_{ss}	solid discharge, as suspended load, by volume	$Q_{ss} = q_{ss}\, b$
19	Q_s	solid discharge, as total load, by volume	$Q_s = Q_{sb} + Q_{ss}$
20	G_s	solid discharge, as total load, by mass	$G_s = Q_s\, \rho_s$
21	G_s	solid discharge, as total load, by weight	$G_s = Q_s\, \rho_s\, g$

2° The formula of *Graf* et *Acaroglu* (1968), which allows the calculation of the total load transported by the flow, is given by :

$$\Phi_A = f(\Psi_A) \tag{6.63}$$

with the parameter of transport :

$$\Phi_A = \frac{C_s \, U R_h}{\sqrt{(s_s-1)g d_{50}^3}} \tag{6.62}$$

and the parameter of shear intensity :

$$\Psi_A = \frac{(s_s-1) \, d_{50}}{S_e \, R_h} \tag{6.61}$$

where the equivalent diameter is taken as (see Table 6.3) :

$d \equiv d_{50} = 0.00032$ [m].

It is to be noted, that R_h is the total hydraulic radius and $C_s = q_s/q$ is the average concentration by volume. The functional relationship is evaluated according to eq. 6.63.

As in the previous case, the calculations are programmed on a microcomputer using a spreadsheet program. The computation sheet prepared in this way is presented in Table Ex. 6.F.3.

3° The formula of *Ackers* et *White* (1973), which allows the calculation of average concentration, C_s, by volume is given by :

$$C_s = G_{gr} \frac{d_{35}}{h_m} \left(\frac{U}{u_*}\right)^{n_w} \tag{6.66}$$

where the equivalent diameter is taken as (see Table 6.3) :

$d \equiv d_{35} = 0.00029$ [m]

The sediment-transport parameter is calculated as :

$$G_{gr} = C_w \left(\frac{F_{gr}}{A_w} - 1\right)^{m_w} \tag{6.65}$$

with the mobility parameter defined as :

$$F_{gr} = \frac{u_*^{n_w}}{\sqrt{(s_s-1) \, g \, d_{35}}} \left[\frac{U}{\sqrt{32} \log (10 h_m/d_{35})}\right]^{(1-n_w)} \tag{6.64}$$

The dimensionless diameter for $d \equiv d_{35}$ is determined using :

Table Ex.6.F.3

	Computation sheet for determining the *stage*—**solid**-discharge curve, using the method of *Graf* et *Acaroglu* (1968)								
	$S_f = 0.0005$ [-] $\rho = 999.1$ [kg/m³]					$d_{50} = 0.00032$ [m] $\rho_s = 2650$ [kg/m³]			
1	2	3	4	5	6	7	8	9	10
h	R_h	U	Q	Ψ_A	Φ_A	C_s	Q_s	G_s	G_s
[m]	[m]	[m/s]	[m³/s]	[ ]	[ ]	[-]	[m³/s]	[kg/s]	[N/s]
0.29	0.28	0.16	4.2	3.718	0.380	1.91E–04	7.95E–04	2	21
0.44	0.44	0.29	11.7	2.401	1.143	2.06E–04	2.41E–03	6	63
0.65	0.65	0.45	26.5	1.639	2.990	2.39E–04	6.33E–03	17	165
0.79	0.78	0.58	41.4	1.356	4.821	2.48E–04	1.03E–02	27	266
0.89	0.88	0.69	55.8	1.203	6.522	2.49E–04	1.39E–02	37	362
1.09	1.07	1.05	103.7	0.992	10.615	2.20E–04	2.28E–02	60	592
1.29	1.26	1.33	157.4	0.839	16.163	2.22E–04	3.49E–02	92	907
1.32	1.29	1.58	191.3	0.821	17.091	1.93E–04	3.69E–02	98	960
1.41	1.37	1.81	232.4	0.773	19.864	1.85E–04	4.30E–02	114	1118
1.57	1.52	2.06	297.3	0.694	26.121	1.91E–04	5.69E–02	151	1478
1.76	1.70	2.30	371.6	0.622	34.445	2.03E–04	7.54E–02	200	1960
2.23	2.13	2.72	559.3	0.496	60.791	2.41E–04	1.35E–01	358	3507
2.71	2.57	3.11	780.9	0.411	97.757	2.82E–04	2.20E–01	583	5721
3.19	3.00	3.46	1026.7	0.353	143.802	3.20E–04	3.28E–01	870	8531

col.	symbol	explanations	expression
1	h	flow depth (see Table Ex.6.F.1)	
2	R_h	total hydraulic radius (see Table Ex.6.F.1)	
3	U	average velocity (see Table Ex.6.F.1)	
4	Q	liquid discharge (voir Tableau Ex.6.F.1)	
5	Ψ_A	shear-stress intensity parameter, eq. 6.61,	$\Psi_A = \dfrac{(s_s - 1)\, d_{50}}{S_f\, R_h}$
6	Φ_A	transport parameter, eq. 6.63,	$\Phi_A = 10.39\, \Psi_A^{-2.52}$
7	C_s	concentration by volume in the section, eq. 6.62,	$C_s = \Phi_A \dfrac{\sqrt{(s_s - 1)\, g\, d_{50}^3}}{U\, R_h}$
8	Q_s	solid discharge, as total load, by volume	$Q_s = C_s\, Q$
9	G_s	solid discharge, as total load, by mass	$G_s = Q_s\, \rho_s$
10	G_s	solid discharge, as total load, by weight	$G_s = Q_s\, g\, \rho_s$

Table Ex.6.F.4

Computation sheet for determining the *stage*—**solid**-discharge curve, using the method of *Ackers* et *White* (1973)

$\rho = 999.1$ [kg/m³] $\nu = 1.186 \times 10^{-6}$ [m²/s]
$\rho_s = 2650$ [kg/m³] $d_{35} = 0.00029$ [m] $\Rightarrow$ $d_* = (g(s_s - 1)/\nu^2)^{1/3} d_{35} = 6.536$
$n_w = 1 - 0.56 \log d_* = 0.5434$ $m_w = 9.66/d_* + 1.34 = 2.8180$
$A_w = 0.23/\sqrt{d_*} + 0.14 = 0.2300$ $C_w = 10^{(2.86 \log d_* - (\log d_*)^2 - 3.53)} = 0.0137$

1	2	3	4	5	6	7	8	9	10
h	u_*	U	Q	F_{gr}	G_{gr}	C_s	Q_s	G_s	G_s
[m]	[m/s]	[m/s]	[m³/s]	[-]	[-]	[-]	[m³/s]	[kg/s]	[N/s]
0.29	0.04	0.16	4.2	0.256	2.979E−05	6.69E−08	2.79E−07	0	0
0.44	0.05	0.29	11.7	0.369	3.316E−03	5.85E−06	6.83E−05	0	2
0.65	0.06	0.45	26.5	0.490	1.937E−02	2.65E−05	7.04E−04	2	18
0.79	0.06	0.58	41.4	0.573	4.242E−02	5.22E−05	2.16E−03	6	56
0.89	0.07	0.69	55.8	0.638	6.923E−02	8.03E−05	4.48E−03	12	117
1.09	0.07	1.05	103.7	0.808	1.845E−01	2.10E−04	2.17E−02	58	565
1.29	0.08	1.33	157.4	0.939	3.268E−01	3.41E−04	5.37E−02	142	1397
1.32	0.08	1.58	191.3	1.020	4.446E−01	4.95E−04	9.48E−02	251	2463
1.41	0.08	1.81	232.4	1.099	5.808E−01	6.44E−04	1.50E−01	397	3891
1.57	0.09	2.06	297.3	1.197	7.845E−01	8.11E−04	2.41E−01	639	6269
1.76	0.09	2.30	371.6	1.289	1.014E+00	9.64E−04	3.58E−01	949	9314
2.23	0.10	2.72	559.3	1.467	1.570E+00	1.22E−03	6.81E−01	1805	17707
2.71	0.11	3.11	780.9	1.626	2.210E+00	1.44E−03	1.12E+00	2970	29133
3.19	0.12	3.46	1026.7	1.769	2.906E+00	1.63E−03	1.68E+00	4440	43559

col.	symbol	explanations	expression
1	h	flow depth (see Table Ex.6.F.1)	$h \cong h_m = S/B$
2	u_*	total shear velocity (see Table Ex.6.F.1)	
3	U	average velocity (see Table Ex.6.F.1)	
4	Q	liquid discharge (see Table Ex.6.F.1)	
5	F_{gr}	parameter of mobility, eq. 6.64,	$F_{gr} = \dfrac{u_*^{n_w}}{\sqrt{(s_s-1)gd_{35}}} \left[\dfrac{U}{\sqrt{32}\log(10h/d_{35})}\right]^{(1-n_w)}$
6	G_{gr}	transport parameter, eq. 6.65,	$G_{gr} = C_w \left(\dfrac{F_{gr}}{A_w} - 1\right)^{m_w}$
7	C_s	concentration by volume, eq. 6.66,	$C_s = G_{gr} \dfrac{d_{35}}{h} \left(\dfrac{U}{u_*}\right)^{n_w}$
8	Q_s	solid discharge, as total load, by volume	$Q_s = C_s Q$
9	G_s	solid discharge, as total load, by mass	$G_s = Q_s \rho_s$
10	G_s	solid discharge, as total load, by weight	$G_s = Q_s g \rho_s$

TRANSPORT OF SEDIMENTS 463

$$d_* = d_{35}\left((s_s-1)\frac{g}{v^2}\right)^{1/3} \cong 6.5 \qquad (6.26)$$

which allows calculation of the coefficients, n_w, m_w, A_w and C_w (see point 6.5.2.4°).

Again, the calculations are programmed on a microcomputer using a spreadsheet program. The computation sheet is presented in Table Ex. 6.F.4.

The Fig. Ex. 6.F.2 gives the *stage—solid*-discharge curves — supplemented by the *stage—liquid*-discharge curve — for the **total load** calculated using the total-load relations of (1) *Einstein*, (2) *Graf* et *Acaroglu* et 3) *Ackers* et *White* as well as for the **bed load** calculated using the bed-load relation of (1) *Einstein*. (*Einstein*'s method, 1950, is an indirect method, thus it allows the evaluation of the bed load transport automatically.).

The Fig. Ex. 6.F.2 shows that the three methods used for the calculations do not give the same values for the solid discharge, Q_s. It is important however to remind that the formulae for the sediment transport can only give the engineer an idea about the order of magnitude of the solid discharge that one should reasonably expect in a particular flow situation. It should also be clear that the sediment-transport capacity (voir sect. 6.1.4) has been calculated.

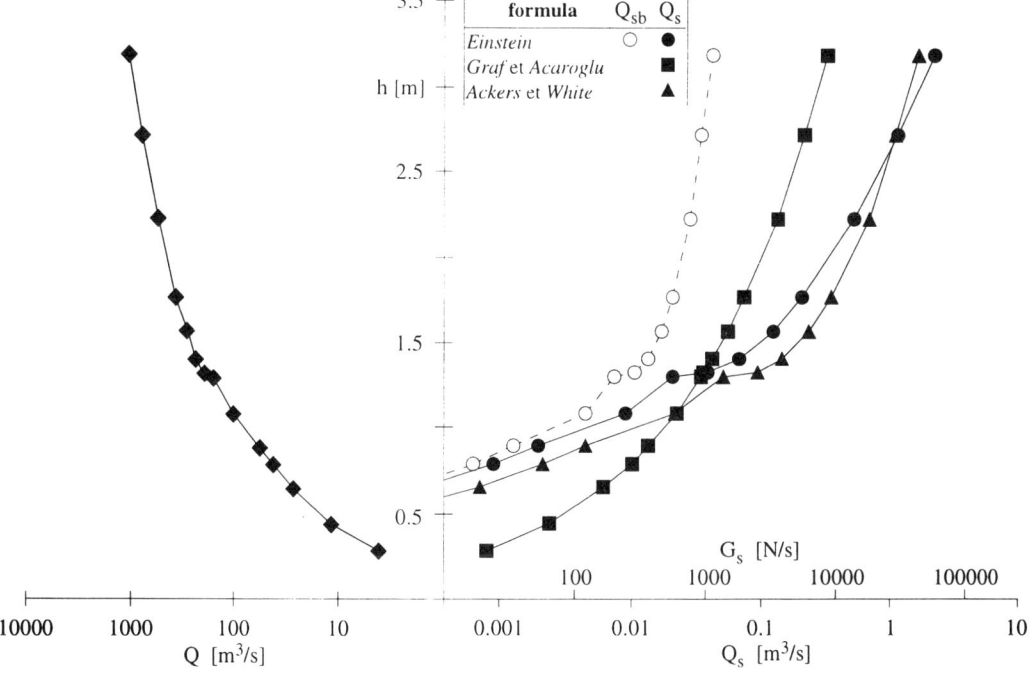

Fig. Ex.6.F.2 *Stage—liquid*-discharge and *stage—solid*-discharge curves.

6.6.2 Problems, unsolved

Ex. 6.1

A long channel of rectangular cross section has a bed slope of $S_f = 0.0004$ [-]. The channel bed is composed of a near-uniform granulate of $d_{50} = 0.5$ [mm] with a porosity of $p = 0.3$ [-]. The normal flow depth was measured as $h = h_n = 2.10$ [m]. This channel enters a lake ; at the juncture the water levels of the channel and of the lake are the same.

The water level in the lake is now *lowered* by $\Delta h = 0.10$ [m]. Determine, at what time the channel bed will be lowered by 90 % and by 50 % of Δh and this at two stations, situated at 1.5 [km] and at 20.0 [km] upstream of the juncture.

Ex. 6.2

In the channel, described in Ex. 6.1, the fixed point at the juncture will be *raised* by $\Delta h = 0.20$ [m]. Determine the temporal variation of the channel bed at a station, being situated 20.0 [km] upstream of the juncture.

Ex. 6.3

The unit discharge in a river is $q = 5$ [m²/s]. The bed has a slope of $S_f = 0.0005$ [-] and the porosity is $p = 0$ [-] ; its granulate is given as $d_{50} = 0.4$ [mm].

Well downstream, this river enters a reservoir, where the water depth is kept constant at a height of $H = 5$ [m]. Determine the aggradation one may expect in the reservoir after 250 [h] and 500 [h].

Ex. 6.4

The irrigation channel "Sivan" must be controlled for 81 days per year to deliver a constant discharge of 10 [m³/s] (during the remaining days of the year the discharge will be less). This rectangular channel, having a width of $B = 5.0$ [m], has a mobile bed with a uniform granulate of $d_{50} = 5$ [mm] ; the bed slope is $S_f = 0.0003$ [-].

i) Calculate the solid discharge, which is annually transported.
ii) Study a proposal, which envisions that a spillway— blocking the flow at a water depth being three times the normal flow depth — is installed at the downstream of the channel. What will the sediment deposition amount to and where will it take place ?

Ex. 6.5

A reach of a river of length $L = 38$ [km] — being indicative of the Bas-Rhône — conveys a constant discharge of $Q = 4000$ [m³/s]. In this reach the cross sections may be approximated by rectangular ones having a width of $B = 250$ [m] and having a constant bed slope of $S_f = 0.0007$ [-]. The roughness coefficient was estimated as being $K_s = 36$ [m$^{1/3}$/s] ; the diameter of the rather uniform granulate is $d_{50} = 27.4$ [mm].

TRANSPORT OF SEDIMENTS

It is envisioned to build a system of weirs, which would raise the water level by 10 [m], thus to H = 10 [m] + h_n. Study the evolution of bed level for a period of two years. Investigate the sediment-transport problem in the reach behind the weirs.

Ex. 6.6

The discharge in a channel, having a bed slope of $S_f = 0.00027$ [-] is $Q = 100$ [m³/s] ; the width at the channel bed is b = 46 [m] and the side slopes are 2 / 1 . Flow in this channel is uniform and the temperature of the water is $T = 15$ [C°]. Samples of the bed material have been evaluated ; the granulometry is given in the following table and its density is $s_s = 2.65$ [-].

median diameter [mm]	granulometric fraction [%]
0.088	5
0.177	22
0.354	37
0.707	31
1.414	5

Calculate the total-load transport, Q_s, by making use of different available formulae. This is to be done :

i) for each individual granulometric fraction,
ii) for all fractions together, taking an equivalent diameter.

Ex. 6.7

For the channel studied in Ex. 6.6, verify if the transport of sediments is influenced by the water temperature ; the lowest and highest temperatures expected are respectively, $T = 10$ [C°] and $T = 20$ [C°]. Make the calculations using an equivalent diameter.

Ex. 6.8

For the channel studied in Ex. 6.6, determine the diameter of the armouring and the flow depth for which the armour layer turns unstable.

7. TURBIDITY CURRENTS

In this chapter will be exposed our knowledge about turbidity currents, being a special case of density currents. These currents are caused essentially by density differences resulting from the presence of sediments in the current. A turbidity current is a bottom current, where the entrainment of sediments plays an important role; however there is also entrainment of ambient water from above the current. Such a current can be of high velocity and become auto-accelerative, if sediment entrainment continues.

The chapter begins by describing the current, being made up of a front followed by a body. Subsequently the hydrodynamic equations with the entrainment coefficients will be developed. The evolution of the current gives the characteristic form of the interface between the current and the ambient layer of water.

Finally, the front of the current will be investigated. In the problem which follows, two different turbidity currents and one density current illustrate a numerical application of our knowledge.

```
                    TABLE OF CONTENTS

    7.1     GENERALITIES
            7.1.1   Description of Current
            7.1.2   Plunge Point

    7.2     HYDRODYNAMIC EQUATIONS

    7.3     INTERFACE-PROFILE CURVES

    7.4     ENTRAINMENT COEFFICIENTS

    7.5     FRONT OF CURRENT

    7.6     DISTRIBUTION OF VELOCITY
            AND OF CONCENTRATION

    7.7     EXERCISES
            7.7.1   Problems, solved
            7.7.2   Problems, unsolved
```

7.1 GENERALITIES

7.1.1 Description of Current (see Fig. 7.1)

1° A density or gravity current is a (two-phase) flow of a fluid of density ρ_t, which is caused essentially by the influence of a density difference, $\Delta\rho$, on the gravity, g. It is as if the gravity were reduced by the ratio of $\Delta\rho/\rho_a$. The reduced gravity — which is the driving force of a density current — is expressed as :

$$g' = g\left(\frac{\rho_t - \rho_a}{\rho_a}\right) = g\frac{\Delta\rho}{\rho_a} \qquad (7.1)$$

where ρ_t is the average density of the current and ρ_a the one of the ambient fluid.

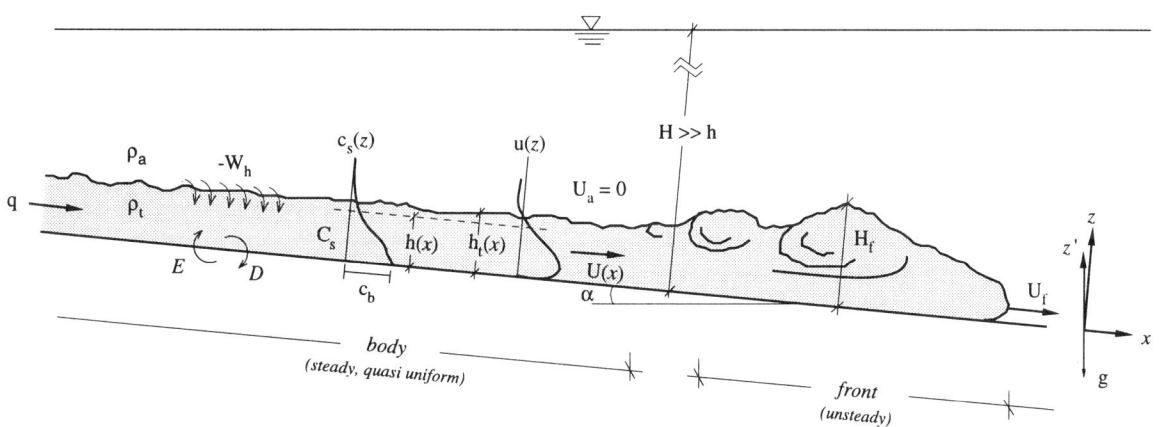

Fig. 7.1 Scheme of a turbidity current.

2° The cause of such a density variation, $\Delta\rho$, — taking water as the ambient fluid — could be a difference in :

i) temperature : $\Delta\rho \cong 2$ [kg/m³],

ii) salinity : $\Delta\rho \cong 20$ [kg/m³],

iii) turbidity : $\Delta\rho \cong 20$ à 130 (200) [kg/m³].

3° A *turbidity current* is thus a density current where the heavy (turbulent) fluid is a mixture of light ambient fluid of density ρ_a, and of granular (non-cohesive) material of density ρ_s, in suspension. Such a current must generate enough turbulence to carry the granular material in suspension.

The average density of a turbidity current is given by :

$$\rho_t = C_s\rho_s + (1-C_s)\rho_a = \rho_a + (\rho_s - \rho_a) C_s \qquad (7.2)$$

where C_s is the volume concentration of the granular material (see eq. 7.11), averaged over the height of the current (see Fig. 7.1); if $\rho_t > \rho_a$, the turbidity current is a bottom current.

The average reduced gravity, eq. 7.1, is expressed by :

$$g' = g\left(\frac{\rho_s - \rho_a}{\rho_a}\right) C_s = g R C_s \qquad (7.3)$$

and the local reduced gravity by :

$$g'_z = g R c_s \qquad (7.3a)$$

where c_s is the local concentration within the current at a height, z, from the bottom. R is the specific density of the submerged granular material.

4° If the suspension of sediments in the current is sufficiently diluted, $(\Delta\rho/\rho_a) \ll 0.1$, the Boussinesq approximation (see *Turner*, 1973, p. 9) is possible; this implies :

- $\dfrac{\Delta\rho}{\rho_a} \cong 0,\qquad$ before the inertia terms, thus $\rho_t \cong \rho_a$,

- $g\dfrac{\Delta\rho}{\rho_a} \equiv g' \neq 0,\qquad$ before the gravity terms, g.

5° For a parametrisation of a density current, the Froude number (see *Graf & Altinakar*, 1991, p. 295) — the ratio of inertia to reduced gravity forces — is written as :

$$Fr_D = \frac{U}{\sqrt{g'h\,\cos\alpha}} \qquad (7.4)$$

called also the densimetric Froude number. However, the following form (see *Turner*, 1973, p. 12 and p. 179) is frequently used :

$$\frac{1}{Fr_D^2} = \frac{g'h\,\cos\alpha}{U^2} = Ri \qquad (7.5)$$

which is the definition of the *global Richardson number*. For both of these numbers, one takes the velocity, U, averaged over the height, h, of the turbidity current (see eqs. 7.9 and 7.10) and the angle, α, of the bottom slope.

TURBIDITY CURRENTS

The local or gradient Richardson number is defined (see *Turner*, 1973) by :

$$Ri_z = - g \frac{\partial \rho}{\partial z} / \rho \left(\frac{\partial u}{\partial z}\right)^2 \qquad (7.5a)$$

where $\partial \rho \equiv d\rho$ is the density difference between two layers over a distance dz.

As shall be seen later (see sect. 7.2), the movement of a turbidity current depends on the entrainment of ambient fluid which is parametrised by the entrainment coefficient, E_w, being itself dependent on the Richardson number :

$$E_w = f(Ri) \qquad (7.6)$$

6° The reduced sediment flux per unit width is defined as :

$$B = g' h U = g R(C_s U h) = g' q \qquad (7.7)$$

One distinguishes now :

i) the *conservative* turbidity currents:

$$\frac{dB}{dx} = 0 \qquad (7.8)$$

ii) the non-*conservative* turbidity currents :

$$\frac{dB}{dx} \neq 0 \qquad (7.8a)$$

where the variation of the reduced sediment flux, B, is due to an erosion of the bed and/or a deposition of suspended material on the bed; q is the unit discharge of the current.

Turbidity currents are often non-conservative currents. Density currents are always conservative currents.

7° A turbidity current (see Fig. 7.1) is made up of a *front* or head advancing into the ambient fluid, being followed by the *body*.

The driving force for the front is essentially the pressure gradient, being due to a density difference between the front and the ambient fluid. The flow is unsteady.

The driving force for the body is the gravitational force of the heavier fluid. The flow is often considered to be a steady one.

The duration of a turbidity current, thus its length, will depend upon the incoming reduced sediment flux, B.

8° The interface between the turbidity current and the ambient fluid (see Fig. 7.1) is usually not easy to distinguish.

For this reason, one defines an height, h, and a velocity, U, using the integral scales of the current (see *Turner*, 1973, p. 179), such as :

$$Uh = \int_0^\infty u \, dz = \int_0^{h_t} u \, dz = \overline{U} h_t = q \qquad (7.9)$$

$$U^2 h = \int_0^\infty u^2 \, dz = \int_0^{h_t} u^2 \, dz = \beta_u \overline{U}^2 h_t \qquad (7.10)$$

where $u(z)$ is the point velocity, h_t is the height where the velocity, u, is zero and $\overline{U}$ is the average velocity of the current; β_u is a coefficient (Boussinesq) of velocity distribution.

The vertical variation of the concentration, c_s, is a gradual one, notably at the interface. The average concentration, C_s, is defined by :

$$C_s Uh = \int_0^\infty (uc_s) \, dz = \int_0^{h_t} (uc_s) \, dz = C_s \overline{U} h_t \qquad (7.11)$$

Across the interface there is entrainment (see eq. 7.6) of the ambient fluid.

9° It appears interesting to compare a flow in a channel with a free surface with a flow of a turbidity current into an ambient fluid :

i) For a flow with a free surface, where the (well-defined) interface separates the air and the water, one writes (see eq. 7.1) :

$$g \left(\frac{\rho_{eau} - \rho_{air}}{\rho_{eau}} \right) \cong g \qquad \text{since} \qquad \rho_{eau} \gg \rho_{air},$$

here the gravity, g, plays the major role.

ii) For a flow of a turbidity current, where the (weakly-defined) interface separates the heavy and the light fluid, one writes :

$$g \left(\frac{\rho_t - \rho_a}{\rho_a} \right) = g' \qquad \text{since} \qquad \rho_t \cong \rho_a \qquad (7.1)$$

here the reduced gravity, g', plays the major role.

10° Turbidity currents have been encountered in watercourses and reservoirs. Such currents displace themselves with often great velocity, being independent of the ambient velocity.

Turbidity currents are frequently caused by large volumes of sediments (earth slides) entering the channel or by extreme floods with large sediment transport in a tributary which, when entering the main channel, keeps its identity and remains close to the bed — due to its density, ρ_t — and gradually mixes with the flow. Some prominent turbidity currents have been observed in large streams such as the Nile, the Colorado and the Mississippi.

11° Turbidity currents form themselves also in the air (atmosphere) such as avalanches, lava flows, earth slides or dust storms.

7.1.2 Plunge Point

1° If a watercourse, $U \neq 0$, which transports large quantities of sediments, $C_s \neq 0$, enters a stagnant reservoir, $U_a = 0$, of reasonably clear water, $C_s = 0$, a plunge zone (often clearly visible) forms itself. Subsequently, it is possible that — after a zone of transition — a turbidity current, composed of fine particles in suspension, establishes itself on the bottom of the reservoir (see Fig. 7.2). The larger particles deposit immediately at the beginning of the reservoir by forming a delta.

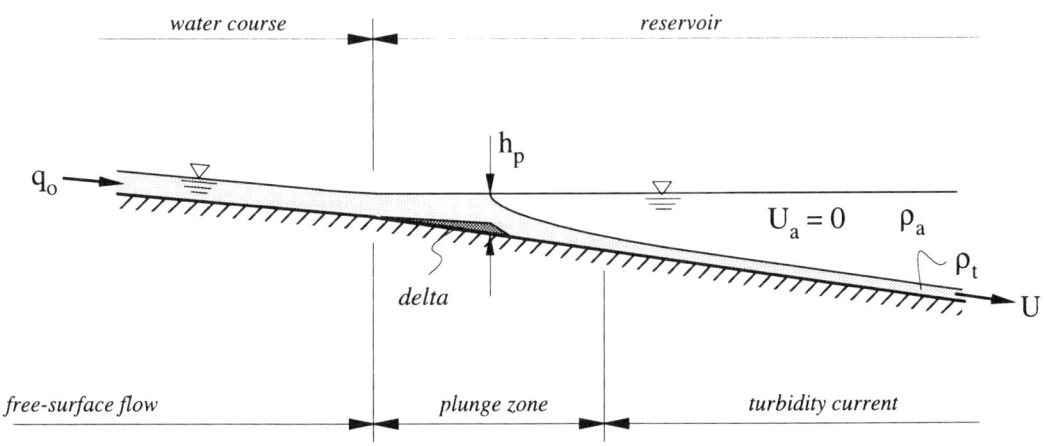

Fig. 7.2 Scheme of the plunge zone.

2° The water depth, h_p, at the plunge point can be calculated by considering that the momentum of the flow (in channel and reservoir) is conserved. An approximate relation (see *Akiyama* et *Stefan*, 1985) is given with the Froude number at the plunge point :

$$\text{Fr}_p = \frac{q_o/h_p}{(g'h_p)^{1/2}} \cong 0.68$$

being valid for channel slopes, $0.017 < S_f < 0.123$; q_o is the unit discharge of the entering current.

3° A turbidity current, when it passes along a reservoir, lake or ocean, has a tendency to deposit its granular material, causing rather important sedimentation (see *Graf, 1983*). Spectacular sedimentation was observed behind Hoover dam and Elephant Butte dam, but also in the delta formed by the Rhone river entering Lake Geneva or in the submarine Scripps Canyon of the Pacific Ocean.

However, turbidity currents can also be beneficial for the life of artificial reservoirs. During floods, the deposited sediments can possibly create turbidity currents, which in turn transport the sediments downstream towards the dam. By clever manipulation of the bottom outlets, turbidity currents can be used to evacuate the sediments accumulated in the reservoir.

7.2 HYDRODYNAMIC EQUATIONS

1° Consider the body of a turbidity current, two-dimensional and plane, with the flow, $\vec{V}(u, 0, w)$, being turbulent and incompressible (see Fig. 7.1 and Fig. 7.3). The height, h, the velocity, U, and the concentration, C_s, are average values, defined by the integral scales (see eqs. 7.9, 7.10 and 7.11). The current moves in the longitudinal direction, x, over a bottom slope, S_f, with an angle, α, under a deep layer, $H \gg h$, of ambient stagnant fluid of density, ρ_a, being slightly smaller than the density of the turbidity current, $\rho_t > \rho_a$.

The flow is well established, continuous, steady and gradually varied. The equation of continuity and of motion for the fluid phase (mixture of water/sediment) and for the solid phase, may be established (see Fig. 7.3). This current, being relatively thin, $h \ll H$, is taken to be a boundary-layer flow, where the conditions of $u \gg w$ and $\partial/\partial z \gg \partial/\partial x$ are valid.

2° The *equation of continuity for the fluid phase* is written as :

$$\frac{\partial u}{\partial x} + \frac{\partial w}{\partial z} = 0 \qquad (7.12)$$

where u and w are (time) averaged (point) velocities in the x and z-directions.

After integration, over the depth, $0 < z < h_t$, one writes :

$$\int_0^{h_t} \frac{\partial u}{\partial x} dz + \int_0^{h_t} \frac{\partial w}{\partial z} dz = \left(\frac{\partial}{\partial x}\int_0^{h_t} u \, dz - u_{h_t}\frac{\partial h_t}{\partial x}\right) + w_{h_t} - w_b = 0$$

where u_{h_t} and w_{h_t} are the velocity components at the interface. By definition, the horizontal velocity at the interface and the vertical velocity at the bed are zero: $u_{h_t} = 0$ and $w_b = 0$. The vertical velocity at the interface, $w_{h_t} = W_h$, is defined as the velocity of entrainment of the ambient fluid into the current. Making use of the definition given with eq. 7.9, one obtains:

$$\frac{\partial}{\partial x}(Uh) = -W_h \tag{7.13}$$

The entrainment velocity, $-W_h$, is assumed (see *Turner*, 1973, p. 179) to be proportional to the velocity of the turbidity current, U, such that:

$$-W_h = E_w U \tag{7.14}$$

where the constant of proportionality, E_w, is the entrainment coefficient of ambient fluid (see eq. 7.6), which in turn depends upon the global Richardson number, Ri.

The entrainment of ambient fluid can be demonstrated with the Bernoulli equation. In the ambient fluid, where is no motion, the static pressure is larger than pressure in the current with motion. The resulting gradient of pressure will cause fluid penetration from the ambient fluid into the turbidity current.

3° The *equation of continuity for the solid phase* is given by the equation of diffusion of granular material (see *Graf*, 1971, chap. 8.3) or:

$$\frac{\partial(uc_s)}{\partial x} + \frac{\partial(wc_s)}{\partial z} = v_{ss}\frac{\partial c_s}{\partial z} + \varepsilon_s\frac{\partial^2 c_s}{\partial z^2} \tag{7.15}$$

where $v_{ss} (\cong v_{ss} \cos\alpha)$ is the settling velocity and ε_s is the diffusion coefficient of granular material. The diffusion term can be expressed by Elder's relation:

$$\varepsilon_s \frac{\partial^2 c_s}{\partial z^2} \cong -\frac{\partial}{\partial z}(\overline{c'_s w'})$$

where $(\overline{c'_s w'})$ is the Reynolds flux of the solid phase (sediments).

After integration, over the depth, $0 < z < h_t$, one writes:

$$\frac{\partial}{\partial x}\int_0^{h_t}(uc_s) \, dz = -v_{ss} c_s\big|_{z=b} + (\overline{c'_s w'})\big|_{z=b}$$

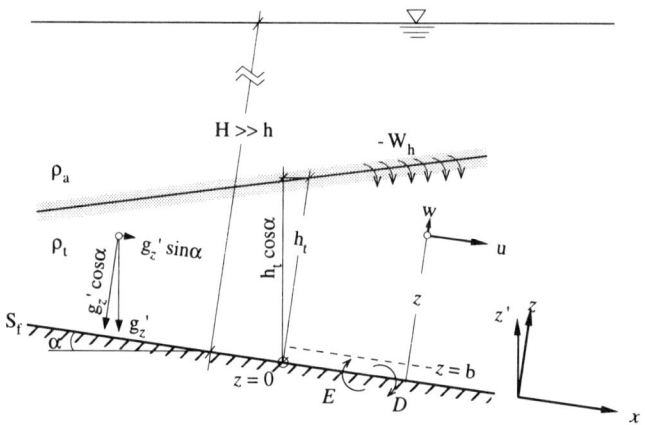

Fig. 7.3 Scheme of the body of a turbidity current.

The values are evaluated at a distance, $z = b$, close to the bed; at a distance, $z = h_t$, they are zero.

It has been proposed (see *Parker* et al., 1987) that :

$$(\overline{c'_s w'})|_b = v_{ss} E_s = E(b) \tag{7.16}$$

represents the erosion of sediments, with E_s as the entrainment coefficient of sediments from the bed, and that :

$$(v_{ss} c_s)|_b = v_{ss} c_b = D(b) \tag{7.17}$$

represents the deposition of sediments on the bed.

One obtains now, utilizing the definition of eq. 7.11, the following :

$$\frac{\partial (C_s U h)}{\partial x} = v_{ss}(E_s - c_b) = E(b) - D(b) \tag{7.18}$$

Consequently (see eqs. 7.7 and 7.8), if :

i) $E > D$:   the turbidity current is non-conservative and erosive;

ii) $E < D$:   the turbidity current is non-conservative and depositive;

iii) $E = D$:   the turbidity current is conservative and/or in equilibrium.

The criterion of autosuspension advanced by *Bagnold*, written as :

$$\frac{U \sin\alpha}{v_{ss}} > 1$$

gives also an information (see *Middleton*, 1984) about existence of turbidity current.

TURBIDITY CURRENTS

A turbidity current can be auto-generative : after its initiation, it may develop sufficient velocity to entrain more and more sediments; its reduced sediment flux, B, increases consequently and produces movement.

4° The *equation of motion of the turbidity current* — the equation of continuity, eq. 7.12, and the conditions of boundary-layer flow will be used — can be written as :

$$\frac{\partial u^2}{\partial x} + \frac{\partial (uw)}{\partial z} = -\frac{1}{\rho_t}\frac{\partial}{\partial x}(p_t + \rho_t g z') + \frac{1}{\rho_t}\frac{\partial \tau_{zx}}{\partial z} \qquad (7.19)$$

where the shear stress for turbulent flow is expressed as :

$$\tau_{zx} = -\rho_t \overline{(u'w')} \qquad (7.20)$$

The pressure in the fluid of the turbidity current, p_t, can be decomposed in a pressure, p_a, of the ambient fluid and a supplementary one, $p_s = p_t - p_a$, which is due to the presence of the sediments. Subsequently, one assumes the vertical pressure distribution, due to the ambient fluid, to be hydrostatic, $(p_a + \rho_a g z') = $ Cte, and one may write :

$$\frac{\partial}{\partial x}(p_t + \rho_t g z') = \frac{\partial}{\partial x}(p_a + \rho_a g z') + \frac{\partial}{\partial x}(p_s + \Delta\rho g z') = \frac{\partial p_s}{\partial x} + \frac{\partial(\rho_t g_z' z')}{\partial x} =$$

$$= \rho_t g R \cos\alpha \, \frac{\partial}{\partial x} \int_z^H c_s dz - \rho_t g R c_s \sin\alpha \qquad (7.21)$$

where $gRc_s = g'_z$ (see eq. 7.3a), $z' = z \cos\alpha$, $(-dz'/dx) = \sin\alpha$ and $\Delta\rho g = \rho_a g' \cong \rho_t g'$.

After integration of eq. 7.19, over the depth, $0 < z < h_t$, one obtains :

$$\frac{\partial}{\partial x}\int_0^{h_t} u^2 dz - u_{h_t}^2 \frac{\partial h_t}{\partial x} + (u_{h_t}w_{h_t} - u_b w_b) =$$

$$= -gR\cos\alpha \, \frac{\partial}{\partial x}\int_0^{h_t} c_s(H-z)\,dz + gR\sin\alpha \int_0^{h_t} c_s dz - u_{*b}^2 \qquad (7.22)$$

By definition, the horizontal velocity at the interface and the vertical velocity at the bed are zero : $u_{h_t} = 0$ and $w_b = 0$. The friction velocity, u_{*b}, close to the bed is expressed by :

$$\int_b^{h_t}\frac{\partial \tau_{zx}}{\partial z}dz = -\tau_{zx}\big|_{z=b} + \tau_{zx}\big|_{z=h_t} \cong -\tau_{zx}\big|_{z=b} + 0 = -\rho_t u_{*b}^2 \qquad (7.23)$$

Subsequently, utilizing the integral scales (see eqs. 7.9, 7.10 and 7.11), the equation of motion can be written as:

$$\frac{\partial(U^2h)}{\partial x} = -\frac{1}{2}\cos\alpha \frac{\partial(S_1 g'h^2)}{\partial x} + S_2 g'h \sin\alpha - u_{*b}^2 \qquad (7.24)$$

where $g' = g R C_s$ (see eq. 7.3). The coefficients of the concentration profile, S_1 and S_2, are defined by:

$$S_1 = \frac{2}{g'h^2} \int_0^{h_t} (gRc_s)(H-z)\,dz$$

$$S_2 = \frac{1}{g'h} \int_0^{h_t} (gRc_s)\,dz \qquad (7.25)$$

The hypothesis that $S_1 \cong 1$ and $S_2 \cong 1$ is frequently assumed (see *Parker* et al., 1987 and *Altinakar* et al., 1993) being a rather good approximation.

The right-hand terms of eq. 7.24 represent the pressure force on the turbidity current, being due to the depth variation, the reduced gravity force, which accelerates the currents, and the friction force on the bed. For the case when the pressure and resistance forces are negligible, a simple form of the equation of motion, eq. 7.24, is now given by:

$$\frac{\partial(U^2h)}{\partial x} = g'h \sin\alpha \qquad (7.26)$$

5° In summary, the equations, vertically integrated over the body of the turbidity current, being one-dimensional and steady but non-conservative, and moving on a slope, $S_f = \sin\alpha$, are:

$$\frac{d}{dx}(Uh) = E_w U \qquad (7.13a)$$

$$\frac{d}{dx}(C_s Uh) = v_{ss}(E_s - c_b) \qquad (7.18)$$

$$\frac{d}{dx}(U^2h) = -\frac{1}{2}gR\cos\alpha \frac{d}{dx}(C_s h^2) + (gR C_s h)\sin\alpha - u_{*b}^2 \qquad (7.24a)$$

The equations being integrated over the vertical, are written with ordinary derivatives, thus replacing the partial derivatives.

TURBIDITY CURRENTS

There are three unknowns, h, U and C_s, and three equations, eqs. 7.13a, 7.18 and 7.24a. The empirical relations for the parameters, E_w, E_s, v_{ss}, c_b and u_{*b}, must still be developed.

Note, that for a conservative turbidity current, the system of equations, eqs. 7.13a, 7.18 and 7.24a, remains valid, but the underlined term in eq. 7.18 will be zero.

This system of equations can still be supplemented with an equation for the turbulent energy (see *Parker et al.*, 1987).

7.3 INTERFACE-PROFILE CURVES

1° The evolution of the turbidity current, dh/dx, gives the form of the interface between the current, h, and the deep layer, H, of the ambient fluid (see Fig. 7.1).

By using the equations of continuity, eq. 7.13a and eq. 7.18, together with the equation of motion, eq. 7.24a, one can obtain:

$$\frac{dh}{dx} = \frac{1}{(1-Ri)} \left\{ \frac{1}{2}(4-Ri) E_w + \frac{1}{2} Ri \frac{v_{ss}}{UC_s}(E_s - c_b) - Ri\, tg\alpha + \left(\frac{u_{*b}}{U}\right)^2 \right\} \quad (7.27)$$

with the global Richardson number being:

$$Ri = \frac{g'h\cos\alpha}{U^2} = \frac{gRC_s h\cos\alpha}{U^2} \quad (7.5)$$

2° The Richardson number, Ri, allows also a parametrisation of the reduced sediment flux, B (see eq. 7.7); this is given as:

$$Ri = \frac{g'h\, U\cos\alpha}{U^3} = \frac{B\cos\alpha}{U^3} \quad (7.5b)$$

The variation of the Richardson number, Ri, thus of the reduced sediment flux, B, is thus obtained as being:

$$\frac{h}{3Ri}\frac{dRi}{dx} = \frac{1}{(1-Ri)}\left\{ \left[E_w + \frac{1}{3}\frac{v_{ss}}{UC_s}(E_s - c_b)\right]\frac{1}{2}(2+Ri) - Ri\, tg\alpha + \left(\frac{u_{*b}}{U}\right)^2 \right\} \quad (7.28)$$

3° Note, that for a conservative turbidity current, these relations, eq. 7.27 and eq. 7.28, remain valid (see *Turner*, 1973, p. 180), but the underlined terms will be zero.

Such a conservative turbidity current is in *equilibrium*, if the Richardson number is independent of the distance, x; or $dRi/dx = 0$. Consequently, the solution between these relations, eq. 7.27 and eq. 7.28, gives:

$$\frac{dh}{dx} = E_w$$

which upon integration, describes a linear growth of the current:

$$h = E_w (x-x_o) + h_o$$

where the index zero, o, refers itself to the beginning point. The velocity of the current can now be obtained using eq. 7.5b, or:

$$U = \left(\frac{B \cos\alpha}{Ri}\right)^{1/3} = \left(\frac{g'q \cos\alpha}{Ri}\right)^{1/3}$$

4° For a conservative turbidity current over a weak slope, thus without entrainment of ambient fluid, $E_w \cong 0$, the eq. 7.27 becomes:

$$\frac{dh}{dx} = \frac{(u_{*b}/U)^2 - Ri\, S_f}{(1 - Ri)} \tag{7.27a}$$

This relation may be compared with the equations of the water-surface profile curves for non-uniform flow in open channels (see eq. 4.8), if one assumes that:

- the gravity replaces the reduced gravity: $g' \Rightarrow g$;

- the densimetric Froude number becomes the Froude number (see eq. 7.5):

$$Ri = \frac{1}{Fr_D^2} \Rightarrow \frac{1}{Fr^2}$$

- the friction coefficient is expressed (see eq. 3.8) by:

$$\left(\frac{u_{*b}}{U}\right)^2 = \frac{g}{C^2}$$

where C is the Chézy coefficient.

5° Thus it is evident that there is an analogy between (internal) flow of the turbidity current and (external) flow of free-surface channel flow.

Consequently, all information developed for free-surface open-channel flow in non-uniform state (see chap. 4) will remain valid — if the reduced gravity, g', and the entrainment process, E_w and E_s, are respected — for the calculations of the interface of an internal hydraulic jump, of a flow over sills or other obstructions (see *Turner*, 1973, p. 64).

Furthermore, a distinction can be made between subcritical flow, where $Ri > 1$ or $Fr_D < 1$, and supercritical flow, where $Ri < 1$ or $Fr_D > 1$ (see *Turner*, 1973, p. 181 and p. 64).

6° Under certain conditions, the non-uniform flow, described by eq. 7.27, can be approximated by *uniform flow* ; thus $(dh/dx) = 0$. Here, it must be assumed that the turbidity current be conservative (the underlined terms in eq. 7.27 are zero) and the fluid entrainment is negligible ($E_w \cong 0$, valid for small slopes). Thus eq. 7.27a can be expressed by :

$$0 = - Ri\, tg\alpha + \left(\frac{u_{*b}}{U}\right)^2 \tag{7.29}$$

With the use of the definition of the Richardson number, eq. 7.5, and by taking $S_f \cong \sin\alpha$, one obtains :

$$g'h\, S_f = u_{*b}^2 \tag{7.29a}$$

Expressing the friction velocity, u_{*b}, with the friction coefficient (see eq. 3.8), one gets the velocity of the current (see eq. 7.9) as :

$$U = \sqrt{8/f_{CT}}\, \sqrt{g'h\, S_f} \tag{7.30}$$

This relation can be compared to the Weisbach-Darcy equation (see eq. 3.10) established for open-channel flow.

7° The friction coefficient, f_{CT}, for the uniform motion of a turbidity current, can be determined (see *Graf*, 1983) as follows :

 i) as a first approximation, one may take the friction coefficient for free-surface open-channel flow (see sect. 3.2), or $f_{CT} \equiv f$; herewith one neglects completely the friction of the turbidity current at the interface;

 ii) according to Harleman, considering that the turbidity current is delayed by friction at the bed and equally at the interface, one takes :

$$f_{CT} = f(1 + \alpha_H)$$

where f is the friction coefficient given with the Moody-Stanton diagram (see *Graf & Altinakar*, 1991, p. 438) and $\alpha_H \cong 0.43$ being valid for turbulent flow.

7.4 ENTRAINMENT COEFFICIENTS

1° In the three equations, eqs. 7.13a, 7.18 and 7.24a, which describe the dynamics of a turbidity current, it is still necessary to specify the entrainment parameters, E_w, $v_{ss}(E_s - c_b)$, and the friction velocity, u_{*b}. These parameters are given with empirical relations.

2° In the equation of continuity for the fluid phase, eq. 7.13, it still is necessary to find a value for the entrainment coefficient, E_w, of the ambient fluid.

Many experiments (notably in the laboratory) with different types of density and turbidity currents for large Reynolds numbers have been done and are summarised in Fig. 7.4 (see *Turner*, 1973, p. 182 and *Altinakar* et al., 1993). The coefficient, E_w, related to the global Richardson number, Ri, is given by the following empirical relationship (see *Parker* et al., 1987), obtained by regression:

$$E_w = 0.075 \, (1 + 718 \, Ri^{2.4})^{-0.5} \qquad (7.31)$$

where the experimental dispersion is rather large. Note, the rapid decrease of E_w with increasing Richardson numbers, Ri; the slope of the bed plays also an important part.

An indicative relationship can also be obtained (see *Turner*, 1973, p. 180), when putting together eqs. 7.13 and 7.14 with eq. 7.26, such as :

$$E_w = Ri \, tg \, \alpha$$

being valid for large slopes, $\alpha > 12°$, when the current is in equilibrium.

On the Fig. 7.4, one can observe :

i) if $Ri \to 0$, $E_w \cong 0.075$

such a value is obtained for an impulse jet where the effect of the reduced gravity is negligible, and flow is without stratification.

ii) if $Ri \simeq 1$, $E_w \simeq 0.003$

such a value is obtained if the entrainment of ambient fluid is negligible, notably for $Ri > 1$, when the flow is subcritical at weak slopes of the bed.

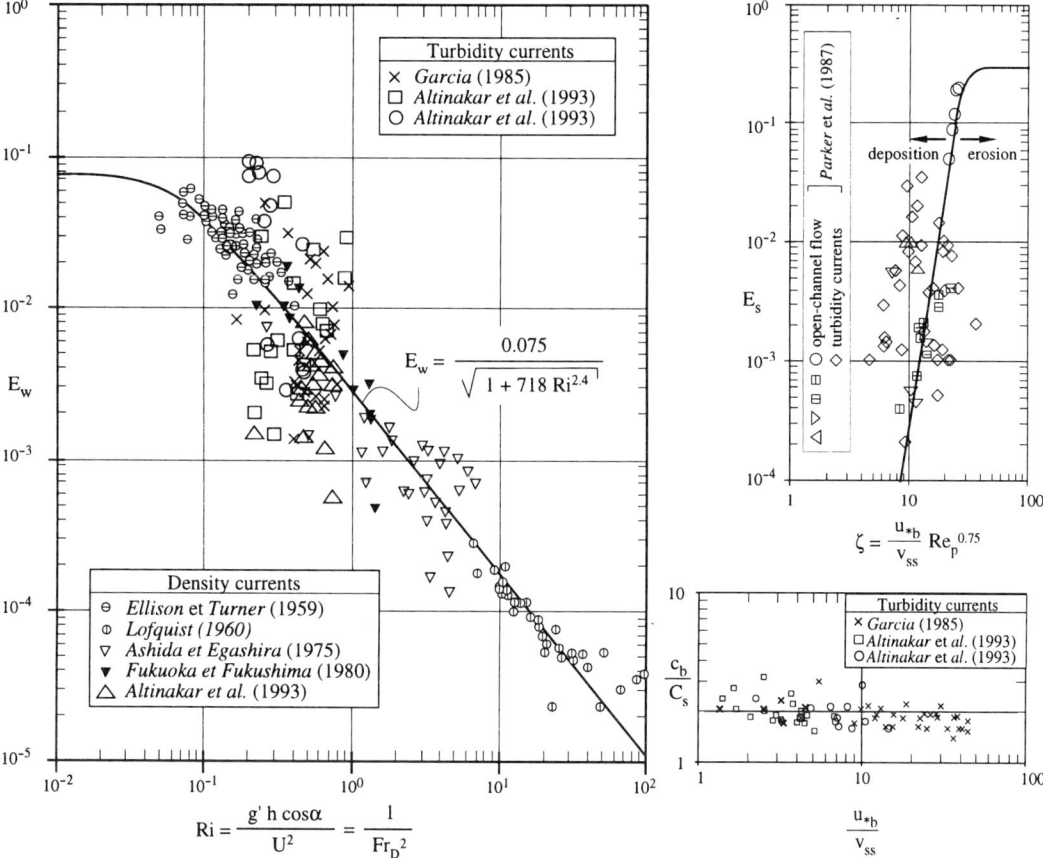

Fig. 7.4 Entrainment coefficient, $E_w(Ri)$, of ambient fluid and $E_s(\zeta)$, of sediments; reference concentration, c_b.

iii) for $Ri > 0.05$

a simple relationship (see *Egashira* et *Ashida*, 1980) can be used :

$$E_w = 0.0015 \ Ri^{-1} \tag{7.31a}$$

Across a (stable) interface, if $Ri > 1$, there is almost no mixing, but *internal gravity waves* can form themselves being due to small perturbations (see *Turner*, 1973, p. 14).

If these are *long* internal waves of small amplitude, the celerity is given by :

$$c_i^2 = g'h$$

This relation can be compared to the celerity, $c > c_i$, of the surface wave (see eq. 2.27). The vertical displacement, η_i, of the internal wave is larger than the one, η, of the surface wave, and this by a factor of :

$$\frac{\eta}{\eta_i} \cong O\left(\frac{\Delta\rho}{\rho_a}\right)$$

These internal waves can break and thus will contribute somewhat to the mixing (entrainment), if the criterion of *Keulegan* (see *Turner*, 1973, p. 108) is fulfilled :

$$\frac{Ri}{Re} = \frac{\nu\, g'}{U^3} < (0.18)^3$$

and the flow of the turbidity current stays turbulent.

3° In the equation of continuity for the solid phase, eq. 7.18, it is still necessary to find a value for the entrainment coefficient, $v_{ss}(E_s - c_b)$, of the sediments.

The experiments, performed in open-channel flows with sediment transport (see *Akiyama* et *Stefan*, 1985), and in flows of turbidity currents (see *Altinakar* et al., 1993), show that the following empirical relation, proposed by *Parker* et al. (1987), can be used (see Fig. 7.4) :

$$E_s = \frac{3 \cdot 10^{-11}\, \zeta^7}{1 + 10^{-10}\, \zeta^7} \tag{7.32}$$

Using a definition of a particle Reynolds number, Re_p, one defines :

$$\zeta = \frac{u_{*b}}{v_{ss}} Re_p^{0.75} \qquad \text{with} \qquad Re_p = \frac{d_{50}\sqrt{gRd_{50}}}{\nu}$$

This relation, eq. 7.32, is rather steep and arrives at a maximum constant value where the entrainment coefficient is about $E_s \cong 0.3$.

4° The reference concentration, c_b, being evaluated close to the bed, $b \cong 0.05\, h_t$, is given (see *Graf*, 1971, p. 173) by :

$$\frac{c_b}{C_s} = f\left(\frac{u_{*b}}{v_{ss}}\right)$$

From experiments with turbidity currents (see *Altinakar* et al., 1993 and *Parker* et al., 1987), it is found (see Fig. 7.4) that :

TURBIDITY CURRENTS

$$\frac{c_b}{C_s} \cong 2 \tag{7.33}$$

This value remains more or less constant for $1 < u_{*b}/v_{ss} < 50$; C_s is the average concentration in the cross section.

5°   The settling velocity, v_{ss}, can be calculated (see Fig. 6.10) with different methods described in the literature (see *Graf*, 1971, chap. 4) ; it can be expressed as :

$$v_{ss} = \sqrt{\frac{4}{3} g \frac{\rho_s - \rho}{\rho} \frac{1}{C_D} d} \tag{7.34}$$

where C_D is the particle's drag coefficient (see *Graf & Altinakar*, 1991, p. 388) and d is the particle diameter.

6°   The friction velocity, u_{*b}, on the bed is given (see sect. 3.1.3) by :

$$u_{*b}^2 = \frac{\tau_o}{\rho} = \left(\frac{f}{8}\right) U^2 \tag{7.35}$$

where f is the friction coefficient of Weisbach-Darcy, given with eq. 3.13. Also used can be the friction coefficient of Chézy or of Manning, or any other one (see sect. 3.2).

7.5   FRONT OF CURRENT

1°   The flow of the *body* of the turbidity current is quasi-uniform and steady, while the one of the preceding *front* is usually unsteady (see Fig. 7.1).

It is with the front that the turbidity current displaces the ambient fluid and penetrates into it. The driving force is essentially the gradient of the pressure, resulting from the density difference, $\Delta\rho$, between the front, ρ_t, and the ambient fluid, ρ_a.

The front has a characteristic shape (see Fig. 7.5); its height, H_f, is the highest point, being well defined. The front — which is rather irregular — is a region where the mixing of ambient fluid into the current is very active. The velocity, U_f, of the front is the celerity of the advancement of the front. The front has a nose which is situated, h_f, slightly above the bed.

2°   The velocity of the front, U_f, of the turbidity current can be calculated by using simple hydraulic considerations (see *Turner*, 1973, p. 73). It is assumed that the turbidity current is quasi-uniform, moving on an horizontal slope without friction and that mixing (entrainment) can be neglected. The difference in hydraulic pressure — far before and after the front — is given by : $g(\rho_t-\rho_a)h$. This produces a dynamic pressure at the stagnation point of : $1/2\, \rho_t U_f^2$. Thus one may write — assuming that $\rho_t \simeq \rho_a$ — :

$$U_f = \sqrt{2\left(\frac{\rho_t-\rho_a}{\rho_a}\right)gh} = \sqrt{2g'h} \qquad (7.36)$$

3° Using various experiments from turbidity currents — but also from density currents — for a large range of slopes and of roughness, a simple and useful relationship was proposed (see *Altinakar* et al., 1990), such as :

$$U_f = 0.75\sqrt{g'H_f} \qquad (7.37)$$

where H_f is the height of front of the current.

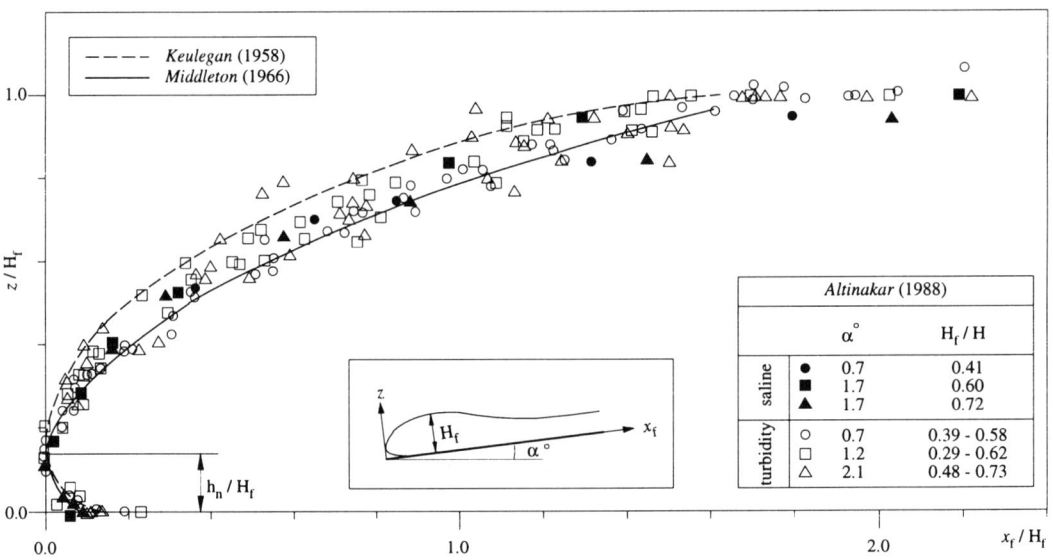

Fig. 7.5 Front of turbidity current; dimensionless height, H_f, of the front.

4° A comparison between the velocity of the body and the one of the front — assuming the flow is uniform — can be done by using eq. 7.30 and eq. 7.36 :

$$\frac{U_f}{U} = \frac{\sqrt{2g'h}}{\sqrt{8/f_{CT}}\sqrt{g'hS_f}} = \frac{\sqrt{f_{CT}}}{\sqrt{4\, S_f}}$$

Taking a reasonable value for the friction coefficient — for example $f = 0.03$ — one obtains (see point 7.3.7°) : $f_{CT} = f(1 + 0.43) \simeq 0.04$. It is now evident that — for slopes $S_f \geq 0.01$ or $\alpha \geq 0.7°$ — the velocity of the front remains always smaller than the one of the body, $U_f < U$. This inequality grows, if the slope, S_f, increases (see Fig. 7.6). However, in order to maintain the flow continuity, the height of the front will always be larger than the one of the body, $H_f > h$.

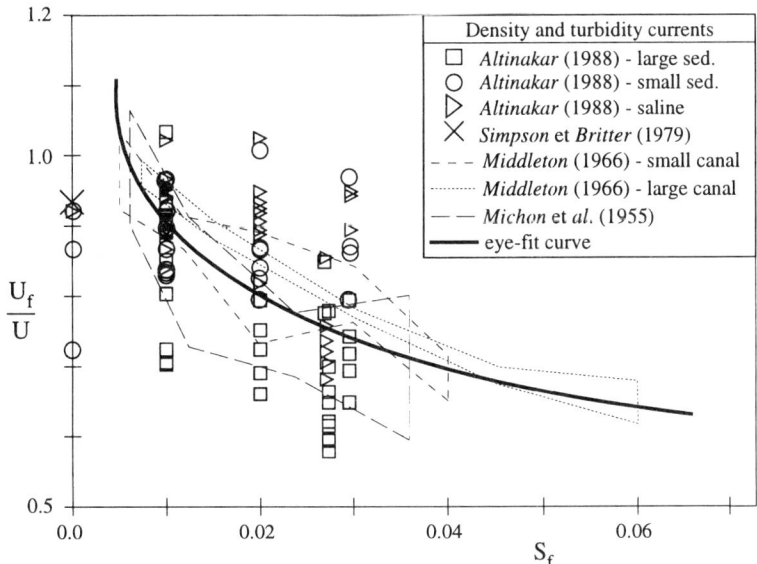

Fig. 7.6 Ratio of velocity of front to velocity of body, U_f/U, as function of bottom slope, S_f.

5°   Furthermore, it has been observed (see *Britter* et *Linden*, 1980 and *Altinakar et al.*, 1990) that velocity of the front, U_f, remains more or less constant and independent of the distance covered by the turbidity current. This implies that the reduced gravity compensates the frictional force (see eq. 7.24). However, for turbidity currents over small slopes, $\alpha \leq 0.5°$, and for depositing ones, the front of the current decelerates slightly. The height of the front, H_f, increases with the distance covered; this is attributed to the entrainment of the ambient fluid.

6°   The velocity of the front, U_f, can also be related to the reduced sediment flux, $B_o = g'_o q_o$, of the entering current and to the bed slope, α, or :

$$U_f = (g'_o q_o)^{1/3} f(\alpha) \tag{7.38}$$

The roughness of the bed, f_{CT}, and the entrainment coefficient, E_w, as well as the Reynolds number, $Re = Uh/\nu$, could also play a role.

In order to determine this relation, eq. 7.38, experimental data from turbidity and density currents can be used (see Fig. 7.7). While the dispersion of these data is rather large, the following tendencies can be seen :

i)   for large slopes (see *Britter* et *Linden*, 1980) :

$$5° \leq \alpha \leq 90° \quad \Rightarrow \quad \frac{U_f}{B_o^{1/3}} = 1.5 \pm 0.2$$

the dimensionless velocity stays essentially constant;

ii)   for small slopes (see *Altinakar* et al., 1990) :

$$\alpha < 5° \quad \Rightarrow \quad 0.7 < \frac{U_f}{B_o^{1/3}} \leq 1.5$$

the dimensionless velocity varies linearly.

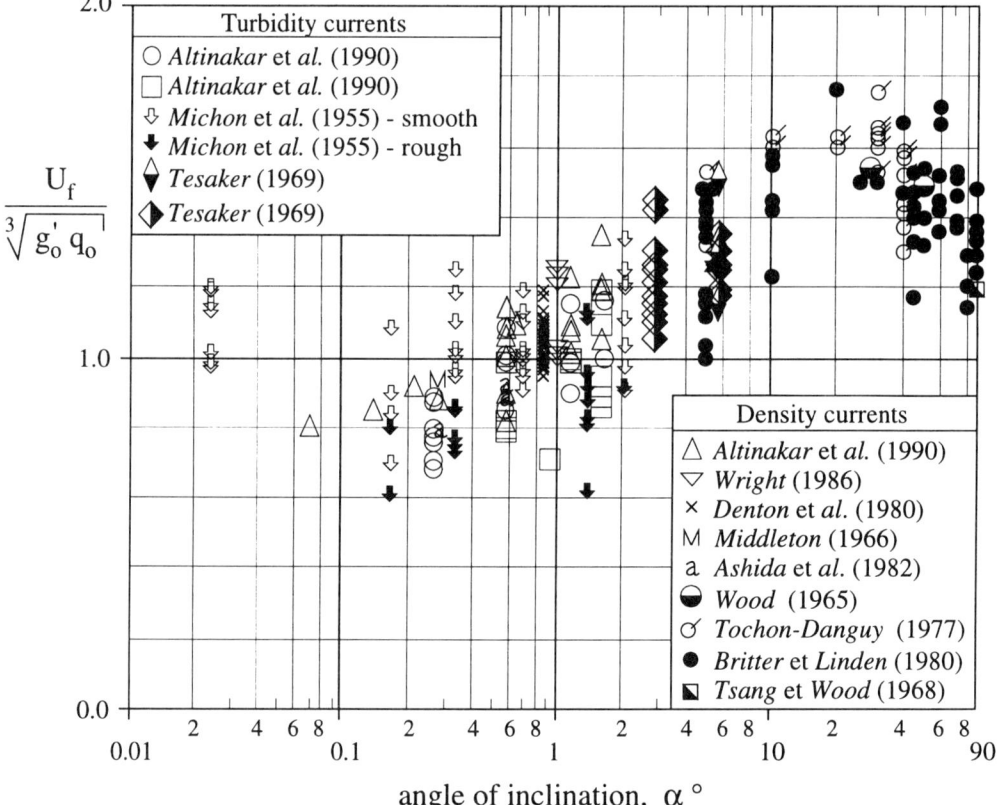

Fig. 7.7 Dimensionless velocity of front, $U_f/B_o^{1/3}$, as function of the bottom slope, α.

7.6 DISTRIBUTION OF VELOCITY AND OF CONCENTRATION

1° The body of a turbidity current may be assimilated to a *wall jet*, being turbulent and established; one may distinguish two regions : the wall and the jet region. The height, h_m, where the maximum velocity, $u = U_m$, is measured, separates these regions (see Fig. 7.8) :

   i) in the wall region, $z < h_m$, turbulence is created at the wall (bottom); entrainment of sediments, E_s, can take place ;

   ii) in the (free) jet region, $z > h_m$, turbulence is created by friction and by entrainment, E_w, of ambient fluid.

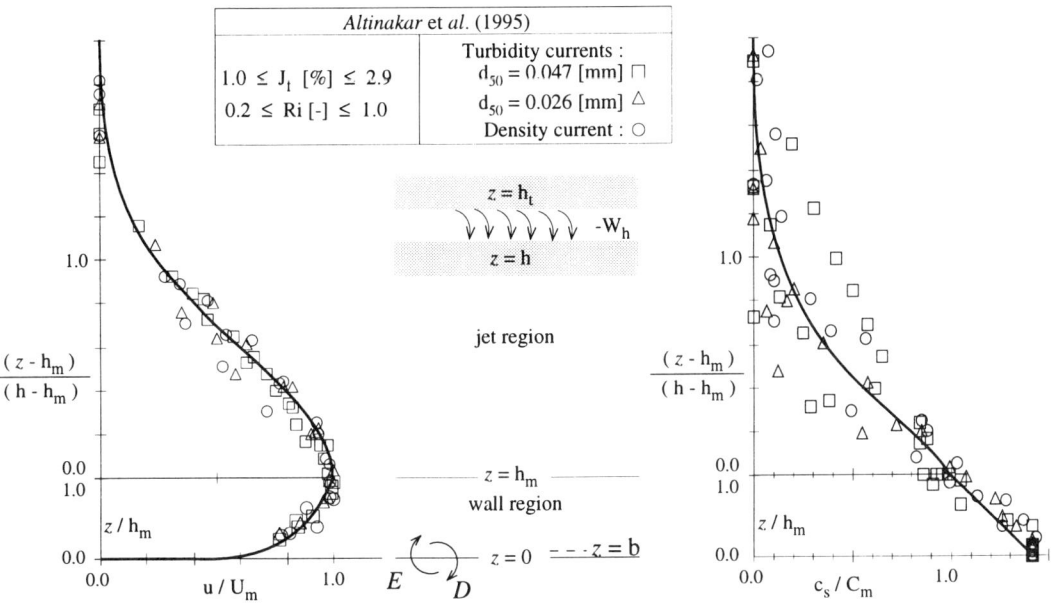

Fig. 7.8 Profiles of velocity, u(z), and of concentration, $c_s(z)$, measured for different types of gravity currents.

2° The distribution of the local velocity, u, of a turbidity current (see Fig. 7.8) in each of these regions is given as follows :

   i) in the wall region, $z < h_m$, it is logarithmic :

$$\frac{u(z)}{u_*} = \frac{1}{\kappa} \ln(z) + \text{Cte} \qquad (2.50)$$

   but may also be expressed by an empirical relation :

$$\frac{u(z)}{U_m} = \left(\frac{z}{h_m}\right)^n \qquad (2.50a)$$

   where one may take n = 1/6 (see *Altinakar et al.*, 1996).

ii) in the jet region, $z > h_m$, it is gaussian (see *Hug*, 1975, p. 270):

$$\frac{u(z)}{U_m} = \exp\left[-\alpha_c \left(\frac{z - h_m}{h - h_m}\right)^2\right] \qquad (7.39)$$

where one may take $\alpha_c \cong 1.4$ (see *Altinakar* et al., 1996).

Experimentally it was found (see *Altinakar* et al., 1996) that:

$$\frac{h_m}{h} \approx 0.3 \quad ; \quad \frac{U_m}{U} \approx 1.3 \quad ; \quad \frac{h_t}{h} \approx 1.3 \quad ; \quad \frac{U}{\overline{U}} \approx 1.3 \qquad (7.40)$$

where h and U are the height and the velocity, as defined by the integral scales (see eqs. 7.9 and 7.10); h_t is the height where the point velocity, u, is zero and $\overline{U}$ is the average velocity of the current.

3° The distribution of the local concentration, c_s, of the turbidity current (see Fig. 7.8) in each of the regions is given as follows:

i) in the wall region, $z < h_m$, it is given (see *Graf*, 1971, p. 189 or sect. 6.4.2) by:

$$\frac{c_s(z)}{c_b} = \left(\frac{h_m - z}{z} \frac{b}{h_m - b}\right)^{\zeta} \qquad (6.51)$$

where c_b is a reference concentration evaluated at a distance rather close to the bed, $b \cong 0.05\, h_t$. The exponent is, according to *Rouse*, given by $\zeta = v_{ss}/\kappa u_{*b}$. If this value is large, the concentration of the granulate will accumulate on the bottom. It is also possible to approximate the distribution by a linear relation:

$$\frac{c_s(z) - C_m}{c_b - C_m} = \frac{h_m - z}{h_m} \qquad (7.41)$$

ii) in the jet region, $z > h_m$, it is gaussian (see *Hug*, 1975, p. 270):

$$\frac{c_s(z)}{C_m} = \exp\left[-\beta_c \left(\frac{z - h_m}{h - h_m}\right)^{4/3}\right] \qquad (7.42)$$

where one takes $1.7 \leq \beta_c \leq 4.1$, a value which increases with the distance, x, (see *Altinakar* et al., 1996); C_m is the concentration evaluated at $z = h_m$.

Experimentally it was found (see *Altinakar* et al., 1996) that:

$$\frac{c_b}{C_s} \approx 2 \quad ; \quad \frac{c_b}{C_m} \approx 1.4 \qquad (7.43)$$

where C_s is the concentration defined by the integral scale (see eq. 7.11).

A preliminary version of sects 7.1 to 7.5 was published under the title:
Graf, W.H. et M.S. Altinakar (1995): *Courants de Turbidité*.
La Houille blanche, year 50, N°7, Paris

TURBIDITY CURRENTS

7.7 EXERCISES

7.7.1 Problems, solved

Ex. 7.A

Following a heavy rainfall causing a strong erosion in its catchment, the river *Lentier* is carrying a considerable amount of suspended sediments. This river flows into the reservoir *Cloison*, whose bed is made up of the same sediments carried by the river. The reservoir bed has a slope of $\alpha = 5°$ and a friction coefficient of $f = 0.032$ [-]. The muddy river water forms a turbidity current in the reservoir. At a certain distance downstream from the plunge point, where the reservoir depth is $H_o = 10$ [m], the following observations (measurements) are made :

- height of the current : $h_o = 1.0$ [m]
- unit discharge of the current : $q_o = 1.0$ [m²/s]
- turbidity of the current : $\Delta\rho_o = 35$ [kg/m³]
- density of the ambient water : $\rho_a \cong 1000$ [kg/m³] at $T = 10$ [°C]
- sediments (quartz) : $d_{50} = 0.05$ [mm], $\rho_s \cong 2650$ [kg/m³]

i) Predict the variation of the current depth, h(x), the current speed, U(x), and the average concentration, $C_s(x)$, along the reservoir, from the measuring station to a section $L = 1000$ [m] downstream.

ii) Estimate the water depth, h_p, at the plunge point.

iii) Repeat the same calculations considering a sediment with $d_{50} = 0.10$ [mm]; all the other parameters of the problem remain the same. Compare the results.

iv) Repeat the same calculations considering a fixed bed and a sediment whose settling velocity can be neglected. All the other parameters of the problem remain the same. Compare the results.

v) Estimate the distribution of the velocity, u(z), and of the concentration, $c_s(z)$, at the stations located at $x = 0$ [m], $x = 500$ [m] and $x = 1000$ [m], for the case with the sediments of $d_{50} = 0.05$ [mm].

SOLUTION :

i) Determination of the variation of the current depth

a) *Equations for the body of a turbidity current*
   In sect. 7.3, by combining the continuity equations, eq. 7.13a and eq. 7.18, with the equation of motion, eq. 7.24a, we have obtained a set of two differential equations expressing the variation of the current depth, h(x), eq. 7.27, and of the overall Richardson number, Ri(x), eq. 7.28, as a function of the distance :

$$\frac{dh}{dx} = \frac{1}{(1-\text{Ri})} \left\{ \frac{1}{2}(4-\text{Ri})E_w + \frac{1}{2}\text{Ri}\frac{v_{ss}}{UC_s}(E_s - c_b) - \text{Ri tg}\alpha + \left(\frac{u_{*b}}{U}\right)^2 \right\}$$

$$\frac{h}{3\text{Ri}}\frac{d\text{Ri}}{dx} = \frac{1}{(1-\text{Ri})} \left\{ \left[E_w + \frac{1}{3}\frac{v_{ss}}{UC_s}(E_s - c_b)\right]\frac{1}{2}(2+\text{Ri}) - \text{Ri tg}\alpha + \left(\frac{u_{*b}}{U}\right)^2 \right\}$$

By using eq. 7.18, one can also derive a third differential equation describing the variation of the volume concentration of the granular material, $C_s(x)$:

$$\frac{d}{dx}(C_s Uh) = Uh \frac{d}{dx}(C_s) + C_s \frac{d}{dx}(Uh) = v_{ss}(E_s - c_b) \tag{7.18}$$

Using eq. 7.13a, one can write :

$$Uh \frac{d}{dx}(C_s) + C_s (E_w U) = v_{ss}(E_s - c_b)$$

thus one obtains :

$$\frac{dC_s}{dx} = \frac{1}{Uh}\left[v_{ss}(E_s - c_b) - C_s E_w U\right] \tag{7.18a}$$

However, supplementary relations for E_w, u_{*b}, v_{ss}, E_s and c_b are still necessary for closing the system of equations. These relations are the following :

- the entrainment coefficient of ambient water (see Fig. 7.4) :

$$E_w = 0.075 \, (1 + 718 \, Ri^{2.4})^{-0.5} \tag{7.31}$$

- the shear velocity at the bed :

$$u_{*b}^2 = \frac{\tau_o}{\rho} = (\frac{f}{8}) U^2 \tag{7.35}$$

- the settling velocity of the particles; for very fine particles the Stokes equation (see *Graf*, 1971, p.43) can be used to calculate the settling velocity :

$$v_{ss} = \frac{d^2}{18 \, \nu} g \frac{\rho_s - \rho_a}{\rho_a} = \frac{d^2 \, g \, R}{18 \, \nu}$$

- the entrainment coefficient of sediments (see Fig. 7.4) :

$$E_s = \frac{3 \cdot 10^{-11} \, \zeta^7}{1 + 10^{-10} \, \zeta^7} \quad \text{where} \quad \zeta = \frac{u_{*b}}{v_{ss}} \left(\frac{d_{50}\sqrt{gRd_{50}}}{\nu}\right)^{0.75} \tag{7.32}$$

- the reference concentration (see Fig. 7.4) :

$$\frac{c_b}{C_s} \cong 2 \tag{7.33}$$

The system of three differential equations, dh/dx, dRi/dx and dC_s/dx, can be solved simultaneously using numerical methods.

TURBIDITY CURRENTS

b) *Numerical methods*

The three equations, eqs. 7.27, 7.28 and 7.18a, are ordinary differential equations of the type :

$$\frac{dy}{dx} = f(x,y)$$

Different numerical methods are available for solving ordinary differential equations. A simple and direct one is the *method of Euler*, where one writes the differential equation directly as an equation of finite differences :

$$\frac{\Delta y}{\Delta x} \cong f(x,y) \quad \Rightarrow \quad \Delta y = \Delta x \, f(x,y)$$

If for a given value of x the value of the dependent variable, y_x, is known, then the next value of the dependent variable, $y_{x+\Delta x}$, for $(x+\Delta x)$ can be calculated from the following expression :

$$y_{x+\Delta x} = y_x + \Delta x \, \frac{\Delta y}{\Delta x} = y_x + \Delta x \, f(x,y_x)$$

This procedure is illustrated in Fig. Ex.7.A.1a. The method of Euler is non symmetrical, i.e.: the solution advances based only on the derivative, dy/dx, being the slope at the preceding point. In the practice, the use of Euler's method, which is a first-order method (i.e.: the error is of second order), is not recommended. It is neither stable nor sufficiently precise.

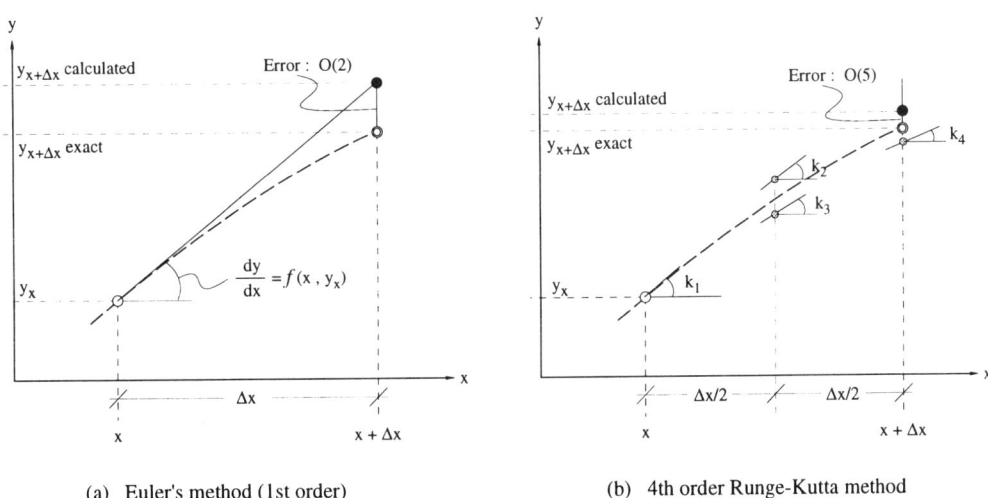

(a) Euler's method (1st order)   (b) 4th order Runge-Kutta method

Fig. Ex.7.A.1 Numerical methods for solving ordinary differential equations.

There exist other methods, being based on a similar principle, which allow to obtain solutions with a higher order. Among those methods, the *fourth-order*

Runge-Kutta method is the most popular one. The formulas for this simple and elegant method are given below, without going into their details :

$$y_{x+\Delta x} = y_x + \frac{\Delta x}{6}(k_1 + 2k_2 + 2k_3 + k_4)$$

where y_x : value of the dependent variable y at x
  $y_{x+\Delta x}$: value of the dependent variable y at $(x + \Delta x)$
  $k_1 = f(x, y_x)$
  $k_2 = f(x + \frac{\Delta x}{2}, y_x + \frac{\Delta x}{2} k_1)$
  $k_3 = f(x + \frac{\Delta x}{2}, y_x + \frac{\Delta x}{2} k_2)$
  $k_4 = f(x + \Delta x, y_x + \Delta x\, k_3)$

The graphical representation of this method is given in Fig. Ex.7.A.1b. In the fourth-order Runge-Kutta method, the right-hand side of the differential equation is evaluated 4 times. Each evaluation is a first-order evaluation but, the coefficients of the error terms are different. By combining these individual evaluations in a particular way the lower-order error terms are canceled out, yielding a fourth-order final result (i.e.: the error is of fifth order).

c) *Definition of the solution domain and the boundary conditions*
The system of ordinary differential equations, eqs. 7.27, 7.28 and 7.18a, describes a one-dimensional turbidity current moving along the x-axis. It is to be noted that a second dimension, z, is represented by taking depth-averaged values, h, U and C_s, defined by the integral scales, eqs. 7.9, 7.10 and 7.11. This system of equations can only be solved if the conditions, namely the values of the three independent variables, h, U and C_s, are known at a given distance, x, forming the upstream boundary of the solution domain. In the present case, the observed values of h_o, U_o and C_{so} at $x = x_o = 0$ will be used.

The fourth-order Runge-Kutta method will be used to solve the system of equations, eqs. 7.27, 7.28 and 7.18a, simultaneously. Since a one-dimensional turbidity current is considered, the solution domain can be represented as a line coinciding with the axis of the reservoir bed. The origin of the coordinate system, x-z, is placed at the measuring station, $x_o = 0$ (see Fig. Ex.7.A.6). The x-axis which coincides with the reservoir bed is divided into ND sections each having a length of Δx (variable DX). The number of nodes, or stations, obtained in this way are NS = ND + 1.

Since we are dealing with equations integrated over the depth, the z-axis, coinciding with the reservoir bed, does not enter into the picture.

TURBIDITY CURRENTS

The numerical solution starts at the measuring station, $x_o = 0$, and progresses towards the downstream, positive x-direction, with a constant step length of Δx. Attention must be paid to the fact that, since the solution is being advanced towards the downstream, the information can only be transmitted from upstream to downstream and not in the opposite direction. *The turbidity current must therefore be in supercritical regime*, $Ri < 1$ or $Fr_D > 1$.

The depth of the current at the measuring section is given directly: $h_o = 1.0$ [m].

The concentration of the suspended sediments, C_{so}, at the measuring section can be calculated from eq. 7.2, using the measured turbidity, $\Delta\rho_o$; one obtains:

$$\Delta\rho_o = (\rho_t - \rho_a)_o = (\rho_s - \rho_a) C_{so} \Rightarrow C_{so} = \frac{35.0}{2650.0 - 1000.0} = 0.0212 \text{ [-]}$$

The specific density of the submerged granular material is:

$$R = \frac{\rho_s - \rho_a}{\rho_a} = \frac{2650.0 - 1000.0}{1000.0} = 1.65 \text{ [-]}$$

The velocity of the current is:

$$U_o = \frac{q_o}{h_o} = \frac{1.0}{1.0} = 1.0 \text{ [m/s]}$$

The reduced sediment flux at the measuring station is calculated by using eq. 7.7:

$$B_o = g R (C_{so} U_o h_o) = 9.81 \times 1.65 \times (0.0212 \times 1.0 \times 1.0) = 0.343 \text{ [m}^3/\text{s}^3]$$

The Richardson number is calculated with the eq. 7.5:

$$Ri_o = \frac{gRC_{so} h_o \cos\alpha}{U_o^2} = \frac{9.81 \times 1.65 \times 0.0212 \times 1.0 \times \cos(5°)}{1.0^2} = 0.342 \text{ [-]}$$

Since $Ri_o = 0.342 < 1$, the turbidity current under consideration is supercritical. The solution should therefore proceed *from upstream to downstream*.

d) *Program GRAVIT*
A program in FORTAN IV language has been written to solve simultaneously the system of ordinary differential equations, eqs. 7.27, 7.28 and 7.18a (supplemented by the additional relations for E_w, u_{*b}, v_{ss}, E_s and c_b, and the boundary conditions: $h_o = 1.0$ [m], $C_{so} = 0.0212$ [-] and $Ri_o = 0.342$ [-]), by using the fourth-order Runge-Kutta method.

The program is written in standard FORTAN and can be run on most of the personal computers. Although a basic knowledge of computers and of programming in languages like FORTRAN, BASIC or PASCAL can be certainly helpful in understanding the rest of the exercise, it is not essential. Extreme care has been given to make the general programming techniques understandable to everybody, even to those who lack any experience in programming.

The program code is written in a pedagogical style and does not have the pretention of being optimized. Numerous comments inserted in the code explain the flow of the program almost step by step. As far as possible, the variable names are chosen to recall the notation used in the text. An exhaustive list of variables together with the type of variables and their explanations, are provided at the beginning of the main program and the related subprograms.

The program has a *modular* structure. It is composed of a main program and several subprograms, each one accomplishes a specific pre-defined task. The flowchart presented in Fig. Ex.7.A.2, shows the relations between different program units. We advise the reader to follow the logic of the program directly from the code given in Figs. Ex.A.9. The specific tasks carried out by each subprogram are described below in detail (see Figs. Ex.A.2 and 9).

The main program GRAVIT controls the flow of the entire program by calling five subroutine subprograms: DREAD, DIVPAR, TITLES, RWRITE and RK4. A sixth subroutine subprogram is declared as "EXTERNAL", to be transmitted to the subroutine subprogram RK4. After finishing the calculations for a station, the main program verifies if the current is still supercritical, Ri < 1. If this is the case, the program starts calculating the values of the independent variables at the next station. This procedure is repeated until the calculations are done for all the stations. If at a station the flow is not supercritical, the program prints an error message and stops.

- The subroutine subprogram DREAD reads the program data by questioning the user. The data are read in four groups :
  - *Physical characteristics* (slope of the bed, friction coefficient, mean diameter and density of the sediments, density and dynamic viscosity of the ambient fluid),
  - *Boundary conditions* (depth and discharge of current, concentration of granular material),
  - *Information on the solution domain* (coordinates of the upstream and downstream end of the solution domain, step length),
  - *Information on the printing of the results* (frequency of the printing of the results and the name of the output file),

  The subroutine subprogram DREAD also calculates some of the parameters to be used later in the program, namely R, Re_p, v_{ss}, Ri_o, the number of sections (ND) and the number of stations (NS). It also verifies whether $Ri_o < 1$.

- After the reading of the program data by the subroutine subprogram DREAD and after the calculation of h, Ri and C_s at a station, the subroutine subprogram DIVPAR is called upon by the main program for calculating the other parameters, g', ρ_t, c_b, U, Q, B and u_{*b}, at that station. The entrainment coefficients, E_s and E_w, are calculated by calling the function subprograms SEDENT and WATENT, respectively.

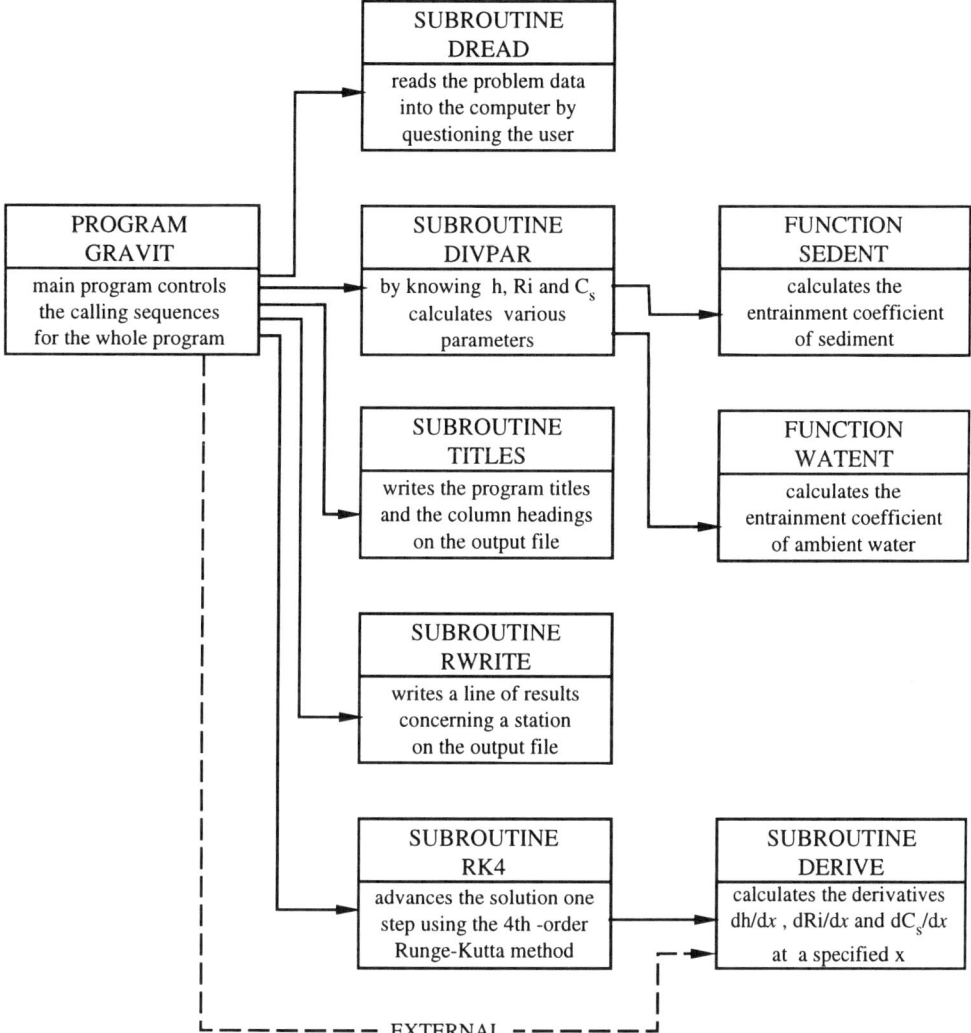

Fig. Ex.7.A.2 Flowchart of the program GRAVIT.

- The subroutine subprogram TITLES is called only once at the very beginning for writing the problem titles and the column headings on the output file.

- The subroutine subprogram RWRITE writes one line of results, concerning a station, on the output file.

- The subroutine subprogram RK4 advances the solution of the three differential equations by one step using the fourth-order Runge-Kutta method. It calls the subroutine subprogram DERIVE to calculate k_1, k_2, k_3 and k_4 for each one of the three differential equations.

- The subroutine subprogram DERIVE calculates the derivatives dh/dx, dRi/dx and dC_s/dx by using the values of h, Ri and C_s sent by the subprogram RK4.

- The function subprogram SEDENT calculates the sediment-entrainment coefficient.

- The function subprogram WATENT calculates the ambient water-entrainment coefficient.

e) *Using the program GRAVIT for solving the problem*
The source code of the program GRAVIT presented in the Fig. Ex.7.A.9 is first *compiled* and *linked* to obtain an executable code. A FORTRAN compiler is of course necessary to do these operations on your computer. The reader should consult the manuals of his computer to learn the exact procedure to follow.

The user feeds the program data into the computer interactively by answering the questions asked by the program. Numerous comments has been introduced into the conversation to remind the user of certain important points treated in chap. 7 and to guide the user in making his choices. The program also checks some of the likely errors in data introduction and warns the user. In case of an error, the question is repeated until the user answers correctly.

The dialog between the user and the computer for solving the present problem is presented in Fig. Ex.7.A.3. The values typed in by the user are highlighted by a white background. They are followed by a sign representing the RETURN (CR or ENTER on some computers) key on the keyboard. The text of the dialog is considered to be sufficiently self-explanatory and will not be commented any further. Nevertheless, there are two important points that need to be clarified :

The first point concerns the step length, being the distance between the stations to be used in the calculations. A too small step size increases computation time. A too big step length can cause a numerical instability in the solution. For the present case, a constant step length of $\Delta x = 0.05$ [m] has been used. The length of the solution domain being TL = 1000.0 [m], the program calculates the independent variables at NS = 20001 stations. The choice of a constant step length, Δx (DX) , has been deliberately adopted here to keep the programming as simple as possible. [There exists however algorithms allowing one to write programs with adaptive step length. In such a program the step length can be automatically adapted to achieve an accuracy desired by the user (see *Press* et al., "Numerical Recipes in Fortran", Cambridge University Press, 1989, p. 607-614).]

The second point concerns the frequency of printing the results on the output file. Obviously, writing all the calculated values (current depth, current speed, discharge, etc.) at all the stations, namely at every $\Delta x = 5.0$ [cm], is not very practical. For this reason the program asks the user to specify the frequency with which the results will be written on the output file. In the present case a printing frequency of NPP = 500 is adopted. The results are therefore printed at every 25.0 [m].

```
            PROGRAM FOR SOLVING THE SYSTEM OF DIFFERENTIAL EQUATIONS
            GOVERNING THE NON-UNIFORM FLOW OF ONE-DIMENSIONAL, STEADY,
              CONSERVATIVE (DENSITY) OR NON CONSERVATIVE (TURBIDITY),
        SUPERCRITICAL GRAVITY CURRENTS USING THE 4TH ORDER RUNGE-KUTTA METHOD.

        NUMERICAL METHOD = 4TH ORDER RUNGE-KUTTA.
        UNIT SYSTEM      = SI

        NOTE : THE CALCULATIONS ARE DONE FOR A UNIT WIDTH.


        READING PHYSICAL CHARACTERISTICS :

        TYPE OF GRAVITY CURRENT ?
        1- CONSERVATIVE CURRENT (DENSITY CURRENT)
        2- NON-CONSERVATIVE CURRENT (TURBIDITY CURRENT)
        YOUR CHOICE (1 OU 2) ? = 2         (RETURN)

        Inclination angle of bed,           ALPHA    (degree)? = 5        (RETURN)
        Friction coefficient,               CF       (-)     ? = 0.032    (RETURN)
        Average diameter of sediments,      D50      (mm)    ? = 0.05     (RETURN)
        Density of sediments,               ROS      (kg/m3) ? = 2650     (RETURN)
        Density of ambient fluid,           ROA      (kg/m3) ? = 1000     (RETURN)
        Kinematic viscosity of ambient fluid, NU     (m2/s)  ? = 1.3E-6   (RETURN)

        READING BOUNDARY CONDITIONS :
        Values known at first station :
        Current depth,                      H0       (m)     ? = 1        (RETURN)
        Unit discharge,                     Q0       (m2/s)  ? = 1        (RETURN)
        Sediment concentration by volume,   CS0      (-)     ? = 0.0212   (RETURN)

        PARAMETERS CALCULATED USING THE GIVEN DATA :
        Submerged specific density of particles, R   (-)       = 1.650
        Particle Reynolds number,           REP      (-)       =   1.09
        Settling velocity of particles,     VSS      (m/s)     =  .001729
        Richardson number for 1st station,  Rio      (-)       =  .342

        READING INFORMATIONS ON THE REACH (COMPUTATIONAL DOMAIN) :
        x-coordinate of first station,      X1       (m)     ? = 0        (RETURN)
        x-coordinate of last station,       XF       (m)     ? = 1000     (RETURN)

        Length of computational reach,      TL       (m)       = 1000.00

        Now you have to specify the step length in x-direction.
        A too large distance between the stations may lead to erroneous
        results. We advise you to specify a step length of 0.05 to 0.10 H0.

        Step length in x-direction,         DX       (m)     ? = 0.05     (RETURN)
        Number of divisions in reach,       ND       (m)       =  20000.
        Number of stations,                 NS                 =  20001.

        Print frequency (number of steps) ? NPP                = 500      (RETURN)
        Name of output file (max. 40 char.) ?                  = SED01    (RETURN)

        NORMAL END OF THE PROGRAM.

        PRESS ENTER TO END THE PROGRAM

        FORTRAN STOP
```

Fig. Ex.7.A.3 Dialog for introducing the problem's data into the computer.

RESULTS OF THE CALCULATIONS WITH THE PROGRAM <GRAVIT>
SOLUTION OF THE SYSTEM OF DIFFERENTIAL EQUATIONS GOVERNING THE NON-UNIFORM FLOW OF ONE-DIMENSIONAL, STEADY,
CONSERVATIVE OR NON CONSERVATIVE GRAVITY CURRENTS USING THE 4TH ORDER RUNGE-KUTTA METHOD.
THE CURRENT HAS A SUPERCRITICAL REGIME. THE SOLUTION IS THEREFORE ADVANCED FROM UPSTREAM TO DOWNSTREAM.

NON CONSERVATIVE GRAVITY CURRENT (TURBIDITY CURRENT)
PHYSICAL CHARACTERISTICS :

Inclination angle of bed,	ALPHA (degre)	=	5.000
Friction coefficient,	CF (-)	=	.03200
Mean diameter of sediments,	D50 (m)	=	.00005
Density of sediments,	ROS (kg/m3)	=	2650.00
Density of ambient fluid,	ROA (kg/m3)	=	1000.00

Kinematic visc. of ambient fluid,   NU   (m2/s) = 0.130E-05

USEFUL PARAMETERS CALCULATED USING GIVEN DATA :

Submerged specific density,	R (-)	=	1.6500
Particle Reynolds number,	REP (-)	=	1.09
Settling velocity of particles,	VSS (m/s)	=	.00173

BOUNDARY CONDITIONS :
Values known at first station :

Current depth,	H0 (m)	=	1.00
Unit discharge of current,	Q0 (m2/s)	=	1.000
Sediment concentration by volume,	CS0 (-)	=	.0212

INFORMATIONS ON COMPUTATIONAL REACH (DOMAIN) :

x-coordinate of first stn,	X1 (m)	=	.00
x-coordinate of last stn,	XF (m)	=	1000.00
Computational reach length,	TL (m)	=	1000.00
Step length in x-direction,	DX (m)	=	.050
Number of divisions of the reach,	ND	=	20000.
Number of stations,	NS	=	20001.

The results are written every (NPP =) 500 steps (stations).

STN	X (m)	H (m)	RI (-)	CS (-)	U (m/s)	B (m3/s3)	Q (m2/s)	GR (m/s2)	CB (-)	ROT (kg/m3)	USTAR (m/s)	EW (-)	ES (-)
1.	.00	1.00	.342	.0212	1.00	.34	1.00	.3432	.0424	1034.98	.06	.01006	.28005
500.	24.95	1.09	.280	.0237	1.22	.51	1.34	.3844	.0475	1039.18	.08	.01271	.29487
1000.	49.95	1.29	.274	.0242	1.36	.69	1.75	.3920	.0484	1039.96	.09	.01304	.29751
1500.	74.95	1.51	.271	.0240	1.47	.86	2.22	.3888	.0480	1039.63	.09	.01319	.29856
2000.	99.95	1.74	.270	.0236	1.57	1.04	2.72	.3816	.0472	1038.90	.10	.01327	.29907
2500.	124.95	1.97	.269	.0231	1.65	1.22	3.26	.3732	.0461	1038.04	.10	.01333	.29936
3000.	149.95	2.21	.268	.0225	1.73	1.39	3.82	.3646	.0451	1037.17	.11	.01336	.29954
3500.	174.96	2.45	.268	.0220	1.80	1.57	4.41	.3563	.0440	1036.32	.11	.01339	.29965
4000.	199.96	2.69	.267	.0215	1.87	1.75	5.03	.3484	.0430	1035.51	.12	.01341	.29973
4500.	224.96	2.93	.267	.0211	1.93	1.93	5.66	.3410	.0421	1034.76	.12	.01342	.29979

continued on the next page

Fig. Ex.7.A.4   Output file of the program GRAVIT.

5000.	249.96	3.18	.267	.0206	1.99	2.11	6.32	.3340	.0413	1034.05	.13	.01343	.29983
5500.	274.96	3.42	.267	.0202	2.05	2.29	7.00	.3276	.0405	1033.39	.13	.01344	.29986
6000.	299.95	3.67	.267	.0199	2.10	2.47	7.70	.3216	.0397	1032.78	.13	.01345	.29988
6500.	324.94	3.91	.267	.0195	2.15	2.66	8.41	.3159	.0390	1032.20	.14	.01345	.29990
7000.	349.94	4.16	.267	.0192	2.20	2.84	9.14	.3106	.0384	1031.67	.14	.01346	.29991
7500.	374.93	4.41	.267	.0189	2.24	3.02	9.89	.3057	.0378	1031.16	.14	.01346	.29993
8000.	399.93	4.65	.266	.0186	2.29	3.21	10.65	.3010	.0372	1030.69	.14	.01347	.29994
8500.	424.92	4.90	.266	.0183	2.33	3.39	11.43	.2966	.0367	1030.24	.15	.01347	.29994
9000.	449.91	5.15	.266	.0181	2.37	3.57	12.22	.2925	.0361	1029.82	.15	.01347	.29995
9500.	474.91	5.40	.266	.0178	2.41	3.76	13.03	.2886	.0357	1029.42	.15	.01348	.29996
10000.	499.90	5.64	.266	.0176	2.45	3.94	13.85	.2849	.0352	1029.04	.16	.01348	.29996
10500.	524.90	5.89	.266	.0174	2.49	4.13	14.68	.2813	.0348	1028.68	.16	.01348	.29996
11000.	549.89	6.14	.266	.0172	2.53	4.32	15.53	.2780	.0343	1028.33	.16	.01348	.29997
11500.	574.88	6.39	.266	.0170	2.56	4.50	16.39	.2747	.0339	1028.01	.16	.01349	.29997
12000.	599.88	6.64	.266	.0168	2.60	4.69	17.26	.2717	.0336	1027.70	.16	.01349	.29997
12500.	624.87	6.89	.266	.0166	2.63	4.87	18.14	.2688	.0332	1027.40	.17	.01349	.29998
13000.	649.86	7.14	.266	.0164	2.67	5.06	19.03	.2660	.0329	1027.11	.17	.01349	.29998
13500.	674.86	7.39	.266	.0163	2.70	5.25	19.94	.2633	.0325	1026.84	.17	.01349	.29998
14000.	699.85	7.64	.266	.0161	2.73	5.44	20.85	.2607	.0322	1026.57	.17	.01349	.29998
14500.	724.85	7.89	.266	.0160	2.76	5.62	21.78	.2582	.0319	1026.32	.17	.01349	.29998
15000.	749.84	8.13	.266	.0158	2.79	5.81	22.71	.2558	.0316	1026.08	.18	.01349	.29998
15500.	774.83	8.38	.266	.0157	2.82	6.00	23.66	.2535	.0313	1025.84	.18	.01349	.29998
16000.	799.83	8.63	.266	.0155	2.85	6.19	24.62	.2513	.0311	1025.62	.18	.01350	.29999
16500.	824.82	8.88	.266	.0154	2.88	6.37	25.58	.2492	.0308	1025.40	.18	.01350	.29999
17000.	849.82	9.13	.266	.0153	2.91	6.56	26.56	.2471	.0305	1025.19	.18	.01350	.29999
17500.	874.81	9.38	.266	.0151	2.94	6.75	27.55	.2451	.0303	1024.99	.19	.01350	.29999
18000.	899.80	9.63	.266	.0150	2.96	6.94	28.54	.2432	.0300	1024.79	.19	.01350	.29999
18500.	924.80	9.88	.266	.0149	2.99	7.13	29.55	.2413	.0298	1024.60	.19	.01350	.29999
19000.	949.79	10.13	.266	.0148	3.02	7.32	30.56	.2395	.0296	1024.41	.19	.01350	.29999
19500.	974.79	10.38	.266	.0147	3.04	7.51	31.58	.2377	.0294	1024.23	.19	.01350	.29999
20000.	999.78	10.63	.266	.0146	3.07	7.70	32.61	.2360	.0292	1024.06	.19	.01350	.29999
20001.	999.83	10.63	.266	.0146	3.07	7.70	32.62	.2360	.0292	1024.06	.19	.01350	.29999

Fig. Ex.7.A.4   Output file of the program GRAVIT (end).

f) *Analysis of results and some remarks*

The output file containing the solution of the present problem is presented in Fig. Ex.7.A.4. The calculated parameters, such as Richardson number (Ri = RI), concentration by volume of sediments (C_s = CS), speed of the current (U), sediment flux (B), discharge per unit width (q = Q), reduced gravitational acceleration (g' = GR), concentration by volume of the sediments near the bed (c_b = CB), density of the current (ρ_t = ROT), bed-shear stress (u_{*b} = USTAR), entrainment coefficient of the ambient fluid (E_w = EW) and sediment entrainment coefficient (E_s = ES), are listed as functions of the distance (x = X) for the solution domain extending from the measuring section at $x = x_o = 0$ [m] to a station at x = L = 1000 [m] (a total of 42 stations separated by a distance of 25 [m]). In the first part of the problem statement we were asked to predict the variation of the current depth, h(x), the current speed, U(x), and the average concentration, $C_s(x)$, along the reservoir, from the measuring station to a downstream section at L = 1000 [m]; the output file (see Fig. Ex.7.A.4) is therefore the answer for this part of the problem.

By using these calculated values, one can also plot the variation of various parameters of the current as functions of the distance. Fig. Ex.7.A.5 shows the variation of the dimensionless parameters, h/h_o, Ri/Ri_o, C_s/C_{so} and U/U_o, as functions of the dimensionless distance, x/h_o.

The Fig. Ex.7.A.6 is a scaled representation of the turbidity current in the reservoir. It is to be noted that the vertical scale was exaggerated to make visible the details of the turbidity current.

By examining the contents of the output file (see Fig. Ex.7.A.4) as well as these two figures (see Figs. Ex.7.A.5 and Ex.7.A.6), the following observations can be made:

- It is admitted that the current is in *equilibrium* if the Richardson number is independent of the distance, x, or dRi/dx = 0 (see point 7.3.3 °). The output file (see Fig. Ex.7.A.4) shows that the Richardson number, which initially has a value of Ri_o = 0.342, decreases rapidly up to a distance of $x' \cong 125$ [m]. From this section downstream the rate of decrease of the Richardson number is considerably smaller and finally in the reach downstream of the section $x' \cong 375$ [m], the Richardson number reaches a constant value of Ri = 0.266. For the reach downstream of the section at $x' \cong 125$ [m] one can combine eq. 7.27 and eq. 7.28 to obtain :

dh/dx = E_w

which after integration gives a *linear* expression for the growth of the current:

h = $h_{x'}$ + E_w (x – x ') = 1.97 + E_w (x – 125)

being clearly evident in Fig. Ex.7.A.5.

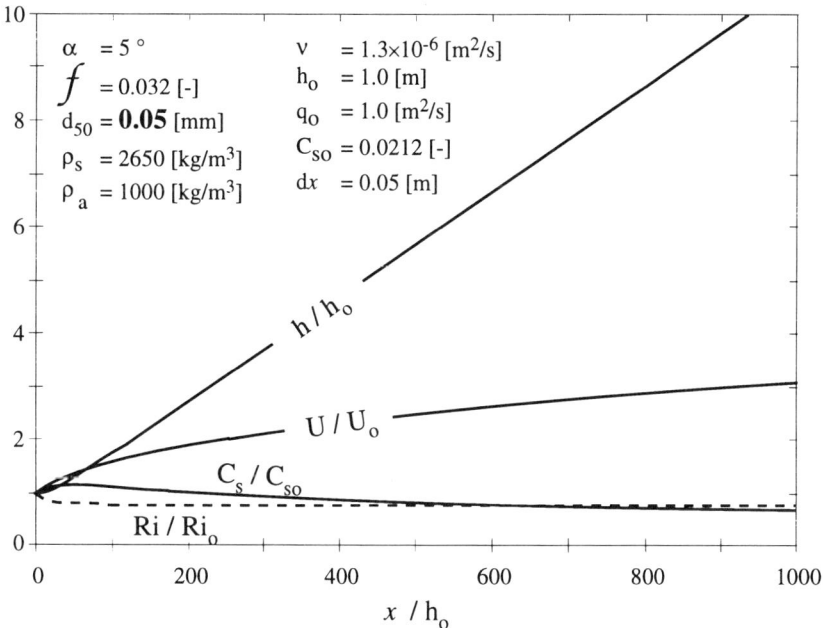

Fig. Ex.7.A.5 Dimensionless curves for the variation of the depth, h, the velocity, U, the sediment concentration by volume, C_s, and the Richardson number, Ri, for a turbidity current with a sediment size of d_{50} = 0.05 [mm].

- The current velocity increases from U_o = 1.00 [m/s] to U_{1000} = 3.07 [m/s] whereas the sediment concentration by volume decreases from C_{s_o} = 0.021 [-] to $C_{s_{1000}}$ = 0.015 [-].

- The reduced sediment flux of the current, B = gR (C_sUh), increases almost linearly from B_o = 0.34 [m³/s³] to B_{1000} = 7.07 [m³/s³] because of the sediment entrainment from the reservoir bed. This type of turbidity current, which is non-conservative, is called *accelerating supercritical current* (With a different set of parameters one can also simulate a turbidity current in which the reduced sediment flux decreases. This type of current is called *decelerating supercritical current*).

The decrease in the concentration by volume, C_s, while the reduced sediment flux, B = gR (C_sUh), is increasing may, at a first sight, seem contradictory and it may give the impression that the quantity of the sediments transported is decreasing from one section to the other. One should however not forget that the depth, h, and the velocity, U, are both increasing in the direction of the current. Therefore, although the percentage of the volume of sediments with respect to the total volume of the current, i.e.: C_s, decreases from one section to the other, the reduced sediment flux, B, increases because of the sediment entrainment from the bottom.

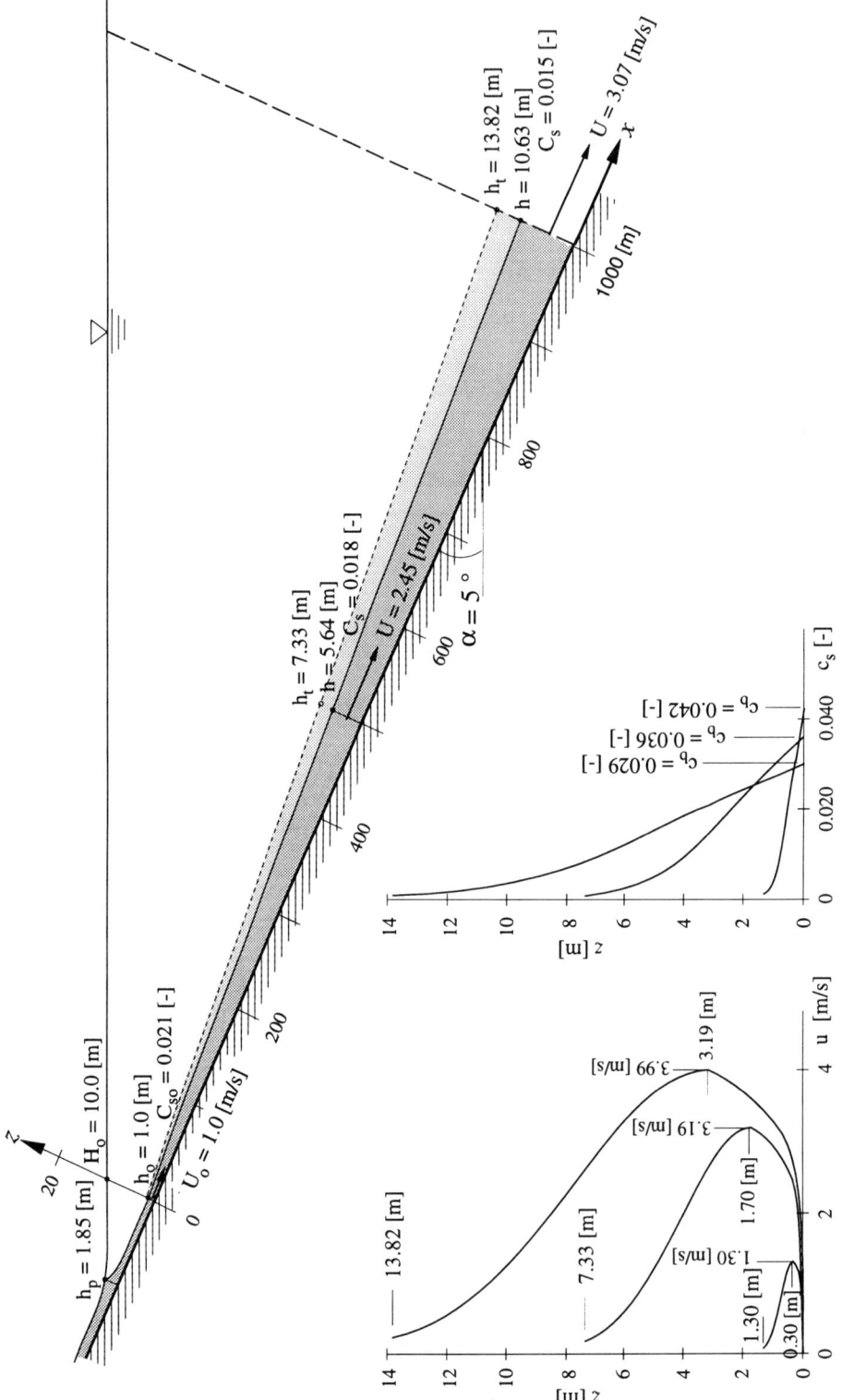

Fig. Ex.7.A.6 A scaled representation of the turbidity current ($d_{50} = 0.05$ [mm]) flowing into the reservoir.

TURBIDITY CURNTS

- The sediment entrainment coefficient increases rapidly to a maximum value, $E_s \cong 0.3$ [-], (see Fig. Ex.7.A.4). It is known that (see Fig. 7.4 and eq. 7.32) this is the maximum value for the sediment entrainment coefficient, E_s, if $\zeta > 45$. The program does not print the values of this parameter, ζ, on the output file but these values can be readily calculated by knowing the characteristics of the sediment, $d_{50} = 0.05$ [mm], $\rho_s \cong 2650$ [kg/m^3] and $v_{ss} = 0.173 \times 10^{-2}$ [m/s], and by using the eq. 7.32 :

$$\zeta = \frac{u_{*b}}{v_{ss}} \left(\frac{d_{50}\sqrt{gRd_{50}}}{\nu} \right)^{0.75} = 618.4 \, u_{*b}$$

For bed-shear stress velocities of $u_{*b} > 0.073$ [m/s], this parameter, ζ, becomes larger than 45 and the sediment entrainment coefficient reaches its maximum value, $E_s \cong 0.3$ [-] ; this is in fact the case from the second station downstream (see Fig. Ex.7.A.4).

ii) According to *Akiyama* et *Stefan* (1985), the densimetric Froude number at the plunge-point is given by (see sect. 7.1.2) :

$$Fr_p = \frac{q_o/h_p}{(g_o'h_p)^{1/2}} \cong 0.68$$

In the present case, the discharge per unit width, $q_o = 1.0$ [m^2/s], is given directly. By using the definition given by the eq. 7.1 and by assuming that the turbidity at the plunge point is the same as the one at the first station, the reduced gravity can be calculated as follows :

$$g_o' = g\left(\frac{\rho_t - \rho_a}{\rho_a}\right)_o = g\frac{\Delta\rho_o}{\rho_a} = 9.81 \frac{35.0}{1000.0} = 0.343 \; [\text{m/s}^2]$$

By substituting this value in the above expression, one obtains an estimation for the *water depth at the plunge point* ; being :

$$h_p = \left(\frac{q_o}{0.68 \sqrt{g_o'}} \right)^{2/3} = \left(\frac{1}{0.68 \sqrt{0.343}} \right)^{2/3} = 1.85 \; [\text{m}]$$

iii) The calculations for the second turbidity current was done again by running the same program with a particle diameter of $d_{50} = 0.10$ [mm]. Neither the dialog with the computer nor the output file are given here to encourage the reader to do this by himself.

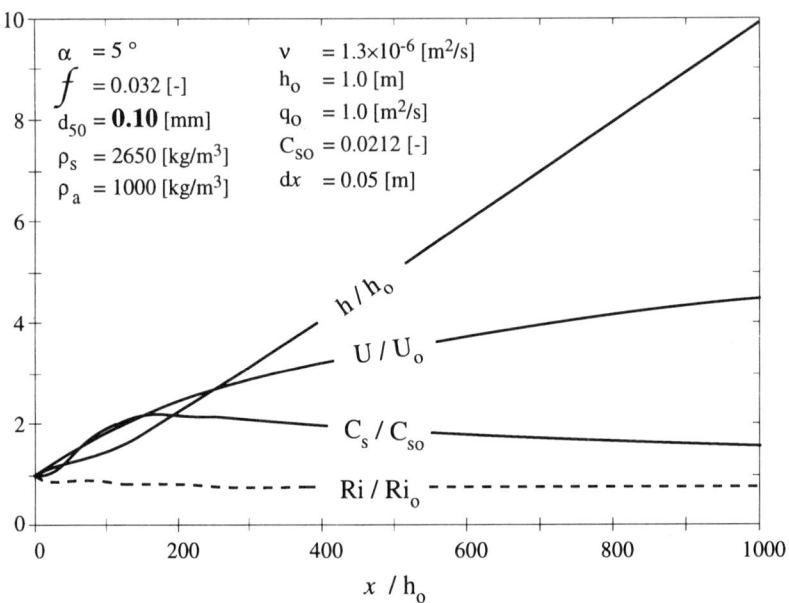

Fig. Ex.7.A.7 Dimensionless curves for the variation of the depth, h, the velocity, U, the sediment concentration by volume, C_s, and the Richardson number, Ri, for a turbidity current with a sediment size of d_{50} = 0.10 [mm].

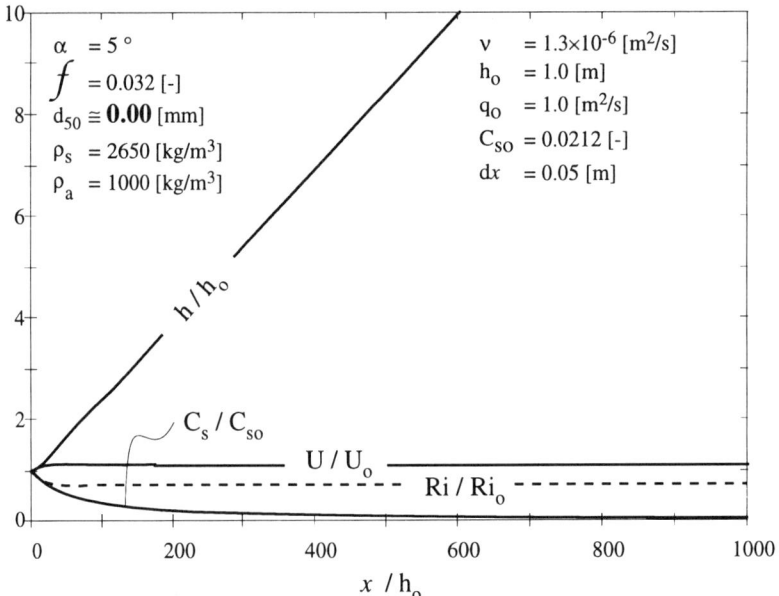

Fig. Ex.A.7.8 Dimensionless curves for the variation of the depth, h, the velocity, U, the sediment concentration by volume, C_s, and the Richardson number, Ri, for a density current (d_{50} = 0.00 [mm]).

TURBIDITY CURRENTS 507

The Fig. Ex.7.A.7 shows the variation of the dimensionless parameters, h/h_o, Ri/Ri_o, C_s/C_{s_o} and U/U_o, as functions of the dimensionless distance, x/h_o. By comparing this figure with Fig. Ex.7.A.5 one can make the following observations :

- The growth rate of the current depth, dh/dx, for the turbidity current with $d_{50} = 0.10$ [mm] is smaller than the one of the preceding turbidity current with $d_{50} = 0.05$ [mm]. The current depth increases linearly from $x \cong 250$ [m] downstream.

- In the present case the velocity, U, and the concentration by volume, C_s, increase faster than they do in the preceding current.

- The present turbidity current, being also non conservative, is also an accelerating, supercritical current. As a matter of fact, the output file will show that its reduced sediment flux decreases more rapidly (from $B_o = 0.34$ [m³/s³] to $B_{1000} = 23.95$ [m³/s³]) than for the preceding current.

iv) If the sediment entrainment at the bottom is neglected because of the given fixed-bed condition, the current can be modeled as a density current. The program must therefore be run by answering "1" to the first question asked by the computer concerning the choice of the type of gravity current (see Fig. Ex.7.A.3). The program now does not ask any questions concerning the sediment characteristics. Neither the dialog with the computer, nor the output file are given here to encourage the reader to do this by himself.

The Fig. Ex.7.A.8 shows the variation of the dimensionless parameters, h/h_o, Ri/Ri_o, C_s/C_{s_o} and U/U_o, as functions of the dimensionless distance, x/h_o. By comparing this figure with Fig. Ex.7.A.5 and Fig. Ex.7.A.7 one can make the following observations :

- The Richardson number, Ri, becomes rapidly constant. It is well known that a density current reaches rapidly a *state of equilibrium* (see point 7.3.3 °).

- The current depth, h, decreases (almost) linearly from the first meters downstream. The rate of growth of the current depth, dh/dx, for the density current is considerably higher than the ones for both turbidity currents.

- The current velocity, U, remains practically constant over the entire reach.

- Since there is no sediment entrainment at the bottom, the volume concentration, C_s, decreases in the direction of the flow.

- The reduced sediment flux, $B_o = B_{1000} = 0.34$ [m³/s³], remains constant; the gravity current is therefore a conservative current (see eq. 7.8).

v) Prior to the determination of the velocity profile, u(z), and the concentration profile, $c_s(z)$, it will be useful to summarize the integral scales, h, U and C_s, previously calculated at the three stations (see Fig. Ex.7.A.6).

Station	h [m]	U [m/s]	C_s [-]	h_m [m]	h_t [m]	U_m [m/s]	c_b [-]	C_m [-]
0 [m]	1.00	1.00	0.021	0.30	1.30	1.30	0.042	0.030
500 [m]	5.64	2.45	0.018	1.70	7.33	3.19	0.036	0.026
1000 [m]	10.63	3.07	0.015	3.19	13.82	3.99	0.029	0.021

Other useful parameters, h_m, U_m, h_t, c_b, C_m, were calculated using the following empirical relationships:

$$\frac{h_m}{h} \approx 0.3 \quad ; \quad \frac{U_m}{U} \approx 1.3 \quad ; \quad \frac{h_t}{h} \approx 1.3 \quad ; \tag{7.40}$$

as well as:

$$\frac{c_b}{C_s} \approx 2 \quad ; \quad \frac{C_m}{c_b} \approx 0.7 \quad . \tag{7.43}$$

The velocity profile is then calculated using:

$$z < h_m : \qquad \frac{u(z)}{U_m} = \left(\frac{z}{h_m}\right)^{1/6} \tag{2.50a}$$

$$z > h_m : \qquad \frac{u(z)}{U_m} = \exp\left[-1.4\left(\frac{z - h_m}{h - h_m}\right)^2\right] \tag{7.39}$$

Subsequently the concentration profile is calculated using:

$$z < h_m : \qquad \frac{c_s(z) - C_m}{c_b - C_m} = \frac{h_m - z}{h_m} \tag{7.41}$$

$$z > h_m : \qquad \frac{c_s(z)}{C_m} = \exp\left[-2.0\left(\frac{z - h_m}{h - h_m}\right)^{1.3}\right] \tag{7.42}$$

The graphical representation of these profiles at the three stations is given in Fig. Ex.7.A.6. This figure shows the variation of parameters h_t, U_m, c_b and the profiles of velocity, u(z), and of concentration, $c_s(z)$, from one station to another.

```
      PROGRAM GRAVIT
C***********************************************************
C
C   MAIN PROGRAM FOR SOLVING THE SYSTEM OF DIFFERENTIAL EQUATIONS
C   GOVERNING THE NON-UNIFORM FLOW OF ONE-DIMENSIONAL, STEADY,
C   CONSERVATIVE OR NON-CONSERVATIVE GRAVITY CURRENTS BY USING THE
C   4TH ORDER RUNGE-KUTTA METHOD.
C
C   THE SOLUTION IS ADVANCED FROM UPSTREAM TO DOWNSTREAM. THE GRAVITY
C   CURRENT MUST THEREFORE HAS A SUPERCRITICAL REGIME, Ri < 1.
C
C***********************************************************
C   ENTRAINMENT
C   LIST OF VARIABLES DEFINED GLOBALLY FOR MAIN PROGRAM AND THE
C   SUBROUTINES. VARIABLES DEFINED LOCALLY IN SUBROUTINES ARE LISTED AT
C   THE BEGINNING OF EACH SUBROUTINE.
C
C   TYPE  NAME     DIMEN.    EXPLANATIONS
C   ====  ====     ======    ============
C   R*4   ALPHA              = INCLINATION ANGLE OF RESERVOIR BED
C   R*4   B                  = SEDIMENT FLUX AT A STATION
C   R*4   CB                 = NEAR-BED CONCENTRATION BY VOLUME
C   R*4   CBCS               = RATIO OF NEAR-BED CONCENTRATION BY VOLUME, cb,
C                              TO AVERAGE CONCENTRATION BY VOLUME, c :
C                              CBCS = ro = cb / Cs = 2.0
C   R*4   CF                 = FRICTION COEFFICIENT USED IN CALCULATING USTAR
C   LOG   CFLAG              = TYPE OF GRAVITY CURRENT
C                              CFLAG =.TRUE.   ---> NON-CONSERVATIVE CURRENT
C                              CFLAG =.FALSE.  ---> CONSERVATIVE CURRENT
C   R*4   CS0                = CONCENTRATION BY VOLUME OF SEDIMENTS AT THE
C                              STATION 0
C   R*4   D50                = MEAN DIAMETER OF SEDIMENTS
C   R*4   DX                 = STEP SIZE IN X-DIRECTION
C   R*4   ES                 = ENTRAINMENT COEFFICIENT FOR SEDIMENTS
C   R*4   EW                 = ENTRAINMENT COEFFICIENT FOR AMBIENT FLUID
C   C*40  FICHS              = OUTPUT FILE NAME
C   R*4   G                  = GRAVITATIONAL ACCELERATION, g
C   R*4   GR                 = REDUCED GRAVITATIONAL ACCELERATION
C   R*4   H0                 = CURRENT DEPTH AT STATION 0
C   I*4   JP                 = NUMB. OF STATIONS FOR WHICH THE CALCULATIONS ARE
C                              DONE SINCE THE LAST WRITING ON THE OUTPUT FILE
C   I*4   ND                 = NUMBER OF INTERVALS IN THE REACH (COMPUTATION
C                              DOMAIN)
C   I*4   NPP                = FREQUENCY WITH WHICH THE RESULTS ARE WRITTEN
C                              ON OUTPUT FILE
C   I*4   NEQ                = NUMBER OF EQUATIONS (AND/OR NUMBER OF DEPENDENT
C                              VARIABLES) TO BE SOLVED
C   I*4   NOUT               = UNIT NUMBER FOR OUTPUT FILE
C   I*4   NS                 = NUMBER OF STATIONS
C   R*4   NU                 = KINEMATIC VISCOSITY OF AMBIENT FLUID
C   R*4   Q                  = CURRENT DISCHARGE AT A STATION
C   R*4   Q0                 = CURRENT DISCHARGE AT STATION 0
C   C*1   QUIT               = TO BE READ FOR STOPPING THE PROGRAM
C   R*4   PI                 = NUMBER PI
C   R*4   R                  = SPECIFIC DENSITY OF SEDIMENTS
C   R*4   REP                = PARTICLE REYNOLDS NUMBER
C   R*4   RILIM              = LIMITING VALUE OF RICHARDSON NUMBER ABOVE WHICH
C                              THE REGIME OF CURRENT BECOMES CRITICAL AND
C                              SUBCRITICAL.
C                              IT IS ADMITTED THAT : RILIM = 0.95
C   R*4   ROA                = DENSITY OF AMBIENT FLUID
C   R*4   ROS                = DENSITY OF SEDIMENTS
C   R*4   ROT                = DENSITY OF CURRENT
C   R*4   STN                = VARIABLE USED TO COUNT THE NUMBER OF STATIONS
C                              FOR WHICH THE SOLUTION IS ALREADY OBTAINED
C   R*4   TL                 = TOTAL LENGTH OF THE REACH (SOLUTION DOMAIN)
C   R*4   U                  = AVERAGE VELOCITY OF CURRENT AT A STATION
C   R*4   USTAR              = BED-SHEAR VELOCITY AT A STATION
C   R*4   VSS                = SETTLING VELOCITY OF PARTICLES
C   R*4   X1                 = X-COORDINATE OF FIRST STATION
C   R*4   XF                 = X-COORDINATE OF LAST STATION
C   R*4   Y       (NEQ)      = VECTOR CONTAINING THE VALUES OF THE DEPENDENT
C                              VARIABLES
C                              Y(1) = CURRENT DEPTH, h
C                              Y(2) = RICHARDSON NUMBER, Ri
C                              Y(3) = SEDIMENT CONCENTRATION BY VOLUME, Cs
C***********************************************************
C   PARAMETERS
      PARAMETER ( NOUT = 10 , NEQ = 3 , RILIM = 0.95 )
C
C   LABELED COMMON BLOCKS (SHARED VARIABLES)
      COMMON /VARIA/ X , DX , Y
      COMMON /DONNE/ ALPHA , CF , D50 , ROS , ROA , NU , R , REP , VSS
      COMMON /LIMIT/ X1 , XF , ND , NS , NPP
      COMMON /DIVP/ GR , ROT , CB , U , Q , B , USTAR , EW , ES
      COMMON /CONST/ CBCS , PI , G , CFLAG
C
C   THE MAIN PROGRAM "GRAVIT" SENDS THE NAME OF THE SUBROUTINE SUBPROGRAM
C   "DERIVE" TO THE SUBROUTINE SUBPROGRAM "RK4" IN THE ARGUMENT LIST.
C   IN FORTRAN, THE SUBROUTINE WHOSE NAME APPEARS IN AN ARGUMENT LIST MUST
C   BE DECLARED AS EXTERNAL.
      EXTERNAL DERIVE
C
C   DECLARATION OF VARIABLES
      CHARACTER*1   QUIT
      CHARACTER*40  FICHS
      LOGICAL       CFLAG
      REAL          NU , ND , NS
      DIMENSION     Y(NEQ) , YP(NEQ)
C
C   INITIALISATIONS
      CBCS = 2.0
      PI   = 3.1415927
      G    = 9.81
C
C   READ THE PROBLEM DATA
      CALL DREAD( NOUT , FICHS )
C
C   CALCULATE VARIOUS USEFUL PARAMETERS AT STATION 0 FOR WHICH
C   VALUES OF DEPENDENT VARIABLES ARE GIVEN BY THE USER AS BOUNDARY
C   CONDITIONS.
      CALL DIVPAR
C
C   WRITE TITLES AND COLUMN HEADINGS ON OUTPUT FILE
      CALL TITLES( NOUT )
C
C   WRITE A LINE OF RESULTS FOR FIRST STATION
      STN = 1
      JP  = 1
      CALL RWRITE( STN , CFLAG , NOUT )
C
C   CALCULATION LOOP
 10   STN = STN + 1
      JP  = JP + 1
      IF(STN.GT.NS) GO TO 999
C
C   ADVANCE THE SOLUTION 1 STEP FOR ALL EQUATIONS
      CALL RK4( X , Y , DX , YP , DERIVE )
C
C   PREPARE FOR WRITING THE RESULTS AT RUNNING STATION ON OUTPUT
C   FILE
```

Fig.Ex.7.A.9 Program GRAVIT (continued)

```fortran
      X = X + DX
      DO 50 I = 1,NEQ
      Y(I) = YP(I)
50    CONTINUE
C
C CALCULATE VARIOUS USEFUL PARAMETERS AT THE RUNNING STATION FOR
C WHICH THE DEPENDENT VARIABLES HAVE JUST BEEN CALCULATED
      CALL DIVPAR
C
C WRITE THE RESULTS AT THE RUNNING STATION ON THE OUTPUT FILE
      IF(JP.EQ.NPP.OR.STN.EQ.NS.OR.Y(2).GT.RILIM)THEN
          CALL RWRITE( STN , CFLAG , NOUT )
          JP = 0
      ENDIF
C
C VERIFY IF THE FLOW REGIME OF THE CURRENT IS STILL SUPERCRITICAL
      IF(Y(2).GE.RILIM) THEN
          WRITE(*,20) RILIM , JP , X
          WRITE(NOUT,20) RILIM , JP , X
20        FORMAT(' ATTENTION !!..... ABNORMAL END OF THE PROGRAM.'/
     1           ' RICHARDSON NUMBER, Ri, IS GREATER THAN ',F7.3/
     2           ' CALCULATIONS CAN NO LONGER PROCEED FROM',
     3           ' UPSTREAM TO DOWNSTREAM.'/
     4           ' THE FORMATION OF AN INTERNAL JUMP IS PROBABLE.'/
     5           ' CALCULATION STEP NO.    JP = ',I8/
     6           ' DISTANCE,                X = ',F9.2,' [m]')
          STOP
      ENDIF
C
      GO TO 10
C
999   WRITE(*,1000)
1000  FORMAT(/ ' NORMAL END OF THE PROGRAM.'/
     1         '  PRESS ENTER TO END THE PROGRAM.'/)
      READ(*,1001)QUIT
1001  FORMAT(A)
      STOP
      END
C
C***************************************************************
C
      SUBROUTINE DREAD( NOUT , FICHS )
C
C***************************************************************
C
C SUBROUTINE SUBPROGRAM TO READ THE PROBLEM DATA INTERACTIVELY.
C
C***************************************************************
C
C LIST OF VARIABLES LOCALLY DEFINED FOR SUBROUTINE SUBPROGRAM 'DREAD'
C
C TYPE NAME    DIMEN.   EXPLANATIONS
C ==== ====    ======   ============
C I*4  CHOICE           = ANSWER TO A MULTIPLE-CHOICE QUESTION
C
C SEE THE BEGINNING OF THE MAIN PROGRAM FOR A LIST OF SIMPLE VARIABLES,
C VECTORS AND MATRICES DEFINED GLOBALLY.
C
C***************************************************************
C
C PARAMETERS
      PARAMETER ( NEQ = 3 )
C
C COMMON BLOCKS
      COMMON /VARIA/ X , DX , Y
      COMMON /DONNE/ ALPHA , CF , D50 , ROS , ROA , NU , R , REP , VSS
      COMMON /LIMIT/ X1 , XF , ND , NS , NPP
      COMMON /CONST/ CBCS , PI , G , CFLAG
C
C DECLARATION OF VARIABLES
      LOGICAL      CFLAG
      INTEGER      CHOICE
      REAL         NU , ND , NS
      CHARACTER*40 FICHS
      DIMENSION    Y(NEQ)
C
C WRITE THE TITLE OF THE PROGRAM ON THE SCREEN
      WRITE(*,10)
10    FORMAT(//'          PROGRAM FOR SOLVING THE SYSTEM OF DIFFERENTIAL',
     1        /'          EQUATIONS.'/
     2        /'       GOVERNING THE NON-UNIFORM FLOW OF ONE-DIMENSIONAL, STEADY,',
     3        /'       CONSERVATIVE (DENSITY) OR NON CONSERVATIVE (TURBIDITY)',
     4        /'       SUPERCRITICAL GRAVITY CURRENTS USING THE 4TH ORDER',
     5        /'       RUNGE-KUTTA METHOD.'//
     6        /'       NUMERICAL METHOD = 4TH ORDER RUNGE-KUTTA.'/
     7        /'       UNIT SYSTEM      = SI '/
     8        /'       NOTE : THE CALCULATIONS ARE DONE FOR A UNIT WIDTH.'///)
C
C READ PHYSICAL CHARACTERISTICS
      WRITE(*,90)
90    FORMAT(/' READING PHYSICAL CHARACTERISTICS :'/)
C
95    WRITE(*,96)
96    FORMAT(' TYPE OF GRAVITY CURRENT ?'/
     1       '    1 - CONSERVATIVE CURRENT (DENSITY CURRENT)'/
     2       '    2 - NON-CONSERVATIVE CURRENT (TURBIDITY CURRENT)'/
     3       '    YOUR CHOICE (1 OU 2) ? = ',$)
      READ(*,*,ERR=95)CHOICE
      IF(CHOICE.EQ.1)THEN
          CFLAG=.FALSE.
      ELSE IF(CHOICE.EQ.2)THEN
          CFLAG=.TRUE.
      ELSE
          WRITE(*,97)
97        FORMAT(' ERROR !.... NON-EXISTING CHOICE !'//)
          GO TO 95
      ENDIF
C
100   WRITE(*,110)
110   FORMAT(/' Inclination angle of bed,               ALPHA    (d'
     1       ,'egree) ? = ',$)
      READ(*,*,ERR=100)ALPHA
C
120   WRITE(*,130)
130   FORMAT(' Friction coefficient,                   CF       (-)'
     1       ,' ? = ',$)
      READ(*,*,ERR=120)CF
C
      IF(CFLAG)THEN
140       WRITE(*,150)
150       FORMAT(' Average diameter of sediments,          D50      (mm'
     1           ,')'/,' ? = ',$)
          READ(*,*,ERR=140)D50
C
160       WRITE(*,170)
170       FORMAT(' Density of sediments,                   ROS      (kg'
     1           ,'/m3) ? = ',$)
          READ(*,*,ERR=160)ROS
      ELSE
162       WRITE(*,172)
172       FORMAT(' Density of dissolved substance,         ROS      (kg'
     1           ,'/m3) ? = ',$)
          READ(*,*,ERR=162)ROS
      ENDIF
C
180   WRITE(*,190)
```

Fig.Ex.7.A.9 Program GRAVIT (continued)

```
  190   FORMAT(' Density of ambient fluid,                    ROA       (kg'
       1  '/m3) ? = ',$)
        READ(*,*,ERR=180)ROA
  C
  200   WRITE(*,210)
  210   FORMAT(' Kinematic viscosity of ambient fluid,        NU        (m2'
       1  '/s) ? = ',$)
        READ(*,*,ERR=200)NU
  C
  C READ BOUNDARY CONDITIONS
        WRITE(*,215)
  215   FORMAT(// ' READING BOUNDARY CONDITIONS :'/
       1  ' Values known at first station :')
  C
  220   WRITE(*,230)
  230   FORMAT(' Current depth,                               H0        (m)'
       1  ' ? = ',$)
        READ(*,*,ERR=220)H0
  C
  240   WRITE(*,250)
  250   FORMAT(' Unit discharge,                              Q0        (m2'
       1  '/s) ? = ',$)
        READ(*,*,ERR=240)Q0
  C
  260   IF(CFLAG)THEN
          WRITE(*,270)
  270     FORMAT(' Sediment concentration by volume,          CS0       (-)'
       1    ' ? = ',$)
        ELSE
          WRITE(*,271)
  271     FORMAT(' Dissolved substance concent. by vol.,      CS0       (-)'
       1    ' ? = ',$)
        ENDIF
        READ(*,*,ERR=260)CS0
  C
  C UNIT CONVERSION
        D50   = D50 * 0.001
        ALPHA = ALPHA * PI / 180
  C
  C CALCULATION OF VARIOUS USEFUL PARAMETERS
        R    = (ROS - ROA) / ROA
        IF(CFLAG)THEN
          REP = D50 * SQRT(G * R * D50) / NU
          VSS = D50**2 * G * R / (18.0 * NU)
        ELSE
          REP = 0
          VSS = 0
        ENDIF
  C
  C VALUES OF THE THREE DEPENDENT VARIABLES AT THE STATION 0
        X    = X1
        Y(1) = H0
        Y(2) = G * R * CS0 * H0 * COS(ALPHA) / (Q0 / H0)**2
        Y(3) = CS0
  C
  C WRITE THE CALCULATED PARAMETERS ON THE SCREEN
        IF(CFLAG)THEN
          WRITE(*,280)R,REP,VSS,Y(2)
  280     FORMAT(// ' PARAMETERS CALCULATED USING THE GIVEN DATA :'/
       1    ' Submerged specific density of particles,   R         (-)'
       2    ' = ',F5.3/
       3    ' Particle Reynolds number,                  REP       (-)'
       4    ' = ',F6.2/
       5    ' Settling velocity of particles,            VSS       (m/'
       6    's) = ',F8.6/
       7    ' Richardson number for 1st station,         Rio       (-)'
       8    ' = ',F6.3/)
          WRITE(*,281)Y(2)
  281     FORMAT(// ' PARAMETERS CALCULATED USING THE GIVEN DATA :'/
       1    ' Richardson number for the 1st station,     Rio       (-)'
       2    ' = ',F6.3/)
        ENDIF
  C
  C VERIFY IF THE CURRENT IS SUPERCRITICAL ( Ri < 1 )
        IF(Y(3).GE.1)THEN
  284     WRITE(*,285)
  285     FORMAT(' ERROR !.. THE CURRENT IS NOT SUPERCRITICAL ! (Ri >= 1'
       1    ')'/'                     YOU HAVE TWO POSSIBILITIES :'/
       2    '                     1- RESTART WITH THE DATA ENTRY'/
       3    '                     2- STOP THE PROGRAM'//
       4    '                     WHAT IS YOUR CHOICE (1 ou 2) ? = ',$)
          READ(*,*,ERR=284)CHOICE
          IF(CHOICE.EQ.2) STOP
          GO TO 5
        ENDIF
  C
  C BEGINNING AND END OF THE REACH (COMPUTATIONAL DOMAIN)
        WRITE(*,290)
  290   FORMAT(' READING INFORMATIONS ON THE REACH (COMPUTATIONAL',
       1  ' DOMAIN) :')
  C
  300   WRITE(*,310)
  310   FORMAT(' x-coordinate of first station,               X1        (m)'
       1  ' ? = ',$)
        READ(*,*,ERR=300)X1
  C
  320   WRITE(*,330)
  330   FORMAT(' x-coordinate of last station,                XF        (m)'
       1  ' ? = ',$)
        READ(*,*,ERR=320)XF
  C
  335   WRITE(*,335)XF-X1
        FORMAT(/' Length of computational reach,             TL        (m)'
       1  ' = ',F9.2//
       2  ' Now you have to specify the step length in x-direction.'/
       3  ' A too large distance between the stations may lead to'/
       4  ' erroneous'/
       5  ' results. We advise you to specify a step length of 0.05 to'/
       6  ' 0.10 H0.'//)
  340   WRITE(*,350)
  350   FORMAT(' Step length in x-direction,                  DX        (m)'
       1  ' ? = ',$)
        READ(*,*,ERR=340)DX
  C
        ND = (XF -X1) / DX
        NS = ND + 1
  C
  400   WRITE(*,400) ND, NS
  400   FORMAT(' Number of divisions in reach,                ND'
       1  ' = ',F10.0/
       2  ' Number of stations,                          NS'
       3  ' = ',F10.0/)
  C
  450   WRITE(*,460)
  460   FORMAT(' Print frequency (number of steps) ?          NPP'
       1  ' = ',$)
        READ(*,*,ERR=450)NPP
  C
  C READ THE NAME OF THE OUTPUT FILE
  500   WRITE(*,510)
```

Fig.Ex.7.A.9 Program GRAVIT (continued)

```
510   FORMAT(' Name of output file (max. 40 char.) ?
     1',      = ',$)
      READ(*,520)FICHS
520   FORMAT(A)
      OPEN( UNIT = NOUT , FILE = FICHS , STATUS = 'NEW' , ERR = 500 )
C
      RETURN
      END
      SUBROUTINE DIVPAR
C****************************************************************
C     CALCULATE THE VARIOUS PARAMETERS FROM THE THREE INDEPENDENT  *
C     VARIABLES                                                    *
C****************************************************************
C
C     NO LOCALLY DEFINED VARIABLES.
C     SEE THE BEGINNING OF THE MAIN PROGRAM FOR A LIST OF SIMPLE VARIABLES,
C     VECTORS AND MATRICES DEFINED GLOBALLY.
C
C     PARAMETERS
      PARAMETER ( NEQ = 3 )
C
C     COMMON BLOCKS
      COMMON /VARIA/  X , DX , Y
      COMMON /DONNE/  ALPHA , CF , D50 , ROS , ROA , NU , R , REP , VSS
      COMMON /DIVP/   GR , ROT , CB , U , Q , B , USTAR , EW , ES
      COMMON /CONST/  CBCS , PI , G , CFLAG
C
C     DECLARATION OF VARIABLES
      LOGICAL    CFLAG
      REAL       NU
      DIMENSION  Y(NEQ)
C
C     CALCULATION OF VARIOUS PARAMETERS
      GR   = G * R * Y(3)
      ROT  = ROA + Y(3) * (ROS - ROA)
      U    = SQRT( G * R * Y(3) ) * Y(1) * COS(ALPHA) / Y(2) )
      Q    = U * Y(1)
      B    = GR * U * Y(1)
      USTAR = SQRT( CF * U**2 / 8.0 )
      EW   = WATENV(Y(2))
      IF(CFLAG)THEN
          CB = CBCS * Y(3)
          ES = SEDENT(USTAR,VSS,REP)
      ELSE
          CB = 0.0
          ES = 0.0
      ENDIF
C
      RETURN
      END
      SUBROUTINE RK4( X , Y , DX , YP , DERIVE )
C****************************************************************
C     CALCULATE THE DEPENDENT VARIABLES AT THE STATION X+DX       *
C****************************************************************
C
C     LIST OF VARIABLES LOCALLY DEFINED OF THE SUBROUTINE SUBPROGRAM "RK4"
C
C TYPE   NAME   DIMEN.    EXPLANATIONS
C =====  =====  =====     ==============================================
C R*4    K1     (NEQ)     = K1 IN THE METHOD OF 4TH ORDER RUNGE-KUTTA
C R*4    K2     (NEQ)     = K2 IN THE METHOD OF 4TH ORDER RUNGE-KUTTA
C R*4    K3     (NEQ)     = K3 IN THE METHOD OF 4TH ORDER RUNGE-KUTTA
C R*4    K4     (NEQ)     = K4 IN THE METHOD OF 4TH ORDER RUNGE-KUTTA
C R*4    YP     (NEQ)     = K4 IN THE METHOD OF 4TH ORDER RUNGE-KUTTA
C R*4    YT     (NEQ)     = TEMPORARY VALUE OF Y AT (X+DX/2) AND/OR A (X+DX)
C                           USED FOR THE CALCULATION OF K2, K3 ET K4
C R*4    DX2              = DX*0.5
C R*4    DX6              = DX/6
C R*4    XM               = VALUE OF X AT A HALF STEP (X+DX2) AWAY
C R*4    XP               = VALUE OF X AT A FULL STEP (X+DX) AWAY
C
C     SEE THE BEGINNING OF THE MAIN PROGRAM FOR A LIST OF SIMPLE VARIABLES,
C     VECTORS AND MATRICES DEFINED GLOBALLY.
C
C     ATTENTION : THE NAME OF THE SUBROUTINE SUBPROGRAM "DERIVE" IS SENT BY
C                 THE MAIN PROGRAM IN THE ARGUMENT LIST.
C
C     PARAMETERS
      PARAMETER ( NEQ = 3 )
C
C     DECLARATION OF VARIABLES
      REAL K1(NEQ) , K2(NEQ) , K3(NEQ) , K4(NEQ)
      DIMENSION Y(NEQ) , YP(NEQ) , YT(NEQ)
C
      DX2=DX*0.5
      DX6=DX/6.0
      XM=X+DX2
      XP=X+DX
C
C     CALCULATE K1
      CALL DERIVE( X  , Y  , K1 )
C
C     CALCULATE K2
      DO 10 I=1,NEQ
          YT(I) = Y(I) + DX2 * K1(I)
10    CONTINUE
      CALL DERIVE( XM , YT , K2 )
C
C     CALCULATE K3
      DO 20 I=1,NEQ
          YT(I) = Y(I) + DX2 * K2(I)
20    CONTINUE
      CALL DERIVE( XM , YT , K3 )
C
C     CALCULATE K4
      DO 30 I=1,NEQ
          YT(I) = Y(I) + DX * K3(I)
30    CONTINUE
      CALL DERIVE( XP , YT , K4 )
C
      DO 40 I=1,NEQ
          YP(I) = Y(I) + DX6 * (K1(I) + 2*K2(I) + 2*K3(I) + K4(I))
40    CONTINUE
      RETURN
      END
```

Fig.Ex.7.A.9   Program GRAVIT (continued)

```fortran
      SUBROUTINE DERIVE( X , Y , DYDX )
C***********************************************************************
C
C     SUBROUTINE SUBPROGRAM FOR CALCULATING THE DERIVATIVES OF NEQ
C     EQUATIONS.
C
C***********************************************************************
C
C     LIST OF VARIABLES LOCALLY DEFINED FOR THE SUBROUTINE SUBPROGRAM
C     "DERIVE"
C
C     TYPE NAME     DIMEN.     EXPLANATIONS
C     ==== ====     ======     ============
C     R*4  DYDX     (NEQ)    = VECTOR CONTAINING THE VALUES OF THE DERIVATIVES
C                              OF NEQ EQUATIONS CALCULATED AT X BY SUBSTITUTING
C                              Y AS THE VALUES OF DEPENDENT VARIABLES
C                              DYDX(1) = dh/dx
C                              DYDX(2) = dRi/dx
C                              DYDX(3) = dCs/dx
C     R*4  X                 = THE VALUE OF THE INDEPENDENT VARIABLE SENT BY
C                              THE SUBPROGRAM RK4 (IT REPLACES SUCCESSIVELY
C                              X, XM AND XP)
C     R*4  Y        (NEQ)    = THE VALUES OF THE DEPENDENT VARIABLES SENT BY
C                              THE SUBPROGRAM RK4 (IT REPLACES SUCCESSIVELY
C                              Y AND YT)
C
C     SEE THE BEGINNING OF THE MAIN PROGRAM FOR A LIST OF SIMPLE VARIABLES,
C     VECTORS AND MATRICES DEFINED GLOBALLY.
C
C***********************************************************************
C
C     PARAMETRES
      PARAMETER ( NEQ = 3 )
C
C     COMMON BLOCKS
      COMMON /DONNE/ ALPHA , CF , D50 , ROS , ROA , NU , R , REP , VSS
      COMMON /DIVP/ GR , ROT , CB , U , Q , B , USTAR , EW , ES
      COMMON /CONST/ CBCS , PI , G , CFLAG
C
C     DECLARATION OF VARIABLES
      LOGICAL CFLAG
      REAL NU
      DIMENSION Y(NEQ) , DYDX(NEQ)
C
C     CALCULATION OF dh/dX
      DYDX(1) = (4.0 - Y(2)) * EW / 2.0
      DYDX(1) = DYDX(1) + Y(2) * VSS * (ES-CB) / (2*U*Y(3))
      DYDX(1) = DYDX(1) - Y(2) * TAN(ALPHA)
      DYDX(1) = DYDX(1) + (USTAR/U)**2
      DYDX(1) = DYDX(1) / (1 - Y(2))
C
C     CALCULATION OF dRi/dX
      DYDX(2) = EW * VSS * (ES-CB) / (3*U*Y(3))) * (2 + Y(2)) / 2.0
      DYDX(2) = DYDX(2) - Y(2) * TAN(ALPHA)
      DYDX(2) = DYDX(2) + (USTAR/U)**2
      DYDX(2) = DYDX(2) * 3 * Y(2) / (Y(1) * (1-Y(2)))
C
C     CALCULATION OF dCs/dX
      DYDX(3) = VSS * (ES - CB) - Y(3) * EW * U
      DYDX(3) = DYDX(3) / (U * Y(1))
      RETURN
      END
C***********************************************************************
      FUNCTION SEDENT(USTAR,VSS,REP)
C***********************************************************************
C
C     FUNCTION SUBPROGRAM TO CALCULATE THE ENTRAINMENT COEFFICIENT FOR
C     SEDIMENTS.
C
C***********************************************************************
C
C     LIST OF VARIABLES LOCALLY DEFINED FOR THE SUBROUTINE SUBPROGRAM
C     "SEDENT"
C
C     TYPE NAME     DIMEN.     EXPLANATIONS
C     ==== ====     ======     ============
C     R*4  KSI                = PARAMETER IN EQ. 7.32
C     R*4  SEDENT             = ENTRAINMENT COEFFICIENT FOR SEDIMENTS
C
C     SEE THE BEGINNING OF THE MAIN PROGRAM FOR A LIST OF SIMPLE VARIABLES,
C     VECTORS AND MATRICES DEFINED GLOBALLY.
C
C***********************************************************************
C
C     DECLARATION OF VARIABLES
      REAL KSI
C
C     CALCULATION OF THE ENTRAINMENT COEFFICIENT FOR SEDIMENTS
      KSI = USTAR * REP**0.75 / VSS
      SEDENT = 3.0E-11 * KSI**7 / (1 + 1.0E-10 * KSI**7)
      RETURN
      END
C***********************************************************************
      FUNCTION WATENT(RI)
C***********************************************************************
C
C     FUNCTION SUBPROGRAM TO CALCULATE THE ENTRAINMENT COEFFICIENT FOR
C     AMBIENT FLUID.
C
C***********************************************************************
C
C     LIST OF VARIABLES LOCALLY DEFINED FOR THE SUBROUTINE SUBPROGRAM
C     "WATENT"
C
C     TYPE NAME     DIMEN.     EXPLANATIONS
C     ==== ====     ======     ============
C     R*4  WATENT             = ENTRAINMENT COEFFICIENT FOR AMBIENT FLUID
C     R*4  RI                 = RICHARDSON NUMBER
C
C     SEE THE BEGINNING OF THE MAIN PROGRAM FOR A LIST OF SIMPLE VARIABLES,
C     VECTORS AND MATRICES DEFINED GLOBALLY.
C
C***********************************************************************
C
C     CALCULATION OF THE ENTRAINMENT COEFFICIENT FOR AMBIENT FLUID
      WATENT = 0.075 / SQRT(- + 718.0 * RI**2.4)
      RETURN
      END
C***********************************************************************
      SUBROUTINE TITLES( NOUT )
C***********************************************************************
C
C     SUBROUTINE SUBPROGRAM FOR WRITING TITLES AND COLUMN HEADINGS ON THE
C     OUTPUT FILE.
C
C***********************************************************************
C
C     LIST OF VARIABLES LOCALLY DEFINED FOR THE SUBROUTINE SUBPROGRAM
C     "TITLES"
```

Fig.Ex.7.A.9   Program GRAVIT (continued)

```
C   TYPE NAME     DIMEN.      EXPLANATIONS
C   ==== ====     ======      ============
C   R*4  TITLE    (20)    = ARRAY CONTAINING THE LIGNES OF THE OUTPUT TITLE
C
C   SEE THE BEGINNING OF THE MAIN PROGRAM FOR A LIST OF SIMPLE VARIABLES,
C   VECTORS AND MATRICES DEFINED GLOBALLY.
C*********************************************************************
C
C   PARAMETERS
      PARAMETER ( NEQ = 3 )
C   COMMON BLOCKS
      COMMON /VARIA/ X , DX , Y
      COMMON /DONNE/ ALPHA , CF , D50 , ROS , ROA , NU , R , REP , VSS
      COMMON /LIMIT/ X1 , XF , ND , NS , NPP
      COMMON /DIVP/ GR , ROT , CB , U , Q , B , USTAR , EW , ES
      COMMON /CONST/ CBCS , PI , G , CFLAG
C   DECLARATION OF VARIABLES
      LOGICAL         CFLAG
      REAL            NU , ND , NS
      CHARACTER*132 TITLE(20)
      DIMENSION Y(NEQ)
C
C   INITIALISATION OF THE TITLE
      DATA TITLE(1)/'                                 RESULTS OF
     1THE CALCULATIONS WITH THE PROGRAM <GRAVIT>'/
      DATA TITLE(2)/'                    SOLUTION OF THE SYSTEM OF DIFFERENTIA
     1L EQUATIONS GOVERNING THE NON-UNIFORM FLOW OF ONE-DIMENSIONAL, STE
     2ADY,'/
      DATA TITLE(3)/'           CONSERVATIVE OR NON CONSERVAT
     1IVE GRAVITY CURRENTS USING THE 4TH ORDER RUNGE-KUTTA METHOD.'/
      DATA TITLE(4)/'          THE CURRENT HAS A SUPERCRITICAL REGI
     1ME. THE SOLUTION IS THEREFORE ADVANCED FROM UPSTREAM TO DOWNSTREAM
     2.'/
      DATA TITLE(5)/'                      '/
      DATA TITLE(6)/' BOUNDARY CONDITIONS :'/
      DATA TITLE(7)/' PHYSICAL CHARACTERISTICS :
     1            Values known at first station :'/
      DATA TITLE(8)/'   Inclination angle of the bed,    ALPHA (degree)
     1= x.xxx                              Current depth,             H0         (m)
     2= x.xx  '/
      DATA TITLE(9)/'   Friction coefficient,            CF         (-)
     1= x.xxxxx                            Unit discharge of current,  Q0        (m2/s)
     2 = '/
      DATA TITLE(10)/'  Mean diameter of the sediments,  D50        (m)
     1= x.xxxxx                            Sediment concentration by volume, CS0 (-)
     2= xxxx.xx '/
      DATA TITLE(11)/'  Density of sediments,            ROS        (kg/m3)
     1= xxxx.xx                            Density of ambient fluid,   ROA       (kg/m3)
     2/'
      DATA TITLE(12)/'  Kinematic visc. of ambient fluid, NU        (m2/s)
     1= x.xxxExxxx                          x-coordinate of first stn,  X1        (m)
     2= '/
      DATA TITLE(13)/'                                   INFORMATIONS ON COMPUTATIONAL REACH (DOMAIN) :'
     1                                     x-coordinate of last stn,   XF        (m)
     2= '/
      DATA TITLE(14)/' USEFUL PARAMETERS CALCULATED USING THE GIVEN DATA
     1 :                                    Computational reach length, TL        (m)
     2= '/
      DATA TITLE(15)/'  Submerged specific density,      R          (-)
     1= x.xxxx                              Step length in x-direction, DX        (m)
     2= '/
      DATA TITLE(16)/'  Particle Reynolds number,        REP
     1= xxx.xx                              Number of divisions of the reach, ND
     2= '/
      DATA TITLE(17)/'  Settling velocity of particles,  VSS        (m/s)
     1= x.xxxxx                            Number of stations,          NS
     2= '/
      DATA TITLE(18)/'                                                   The results a
     1re written every (NPP =  )    steps (stations).'/
      DATA TITLE(19)/'                                                  X    H     RI    CS   U
     1       B       Q        GR       ROT   USTAR     EW     E
     2S'/
      DATA TITLE(20)/'                                                 STN     (m)   (-)    (-)  (m/s
     1) (m3/s)   (m2/s)   (m/s2)   (-)   (kg/m3)   (m/s)   (-
     1)'/
C
C   FILL IN THE INFORMATIONS IN THE TITLE TEXT
      IF(CFLAG)THEN
         TITLE(5)(2:58)='NON CONSERVATIVE GRAVITY CURRENT (TURBIDITY CURR
     1ENT)'
      ELSE
         TITLE(5)(2:58)='CONSERVATIVE GRAVITY CURRENT (DENSITY CURRENT)'
      ENDIF
C
      WRITE(TITLE(7)(53:58),'(F6.3)')ALPHA*180.0/PI
      WRITE(TITLE(7)(120:124),'(F5.2)')Y(1)
C
      WRITE(TITLE(8)(53:59),'(F7.5)')CF
      WRITE(TITLE(8)(120:126),'(F7.3)')Q
C
      IF(CFLAG)THEN
         WRITE(TITLE(9)(53:59),'(F7.5)')D50
      ELSE
         TITLE(9)(1:69)     = '
     1                   '
         TITLE(9)(70:119) = 'Conc. by vol. of dissolved subst., CS0      (-)
     1 = '
      ENDIF
      WRITE(TITLE(9)(120:125),'(F6.4)')Y(3)
C
      IF(.NOT.CFLAG)THEN
         TITLE(10)(1:69) = '      Density of dissolved substances,      ROS       (kg
     1/m3) = '
      ENDIF
      WRITE(TITLE(10)(53:59),'(F7.2)')ROS
C
      WRITE(TITLE(11)(53:59),'(F7.2)')ROA
      WRITE(TITLE(12)(53:61),'(E9.3)')NU
      WRITE(TITLE(12)(120:128),'(F9.2)')X1
C
      WRITE(TITLE(13)(120:128),'(F9.2)')XF
C
      WRITE(TITLE(14)(120:128),'(F9.2)')XF-X1
C
      WRITE(TITLE(15)(53:58),'(F6.4)')R
      WRITE(TITLE(15)(120:126),'(F7.3)')DX
C
      IF(CFLAG)THEN
         WRITE(TITLE(16)(53:58),'(F6.2)')REP
      ELSE
         TITLE(16)(1:69) = ' '
      ENDIF
      WRITE(TITLE(16)(120:127),'(F8.0)')ND
C
```

Fig.Ex.7.A.9 Program GRAVIT (continued)

```
      IF(CFLAG)THEN
         WRITE(TITLE(17)(53:59),'(F7.5)')VSS
      ELSE
         TITLE(17)(1:69) = ' '
      ENDIF
      WRITE(TITLE(17)(120:127),'(F8.0)')NS
C
      WRITE(TITLE(18)(77:80),'(I4)')NPP
C
C     WRITE THE TITLE ON THE OUTPUT FILE
      DO 10 I = 1, 4
10       WRITE(NOUT,20)TITLE(I)
20    FORMAT(A)
      WRITE(NOUT,*)
C
      DO 30 I = 5, 17
30       WRITE(NOUT,20)TITLE(I)
      WRITE(NOUT,*)
C
      WRITE(NOUT,20)TITLE(18)
      WRITE(NOUT,*)
      WRITE(NOUT,*)
C
      DO 40 I = 19, 20
40       WRITE(NOUT,20)TITLE(I)
      RETURN
      END

C*****************************************************************
      SUBROUTINE RWRITE( STN , CFLAG , NOUT )
C*****************************************************************
C*                                                               *
C* THIS SUBROUTINE SUBPROGRAM WRITES A LINE OF RESULTS ON THE OUTPUT *
C* FILE                                                          *
C*                                                               *
C*****************************************************************
C
C     NO LOCALLY DEFINED VARIABLES.
C     SEE THE BEGINNING OF THE MAIN PROGRAM FOR A LIST OF SIMPLE VARIABLES,
C     VECTORS AND MATRICES DEFINED GLOBALLY.
C
C     PARAMETERS
      PARAMETER ( NEQ = 3 )
C
C     COMMON BLOCKS
      COMMON /VARIA/ X  , DX , Y
      COMMON /DIVP/  GR , ROT , CB , U , Q , B , USTAR , EW , ES
C
C     DECLARATION OF VARIABLES
      LOGICAL       CFLAG
      DIMENSION     Y(NEQ)
C
C     WRITE A LINE OF RESULTS
      IF(CFLAG) THEN
         WRITE(NOUT,20)STN, X , (Y(I),I=1,NEQ) , U , B , Q , GR , CB ,
     1       ROT , USTAR , EW , ES
20       FORMAT(F8.0,2X,F9.2,2X,F6.3,2X,F6.4,2X,F5.2,2X,F9.2,2X,
     1       F6.2,2X,F6.4,2X,F6.4,2X,F7.2,2X,F5.2,2X,F7.5,2X,F7.5)
      ELSE
         WRITE(NOUT,30)STN, X , (Y(I),I=1,NEQ) , U , B , Q , GR ,
     1       ROT , USTAR , EW
30       FORMAT(F8.0,2X,F9.2,2X,F6.3,2X,F6.4,2X,F5.2,2X,F9.2,2X,
     1       F6.2,2X,F6.4,2X,F6.4,8X,F7.2,2X,F5.2,2X,F7.5)
      ENDIF
      RETURN
      END
```

Fig.Ex.7.A.9   Program GRAVIT (end)

7.7.2 Problems, unsolved

Ex. 7.1
Study the turbidity current of Ex. 7.A, by varying the slope of the bed, taking $\alpha_1 = 0.1°$, $\alpha_2 = 1°$, $\alpha_3 = 4°$, $\alpha_4 = 8°$ and $\alpha_5 = 10°$. What is the influence of a slope variation on the evolution of the relative depth, h/h_o ?

Ex. 7.2
For the turbidity current described in Ex. 7.A, its turbidity, $\Delta\rho_o$, is now doubled. Make a comparison of the two currents.

Ex. 7.3
Study a turbidity current with the same data set as for Ex. 7.A, but this time the unit discharge is doubled.

Ex. 7.4
For the turbidity current given in Ex. 7.A (*iii*), determine the evolution of the distribution of velocity and of concentration.

Ex. 7.5
Show, whether it is correct that the Richardson number is independent of the distance, for a current in equilibrium.

8. TRANSPORT AND MIXING OF MATTER

In this chapter, the transport of a physical quantity (mass, heat and momentum) in the fluid and by the fluid will be investigated. Considered will be transport by diffusion and by convection.

The theoretical concepts of transport are developed. The different kinds of diffusion, namely diffusion of mass, of temperature and of momentum are presented. Subsequently, the balance equations in laminar and turbulent regime are established. It is evident that the coefficient of diffusion plays an important role.

Some solutions to the equation of pure diffusion and of convection-diffusion in laminar flow are presented, and this for an instantaneously and a continuously introduced source.

Of particular importance is the case of convection-diffusion in turbulent flow. The coefficients of turbulent diffusivity and of dispersivity are presented.

The longitudinal dispersion as encountered in channel flow is of great practical importance. Solutions will be presented for the cases where a passive and an active substance is introduced instantaneously or for a limited time or continuously.

Also investigated will be the dispersion as well as the transport with reaction. The later one is used to estimate the self-cleaning capacity of a waterway receiving wastewater.

TABLE OF CONTENTS

8.1 THEORETICAL CONSIDERATIONS
    8.1.1 Diffusion
    8.1.2 Convection-Diffusion
    8.1.3 Turbulent Convection-Diffusion
    8.1.4 The Diffusivity

8.2 DIFFUSION
    8.2.1 Instantaneous Source
    8.2.2 Continuous Source

8.3 CONVECTION-DIFFUSION IN LAMINAIR REGIME
    8.3.1 Instantaneous Source
    8.3.2 Continuous Source

8.4 CONVECTION-DIFFUSION IN TURBULENT REGIME
    8.4.1 Diffusivity
    8.4.2 Transversal Diffusion
    8.4.3 Dispersivity
    8.4.4 Longitudinal Dispersion
    8.4.5 Dispersion with Reaction
    8.4.6 Transport with Reaction

8.5 EXERCISES
    8.5.1 Problems, solved
    8.5.2 Problems, unsolved

8.1 THEORETICAL CONSIDERATIONS

1° Studied will be the transport (or transfer) of a "physical quantity" in the fluid *by* the fluid (mixture), such as the mass of a conservative substance (salt, colour or waste), the heat, the momentum, but also solid particles in suspension.

2° Two processes of transport, one by *diffusion* and one by *convection*, shall be investigated.

8.1.1 Diffusion

1° The transport by molecular diffusion (in a motionless fluid) is given by a phenomenological law.

Suppose (see *Padet*, 1991, p. 37) that an extensive property (scalar or vector), being the physical quantity, has a local volumic density, $c_f(x,y,z;t)$. The vector of the density flux, q_f, being proportional to the gradient of the density, is directed in a (negative) sense of decreasing densities and given by :

$$\vec{q_f} = -k \, \overrightarrow{\text{grad}} \, c_f$$

$$q_{f_i} = -k \frac{\partial c_f}{\partial x_i} \qquad (8.1)$$

where k is a constant, the coefficient of diffusion or in short the (molecular) *diffusivity*; it is an intrinsic property of the fluid and depends on temperature and on pressure. The transport by diffusion corresponds to a source at the surface in the balance equation, eq. 8.5. Diffusion is an irreversible process and is responsible for energy dissipation within the fluid.

Transport by diffusion is shown in Fig. 8.1a. A point source of an extensive property, $c_f(x,y;t)$, grows in forming a cloud, increasing in size but decreasing in density.

There are different types of diffusion : diffusion of mass, diffusion of heat and diffusion of momentum.

2° *Diffusion of mass* :

For the transport by diffusion of a passive substance, A, in a fluid — the extensise property is here associated to the concentration of the substance, $c_f \equiv c$, — the vector of the density flux is given by the *law of Fick* :

$$\vec{q_m} = -\varepsilon_m \, \overrightarrow{\text{grad}} \, c$$

$$q_{m_i} = -\varepsilon_m \frac{\partial c}{\partial x_i} \qquad (8.2)$$

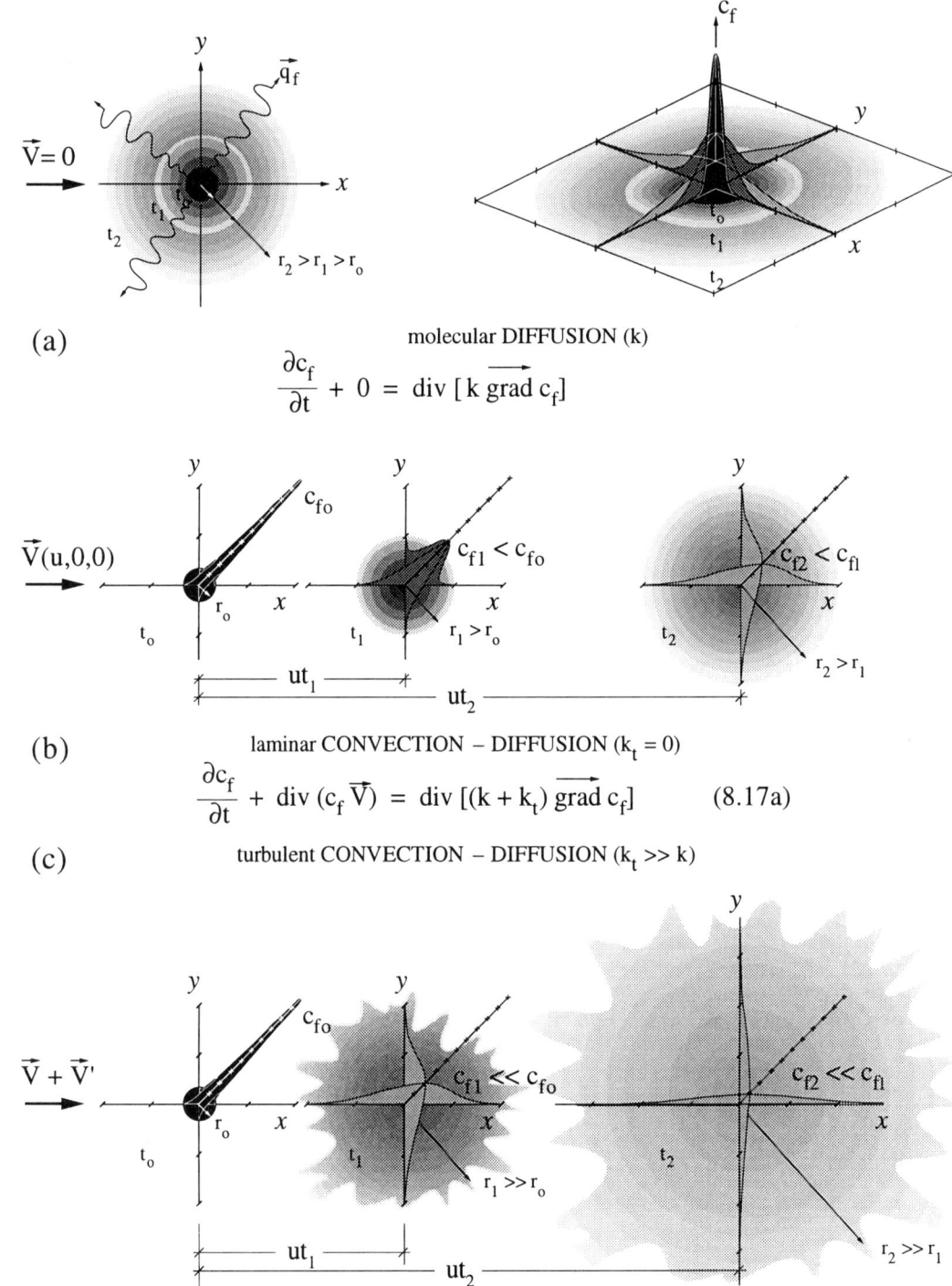

Fig. 8.1 Schematic representation of the *diffusion* and of the *convection-diffusion* in laminar and turbulent regime for a physical property, $c_f(x,y;t)$.

TRANSPORT AND MIXING OF MATTER

where c is the (local) concentration, defined by the mass of the substance per volume of the fluid (mixture), in units of [kg/m³]; one replaces $c \to (c - C_o)$, if the medium has an initial homogeneous concentration, C_o. Here, ε_m is the *mass (molecular) diffusivity* of the substance, A, in the mixture, expressed in units of $[L^2/T]$; it depends on the chemical composition and the molecular concentration of the substance (see *Bird* et al., 1960, chap. 16).

The law of Fick, eq. 8.2, states that a substance in a mixture tends to equalise its distribution; that is, it flows from zones of high to zones of low concentration.

3° *Diffusion of heat*:

For the transport by diffusion (conduction) of heat — the extensive property is here associated to the thermal energy (enthalpy), $c_f \equiv \rho C_p T$, — the vector of the density flux is given by the *law of Fourier*:

$$\vec{q}_h = - a_h \, \overrightarrow{\text{grad}} \, (\rho C_p T)$$

(8.3)

$$q_{h_i} = - \rho a_h C_p \frac{\partial T}{\partial x_i} \qquad (\rho C_p = \text{Cte})$$

where T is the (local) temperature and ρ is the density of the mixture; one replaces $T \to (T - T_o)$, if the medium has an initial homogeneous temperature, T_o. Here, a_h is the *thermal diffusivity*, expressed in units of $[L^2/T]$; $\rho a_h C_p = \lambda$ is the thermal conductivity of the fluid and C_p is the specific heat at constant pressure. For water at $T = 20$ [C°] one takes: $\lambda = 0.60$ [W/(mK)] and $C_p = 4182$ [J/(kgK)].

The law of Fourier, eq. 8.3, states that the temperature tends to equalise its distribution; that is, heat always flows from high to low temperatures. Not to be overlooked is the analogy between the law of Fourier with the one of Fick, eq. 8.2.

4° *Diffusion of momentum*:

For the transfer of momentum — the extensive property is here associated to the momentum, $\vec{c}_f \equiv \rho \vec{V}$, — the tensor of the density flux is given by the *law of Newton*:

$$\overset{=}{q}_{mt} = - \nu \, \overrightarrow{\text{grad}} \, (\rho \vec{V})$$

(8.4)

$$q_{mt_{ij}} = - \rho \nu \frac{\partial V_i}{\partial x_j} = \tau_{ij} \qquad (\rho = \text{Cte})$$

where $\vec{V}(u,v,w)$ is the (local) velocity of the fluid and τ_{ij} is the shear stress. Here, ν is *momentum diffusivity*, being nothing else but the *kinematic (molecular) viscosity*, expressed in units of $[L^2/T]$; ρ is the density of the mixture.

The transfer of momentum caused by the forces of viscosity is thus considered as a mechanism of diffusion.

5° In summary, there exists a formal analogy between the molecular diffusivity, ε_m, the thermal diffusivity, a_h, and the kinematic viscosity, ν, (see *Bird et al.*, 1960, p. 503). Each of the three different cases of transport by diffusion can be described by a kinematic property of the fluid, namely the diffusivities, ε_m, a_h and ν, whose units, $[L^2/T]$, are the same. Some indicative values in an aqueous medium are given in Table 8.1.

Table 8.1 Coefficients of molecular (laminar) and turbulent *diffusivity* in an aqueous medium; the values are only indicative, expressed in order of magnitude.

Diffusivity		*molecular*	*turbulent*	
[cm^2/s]		(at 20 [°C])	longitudinal	vertical
Mass diffusivity	: $\varepsilon_m, \varepsilon_t$	~ 10^{-5}	$O(10^5)$	$O(10^3)$
Thermal diffusivity	: a_h, a_{ht}	~ 10^{-3}	$O(10^5)$	$O(10^3)$
Viscosity	: ν, ν_t	10^{-2}	$O(10^5)$	$O(10^3)$

6° The ratio between the diffusivities gives dimensionless numbers, such as :

i) the *Prandtl* number :

$$\text{Pr} = \frac{\nu}{a_h}$$

for water at $T = 20$ [°C] : Pr $\cong$ 7
for cold water : Pr $\cong$ 13 ;

ii) the *Schmidt* number :

$$\text{Sc} = \frac{\nu}{\varepsilon_m}$$

for an aqueous solution at $T = 20$ [°C] : $10^2 < \text{Sc} < 10^3$,

depending on the molecular weight of the substance, A, in the mixture.

8.1.2 Convection–Diffusion

1° The law of diffusion, eq. 8.1, provides a relationship between the spatial distribution of an extensive property and the density flux. Yet another relationship can be obtained by the application of the principle of conservation of mass. An equation of balance or budget is thus obtained.

2° Consider (see Fig. 8.2) a mass (domain) of fluid, D, bounded by a fixed surface, S, across which an extensive property, $c_f(x,y,z;t)$, is being transported by *diffusion* as well as by fluid motion having a velocity, $\vec{V}$; the latter transport is called *convection*.

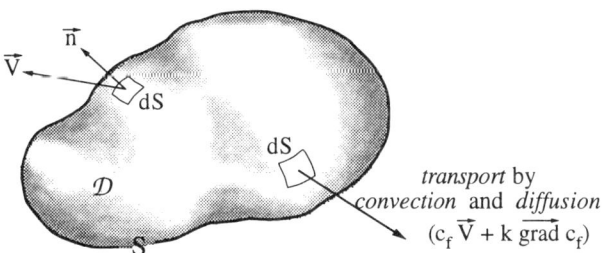

Fig. 8.2 Mass of fluid, D, bounded by a surface, S, submitted to a balance for an extensive property, c_f.

The resulting general balance equation written in *integral* form (see *Padet*, 1991, p. 22) is :

$$\int_D (\partial c_f/\partial t)\, d\delta + \int_D \text{div}(c_f \vec{V})\, d\delta = - \int_D \text{div}\,\vec{q_f}\, d\delta + \int_D q_e\, d\delta \qquad (8.5a)$$

and written in *local* form (see *Daily* et *Harleman*, 1966, p. 48) it is :

$$\frac{\partial c_f}{\partial t} + \text{div}(c_f \vec{V}) = - \text{div}\,\vec{q_f} + q_e \qquad (8.5)$$

where $\vec{V}(u,v,w)$ is the vector of the local velocity. $\vec{q_f}$ is the flow from (external) sources per unity of surface — here enters the transport by diffusion, eq. 8.1, — and q_e is the flow from (internal) sources per unity of volume. The term $\text{div}(c_f\vec{V})$ represents the convection of the physical quantity.

Transport by diffusion and by convection is shown in Fig. 8.1b. A point source of an extensive property, $c_f(x,y;t)$, is diffused (see Fig. 8.1a) but also convected with a velocity, $\vec{V}(u,0,0)$.

The different physical quantities, c_f, shall now be considered; in this way one obtains the balance of mass, of heat and of momentum.

3° *Balance of mass* :

To establish the balance of mass, one writes eq. 8.5 as follows :

$$\frac{\partial c}{\partial t} + \text{div}(c\vec{V}) = -\text{div}\,\vec{q_m} + q_e \tag{8.6a}$$

where $c_f = c$ is the concentration of the substance, A, in the mixture; $\vec{q_m}$ is the vector of the density flux expressed by the *law of Fick*, eq. 8.2, and q_e is the rate of local production of the substance. For the incompressible mixture, $\rho = \text{Cte}$ and $\text{div}\,\vec{V} = 0$, one obtains (see *Padet*, 1991, p. 40) :

$$\frac{\partial c}{\partial t} + \vec{V} \cdot \overrightarrow{\text{grad}}\,c = \text{div}(\varepsilon_m \overrightarrow{\text{grad}}\,c) \tag{8.6}$$

where c is the local concentration.

The term of the (internal) sources, q_e, — this term will include the phenomenon of chemical reaction or of phase change — is in eq. 8.6 omitted, being considered small with respect to the term of diffusion.

This equation, eq. 8.6, is the *convection-diffusion equation* of mass.

4° *Balance of heat* :

The balance of heat (see *Padet*, 1991, p. 40) is expressed in an analogous way to the balance of mass, eq. 8.6, using the *law of Fourier*, eq. 8.3, or :

$$\frac{\partial T}{\partial t} + \vec{V} \cdot \overrightarrow{\text{grad}}\,T = \text{div}(a_h \overrightarrow{\text{grad}}\,T) \tag{8.7}$$

where it was assumed that $\rho = \text{Cte}$ and $C_p = \text{Cte}$. It is here considered that heat is a conservative physical quantity; there is no heat transfer across the bed and the banks of the channel.

This equation, eq. 8.7, is the *convection-conduction equation* of heat transfer.

5° *Balance of momentum* :

The balance of momentum (see *Padet*, 1991, p. 40) is expressed in the same way, using the *law of Newton*, eq. 8.4, or :

$$\frac{\partial V_j}{\partial t} + \vec{V} \cdot \overrightarrow{\text{grad}}\,V_j = \text{div}(\nu\,\overrightarrow{\text{grad}}\,V_j) \tag{8.8}$$

written for an incompressible mixture, $\rho = \text{Cte}$ and $\text{div}\,\vec{V} = 0$; V_j is the component in the j-direction of the velocity, $\vec{V}$. It is to be kept in mind that the transfer of momentum resulting from the friction (viscosity) forces is also a mechanism of diffusion.

TRANSPORT AND MIXING OF MATTER

In eq. 8.8, the term of pressure, p, and of gravity, g, — both are forces applied on the system — have been omitted. However, if they are also included, one obtains:

$$\frac{\partial V_j}{\partial t} + \vec{V} \cdot \overrightarrow{\text{grad}} V_j = -\frac{1}{\rho} \frac{\partial p^*}{\partial x_j} + \text{div}(v\,\overrightarrow{\text{grad}} V_j) \tag{8.9}$$

where $p^* = p + \rho g z$ is the driving pressure.

This equation, eq. 8.9, is the *Navier-Stokes equation*, valid for any type of flow, but in particular for laminar flow (see *Schlichting*, 1979).

6° *Pure* diffusion exists if the velocity vector is zero, $\vec{V} = 0$, in eq. 8.6 and eq. 8.7; in eq. 8.8, there is no diffusion possible.

If furthermore the coefficients of diffusivity, ε_m and a_h, remain constant, one can write for the:

i) *balance of mass* (see eq. 8.6):

$$\frac{\partial c}{\partial t} = \varepsilon_m \, \text{div}\,(\overrightarrow{\text{grad}}\, c) \tag{8.10}$$

and, expressed in Cartesian coordinates:

$$\frac{\partial c}{\partial t} = \varepsilon_m \left(\frac{\partial^2 c}{\partial x^2} + \frac{\partial^2 c}{\partial y^2} + \frac{\partial^2 c}{\partial z^2}\right) \tag{8.10a}$$

where c is the local concentration.

ii) *balance of heat* (see eq. 8.7):

$$\frac{\partial T}{\partial t} = a_h \, \text{div}\,(\overrightarrow{\text{grad}}\, T) \tag{8.11}$$

and, expressed in Cartesian coordinates:

$$\frac{\partial T}{\partial t} = a_h \left(\frac{\partial^2 T}{\partial x^2} + \frac{\partial^2 T}{\partial y^2} + \frac{\partial^2 T}{\partial z^2}\right) \tag{8.11a}$$

where *T* is the local temperature.

These equations, eq. 8.10 and eq. 8.11, have the classic form of the *diffusion equation* with partial derivatives; they are parabolic equations having constant coefficients.

8.1.3 Turbulent Convection-Diffusion

1° For laminar flow with molecular diffusion, eq. 8.1, the equation of balance in local form, eq. 8.5, is given by:

$$\frac{\partial c_f}{\partial t} + \text{div}(c_f \vec{V}) = + \text{div}(k \, \overrightarrow{\text{grad}} \, c_f) + q_e \qquad (8.12)$$

where c_f is a physical quantity (an extensive property) and k is a constant, the molecular diffusivity.

2° When flow is turbulent, all instantaneous physical quantities, c_f, must be expressed by a time-averaged value, $\overline{c_f} \equiv c_f$ (the bar shall be deleted to simplify writing), plus a value due to the random fluctuations, c_f', whose average value is $\overline{c_f'} = 0$; namely:

$$c_f \rightarrow c_f + c_f' \qquad (8.13)$$

For the physical quantities under consideration, this gives:

$$c \rightarrow c + c' \quad ; \quad T \rightarrow T + T' \quad ; \quad \vec{V} \rightarrow \vec{V} + \vec{V}'$$
$$(u \rightarrow u + u', \text{ etc.})$$

3° The above balance equation, eq. 8.12, for turbulent flow with turbulent diffusion, reads now:

$$\frac{\partial (c_f + c_f')}{\partial t} + \text{div}\left[(c_f + c_f')(\vec{V} + \vec{V}')\right] = \text{div}\left[k \, \overrightarrow{\text{grad}}(c_f + c_f')\right] + q_e \qquad (8.14)$$

Using the definition for the instantaneous terms, eq. 8.13, and their statistical properties (see *Padet*, 1991, p. 136) one writes eq. 8.14 as:

$$\frac{\partial c_f}{\partial t} + \text{div}(c_f \vec{V} + \overline{c_f' \vec{V}'}) = \text{div}(k \, \overrightarrow{\text{grad}} \, c_f) + q_e \qquad (8.15a)$$

or more commonly as:

$$\frac{\partial c_f}{\partial t} + \text{div}(c_f \vec{V}) = \text{div}\left[(k \, \overrightarrow{\text{grad}} \, c_f) - \overline{c_f' \vec{V}'}\right] + q_e \qquad (8.15)$$

This is the equation of balance in turbulent regime for the average values, $\overline{c_f} \equiv c_f$. Comparing eq. 8.15 with eq. 8.12, a new term, $\text{div}(\overline{c_f' \vec{V}'})$, appears, which is due to the fluctuations of $\vec{V}'$ and c_f'.

The transport by molecular and by turbulent diffusion and by convection is shown with Fig. 8.1c. A point source of an extensive property, c_f, is diffused

TRANSPORT AND MIXING OF MATTER 527

(see Fig. 8.1a), but also convected with a velocity, $\vec{V}(u,0,0)$, (see Fig. 8.1b) and its turbulent fluctuations, $\vec{V'}(u',v',w')$.

4° The expression, $\overline{(c_f'\vec{V'})}$, which is a covariance, called the turbulent flux of Reynolds, can be associated to the turbulent diffusion; this is in analogy with the other term in the brackets of eq. 8.15, which represents the molecular diffusion. Commonly (see *Hinze*, 1959, p. 21 and p. 25) used is the following expression :

$$-\overline{(c_f'\vec{V'})} = k_t \overrightarrow{\text{grad}}\, c_f \quad (8.16)$$

where k_t is the turbulent diffusivity of the physical quantity, c_f, which depends on the local (statistical) properties of the turbulent flow, but does not depend on the fluid itself. This diffusivity can be a scalar, k_t, or a diagonal tensor, $\vec{k_t}$ (k_{tx}, k_{ty}, k_{tz}), where one takes in account different values for the different (corresponding) directions, x, y et z.

5° The equation of balance in turbulent regime, eq. 8.15, can thus be written as :

$$\frac{\partial c_f}{\partial t} + \text{div}(c_f\, \vec{V}) = \text{div}\left[(k+k_t)\, \overrightarrow{\text{grad}}\, c_f\right] + q_e \quad (8.17)$$

For turbulent flow, the turbulent diffusivity, k_t, is much larger than the molecular diffusivity, $k_t \gg k$; thus one may postulate :

$(k + k_t) \approx k_t$

The turbulence in the flow is thus responsible for a strong diffusion of the physical quantity, c_f.

The different physical quantities, c_f, can now be considered; in this way one obtains expressions for the balance of mass, of heat and of momentum in turbulent regime. The term of the internal source, q_e, shall be omitted from most of the further considerations.

6° *Balance of mass* :

For an incompressible mixture, $\rho = \text{Cte}$ and $\text{div}\, \vec{V} = 0$, one may write eq. 8.17 (see eq. 8.6) as follows :

$$\frac{\partial c}{\partial t} + \vec{V} \cdot \overrightarrow{\text{grad}}\, c = \text{div}\left[(\varepsilon_m + \varepsilon_t)\, \overrightarrow{\text{grad}}\, c\right] \quad (8.18)$$

The turbulent flux of mass per unit surface (see eq. 8.16) is expressed by :

$$-\overline{(c'\vec{V'})} = \varepsilon_t\, \overrightarrow{\text{grad}}\, c \quad (8.18a)$$

where c is time-averaged concentration and c' stands for the concentration fluctuations. ε_t is the turbulent mass diffusivity of the substance, A, in the mixture; in general, $\varepsilon_t \gg \varepsilon_m$. The diffusivity, ε_t, is taken to be a scalar; often the diffusivity depends on the directions, x, y and z, and is a diagonal tensor, $\vec{\varepsilon}_t$.

This equation, eq. 8.18, is the *convection-diffusion equation* of mass for time-averaged values in turbulent regime. This equation is a basic one for analysis of many problems of mass transport in water courses. Solutions to this equation are difficult to obtain; this is the reason to see if simplification can be made (see sect. 8.4).

In Cartesian coordinates, eq. 8.18 is written as :

$$\frac{\partial c}{\partial t} + u \frac{\partial c}{\partial x} + v \frac{\partial c}{\partial y} + w \frac{\partial c}{\partial z} = \varepsilon_m \left(\frac{\partial^2 c}{\partial x^2} + \frac{\partial^2 c}{\partial y^2} + \frac{\partial^2 c}{\partial z^2} \right) +$$
$$+ \frac{\partial}{\partial x}(\varepsilon_{tx} \frac{\partial c}{\partial x}) + \frac{\partial}{\partial y}(\varepsilon_{ty} \frac{\partial c}{\partial y}) + \frac{\partial}{\partial z}(\varepsilon_{tz} \frac{\partial c}{\partial z}) \quad (8.19)$$

where one takes (see eq. 8.18a) :

$$-\varepsilon_{tx} \frac{\partial c}{\partial x} = \overline{(u'c')} \quad , \quad -\varepsilon_{ty} \frac{\partial c}{\partial y} = \overline{(v'c')} \quad , \quad -\varepsilon_{tz} \frac{\partial c}{\partial z} = \overline{(w'c')} \quad (8.19a)$$

The molecular mass diffusivity, being a property of the fluid, remains constant, ε_m = Cte. The turbulent mass diffusivity, $\vec{\varepsilon}_t(\varepsilon_{tx}, \varepsilon_{ty}, \varepsilon_{tz})$ — here a diagonal tensor — depends on the local properties of the flow; for an isotropic turbulence, their values are the same, $\varepsilon_{tx} = \varepsilon_{ty} = \varepsilon_{tz}$.

Evidently, eq. 8.19 becomes eq. 8.10a, if there is no motion, $\vec{V} + \vec{V}' = 0$; this is the case for pure diffusion.

7° *Balance of heat* :

For the temperature, eq. 8.17 (see eq. 8.7) can be written in an analogous way, or :

$$\frac{\partial T}{\partial t} + \vec{V} \cdot \overrightarrow{\text{grad}}\, T = \text{div}\left[(a_h + a_{ht}) \overrightarrow{\text{grad}}\, T\right] \quad (8.20)$$

where it is admitted that

$$-\overline{(T'\vec{V}')} = a_{ht} \overrightarrow{\text{grad}}\, T \quad (8.20a)$$

T are the fluctuations of the temperature, T. a_{ht} is the thermal turbulent diffusivity; in general $a_{ht} \gg a_h$.

This equation, eq. 8.20, is the *convection-conduction equation* of heat transfer for time-averaged values in turbulent regime.

TRANSPORT AND MIXING OF MATTER   529

8° *Balance of momentum* :

For the momentum, eq. 8.17 (see eq. 8.9) can be written in an analogous way (see *Padet*, 1991, p. 108), or :

$$\frac{\partial V_j}{\partial t} + \vec{V} \cdot \overrightarrow{\text{grad}}\, V_j = -\frac{1}{\rho}\frac{\partial p^*}{\partial x_j} + \text{div}\left[(\nu + \nu_{tj})\, \overrightarrow{\text{grad}}\, V_j\right] \quad (8.21)$$

where the Reynolds stresses are expressed as :

$$-\rho\, \overline{(V_j'\vec{V}')} = \rho \nu_{tj}\, \overrightarrow{\text{grad}}\, V_j \quad (8.21a)$$

with V_j' being the fluctuations of the velocity, V_j, in the j-direction. ν_{tj} is the turbulent viscosity in this direction; in general $\nu_{tj} \gg \nu$.

This equation, eq. 8.21, is the *Reynolds equation*, valid for turbulent flow (see *Hinze*, 1959, p. 19).

For steady and two-dimensional, $\vec{V}(u,0,w)$, flow in a channel, where one may assume that $u \gg w$, the above equation reads :

$$u\frac{\partial u}{\partial x} + w\frac{\partial u}{\partial z} = -\frac{1}{\rho}\frac{\partial p^*}{\partial x} + \nu\frac{\partial^2 u}{\partial z^2} - \frac{\partial}{\partial x}\overline{(u'u')} - \frac{\partial}{\partial z}\overline{(u'w')} \quad (8.22)$$

admitting that :

$$\nu_{tx}\frac{\partial u}{\partial x} = -\overline{(u'u')} \quad ; \quad \nu_{tz}\frac{\partial u}{\partial z} = -\overline{(u'w')}$$

This equation, eq. 8.22, was previously (see sect. 2.5.1) developed.

8.1.4 The Diffusivity

1° Transport by *molecular diffusion* in a fluid, either motionless or in laminar flow, was described by eq. 8.1. The molecular diffusivity, k , is a scalar and is constant, depending on the fluid itself and not on the fluid motion and direction, x, y and z.

For the different physical quantities, one has :

- mass diffusivity   : ε_m
- thermal diffusivity   : a_h
- viscosity   : ν

Indicative values are given in Table 8.1.

2° Transport by *turbulent diffusion* in a fluid with turbulent flow, was described by eq. 8.17. The turbulent diffusivity, $\vec{k_t}$, is a diagonal tensor or a scalar, depending on the fluid motion and direction, *x, y* and *z*, and not on the fluid itself.

For the different physical quantities, one has :

- mass diffusivity : $\vec{\varepsilon_t}\,(\varepsilon_{tx}, \varepsilon_{ty}, \varepsilon_{tz})$
- thermal diffusivity : $\vec{a_{ht}} \equiv \vec{a_{ht}}'$
- viscosity : $\vec{v_t}\,(v_{tx}, v_{ty}, v_{tz})$

In general, turbulent diffusivities have much larger values than molecular diffusivities, or :

$$\vec{k_t} \gg k$$

3° A parametrization of the diffusivity, k_t, will depend on the scales of the turbulent motion. Applying dimensional arguments — using eq. 8.10a —, it is possible to show that the diffusivity is related to a characteristic length and velocity, or :

$$k_t \approx L_c \cdot U_c \qquad (8.23)$$

The choice of these characteristic dimensions, L_c et U_c, as well as the determination of the constant of proportionality, must be obtained from relevant experiments.

4° An estimation by order of magnitude, O, of the diffusivity may be envisioned; here it is done for flow in a channel :

i) Admitted are the following (typical) scales :

- characteristic longitudinal length, L_L : $O(10^1$ [m])
- characteristic vertical length, L_V : $O(10^0$ [m])
- characteristic longitudinal velocity, V_L : $O(10^0$ [m/s])

ii) Used is the equation of continuity, eq. 2.35a :

$$\frac{\partial u}{\partial x} + \frac{\partial w}{\partial z} = 0$$

$$\frac{V_L}{L_L} \sim \frac{V_V}{L_V}$$

where one sees that the order of magnitude of the two terms has to be the same; thus it follows that :

$$V_V \sim \frac{L_V}{L_L} V_L$$

TRANSPORT AND MIXING OF MATTER

Subsequently, one obtains :

- characteristic vertical velocity, $\quad V_V : O(10^{-1}\ [m/s])$

iii) Used is the (simplified) equation of motion in the x-direction, eq. 8.22 :

$$u\frac{\partial u}{\partial x} + w\frac{\partial u}{\partial z} = -\frac{1}{\rho}\frac{\partial p^*}{\partial x} + v_{tx}\frac{\partial^2 u}{\partial x^2} + v_{tz}\frac{\partial^2 u}{\partial z^2}$$

$$V_L\frac{V_L}{L_L} \sim V_V\frac{V_L}{L_V} = \qquad v_{tx}\frac{V_L}{L_L^2} \quad v_{tz}\frac{V_L}{L_V^2}$$

where one sees that the order of magnitude of the terms due to the inertia force and the ones due to the diffusion — admitted is $v_t = \varepsilon_t$ (see eq. 8.26) — has to be the same; thus it follows that :

$$\frac{V_L^2}{L_L} \sim \varepsilon_{tx}\frac{V_L}{L_L^2} \quad \text{and} \quad \frac{V_L^2}{L_L} \sim \varepsilon_{tz}\frac{V_L}{L_V^2}$$

Subsequently (see also eq. 8.23) one writes :

$$\varepsilon_{tx} \sim V_L L_L \quad \text{and} \quad \varepsilon_{tz} \sim V_L\frac{L_V^2}{L_L} \sim \varepsilon_{tx}\frac{L_V^2}{L_L^2} \qquad (8.24)$$

and one obtains :

- characteristic longitudinal diffusivity, $\quad \varepsilon_{tx} : O(10^1\ [m^2/s])$
- characteristic vertical diffusivity, $\quad \varepsilon_{tz} : O(10^{-1}\ [m^2/s])$

iv) Above indicates that the longitudinal diffusivity is far more important than the vertical diffusivity, for two-dimensional flow in open channels, or :

$$\varepsilon_{tx} \sim 10^2\ \varepsilon_{tz} \qquad (8.25)$$

Needless to say, the exact values for the diffusivity are much linked to the characteristic scales of the flow, L_c et U_c.

For geophysical scales (oceans and lakes), one obtains (see *Graf*, 1988, p. 4) the following :

$$\varepsilon_{tx} : O(10^3\ [m^2/s]) \quad \text{and} \quad \varepsilon_{tz} : O(10^{-3}\ [m^2/s])$$

5° According to the *analogy of Reynolds* (see *Taylor*, 1954, p. 451), — an hypothesis assuming that turbulent transfer of mass, of heat and of momentum are exactly analogous — one postulates :

$$\varepsilon_t = a_{ht} = v_t \tag{8.26}$$

This has been experimentally verified for turbulent flow close to the wall, but one may assume that this analogy is also reasonably valid for flow in channels (see *Jobson* et *Yotsukura*, 1973).

Consequently, in turbulent regime the *Prandtl* and the *Schmidt* numbers (see point 8.1.2,6°) are unity :

$$Pr_t = \frac{v_t}{a_{ht}} \approx 1 \quad , \quad Sc_t = \frac{v_t}{\varepsilon_t} \approx 1$$

6° Some indicative values for turbulent diffusivities — taking into account the analogy of Reynolds and the estimation by order of magnitude — are given in Table 8.1. These values must be taken with the greatest caution : the order of magnitude might well vary with the considered scale.

8.2 DIFFUSION

1° *Pure* diffusion exists if in eq. 8.5, the velocity vector is zero, $\vec{V}(u,v,w) = 0$.

The *balance of mass*, in Cartesian coordinates, is written (see eq. 8.10) :

$$\frac{\partial c}{\partial t} = \varepsilon_m \left(\frac{\partial^2 c}{\partial x^2} + \frac{\partial^2 c}{\partial y^2} + \frac{\partial^2 c}{\partial z^2} \right) \tag{8.10a}$$

where c is the local concentration.

The *balance of heat*, in Cartesian coordinates, is written (see eq. 8.11) :

$$\frac{\partial T}{\partial t} = a_h \left(\frac{\partial^2 T}{\partial x^2} + \frac{\partial^2 T}{\partial y^2} + \frac{\partial^2 T}{\partial z^2} \right) \tag{8.11a}$$

where *T* is the local temperature.

Here it was assumed that the coefficients of diffusivity, ε_m and a_h, are constant in an homogeneous aqueous medium.

TRANSPORT AND MIXING OF MATTER

2° These two equations, eq. 8.10a and eq. 8.11a, are diffusion equations; they are mathematically identical.

Presented will be only solutions for the diffusion of mass, c ; solutions for the diffusion of temperature, T, are analogous, by simply replacing c by T and ε_m by a_h.

3° Some analytical solutions to eq. 8.10a for simple problems, but useful ones for the hydraulic engineer, will hereafter be presented.

For more complex problems (see *Crank*, 1956 and 1989 or *Csanady*, 1973), there exist solutions, using analytical or numerical methods.

8.2.1 Instantaneous Source

1° Considered will be the one-dimensional diffusion in the x-direction — thus independent of y and z — in an infinite medium without motion, $\vec{V} = 0$.

If M_0 is the total mass of the substance, being introduced (injected) instantaneously into the aqueous medium, $\mathcal{M}_1$ is defined as the mass introduced per unit surface, S, or :

$$\mathcal{M}_1 = \frac{M_0}{S} = \int_{-\infty}^{+\infty} c(x,t) \, dx = \int_{-\infty}^{+\infty} c(x,0) \, dx \tag{8.27}$$

Thus it is an instantaneous *surface* (plane) source, M_0/S, in units of [kg/m²].

2° This mass diffuses according to the equation of pure, one-dimensional diffusion :

$$\frac{\partial c}{\partial t} = \varepsilon_m \frac{\partial^2 c}{\partial x^2} \tag{8.10b}$$

3° A solution to eq. 8.10b is possible, if the following boundary and initial conditions are applied :

$$c(\pm\infty, t) = 0 \qquad c(x,0) = \mathcal{M}_1 \delta(x)$$

where $\delta(x)$ is the delta function of Dirac, which is everywhere zero, but at $x = 0$; it is given by :

$$\int_{-\infty}^{+\infty} \delta(x) \, dx = 1$$

The total mass of the substance, M_0, must be conserved at all times, thus:

$$\int_{-\infty}^{+\infty} c(x,t)dx = \int_{-\infty}^{+\infty} c(x,0)dx = M_1 \int_{-\infty}^{+\infty} \delta(x)dx = M_1$$

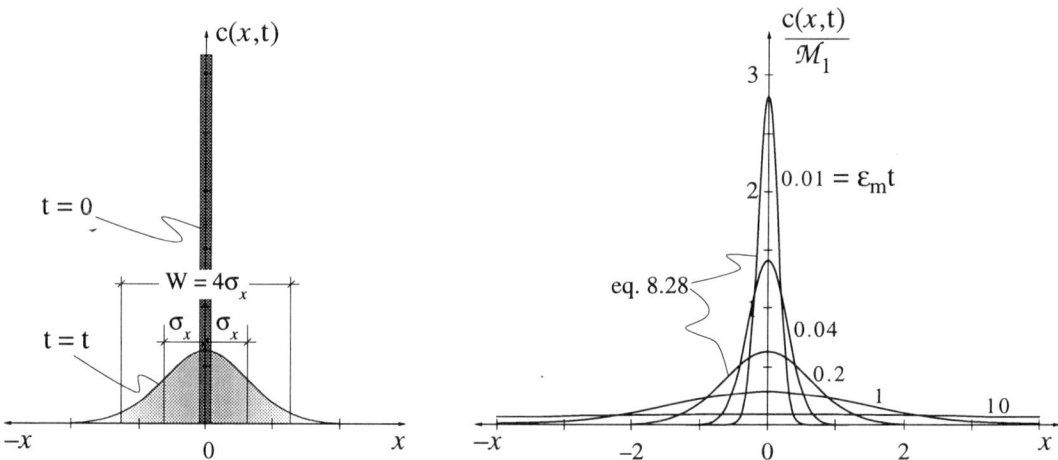

Fig. 8.3 Evolution of the distribution of concentration, $c(x,t)$, for a mass, M_0, injected instantaneously at $x = 0$ into a stagnant medium, $\vec{V} = 0$.

4° With these conditions, the solution to eq. 8.10b is given (see *Csanady*, 1973, p. 8 and *Crank*, 1989, p.12) by:

$$c(x,t) = \frac{M_1}{\sqrt{4\pi \, \varepsilon_m \, t}} \exp\left(-\frac{x^2}{4 \, \varepsilon_m \, t}\right) \tag{8.28}$$

This equation, eq. 8.28, describes the diffusion of a mass, M_0, which is initially introduced punctually and which spreads according to a Gaussian or normal curve (of bell shape), being symmetrical in x (see Fig. 8.3). The maximum concentration, which remains always at the origin, $x = 0$, decreases with time.

Making use of the definition of the variance, σ_x^2, this equation, eq. 8.28, can be written as:

$$c(x,t) = \frac{M_1}{\sqrt{2\pi} \, \sigma_x} \exp\left(-\frac{x^2}{2\sigma_x^2}\right) \tag{8.28a}$$

5° For a normal distribution, eq. 8.28, the variance of the distribution of a mass in the x-direction, is given by :

$$\sigma_x^2(t) = 2\,\varepsilon_m\,t \qquad (8.29)$$

This represents a measure (scale) of the spreading of the mass, namely the width of the diffusing cloud. If one takes a spread of :

$$W = (2 \times 1.96)\,\sigma_x \approx 4\sigma_x$$

one considers about 95% of the area under the normal curve of the concentration. This dimension, W, is used as being indicative of the *size of the cloud*.

With eq. 8.29, it is evident that the standard deviation increases with the root of the time :

$$\sigma_x \sim \sqrt{t}$$

whereas the (maximum) concentration at the origin, $x = 0$, (see eq. 8.28a) decreases with the root of the time :

$$c(x,t) \sim \frac{1}{\sigma_x} \sim \frac{1}{\sqrt{t}}$$

The definition of the variance can be used (see *Fischer et al.*, 1979, p. 41) to deduce the diffusivity :

$$\varepsilon_m = \frac{1}{2}\frac{d\sigma_x^2}{dt} = \frac{1}{2}\frac{\sigma_x^2(t_2) - \sigma_x^2(t_1)}{(t_2 - t_1)} \qquad (8.29a)$$

Consequently, the measure of the spatial variances, $\sigma_x^2(t_1)$ and $\sigma_x^2(t_2)$, at two different moments, t_2 and t_1, but remaining always at the same station, x, allows the determination of the diffusivity, ε_m.

6° Two-dimensional diffusion (see Fig. 8.1) in the x and y-directions — thus independent of z — may be considered (see *Csanady*, 1973, p. 14) as the product of one-dimensional diffusion in two directions, x and y. It is given by eq. 8.10a, whose solution is :

$$c(x,y;t) = \left[\frac{M_x}{\sqrt{2\pi}\,\sigma_x}\exp\left(-\frac{x^2}{2\sigma_x^2}\right)\right]\left[\frac{M_y}{\sqrt{2\pi}\,\sigma_y}\exp\left(-\frac{y^2}{2\sigma_y^2}\right)\right]$$

and written for an aqueous homogeneous medium, $\sigma_x = \sigma_y = \sigma$:

$$c(x,y;t) = \frac{M_2}{(\sqrt{2\pi}\,\sigma)^2} \exp\left[-\frac{(x^2+y^2)}{2\sigma^2}\right] \qquad (8.30)$$

with the variance given, as above, by :

$$\sigma^2(t) = 2\,\varepsilon_m\,t \qquad (8.29)$$

If M_0 is the total mass of the substance, being introduced instantaneously into the aqueous medium, M_2 is defined as the mass introduced per unit of length, L, or :

$$M_2 = \frac{M_0}{L} = \int_{-\infty}^{+\infty}\int_{-\infty}^{+\infty} c(x,y;t)\,dx\,dy = \int_{-\infty}^{+\infty}\int_{-\infty}^{+\infty} c(x,y;0)\,dx\,dy \qquad (8.31)$$

This is thus an instantaneous *length* (line) source, M_0/L, in units of [kg/m].

With eq. 8.30, one remarks that the concentration spreads in form of circles, $r^2 = x^2+y^2$, whose center remains always at the origin. A cut along any diameter yields a concentration profile, as has already been seen for one-dimensional diffusion (see Fig. 8.3).

7° Three-dimensional diffusion — in x, y and z-directions — may be considered (see *Csanady*, 1973, p. 15) as the product of one-dimensional diffusion in these three directions, x, y and z (see eq. 8.10a), or :

$$c(x,y,z;t) = \frac{M_3}{(\sqrt{2\pi}\,\sigma)^3} \exp\left(-\frac{r^2}{2\sigma^2}\right) \qquad (8.32)$$

with:

$$M_3 = M_0 = \int_{-\infty}^{+\infty}\int_{-\infty}^{+\infty}\int_{-\infty}^{+\infty} c(x,y,z;t)\,dx\,dy\,dz$$

where $M_3 = M_0$ is the instantaneous *point* source, in units of [kg]; σ^2 is the variance, eq. 8.29, and $r^2 = x^2+y^2+z^2$ is the radius of a sphere, being the distance from the origin.

With eq. 8.32, it is seen that the concentration at the origin, r = 0, decreases with the time :

$$c(x,y,z;t) \propto \frac{1}{\sigma^3} \propto t^{-3/2}$$

This decrease is faster than the one for one-dimensional diffusion, where $c(x,t) \propto t^{-1/2}$, or for two-dimensional diffusion, where $c(x,y;t) \propto t^{-2/2}$.

8° The solutions for an instantaneous source, eq. 8.27, eq. 8.31 and eq. 8.32, are of great importance. Since the equation of diffusion, eqs 8.6 and 8.10b, is linear, its solutions can be superposed ; in this way more complex problems can be solved. This is the principle of superposition (see *Rutherford*, 1994, p. 11 and p. 325).

9° Until now, it was considered that the aqueous medium is infinitely large; the mass can spread infinitely by diffusion.

If there are boundaries (walls), the medium will be limited (bounded) and diffusion encounters obstacles ; the diffusing mass will be reflected.

Take an (impermeable) wall of reflection, positioned at a distance, $x = -L_p$, from the source of the injected mass (see Fig. 8.4). An (identical) mirror-image source is created at a distance of $x = -2L_p$. The combined concentrations of the two sources is given by the sum of the solutions (see *Csanady*, 1973, p. 14); for a one-dimensional diffusion (see eq. 8.28a), one has :

$$c(x,t) = \frac{\mathcal{M}_1}{\sqrt{2\pi}\,\sigma_x} \left\{ \exp\left[-\frac{x^2}{2\sigma_x^2}\right] + \exp\left[-\frac{(x+2L_p)^2}{2\sigma_x^2}\right] \right\} \qquad (8.28b)$$

The concentration, c_p, at the wall itself is :

$$c_p = \frac{2\mathcal{M}_1}{\sqrt{2\pi}\,\sigma_x} \exp\left(-\frac{L_p^2}{2\sigma_x^2}\right)$$

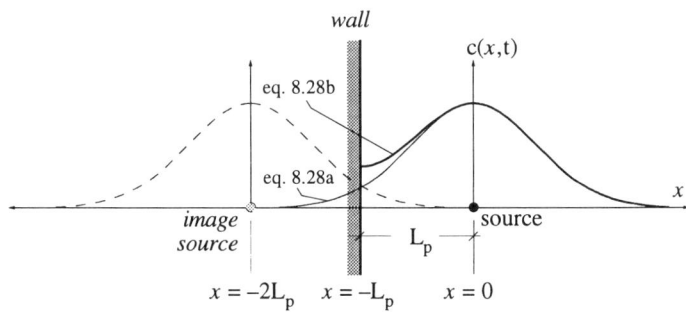

Fig. 8.4 Distribution of the concentration, $c(x,t)$, for a mass, M_0, injected at $x = 0$, and in presence of a reflecting wall at $x = -L_p$.

8.2.2 Continuous Source

1° Considered will be the one-dimensional diffusion in a stagnant medium, $\vec{V} = 0$.

At a certain station, $x = 0$, a constant concentration, $c_o = $ Cte, is introduced (emitted) continuously.

2° This initial and constant concentration will diffuse according to the equation of one-dimensional, pure diffusion, eq. 8.10a, given by :

$$\frac{\partial c}{\partial t} = \varepsilon_m \frac{\partial^2 c}{\partial x^2} \tag{8.10b}$$

3° A solution to eq. 8.10b is possible, if the following boundary and initial conditions are applied :

$c(0, t \geq 0) = c_o$ $\qquad c(x > 0, 0) = 0$

$c(\infty, t \geq 0) = 0$

where the concentration at the origin remains constant, c_o = Cte, at all times.

4° With these conditions, the solution to eq. 8.10b is (see *Crank*, 1989, p. 20 or *Daily* et *Harleman*, 1966, p. 434) given by :

$$c(x,t) = c_o \, erfc\left(\frac{x}{\sqrt{4\varepsilon_m t}}\right) \tag{8.33}$$

where the complementary error function is defined as :

$$erfc(Y) = \frac{2}{\sqrt{\pi}} \int_Y^\infty e^{-\xi^2} d\xi \tag{8.34}$$

to be found in mathematical tables (see Ex. 6.A).

This equation, eq. 8.33, describes the one-dimensional, pure diffusion of constant concentration at the origin, $x = 0$, which spreads (see Fig. 8.5) according to an *erfc* curve, eq. 8.34.

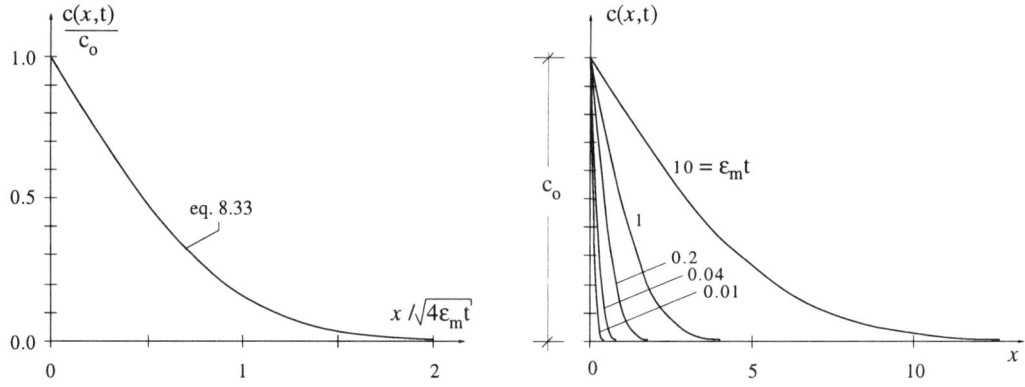

Fig. 8.5 Evolution of the distribution of the concentration, $c(x,t)$ for a constant concentration, c_o, introduced into a stagnant medium.

TRANSPORT AND MIXING OF MATTER   539

8.3 CONVECTION-DIFFUSION IN LAMINAR REGIME

1° Convection-diffusion is to be observed if, in eq. 8.5, the velocity vector of translation is unequal to zero, $\vec{V}(u,v,w) \neq 0$.

The *balance of mass,* expressed in Cartesian coordinates, is given (see eq. 8.6) by :

$$\frac{\partial c}{\partial t} + u\frac{\partial c}{\partial x} + v\frac{\partial c}{\partial y} + w\frac{\partial c}{\partial z} = \varepsilon_m \left(\frac{\partial^2 c}{\partial x^2} + \frac{\partial^2 c}{\partial y^2} + \frac{\partial^2 c}{\partial z^2} \right) \tag{8.6b}$$

where c is the local concentration; the diffusivity, ε_m, is constant.

The *balance of heat,* expressed in Cartesian coordinates, is given (see eq. 8.7) by :

$$\frac{\partial T}{\partial t} + u\frac{\partial T}{\partial x} + v\frac{\partial T}{\partial y} + w\frac{\partial T}{\partial z} = a_h \left(\frac{\partial^2 T}{\partial x^2} + \frac{\partial^2 T}{\partial y^2} + \frac{\partial^2 T}{\partial z^2} \right) \tag{8.7a}$$

where T is the local temperature; the thermal diffusivity, a_h, is constant.

2° These two equations, eq. 8.6b and eq. 8.7a, are equations of convection-diffusion; they are mathematically identical. Presented will be only solutions for diffusion of mass, c ; solutions for diffusion of temperature, T, are analogous, by simply replacing c by T and ε_m by a_h.

3° Some analytical solutions to eq. 8.6b for simple problems will be presented herewith. Using analytical or numerical methods (see *Crank*, 1989, or *Csanady*, 1973), solutions for more complex problems are possible.

8.3.1 Instantaneous Source

1° In the case of pure diffusion (see sect. 8.2.1), the mass diffuses in a stagnant medium, $\vec{V} = 0$. In the case of pure convection, the mass displaces itself with a velocity of translation, $\vec{V}(u,v,w) \neq 0$.

When the medium moves at a constant, uniform velocity, $\vec{V}(u,0,0)$, one can imagine that the coordinate system moves with this velocity. Hence applying a coordinate transformation, $x' = x - ut$, the solution to the equation of three-dimensional convection-diffusion, eq. 8.6b, is (see *Csanady*, 1973, p. 17) given by

$$c(x,y,z;t) = \frac{\mathcal{M}_3}{\left(\sqrt{2\pi}\,\sigma\right)^3} \exp\left[-\frac{\left((x-ut)^2 + y^2 + z^2\right)}{2\sigma^2} \right] \tag{8.35}$$

where $\sigma^2 = (2\,\varepsilon_m\,t)$ is the variance (see eq. 8.29).

In the absence of velocity, $\vec{V} = 0$, eq. 8.35 becomes evidently eq. 8.32.

2° For one-dimensional convection-diffusion in the x-direction, in a channel with a weak velocity, $u \equiv U$, being uniformly distributed over the flow depth, the above equation, eq. 8.35, can be written as :

$$C(x,t) = \frac{M_1}{\sqrt{4\pi\, \varepsilon_m t}} \exp\left[-\frac{(x - Ut)^2}{4\, \varepsilon_m t}\right] \qquad (8.36)$$

where C is the average concentration in a section, S, of the channel.

The total mass of the substance, M_0, introduced instantaneously and uniformly over the section, S, defines :

$$M_1 = \frac{M_0}{S} = \int_{-\infty}^{+\infty} C(x,t)\, dx = \int_{-\infty}^{+\infty} C(x,0)\, dx \qquad (8.27a)$$

3° All remarks made for pure diffusion (see sect. 8.2.1) remain valid; the velocity of translation, u or U, is taken into account by coordinate transformation, $x' = x - ut$ or $x' = x - Ut$ (see Fig. 8.3 and Fig. 8.6).

The mass, M_0, displaces itself with the velocity of translation, U, and at the same time it spreads out according to the normal curve (see Fig. 8.6). The maximum concentration, C_{max}, is propagated with the velocity and it decreases with time.

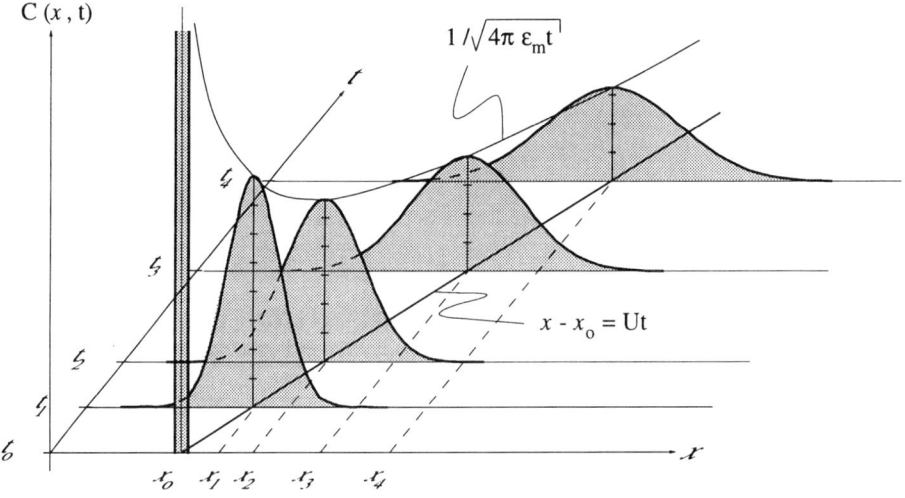

Fig. 8.6 Evolution of the concentration, $C(x,t)$, for a mass, M_0, injected instantaneously at $x = x_0$ into a medium in motion, U.

TRANSPORT AND MIXING OF MATTER

8.3.2 Continuous Source

1° Considered will be the one-dimensional convection-diffusion in a medium moving with a non-zero velocity, $\vec{V}(u,0,0) \neq 0$.

At a certain station, $x = 0$, an average concentration is introduced in a continuous and constant way, $C_o = Cte$.

The average velocity, U, being weak (without distribution over the flow depth) transports the average concentration, C, and diffusion takes place at the same time.

2° The solution to the one-dimensional convection-diffusion equation (see eq. 8.6b) is (see *Daily* et *Harleman*, 1966, p. 434) given by :

$$C(x,t) = \frac{C_o}{2}\left[\exp\left(\frac{Ux}{\varepsilon_m}\right) erfc\left(\frac{x+Ut}{\sqrt{4\varepsilon_m t}}\right) + erfc\left(\frac{x-Ut}{\sqrt{4\varepsilon_m t}}\right)\right] \qquad (8.37)$$

In the absence of velocity, U = 0, eq. 8.37 becomes evidently eq. 8.33.

The evolution of the concentration, $C(x,t)$, is shown at Fig. 8.7. Note that the concentration of the value $C_o/2$ displaces itself with the velocity of the flow, U.

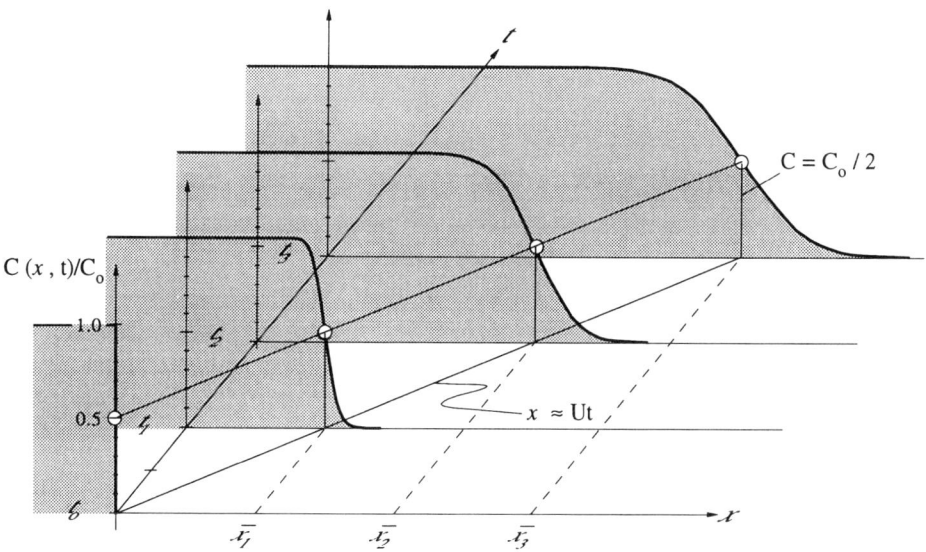

Fig. 8.7 Evolution of the concentration, $C(x,t)$, for a concentration, C_o, introduced continuously into a flow with an average velocity, U.

8.4 CONVECTION-DIFFUSION IN TURBULENT REGIME

1° The convection-diffusion equation of mass for time-averaged values in turbulent regime, written in Cartesian coordinates, was given with eq. 8.19. For two-dimensional flow, $\vec{V}(u,0,w)$, in a channel where one may take $u \gg w$, one writes :

$$\frac{\partial c}{\partial t} + \frac{\partial uc}{\partial x} = \frac{\partial}{\partial x}\left(\varepsilon_{tx}\frac{\partial c}{\partial x}\right) + \frac{\partial}{\partial z}\left(\varepsilon_{tz}\frac{\partial c}{\partial z}\right) \tag{8.38}$$

where $\vec{V}$ is the local velocity and c is the local concentration. $\vec{\varepsilon}_t$ is the turbulent diffusivity of mass, being much larger than the molecular one, $\vec{\varepsilon}_t \gg \varepsilon_m$; the latter can thus be often neglected.

The balance of heat is expressed by :

$$\frac{\partial T}{\partial t} + \frac{\partial uT}{\partial x} = \frac{\partial}{\partial x}\left(a_{ht}\frac{\partial T}{\partial x}\right) + \frac{\partial}{\partial z}\left(a_{ht}\frac{\partial T}{\partial z}\right) \tag{8.38a}$$

where, according to the analogy of Reynolds (see eq. 8.26), one assumes that $\varepsilon_t = a_{ht}$ (see *Jobson* et *Yotsukura*, 1973).

These two equations, eq. 8.38 and eq. 8.38a, are the *equations of convection-diffusion* ; they are mathematically identical.

Presented will only be solutions for diffusion of mass, c ; solutions for diffusion of temperature, *T,* are analogous, by simply replacing c by *T*.

2° In turbulent regime, the flow in a channel has a pronounced vertical velocity distribution; this has to be accounted for.

By integration of eq. 8.38 over the flow depth, h, (see *Rutherford*, 1994, pp. 38-41), one obtains :

$$\frac{\partial C}{\partial t} + \frac{\partial}{\partial x}(UC) + \frac{\partial}{\partial x}\left(\overline{U'C'}\right) = \frac{\partial}{\partial x}\left(\overline{\varepsilon_{tx}}\frac{\partial C}{\partial x}\right) + \frac{\partial}{\partial z}\left(\overline{\varepsilon_{tz}}\frac{\partial C}{\partial z}\right) \tag{8.39}$$

The depth-averaged values are defined as :

$$\int_0^h u \, dz = U h \qquad \int_0^h c \, dz = C h$$

TRANSPORT AND MIXING OF MATTER

$$\int_0^h uc\, dz = \overline{uc}\, h = \left(UC + \overline{U'C'}\right) h \tag{8.40}$$

$$\int_0^h \varepsilon_{ti} \frac{\partial c}{\partial x_i}\, dz \cong \left(\overline{\varepsilon_{ti}} \frac{\partial C}{\partial x_i}\right) h$$

where U is the average velocity and C is the average concentration. U' = u − U and C' = c − C represent, for a given profile, the deviations of u(z) and of c(z) from their respective average values, U and C, (see Fig. 8.10). $\overline{\varepsilon_{ti}}$ are the depth-averaged turbulent diffusivities.

The term, $\partial(\overline{U'C'}) / \partial x$, stands for the longitudinal dispersivity — also called the differential diffusion — which is due to the vertical velocity distribution.

3° If one assumes now that the diffusion in the vertical z-direction and the transversal y-direction are completely achieved — this is the case after a certain time, $t_y > \xi_y\, h^2/\varepsilon_{ty}$ (see eq. 8.56a) — one obtains (see eq. 8.39) the one-dimensional equation of convection-dispersion for turbulent uniform flow, or :

$$\frac{\partial C}{\partial t} + U\frac{\partial C}{\partial x} + \frac{\partial}{\partial x}\left(\overline{U'C'}\right) = \frac{\partial}{\partial x}\left(\overline{\varepsilon_{tx}} \frac{\partial C}{\partial x}\right) \tag{8.41}$$

4° The terms in the above equations, namely the :

   − diffusivity : ε_{ti} et $\overline{\varepsilon_{ti}}$

   − dispersivity : $\partial(\overline{U'C'})/ \partial x$

shall be discussed further on.

8.4.1 Diffusivity

1° With the *theorem of Taylor* (see *Hinze*, 1959, p. 47), one states a relationship, which links a statistical value of the turbulence, namely the correlation coefficient, $R_L(\tau)$, to a characteristic value of the diffusion, namely the variance, $\sigma^2(t)$, or :

$$\sigma^2(t) = 2\overline{u'^2} \int_0^t (t-\tau)\, R_L(\tau)\, d\tau \tag{8.42}$$

where σ^2 is the variance of the displacement of particles and $\overline{u'^2}$ is the mean-square of the velocity fluctuations. The Lagrangian correlation coefficient (see *Graf & Altinakar*, 1991, p. 260) is given by:

$$R_L(\tau) = \frac{\overline{u'(t) \cdot u'(t + \tau)}}{\overline{u'^2}} \qquad (8.43)$$

It represents a quantification of the homogeneous and stationary turbulence; furthermore it permits to express the macro scale, Λ_L, and the time scale, T_L, (see *Hinze*, 1959, p. 48), such as:

$$\Lambda_L = \sqrt{\overline{u'^2}} \int_0^\infty R_L(\tau)d\tau = \sqrt{\overline{u'^2}} \cdot T_L \qquad (8.44)$$

$$T_L = \int_0^\infty R_L(\tau)d\tau \qquad (8.45)$$

The above relationship, eq. 8.42, gives a Lagrangian description — followed is a moving particle — of the phenomenon of turbulent diffusion in turbulent flow, being *homogeneous* and stationary.

Since the correlation coefficient, eq. 8.43, is rarely known, the variance cannot be calculated with eq. 8.42. Nevertheless, this equation reveals something about the time-rate of growth of the mixing (cloud).

2° Considering eq. 8.42, two limiting cases are of interest:

   i) At the beginning, for very *short* periods of time, $t \ll T_L$, the correlation coefficient becomes:

   $$R_L(0) \cong 1 \qquad (8.46a)$$

   and, consequently (see *Hinze*, 1959, p. 48), the above relationship, eq. 8.42, yields:

   $$\sigma^2(t) = \overline{u'^2} \cdot t^2 \qquad (8.46)$$

   Diffusion, namely the cloud, grows with time; it depends on the turbulence.

ii) For very *long* periods of time, $t \gg T_L$, the correlation coefficient becomes:

$$R_L(\infty) \cong 0 \tag{8.47a}$$

and consequently (see *Hinze*, 1959, p. 48), the above relationship, eq. 8.42, yields:

$$\sigma^2(t) = 2\overline{u'^2} \cdot T_L \cdot t \tag{8.47}$$

Diffusion, namely the cloud, grows with the square root of time; it depends on the turbulence.

3° It can be shown (see *Hinze*, 1959, p. 311) that turbulent diffusion is well described by the diffusion equation, eq. 8.6, where the turbulent diffusivity is given by:

$$\varepsilon_t = \frac{1}{2} \frac{d\sigma^2}{dt} \tag{8.48}$$

This relation is thus analogous with the one for molecular diffusion, eq. 8.29a.

Using the expression for the variance, eq. 8.42, eq. 8.48 can be written as:

$$\varepsilon_t = \overline{u'^2} \int_0^t R_L(\tau) \, d\tau$$

For the two limiting cases, the above relationship reduces to:

i) for $t \ll T_L$ with $R_L \cong 1$:

$$\varepsilon_t = \overline{u'^2} \cdot t \tag{8.49}$$

ii) for $t \gg T_L$ with $R_L \cong 0$:

$$\varepsilon_t = \overline{u'^2} \int_0^\infty R_L(\tau) \, d\tau = \overline{u'^2} \cdot T_L = \sqrt{\overline{u'^2}} \cdot \Lambda_L \tag{8.50}$$

An estimation for the value of the diffusivity shall be presented later (see point 8.4.1,5 °).

4° The time evolution of the variance, σ^2, and of the diffusivity, ε_t, are summarized in Table 8.2.

Table 8.2 Diffusivity, ε_t, and variance, σ^2, in relation to time.

	ε_t	σ^2
$t \ll T_L$	$\varepsilon_t = \overline{u'^2} \cdot t$	$\sigma^2 \propto t^2$
$t < T_L$	$\varepsilon_t \propto \sigma^{4/3}$	$\sigma^2 \propto t^3$
$t \gg T_L$	$\varepsilon_t = \overline{u'^2} \cdot T_L$	$\sigma^2 \propto t^1$

The following observations can be made :

i) $t \ll T_L$: The diffusivity increases with time (this shows an important difference from the molecular diffusivity; the latter remains constant). The diffusion of the growing cloud, being still of small size, is achieved by the turbulent eddies of the same (small) size; the turbulent eddies of larger sizes transport the small cloud and barely participate in the diffusion.

ii) $t \gg T_L$: The diffusivity is independent of time (this shows a clear similarity to the molecular diffusivity; both remain constant). The diffusion of the well-developed cloud, being already of large size — being of the order of the spatial geometry — is achieved by the turbulent eddies of large size.

iii) $t < T_L$: The diffusion of the growing cloud, being of medium size, is achieved by turbulent eddies in the inertial range. The diffusivity (see *Fischer* et al., 1979, p. 75) is given by :

$$\varepsilon_t = \alpha_f \sigma^{4/3} \qquad (8.51)$$

This relation was obtained with dimensional considerations by Batchlor (1952) and with observations by Richardson (1926). The value of the constant falls in a range of $0.002 < \alpha_f \; [\text{cm}^{2/3}/\text{s}] < 0.01$.

The diffusion, namely the cloud, grows according to :

$$\sigma^2 \propto t^3$$

From above it is evident, that the turbulent diffusion is essentially produced by those turbulent eddies which have sizes comparable to the size of the diffusing cloud (note that the spectrum of the turbulence contains all possible sizes of turbulent eddies).

TRANSPORT AND MIXING OF MATTER

5° An estimation of the value of the turbulent diffusivity is of great practical importance. Earlier (see sect. 8.1.4) it has been shown, that the diffusivity is related to a characteristic length and to a characteristic velocity :

$$\varepsilon_t \approx L_c \cdot U_c \tag{8.23a}$$

For flow in a channel, having a flow depth, h, and a shear velocity, u_*, one may postulate :

$$\varepsilon_t \approx h \cdot u_* \tag{8.52}$$

The turbulent diffusivity is usually a diagonal tensor, $\vec{\varepsilon}_t$ ($\varepsilon_{tx}, \varepsilon_{ty}, \varepsilon_{tz}$). Consequently, one should distinguish a longitudinal, ε_{tx}, a transversal, ε_{ty}, and a vertical, ε_{tz}, diffusivity.

6° *Vertical Diffusivity*, $\overline{\varepsilon_{tz}}$ (see point 6.4.2,6°) :

The vertical diffusion is due to the turbulence generated by friction at the bed of the channel.

By assuming that the vertical distribution of the velocity is logarithmic (see sect. 2.5.2) and that the analogy of Reynolds (see eq. 8.26) remains valid, the vertical diffusivity can be expressed (see *Graf*, 1971, p. 189) as :

$$\varepsilon_{tz} = \kappa u_* \frac{z}{h} (h-z) \tag{6.49}$$

A depth-averaged value is subsequently obtained :

$$\overline{\varepsilon_{tz}} = \frac{\kappa u_*}{h^2} \int_0^h (hz - z^2)\, dz \tag{8.53a}$$

and, taking $\kappa = 0.4$, one gets :

$$\overline{\varepsilon_{tz}} = 0.067\, (h\, u_*) \tag{8.53}$$

which is verified (see *Rutherford*, 1994, p. 61) by experimental studies.

An estimation, expressed in order of magnitude, may be done:

$$\overline{\varepsilon_{tz}} = O(0.1 \times 100 \times 10) = O(10^2 \, [\text{cm}^2/\text{s}])$$

where it was admitted that $u_* \cong 0.1U$.

A similar value, already estimated (see sect. 8.1.4), is found in Table 8.1.

The vertical diffusion (mixing) extends over practically the entire flow depth, h, and this after a certain distance, L_z, or a certain time, t_z, counted from the introduction of the substance (see *Fischer et al.*, 1979, p. 13, 113); or :

$$t_z = \xi_z \frac{h^2}{\overline{\varepsilon_{tz}}} \quad \text{or} \quad L_z = \xi_z U \frac{h^2}{\overline{\varepsilon_{tz}}} \tag{8.54}$$

with $\xi_z \cong 0.4$ (0.1), if the point of introduction of the substance is situated on the bed (or at mid-depth) of the channel (see *Rutherford*, 1994, p. 69 and Fig. 8.B.1 in Ex. 8.B). By using eq. 8.53, one can also write :

$$t_z \cong 6 \frac{h}{u_*} \quad \text{or} \quad L_z \cong 6U \frac{h}{u_*} \tag{8.54a}$$

If one takes $u_* \cong 0.1U$, this distance, L_z, expressed in order of magnitude, can be estimated as being :

$$L_z = O(60 \, h) \tag{8.54b}$$

Different scenarios of vertical convection-diffusion have been investigated (see *Rutherford*, 1994, pp. 62-92).

7° *Transversal Diffusivity*, $\overline{\varepsilon_{ty}}$:

It is not possible to present a theoretical relationship for the determination of the transversal diffusivity, $\overline{\varepsilon_{ty}}$. Relying on experiments in laboratory channels (see *Fischer et al.*, 1979, p. 107), the following relationship was proposed:

$$\overline{\varepsilon_{ty}} \cong 0.15 \, (h \, u_*) \tag{8.55}$$

For irregular waterways with weak meanders (see *Fischer et al.*, 1979, p. 112), the following approximate relation was deduced :

$$\overline{\varepsilon_{ty}} \cong 0.6 \, (h \, u_*) \tag{8.55a}$$

TRANSPORT AND MIXING OF MATTER

The transversal diffusion (mixing) extends over practically the entire channel width, B, and this after a certain distance, L_y, or a certain time, t_y; or :

$$t_y = \xi_y \frac{B^2}{\overline{\varepsilon_{ty}}} \quad \text{or} \quad L_y = \xi_y U \frac{B^2}{\overline{\varepsilon_{ty}}} \tag{8.56}$$

with $\xi_y \cong 0.5$ (0.1), if the point of introduction of the substance is situated at one of the lateral walls (or in the center) of the channel (see Fig. 8.9 and *Rutherford*, 1994, p. 120). By using eq. 8.55a, one can also write :

$$t_y \cong 0.8 \frac{B^2}{h\, u_*} \quad \text{or} \quad L_y \cong 0.8 U \frac{B^2}{h\, u_*} \tag{8.56a}$$

If one takes $u_* \cong 0.1U$ and $(B/h) \cong 50$, this distance, L_y, can be estimated, expressed in order of magnitude, as being :

$$L_y = O\,(400\,B) \tag{8.56b}$$

Subsequently, one can estimate, expressed in order of magnitude, the relative value of these times and distances, — by taking $(B/h) \cong 50$ —; or:

$$\frac{t_y}{t_z} \cong O\,(300) \quad \text{or} \quad \frac{L_y}{L_z} \cong O\left(\frac{400\,B}{60\,h}\right) \cong O(300)$$

Thus it is shown, that the necessary distance for total transversal mixing, L_y, is about 300 times longer than the one for total vertical mixing, L_z. This is the reason why one often admits that vertical mixing is achieved almost instantaneously (see *Fischer* et al., 1979, p. 113).

Different scenarios of transversal convection-diffusion have been investigated (see sect. 8.4.2, *Rutherford*, 1994, pp 95-171 and *Fischer* et al., 1979, pp. 107-124).

8° *Longitudinal Diffusivity*, $\overline{\varepsilon_{tx}}$:

It is not possible to present a theoretical relationship for the determination of the longitudinal diffusivity, $\overline{\varepsilon_{tx}}$. Relying on experiments in laboratory channels (see *Elder*, 1959), the following relationship was proposed :

$$\overline{\varepsilon_{tx}} \cong 0.23\,(h\, u_*) \tag{8.57}$$

The longitudinal diffusion by turbulence is commonly considered of minor importance, since the longitudinal dispersion, K_x', which is due to the velocity distribution, usually dominates (see sect. 8.4.3).

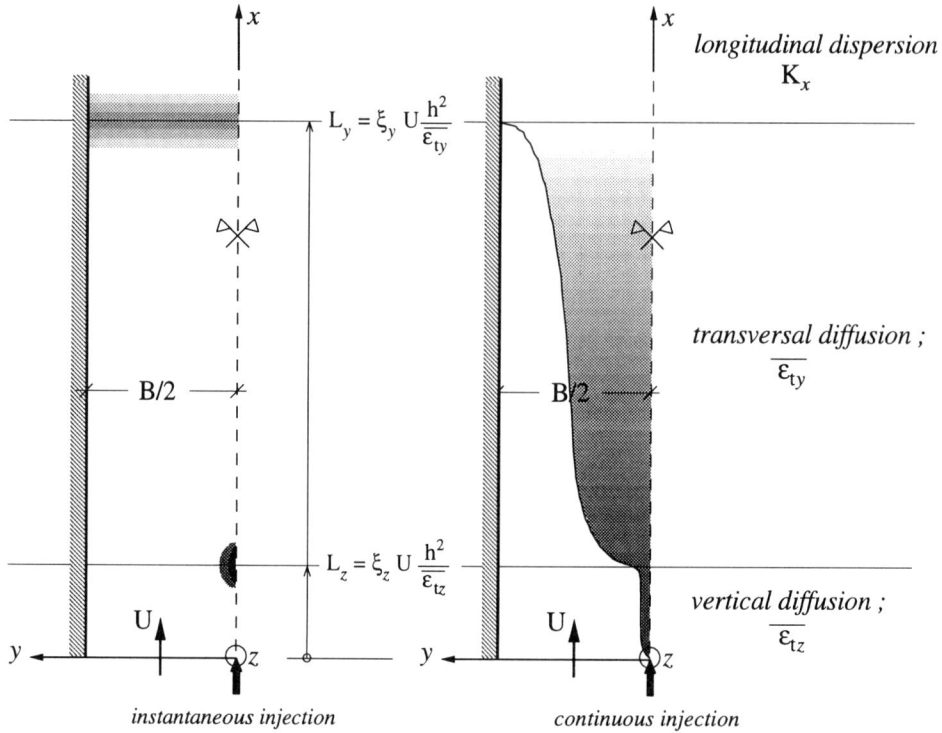

Fig. 8.8 Scheme of the convection-diffusion after an instantaneous and continuous injection; the distances, L_z and L_y, of total mixing are shown.

9° The Fig. 8.8 summarizes schematically the distances of total mixing in the vertical, L_z, and in the transversal, L_y, for an instantaneous and continued injection. For larger distances, $L > L_y$, the longitudinal dispersion is encountered (see sect. 8.4.4). Consequently, three zones can be distinguished :

- The *vertical* convection-*diffusion* is a phenomenon which is encountered in a zone close to the source; the *near-field zone of mixing*. This zone extends itself to a distance, L_z, given by eq. 8.54a. Note, that in this zone, the initial momentum with which the substance is injected can be of importance.

- This near-field zone is followed by a zone, where the *transversal* convection-*diffusion* develops itself ; the *mid-field zone of mixing*. This zone extends itself to a distance, L_y, given by eq. 8.56.

- This mid-field zone is then followed by a zone, where the *longitudinal dispersion* takes place; the *far-field zone of mixing*. The concentration of the source is distributed over the entire cross section of the flow (see sect. 8.4.3); the dispersion will tend to erase any longitudinal concentration variation.

The values, expressed in order of magnitude, for the distances of total mixing, L_z (see eq. 8.54a) and L_y (see eq. 8.56a), give approximate informations, which are rather useful for the hydraulic engineer.

8.4.2 Transversal Diffusion

1° The equation of convection-diffusion in turbulent regime was given with eq. 8.19. Neglecting molecular diffusion, admitting the flow as one-dimensional of constant velocity, $\vec{V}(u,0,0) = U$, and taking the diffusivities, $\varepsilon_{ti} = \overline{\varepsilon_{ti}}$, as being constant, this equation, eq. 8.19, reads :

$$\frac{\partial c}{\partial t} + U \frac{\partial c}{\partial x} = \varepsilon_{tx} \frac{\partial^2 c}{\partial x^2} + \varepsilon_{ty} \frac{\partial^2 c}{\partial y^2} + \varepsilon_{tz} \frac{\partial^2 c}{\partial z^2} \tag{8.58}$$

Considered will be that the diffusion of the source is steady, $\partial c/\partial t = 0$, that the longitudinal diffusion is negligible, $(\varepsilon_{tx}) \partial^2 c/\partial x^2 = 0$, and that the vertical diffusion is almost completely achieved, $(\varepsilon_{tz}) \partial^2 c/\partial z^2 = 0$. The above equation, eq. 8.58, can now be written as :

$$U \frac{\partial C}{\partial x} = \varepsilon_{ty} \frac{\partial^2 C}{\partial y^2} \tag{8.59}$$

where C is the depth-averaged concentration.

This equation, eq. 8.59, is the *equation of transversal convection-diffusion*. Since the above assumptions are rather restrictive, this model is of only indicative interest to the hydraulic engineer.

2° The solution (see *Rutherford*, 1994, p. 118) for a medium of infinite extent, B, is :

$$C_u(x,y) = \frac{G_0}{h \sqrt{4\pi \varepsilon_{ty} xU}} \exp\left(-\frac{y^2 U}{4 \varepsilon_{ty} x}\right) \tag{8.60}$$

where $G_0 = M_0/t$ is the mass-flow of a linear source, in units of [kg/s], distributed over the flow depth, h. A comparison with eq. 8.28 is recommended.

3° For a finite (bounded) medium of width, B, the following boundary conditions must be applied :

$$\frac{\partial C}{\partial y} = 0 \qquad \text{at:} \quad y = 0 \quad \text{and} \quad y = B$$

Consequently, the lateral walls become (impermeable) walls of reflection (see point 8.2.1,8°). These boundary conditions can be fulfilled by using the method of images, $n = 1,2 \ldots N$. A solution for a bounded flow can be given (see *Rutherford*, 1994, p. 118) by :

$$C(x,y) = C_u(x,y+y_o) + C_u(x,y-y_o) + \sum_{n=1}^{N} C_u(x, 2nB \pm y \pm y_o) \qquad (8.61)$$

where C and C_u are the concentrations in respectively bounded and unbounded medium. B is the width of the channel and y_o is the transversal position, where the source is injected.

4° A solution, eq. 8.61, in graphical form (see *Rutherford*, 1994, pp. 119-121) is of great practical importance. Fig. 8.9 shows the results for a source positioned in the center, $y^* = y/B = 0.5$, and at one of the lateral walls, $y^* = y/B = 0.0$, of the channel. This solution is also valid for a source positioned at $y^* = 1.0$, namely at the other lateral wall, since the boundary conditions are symmetrical.

The variables in eq. 8.61, are here expressed as dimensionless variables, or :

- the transversal distance : $\quad y^* = y \dfrac{1}{B}$
- the longitudinal distance : $\quad x^* = x \dfrac{\varepsilon_{ty}}{UB^2}$
- the concentration : $\quad c^* = C \dfrac{1}{C_M}$

where C is the depth-averaged concentration; $C_M = G_o/Q$ is the final concentration after the transversal diffusion is completely achieved over the entire width, B, of the channel. Q is the total flow in the channel and $G_o = (M_o/t)$ is the mass-flow of the injected substance. The velocity, u = U, and the transversal diffusivity, $\varepsilon_{ty} = \overline{\varepsilon_{ty}}$, have been assumed as being constant over the entire section of the channel.

5° Some interesting conclusions of this solution, eq. 8.61, presented in Fig. 8.9, can be drawn :

- For $x^* \cong 0.5$ (0.1) : with an injection of the substance at a lateral wall (at the center), the transversal diffusion is completed at $0.95 < c^* < 1.05$. One may thus consider that the concentration, $C \approx C_M$, is almost uniform over the entire width. It should be remarked that these values, $x^* \Rightarrow \xi_y$, are used in eq. 8.56.
- For $x^* \cong 0.06$: with the injection of the substance at a lateral wall (at the center), the transversal diffusion is reached at $0.10 < c^*$ ($0.75 < c^*$) at the opposite lateral wall.
- To obtain the transversal diffusion (see eq. 8.56) rather fast — in a short distance or time — the injection of the substance should be done at the center of the channel.

6° Solutions, eq. 8.61, in graphical form for an introduction of the substance at $y^* = 0.25$ and $y^* = 0.33$ are found elsewhere (see *Rutherford*, 1994, p. 120).

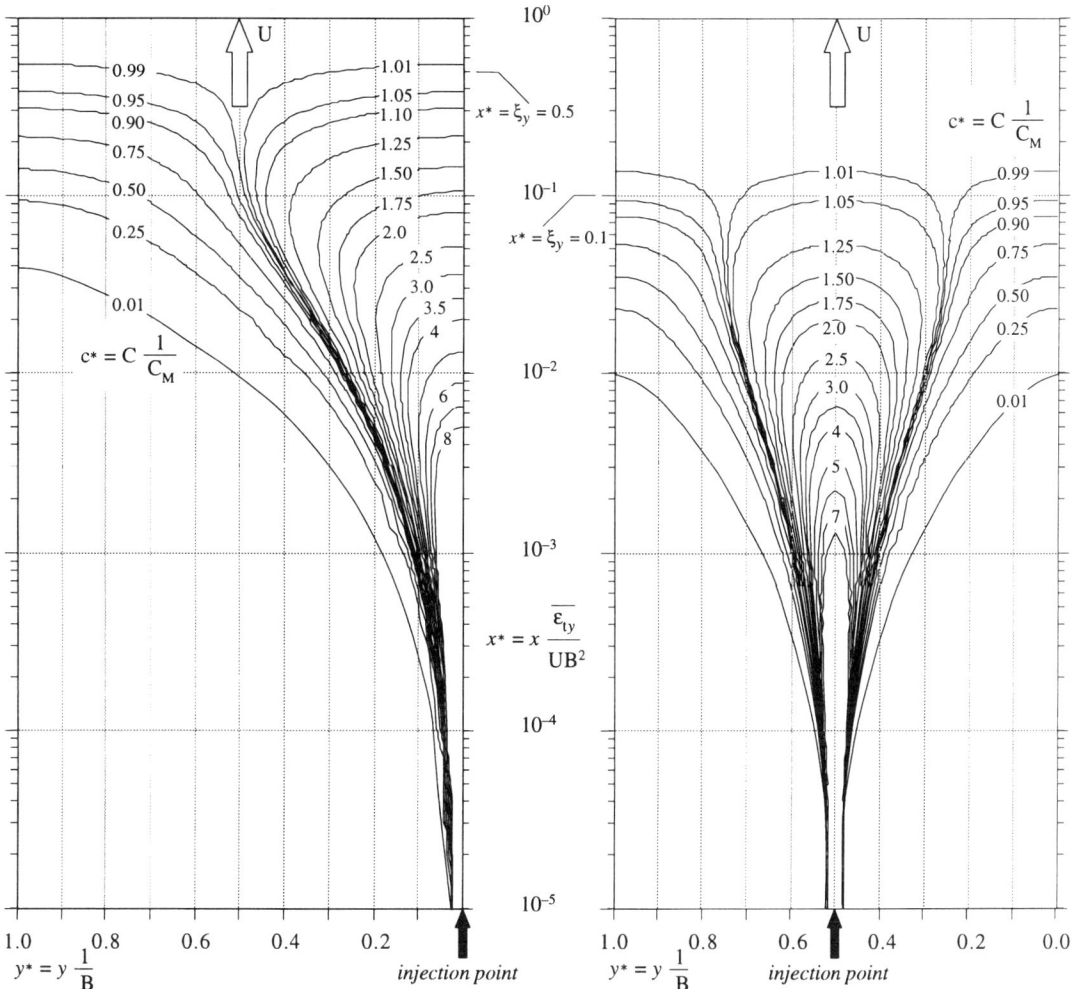

Fig. 8.9 Evolution of the concentration distribution, $c^*(x^*,y^*)$, for a source positioned at the center, $y^* = 0.5$, and at a lateral wall, $y^* = 0.0$, of the channel.

7° The zone where the transversal convection-diffusion develops itself, the *mid-field zone of mixing*, is generally of a rather long distance, namely:

$$L_y = \xi_y U \frac{B^2}{\overline{\varepsilon_{ty}}} \tag{8.56}$$

For flow in a channel, whose parameters, U, B and $\overline{\varepsilon_{ty}}$, are given in advance, it might be advantageous that the value of the coefficient, ξ_y, were small. If the introduction of the substance is situated as indicated below, this coefficient (see *Rutherford*, 1994, p. 141) will be :

- at one of the lateral walls, at $y^* = 0.0$ or $y^* = 1.0$: $\xi_y \cong 0.50$
- at the center of the channel, at $y^* = 0.5$: $\xi_y \cong 0.10$
- at two points, situated at $y^* = 0.4$ and $y^* = 0.6$: $\xi_y \cong 0.07$
- at many points (multipoint diffuser) arranged over the entire width : $\xi_y \leq 0.07$

From above, one sees clearly that by an appropriate choice of the point(s) of introduction of the substance, the distance of complete transversal diffusion, L_y, can be considerably reduced.

It should be remembered that this distance — for $\xi_y \cong 0.50$ — expressed in order of magnitude, is :

$$L_y = O(400\,B) \tag{8.56b}$$

thus being rather long. This distance would be reduced to $15\% \times L_y$, if an introduction with a multipoint diffuser, namely with $\xi_y \leq 0.07$, is chosen.

8°   For more complex problems — the source and/or the flow being unsteady, the flow depth, the width and/or the diffusivity are not constant — numerical methods should be used (see *Rutherford*, 1994, pp. 125-155).

9°   Different methods exist (see *Rutherford*, 1994, pp. 245-269), which allow to determine the transversal diffusivity, ε_{ty}.

8.4.3 Dispersivity

1°   The equation of convection-diffusion for turbulent, uniform flow, given by :

$$\frac{\partial C}{\partial t} + U\frac{\partial C}{\partial x} + \frac{\partial}{\partial x}\left(\overline{U'C'}\right) = \frac{\partial}{\partial x}\left(\varepsilon_{tx}\frac{\partial C}{\partial x}\right) \tag{8.41}$$

includes a term of diffusion (see sect. 8.4.1) as well as one of dispersion. The term of longitudinal dispersion, $\partial(\overline{U'C'})/\partial x$, results from the velocity distribution of the longitudinal convective transport.

2°   In a given cross section, the velocity, $u(z)$, and the concentration, $c(z)$, show a vertical distribution (see Fig. 8.10). Their average values, U et C, are defined by eq. 8.40. U' and C' are the deviations of the values, $u(z)$ and $c(z)$, with respect to the values, U and C.

It has been proposed by Taylor (see *Fischer* et al., 1979 p. 86) that the quantity $(\overline{U'C'})$ be proportional to the gradient of the average concentration, C, or :

$$-(\overline{U'C'}) = -\frac{1}{h}\int_0^h U'C'\,dz = K_x'\frac{\partial C}{\partial \xi} \tag{8.62}$$

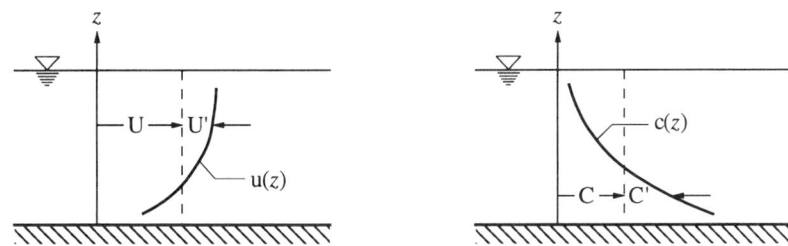

Fig. 8.10 Distribution of the velocity, u(z), and of the concentration, c(z).

where $\xi = x - Ut$ is the abscissa of an observer moving downstream at the average velocity, U. The analogy of this expression, eq. 8.62, with the one for turbulent, eq. 8.16, and molecular, eq. 8.1, diffusion should be remarked.

The product $\overline{(U'C')}$ represents an average value over the cross section, whose deviations (of the velocity and of the concentration with respect to their average values) are more important than the turbulent fluctuations. This explains why longitudinal dispersion conceals turbulent diffusion, which in turn conceals molecular diffusion, $K_x' \gg \varepsilon_{tx} \gg \varepsilon_m$; this was well demonstrated by experiments (see eq. 8.67).

3°   The expression of Taylor's model, eq. 8.41, can now be written as :

$$\frac{\partial C}{\partial t} + U\frac{\partial C}{\partial x} = \frac{\partial}{\partial x}\left[\left(\overline{\varepsilon_{tx}} + K_x'\right)\frac{\partial C}{\partial x}\right] \tag{8.63}$$

where K_x' and $\overline{\varepsilon_{tx}} + K_x' = K_x$ are the coefficients of longitudinal dispersion or, in short, the turbulent *dispersivity*.

For uniform, steady flow, the dispersivity, K_x, remains constant, thus :

$$\frac{\partial C}{\partial t} + U\frac{\partial C}{\partial x} = K_x\frac{\partial^2 C}{\partial x^2} \tag{8.64}$$

This equation is known as the *equation of one-dimensional dispersion* ; it is very useful for the analysis of dispersion/diffusion in environmental (turbulent) flows, such as in waterways.

However, it must be recalled, that this equation, eq. 8.64, becomes only valid after a certain time period or distance after introduction of the substance, given by :

$$t_y = \xi_y U \frac{B^2}{\overline{\varepsilon_{ty}}} \quad \text{or} \quad L_y = \xi_y \frac{B^2}{\overline{\varepsilon_{ty}}} \tag{8.56}$$

when the vertical and the transversal diffusion have already been achieved.

4° The coefficient of dispersion, K_x', is in general an unknown, yet to be determined. It has been shown (see *Fischer* et al., 1979, p. 84 and 91) that the convection-diffusion equation, eq. 8.38, can be simplified to a form of :

$$U' \frac{\partial C}{\partial \xi} = \frac{\partial}{\partial z} (\varepsilon_{tz} \frac{\partial C'}{\partial z})$$

If it is admitted, that $(\partial C/\partial \xi)$ becomes constant after a certain time, one can double integrate the above equation to obtain :

$$C' = \frac{\partial C}{\partial \xi} \int_0^z (\frac{1}{\varepsilon_{tz}} \int_0^{z'} U' dz'') dz'$$

Introducing this value into the integral of eq. 8.62, one obtains (see *Fischer* et al., 1979, p. 91) the following expression for the dispersivity :

$$K_x' = -\frac{1}{h} \int_0^h U' \int_0^z \frac{1}{\varepsilon_{tz}} \int_0^{z'} U' dz'' \, dz' \, dz \qquad (8.65)$$

This relation, eq. 8.65, as well as the one given by eq. 8.70, can be used to establish experimentally the different formulas of the dispersivity, K_x' :

– For flow in a pipe of radius r — the hydraulic radius is $r/2 = R_h$ — the following expression (see *Taylor*, 1954, p. 453) was obtained :

$$K_x' = 10.06 \, (r \, u_*) = 20.1 \, (R_h \, u_*) \qquad (8.66)$$

– For flow in a channel, being rectangular and very wide — with $R_h \approx h$ — where the velocity is distributed logarithmically in the vertical, the following expression (see *Elder*, 1959) was given :

$$K_x' = 5.86 \, (h \, u_*)$$

By taking into account the longitudinal diffusion, $\overline{\varepsilon_{tx}}$ (see eq. 8.57), the dispersivity is :

$$K_x = K_x' + \overline{\varepsilon_{tx}} = (5.86 + 0.23) \, (h \, u_*)$$

where it is to be seen that the contribution of the diffusion can usually be (and is) ignored; thus :

$$K_x \cong K_x' \cong 6.0 \, (h \, u_*) \qquad (8.67)$$

TRANSPORT AND MIXING OF MATTER

– For flow in waterways, where there exists a velocity distribution in both the vertical and transversal directions (see *Fischer* et al., 1979, p. 125), the range of measured values is :

$$140 < \frac{K_x}{h\, u_*} < 500 \tag{8.68}$$

For the Missouri river, some very large values of 7500 have been reported.

– For flow in laboratory channels, 2 < B/h < 20, and in waterways, 10 < B/h < 100, the influence of the width/depth ratio was evaluated (see *Iwasa* et *Aya*, 1991, p. 509), giving :

$$\frac{K_x}{h\, u_*} = 2.0 \left(\frac{B}{h}\right)^{1.5} \tag{8.68a}$$

being valid for $6 < K_x/h u_* < 2000$.

– For natural waterways, the following approximate value (see *Fischer* et al., 1979, p. 136) was proposed :

$$K_x = 0.011 \frac{B^2\, U^2}{h\, u_*} \tag{8.69}$$

High values of K_x are often measured when the flow passes over large geometrical irregularities; regions of stagnant dead-water zones are encountered, which participate only marginally in the mixing process (see *Rutherford*, 1994, p. 190).

The large variability of the dispersivity, K_x, is at least partially due to the fact that the hypothesis used to obtain Taylor's model, eq. 8.63, are not always satisfied.

Consequently the chosen value for the dispersivity, K_x, must always be considered as an approximate value.

5° A method to calculate the dispersivity, K_x, which does not require measurements of the velocities, eq. 8.65, consists in measuring the variances of the cloud of the substance in uniform flow. The variance and the dispersivity are related (see *Rutherford*, 1994, p. 269 et p. 189) according to :

$$K_x = \frac{1}{2}\frac{d\sigma_x^2}{dt} = \frac{1}{2}\frac{\sigma_x^2(t_2) - \sigma_x^2(t_1)}{t_2 - t_1} \qquad (8.70)$$

where $\sigma_x^2(t_1)$ and $\sigma_x^2(t_2)$ are the longitudinal variances of the curves, $C(x)$, at two different time periods, t_2 et t_1. It often proves difficult to measure these variances. However, it was shown (see *Fischer*, 1966), that $\Delta\sigma_x^2 = U^2\Delta\sigma_t^2$; consequently, eq. 8.70 can also be expressed as :

$$K_x = \frac{1}{2}U^2\frac{\sigma_t^2(x_2) - \sigma_t^2(x_1)}{\bar{t}_2 - \bar{t}_1} = \frac{1}{2}U^3\frac{\sigma_t^2(x_2) - \sigma_t^2(x_1)}{x_2 - x_1} \qquad (8.70a)$$

$\sigma_t^2(x_2)$ and $\sigma_t^2(x_1)$ are the time variances of the curves, $C(t)$, at two different stations, x_2 and x_1. U is the average velocity of the uniform flow; $\bar{t}_1$ and $\bar{t}_2$ are the mean times of the passage of the cloud at each of the stations. However, it should be pointed out, that both sampling stations must be sufficiently downstream (see eq. 8.56a) from the injection of the substance, in order to assure a constant dispersivity, K_x , (see eq. 8.64). Furthermore, one should also ascertain that the influence of side walls and of dead-water zones is negligible.

Instead of measuring the variance of the cloud, σ_x^2, one could also measure the width of the cloud, $W = 4\sigma_x$, and put these values into eq. 8.70.

Remark that this relationship, eq. 8.70, is analogous to the one for laminar, eq. 8.29a, and for turbulent, eq. 8.48, diffusion.

8.4.4 Longitudinal Dispersion

1° The equation of one-dimensional convection-dispersion was given with :

$$\frac{\partial C}{\partial t} + U\frac{\partial C}{\partial x} = K_x\frac{\partial^2 C}{\partial x^2} \qquad (8.64)$$

where U and C are the average velocity and concentration in the cross section. The coefficient of dispersion — also constant over the cross section — is given (see point 8.4.3,4°), for example, by :

$$K_x \cong 6.0\,(h\,u_*) \qquad (8.67)$$

or by one of the other formulas, eq. 8.68 and eq. 8.69, presented previously.

However, the use of this equation, eq. 8.64, is only possible :

i)   if the injected substance has become well mixed over the entire cross section;

ii)   at, or after, a certain distance or time (see Fig. 8.8) after injection of the substance, namely in the *far-field zone of mixing*, given as :

$$L_y = \xi_y U \frac{B^2}{\varepsilon_{ty}} \tag{8.56}$$

As shown earlier (see eq. 8.56b), the distance, expressed in order of magnitude as being $L_y = O\,(400\,B)$, can be rather long.

2°   The dispersivity, K_x, is independent of time; thus it is a constant. This shows a clear similarity with the turbulent diffusivity, ε_t, which is also constant for large time periods, $t \gg T_L$ (see point 8.4.1,4ii°), but also with the molecular diffusivity, ε_m, which also remains constant.

Solutions to the equation of convection-dispersion in turbulent regime, eq. 8.64, will consequently be similar to solutions to the equation of convection-diffusion in laminar regime, eq. 8.6b (see sect. 8.3).

3°   Such a simplified model, eq. 8.64, can have severe limitations. The assumptions leading to the expression of Taylor's model, eq. 8.41 and eq. 8.64, are not always fulfilled; notably, the one-dimensionality of uniform flow or the distribution of the velocity and of the diffusivity. Such are some of the reasons why one should rather work with the more complete equation, eq. 8.19. However, in such a case, analytical solutions are often not available and use should be made of numerical methods (see *Crank*, 1989, chap. 8).

The numerical methods use in general an explicit or implicit scheme of finite differences (see sect. 5.2). However, particular care must be taken in using these methods, since they may give rise to numerical (artificial) dispersion/diffusion (see *Cunge et al.*, 1980, p. 320). The latter can sometimes be stronger than the physical dispersion/diffusion and may thus render all calculations rather useless.

–   The explicit method (see sect. 5.2.3) has been used by *Chen* et al. (1975, p.353), *Fischer* et al. (1979, p. 285), *Cunge* et al. (1980, p. 331), *Sauvaget* (1985), *Abbott* et *Basco* (1989, chap. 5) and *Crank* (1989, p. 141).

–   The implicit method (see sect. 5.2.4) has been used by *Sauvaget* (1985, p. 54) and *Abbott* et *Basco* (1989, chap. 5).

4° Some analytical solutions to eq. 8.64 shall now be presented, which are often rather useful for the hydraulic engineer; they can also be used to check the numerical results.

   i) The total mass of the substance, M_0, is introduced (or arrives) *instantaneously*, being uniformly distributed over the entire cross section, S, of the channel (see Fig. 8.11a).

   A solution (see *Daily* et *Harleman*, 1966, p. 436) to eq. 8.64, for all time instants, t, is :

$$C(x,t) = \frac{\mathcal{M}_1}{\sqrt{4\pi K_x t}} \exp\left[-\frac{(x-Ut)^2}{4 K_x t}\right] \quad (8.71)$$

   where $\mathcal{M}_1 = M_0/S$, is an instantaneous *surface* source, in units of [kg/m²], defined before with eq. 8.27.

   This solution should be compared with eq. 8.36 (see Fig. 8.6); the dispersivity, K_x, has replaced the molecular diffusivity, ε_m.

   For this normal distribution of the concentration, eq. 8.71, the variance is given by :

$$\sigma_x^2(t) = 2 K_x t \quad (8.72)$$

   The matter, M_0, is convected downstream with the average velocity, U, and spreads according to a Gaussian curve (see Fig. 8.6), being symmetrical in the x-direction at any given time. The maximum concentration, C_{max}, moves with a speed of $U = x/t$; its magnitude decreases with time or distance :

$$C_{max}(t) = \frac{\mathcal{M}_1}{\sqrt{4\pi K_x t}} \times 1 = \frac{\mathcal{M}_1}{\sqrt{4\pi K_x x/U}} \quad (8.73)$$

   Knowledge of the time evolution, given with eq. 8.73, can be used to determine the dispersivity, K_x. From a plot of $C_{max}(t) = f(1/\sqrt{t})$, the slope gives directly (by regression) the value for the diffusivity, K_x ; the instantaneous surface source, M_0/S, must of course be known.

   ii) The total mass of the substance, M_0, is introduced (or arrives) uniformly distributed over the entire cross section, S, of the channel during a (finite, variable) *time-limited period*, T (see Fig. 8.11b).

   Since the governing equation, eq. 8.64, is linear, its solutions, such as the one given with eq. 8.71, can be added together to obtain a solution for complex problems; this is the principle of superposition (see *Rutherford*, 1994, p. 210 and p. 325).

TRANSPORT AND MIXING OF MATTER

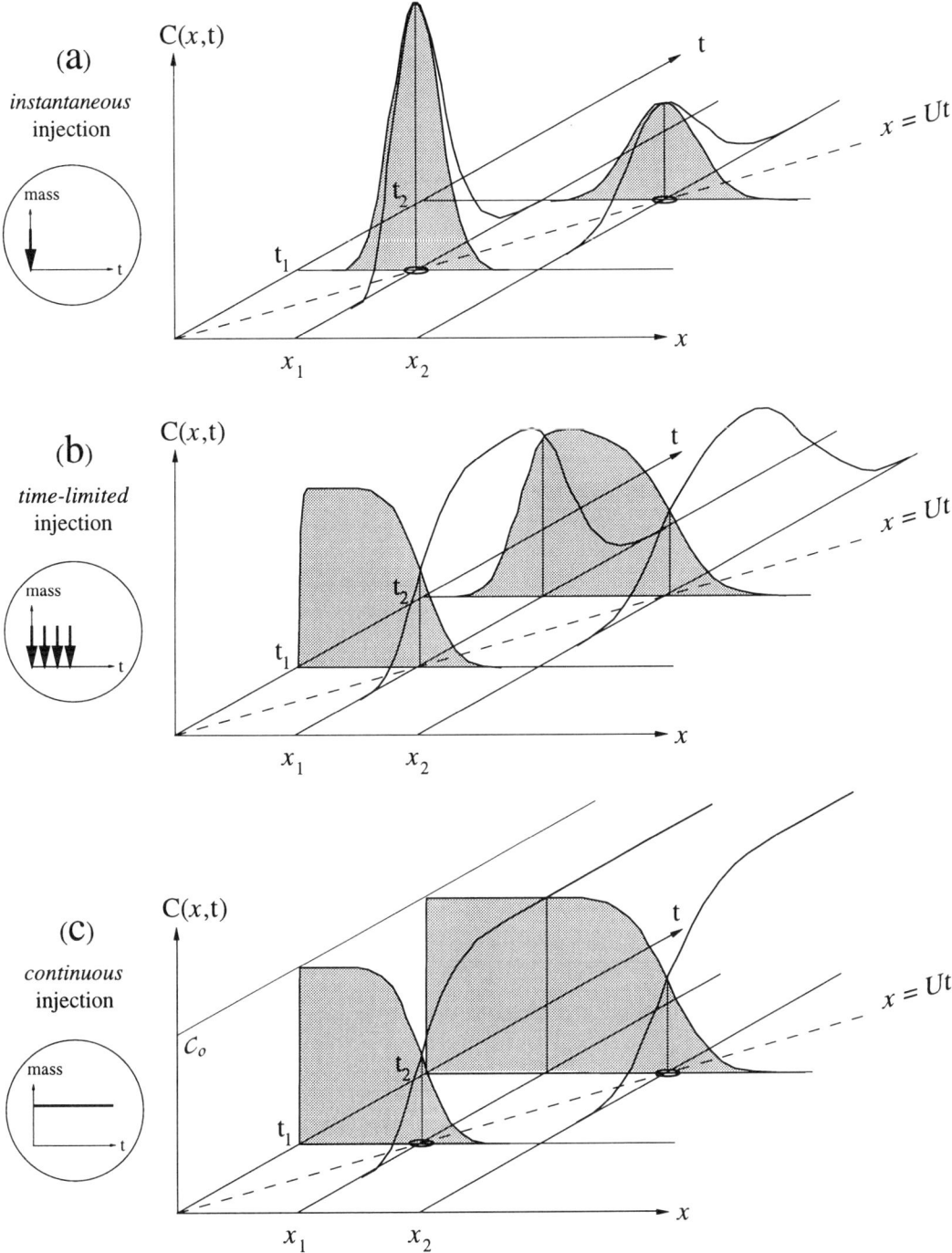

Fig. 8.11 Evolution of the concentration, C(x,t), introduced uniformly over the cross section :
a) *instantaneously*, b) *time-limited* and c) *continuously*.

The injection of the total mass of the substance during a finite time period, M_0/T, can be simulated (see *Fischer* et al., 1979, p. 45) by a series of successive injections of small masses, $(M_0/T)\Delta\tau = m_i$, during a time period, $\Delta\tau$, being infinitesimally small.

Each *mini*-injection will create a cloud which develops according to the normal distribution (see eq. 8.71):

$$\Delta C_i(x,t) = \frac{m_i}{S\sqrt{4\pi K_x(t-\tau_i)}} \exp\left[-\frac{[x-U(t-\tau_i)]^2}{4K_x(t-\tau_i)}\right] \quad (8.74)$$

Finally, the total concentration is the sum of the concentrations, resulting from the individual mini-injections, or:

$$C(x,t) = \sum_{i=1}^{n} \Delta C_i(x,t) = \frac{1}{S\sqrt{4\pi K_x}} \sum_{i=1}^{n} \frac{m_i}{\sqrt{t-\tau_i}} \exp\left[-\frac{[x-U(t-\tau_i)]^2}{4K_x(t-\tau_i)}\right] \quad (8.75)$$

iii) At a certain station, $x = 0$, the concentration is introduced (or arrives) *continuously* in a permanent and constant way, $C_0 = $ Cte, being uniformly distributed over the entire cross section, S, of the channel (see Fig. 8.11c).

A solution (see *Daily* et *Harleman*, 1966, p. 434) to eq. 8.64, for all time instants, t, is:

$$C(x,t) = \frac{C_0}{2}\left[\exp\left(\frac{Ux}{K_x}\right) erfc\left(\frac{x+Ut}{\sqrt{4K_x t}}\right) + erfc\left(\frac{x-Ut}{\sqrt{4K_x t}}\right)\right] \quad (8.76)$$

This solution should be compared with eq. 8.37 (see Fig. 8.7); the dispersivity, K_x, has replaced the diffusivity, ε_m.

In case of a steady state, when $t \to \infty$, this solution, eq. 8.76, becomes:

– if $U(x)$ is positive: $\quad \dfrac{C}{C_0} = 1 \quad (8.76a)$

– if $U(x)$ is negative: $\quad \dfrac{C}{C_0} = \exp\left(-\dfrac{Ux}{K_x}\right) \quad (8.76b)$

Note the following: $erfc(+\infty) = 0$ and $erfc(-\infty) = 2$.

TRANSPORT AND MIXING OF MATTER

5° It is rather obvious, that the mixing potential of dispersion (diffusion) in turbulent regime is much larger than the one of molecular diffusion in laminar regime; this is due to the importance of the value of the dispersivity, $K_x \gg \varepsilon_{tx} \gg \varepsilon_m$ (see Table 8.1), which conceals the one of diffusion.

8.4.5 Dispersion with Reaction

1° The general balance equation, written in local form, for a physical quantity, c_f, was given as:

$$\frac{\partial c_f}{\partial t} + \text{div}(c_f \vec{V}) = -\text{div}\,\vec{q_f} + q_e \tag{8.5}$$

where q_e is the flow from internal sources per unity of volume in the domain, D. To establish the *balance of mass*, this equation was written as:

$$\frac{\partial c}{\partial t} + \text{div}(c\vec{V}) = -\text{div}(\varepsilon_m \overrightarrow{\text{grad}}\,c) + q_e \tag{8.6a}$$

where $c_f \equiv c$ is the local concentration of the substance in the mixture. The term of the internal sources, q_e — which might include a term due to a chemical or biological reaction, due to a phase change or due to a transfer across boundaries — was omitted, being considered small with respect to the term of diffusion. Thus, up till now, this term of internal sources, q_e, was never taken into account. This seems reasonable for many transport phenomena encountered in fluvial hydraulics.

2° Nevertheless a certain number of problems require, that this term of internal sources, q_e, should also be considered. This is notably the case when the substance is (re)active, thus non-conservative; thus, the mixture may be subject to transformation of its composition.

3° The equation of convection-dispersion, eq. 8.64, is now written as:

$$\frac{\partial C}{\partial t} + U\frac{\partial C}{\partial x} = K_x \frac{\partial^2 C}{\partial x^2} \pm (k_r C) \tag{8.77}$$

where $(k_r C)$ is the term due to a first-order *reaction*, expressing the rate at which the active substance increases (+) or decreases (−). k_r is a coefficient (constant) of reaction of the substance in the mixture, given in units of [1/T]. U and C are the average velocity and concentration in the cross section.

It should here be recalled, that this one-dimensional equation, eq. 8.77, is only valid, if the concentration of the substance, C, is already uniformly distributed over the entire cross section.

4° Some analytical solutions to this equation, eq. 8.77, which are often useful for hydraulic engineers, shall now be presented.

    i) The total mass of an active substance, M_0, is introduced uniformly over the entire cross section, S, in an *instantaneous* way (see Fig. 8.11a).

The solution (see *Metcalf*, 1972, p. 688) to eq. 8.77 is given by :

$$C(x,t) = \frac{M_0}{S\sqrt{4\pi K_x t}} \exp\left[-\frac{(x-Ut)^2}{4 K_x t} \pm k_r t\right] \quad (8.78)$$

where k_r is the coefficient of reaction of the active substance.

In the absence of reaction of the substance, $k_r = 0$, this equation, eq. 8.78, becomes obviously eq. 8.71.

The matter, M_0, is convected downstream with the average velocity, U, spreads according to the Gaussian curve (see Fig. 8.6) but increases (+) or decreases (–) slowly in an exponential way due to reaction.

The total mass of the substance, M_0, does not conserve itself (see eq. 8.27); there is a weak, continuous variation of the concentration according to :

$$\int_{-\infty}^{+\infty} C(x,t)dx = \exp(\pm k_r t)$$

    ii) The concentration of an active substance, C_0, is introduced uniformly over the entire cross section, S, in a *permanent* and constant way. Admitted is the boundary condition of $C \to 0$ at $x \to \infty$ and a steady state for $t \to \infty$.

The solution (see *Fischer* et al., 1979, p. 146) to eq. 8.77, is given by :

$$\frac{C(x)}{C_0} = \exp\left\{-\frac{k_r x}{U}\left[\frac{2}{\alpha_r}\left(\sqrt{\alpha_r + 1} - 1\right)\right]\right\}$$

$$= \exp\left\{-\frac{Ux}{2K_x}\left(\sqrt{\alpha_r + 1} - 1\right)\right\} \quad (8.79)$$

where $\alpha_r = (4 K_x k_r / U^2)$ is a dimensionless coefficient of reaction-dispersion; $(-k_r)$ is here taken for a decaying substance.

TRANSPORT AND MIXING OF MATTER   565

For small values, $\alpha_r \ll 1$, when $\sqrt{\alpha_r + 1} \approx 1 + \alpha_r/2$, the expression in brackets tends towards unity, $[\cdot/\cdot] \to 1$, — this means that the term due to dispersion in eq. 8.77 is neglected — and the above relation, eq. 8.79, becomes :

$$\frac{C(x)}{C_0} = \exp\left(-\frac{k_r x}{U}\right) \tag{8.80}$$

The solution is the one for a simple first-order (decay) reaction. In the absence of reaction, $k_r = 0$, the above relation, eq. 8.79, becomes :

$$\frac{C}{C_0} = 1 \tag{8.76a}$$

The evolution of the concentration, according to eq. 8.79, is shown in Fig. 8.12 ; the concentration of the active substance decreases in an exponential way.

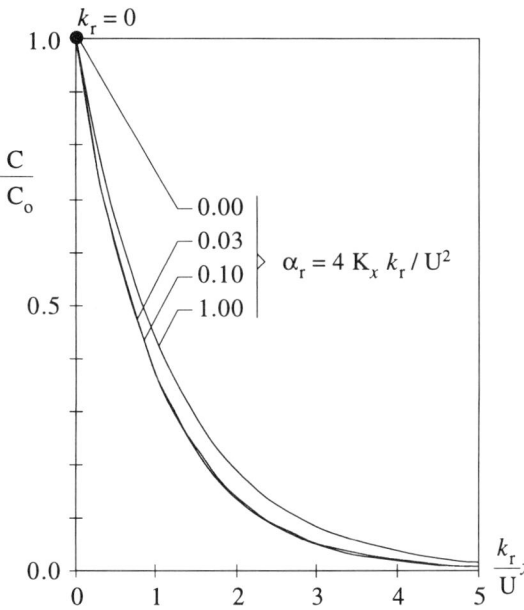

Fig. 8.12 Evolution of the distribution of the concentration, C/C_0, for convection-dispersion with reaction.

5° The coefficient (constant) of reaction of a substance in the mixture, k_r, depends on the material (wastewater) itself and on the temperature.

Values of this coefficient, k_r, for the rate of decrease (decay or decomposition) are given (see *Metcalf*, 1972, p. 245) as :

$$0.05 < k_r \text{ [days}^{-1}\text{]} < 0.30 \ (2.0) \tag{8.81}$$

or, written in a dimensionless form :

$$\alpha_r = 4 \, k_r \, K_x / U^2 < 0.4 \tag{8.82}$$

These values are rather representative, notably for the biochemical oxygen demand, BOD, which is used as a global parameter of the organic substances in (treated or untreated) wastewater.

8.4.6 Transport with Reaction

1° Consider the transport of a continuous source under steady conditions, $\partial C/\partial t = 0$, and assume that the longitudinal dispersion (mixing) is negligible, $K_x(\partial^2 C/\partial x^2) = 0$. Consequently, the equation of convection-dispersion with reaction, eq. 8.77, becomes :

$$U \frac{\partial C}{\partial x} = (\pm k_r C) \tag{8.83}$$

where U and C are the average velocity and concentration in the cross section. This equation, eq. 8.83, is an important equation for modelling the water quality in waterways.

2° Integration of eq. 8.83, by separation of the variables, yields :

$$\frac{C(x)}{C_o} = e^{\pm k_r(x/U)} \tag{8.84}$$

with the reference concentration $C_o = (G_0/Q)$ taken at a section of $x = 0$; $G_0 = (M_0/t)$ is the total mass flow of the substance and Q is the total discharge of the mixture.

It should be noted, that eq. 8.84 could have been obtained directly from eq. 8.79, by assuming that $\alpha_r \ll 1$ has a very small value.

For a decaying substance, one writes :

$$\frac{C(x)}{C_o} = e^{-k_r(x/U)} \tag{8.80}$$

TRANSPORT AND MIXING OF MATTER

This relation, eq. 8.80, can also be used for the determination of the coefficient of reaction, k_r, such as :

$$k_r = -\frac{U}{x} \ln \frac{C(x)}{C_o(x=0)}$$

by measuring the average velocity, U, and concentrations, $C(x)$ and $C_o(x=0)$.

3° The equation of transport with reaction, eq. 8.83, can well be used to study the self-cleaning (purifying) capacity of a waterway receiving wastewater.

The water in a waterway has a certain capacity to absorb organic matter (pollutants) : it is the micro-organisms which do the cleaning since they take care of the degradation of certain non-conservative pollutants of organic origin. In order to fulfil this degradation, the main source of energy for these organisms comes from the available dissolved oxygen. Consequently : the *budget* (balance) *of the dissolved oxygen* in a channel must be established. One expresses the total variation of the concentration of dissolved oxygen in water, $C_{(OD)}$, by the following (see *Metcalf*, 1972, p. 683) differential equation :

$$\frac{dC_{(OD)}}{dt} = R + P - (k_r C_{(BOD)} + F + A) \tag{8.85}$$

The *sources* of oxygen are : the re-oxygenation across the water surface, R, and the photosynthesis, P.

The *sinks* (consumption) of oxygen are : the biochemical oxygenation (oxidation) of organic matter, $C_{(BOD)}$, the decomposition of sludge deposits, F, and the respiration of aquatic plants, A.

An evaluation — by measuring *in situ* — of certain parameters, P, F and A, is a difficult task ; therefore these parameters are usually considered to be not of importance. Furthermore, being concerned with steady-state conditions, $\partial C/\partial t = 0$, the above relationship, eq. 8.85, is now written as :

$$U \frac{\partial C_{(OD)}}{\partial x} = R - k_r C_{(BOD)} \tag{8.86}$$

The terms in this equation,

– re-oxygenation : R
– biochemical oxygenation : $C_{(BOD)}$

shall be elaborated in what follows.

The content of dissolved oxygen, $C_{(OD)}$, in water is usually considered as an important criterion for the quality of water.

In case of absence of dissolved oxygen, the anaerobic micro-organisms take over ; they take the necessary oxygen from such substances, as nitrates and sulphates, and cause a disagreeable smell.

4° *Biochemical Oxygenation* :

The equation using first-order reaction kinetics, eq. 8.83, can be used :

$$U \frac{\partial C_{(BOD)}}{\partial x} = - k_r \, C_{(BOD)} \tag{8.87}$$

to study the evolution of the concentration of the *biochemical oxygen demand*, written abbreviated as BOD, (see *Metcalf*, 1972, p. 241).

i) The BOD is commonly employed as parameter of the organic (biodegradable) content in both wastewater and clear water.

ii) The BOD is a measure of the quantity of dissolved oxygen, which is absorbed by the micro-organisms to take care of the degradation of organic matter (pollutants) during a certain time.

iii) The BOD, being a global parameter for the organic matter, reveals itself as an important parameter for the sanitary engineer ; it is used as a criterion for pollution in general and for the conception of treatment facilities of wastewater.

The solution to the above equation of transport, eq. 8.87, can be taken to describe the evolution (degradation) of the concentration of BOD, $C_{(BOD)}$, still *not satisfied* — thus remaining in the water — or :

$$C_{(BOD)}(x) = C_{0(BOD)} \, [e^{-k_r(x/U)}] \tag{8.88}$$

where $C_{0(BOD)}$ is the total or potential BOD, initially present at $x = 0$. k_r is the reaction (degradation) coefficient of the BOD in the mixture, given with the relation of eq. 8.81; some indicative values (see *Fair* et al. 1968, p. 33/16) are given in Table 8.3.

For the BOD, which is *satisfied* — thus consumed in the water — after a distance, x, one writes :

$$C^*_{(BOD)}(x) = C_{0(BOD)} - C_{(BOD)}(x) = C_{0(BOD)}[1 - e^{-k_r(x/U)}] \tag{8.89}$$

Table 8.3 Reaction (degradation) coefficient of organic matter, contained in water and wastewater.

	k_r [d^{-1}]	$C_{0(BOD)}$ [mg/l]
strong wastewater	0.39	250
weak wastewater	0.35	150
secondary effluent	0.12 – 0.23	15 à 75
tap water	< 0.12	< 1

If one speaks of BOD, without any particular specification, one implies :

$$BOD_5 \equiv C_5^*{}_{(BOD)}$$

which is the biochemical demand of *satisfied* (consumed) oxygen during a period of T = 5 [d], at a temperature of $T = 20$ [°C] in obscurity, when the oxidation of the matter is completed at 60-70 % (see *Metcalf*, 1972, p. 243). This is formulated (see eq. 8.89) as follows :

$$C_5^*{}_{(BOD)} = C_{0(BOD)} - C_{5(BOD)} = C_{0(BOD)} [1 - e^{-k_r(5)}] \tag{8.90}$$

The BOD_5 is in general measured by a test on a water sample from the waterway and subsequently incubated in the laboratory during 5 days (see *Metcalf*, 1972, p. 241). This represents an index, apparently not precise but rather representative, of the consumption of the oxygen, which is necessary for the degradation of an organic matter.

These relations, eq. 8.88 and eq. 8.89, are shown in Fig. 8.13, together with the value obtained with the relation of eq. 8.90. [These curves represent only the carbonaceous demand, to which must still be added — after > 10 days— the nitrogenous demand (see *Metcalf*, 1972, p. 246) which has the same tendency].

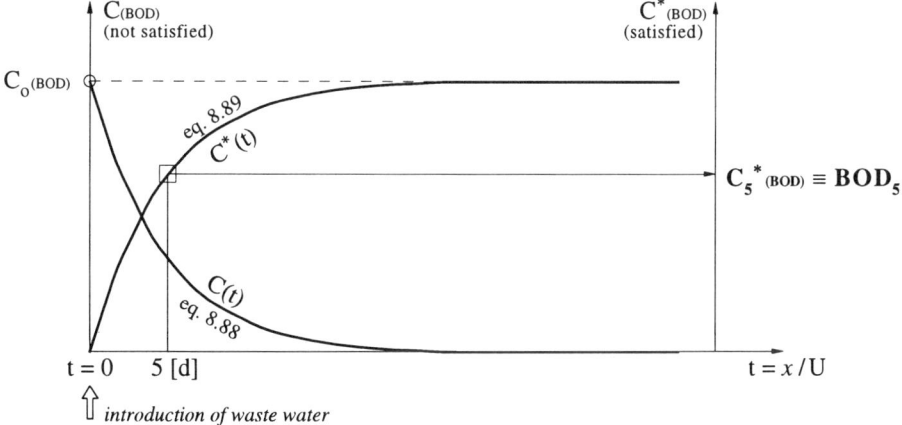

Fig. 8.13 Evolution of the concentration of BOD not satisfied, $C_{(BOD)}$, and of BOD satisfied, $C^*{}_{(BOD)}$, and the value of $C_5^*{}_{(BOD)}$.

5° *Reoxygenation*:

The (re)oxygenation (reaeration) of the water in a waterway is essentially obtained through the gaseous transfer from the atmosphere into the water ; wind action, waves and constant renewal at the surface water will contribute. Such a natural oxygenation can be complemented by different types of artificial oxygenation. The rate of oxygenation will also depend on the oxygen deficit in the water.

The above equation, eq. 8.83, can be used to study (see *Fair* et al., 1968, p, 23/11 and *Metcalf*, 1972, p. 680) the rate of transfer of atmospheric oxygen, or :

$$-U \frac{\partial D_{(OD)}}{\partial x} = U \frac{\partial C_{(OD)}}{\partial x} = k_a (C_{s(OD)} - C_{(OD)}) = k_a D_{(OD)} \tag{8.91}$$

where $C_{(OD)}$ is the concentration of dissolved oxygen and $C_{s(OD)}$ is the saturation concentration of dissolved oxygen. The term, $D_{(OD)} = (C_{s(OD)} - C_{(OD)})$, is called the dissolved-oxygen deficit. k_a is a coefficient (constant) of reoxygenation, which depends — such as does the value of $C_{s(OD)}$ — on the temperature of the water. Integration of the above equation, eq. 8.91, between the limits, $C_{o(OD)}$ at $x = 0$ and $C_{(OD)}$ at $x = x$, yields :

$$(C_{(OD)} - C_{o(OD)}) = (C_{s(OD)} - C_{o(OD)}) [1 - e^{-k_a(x/U)}] \tag{8.92}$$

Expressed in terms of oxygen deficit (see *Fair* et al., 1968, p. 33/21) it reads :

$$D_{(OD)} = D_{o(OD)} [e^{-k_a(x/U)}] \tag{8.92a}$$

where $D_{o(OD)}$ is the initial dissolved-oxygen deficit at $x = 0$.

The saturation concentration, $C_{s(OD)}$ [mg/l], is a function which depends (decreasingly) on the temperature, T [°C]; it can be calculated (see *Fair* et al., 1968, p. 23/10) with the following empirical relation :

$$C_{s(OD)} \approx 14.65 - 0.41\ (T) + 0.008\ (T^2) \tag{8.93}$$

being valid for an atmospheric pressure at 760 [mm] of mercury.

The coefficient of reoxygenation, k_a, depends on the volume of water participating in the transfer as well as on the water surface being in contact with the atmosphere. Some empirical formulae have been developed ; here the one proposed by *Fair* et al. (1968, p. 33/21) is given :

$$k_a \cong 5\ U/R_h^{5/3}\ [d^{-1}] \tag{8.94}$$

where U [m/s] is the average velocity and R_h [m] is the average hydraulic radius of the channel.

6° The mass *balance (budget) for the dissolved oxygen* in steady state, in a waterway, was given with eq. 8.86. Using the expressions for reoxygenation, eq. 8.91, and for desoxygenation, eq. 8.87, one can now write :

$$U \frac{\partial C_{(OD)}}{\partial x} = k_a (C_{s(OD)} - C_{(OD)}) - k_r C_{(BOD)} \tag{8.95}$$

Expressed in terms of oxygen deficit, $D_{(OD)}$, it reads :

$$U \frac{\partial D_{(OD)}}{\partial x} = -k_a D_{(OD)} + k_r C_{(BOD)} \tag{8.95a}$$

where $D_{(OD)} = C_{s(OD)} - C_{(OD)} = D_{0(OD)} [e^{-k_a(x/U)}]$ (8.92a)

$C_{(BOD)} = C_{0(BOD)} [e^{-k_r(x/U)}]$ (8.88)

Integration of the above equation, eq. 8.95a — between the limits of $D_{0(OD)}$ at $x = 0$ when $C_{(BOD)} = C_{0(BOD)}$ and of $D_{(OD)}$ at $x = x$ — gives the following relation (see *Metcalf*, 1972, p.684) :

$$D_{(OD)} = \frac{k_r}{k_a - k_r} C_{0(BOD)} [e^{-k_r(x/U)} - e^{-k_a(x/U)}] + D_{0(OD)} [e^{-k_a(x/U)}] \tag{8.96}$$

The evolution of the oxygen deficit, $D_{(OD)}$, is obtained here by addition of the curves of deoxygenation and of reoxygenation (see Fig. 8.14). The resulting curve, eq. 8.96, is known as the *dissolved-oxygen sac* curve ; it has been proposed in a classical study by Streeter et Phelps (see *Fair et al.*, 1968, p. 33/22) in 1925. It must be cautioned, however, that in the light of the many assumptions made, the model, eq. 8.95, and its solution, eq. 8.96, are of limited use, since they neglect some important phenomena.

The dissolved-oxygen sac curve can be interpreted as follows :

i) Initially the clean water in the channel is saturated or almost saturated with dissolved oxygen, $C_{(OD)} \cong C_{s(OD)}$; the oxygen deficit is thus zero, $D_{(OD)} \cong 0$.

ii) Upon introduction of wastewater, a biochemical decomposition of the wastewater with dissolved oxygen, $C_{(OD)}$, becomes active, according to eq. 8.88. An oxygen deficit, $D_{(OD)}$, establishes itself.

iii) However this oxygen deficit, $D_{(OD)}$, is continuously made good by the reoxygenation of the mixture by gaseous transfer, according to eq. 8.91.

iv) After a certain distance (time), being often very large, the water in the channel together with the wastewater, become again more or less saturated with dissolved oxygen, $C_{(OD)} \cong C_{s(OD)}$; the oxygen deficit is reduced back to zero, $D_{(OD)} \cong 0$.

The above equation, eq. 8.96, can also be used for the determination of the highest (critical) oxygen deficit, $D_{K(OD)}$. This is obtained by taking, in eq. 8.95, $\partial C_{(OD)}/\partial x = 0$; thus :

$$D_{K(OD)} = \frac{k_r}{k_a} C_{(BOD)} = \frac{k_r}{k_a} C_{o(BOD)} [e^{-k_r(x/U)}] \tag{8.97}$$

This critical state in a channel, does not arrive at the point of introduction of the wastewater, but at a rather long distance further downstream (see *Metcalf*, 1972, p. 685) ; it can be calculated using eq. 8.96 and taking $\partial D_{(OD)}/\partial x = 0$; one obtains :

$$x_K = \frac{U}{k_a - k_r} \ln \frac{k_a}{k_r} [1 - \frac{D_{o(OD)}(k_a - k_r)}{k_r C_{o(BOD)}}] \tag{8.98}$$

Subsequently, the time of arrival of $D_{K(OD)}$ is given by : $t_K = (x_K/U)$.

One can now also calculate the *minimal* dissolved-oxygen content in the channel :

$$C_{K(OD)} = C_{s(OD)} - D_{K(OD)} \tag{8.99}$$

a value which should not fall below a certain value, being 6 [mg/l] according to Swiss authorities. The value of the dissolved-oxygen content is generally used as a criterion for the quality of the water.

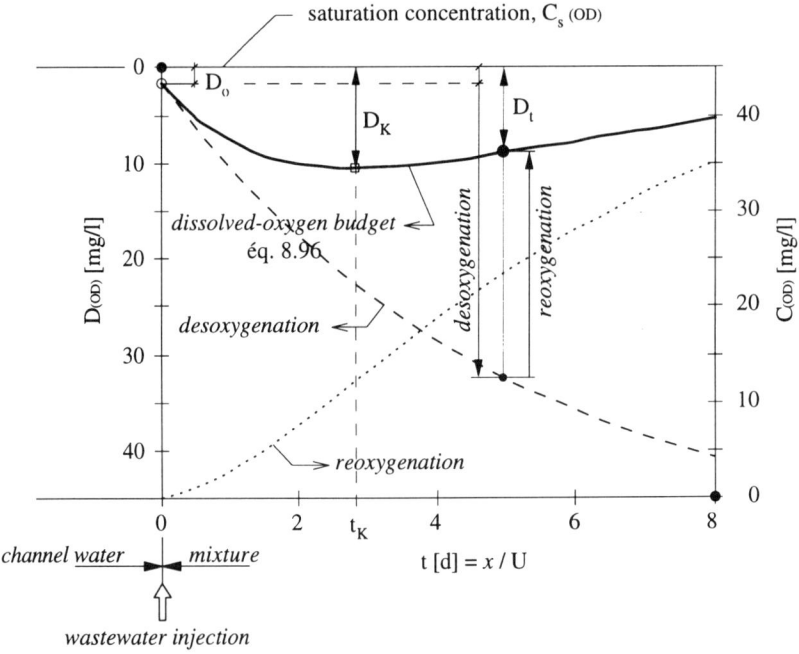

Fig. 8.14 Evolution of the oxygen deficit and of the concentration of dissolved oxygen, $C_{(OD)}$ and $D_{(OD)}$; the dissolved-oxygen sac curve.

8.5 EXERCISES

8.5.1 Problems, solved

Ex. 8.A

A rectangular channel of width B = 2.0 [m], is filled with (stagnant) water, $\vec{V} = 0$, up to a depth of h = 1.0 [m]. A certain quantity, M_0 = 2 [kg], of common salt (NaCl), released into the water at the station x = 0.0 [m], spreads immediately over the entire cross section of the channel.

Determine the evolution in time : *i)* of the longitudinal spreading of the salt and *ii)* of its maximum concentration. Plot afterwards : *iii)* the concentration distribution for the time $t_R = 10^7$ [s].

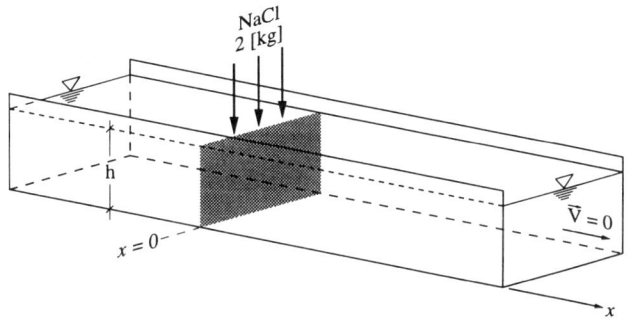

SOLUTION :

When there is no movement and no turbulence the diffusion is molecular. Therefore, the present problem is one of a one-dimensional *pure diffusion* in the x-direction from an instantaneous surface source (see sect. 8.2.1). Spreading of the cross-sectional average concentration can be calculated using eq. 8.28a :

$$C(x,t) = \frac{M_1}{\sqrt{2\pi}\,\sigma_x} \exp\left(-\frac{x^2}{2\sigma_x^2}\right)$$

with the variance, σ_x^2, given by :

$$\sigma_x^2(t) = 2\,\varepsilon_m\,t \tag{8.29}$$

The molecular diffusivity of the salt (NaCl) in water (see *Csanady*, 1973, p. 5) is :

$$\varepsilon_m = 1.24 \times 10^{-5} \text{ [cm}^2\text{/s]}$$

i) The longitudinal spreading of the matter, expressed by the variance, σ_x^2, is calculated as follows :

$$\sigma_x^2 = 2\,(1.24 \times 10^{-5})\,t$$

The width of the cloud is approximately given by :

W = $4\sigma_x$

The following table shows the time evolution of the standard deviation, σ_x and $4\sigma_x$, which increases proportionally to the square-root of the time, $\sqrt{t}$.

It can be seen that the time evolution is very slow; the cloud will have a width of W = 0.63 [m] after t_R = 10^7[s] = 115.7[d].

t	0[s]	10^5[s] = 27.77[h]	10^7[s] = 115.7[d]	10^9[s] = 32[year]
σ_x[m]	0	0.016	0.157	1.575
W[m]	0	0.063	0.630	6.300
C_{max}[kg/m³]	[∞]	25.53	2.53	0.25

ii) The maximum concentration, which remains at the origin, x = 0, is calculated using eq. 8.28a, or :

$$C_{max}(t) = C(0,t) = \frac{M_1}{\sqrt{2\pi}\ \sigma_x} \times 1$$

with the instantaneous surface source given by the eq. 8.27 :

$$M_1 = \frac{M_0}{S} = \frac{M_0}{hB} = \frac{2}{1 \times 2} = 1[kg/m^2]$$

This allows to write :

$$C_{max}(t) = \frac{1}{\sqrt{2\pi}\ \sigma_x} = \frac{0.399}{\sigma_x}\ [kg/m^3]$$

The above table shows the time evolution of the maximum concentration, C_{max}, which varies with the inverse of the standard deviation, $1/\sigma_x$, namely with $1/\sqrt{t}$.

The concentration, C, is taken here as the concentration of the mass of the substance in a given volume of mixture, defined (see point 8.1.1. 2°) as :

$$C = \frac{\text{mass of substance}}{\text{volume of mixture}}\ \left[\frac{kg}{m^3}\right]$$

Another type of concentration, the concentration by volume, defined by :

$$C_v = \frac{\text{volume of substance}}{\text{volume of mixture}}\ \left[\frac{m^3}{m^3}\right]$$

can be calculated — here taken as the maximum concentration : $C_{max} = 2.53$ [kg/m³] and the density of the salt : $\rho_s = 2200$ [kg/m³] — as follows :

$$C_v = \frac{1}{\rho_s} C = \frac{1}{2200} 2.53 = 1.15 \times 10^{-3} \text{ [-]}$$

The density of the salt/water mixture — using the same values — is :

$$\rho_m = \rho_{H_2O} + (\rho_s - \rho_{H_2O})C_v = 1000 + (2200 - 1000)\, 1.15 \times 10^{-3} = 1001.4 \text{ [kg/m³]}$$

from which one can see that : $\rho_m \approx \rho_{H_2O} = \rho$.

iii) The concentration distribution is given by eq. 8.28a, or :

$$C(x,t) = \frac{M_1}{\sqrt{2\pi}\,\sigma_x} \exp\left(-\frac{x^2}{2\sigma_x^2}\right)$$

For the time $t_R = 10^7$[s], when the standard deviation is $\sigma_x = 0.157$[m], and for $M_1 = 1$[kg/m²], this relationship renders :

$$C(x,t_R) = \frac{1}{\sqrt{2\pi}\,0.157} \exp\left(-\frac{x^2}{2 \times 0.157^2}\right) = 2.54 \exp\left(-\frac{x^2}{0.049}\right)$$

By substituting different values for *x*, one obtains the concentration distribution as shown in the figure below.

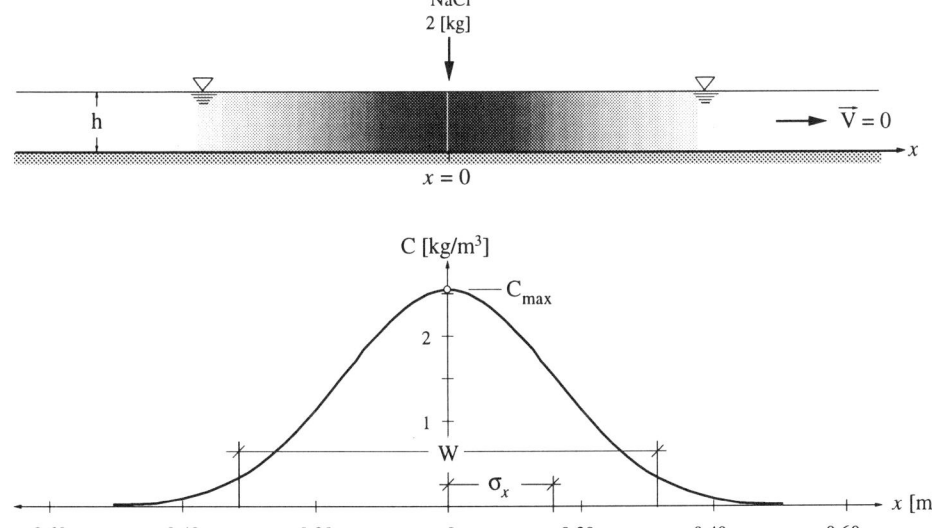

Ex. 8.B

In a rectangular channel the flow is controlled and has a constant discharge of $Q = 103.7$ [m³/s]. The following hydraulic parameters are measured: $U = 1.05$ [m/s], $A = 98.8$ [m²], $h = 1.09$ [m], $B = 90.6$ [m], $u_* = 0.07$ [m/s] and $S_f = 0.0005$ [–]; the uniform flow in the channel is subcritical and turbulent. The water is at a temperature of $T = 10$ [°C] and has a salt content of $C = 1$ [kg/m³].

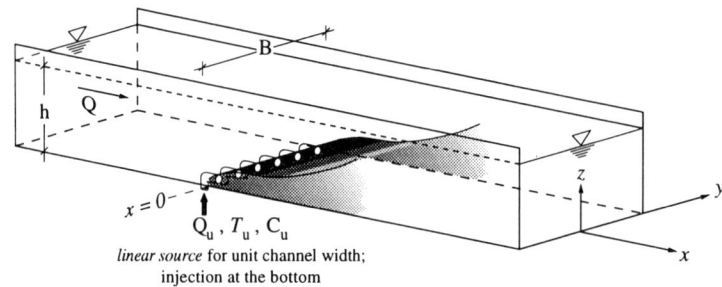

linear source for unit channel width; injection at the bottom

This channel receives wastewater from a nearby treatment plant. A continuous discharge, $Q_u = 0.5$ [m³/s], of wastewater is released uniformly over the entire width, B, of the channel by means of a multipoint-diffuser system installed at its bottom. The wastewater has a temperature of $T_u = 20$ [°C] and a salt content of $C_u = 30$ [kg/m³].

Study the *vertical* diffusion of the injected wastewater.

i) Derive the hydraulic relationships for the vertical diffusion.
ii) Determine the distance required for the vertical diffusion to be completed over the entire flow depth, h.
iii) Find the distance from the source of the point at which a concentration, being equal to the half of the depth-averaged final concentration, $c = C_M/2$, appears at the water surface, h.
iv) Calculate and plot the evolution of the salt content and of the temperature as a function of the distance.
v) Propose a method for measuring the vertical diffusivity, ε_{tz}.

SOLUTION :

i) Hydraulics of the vertical diffusion from a source located at the bottom of a channel, where the flow is uniform, turbulent and one-dimensional.

Neglecting the molecular diffusion and assuming that the velocity, $\vec{V}$ $(u,0,0) = U$, and also the diffusivities, ε_{ti}, are constant, the convection-diffusion equation in turbulent regime, eq. 8.19, is written as:

$$\frac{\partial c}{\partial t} + U \frac{\partial c}{\partial x} = \varepsilon_{tx} \frac{\partial^2 c}{\partial x^2} + \varepsilon_{ty} \frac{\partial^2 c}{\partial y^2} + \varepsilon_{tz} \frac{\partial^2 c}{\partial z^2} \quad (8.58)$$

The diffusion from a steady, $\partial c/\partial t = 0$, continuous source will be considered. It will also be assumed that the longitudinal diffusion is negligible, $(\varepsilon_{tx}) \partial^2 c/\partial x^2 = 0$, and the transversal diffusion is achieved almost instantaneously, $(\varepsilon_{ty}) \partial^2 c/\partial y^2 = 0$. Thus, one can rewrite eq. 8.58 as :

$$U \frac{\partial c}{\partial x} = \varepsilon_{tz} \frac{\partial^2 c}{\partial z^2} \qquad (8.\alpha)$$

The solution (see *Rutherford*, 1994, p.64), for an infinitely large (unbounded) water body is :

$$c_u(x,z) = \frac{G_u}{B \sqrt{4\pi \, \varepsilon_{tz} \, xU}} \exp\left(-\frac{z^2 \, U}{4 \, \varepsilon_{tz} \, x}\right) \qquad (8.\beta)$$

where $G_u = C_u Q_u$ is the injected mass discharge taken over the entire channel width, B. A comparison with eq. 8.28 and eq. 8.60 will be useful.

In the present situation the receiving water body has finite (bounded) dimensions. A new set of boundary conditions must therefore be satisfied, namely :

$$\frac{\partial c}{\partial z} = 0 \qquad \text{at:} \quad z = 0 \quad \text{et} \quad z = h$$

implying that the water surface, $z = h$, and the channel bottom, $z = 0$, are (impervious) reflecting boundaries (see point 8.2.1.8°). Subsequently one sets up image sources, $n = 1,2,...N$, for finding a solution for the case of a water body of finite dimensions by using the following relationship (see *Rutherford*, 1994, p. 64), or :

$$c(x,z) = c_u(x,z+z_0) + c_u(x,z-z_0) + \sum_{n=1}^{N} c_u(x, 2nh \pm z \pm z_0) \qquad (8.\gamma)$$

where c et c_u are the concentrations in finite and infinite water bodies, respectively; h is the flow depth and z_0 is the position of the source on the vertical, z, coordinate axis.

The solution of eq.8.γ is given in Fig. 8.B.1 in graphical form for a source located at $z = z_0 = 0$, that is on the channel bed.

This solution is also valid for a source located at $z = z_0 = h$, namely at the water surface, being due to the symmetrical nature of the boundary conditions. The graphical solutions for different positions of the injection point, $z^* = z_0/h = 0.25$, 0.75, 0.33, 0.67 et 0.50, are given by *Rutherford* (1994, p. 66-67).

In the Fig. 8.B.1 the variables are expressed in a dimensionless form, such as :

- the vertical distance : $z \dfrac{1}{h} = z^*$

- the longitudinal distance : $x \dfrac{\varepsilon_{tz}}{Uh^2} = x^*$

- the concentration : $c \dfrac{1}{C_M} = c^*$

where c is the local concentration and $C_M = G_u/Q$ is the final average concentration (i.e.: after the mixing has been completed); Q is the total discharge in the channel and G_u is the mass discharge of the *injected* substance. The velocity, $u = U$, and the vertical diffusivity, $\varepsilon_{tz} = \overline{\varepsilon_{tz}}$, are considered to be constant in the vertical.

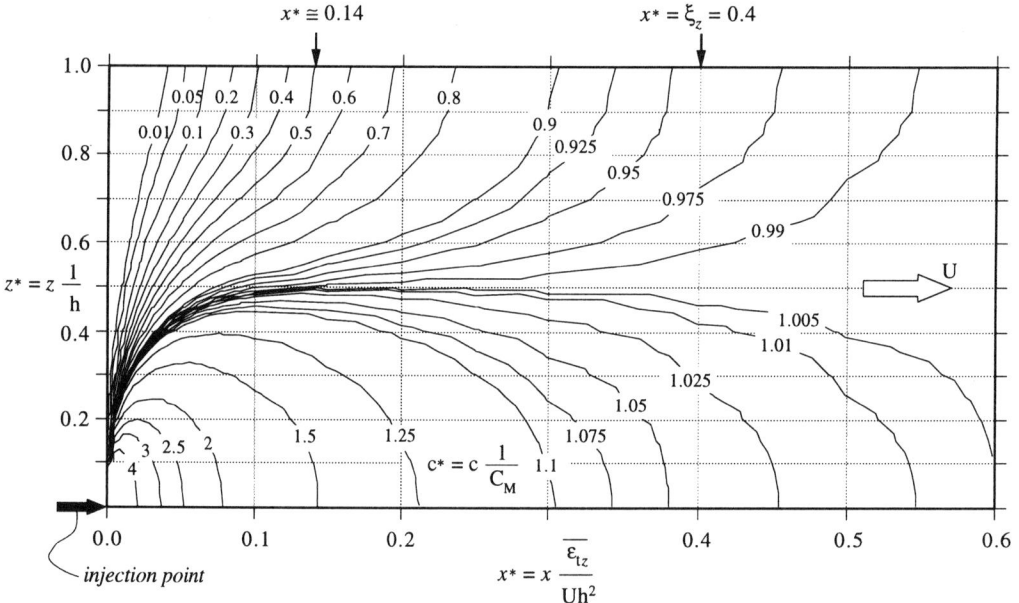

Fig. 8.B.1 Concentration distribution, $c^*(x^*,z^*)$, for the vertical diffusion from a source located at $z^* = 0$.

Some interesting conclusions can be drawn from the analytical solution presented in the Fig. 8.B.1 :

- for $x^* \geq 0.54$: the vertical diffusion is practically completed and the concentration is almost uniform over the entire water depth, h.
- for $x^* \geq 0.4$: the vertical diffusion is completed up to $0.95 < c^* < 1.05$; this is the value used in eq. 8.54.
- for $z^* = 0.0$: the (initial) concentration at the channel bed decreases rapidly.
- for $z^* = 1.0$: the concentration at the water surface increases slowly.

TRANSPORT AND MIXING OF MATTER

ii) Determination of the distance, L_z.

The distance, measured from the point of injection of the substance, can be calculated using :

$$L_z = U t_z = \xi_z \frac{U h^2}{\overline{\varepsilon_{tz}}} \tag{8.54}$$

where $\xi_z \cong 0.4$ for a source located at the channel bed, $z_0/h = 0$. With the vertical diffusivity given by eq. 8.53 :

$$\overline{\varepsilon_{tz}} = 0.067 \, (h \, u_*) = 0.067 \, (1.09 \times 0.07) = 0.0051 \, [\text{m}^2/\text{s}]$$

one can now obtain :

$$L_z = 0.4 \, \frac{1.05 \times 1.09^2}{0.0051} \cong 98 \, [\text{m}]$$

After this distance the concentration has reached 95% of the final average (injected) concentration, C_M ; this is shown in Fig. 8.B.2.

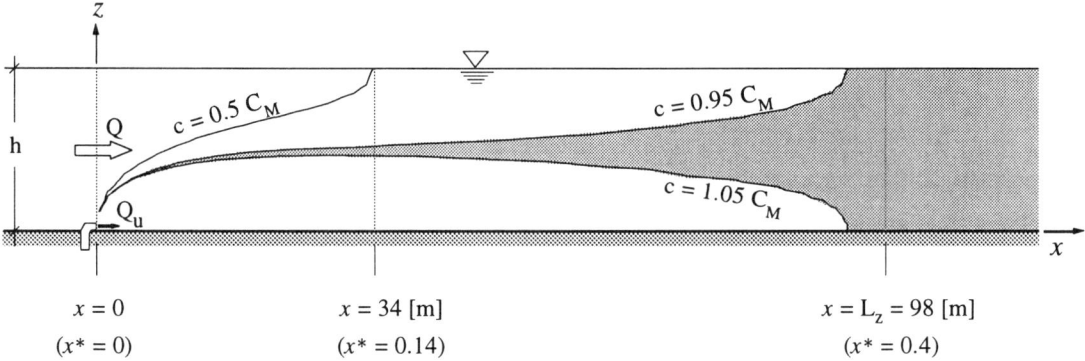

Fig. 8.B.2   Summary of the study of the vertical diffusion.

iii) Determination of the distance at which $C_M/2$ appears at the water surface.

In Fig. 8.B.1, one can see that this dimensionless concentration, $c^* = c/C_M = 0.5$, reaches the water surface, $z^* = 1$, when the dimensionless distance is equal to :

$$x^* = \frac{x \, \overline{\varepsilon_{tz}}}{U h^2} = 0.14$$

Using this value, the distance at which the concentration of $c = C_M/2$ reaches the water surface is calculated as follows :

$$x = 0.14 \frac{U h^2}{\overline{\varepsilon_{tz}}} = 0.14 \frac{1.05 \times 1.09^2}{0.0051} \cong 34 \text{ [m]}$$

Fig. 8.B.2 summarizes the results of the study of the *vertical* diffusion.

iv) Evolution of the salt content and of the temperature.

   a) Salt (content) concentration.

      The final average concentration resulting from the injection of the wastewater into the channel is :

$$C_M = \frac{G + G_u}{Q + Q_u} = \frac{CQ + C_u Q_u}{Q + Q_u} = \frac{1 \times 103.7 + 30.0 \times 0.5}{103.7 + 0.5} = 1.139 \text{ [kg/m}^3\text{]}$$

      One can then obtain, for example for $c^* = 0.5$, a concentration of :

$$c^* = 0.5 = \frac{C_{0.5} - C}{C_M - C}$$

      from which it can be calculated that :

$$C_{0.5} = C + 0.5 (C_M - C) = 1.0 + 0.5 (1.139 - 1.0) = 1.07 \text{ [kg/m}^3\text{]}$$

   b) Temperature.

      Recall (see point 8.4,1°) that the diffusion of a substance, C, and of the temperature, T, are analogous; one simply substitutes C by T. The heat is considered to be a passive physical quantity; this implies that there is no heat transfer either across the water surface or across the bottom and the side-walls of the channel.

      The final average temperature resulting from the injection of the wastewater with a temperature of $T_u = 20$ [°C] into the channel is :

$$T_M = \frac{TQ + T_u Q_u}{Q + Q_u} = \frac{10 \times 103.7 + 20 \times 0.5}{103.7 + 0.5} = 10.05 \text{ [°C]}$$

      One can the obtain, for example for $t^* = 0.5$, a temperature of :

$$t^* = 0.5 = \frac{T_{0.5} - T}{T_M - T}$$

      from which it can be calculated that :

$$T_{0.5} = T + 0.5 (T_M - T) = 10.0 + 0.5 (10.05 - 10.0) = 10.025 \text{ [°C]}$$

TRANSPORT AND MIXING OF MATTER

The longitudinal evolution of the concentration and of the temperature are shown in Fig. 8.B.3.

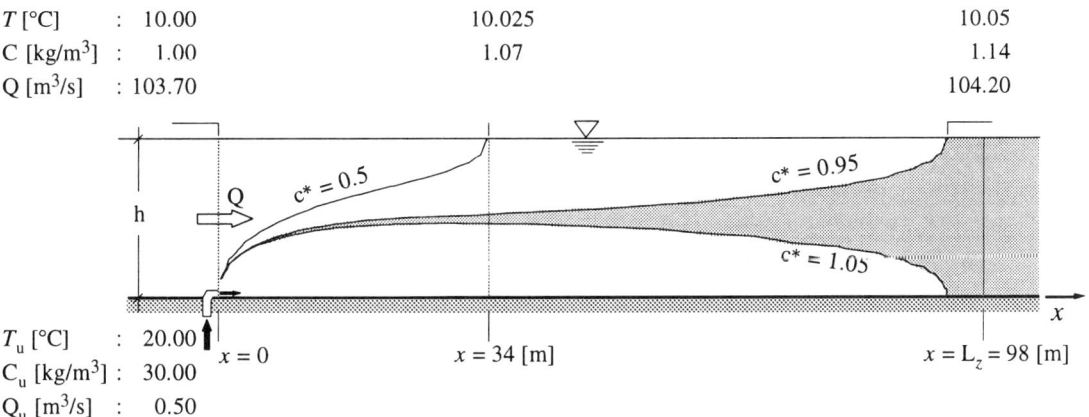

Fig. 8.B.3 Summary of the application of the study of the vertical diffusion.

v) Method for measuring the vertical diffusivity.

The measurements should be made in a uniform flow in an infinite water body where the concentration is given by :

$$c(x,z) = \frac{G_u}{B\sqrt{4\pi \varepsilon_{tz} xU}} \exp\left(-\frac{(z-z_o)^2 U}{4 \varepsilon_{tz} x}\right)$$

The maximum concentration is at the same height as the point of injection, $z = z_o$:

$$c_{max}(x, z = z_o) = \frac{G_u}{B\sqrt{4\pi \varepsilon_{tz} xU}} \times 1 = \frac{G_u}{B\sqrt{4\pi \varepsilon_{tz} U}} \left(\frac{1}{\sqrt{x}}\right)$$

On a plot of $c_{max}(x) = f(1/\sqrt{x})$ the slope of the line gives the diffusivity, ε_{tz}; the parameters, G_u as well as U and B, must be determined previously.

In a given channel where the velocity, U, is known, the measurement of the variation of the maximum concentration, $c_{max}(x)$, allows the determination of the vertical diffusion, ε_{tz}.

Ex. 8.C

A channel designed for a constant discharge of Q = 103.7 [m³/s], has the following hydraulic parameters : U = 1.05 [m/s], h = 1.09 [m], B = 90.6 [m], u_* = 0.07 [m/s] and S_f = 0.0005 [–]. The flow in the channel is uniform, subcritical and turbulent.

Wastewater with a salt content of C_u = 30 [kg/m³] is discharged into this channel at a rate of Q_u = 0.5 [m³/s].

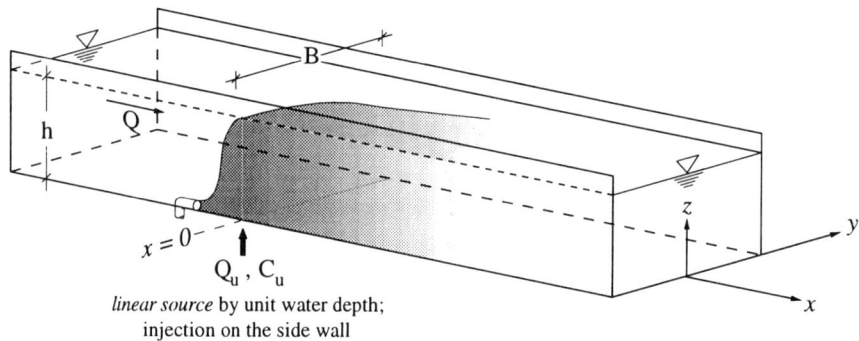

linear source by unit water depth;
injection on the side wall

Study the *transversal* diffusion for the following scenarios where the wastewater is injected either at the center of the channel or at one of the side walls. It shall be admitted that the vertical diffusion takes place almost instantly (see point 8.4.1,7°).

i) Determine the distance necessary to achieve a complete transversal diffusion over the entire width of the channel for both of the injection scenarios.

ii) Calculate and plot the transversal wastewater concentration profile at the stations L_1 = 0.5 [km] and L_2 = 20 [km].

SOLUTION :

i) Determination of the distance, L_y.

Measured from the point of injection of the wastewater, the distance necessary for a complete transversal diffusion over the entire channel width can be calculated with :

$$L_y = \xi_y \, U \, \frac{B^2}{\overline{\varepsilon_{ty}}} \qquad (8.56)$$

where $\xi_y \cong 0.5 \, (0.1)$ for an injection source located at a side wall (at the center) of the channel.

The transversal diffusivity is given by eq. 8.55a :

$$\overline{\varepsilon_{ty}} = 0.6 \, (h \, u_*) = 0.6 \, (1.09 \times 0.07) = 0.046 \, [m^2/s]$$

One can now calculate :

- for a source located at one of the side walls of the channel

$$L_y = 0.5 \, (1.05 \, \frac{90.6^2}{0.046}) \cong 94 \, [km]$$

- for a source located at the center of the channel

$$L_y = 0.1 \, (1.05 \, \frac{90.6^2}{0.046}) \cong 19 \, [km]$$

It can be seen that, in both cases, the distance, L_y, necessary for the transversal diffusion to be almost completely achieved over the entire width of the channel is rather long. To shorten this distance, L_y, one should install a system of multipoint diffusers over the entire channel width, B, giving a better lateral distribution.

One can also see that this distance, L_y, is much longer than the one needed to accomplish a complete vertical diffusion over the entire channel depth, which was calculated before (see Ex. 8.B) as being $L_z \cong 0.098$ [km].

Fig. 8.C.1 presents a summary of the results for the two injection scenarios.

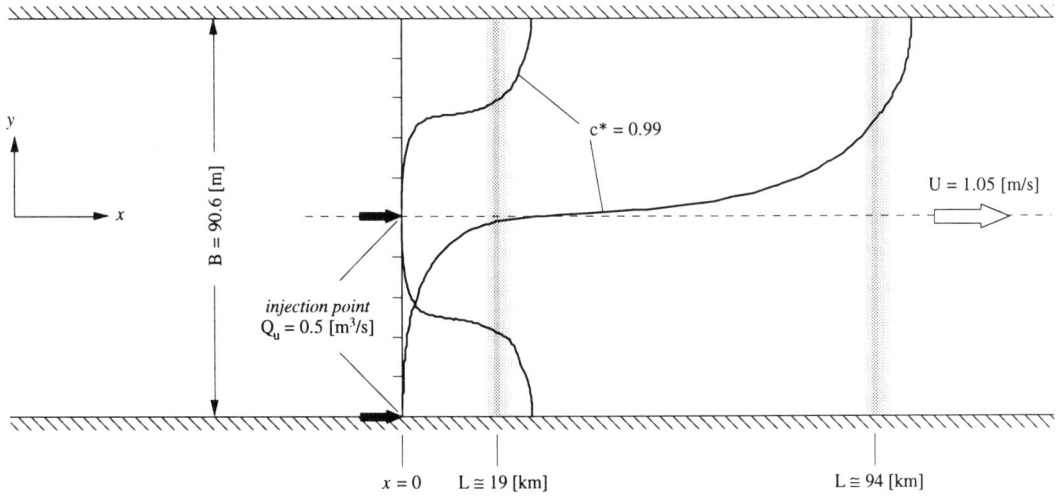

Fig. 8.C.1 The distance, L_y, at which the transversal diffusion is almost completely achieved for both scenarios investigated.

ii) Determination of the transversal concentration profiles at the stations $L_1 = 0.5$ [km] and $L_2 = 20$ [km].

The case we are dealing with here is a problem of uniform and steady transversal convection-diffusion from a linear source (see sect. 8.4.2) for which it was assumed that the vertical diffusion has already been completed.

The solution to this problem is given by eq. 8.61, whose graphical representation, already given in Fig. 8.9, is reproduced here with Fig. 8.C.2 for both of the injection scenarios, namely at the center and at one side wall of the channel.

The dimensionless variables, for the present problem, are :

$$y^* = y\frac{1}{B} = y\frac{1}{90.6}$$

$$x^* = x\frac{\varepsilon_{ty}}{U B^2} = x\frac{0.046}{1.05 \times (90.6)^2} = x\,(5.3 \cdot 10^{-6})$$

$$c^* = C\frac{1}{C_M} = C\frac{Q + Q_u}{G_u} = C\frac{103.7 + 0.5}{30.0 \times 0.5} = \frac{C}{0.144}$$

For the two stations, one can write :

- $L_1 = 500$ [m] $x_1^* = 2.7 \cdot 10^{-3}$
- $L_2 = 20000$ [m] $x_2^* = 1.1 \cdot 10^{-1}$

By using these values, x_1^* et x_2^*, one can draw the transversal distribution of the concentration for the two stations and for each of the two scenarios; this is shown on the Fig. 8.C.2.

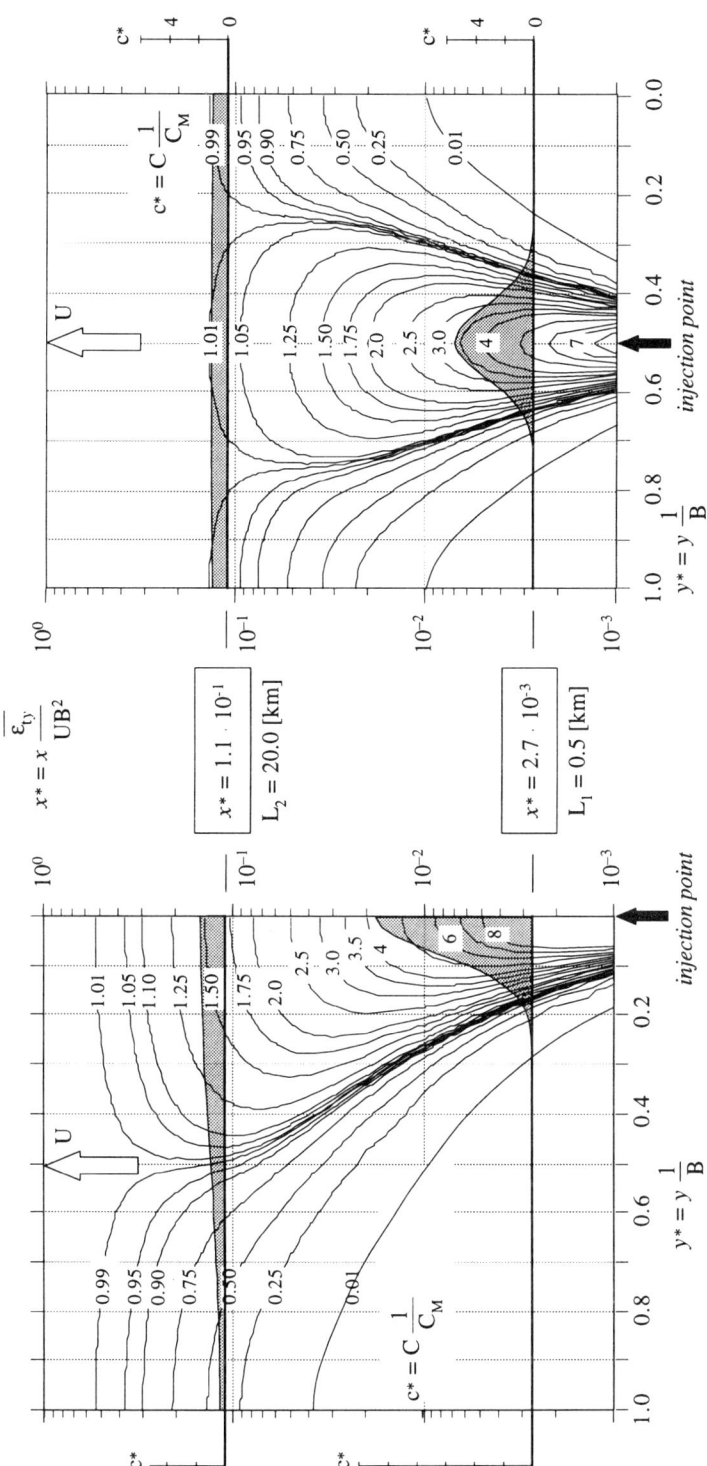

Fig. 8.C.2   Concentration distribution, $c^*(x^*, y^*)$, resulting from transversal diffusion for both of the scenarios studied.

Ex. 8.D

In a large channel the water is flowing with a low average velocity of U = 0.1 [m/s] and the flow can be considered to be one-dimensional. The flow depth is h = 1.0 [m]. The density and the temperature of the water are ρ = 1000 [kg/m³] and T_0 = 15 [°C], respectively.

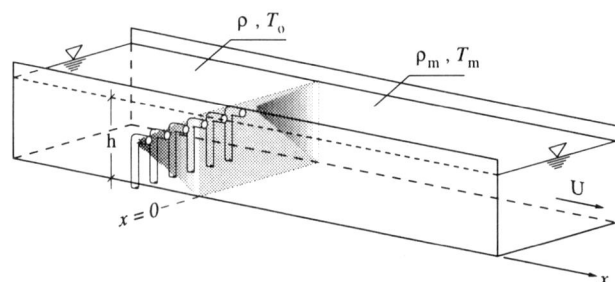

At a certain moment a brine solution is introduced continuously into the flow, and this at a constant rate by a multipoint-diffuser system spreading the mixture uniformly over the entire cross section of the channel almost instantaneously. The density and the temperature of the mixture at the cross section just at the downstream of the diffuser were measured as being ρ_m = 1010 [kg/m³] at T_m = 22 [°C]. The dispersion coefficient was obtained as being K_x = 0.1 [m²/s] by a previous experimental study.

Considering only the longitudinal dispersion in the x-direction, determine : *i)* the concentration distribution and *ii)* the temperature distribution at times t_1 = 1[h] and t_2 = 24 [h] after the start of the injection. *iii)* Subsequently calculate when the mixture with a density of ρ'_m = 1001 [kg/m³] arrives at a station located at 1 [km] downstream of the diffuser system.

SOLUTION :

The present problem is one of one-dimensional *convection-dispersion* from a continuous source (see sect. 8.4.4. 4°*iii*). To determine the flow regime, the Reynolds number (see eq. 1.7) is calculated as:

$$\text{Re}' = \frac{Uh}{\nu} = \frac{0.1 \times 1.0}{1.3 \times 10^{-6}} = 77000 > 2000$$

thus the flow is a *turbulent* one.

The salt content of a mixture with a density of ρ_m = 1010 [kg/m³] can be calculated (see *Handbook of Chemistry and Physics*, 55th Edition, CRC Press, p. D.224) as :

$C_0 \cong 17$ [kg/m³].

i) The concentration distribution for the present problem is given by the following solution to the one-dimensional convection-dispersion equation, eq. 8.64; or :

$$C(x,t) = \frac{C_o}{2}\left[\exp\left(\frac{Ux}{K_x}\right)erfc\left(\frac{x+Ut}{\sqrt{4K_xt}}\right) + erfc\left(\frac{x-Ut}{\sqrt{4K_xt}}\right)\right] \quad (8.76)$$

where C_o is the concentration of the salt introduced continuously at a constant rate.

One should recall that the half concentration, $C = C_o/2$, travels approximately with the flow velocity, U ; the distance traveled being therefore :

– for $\quad t_1 = 1$ [h] : $\quad \overline{x_1} = Ut_1 = (0.1)(3600) = 360$ [m]
– for $\quad t_2 = 24$ [h] : $\quad \overline{x_2} = Ut_2 = (0.1)(24 \times 3600) = 8640$ [m]

		U = 0.1 [m/s] ; K_x = 0.1 [m²/s] ; t_1 = 1 [h] et t_2 = 24 [h]									
	x [m]	0	200	300	340	350	360	370	380	390	420	430
t_1	C/C$_o$	1.00	1.00	0.99	0.78	0.66	0.51	0.37	0.24	0.14	0.01	0.00
	C [kg/m³]	17.0	17.0	16.8	13.3	11.2	8.8	6.2	4.6	2.4	0.2	0.0
	x [m]	0	7880	8400	8540	8590	8640	8670	8730	8800	8940	8980
t_2	C/C$_o$	1.00	1.00	0.97	0.78	0.65	0.50	0.41	0.25	0.11	0.01	0.00
	C [kg/m³]	17.0	17.0	16.5	13.3	11.1	8.5	7.0	4.6	1.9	0.2	0.0

The concentration distribution, C(x,t), at times t_1 and t_2, can be now calculated by substituting different values for the distance, x, into the above equation, eq. 8.76.

The relative concentration calculated for different distances is tabulated above for both instances, t_1 and t_2, and the concentration distributions, $C(x)/C_o$, are plotted Fig. Ex. 8.D.1.

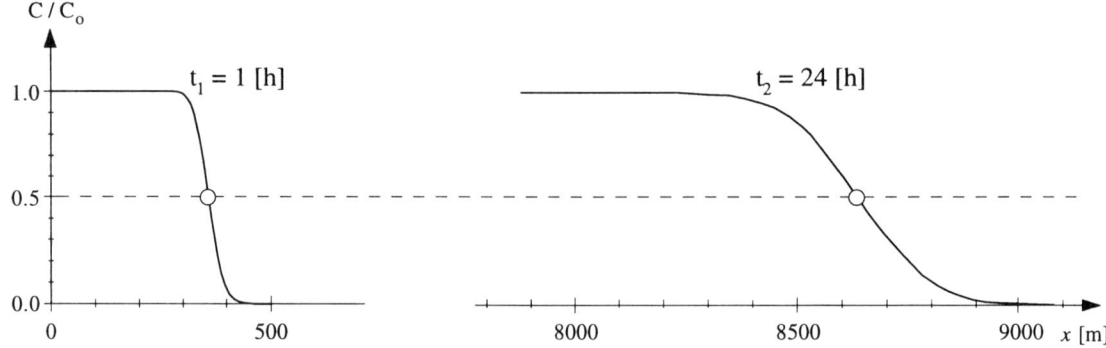

Fig. Ex. 8.D.1 Concentration distributions, $C(x)/C_o$, at $t_1 = 1$ [h] and at $t_2 = 24$ [h].

ii) The temperature distribution for the present problem is given by the following solution to the one-dimensional convection-dispersion equation, eq. 8.64; or :

$$\Delta T(x,t) = \frac{\Delta T_o}{2}\left[\exp\left(\frac{Ux}{a_{ht}}\right) erfc\left(\frac{x + Ut}{\sqrt{4\, a_{ht} t}}\right) + erfc\left(\frac{x - Ut}{\sqrt{4\, a_{ht} t}}\right)\right]$$

where $\Delta T = (T - T_o)$ and $\Delta T_o = (T_m - T_o)$. It will be recalled here (see point 8.4.1°) that the solutions for the diffusion of matter, C, and for the diffusion of temperature, T, are analogous. One can simply replace C by ΔT and one admits that $a_{ht} \equiv K_x$.

		\multicolumn{10}{c}{$U = 0.1$ [m/s] ; $a_{ht} \equiv K_x = 0.1$ [m²/s] ; $t_1 = 1$ [h] et $t_2 = 24$ [h]}
	x [m]	0	200	300	340	350	360	370	380	390	420	430
t_1	$\Delta T/\Delta T_o$	1.00	1.00	0.99	0.78	0.66	0.51	0.37	0.24	0.14	0.01	0.00
	T [°C]	22.0	22.0	21.9	20.5	19.6	18.6	17.6	16.7	16.0	15.1	15.0
	x [m]	0	7880	8400	8540	8590	8640	8670	8730	8800	8940	8980
t_2	$\Delta T/\Delta T_o$	1.00	1.00	0.97	0.78	0.65	0.50	0.41	0.25	0.11	0.01	0.00
	T [°C]	22.0	22.0	21.8	20.5	19.6	18.5	17.9	16.8	15.8	15.1	15.0

The temperature distributions, $\Delta T(x,t)$, at times t_1 and t_2, can be now calculated by substituting different values for the distance, x, in the equation for the temperature.

The temperature calculated for different distances is tabulated above for both instances, t_1 and t_2, and the temperature distributions, $T(x)$, are plotted Fig. Ex. 8.D.2.

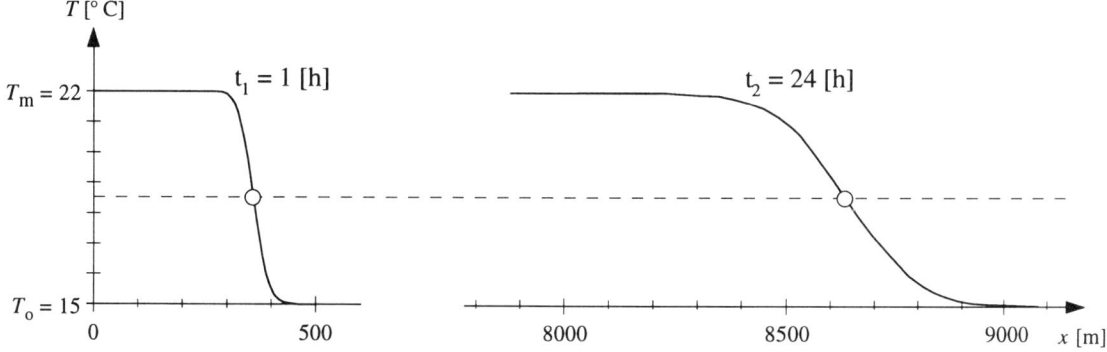

Fig. Ex. 8.D.2 Temperature distributions, $T(x)$, at $t_1 = 1$ [h] and at $t_2 = 24$ [h].

iii) Calculation of the instant, when a mixture with a density of $\rho'_m = 1001$ [kg/m³] arrives at a station located at $x = 1000$ [m] downstream of the injection point.

The salt content of the mixture with a density of $\rho'_m = 1001$ [kg/m³] can be calculated (see *Handbook of Chemistry and Physics*, 55th Edition, CRC Press, p. D.224) as :

$C' \cong 4$ [kg/m³].

By inserting this value in eq. 8.76, one can write :

$$\frac{C'}{C_o} = \frac{4.0}{17} = \frac{1}{2}\left[\exp\left(\frac{0.1 \times 10^3}{1 \times 10^{-1}}\right) erfc\,\frac{10^3 + 0.1 \times t}{\sqrt{4\,(1 \times 10^{-1})\,t}} + erfc\,\frac{10^3 - 0.1 \times t}{\sqrt{4\,(1 \times 10^{-1})\,t}}\right]$$

This expression can be solved by trial and error. It can then be calculated that, $C'/C_o = 0.24$ [-], arrives at this station after $t \cong 9680$[s] $= 2.69$ [h].

Ex. 8.E

The flow in a channel can be considered as turbulent and subcritical. The following hydraulic data have been measured : h = 0.7 [m], B = 6.0 [m], U = 1.0 [m/s] and $u_* = 0.07$ [m/s].

A certain quantity of salt, $M_0 = 4$ [kg], is released into the flow and it spreads itself instantaneously over the entire cross section of the channel.

Study the temporal evolution of the concentration at a distance of $L_R = 1$ [km] downstream from the injection point for the following scenarios :

i) **Scenario A** : the salt is released instantaneously at the station x = 0.0 [m],

ii) **Scenario B** : the salt is released during a time period of T = 8 [min] at the station x = 0.0 [m],

iii) **Scenario C** : 2 [kg] of salt is released simultaneously at two stations located at $x_1 = 0.0$ [m] and $x_2 = 250.0$ [m].

iv) For the above three scenarios calculate the time period during which the salt concentration at $L_R = 1$ [km] is higher than C' = 0.002 [kg/m³].

v) Calculate the concentration distribution at the time $t_R = 1000$ [s] for scenarios A and B.

SOLUTION :

Since the passive substance (salt) spreads after the release, almost instantaneously over the entire cross section neither the vertical nor the transversal diffusion have to be considered. The problem is one of a one-dimensional longitudinal convection-dispersion from a surface source (see sect. 8.4.4. 4°).

The dispersion coefficient can be estimated using eq. 8.68 :

$K_x \approx 200 (h\, u_*) = 200 (0.7 \times 0.07) \cong 9.8$ [m²/s]

i) **Scenario A** : The salt is released into the flow instantaneously. The concentration distribution is given by :

$$C(x,t) = \frac{M_0}{S\sqrt{4\pi\, K_x t}} \exp\left[-\frac{(x-Ut)^2}{4\, K_x t}\right] \qquad (8.71)$$

with the surface source defined by the eq. 8.27 :

$$M_1 = \frac{M_0}{S} = \frac{4}{0.7 \times 6.0} = 0.95 \text{ [kg/m}^2\text{]}$$

TRANSPORT AND MIXING OF MATTER

The temporal evolution of the concentration — with $x \equiv L_R$, $\mathcal{M}_1$, K_x, U, already determined — can be calculated using eq. 8.71 :

$$C(L_R,t) = \frac{0.95}{\sqrt{4\pi \times 9.8 \times t}} \exp\left[-\frac{(1000 - (1.0 \times t))^2}{4 \times 9.8 \times t}\right]$$

by substituting different values for t. This is shown in Fig. 8.E.1, labeled as scenario A; values for a few points are presented in the following table. It is to be noted that the curve, C(t), is slightly non symmetrical.

The temporal evolution of the concentration at L_R = 1000 [m]						
U = 1.0 [m/s]	u_* = 0.07 [m/s]	h = 0.7 [m]	B = 6.0 [m]		M_0 = 4.0 [kg]	K_x = 9.8 [m²/s]

scenario A		scenario B								
A		B_1		B_2		B_3		B_4		B = Σ B_i
$\mathcal{M}_1$ = 0.950 [kg/m²]		$(\mathcal{M}_1)_1$ = 0.238 [kg/m²]		$(\mathcal{M}_1)_2$ = 0.238 [kg/m²]		$(\mathcal{M}_1)_3$ = 0.238 [kg/m²]		$(\mathcal{M}_1)_4$ = 0.238 [kg/m²]		Σ $(\mathcal{M}_1)_i$ = $\mathcal{M}_1$
		τ_1 = 60 [s]		τ_2 = 180 [s]		τ_3 = 300 [s]		τ_4 = 420 [s]		T = 8 [min]
t	C(x,t)	t	t - τ_1	ΔC_1(x,t)	t - τ_2	ΔC_2(x,t)	t - τ_3	ΔC_3(x,t)	t - τ_4	ΔC_4(x,t)	$\sum_{i=1}^{4} \Delta C_i(x,t)$
[s]	[kg/m³]	[s]	[s]	[kg/m³]	[s]	[kg/m³]	[s]	[kg/m³]	[s]	[kg/m³]	[kg/m³]
750	3.74E-04	750	690	2.34E-05	570	2.29E-07	450	3.61E-11	330	0.00E+00	2.36E-05
800	8.47E-04	800	740	7.67E-05	620	2.26E-06	500	2.77E-09	380	0.00E+00	7.90E-05
850	1.50E-03	850	790	1.84E-04	670	1.31E-05	550	7.63E-08	430	4.40E-12	1.97E-04
900	2.15E-03	900	840	3.40E-04	720	4.97E-05	600	9.73E-07	480	5.62E-10	3.91E-04
950	2.60E-03	950	890	5.08E-04	770	1.34E-04	650	6.87E-06	530	2.25E-08	6.49E-04
1000	*2.71E-03*	1000	940	6.35E-04	820	2.73E-04	700	3.05E-05	580	3.80E-07	9.39E-04
1060	2.42E-03	*1060*	1000	*6.78E-04*	880	4.76E-04	760	1.13E-04	640	4.84E-06	1.27E-03
1120	1.85E-03	1120	1060	6.04E-04	940	6.35E-04	820	2.73E-04	700	3.05E-05	1.54E-03
1180	1.24E-03	*1180*	1120	4.62E-04	1000	*6.78E-04*	880	4.76E-04	760	1.13E-04	1.73E-03
1240	7.45E-04	1240	1180	3.10E-04	1060	6.04E-04	940	6.35E-04	820	2.73E-04	*1.82E-03*
1300	4.07E-04	*1300*	1240	1.86E-04	1120	4.62E-04	1000	*6.78E-04*	880	4.76E-04	1.80E-03
1360	2.05E-04	1360	1300	1.02E-04	1180	3.10E-04	1060	6.04E-04	940	6.35E-04	1.65E-03
1420	9.58E-05	*1420*	1360	5.12E-05	1240	1.86E-04	1120	4.62E-04	1000	*6.78E-04*	1.38E-03
1480	4.20E-05	1480	1420	2.39E-05	1300	1.02E-04	1180	3.10E-04	1060	6.04E-04	1.04E-03
1540	1.75E-05	1540	1480	1.05E-05	1360	5.12E-05	1240	1.86E-04	1120	4.62E-04	7.10E-04
1600	6.90E-06	1600	1540	4.36E-06	1420	2.39E-05	1300	1.02E-04	1180	3.10E-04	4.40E-04
1660	2.61E-06	1660	1600	1.72E-06	1480	1.05E-05	1360	5.12E-05	1240	1.86E-04	2.50E-04
1720	9.48E-07	1720	1660	6.52E-07	1540	4.36E-06	1420	2.39E-05	1300	1.02E-04	1.31E-04

592   FLUVIAL HYDRAULICS

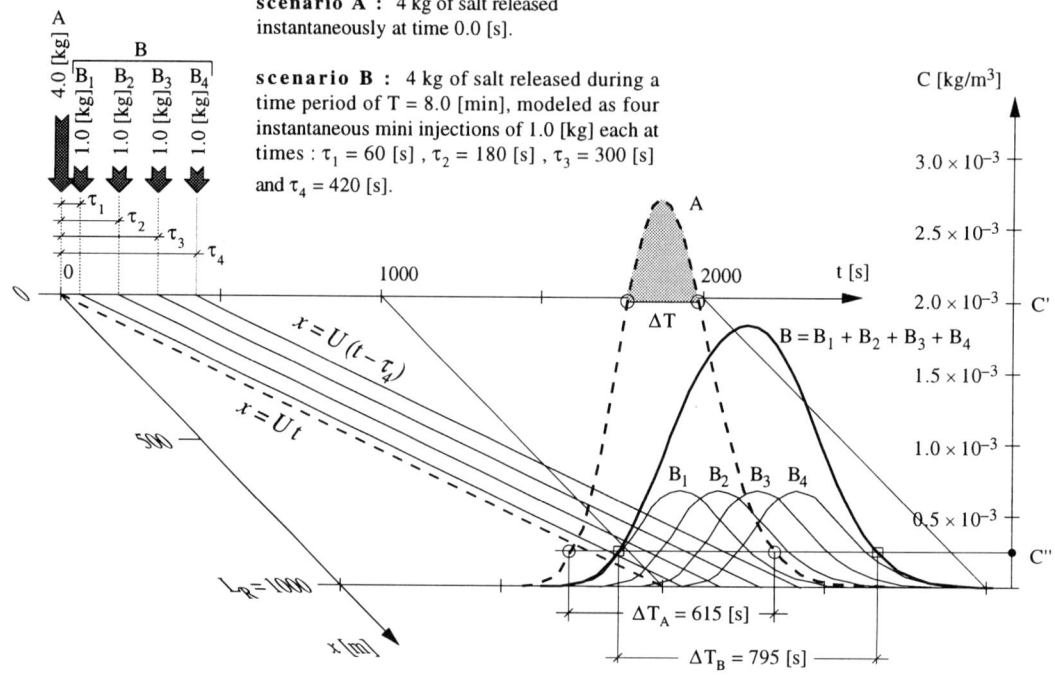

Fig. 8.E.1 Temporal evolution of the concentration for scenarios A and B.

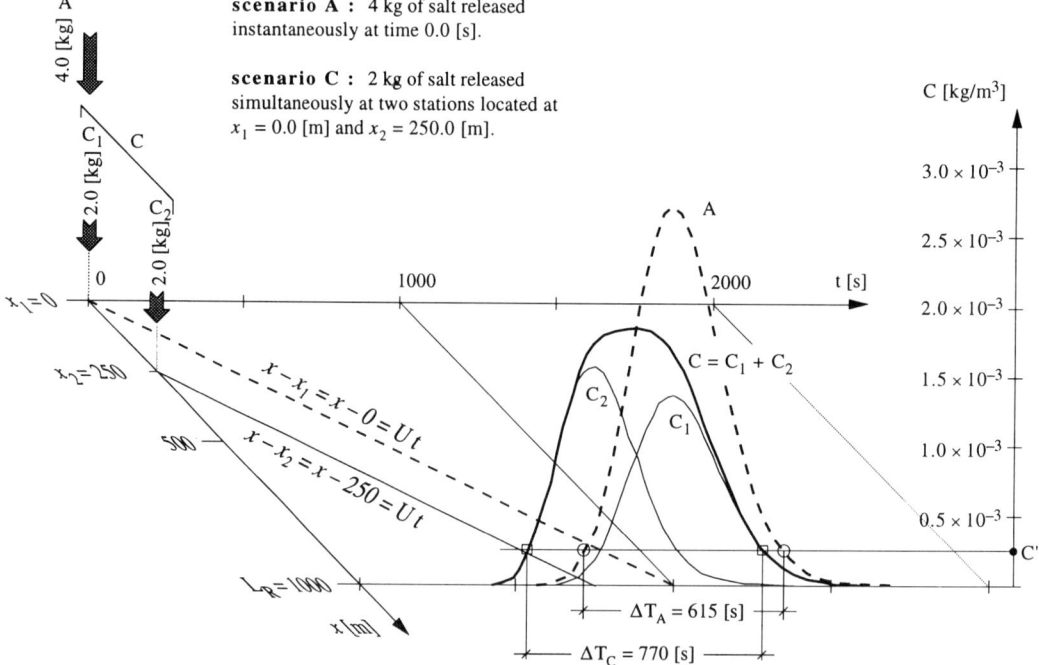

Fig. 8.E.2 Temporal evolution of the concentration for scenarios A and C.

TRANSPORT AND MIXING OF MATTER

ii) **Scenario B** : The salt is released during a finite time period of T = 8 [min].

The continuous injection of 4 [kg] of salt during a time period of T = 8 [min] will be here modeled by four instantaneous mini-injections of m_i = 1 [kg] or $(\mathcal{M}_1)_i$ = 0.238 [kg/m²], taken according to the following time table :

m_1 = 1 [kg]	at	τ_1	=	1 [min]	→	$\Delta C_1(x,t)$	
m_2 = 1 [kg]	at	τ_2	=	3 [min]	→	$\Delta C_2(x,t)$	
m_3 = 1 [kg]	at	τ_3	=	5 [min]	→	$\Delta C_3(x,t)$	
m_4 = 1 [kg]	at	τ_4	=	7 [min]	→	$\Delta C_4(x,t)$	
$\sum_{i=1}^{4} m_i$ = 4 [kg]		after T	=	8 [min]	→	$\sum_{i=1}^{4} \Delta C_i(x,t)$	

The concentration distribution for each one of the four mini-injections, m_1, m_2, m_3 and m_4, is given by :

$$\Delta C_i(L_R, t) = \frac{m_i}{S \sqrt{4\pi K_x (t - \tau_i)}} \exp\left[-\frac{[x - U(t - \tau_i)]^2}{4K_x (t - \tau_i)}\right] \tag{8.74}$$

The four concentration distributions — for the already determined values, $x \equiv L_R$, m_i, S, K_x, U — can be calculated by giving different values to $(t - \tau_i)$. They are then added to obtain :

$$C(L_R, t) = \sum_{i=1}^{n} \Delta C_i(x, t) \tag{8.75}$$

This is shown in Fig. 8.E.1, labeled as scenario B; values for a few points are presented in the above table. It is to be noted that these curves, C(t), are slightly non symmetrical.

Discussion (see Fig. 8.E.1) :

In scenarios A and B, the same total quantity of salt, M_0 = 4 [kg], is released at the same station, x = 0.0 [m], but in a different way.

- In the scenario A, with a single and instantaneous release, the maximum concentration of C_{max} = 2.71 × 10⁻³ [kg/m³] arrives at the station L_R = 1000 [m] after t_R = 1000 [s] .

- In the scenario B, with 4 instantaneous mini-injections during a time period of T = 8 [min] , the maximum concentration of C_{max} ≈ 1.82 × 10⁻³ [kg/m³] — which is less important than the one in the scenario A — arrives after t ≈ 1240 [s] — later than in the scenario A — at the station L_R = 1000 [m]. The same concentration distribution results from each mini-injection, m_i, but they arrive with a time lag of τ_i .

- For a low concentration, such as C" ≈ 0.25 × 10⁻³ [kg/m³], the time period during which this concentration is exceeded is :

  - for the scenario A : $\Delta T_A \approx 1345 - 730 = 615$ [s]
  - for the scenario B : $\Delta T_B \approx 1660 - 865 = 795$ [s]

  It is evident that $\Delta T_B > \Delta T_A$.

iii) **Scenario C** : The total quantity of salt, $M_0 = 4$ [kg], is divided into two equal quantities which are released simultaneously and instantaneously at two stations located at $x_1 = 0.0$ [m] and $x_2 = 250.0$ [m].

The first mini-injection of $m_1 = 2$ [kg] is released at $x_1 = 0.0$ [m] and the second mini-injection of $m_2 = 2$ [kg] is released at $x_2 = 250.0$ [m], both at the same instant, t = 0.0 [s] :

$$\begin{array}{llll}
m_1 = 2 \text{ [kg]} & \text{at} & x_1 = 0.0 \text{ [m]} & \rightarrow \Delta C_1(x,t) \\
m_2 = 2 \text{ [kg]} & \text{at} & x_2 = 250.0 \text{ [m]} & \rightarrow \Delta C_2(x,t) \\
\hline
\sum_{i=1}^{2} m_i = 4 \text{ [kg]} & \text{after} & L_R = 1000.0 \text{ [m]} & \rightarrow \sum_{i=1}^{2} \Delta C_i(x,t)
\end{array}$$

The temporal evolutions of the concentration for each one of the mini-injections, m_1 and m_2, can be calculated using the eq. 8.74, written as :

$$\Delta C_i(x,t) = \frac{m_i}{S\sqrt{4\pi K_x t_i}} \exp\left[-\frac{[(x-x_i) - Ut_i]^2}{4 K_x t_i}\right]$$

The temporal evolution of the concentration — at the station $x = L_R = 1000$ [m] — are calculated separately for each mini-injection by giving different values to t_i. Subsequently they are added to obtain :

$$C(L_R,t) = \sum_{i=1}^{n} \Delta C_i(x,t) \tag{8.75}$$

This is shown in Fig. 8.E.2, labeled as scenario C; values for a few points are presented in the following table. It is to be noted that these curves, C(t), are slightly non symmetrical.

Discussion (see Fig. 8.E.2) :

In the scenarios A and C, the same total quantity of salt, $M_0 = 4$ [kg], is released at the same instant but at different stations.

- In the scenario A, with the single instantaneous release, the maximum concentration of $C_{max} = 2.71 \times 10^{-3}$ [kg/m³] arrives after $t_R = 1000$ [sec] at the station $L_R = 1000$ [m].

The temporal evolution of the concentration at $L_R = 1000$ [m]

$U = 1.0$ [m/s] $u_* = 0.07$ [m/s] $h = 0.7$ [m] $B = 6.0$ [m] $M_o = 4.0$ [kg] $K_x = 9.8$ [m²/s]

scenario A		scenario C			
$x = 1000$ [m]			C_1	C_2	$C = \Sigma C_i$
$\mathcal{M}_1 = 0.950$ [kg/m²]			$(\mathcal{M}_1)_1 = 0.476$ [kg/m²]	$(\mathcal{M}_1)_2 = 0.476$ [kg/m²]	$\Sigma (\mathcal{M}_1)_i = \mathcal{M}_1$
			$x_1 = 0.0$ [m]	$x_2 = 250.0$ [m]	
			$x - x_1 = 1000.0$ [m]	$x - x_2 = 750.0$ [m]	
t [s]	$C(x,t)$ [kg/m³]	t [s]	$\Delta C_1(x-x_1, t)$ [kg/m³]	$\Delta C_2(x-x_2, t)$ [kg/m³]	$\sum_{i=1}^{2} \Delta C_i(x,t)$ [kg/m³]
550	3.05E-07	550	1.53E-07	2.86E-04	2.86E-04
625	1.10E-05	625	5.52E-06	9.07E-04	9.13E-04
700	1.22E-04	700	6.10E-05	1.48E-03	1.54E-03
750	3.74E-04	750	1.87E-04	*1.57E-03*	1.75E-03
850	1.50E-03	850	7.49E-04	1.09E-03	*1.84E-03*
875	1.84E-03	875	9.20E-04	9.20E-04	*1.84E-03*
950	2.60E-03	950	1.30E-03	4.76E-04	1.78E-03
1000	*2.71E-03*	*1000*	*1.36E-03*	2.76E-04	1.63E-03
1075	2.29E-03	1075	1.15E-03	1.07E-04	1.25E-03
1150	1.54E-03	1150	7.68E-04	3.64E-05	8.05E-04
1225	8.54E-04	1225	4.27E-04	1.12E-05	4.38E-04
1300	4.07E-04	1300	2.04E-04	3.14E-06	2.07E-04
1375	1.70E-04	1375	8.52E-05	8.24E-07	8.60E-05
1450	6.39E-05	1450	3.20E-05	2.03E-07	3.22E-05
1525	2.19E-05	1525	1.09E-05	4.76E-08	1.10E-05

- In the scenario C, with 2 mini-injections, the maximum concentration of $C_{max} \approx 1.85 \times 10^{-3}$ [kg/m³] — which is less important than the one in the scenario A — arrives after $t \approx 860$ [s] — earlier than in the scenario A — at the station $L_R = 1000$ [m].

- For a low concentration, such as $C'' \approx 0.25 \times 10^{-3}$ [kg/m³], the time period during which this concentration is exceeded is :

  • for the scenario A : $\Delta T_A \approx 1345 - 730 = \mathit{615}$ [s]
  • for the scenario C : $\Delta T_C \approx 1285 - 515 = \mathit{770}$ [s]

It is evident that $\Delta T_C > \Delta T_A$.

iv) Determination of the time during which the salt concentration exceeds C' = 0.002 [kg/m³].

This concentration, C' = 0.002 [kg/m³], is only exceeded in the case of an instantaneous single release (scenario A), and this for a duration of $\Delta T \approx 220$ [s], as can be seen in Fig. 8.E.1.

v) Study of the concentration distribution at the time, $t_R = 1000$ [s].

Scenario A

The concentration distribution is given by eq. 8.71. With the values, t_R, $\mathcal{M}_1$, K_x, U, already calculated, one can write:

$$C(x, t_R) = \frac{0.95}{\sqrt{4\pi \times 9.8 \times 1000}} \exp\left[-\frac{(x - (1.0 \times 1000))^2}{4 \times 9.8 \times 1000}\right]$$

By substituting different values for x, one obtains the concentration distribution at the time $t_R = 1000$ [s]. This is shown in Fig. 8.E.3, as scenario A. It is to be noted that this time the curve, $C(x)$, is symmetrical.

Scenario B

The concentration distribution for the case of the four mini-injections is given by eq. 8.74. The four concentration distributions — for the values, t_R, m_i, S, K_x, U, already determined — are calculated separately by assigning different values to x and then are added as specified by eq. 8.75. This is shown in Fig. 8.E.3, labeled as scenario B. It is to be noted that all the curves, $C(x)$, are symmetrical.

Discussion (see Fig. 8.E.3)

In both of the scenarios, A and B, the same quantity of salt, $M_0 = 4$ [kg], is released at the same station, $x = 0.0$ [m], but in a different manner.

- In the scenario A, with a single and instantaneous release, the maximum concentration of $C_{max} = 2.71 \times 10^{-3}$ [kg/m³] arrives at the station $x = 1000$ [m] at the time $t_R = 1000$ [s]. This can also be seen in Fig. 8.E.1.

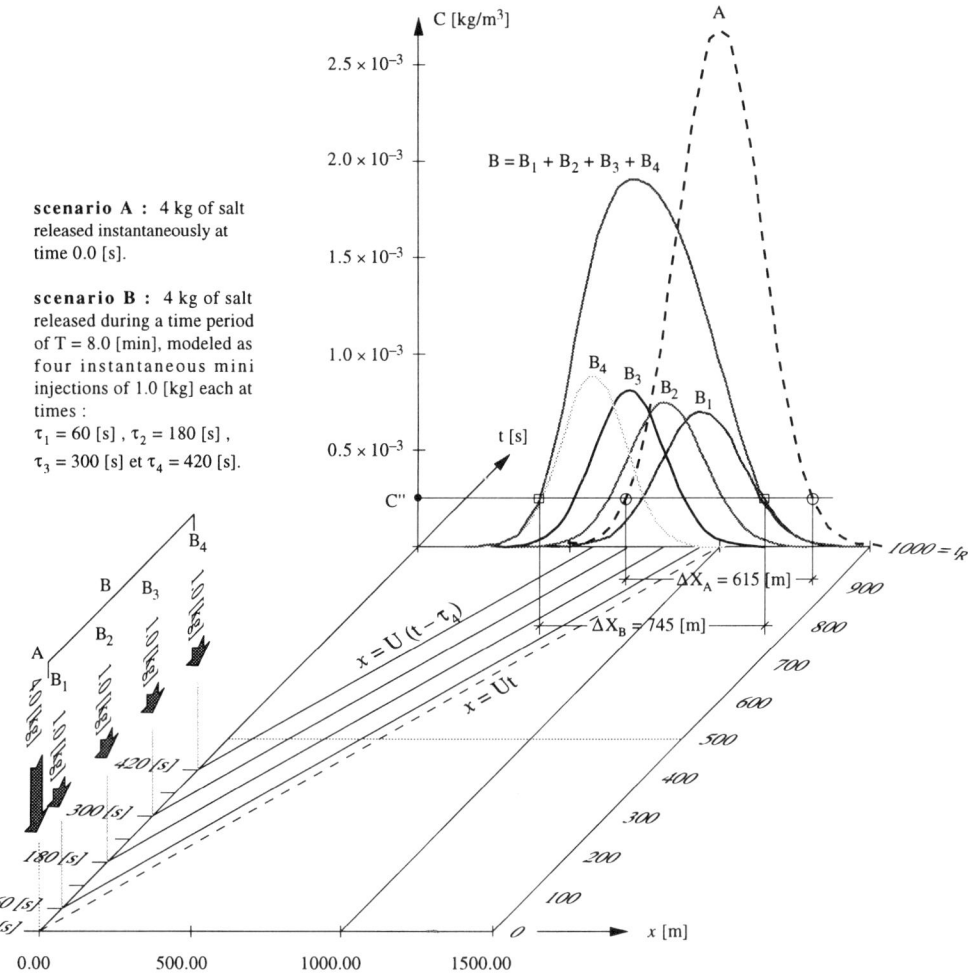

Fig. 8.E.3 Longitudinal concentration distribution for the scenarios A and B.

- In the scenario B, with 4 mini-injections during a time period of T = 8 [min], the maximum concentration of $C_{max} \approx 1.91 \times 10^{-3}$ [kg/m³] — being less important than that in the scenario A — is observed at $x \approx 710$ [m] — at a shorter distance than the one in the scenario A — at the same time $t_R = 1000$ [s]. The maximum values of the concentration distributions of the four mini-injection are decreasing, according to $1/\sqrt{4\pi K_x t}$; they arrive with a time lag of τ_i.

- For a low concentration, such as $C'' \approx 0.25 \times 10^{-3}$ [kg/m³], the time period during which it is exceeded is :

    • for the scenario A : $\Delta X_A = 1305 - 690 = 615$ [m]
    • for the scenario B : $\Delta X_B = 1145 - 400 = 745$ [m]

    It is evident that $\Delta X_B > \Delta X_A$.

598　　　　　　　　　　　　　　　　　　　　　　　　　　　　　　　　　　　　FLUVIAL HYDRAULICS

> Ex. 8.F
>
> The river *Arenne*, corrected long time ago, flows at an almost constant discharge of Q = 1200 [m³/s]. During an accident in a near-by chemical products factory, a quantity of 3600 [kg] of a chemically reactive substance spills into the sewer and this during a period of 12 [h]. This sewer discharges at the bottom of the center of the river. One would like to know when the cloud of this substance arrives at the stations 200 [km] and 330 [km] downstream of the accident.
>
> The available hydraulic parameters are the averaged values, being rather representative of the long downstream reach of the river : the water depth is h = 5.7 [m] and the river's cross section, which can be assumed to be rectangular, has a width of B = 300 [m].

SOLUTION :

i) This is a problem of *convection-diffusion* resulting from a continuous injection of a reactive substance during a finite time period. Three zones (see Fig. 8.8) of mixing should be distinguished :

- the *near-field zone of mixing*, where the vertical convection-diffusion takes place;

- the *mid-field zone of mixing*, where the transversal convection-diffusion takes place;

- the *far-field zone of mixing*, where the mixing is only due to the process of convection-dispersion.

It shall be assumed that the reactive nature of the substance can be ignored for the first two zones, $k_r = 0$, but it should definitely be taken into account, $k_r \neq 0$, in the dispersion zone.

ii) The available hydraulic data are the averaged values being representative of the entire reach downstream of the accident, such as :

    flow discharge :   Q = 1200 [m³/s]
    channel width :   B = 300 [m]
    water depth :   h = 5.7 [m]

The other hydraulic parameters can be calculated :

    flow velocity: $U = \dfrac{Q}{hB} \approx 0.7$ [m/s]

    bed-shear velocity : $u_* \approx 0.1\, U \approx 0.1$ [m/s]

    cross section of the channel : $S = hB \approx 1700$ [m²]

TRANSPORT AND MIXING OF MATTER

iii) The *substance*, $M_0 = 3600$ [kg], is brought into the river during a finite period of time, $T = 12$ [h].

It is proposed to model the injection of the total substance, $M_0 = 3600$ [kg], during a time period of $T = 12$ [h] by 12 instantaneous mini-injections, each of which being $m_i = 300$ [kg] or $(\mathcal{M}_1)_i = 0.5$ [kg/m²], namely:

$$
\begin{array}{llllll}
m_1 & = & 300 \text{ [kg]} & \text{at} & \tau_1 = 0.5 \text{ [h]} & \rightarrow \Delta C_1(x,t) \\
m_2 & = & 300 \text{ [kg]} & \text{at} & \tau_2 = 1.5 \text{ [h]} & \rightarrow \Delta C_2(x,t) \\
.... & & & & ... & ... \\
.... & & & & ... & ... \\
m_{12} & = & 300 \text{ [kg]} & \text{at} & \tau_{12} = 11.5 \text{ [h]} & \rightarrow \Delta C_{12}(x,t) \\
\hline
\sum_{i=1}^{12} m_i & = & 3600 \text{ [kg]} & \text{after } T = 12.0 \text{ [h]} & \rightarrow \sum_{i=1}^{12} \Delta C_i(x,t)
\end{array}
$$

iv) *Vertical diffusion*

The vertical diffusion (mixing) is completed over the entire flow depth, h, after a certain distance, L_z, or after a certain time, t_z, given by:

$$L_z = \xi_z U \frac{h^2}{\varepsilon_{tz}} \qquad\qquad t_z = \xi_z \frac{h^2}{\varepsilon_{tz}} \qquad\qquad (8.54)$$

The vertical diffusivity can be calculated with eq. 8.53 as follows:

$$\overline{\varepsilon_{tz}} = 0.067 \, (h \, u_*) = 0.067 \, (5.7 \times 0.1) \approx 0.04 \, [\text{m}^2/\text{s}]$$

One takes $\xi_z = 0.4$ for the injection of the substance at the bottom of the river (see Fig. 8.B.1).

By using these values one can write:

$$L_z \approx 6 U \frac{h}{u_*} \qquad\qquad t_z \approx 6 \frac{h}{u_*} \qquad\qquad (8.54a)$$

and obtains now:

$$L_z \approx 6 U \frac{h}{u_*} = 6 \times 0.7 \times (5.7/0.1) = 227 \text{ [m]} \approx 0.23 \text{ [km]}$$

$$t_z \approx 6 \frac{h}{u_*} = 6 \times (5.7/0.1) = 342 \text{ [s]} \approx 5.7 \text{ [min]}$$

These are the distance, L_z, and the time, t_z, after which the vertical diffusion is completed.

v) *Transversal diffusion*

The transversal diffusion (mixing) is completed over the entire river width, B, after a certain distance, L_y, or after a certain time, t_y, given by :

$$L_y = \xi_y U \frac{B^2}{\overline{\varepsilon_{ty}}} \qquad t_y = \xi_y \frac{B^2}{\overline{\varepsilon_{ty}}} \qquad (8.56)$$

The transversal diffusivity can be calculated with eq. 8.55a as follows :

$$\overline{\varepsilon_{ty}} = 0.6\,(h\,u_*) = 0.6\,(5.7 \times 0.1) \approx 0.34\ [m^2/s]$$

One takes $\xi_y = 0.1$ for the injection of the substance at the bottom of the river (see Fig. 8.9).

By using these values one can write :

$$L_y \approx 0.17\, U \frac{B^2}{hu_*} \qquad t_y \approx 0.17 \frac{B^2}{hu_*} \qquad (8.56a)$$

and obtains now :

$$L_y \approx 0.17\, U \frac{B^2}{hu_*} = 0.17 \times 0.7 \times (300^2/5.7 \times 0.1) \approx 18789\ [m] \approx 20\ [km]$$

$$t_y \approx 0.17 \frac{B^2}{hu_*} = 0.17 \times (300^2/5.7 \times 0.1) \approx 26842\ [s] \approx 7.5\ [h]$$

These are the distance, L_y, and the time, t_y, after which the transversal diffusion is completed.

vi) *Longitudinal dispersion*

The longitudinal dispersion starts to take place once the vertical and transversal diffusion has distributed the substance over the entire cross section of the river, S (see Fig. 8.F.1).

The equation of convection-dispersion with reaction is given by :

$$\frac{\partial C}{\partial t} + U \frac{\partial C}{\partial x} = K_x \frac{\partial^2 C}{\partial x^2} - (k_r C) \qquad (8.77)$$

For a mass, M_0, arriving uniformly distributed over the cross section, S, during a finite time period, T, each mini-injection, m_i, is going to create its own cloud. The solution to eq. 8.77, given by eq. 8.78, is written as follows for a single mini-injection :

TRANSPORT AND MIXING OF MATTER 601

$$\Delta C_i(x,t) = \frac{m_i}{S\sqrt{4\pi K_x(t-\tau_i)}} \exp\left[-\frac{[x-U(t-\tau_i)]^2}{4 K_x(t-\tau_i)} - k_r(t-\tau_i)\right]$$

Finally, the total concentration is the sum of the concentrations resulting from each mini-injection :

$$C(x,t) = \sum_{i=1}^{n} \Delta C_i(x,t) \tag{8.75}$$

The dispersivity of the flow in such a river can be estimated (see eq. 8.68) as being :

$$\frac{K_x}{h u_*} \cong 435$$

from which one obtains :

$$K_x = 435 (h u_*) = 435 (5.7 \times 0.1) = 248 \ [m^2/s]$$

The reaction coefficient is taken (see eq. 8.81) as being :

$$k_r = 3.0 \times 10^{-6} \ [s^{-1}]$$

Above relationship, eq. 8.75, can be used to determine the time evolution of the concentration at $L_1 = 200$ [km] and $L_2 = 330$ [km]. It is important to note that the dispersion starts only after $L_y \approx 20$ [km].

The time histories of the concentration are calculated for two stations :

- $L_1^* = L_1 - L_y = 200 - 20 = 180$ [km]

The relationship for a single mini-injection is written as :

$$\Delta C_i(L_1^*, t) = \frac{300}{1700\sqrt{4\pi \times 248 \times (t-\tau_i)}} \times$$

$$\times \exp\left[-\frac{[180.000 - 0.7(t-\tau_i)]^2}{4 \times 248 \times (t-\tau_i)} - 3.0 \times 10^{-6} \times (t-\tau_i)\right]$$

By substituting different values for $(t - \tau_i)$, the concentration distribution for a single mini-injection is obtained.

The 12 concentration profiles — for the values, L_1^*, m_i, S, K_x, k_r, U, already determined — can be calculated by giving different values to τ_i. Subsequently they are added, according to the eq. 8.75. This is shown in Fig. 8.F.1.

- $L_2^* = L_2 - L_y = 330 - 20 = 310$ [km]

The calculations for L_1^* described above are repeated for L_2^*.

Discussion (see Fig. 8.F.1)

- At the station $L_1 = 200$ [km], the maximum concentration, $C_{max} = 2.65 \times 10^{-5}$ [kg/m³], arrives after a time of $t_1 = t_1^* + t_y = 3.52$ [d].

At the station $L_2 = 330$ [km], the maximum concentration, $C_{max} = 1.29 \times 10^{-5}$ [kg/m³], arrives after a time of $t_2 = t_2^* + t_y = 5.66$ [d].

The maximum concentration values are decreasing with time.

- For a low concentration, for example $C'' = 0.5 \times 10^{-5}$ [kg/m³], the time during which this concentration is exceeded is :

  • at $L_1 = 200$ [km] $\Delta T_1 = 315200 - 239600 = 75600$ [s]
  • at $L_2 = 330$ [km] $\Delta T_2 = 497000 - 428600 = 68400$ [s]

It can be seen that $\Delta T_1 > \Delta T_2$.

- The concentration distribution, $C(x^*)$, after a time of $t_1 = 3.52$ [d] and of $t_2 = 5.66$ [d] are also shown in Fig. 8.F.1.

TRANSPORT AND MIXING OF MATTER 603

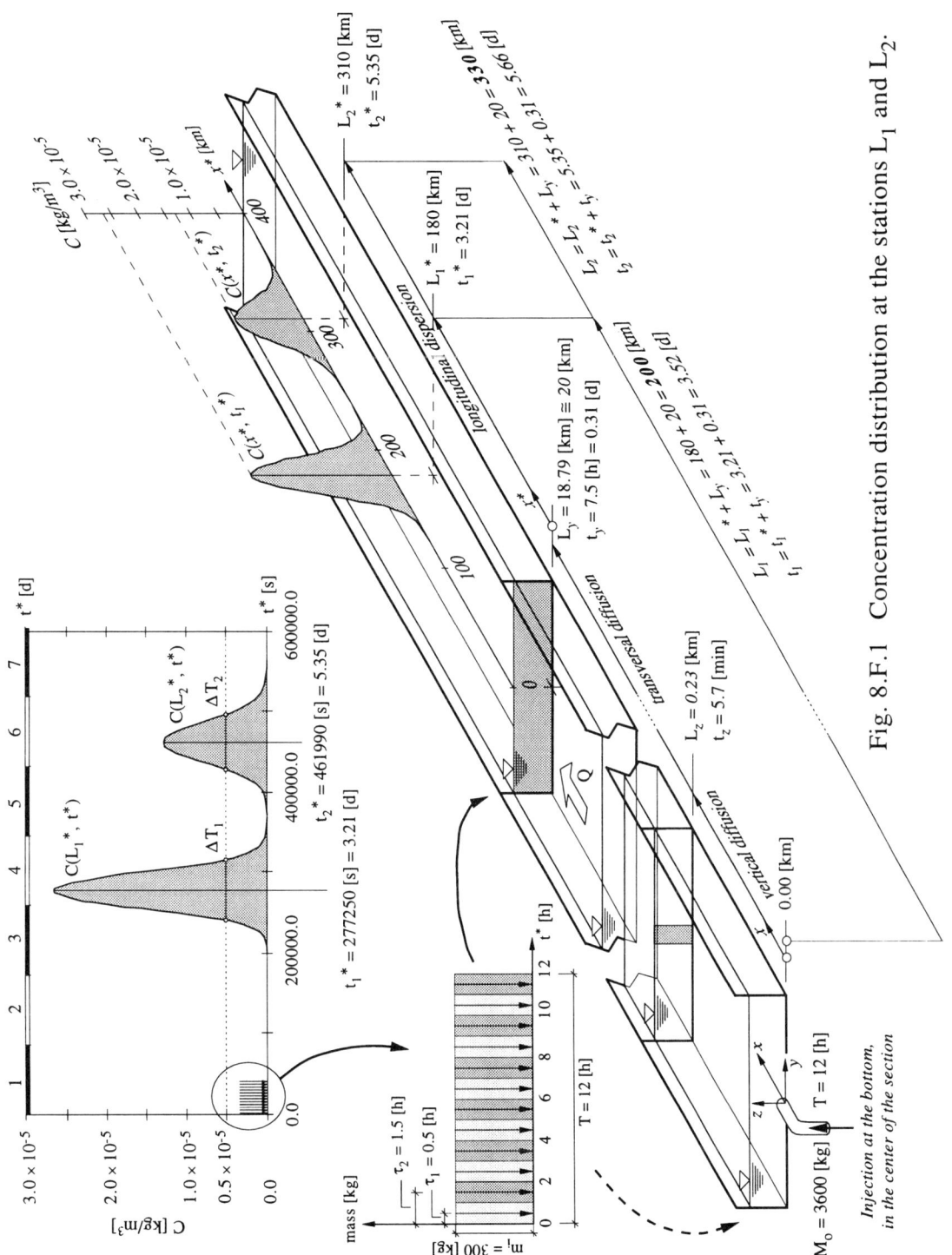

Fig. 8.F.1 Concentration distribution at the stations L_1 and L_2.

Ex. 8.G

A discharge of Q = 8500 [l/s] is flowing in a channel with an average velocity of U = 1 [m/s]. The water is 95 % saturated in dissolved oxygen; the temperature is T = 10 [°C] and the biochemical oxygen demand is (BOD$_5$) = 1 [mg/l].

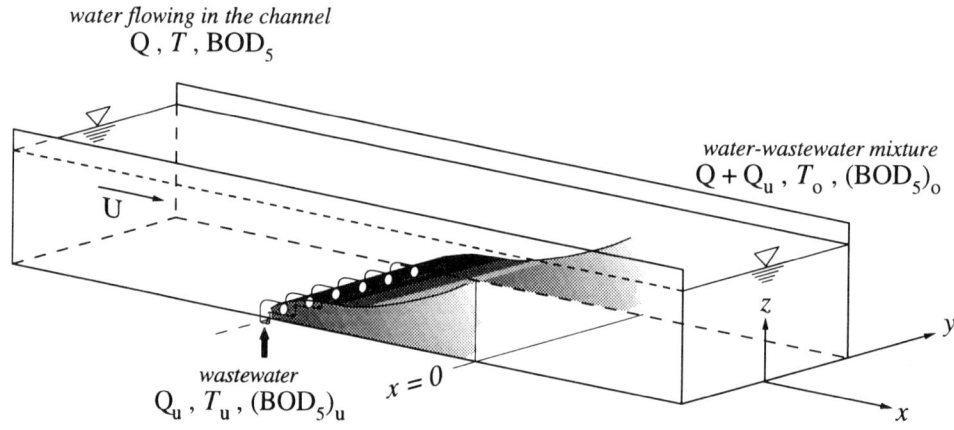

A water-treatment plant discharges wastewater into this channel at a rate of Q_u = 1500 [l/s]; the temperature is T_u = 15 [°C] and the biochemical oxygen demand is $(BOD_5)_u$ = 200 [mg/l]. The wastewater spreads over the entire section almost instantaneously. The coefficients of reaction, k_r = 0.2 [d^{-1}], and of reaeration, k_a = 0.5 [d^{-1}], were determined previously.

i) Determine the maximum oxygen deficit and its arrival time.

ii) Estimate also the value of BOD$_5$ for a sample taken at this location.

SOLUTION :

To determine the dissolved-oxygen deficit for the mixture of water-wastewater, it shall be admitted that the flow in the channel has a low velocity and that dispersion-diffusion can be neglected. The dissolved-oxygen budget, written in terms of the oxygen deficit, was established by eq. 8.95; the solution given by :

$$D_{(OD)} = \frac{k_r}{k_a - k_r} C_{o(BOD)} [e^{-k_r(x/U)} - e^{-k_a(x/U)}] + D_{o(OD)} [e^{-k_a(x/U)}] \quad (8.96)$$

is shown in Fig. Ex. 8.G.1. The different parameters, appearing in eq. 8.96, must be determined.

TRANSPORT AND MIXING OF MATTER 605

i) Determination of the oxygen concentration in the channel just upstream of the point of injection of the wastewater :

– The saturation concentration is calculated with:

$$C_{s(OD)} \approx 14.65 - 0.41\,(T) + 0.008\,(T^2) \tag{8.93}$$

for $T = 10\,[°C]$ it is : $C_{s(OD)} = 11.35\,[mg/l]$

– The concentration of dissolved oxygen is therefore :

$$C_{(OD)} = 0.95\,C_{s(OD)} = 0.95\,(11.35) = 10.78\,[mg/l]$$

ii) Determination of the temperature, T_o, of the dissolved oxygen, $C_{o(OD)}$, and of the biochemical oxygen demand, $(BOD_5)_o$, of the mixture just downstream of the injection point of the wastewater at $x = x_0 = 0$:

$$T_o = \frac{TQ + T_u Q_u}{Q + Q_u} = \frac{10 \times 8500 + 15 \times 1500}{8500 + 1500} = 10.75\,[°C]$$

$$C_{o(OD)} = \frac{C_{(OD)}Q + C_{u(OD)}Q_u}{Q + Q_u} = \frac{10.78 \times 8500 + 0 \times 1500}{8500 + 1500} = 9.17\,[mg/l]$$

$$(BOD_5)_o = \frac{(BOD_5)Q + (BOD_5)_u Q_u}{Q + Q_u} = \frac{1 \times 8500 + 200 \times 1500}{8500 + 1500} = 30.85\,[mg/l]$$

iii) The initial (BOD_5)-concentration, $C_{o(BOD)}$, of the mixture at $x_0 = 0$ can be calculated using :

$$(BOD_5)_o \equiv C_{5(o)}{}^*{}_{(BOD)} = C_{o(BOD)}\,[1 - e^{-k_r(5)}] \tag{8.90}$$

from which one obtains :

$$C_{o(BOD)} = \frac{(BOD_5)_o}{1 - e^{-k_r(5)}} = \frac{30.85}{1 - e^{-0.2 \times 5}} = 48.80\,[mg/l]$$

This value is quite reasonable for a treated wastewater, as can be seen in Table 8.3.

The initial dissolved-oxygen deficit, $D_{o(OD)}$, is given by :

$$D_{o(OD)} = (C_{s(OD)} - C_{o(OD)}) = 11.35 - 9.17 = 2.18\,[mg/l]$$

iv) The (critical) distance, x_K, at which the maximum deficit of dissolved oxygen occurs, can be calculated using :

$$x_K = \frac{U}{k_a - k_r} \ln \frac{k_a}{k_r} [1 - \frac{D_{o(OD)} (k_a - k_r)}{k_r C_{o(BOD)}}] \tag{8.98}$$

Since the arrival time of the maximum dissolved oxygen deficit, t_K, is :

$$t_K = \frac{x_K}{U} = \frac{1}{0.5 - 0.2} \ln \frac{0.5}{0.2} \left[1 - \frac{2.18 (0.5 - 0.2)}{0.2 \times 48.80} \right] = 2.82 \text{ [d]}$$

the distance can now be calculated as being :

$$x_K = U \cdot t_K = 1 \times (2.82 \times 24 \times 3600) = 243.65 \text{ [km]} .$$

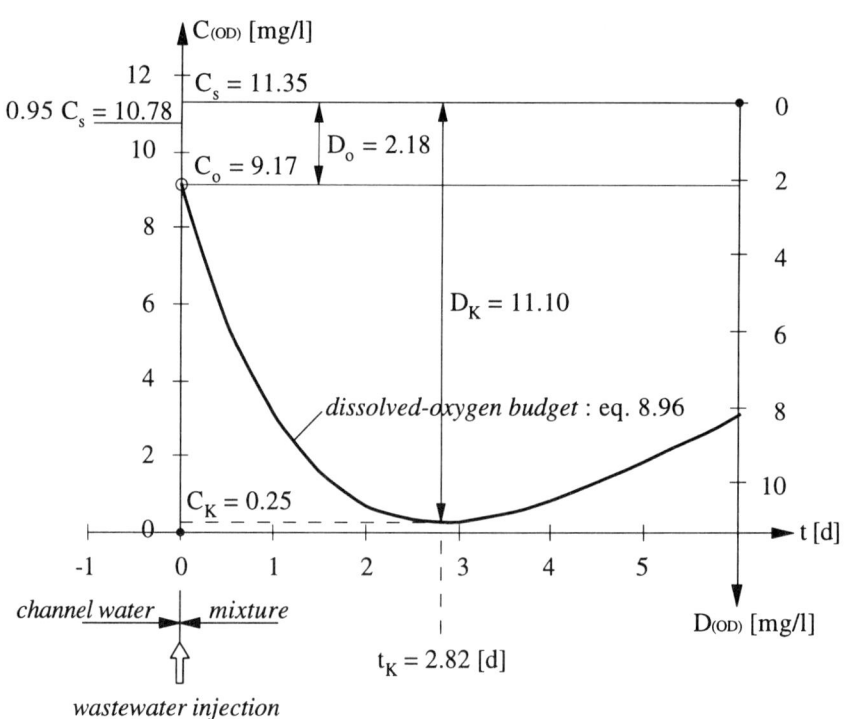

Fig. 8.G.1 Time-history of the concentration and the deficit of dissolved oxygen, $C_{(OD)}$ et $D_{(OD)}$.

v) The maximum (critical) dissolved-oxygen deficit is calculated with :

$$D_{K(OD)} = \frac{k_r}{k_a} C_{(BOD)} = \frac{k_r}{k_a} C_{o(BOD)} [e^{-k_r(x/U)}] \tag{8.97}$$

thus :

$$D_{K(OD)} = \frac{0.2}{0.5} \times 48.80 \times e^{-0.2(2.82)} = \mathit{11.10} \text{ [mg/l]}$$

The dissolved-oxygen concentration can now be calculated as:

$$C_{K(OD)} = C_{s(OD)} - D_{K(OD)} = 11.35 - 11.10 = \mathit{0.25} \text{ [mg/l]}$$

which is a very low value; according to Swiss regulations, the dissolved-oxygen content should not be less than 6 [mg/l].

All these values are indicated in Fig. 8.G.1.

vi) Determination of the value of the (BOD_5) at this critical distance, x_K :

First, the unsatisfied BOD concentration is calculated using eq. 8.88 :

$$C_{(BOD)}(x_K) = C_{o(BOD)} [e^{-k_r(x/U)}] = 48.80 \,[e^{-0.2(2.82)}] = 27.76 \text{ [mg/l]}$$

subsequently, using eq. 8.90, one obtains :

$$(BOD_5)_{x_K} = C_{(BOD)} [1 - e^{-k_r(5)}] = 27.76 \,[1 - e^{-1}] = \mathit{17.55} \text{ [mg/l]}$$

8.5.2 Problems, unsolved

Ex. 8.1
Repeat Ex. 8.A for the following scenario :
Long after a strong storm, the average velocity in the upper layers has become non-existent, U = 0, but the turbulence has not yet died out.

Ex. 8.2
Repeat Ex. 8.A for the following scenario :
The average velocity of the flow is laminar, U = 0.01 [m/s].

Ex. 8.3
Consider the same scenario and the same question as in Ex. 8.A :
This time, the temperature of the water in the channel is $T_o = 15$ [C°], while the one of the water introduced is $T_R = 20$ [C°].

Ex. 8.4
The relationship given with eq. 8.33 is a solution to the equation of pure diffusion. Show mathematically that this is so (see *Bird* et al., 1960, p. 353).

Ex. 8.5
Repeat Ex. 8.B for a scenario, where the channel is designed for a discharge of Q = 144.00 [m³/s], which remains constant.

Ex. 8.6
Repeat Ex. 8.B for a scenario, where the system of multiport diffusers is installed at mid-depth, h/2 = 0.55 [m].

Ex. 8.7
Repeat Ex. 8.B for a scenario, where the system of multiport diffusers is close to the water surface of the channel.

Ex. 8.8
Take the scenario of Ex. 8.B. Determine the zone within which the content of salt will remain above $C_s = 5$ [kg/m³].

Ex. 8.9
Consider the scenario of Ex. 8.C. The water temperature in the channel is $T = 10$ [°C] and the one of the wastewater is $T_u = 25$ [°C]. Determine the temperature at the water surface for the stations L_1 and L_2. Subsequently delimit the zone, where the temperature remains above $T = 15$ [°C] and below $T = 16$ [°C].

TRANSPORT AND MIXING OF MATTER 609

Ex. 8.10
Estimate the dispersivity, K_x, for the different discharges of the channel investigated in Ex. 3.B. Compare these results with the ones of the four rivers studied by *Rutherford* (1966, p. 199 et p. 280).

Ex. 8.11
Repeat Ex. 8.D, using a coefficient of dispersion which was found experimentally as being $K_x = 0.05$ [m²/s].

Ex. 8.12
Consider the scenario of Ex. 8.D :
Now the salt solution has a density of $\rho_m = 1015$ [kg/m³] at $T_m = 12$ [°C].

Ex. 8.13
Repeat Ex. 8.E for the case, where a large quantity of salt, $M_o = 8$ [kg], is released.

Ex. 8.14
Repeat Ex. 8.E for scenarios A and B ; study the temporal evolution of the concentration at a distance, $L_R = 2$ [km], from the salt injection.

Ex. 8.15
The river Arenne has at low water a velocity of $U = 0.6$ [m/s] at a flow depth of $h = 4.5$ [m]. Repeat the study of Ex. 8.F.

Ex. 8.16
Repeat Ex. 8.F, but consider now that the substance is a *passive* one.

Ex. 8.17
Repeat Ex. 8.G, where the wastewater was treated, and the reaction coefficient was taken as $k_r = 0.2$. Now the wastewater remains untreated ; the reaction coefficient is to be taken as $k_r = 0.40$; and the $BOD_5 = 600$ [mg/l].

Ex. 8.18
Repeat Ex. 8.G , for the case where the wastewater discharge is $Q_u = 1000$ [l/s].

9. LOCAL SCOUR

Flow in a channel with a mobile bed is usually accompanied by a transport of sediments; erosion and deposition might be the consequence. Additional erosion of sediments will be caused, where there is a local change in the geometry of the channel or in the flow.

Different kinds of local scour are encountered in fluvial hydraulics. Positioned in the flow, a pier or an abutment will locally alter the flow and cause erosion (deposition) in the vicinity of the obstacle.

Constriction scour is encountered, if the width of a channel is reduced. Flow over or/and under an hydraulic structure has a considerable potential to cause scour at the downstream.

TABLE OF CONTENTS

9.1 GENERAL REMARKS

9.2 PIER SCOUR
    9.2.1 Scour Process
    9.2.2 Flow Pattern
    9.2.3 Functional Relations
    9.2.4 Formulae for Design
    9.2.5 Scour Prevention

9.3 ABUTMENT SCOUR
    9.3.1 Flow Pattern and Scouring
    9.3.2 Functional Relations
    9.3.3 Formulae for Design

9.4 CONSTRICTION SCOUR
    9.4.1 Hydraulic Considerations
    9.4.2 Scour-depth Relations
    9.4.3 Formula for Design

9.5 HYDRAULIC STRUCTURES SCOUR
    9.5.1 Notions
    9.5.2 Flow over a Structure
    9.5.3 Flow under a Structure

9.6 EXERCISES
    9.6.1 Problems, solved
    9.6.2 Problems, unsolved

LOCAL SCOUR

9.1 GENERAL REMARKS

1° Flow in an open channel with a mobile bed is usually accompanied by a transport of sediments (see chap. 6). The latter is a result of the interplay between erosion and deposition of the transported sediments.

2° Any local change in the geometry of the channel or in the flow will produce additional erosion (deposition) of the sediments ; this is called *local scour* (germ. : *Kolk*). One encounters different types of local scour :

i) Pier Scour (see sect. 9.2) and

ii) Abutment Scour (see sect. 9.3) :

A pier or an abutment, positioned in the flow, will *locally* alter the flow in the channel, causing erosion (scour) and possibly deposition in the vicinity of the pier or the abutment.

iii) Constriction Scour (see sect. 9.4) :

A change in the geometry of the channel, like a constriction, will alter the flow in the channel, thus its capacity of transporting sediments ; scour might be the consequence.

iv) Hydraulic Structures Scour (see sect. 9.5) :

Since most hydraulic structures perturb at least locally the flow, erosion (scour) or possibly deposition in the vicinity of the structure might be encountered.

Local scour occurs when the capacity of the flow to erode and to transport the sediments is larger than the capacity to supply (replace) the sediments.

3° Knowledge about the topography of the erosion-deposition of the channel bed is of paramount importance for the hydraulic engineer. Continued scour could undermine the foundation of the structure. Notably, excessive discharges (floods) are liable to produce also severe local scour. Protection (structural) against erosion-deposition might become necessary (see sect. 9.2.5).

4° In the presence of an obstacle (pier, abutment, constriction, etc.) the unidirectional flow in the channel gets to be a three-dimensional one. Hydraulic investigations become rather complex, numerical modelling may be necessary. For such a reason, it became customary to communicate the results of studies in laboratories and in the field dealing with scour, with *dimensionless parameters*.

To decide which of the dimensionless parameters are of importance in scour studies a dimensional analysis must be performed.

5° Two types of scour should be distinguished :
- *clear-water scour* : when the sediments are removed from the scour hole and are not replaced ;
- *sediment-transport (live-bed) scour* : when the scour hole is continuously supplied with sediments from the sediment transport in the channel.

6° Despite the volume of investigations available — most of them come from laboratory studies and only a few from field tests — the hydraulics of local scour is as yet not well established. Consequently the formulae to be presented herein will only offer a guideline for the engineer.

9.2 PIER SCOUR

1° Local scour around a pier is a complex phenomenon resulting from the strong interaction of the three-dimensional turbulent flow field around the pier and the erodible sediment bed.

2° The presence of a (isolated) pier changes the flow pattern, which in turn modifies the surface of the mobile bed (see Fig. 9.1). In the vicinity of the pier, scouring is the consequence. The intensity of scour, thus the scour depth, d_s, will depend on the channel flow, the sediments of the bed and the geometry and alignment of the pier.

9.2.1 Scour process

1° The evolution of the scour depth, d_s, during an increase of the flow velocity, U, — evidenced by *Chabert* et *Engeldinger* in 1956 (see *Shen*, 1971, p. 23.3) — is shown in Fig. 9.2.

Upon reaching a certain flow velocity in the channel, the sediment particles close to the pier begin to move ; scour is initiated. The eroded particles will follow the flow pattern and are carried from the front of the pier towards the downstream. Upon an increase in the flow velocity, more and more particles will get dislodged, forming a scour hole increasing in size and depth. Eventually a maximum scour depth, $(d_s)_{max}$, is attained, which corresponds to a flow velocity being close to the critical velocity, $U \equiv U_{cr}$, for initiation of sediment transport in the channel. For non-uniform sediments, the larger sizes are less likely to be eroded and an armouring layer forms itself in the scour hole.

LOCAL SCOUR 615

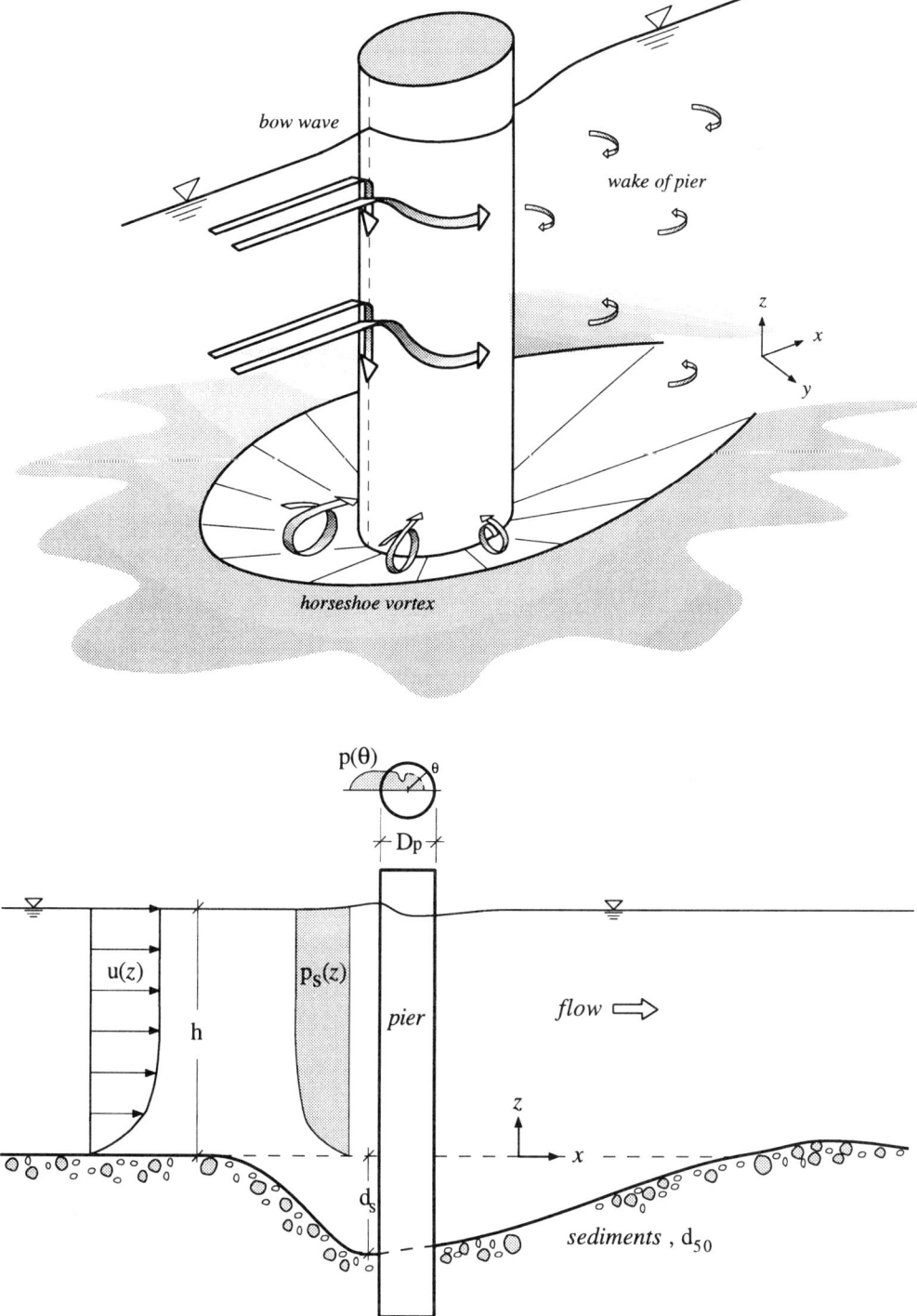

Fig. 9.1 Schema of flow pattern and local scour around a cylindrical pier.

A subsequent further increase in the flow velocity, $U > U_{cr}$, is responsible for a transport of sediments *in* and *out* of the scour hole, but the scour depth remains essentially constant. Thus an average *equilibrium scour depth*, d_s, establishes itself, being slightly smaller than the maximum scour depth, $(d_s)_{max}$.

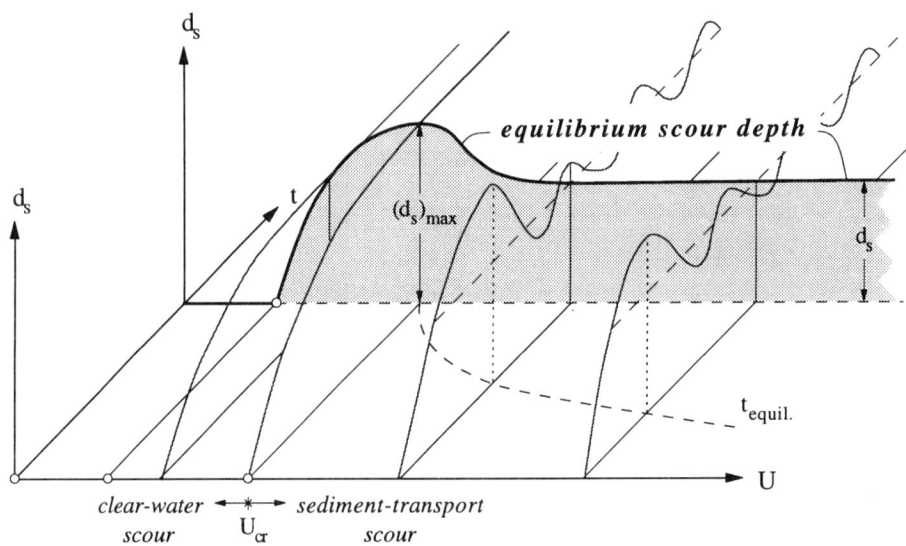

Fig. 9.2   Scheme of evolution of the scour depth, d_s, with respect to flow velocity, U, and time, t.

2°   The equilibrium scour depth, d_s, will usually be reached during high-water flow of a long duration.

In steady flow conditions, the time necessary to attain this equilibrium depth depends on whether the flow is able to transport sediments, $(U/U_{cr}) > 1$, or not, $(U/U_{cr}) < 1$; thus one distinguishes (see Fig. 9.2) :

- *clear-water scour* : the scour depth increases gradually and approaches an asymptotic value, when the capacity of transport *out* of the scour hole is zero (Laboratory experiments have to run continuously for several days before equilibrium conditions are reached).
- *sediment-transport scour* : the scour depth increases rapidly and attains an equilibrium value, when the capacity of sediment transport *out* of the scour hole is equal to the one *into* the scour hole.

LOCAL SCOUR   617

3° Scour is initiated at, or close to, the nose of the pier (Actually, two small lateral scour holes begin to form on the sides of the pier. They rapidly work their way around the pier to meet at the nose of the pier). The scour hole grows in depth and in volume by forming a groove. The upstream portion of the scour hole has the approximate shape of an inverted cone-like surface, stretching around the pier with side slopes about equal to the angle of repose of the sediments. Eroded material is transported — often in form of bursts — towards the rear of the pier, where it may or not be deposited.

The maximum scour depth for a *cylindrical* pier is in general located in front of the pier (see Fig. 9.1) ; for a *rectangular* pier, it is found to occur at the nose of the pier, while for a *streamlined* pier at the sides of the pier.

9.2.2 Flow Pattern

1° Unidirectional flow in a erodible channel which encounters a protruding obstacle, like a pier, becomes three-dimensional. The resulting flow pattern around the pier becomes complex and difficult to assess hydrodynamically.

Results of various studies (see *Shen*, 1971, p. 23/4, *Raudkivi*, 1991, p. 63, and *Graf* et *Yulistiyanto*, 1996) have shown that different components of the flow pattern might play a role in local scouring (see Fig. 9.1).

2° Flow in a channel, being a boundary-layer flow, $u(z)$, approaches the pier and a stagnation pressure, $p_s(z)$, decreasing with depth, establishes itself. This will produce a (weak) pressure gradient along the front of the pier and induce a *downward flow*, namely from high to low velocities. Since there is also a (strong) pressure gradient around the pier, $p(\theta)$, the downstream flow will be laterally diverted. However it is generally agreed upon, that the vertical component of the flow is responsible for the initiation of the scour.

Due to the stagnation pressure in front of the pier (cylinder), the water surface increases, forming a *bow wave* (roller).

If this pressure increase becomes sufficiently strong, the three-dimensional boundary layer undergoes a separation. A *horseshoe-vortex system* forms itself at the base of the pier. This vortex stretches into the downstream direction, diminishing its strength, and is very active in the local scour process.

A trailing *wake-vortex system* is formed in the rear of the pier over the entire flow depth. There the turbulence intensity is increased and consequently erosion and transport of sediments is enhanced.

9.2.3 Functional Relations

1° Considering an isolated single pier in a wide and rectangular open channel, where the flow in unidirectional, uniform and steady, and whose mobile bed is made up of cohesionless sediments, the equilibrium *scour depth*, d_s, depends on :

- the *fluid* : the density, ρ, and viscosity, ν ;
- the *sediments* : the density, ρ_s, and a characteristic diameter, d ;
- the channel *flow* : the flow depth, h, the channel slope, S_f, and the gravity, g ; thus the average velocity, $U = C\sqrt{hS_f}$, or the friction velocity, $u_* = \sqrt{ghS_f}$;
- the geometry of *pier* : a characteristic dimension, D_c ; for a cylindrical pier the diameter of the pier is used, $D_c \equiv D_p$.

This allows to write :

$$d_s = f_1(\rho, \nu\,;\,\rho_s, d\,;\,h, U \text{ or } u_*, g\,;\,D_p) \tag{9.1}$$

2° Using dimensional reasoning the following dimensionless parameters (see *Breusers et al.*, 1977, p. 219) are obtained :

$$\frac{d_s}{D_p} = f_2\left(\frac{u_* d}{\nu},\,\frac{\rho u_*^2}{g(\rho_s - \rho)d},\,\frac{\rho_s}{\rho},\,\frac{h}{D_p},\,\frac{d}{D_p}\right) \tag{9.2}$$

The particle Reynolds number, Re_*, and the dimensionless shear stress, τ_*, are related according to a relation developed by *Shields* (see sect. 3.4.2) such as :

$$\frac{\rho u_*^2}{g(\rho_s - \rho)d} = \tau_* = f_3(Re_*) = f_3\left(\frac{u_* d}{\nu}\right) \tag{9.3}$$

This relation can also be used in expressing the critical condition, when initiation of sediment transport takes place, or :

$$\tau_{*cr} = f_4(Re_*) \tag{9.3a}$$

and consequently

$$\frac{\tau_*}{\tau_{*cr}} = f_5(Re_*) \tag{9.3b}$$

LOCAL SCOUR 619

Considering now the proportionalities (see eq. 9.3) of :

$$\tau_* \propto u_* \propto U$$

and realising that the relative density, ρ_s/ρ, is already used in the definition of τ_* (see eq. 9.3), the above relation, eq. 9.2, becomes :

$$\frac{d_s}{D_p} = f_6\left(\frac{\tau_*}{\tau_{*cr}} \ or \ \frac{u_*}{u_{*cr}} \ or \ \frac{U}{U_{cr}}, \frac{h}{D_p}, \frac{d}{D_p}\right) \qquad (9.4)$$

where the subscript, $_{cr}$, indicates the critical condition, implying the beginning of the sediment transport in the channel (see sect. 3.4.2).

3° This relation, eq. 9.4, can still be generalised to include some dimensionless correction factors such as :

- a coefficient of sediment grading : ξ_g
- a coefficient for pier shape : ξ_s
- a coefficient of angle of approach : ξ_α

A more general functional form for the scour depth reads now :

$$\frac{d_s}{D_p} = f_7\left(\frac{U}{U_{cr}}, \frac{h}{D_p}, \frac{d}{D_p}; \xi_g, \xi_s, \xi_\alpha\right) \qquad (9.5)$$

where D_p is the width (diameter) of the pier.

4° Local scour, parameterised by the scour depth, d_s, is seen to be a complex phenomenon, where the various parameters interact.

The influence of each of the above parameters (see Fig. 9.3) will be examined next by using experimental data from various investigations.

5° Influence of flow velocity

   i)   The influence of the relative flow velocity, U/U_{cr}, on the dimensionless scour depth, d_s/D_p, is shown in Fig. 9.3a (see *Raudkivi*, 1991, p. 76) for a limited series of experiments. The overall tendency for other sediments, d, and different flow depths, h, would be the same.

ii) In Fig. 9.3a, the following zones are to be distinguished :

- $(U/U_{cr}) < 0.5$: there is no local erosion (scour) and no sediment transport in the channel ;

- $0.5 < (U/U_{cr}) < 1.0$: there is active local erosion (scour), but no sediment transport in the channel ; one talks of *clear-water scour* (see Fig. 9.2) ;

- $(U/U_{cr}) > 1.0$: there is almost no net local erosion (scour), but sediment transport in the channel takes place ; one talks of *sediment-transport scour* (see Fig. 9.2) ; (local erosion and deposition are in equilibrium due to the sediment transport in the channel) ; the scour depth might fluctuate in the presence of moving bed forms.

iii) For clear-water scour — $(U/U_{cr}) < 1$ — where the flow velocity is of importance, an approximate relation proposed by *Shen* (1971, p. 23/9) — relating the strength of the horseshoe vortex to the pier Reynolds number — reads :

$$d_s = 0.00022 \left(\frac{UD_p}{\nu}\right)^{0.619} \tag{9.6}$$

For sediment-transport scour — taking $(U/U_{cr}) > 2$ — where the flow velocity is no more of great importance, it is generally admitted that :

$$\frac{d_s}{D_p} \approx 2.0 \text{ to } 2.3 \tag{9.7}$$

remains constant. This has been called the average equilibrium scour depth.

The above relations have been tested for subcritical flow, mainly in laboratory channels. For large Froude numbers, $Fr > 0.8$, the value of eq. 9.7 should be taken as 3.0 (see *Johnson*, 1995). For small Froude numbers, $Fr < 0.2$, this value could be taken as being less than 2.0.

6° Influence of flow depth

i) The influence of the relative flow depth, h/D_p, on the dimensionless scour depth, d_s/D_p, is shown in Fig. 9.3b (see *Breusers et al.*, 1977, p. 242).

LOCAL SCOUR   621

ii) As can be seen in Fig. 9.3b, the various investigations present often conflicting conclusions. Here is a selection of some relations (see *Breusers et al.*, 1977, p. 241) presently in use :

- a relation proposed by *da Cunha* (1970) :

$$\frac{d_s}{D_p} = 1.35 \left(\frac{h}{D_p}\right)^{0.3} \tag{9.8a}$$

- a relation proposed by *Breusers* (1965) :

$$\frac{d_s}{D_p} = 1.4 \tag{9.8b}$$

valid for $h/D_p > 1.3$;

- a relation proposed by *Breusers et al.* (1977) :

$$\frac{d_s}{D_p} = 1.5 \tanh\left(\frac{h}{D_p}\right) \tag{9.8c}$$

whose constant be taken as 2.0 instead of 1.5, in order to be on the safe side.

iii) For shallow flow, when $(h/D_p) < 1.5$, the relative scour depth increases with the flow depth. It is however evident that for $(h/D_p) > 1.5$, the influence of the flow depth has little effect on the relative scour depth, d_s/D_p.

The experimental evidence, compared with the above relations, eqs. 9.8, is indeed not conclusive, as seen in Fig. 9.3b.

7° Influence of sediments

i) The influence of a sediment of uniform diameter, expressed by the relative diameter, D_p/d, on the dimensionless scour depth, d_s/D_p, is shown in Fig. 9.3c (see *Raudkivi*, 1991, p. 69).

ii) As can be seen in Fig. 9.3c, the limited data arrange themselves rather well, notably if $d > 0.7$ [mm]. No relationship was proposed. However it is evident that for $D_p/d > 25$ — when the particle diameter, d, is small compared with the pier diameter, D_p — the influence of the relative diameter, D_p/d, has little effect on the relative scour depth, d_s/D_p.

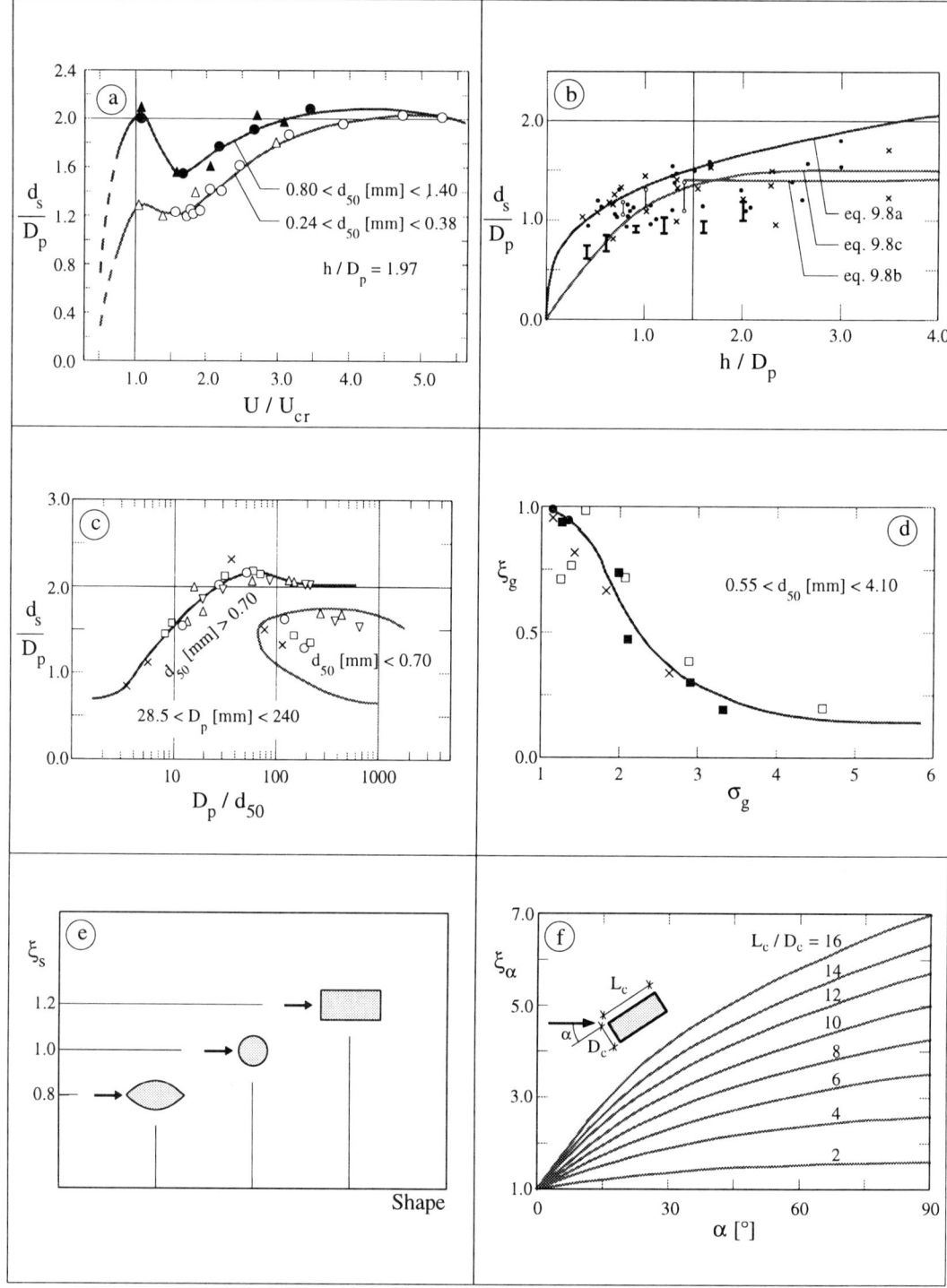

Fig. 9.3 Influence of the different parameters (see eq. 9.5), on the dimensionless scour depth, d_s/D_p, of a pier.

LOCAL SCOUR

iii) The influence of a sediment with a granulometric distribution, expressed with a coefficient of sediment grading, ξ_g, which in turn depends on the geometric standard deviation, $\sigma_g = (d_{84}/d_{16})^{0.5}$, is shown with Fig. 9.3d (see *Raudkivi, 1991, p. 67*).

The coefficient of sediment grading, which is $\xi_g = 1$ for uniform sand, may be considerable reduced, $\xi_g < 1$, for graded sand. This implies that the relative scour depth, d_s/D_p, will also be reduced. This reduction in the scour depth can be partially attributed to the armouring effect in the scour hole.

8° **Influence of pier shape**

i) The influence of the geometry of the pier, expressed with a coefficient of pier shape, ξ_s, is given with Fig. 9.3e (see *Raudkivi, 1991, p. 73*).

ii) Depending on the geometry of the nose of the pier, this coefficient may vary, being for :

- a cylindrical pier : $\xi_s = 1.0$
- a rectangular pier : $\xi_s \cong 1.2$
- a lenticular pier : $\xi_s \cong 0.8$

By streamlining the pier — itself well aligned with the flow — the disturbance of the flow pattern around the pier is reduced.

9° **Influence of pier alignment**

i) The influence of the pier alignment, expressed with the coefficient of the angle of approach of flow, ξ_α, is given with Fig. 9.3f (see *Raudkivi, 1991, p. 73*), using the experimental data of Laursen (1962).

ii) Depending on the pier alignment, α, but also upon the geometry of the pier, this coefficient may vary substantially, being for :

- a cylindrical pier : $\xi_\alpha = 1$
- a rectangular pier : $\xi_\alpha = f(\alpha, L_c/D_c)$

where D_c is the width and L_c is the length of the cross section of the pier. It should be noted, that not the actual width, D_c, counts here, but the projected width, D_{cn}, which increases with the angle of approach, α.

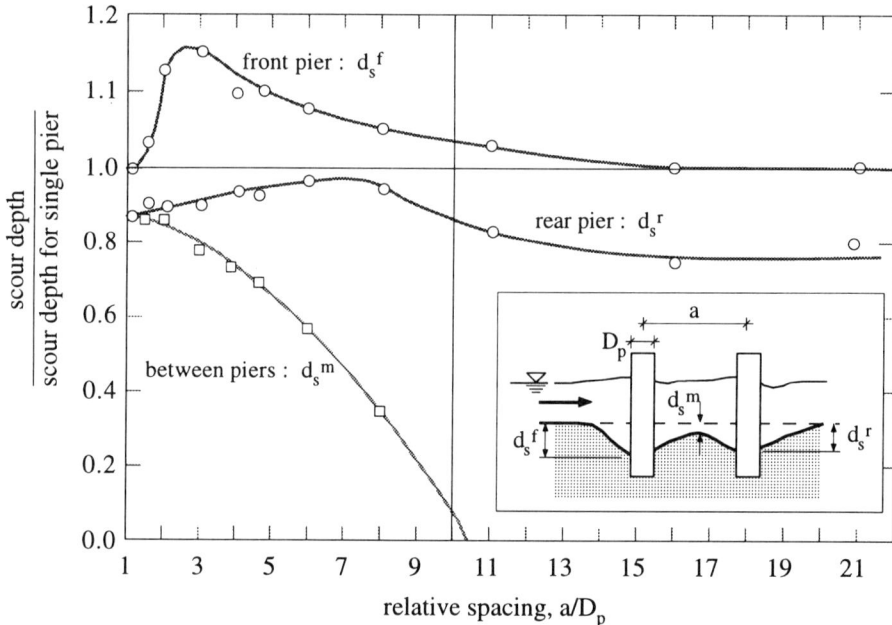

Fig. 9.4 Influence of two piers on the scour depth.

10° Influence amongst piers

i) Until now an isolated, single pier was considered. A group of piers will exhibit a mutual influence on the hydraulic behaviour and on scouring. However, research on this topic is extremely scarce.

ii) Research results on scour at two piers (see *Raudkivi*, 1991, p. 85) are shown in Fig. 9.4. The scour depth, d_s, of the reference pier is obviously influenced by the spacing of the piers ; one observes :

- when the two piers touch each other, $a/D_p = 1$, the scour depth of the front pier is not influenced ;

- when the two piers are separated, the scour depth of the front pier begins to increase but eventually falls off ;

- when the two piers are very separated, $a/D_p \approx 11$, the front pier is not influenced ;

- the scour depth, d_s, of the rear pier is always smaller than the one of the front pier ;

- the scour depth, d_s, *between* the piers diminishes rapidly and is negligible for $a/D_p > 10$.

11° Influence of unsteady flow

Unsteady flow in channels, such as encountered during the passage of floods, of translation waves or of wind waves, complicates the scour process considerably. Such waves may cause considerable pressure fluctuations on the bed, which in turn may help to displace or loosen particle in the scour hole.

Research on the effect of unsteady flow is scarce (see *Raudkivi*, 1991, p. 83) and inconclusive.

9.2.4 Formulae for Design

1° Although it is possible to present a good deal of experimental data, it is rather difficult to propose a scour formula for design. Field data are often lacking.

Theoretical developments, which would help to organise better the data of the experiments, are limited.

2° From the material presented above (see sect. 9.2.3), the scour depth for a single pier, positioned in a wide channel, can be estimated using the following relations :

i) for *clear-water scour* ($U/U_{cr} < 1$) :

- the relation of *Shen* (1971) :

$$d_s = 0.00022 \left(\frac{UD_p}{\nu}\right)^{0.619} \tag{9.6}$$

ii) for *sediment-transport* and *clear-water scour* :

- the relation of *Breusers et al.* (1977, p. 248) :

$$\frac{d_s}{D_p} = g\left(\frac{U}{U_{cr}}\right) \cdot \left(2.0 \tanh \frac{h}{D_p}\right) \cdot \xi_s \cdot \xi_\alpha \tag{9.9}$$

with $g\left(\dfrac{U}{U_{cr}}\right)$
$\begin{cases} = 0 & \text{for } \dfrac{U}{U_{cr}} < 0.5 \\ = \left(2\dfrac{U}{U_{cr}} - 1\right) & \text{for } 0.5 < \dfrac{U}{U_{cr}} < 1.0 \\ = 1 & \text{for } \dfrac{U}{U_{cr}} > 1.0 \end{cases}$

where ξ_s is the coefficient of pier shape and ξ_α is the coefficient of the angle of approach.

iii) for *sediment-transport scour* ($U/U_{cr} > 1$) :

- a relation proposed by *Raudkivi* (1991 p. 88) :

$$\frac{d_s}{D_p} = 2.3 \cdot \xi_\alpha \qquad (9.10)$$

- a relation of the type of eq. 9.5 :

$$\frac{d_s}{D_p} = 2.0 \cdot \xi_g \cdot \xi_s \cdot \xi_\alpha \qquad (9.11)$$

valid for $(h/D_p) > 1.5$ and $(D_p/d_{50}) > 25$.

In the above relations, the scour depth is the average limiting (equilibrium) one, which occurs usually after an extended flooding.

3° Evaluation of field measurements (see *Nordin*, 1989) of 104 cylindrical piers, covering a range of pier diameter, $0.2 < D_p$ [m] < 13.0, of scour depth, $0.17 < d_s$ [m] < 8.5, of flow depth, $0.3 < h$ [m] < 17.1, of approach velocity, $0.35 < U$ [m/s] < 2.72, of Froude number, $0.06 < Fr < 0.86$ and of sediment size, $0.3 < d$ [mm] < 10, showed, that about 10 % of the observed scour depth, d_s, is larger than the design scour depth, using eq. 9.8a. This represents an encouraging result.

A large set of 515 field data — including clear-water and sediment-transport scour — has been recently (see *Johnson*, 1995, p. 628) evaluated. It was found that the above relation, eq. 9.9, appears to work best.

4° All the above relations — in the light of their limitations and complications — must be considered as (conservative) guideline-estimate for hydraulic design.

9.2.5 Scour Prevention

1° An important task for the hydraulic engineer is the concern of the protection against local scour. Prevention of scour can be achieved by diminishing the erosion effect of the downward flow and of the horseshoe vortex.

Different methods have been suggested.

2° Rip-rap apron (see Fig. 9.5a)

A most effective method for remedying local scour is the dumping of stones into the (potential) scour hole.

For the determination of the size of the stones, d [m], a simple working relationship was proposed (see *Breusers et al.*, 1977, p. 249) :

LOCAL SCOUR

$$U_d \cong 2.4 \sqrt{d}$$

where U_d [m/s] is taken to be the cross-sectional average velocity of the approach flow at the design discharge.

The horizontal dimension of the loose-stone rip-rap apron should cover 2 to 3 times the width of the obstacle. Its thickness should be 3 times the stone diameter.

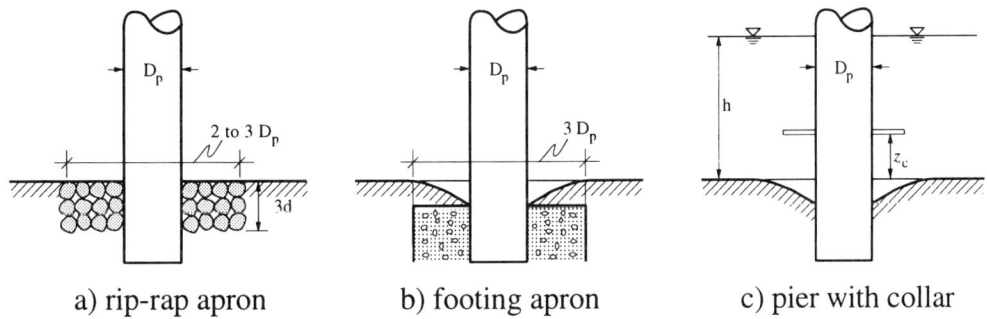

Fig. 9.5   Methods of scour prevention.

3° Footing apron (see Fig. 9.5b)

Footings or foundation blocks installed around the base of the obstacle and set below the bed level, are effective to counteract the downward current.

Pilings and sheet-piles driven below the footing afford a degree of further protection against failure by scour.

4° Pier with collar (see Fig. 9.5c)

The placement of an horizontal ring-formed shield placed on the pier proved to diminish the scour depth. The positioning at $z_c = 0.2\, h$, where h is the flow depth, reduces the scour depth by 50 %. The presence of the collar disturbs the downward current, thus the erosion around the pier.

5° Pile arrangement

A wedge-shaped arrangement of small piles, positioned upstream of the obstacle, can reduce considerably the local erosion. These piles disturb the flow and consequently weaken the horseshoe vortex.

6° Streamlining

If the upstream nose of the pier is streamlined and/or properly aligned to the flow (see Fig. 9.3e and Fig. 9.3f), the scour depth can be reduced. In this way the horseshoe vortex and its effects are diminished.

9.3 ABUTMENT SCOUR

1° The presence of an obstruction, protruding into the flow, such as an abutment, an embankment, a spur dike or a groin — just like the one of a pier (see sect. 9.2) — changes the flow pattern and subsequently modifies the mobile channel bed (see Fig. 9.6).

In the vicinity of the abutment, scour will occur. The intensity of scour will depend upon the channel flow, the sediments of the bed and the geometry of the obstruction.

2° The geometry of the abutment is given by its length, L_A, measured perpendicular to the flow, its shape and its alignment with the flow.

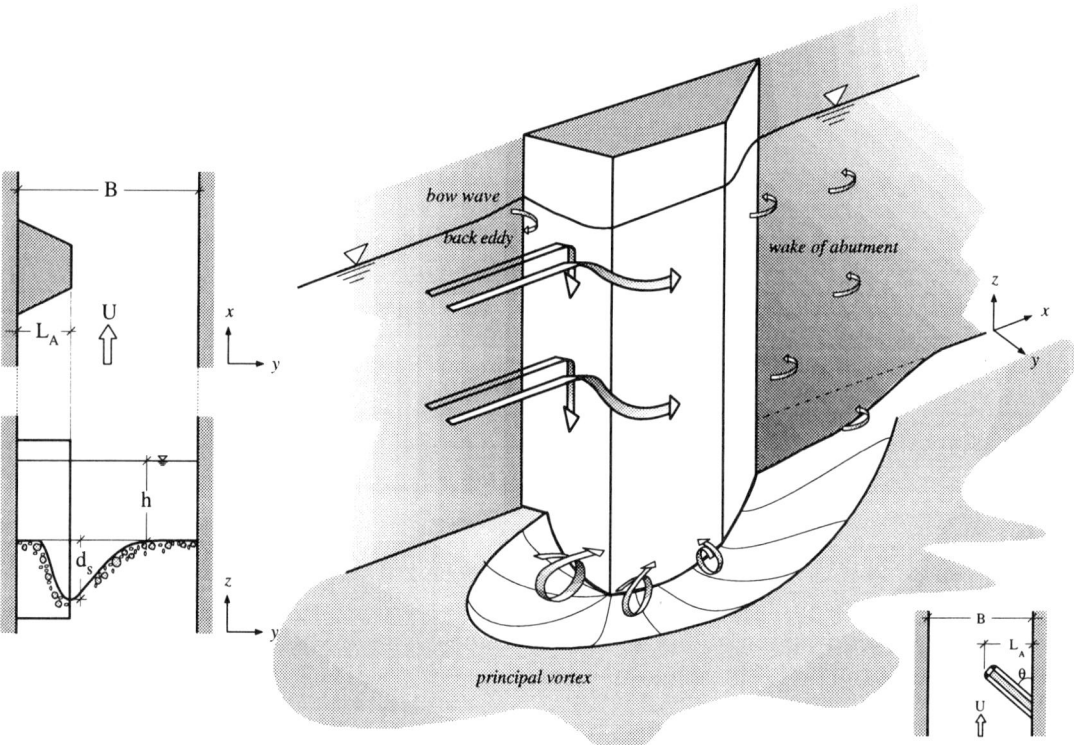

Fig. 9.6 Scheme of flow pattern and local scour around an abutment.

9.3.1 Flow Pattern and Scouring

1° The flow pattern and scouring at abutments are rather similar to the ones around piers (see Fig. 9.6 as well as Fig. 9.1). A *downward flow* produces a *principal* (and a secondary) *vortex*, which is very active in the scour process.

LOCAL SCOUR

2° When the length of the abutment, L_A, is considerable, a back eddy in the dead-water zone establishes itself on the upstream side of the abutment; this may influence the strength of the principal vortex. Furthermore a possible channel constriction, $(B - L_A)/B$, might have to be taken into account (see sect. 9.4).

9.3.2 Functional Relations

1° To obtain functional relations, the same arguments as applied for a single pier (see sect. 9.2.3) can be used.

2° Using dimensional reasoning the following dimensionless parameters are obtained (see eq. 9.4 and eq. 9.5):

$$\frac{d_s}{L_A} = f_8\left(\frac{U}{U_{cr}}, \frac{h}{L_A}, \frac{d}{L_A}; \xi_g, \xi_s, \xi_\alpha\right) \qquad (9.12)$$

where h and U are the flow depth and average velocity of the approach flow, U_{cr} is the velocity at critical erosion condition and d is the average sediment-particle diameter. ξ_g, ξ_s and ξ_α are coefficients of sediment grading, of abutment shape and of angle of approach. Note, that flow depth, h, and abutment length, L_A, could be interchanged in eq. 9.12.

3° Local scour, parametrized by the scour depth, d_s, is seen to be a complex phenomenon, where the various parameters interact. The influence of each of the above parameters will be examined next, using experimental data.

4° Many of the available laboratory data have been systematically evaluated by *Melville* (1992), whose results will be presented herewith.

It must be mentioned, however, that the existing information was obtained from laboratory tests at greatly reduced scales; extrapolation to full-scale conditions may be questionable.

5° Influence of flow velocity:

The influence of the relative flow velocity, U/U_{cr}, on the dimensionless scour depth, d_s/L_A, is shown in Fig. 9.7a (see *Melville*, 1992, p. 624).

In Fig. 9.7a, it is again (see also Fig. 9.3a) apparent that *clear-water scour* takes place at $U/U_{cr} < 1$ and *sediment-transport scour* at $U/U_{cr} > 1$.

The maximum scour depth, $d_{s\,max}$, occurs at flow velocities being identical to the critical one ; this was also observed for scour around piers (see Fig. 9.3a). Subsequently, upon an increase in the flow velocity, the average equilibrium scour depth, d_s, establishes itself, being indistinguishable the same as the maximum one.

For sediment-transport scour, where the flow velocity is of no further importance, it can be admitted that :

$$\frac{d_s}{L_A} \approx 2 \tag{9.13}$$

6° Influence of flow depth :

The influence of the relative flow depth, h/L_A, on the dimensionless scour depth, d_s/L_A, is shown in Fig. 9.7b (see *Melville*, 1992, p. 619), obtained from extensive laboratory data.

As can be seen in Fig. 9.7b , the relative scour depth increases such as :

$$\frac{d_s}{L_A} = 2\sqrt{\frac{h}{L_A}} \qquad \text{for} \quad 0.04 < \frac{h}{L_A} < 1 \tag{9.14}$$

and attains for *short* abutments a limiting value of :

$$\frac{d_s}{L_A} \cong 2 \qquad \text{for} \quad \frac{h}{L_A} > 1 \tag{9.14a}$$

For very *long* abutments one takes :

$$\frac{d_s}{h} \approx 10 \qquad \text{for} \quad \frac{h}{L_A} < 0.04 \tag{9.14b}$$

Consequently, for abutments, which are neither very long nor very short, the scour depth, d_s, is influenced by the flow depth, h, and the abutment length, L_A ; it is given by the above relations, eqs 9.14, which are similar to a relation proposed by Laursen (1958).

7° Influence of sediments :

No successful evaluation of the influence of sediments on the scouring is available.

LOCAL SCOUR

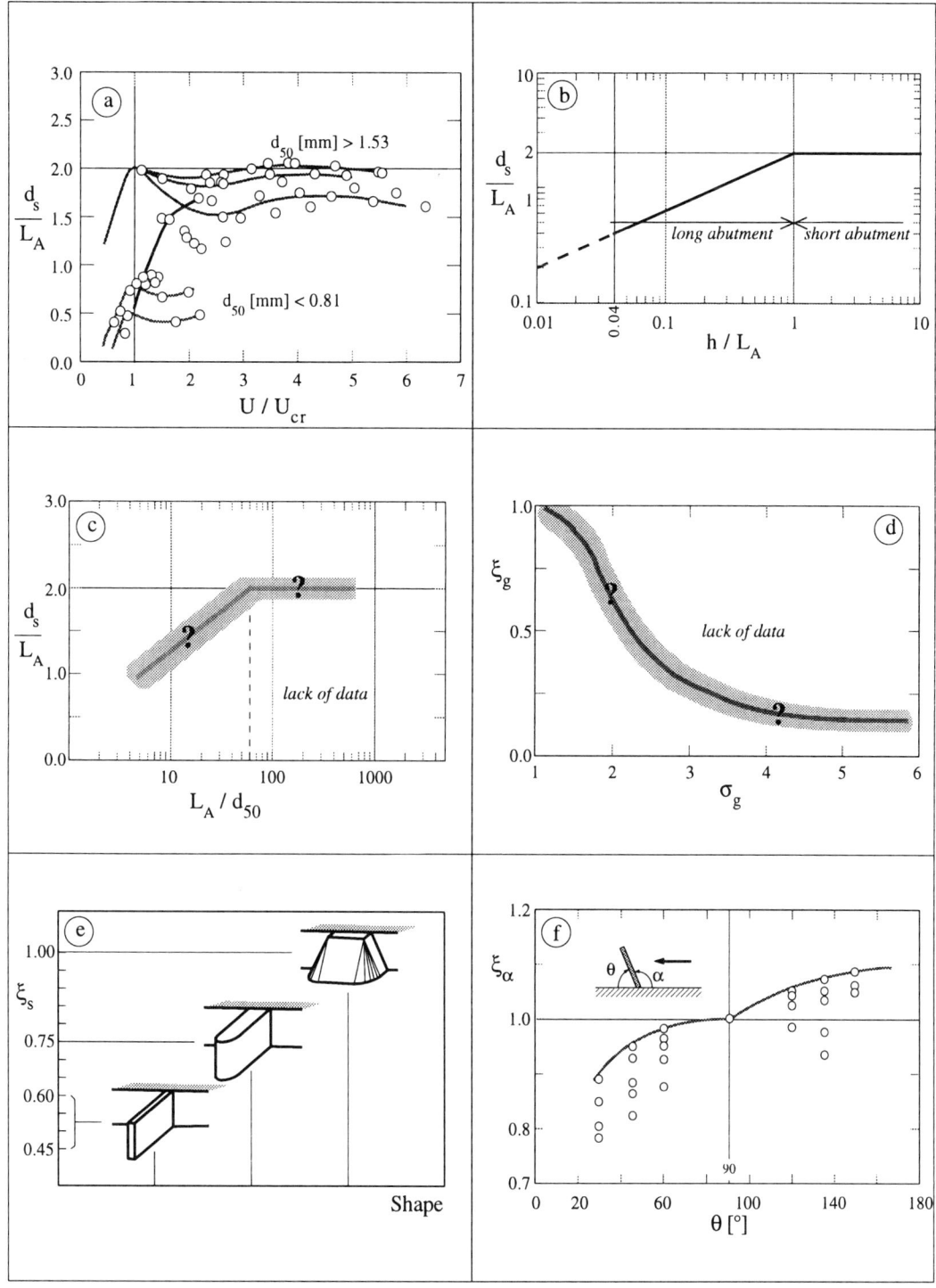

Fig. 9.7 Influence of the different parameters (see eq. 9.12) on the dimensionless scour depth, d_s/L_A, for an abutment.

Using data obtained from pier-scour studies (see Fig. 9.3c), it was deduced (see *Melville*, 1992 p. 620) that for a relative sediment diameter, $L_A/d > 50$, there is little effect on the scour depth (see Fig. 9.7c). Consequently a value of :

$$\frac{d_s}{L_A} \approx 2$$

represents a very conservative value.

For sediments with a grain-size distribution, σ_g, it is reported (see *Melville*, 1992, p. 620) that there is consistently produced a lower scour depth, as was also reported for pier scour (see Fig. 9.3d).

However since no conclusive experiments are available, the coefficient of sediment grading (see Fig. 9.7d) is to be taken as $\xi_g \approx 1$; once again this is a very conservative value.

8° Influence of abutment shape :

The influence of the geometry of the abutment, expressed with a coefficient of shape, ξ_s, is given with Fig. 9.7e (see *Melville*, 1992, p. 617).

Depending on the geometry of the abutment, the coefficient may vary, being for :

- narrow vertical walls : $\xi_s \cong 1.0$
- vertical walls with rounded ends : $\xi_s \cong 0.75$
- spill-through dikes : $0.45 < \xi_s < 0.6$

9° Influence of abutment alignment

The influence of the alignment, expressed with the coefficient of the angle of approach of flow, ξ_α, is given in Fig. 9.7f (see *Melville*, 1992, p. 623).

Depending on the alignment angle of the abutment, α, the scour depth, d_s, increases with an increase in the angle ; the perpendicular aligned abutment, $\alpha = 90°$, serves as a reference. An upper envelope curve delimits the data points.

9.3.3 Formulae for Design

1° A good deal of experimental data are available, but they come exclusively from laboratory measurements. There is a great shortage of field data.

2°   Preliminary design recommendation — to be used with great care — are given, notably valid for sediment-transport scour, as :

$$\frac{d_s}{L_A} = 2.0 \, \xi_s \cdot \xi_\alpha \qquad \text{for } \frac{h}{L_A} > 1 \qquad (9.15a)$$

$$\frac{d_s}{L_A} = 2 \sqrt{\frac{h}{L_A}} \cdot \xi_s \cdot \xi_\alpha \qquad \text{for } \frac{h}{L_A} < 1 \qquad (9.15b)$$

The above relations — in the light of the above limitations and complications — must be considered as (conservative) guideline estimates for hydraulic design.

9.4   CONSTRICTION SCOUR

9.4.1 Hydraulic Considerations

1°   A constriction (contraction) of a watercourse, due to a reduction in width, $B_2 < B_1$, will lead in *fluvial* regime to an increased flow velocity, $U_2 > U_1$, in the reduced cross section. If the channel has a mobile bed, erosion of the bed may be the consequence (see Fig. 9.8).

Considered will be one-dimensional flow in a rectangular channel. It is supposed, that uniform flow exists in both the approach and the constricted cross section. If the reach of the constriction is a *long* and gradual one, the resulting head loss will be negligible.

2°   This problem is one of a channel of variable width (see sect. 4.5.2) and with a variable bed floor (see sect. 4.5.1).

   i)   The equation of continuity for a constant discharge, Q, reads :

$$q_1 B_1 = Q = q_2 B_2 \qquad (9.16)$$

   For a channel of variable width, the unit discharge, q, will be variable.

   ii)   The specific energy, H_s, will be maintained throughout the transition, thus :

$$H_{s_1} = H^*_{s_2} = H_{s_2} - d_s$$

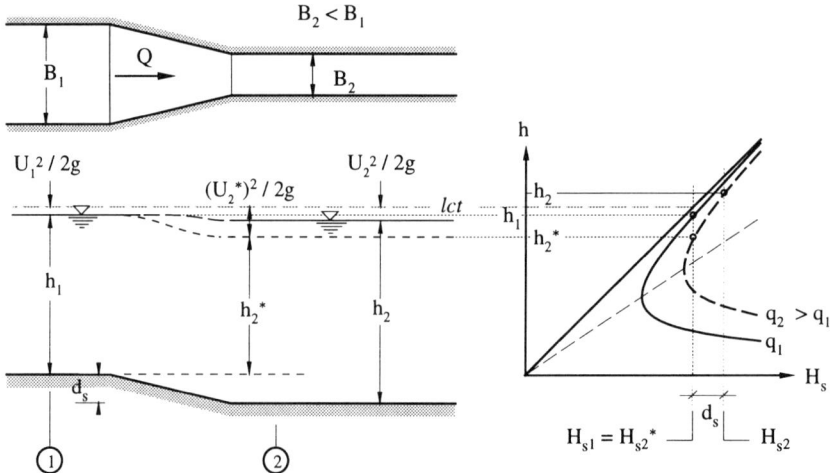

Fig. 9.8 Definition sketch for a long constriction.

A graphical solution is to be followed (see Fig. 9.8) :

- In fluvial regime, a *constriction*, $B_2 < B_1$, causes an increase of the unit discharge, $q_2 > q_1$, and consequently a decrease in the flow depth, $h_2^* < h_1$, and an increase in velocity, $U_2^* > U_1$.

- This increase in flow velocity may cause erosion in the constricted section ; consequently a *drop*, d_s, in the bed floor will appear.

- Downstream of the floor change, the specific energy, H_{s_2}, is changed by the height of the floor change, d_s ; it has now a value of $H_{s_2} = H_{s_2}^* + d_s$, thus a shift of d_s to the right. Consequently, an increase in the flow depth, $h_2 > h_2^*$, and a decrease in the flow velocity, $U_2 < U_2^*$, will install itself.

- Comparing now the hydraulic situation in the approach channel, h_1 and U_1, and in the constricted channel, h_2 and U_2, the following is observed :

    if d_s is small : $h_2' < h_1$; $U_2' > U_1$
    if d_s is large : $h_2' > h_1$; $U_2' \approx U_1$ (see Fig. 9.8)

    This should be kept in mind, when approximating the terms in the equation of energy, eq. 9.20, but also in the interpretation of the values obtained from Fig. 9.9.

3° The equation of continuity, for the water discharge, Q, between the two cross sections, reads :

$$B_1 U_1 h_1 = B_2 U_2 h_2 \tag{9.16a}$$

LOCAL SCOUR

Using the Manning formula (see eq. 3.16) for wide channels, this yields:

$$B_1 \frac{1}{n_1} h_1^{5/3} S_{e_1}^{1/2} = B_2 \frac{1}{n_2} h_2^{5/3} S_{e_2}^{1/2} \qquad (9.17)$$

If the granulate is the same, $d_1 = d_2$, thus the roughness coefficient, $n_1 = n_2$, is also the same, one obtains:

$$B_1 h_1^{5/3} S_{e_1}^{1/2} = B_2 h_2^{5/3} S_{e_2}^{1/2} \qquad (9.18)$$

$S_e \equiv S_f$ may be taken, if the flow is a uniform one.

4° The shear stress in a wide rectangular channel is given by:

$$\tau_o = \rho g h S_e \qquad (9.19)$$

where S_e is the slope of the energy-grade line.

5° The equation of energy, for the two cross sections under consideration (see Fig. 9.8) can be written as:

$$d_s + h_1 + \frac{U_1^2}{2g} = h_2 + \frac{U_2^2}{2g} + h_s + h_r \qquad (9.20)$$

Assuming the head loss due to friction, h_r, and to the constriction, h_s, are negligible, an expression for the scour depth reads:

$$d_s = (h_2 - h_1) + \left(\frac{U_2^2}{2g} - \frac{U_1^2}{2g}\right) \qquad (9.20a)$$

For fluvial flow, it seems often reasonable (see *Laursen*, 1963, p. 97) to assume that the variation in the velocity head is small, which renders:

$$d_s \approx (h_2 - h_1) \quad \text{or} \quad h_2 \approx h_1 + d_s \qquad (9.21)$$

9.4.2 Scour-depth Relations

1° The above relation, eq. 9.18, shall be used to express the relative flow depth:

$$\frac{h_2}{h_1} = \left(\frac{B_1}{B_2}\right)^{3/5} \left(\frac{S_{e_1}}{S_{e_2}}\right)^{3/10} \qquad (9.22)$$

Furthermore the definition of the shear stress, eq. 9.19, yields:

$$\frac{\tau_{o_1}}{\tau_{o_2}} = \frac{\rho g\, h_1\, S_{e_1}}{\rho g\, h_2\, S_{e_2}} \quad \Rightarrow \quad \frac{S_{e_1}}{S_{e_2}} = \frac{\tau_{o_1}\, h_2}{\tau_{o_2}\, h_1} \qquad (9.23)$$

2° Combining the above relations, renders:

$$\frac{h_2}{h_1} = \left(\frac{B_1}{B_2}\right)^{3/5} \left(\frac{\tau_{o_1}\, h_2}{\tau_{o_2}\, h_1}\right)^{3/10}$$

or rewritten:

$$\frac{h_2}{h_1} = \left(\frac{B_1}{B_2}\right)^{6/7} \left(\frac{\tau_{o_1}}{\tau_{o_2}}\right)^{3/7} \qquad (9.24)$$

This relation is plotted in Fig. 9.9, with the relative flow depth, h_2/h_1, as a function of the relative shear stress, τ_{o_1}/τ_{o_2}, for various values of width reduction, B_1/B_2. Here, the following is to be observed:

- for a given relative shear stress, τ_{o_1}/τ_{o_2}, an increase in the width reduction, B_1/B_2, is responsible for an increase in the relative depth, h_2/h_1;

- if the flow depth throughout the entire constriction does not change — the relative depth yields unity, $h_2/h_1 = 1$ — it implies (see eq. 9.24), that:

$$\frac{\tau_{o_1}}{\tau_{o_2}} = \left(\frac{B_2}{B_1}\right)^2 \quad \text{or} \quad \frac{\rho g\, h_1\, S_{e_1}}{\rho g\, h_2\, S_{e_2}} = \frac{S_{e_1}}{S_{e_2}} = \left(\frac{B_2}{B_1}\right)^2 \qquad (9.25)$$

having a limiting value at:

$$\frac{\tau_{o_1}}{\tau_{o_2}} = 1 \quad \text{when} \quad \frac{B_2}{B_1} = 1$$

- for a relative flow depth below unity, $h_2/h_1 < 1$, scour is also possible; the complete equation of energy, eq. 9.20a, must be considered.

3° An interesting relationship can be obtained, assuming that the same granulate, thus the same critical shear stress, $\tau_{cr_1} \equiv \tau_{cr_2}$, exists in the approach and the constricted channel. The above relation, eq. 9.24, can now be rewritten:

$$\frac{h_2}{h_1} = \left(\frac{B_1}{B_2}\right)^{6/7} \left(\frac{\tau_{o_1}/\tau_{cr}}{\tau_{o_2}/\tau_{cr}}\right)^{3/7} \qquad (9.24a)$$

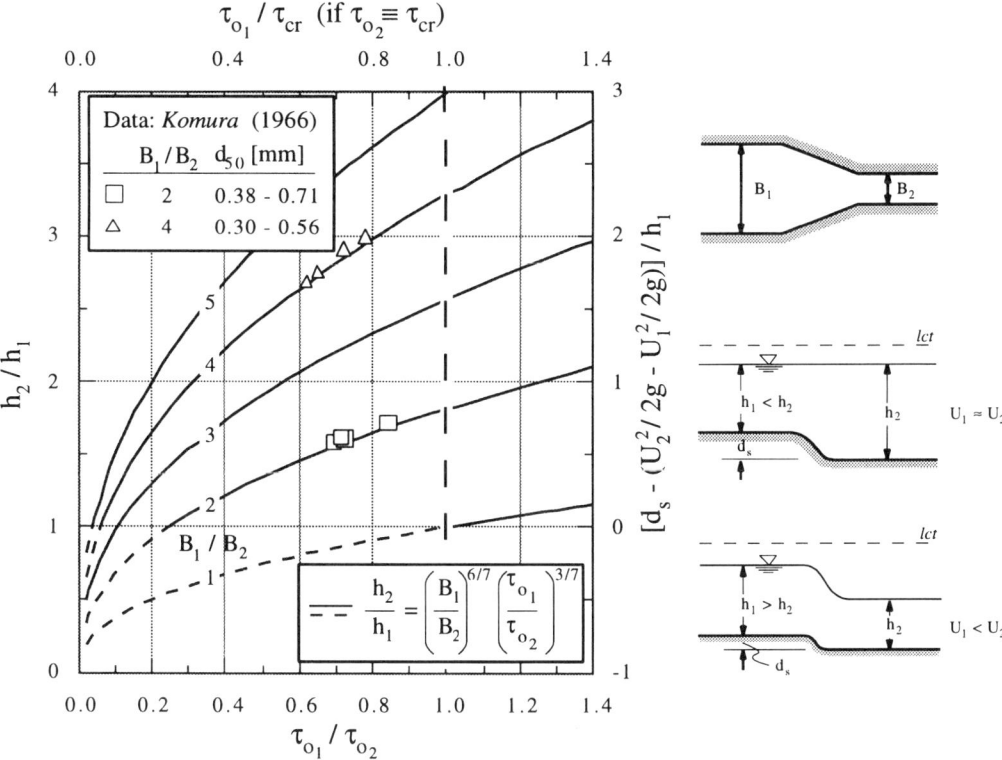

Fig. 9.9 Relative flow depth, h_2/h_1, or scour depth, d_s/h_1, as a function of the relative shear stress, τ_{o_1}/τ_{o_2}, or the relative shear stress, τ_{o_1}/τ_{cr}, (if $\tau_{o_2} \equiv \tau_{cr}$), both for different width reductions, B_1/B_2.

If the shear stress, τ_{o_2}, in the constricted channel is taken to be the critical one, $\tau_{o_2} = \tau_{cr}$, the above relation becomes :

$$\frac{h_2}{h_1} = \left(\frac{B_1}{B_2}\right)^{6/7} \left(\frac{\tau_{o_1}}{\tau_{cr}}\right)^{3/7} \tag{9.26}$$

already presented by *Laursen* (1963, p. 97). This relation is also given with Fig. 9.9, where the following is to be observed :

- for a given relative shear stress, τ_{o_1}/τ_{cr}, an increase in the width reduction, B_1/B_2, is responsible for an increase in the relative depth, h_2/h_1 ;

- for the special case, when the shear stress is the critical one throughout the entire channel, namely, $\tau_{o_1} = \tau_{cr} = \tau_{o_2}$, the above relations, eq. 9.24 or eq. 9.26, reduce to :

$$\frac{h_2}{h_1} = \left(\frac{B_1}{B_2}\right)^{6/7} \tag{9.27}$$

4° Allowing for a variation of the sediment sizes, $d_1 \neq d_2$, the above relation, eq. 9.24, can be written (see *Komura*, 1966, p. 23) as :

$$\frac{h_2}{h_1} = \left(\frac{B_1}{B_2}\right)^{6/7} \left(\frac{\tau_{o_1}}{\tau_{o_2}}\right)^{3/7} \left(\frac{d_2}{d_1}\right)^{1/7} \qquad (9.27a)$$

9.4.3 Formula for Design

1° From the material presented above (see sect. 9.4.2), the scour depth can be estimated using the following relation :

$$\frac{h_2}{h_1} = \left(\frac{B_1}{B_2}\right)^{6/7} \left(\frac{\tau_{o_1}}{\tau_{o_2}}\right)^{3/7} \qquad (9.24)$$

which is given with Fig. 9.9.

The shear stress, τ_{o_1} and τ_{o_2}, defined with eq. 9.19, can be evaluated by means of the friction coefficient, namely according to the relations of Manning, Chézy or Strickler (see sect. 3.2).

2° Due to a lack of laboratory data as well as an almost complete lack of field data, the above relation, eq. 9.24, must be considered as a guide-line estimate for hydraulic design (see *Raudkivi*, 1990, p. 245).

9.5 HYDRAULIC STRUCTURES SCOUR

9.5.1 Notions

1° Weirs or low-head dams as well as underflow gates, are installed in channels and waterways to control and/or measure the discharge (see sect. 4.4.1 and sect. 4.4.3). The flow passes *over*, q_o, and/or *under*, q_u, such an hydraulic structure towards the downstream (see Fig. 9.10).

2° Such a flow, which often appears in form of jets, may have considerable hydraulic potential to scour on the downstream side of the structure ; a scour hole is formed.

3° Since the scour hole develops rather rapidly, the engineer takes interest in the equilibrium scour depth, d_s, and the length of the scour hole, L_s.

4° Scour will not only endanger the stability of the channel bed, but it might also have devastating effects on the hydraulic structure itself.

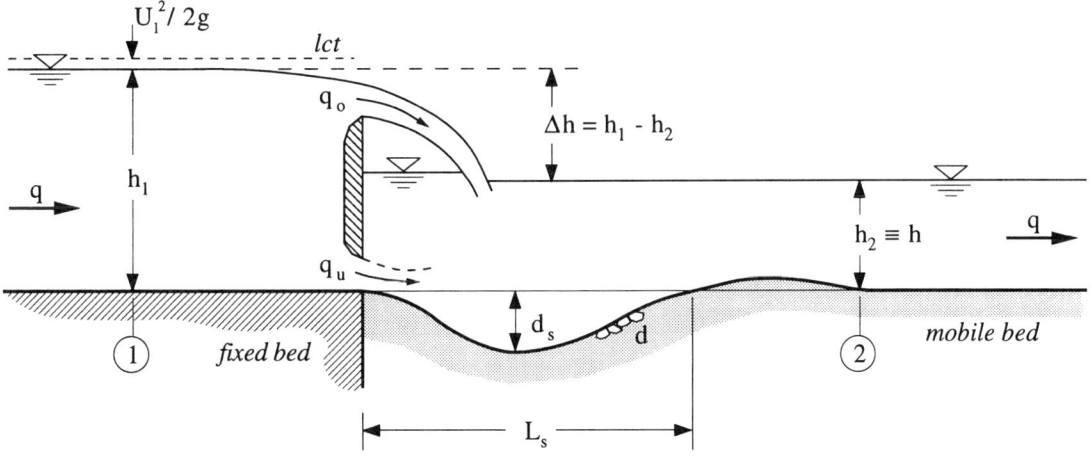

Fig. 9.10 Schema of hydraulic structure scour;
a combined sluice gate allowing overflow and underflow.

5° Considering an hydraulic structure positioned in an open channel, where the flow is unidirectional, uniform and steady, and where the bed is made of cohesionless sediments, the equilibrium *scour depth*, d_s, depends on :

- the *fluid* : the density, ρ, and viscosity, ν ;
- the *sediments* : the density, ρ_s, and a characteristic diameter, d ;
- the channel *flow* (in absence of the structure) :
  the water depth, h, the average velocity, U, and the gravity, g ;
- the interference of the *structure* on the channel flow :
  resulting in a drop in the water depth, $\Delta h = h_1 - h_2$.

This allows to write :

$$d_s = f_1(\rho, \nu \,;\, \rho_s, d \,;\, h, U, g \,;\, \Delta h) \tag{9.25}$$

6° Using dimensional reasoning the following parameters are obtained :

$$\frac{d_s}{d} = f_2\left(\frac{u_* d}{\nu}, \frac{(Uh)^2}{gd^3}, \frac{\rho_s}{\rho}, \frac{h}{d}, \frac{\Delta h}{d}\right) \tag{9.26}$$

Usually it is assumed (see *Kotoulas*, 1967 p. 38) that the particle Reynolds number, Re_*, the relative density, ρ_s/ρ, and the relative roughness, h/d, play a negligible role ; thus :

$$\frac{d_s}{d} = f_3\left(\frac{q^2}{gd^3}, \frac{\Delta h}{d}\right) \tag{9.27}$$

where q = Uh is the unit discharge in the channel.

For flow *over* the structure, the discharge is $q = q_o$;
for flow *under* the structure, the discharge is $q = q_u$;
for flow *over* as well as *under* the structure, the discharge is $q = q_o + q_u$.

7°   Stilling basins are sometimes used to *protect* against destructive scouring. The dimensioning will require knowledge of the possible length, L_s, of the scour hole.

There might still be scouring downstream of the stilling basin (see *Raudkivi*, 1990, p. 283) where rock blocks or rip-rap could be placed as a measure of protection (see sect. 9.2.5).

9.5.2 Flow over a Structure

1°   Scour due to flow *over* an hydraulic (low-head) structure is schematically shown in Fig. 9.11. The scour depth, d_s, is the equilibrium maximum value, which develops rather fast. The unprotected channel bed below the free overfall is eroded. Downstream of the scour hole of a certain length, L_s, may appear a mound of deposition. The free overfall, plunging in form of a jet into the downstream water, forms two vortices, whose erosive power often considerable is responsible for the scour hole.

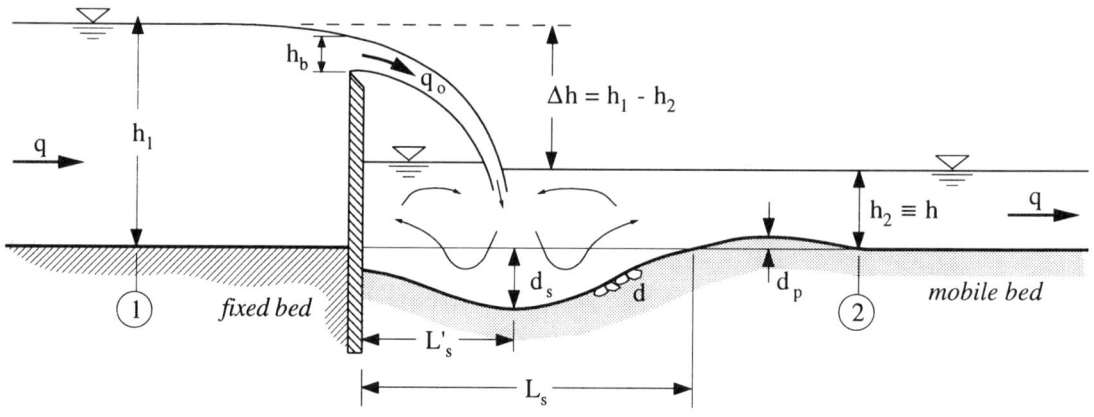

Fig. 9.11 Schema of flow over a structure.

2°   For the prediction of the scour depth, d_s, various formulae have been developed. However, all these prediction formulae are based in small-scale laboratory studies ; upscaling of these in order to predict scour on real structures has to be viewed with caution.

Here is a selection of some relations presently in use.

LOCAL SCOUR 641

3° Early attempts by Schoklitsch (1932) and Veronese (1937) have shown (see *Eggenberger* et *Müller,* 1944), that the scour depth, d_s, cannot be expressed by a simple relation.

4° A limited set of laboratory data, using the similarity of Froude, allowed *Eggenberger* et *Müller* (1944, p. 31 and p. 45) to propose the following relation :

$$h + d_s = w \frac{\Delta h^{0.5} q^{0.6}}{d_{90}^{0.4}} \tag{9.28}$$

where the coefficient is : $w = 1.44 \ [s^{0.6}/m^{0.3}]$, if d_{90} [m]
$w = 22.88 \ [s^{0.6}/m^{0.3}]$, if d_{90} [mm] ;
$w" = 3.6 \ [-]$.

This relation was subsequently generalised to :

$$h + d_s = \frac{w"}{g^{0.3}} \left(\frac{\rho}{\rho_s - \rho}\right)^{4/9} \frac{\Delta h^{0.5} q^{0.6}}{d_{90}^{0.4}} \tag{9.29}$$

where d_{90} [m] is the diameter of the largest particles in the granulometry of the original bed; thus an armouring (see sec. 6.3.4) takes place.

It has been frequently remarked that the above formula, eq. 9.28, overestimates the scour depth.

For the length of the scour hole the following relations are advanced :

$$\frac{L_s}{(h + d_s)} \approx 1.8 \qquad \frac{L'_s}{(h + d_s)} \cong 0.5 \tag{9.30}$$

5° Performing laboratory test and evaluating these data according to the dimensional reasoning (see eq. 9.27) *Kotoulas* (1967, p. 40) proposed a relation for the dimensionless equilibrium scour depth, such as :

$$\frac{h + d_s}{d_{95}} = 1.9 \left(\frac{q^2}{g \, d_{95}^3} \frac{\Delta h}{d_{95}}\right)^{0.35} \tag{9.31}$$

or

$$h + d_s = 1.9 \frac{1}{g^{0.35}} \left(\frac{\Delta h^{0.35} q^{0.7}}{d_{95}^{0.4}}\right)$$

The temporal evolution of the scour depth, $(h + d_s)_t$, was given by:

$$(h + d_s)_t = \left[1.9 \frac{1}{g^{0.35}} \left(\frac{\Delta h^{0.35} q^{0.7}}{d_{95}^{0.4}}\right)\right](1 - e^{-0.55 t^{1/5}}) \qquad (9.32)$$

Equilibrium scour is obtained in the laboratory after 24 [h].

A comparison of eq. 9.31 with eq. 9.28 was done and showed that the scour depth, d_s, obtained with eq. 9.28 is usually 85% larger.

6° Many of the available model and prototype data were analysed by *Mason* et *Arumugam* (1985, p. 232). The following relation was proposed:

$$h + d_s = K_M \frac{1}{g^{0.3}} \left(\frac{\Delta h^y q^x}{d_m^{0.1}}\right) h^{0.15} \qquad (9.33)$$

where d_m [m] is the mean particle size, rather than d_{90} [m]. The coefficient and the exponents are given, being for

model data: $\qquad K_M = 3.27$, $\qquad y = 0.05$
$\qquad\qquad\qquad\qquad\qquad\qquad\qquad\qquad x = 0.6$

prototype and model data: $K_M = 6.42 - 3.10\, \Delta h^{0.1}$, $\quad y = 0.15 + \Delta h/200$
$\qquad\qquad\qquad\qquad\qquad\qquad\qquad\qquad\qquad\qquad\qquad x = 0.6 - \Delta h/300$

This formula is said to be applicable for bed material of $0.001 < d_m$ [m] < 0.028; it can also be applied for all types of rocks, by assuming $d_m = 0.25$ [m].

Since the investigated data set included also data from high-head structures (spillway-chute flip buckets and tunnel outlets), the above relation, eq. 9.33, might be less valuable for low-head structure in fluvial hydraulics.

7° It should be remarked that the above presented formulae, eq. 9.28, eq. 9.29, eq. 9.31 and eq. 9.33, maintain a rather simple combination of the parameters obtained through dimensional reasoning.

Some other formulae, having a complex form – these include the Russian relations (see *Mason* et *Arumugam*, 1985, p. 224) – have not been considered. Also not included are formulae obtained from data of high-head structures (see *Breusers*, 1991, p. 119).

LOCAL SCOUR

9.5.3 Flow under a Structure

1° Scour due to flow *under* an hydraulic structure is schematically shown in Fig. 9.12. The underflow forms a kind of horizontal jet moving towards the downstream, whose erosive power could become considerable.

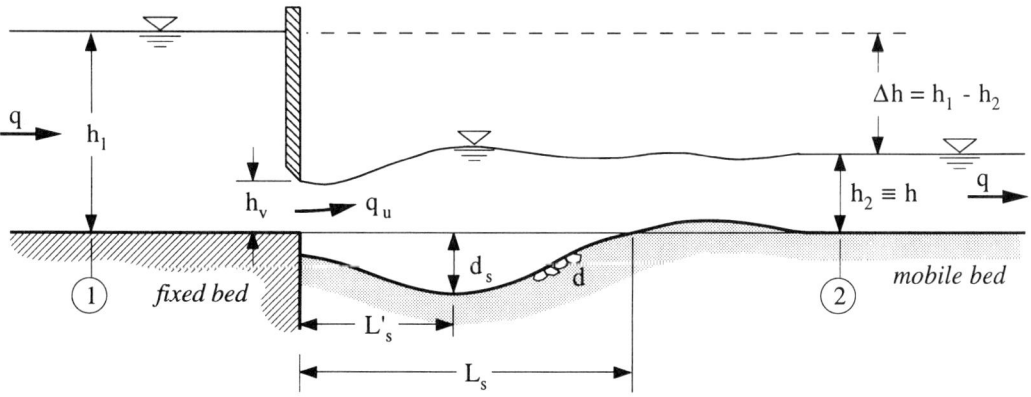

Fig. 9.12 Schema of flow under a structure.

2° For the prediction of the scour depth, d_s, a few formulae have been advanced; all are based on small-scale laboratory studies.

Here is a selection of some relations presently in use.

3° The form of the relation, eq. 9.28, established by *Eggenberger* et *Müller* (1944, p. 40) for overflow, $q = q_o$, was found valid for underflow, $q = q_u$, or :

$$h + d_s = w \frac{\Delta h^{0.5} q^{0.6}}{d_{90}^{0.4}} \qquad (9.34)$$

The coefficient for a submerged jet is $w = 10.35$ [$s^{0.6}/m^{0.3}$] and for a free jet it is $w = 15.40$ [$s^{0.6}/m^{0.3}$], if d_{90} [mm].

For the same flow conditions, q and Δh, and the same sediment, d_{90}, it is seen that scour at underflow produces a smaller scour depth, d_s.

For the length of the scour hole the following relations are advanced :

$$\frac{L_s}{(h + d_s)} \approx 6 \qquad \frac{L'_s}{(h + d_s)} \approx 3 \qquad (9.35)$$

In comparison with the scour hole for overflow (see eq. 9.30) the one for underflow is considerably larger.

4° Reviewing available laboratory data, the following relation was advanced by *Breusers* (1991, p. 107):

$$\frac{d_s}{h_v} = 0.008 \left(\frac{U_u}{u_{*cr}}\right)^2 \qquad (9.36)$$

valid for $(U_u/u_{*cr}) < 100$. Here $U_u = q_u/h_v$ is the underflow velocity and $u_{*cr} = (\tau_{ocr}/\rho)^{1/2}$ is the critical (scour) shear velocity of the bed material.

For the length of the scour hole an estimation is given as:

$$\frac{L_s}{d_s} \approx 5 \text{ to } 7 \qquad (9.37)$$

Above relations are developed for scour by submerged horizontal jets.

5° The scour resulting from a combination of flow *under* and *over* the hydraulic structure (see Fig. 9.10) was investigated by *Eggenberger* et *Müller* (1944, p. 43). The general form of the relation valid for overflow, eq. 9.28, or for underflow, eq. 9.34, was used:

$$h + d_s = w \frac{\Delta h^{0.5} q^{0.6}}{d_{90}} \qquad (9.38)$$

where $q = q_o + q_u$. The coefficient was determined as being:

$$w = 22.88 - \frac{1}{0.0049 R_q^3 - 0.0063 R_q^2 + 0.029 R_q + 0.064} \qquad (9.39)$$

with $R_q = (q_o/q_u)$ as the ratio of the overflow to the underflow unit discharge. Note, for underflow only, when $R_q \ll 1$, $w = 10.35$ [s $^{0.6}$/m$^{0.3}$] and for overflow only, when $R_q \gg 1$, $w = 22.88$ [s $^{0.6}$/m$^{0.3}$].

9.6 EXERCISES

9.6.1 Problems, solved

Ex. 9.A

A *bridge pier* has to be placed into a river estuary. This pier, being well aligned with the flow, is round-nosed having a diameter of $D_p = 18$ [m] and a length of $L_p = 40$ [m].

The water depth in the estuary is $h = 22$ [m]; an approach flow velocity from the tidal current of $U = 0.12$ [m/s] was measured. The bed material is rather fine, having a diameter of $d_{50} = 0.1$ [mm] with $\sigma_g = 1.5$ [-].

Determine the scour depth to be expected.

SOLUTION:

i) *Critical velocity*

The critical velocity, U_{cr}, of erosion will be taken from the diagram of Hjulstrom (see Fig. 3.12):

for $d = 0.1$ [mm] $\Rightarrow$ $U_{cr} \approx 0.2$ [m/s]

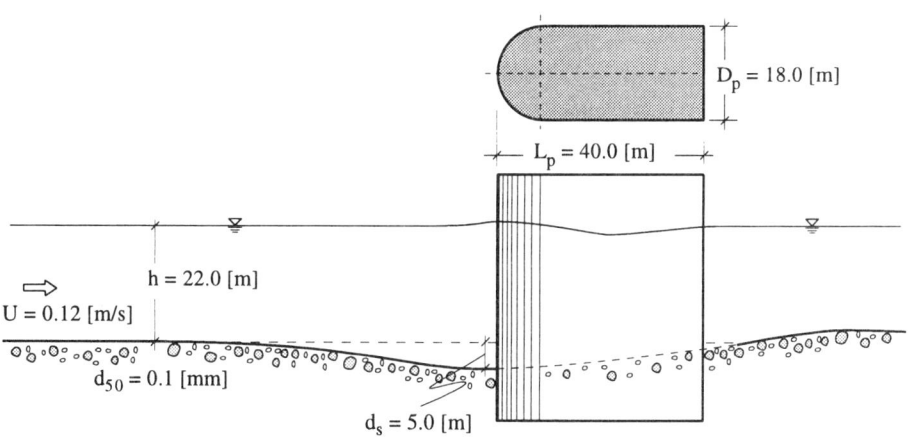

Fig. Ex.9.A.1

ii) *Scour-depth estimation*

The problem is one of clear-water scour, since (see Fig. 9.3a):

$$\frac{U}{U_{cr}} = \frac{0.12}{0.2} = 0.6 \qquad \therefore \frac{U}{U_{cr}} < 1.00$$

The relative flow depth is of no great importance, since (see Fig. 9.3b):

$$\frac{h}{D_p} = \frac{22}{18} = 1.22 \qquad \therefore \frac{h}{D_p} > 1.00$$

The relative grain size is of no further consequence, since (see Fig. 9.3c):

$$\frac{D_p}{d_{50}} = \frac{18}{0.0001} = 1.8 \times 10^5 \qquad \therefore \frac{D_p}{d_{50}} > 100$$

From above, it appears reasonable, that the scour depth can be evaluated, using the following two relations, available for clear-water scour:

— $\qquad d_s = 0.00022 \left(\frac{UD_p}{\nu}\right)^{0.619}$ (9.6)

consequently, one writes:

$$d_s = 2.2 \times 10^{-4} \times \left(\frac{0.12 \times 18}{10^{-6}}\right)^{0.619} = 1.834 \approx \mathbf{1.9 \ [m]}$$

— $\qquad \dfrac{d_s}{D_p} = \left(2 \dfrac{U}{U_{cr}} - 1\right)\left(2.0 \tanh \dfrac{h}{D_p}\right) \xi_s \cdot \xi_\alpha$ (9.9)

valid for: $0.5 < \dfrac{U}{U_{cr}} < 1.0$;

with $\xi_s = 1.0$, for a round pier (see Fig. 9.6e)

with $\xi_\alpha = 1.0$, for an aligned pier (see Fig. 9.6f)

consequently, one writes:

$$\frac{d_s}{D_p} = \left(2 \times \frac{0.12}{0.2} - 1\right)\left(2.0 \tanh \frac{22}{18}\right) \times 1 \times 1$$

$$= (1.2 - 1)(2.0 \times 0.84) = 0.336$$

and one obtains:

$$d_s = 0.336 \ D_p = 0.336 \times 18 = \mathbf{6.05 \ [m]}$$

The scour depth, obtained with the above relations, eq. 9.6 and eq. 9.9, give different results; the more conservative value should be used.

The scour depth to be expected may be taken as:

$$d_s \cong \mathbf{6.0 \ [m]}$$

LOCAL SCOUR 647

Ex. 9.B

A large channel, having a width of B = 90 [m] is designed for a maximum discharge of Q = 1026 [m³/s]. The corresponding depth and hydraulic radius were calculated as being h = 3.19 [m] and R_h = 3.00 [m].

The bottom slope of the channel was measured as being S_f = 0.0005 [-] ; the granulometric analysis gave values of d_{50} = 0.32 [mm] with σ_g = 1.8 and s_s = 2.65 [-].

A bridge which crosses the river will have a round-nosed pier being aligned under an angle of α = 15°. The pier is 1.2 [m] wide and 6.0 [m] long.

Estimate the equilibrium scour depth at the pier.

SOLUTION :

i)   *Hydraulic calculations*

The shear stress of the approach flow shall be calculated using the expression :

$$\tau_o = \rho u_*^2 \tag{3.6}$$

where $u_*^2 = gR_hS_f = 9.81 \times 3.0 \times 0.0005 = 0.0147$ [m²/s²].

Subsequently one obtains :

$$\tau_o = 1000 \times 0.0147 = 14.70 \text{ [N/m}^2\text{]}$$

Furthermore, the dimensionless shear stress is expressed by :

$$\tau_* = \frac{\tau_o}{(\gamma_s-\gamma)d_{50}} = \frac{14.7}{9.81 \times (2650-1000) \times 0.00032} = 2.84 \text{ [-]} \tag{3.38}$$

ii) *Critical condition*

The critical condition will be expressed using the relation of Shields (see Fig. 3.13), given by :

$$\tau_{*cr} = f(d_*) \tag{3.40a}$$

where $\quad d_* = d_{50}\left(\frac{\rho_s-\rho}{\rho}\frac{g}{\nu^2}\right)^{1/3} = 3.2\times10^{-4}\left(\frac{2650-1000}{1000}\frac{9.81}{\left(1.2\times10^{-6}\right)^2}\right)^{1/3}$

d_* = 7.17 [-] $\Rightarrow$ τ_{*cr} = 0.04 [-]

iii) *Scour-depth estimation*

The problem is one of sediment-transport scour, since (see Fig. 9.3a):

$$\frac{\tau_*}{\tau_{*cr}} = \frac{2.84}{0.04} = 71 \qquad \therefore \frac{\tau_*}{\tau_{*cr}} > 1$$

The relative flow depth is of no further consequence, since (see Fig. 9.3b):

$$\frac{h}{D_p} = \frac{3.19}{1.2} = 2.66 \qquad \therefore \frac{h}{D_p} > 1.0$$

The relative grain size is of no further consequence, since (see Fig. 9.3c):

$$\frac{D_p}{d_{50}} = \frac{1.2}{0.00032} = 3750 \qquad \therefore \frac{D_p}{d_{50}} > 100$$

From above it appears reasonable, that the scour depth can be evaluated, using the following two relations, available for sediment-transport scour:

$$- \quad \frac{d_s}{D_p} = 2.3 \cdot \xi_\alpha \qquad (9.10)$$

with $\xi_\alpha = 1.7$, for a rectangular non-aligned pier,
for $\alpha = 15°$ with $L_p/D_p = 5$ (see Fig. 9.3f)

consequently, one obtains:

$$d_s = (2.3 \times 1.7) \times 1.2 = \mathbf{4.7} \text{ [m]}$$

$$- \quad \frac{d_s}{D_p} = 2.0 \cdot \xi_g \cdot \xi_s \cdot \xi_\alpha \qquad (9.11)$$

with $\xi_g = 0.75$ for a graded sediment, $\sigma_g = 1.8$ [-] (see Fig. 9.3d)
$\xi_s = 1.1$, for a round-nosed pier (see Fig. 9.3e)

consequently, one obtains:

$$d_s = 2.0 \cdot (0.75 \times 1.1 \times 1.7) \cdot 1.2 = \mathbf{3.4} \text{ [m]}$$

The scour depth, obtained with the above relations, eq. 9.10 and eq. 9.11, give reasonably similar results.

The relation of eq. 9.10 is a more conservative one, while the relation of eq. 9.11 is all together reasonable in the light of the available research.

LOCAL SCOUR

Ex. 9.C

A spill-through *abutment* is to be constructed in a larger river, which conveys a design discharge of Q = 1600 [m³/s], at a flow depth of h = 4.0 [m].

The river has a width of B = 400 [m] and a bottom slope of S_f = 0.000072 [-]. The sediment in the river was analysed as being d_{50} = 8 [mm] with σ_g = 1.7 [-] and s_s = 2.63 [-].

The abutment to be constructed is inclined by θ = 50° and its extension, measured perpendicular to the flow, should be L_A = 50 [m].

Calculate the equilibrium scour depth. Suggest a possible scour prevention scenario.

SOLUTION :

i) *Hydraulic calculations*

The average flow velocity shall be calculated using the formula of Manning-Strickler :

$$U = K_s R_h^{2/3} S_f^{1/2} \qquad (3.16)$$

where $\qquad K_s = \dfrac{21.1}{d_{50}^{1/6}} = \dfrac{21.1}{(0.008)^{1/6}} = 47.2 \ [m^{1/3}/s] \qquad (3.18)$

$\qquad n = 1/K_s = 0.02 \ [m^{1/3}/s]^{-1}$

and with $\quad R_h^{2/3} \approx h^{2/3} = 4^{2/3} = 2.5 \ [m]$

Subsequently one obtains :

$U = 47.2 \times 2.5 \times (0.000072)^{1/2} = $ **1.0** [m/s]

ii) *Critical velocity*

The critical velocity, U_{cr}, of erosion will be taken from the diagram of Hjulstrom (see Fig. 3.12) :

for d = 8 [mm] $\quad \Rightarrow \quad U_{cr} \approx 0.8$ [m/s]

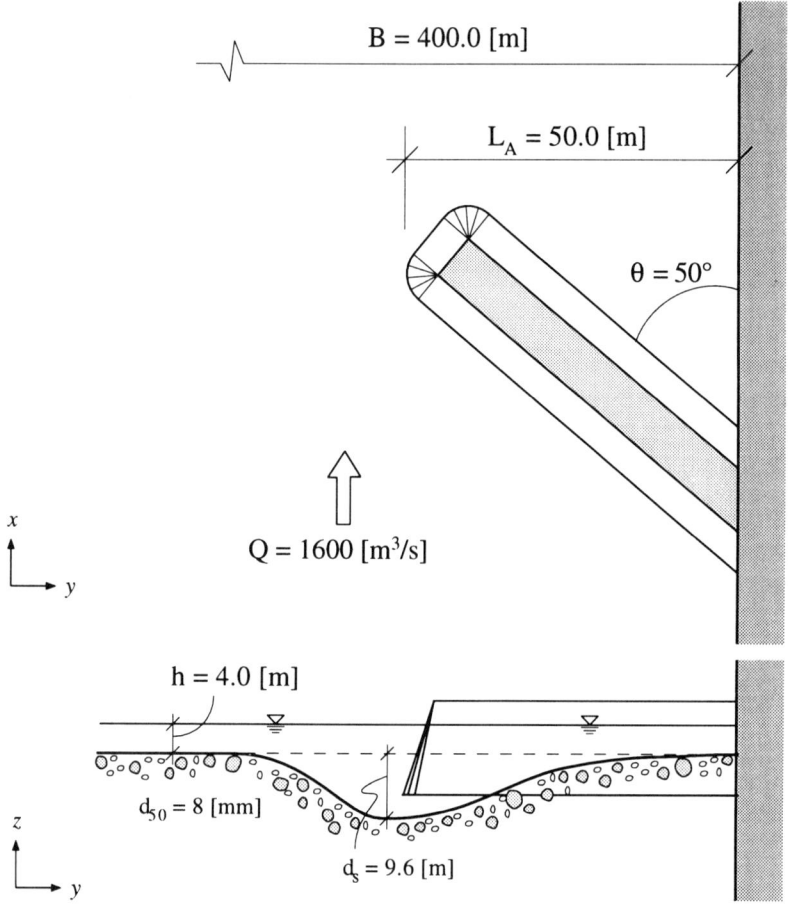

Fig. Ex.9.C.1

iii) *Scour-depth estimation*

The problem is one of sediment-transport scour, since (see Fig. 9.6a):

$$\frac{U}{U_{cr}} = \frac{1.0}{0.8} = 1.25 \qquad \therefore \frac{U}{U_{cr}} > 1.00$$

The abutment is considered to be a "long" one, since (see Fig. 9.6b):

$$\frac{h}{L_A} = \frac{4}{50} = 0.08 \qquad \therefore \frac{h}{L_A} < 1$$

The relative grain size, is of no further consequence, since (see Fig. 9.6c):

$$\frac{L_A}{d_{50}} = \frac{50}{0.008} = 6250 \qquad \therefore \frac{L_A}{d_{50}} > 25$$

LOCAL SCOUR

From above it appears reasonable, that the scour depth can be evaluated, using the following relation :

$$\frac{d_s}{L_A} = 2\sqrt{\frac{h}{L_A}} \cdot \xi_s \cdot \xi_\alpha \qquad (9.15b)$$

with $\xi_s \cong 0.5$, for a spill-through abutment (see Fig. 9.7e)

$\xi_\alpha \cong 0.9$, for the angle of approach, $\theta = 50°$ (see Fig. 9.7f)

consequently one writes :

$$\frac{d_s}{L_A} = 2 \times \left(\frac{4}{50}\right)^{1/2} \times 0.5 \times 0.9 \cong 0.255$$

$$d_s = 0.255 \, L_A = 0.255 \times 50 \cong \mathbf{12.7 \, [m]}$$

This is a rather conservative design estimate, as has been remarked in the development of eq. 9.15b.

In the above relationship, eq. 9.15b, the fact that the sediment is graded, $\sigma_g = 1.7$, has not been taken into account, but $\xi_g = 1$ was taken. Research (see *Raudkivi*, 1991, p. 67 and *Melville*, 1992, p. 623) has shown (see also Fig. 9.3d) that graded sediments produce armouring and a reduction expressed by $\xi_g = 0.75$ can safely be expected. Thus one obtains for the scour depth :

$$d_s = \xi_g \, (0.255 \, L_A) = 0.75 \times (0.255 \times 50) = \mathbf{9.6 \, [m]}$$

This appears to be a more appropriate design estimate.

iv) *Scour prevention*

An effective method to avoid much of the local scour is the dumping of stones into the scour hole.

The size of the stone, d, can be determined from a simple working relationship (see sect. 9.2.5) using :

$$U_d \cong 2.4\sqrt{d}$$

consequently one obtains ($U_d \equiv U = 1$ [m/s]) :

$$d = \left(\frac{U_d}{2.4}\right)^2 = \left(\frac{1.0}{2.4}\right)^2 \approx \mathbf{0.17 \, [m]}$$

Ex. 9.D

A mountain stream was channellised — see Ex. 3.D — in order to flow at the design discharge of Q = 30 [m³/s], essentially free of erosion.

The hydraulic and sedimentological parameters, the following were obtained :

S_{f_1} = 0.01 [-], d_{50} = 50 [mm], s_s = 2.65 [-], n = 0.025 [m$^{-1/3}$s] ;
R_{h_1} = 0.45 [m], h_1 = 0.47 [m], B_1 = 27.5 [m], U_1 = 2.35 [m/s].

The channel width should be reduced, but the part in the constriction should be free of erosion, having a scour depth of no more than d_s = 0.30 [m].

Determine the width in the constricted section of the channel.

SOLUTION :

i) *Hydraulic calculations*

- The shear stress in the approach channel is obtained as :

$$\tau_{o_1} = \rho g\, R_{h_1}\, S_{f_1} = 1000 \times 9.81 \times 0.45 \times 0.01 \cong 44 \text{ [N/m}^2\text{]}$$

- The critical shear stress, τ_{cr}, for the granulate, d_{50}, shall be obtained from Fig. 3.13, where $\tau_* = f(d_*)$.

The dimensionless diameter is :

$$d_* = d_{50}\left(\frac{\rho_s - \rho}{\rho} \cdot \frac{g}{\nu^2}\right)^{1/3} = 0.05\left(\frac{2650 - 1000}{1000} \cdot \frac{9.81}{(1.2 \times 10^{-6})^2}\right)^{1/3} \approx 1120\ [-]$$

The dimensionless shear stress can be obtained from Fig. 3.13, thus :

d_* = 1120 [-] $\Rightarrow$ τ_{*cr} = 0.055 [-]

Subsequently the shear stress is found to be :

$$\tau_{cr} = \tau_{*cr}\, g\, (\rho_s - \rho)\, d_{50} = 0.055 \times 9.81 \times (1650) \times 0.05 \cong 44.5 \text{ [N/m}^2\text{]}$$

Thus it is shown that the approach channel is essentially *free* of erosion, since :

$$\tau_{o_1} \cong \tau_{cr} \cong 45 \text{ [N/m}^2\text{]}$$

LOCAL SCOUR 653

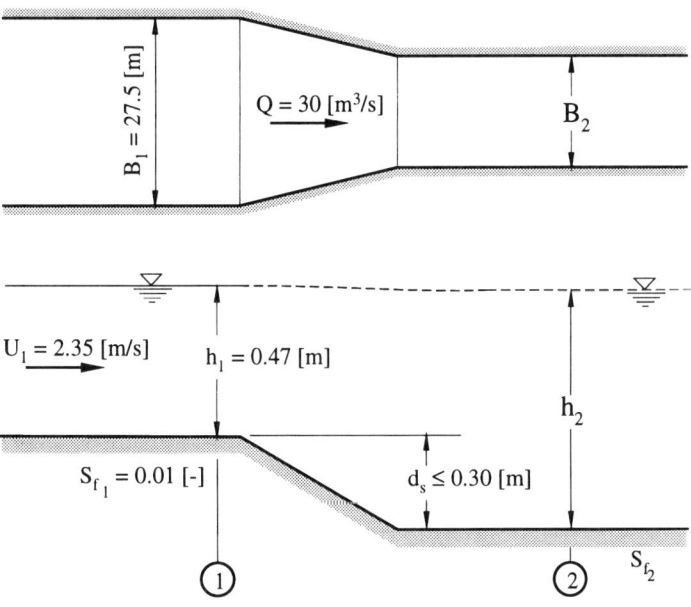

Fig. Ex.9.D.1

ii) *Calculation of scour depth*

- It shall be *assumed*, that the variation of the velocity head is small (see eq. 9.20a), such that :

$$d_{s_1} = (h_2 - h_1) \tag{9.21}$$

Using the "allowable" scour depth of $d_s = 0.30$ [m], one gets :

$$h_2 = h_1 + d_{s_1} = 0.47 + 0.30 = 0.77 \text{ [m]}$$

- Since both, the approach and the constricted section of the channel, are free of erosion, namely $\tau_{o_1} \equiv \tau_{cr}$ and $\tau_{o_2} \equiv \tau_{cr}$, the relative flow depth can be calculated with the following relation :

$$\frac{h_2}{h_1} = \left(\frac{B_1}{B_2}\right)^{6/7} \tag{9.27}$$

This renders :

$$\frac{B_1}{B_2} = \left(\frac{h_2}{h_1}\right)^{7/6} = \left(\frac{0.77}{0.47}\right)^{7/6} = 1.78 \quad \Rightarrow \quad B_2 = 15.5 \text{ [m]}$$

- The assumption of a negligeable variation of velocity head must be controlled. This requires to calculate the velocity heads, or :

$$\frac{U_1^2}{2g} = \frac{\left(\frac{Q}{B_1 h_1}\right)^2}{2g} = \frac{\left(\frac{30}{27.5 \times 0.47}\right)^2}{2 \times 9.81} = 0.28 \, [m]$$

$$\frac{U_2^2}{2g} = \frac{\left(\frac{Q}{B_2 h_2}\right)^2}{2g} = \frac{\left(\frac{30}{15.5 \times 0.77}\right)^2}{2 \times 9.81} = 0.32 \, [m]$$

Using now the complete expression for the scour depth :

$$d_{s_2} = (h_2 - h_1) + \left(\frac{U_2^2}{2g} - \frac{U_1^2}{2g}\right) \qquad (9.20a)$$

One obtains :

$$d_{s_2} = (0.77 - 0.47) + (0.32 - 0.28) = 0.30 + 0.04 = \mathit{0.34} \, [m] \quad > \quad 0.30 \, [m]$$

Consequently the above assumption leads to an unacceptable scour depth, since the "allowable" scour depth of $d_s = 0.30$ [m] is exceeded by 10 %.

- The above calculations must be repeated, *assuming* now a smaller scour depth of $d_{s_3} = 0.25$ [m]. This yields, using eq. 9.21 :

$$h_2 = h_1 + d_{s_3} = 0.47 + 0.25 = 0.72 \, [m]$$

Consequently one obtains, using eq. 9.27 :

$$\frac{B_1}{B_2} = \left(\frac{h_2}{h_1}\right)^{7/6} = \left(\frac{0.72}{0.47}\right)^{7/6} = 1.65 \quad \Rightarrow \quad B_2 = \mathit{16.7} \, [m]$$

The assumption of a negligeable variation of the velocity head must again be controlled, since :

$$\frac{U_1^2}{2g} = 0.28 \, [m]$$

$$\frac{U_2^2}{2g} = \frac{\left(\frac{Q}{B_2 h_2}\right)^2}{2g} = \frac{\left(\frac{30}{16.7 \times 0.77}\right)^2}{2 \times 9.81} = 0.32 \, [m]$$

LOCAL SCOUR

This allows the calculation of the scour depth, eq. 9.20a, as :

$$d_s = (0.72 - 0.47) + (0.32 - 0.28) = 0.25 + 0.04 = 0.29 \text{ [m]} \quad < \quad 0.30 \text{ [m]}$$

Consequently this second assumption leads to an acceptable scour depth, since the "allowable" scour depth of $d_s = 0.30$ [m] is not exceeded.

iii) *Summary*

The channelised mountain stream having the hydraulic characteristics of :

$B_1 = 27.5$ [m], $h_1 = 0.47$ [m] and $U_1 = 2.32$ [m/s]

can be reduced in width such as :

$B_2 = $ *16.7*[m], $h_2 = 0.72$ [m] and $U_2 = 2.50$ [m/s]

The scour depth will be : $d_s = $ *0.29* [m]

The slope of the constricted channel will be :

$$S_{f_2} = \frac{\tau_{o_2}}{\rho g \, R_{h_2}} = \frac{45}{1000 \times 9.81 \times 0.66} = 0.007 \text{ [-]}$$

iv) *Remarks*

- A reasonably similar result could, of course, have been obtained by using directly the Fig. 9.9.

    The entire channel is free of scour, namely :

    $$\tau_{o_1} = \tau_{cr} = \tau_{o_2} \quad \Rightarrow \quad \frac{\tau_{o_1}}{\tau_{o_2}} = 1$$

- The relative depth, h_2/h_1, shall be *approximated* (see eq. 9.21) by :

    $$\frac{h_2}{h_1} \approx \frac{h_1 + d_s}{h_1} = \frac{0.47 + 0.30}{0.47} = \frac{0.77}{0.47} = 1.64$$

- Using Fig. 9.9, one obtains :

    $$\frac{B_1}{B_2} \approx 1.7 \quad \Rightarrow \quad B_2 \approx \textit{16.2} \text{ [m]}$$

- *Note* : no control about the approximation (see eq. 9.21) was performed.

Ex. 9.E

A rectangular channel having a width of $B = 4.0$ [m] carries a discharge of $Q = 6$ [m³/s]. The normal depth has previously (see Ex. 4.D) been determined as being $h_n = 0.82$ [m] and the friction coefficient of Strickler was evaluated as $K_s = 45$ [m$^{1/3}$/s]. A gate was installed in this channel to control the flow. This gate, operating as sluice gate, passes the total discharge as underflow at a gate opening of $h_v = 0.5$ [m]. Water is stored behind the gate at a depth of $h_1 = 2.34$ [m]. The bed behind the gate is a fixed one, while downstream of the gate it is a mobile one.

i) When the water is released under the gate at a high velocity the downstream bed will probably be scoured. What dimensions of the scour hole should one expect ?

ii) Investigate two other scenarios — keeping the same flow depth behind the gate, $h_1 = 2.34$ [m] — to see whether the scour depth could be diminished. The gate shall be operated in a way to produce :
   a) an overflow of $\quad\quad Q = 6$ [m³/s].
   b) an equally divided overflow and underflow of
$$Q = Q_o + Q_u = 3 \text{ [m}^3\text{/s]} + 3 \text{ [m}^3\text{/s]}.$$

SOLUTION :

0) *Determination of the particle size :*

The grain-size distribution of the sediment can be calculated using the relation of Strickler, eq. 3.18, or :

$$K_S = \frac{21.1}{d_{50}^{1/6}} \quad\Rightarrow\quad d_{50} = \left(\frac{21.1}{45}\right)^6 = 0.0106 \text{ [m]}$$

$$K_S = \frac{26}{d_{90}^{1/6}} \quad\Rightarrow\quad d_{90} = \left(\frac{26}{45}\right)^6 = 0.0372 \text{ [m]}$$

i) *Flow under the structure*

The following schema shows the problem with the known parameters.

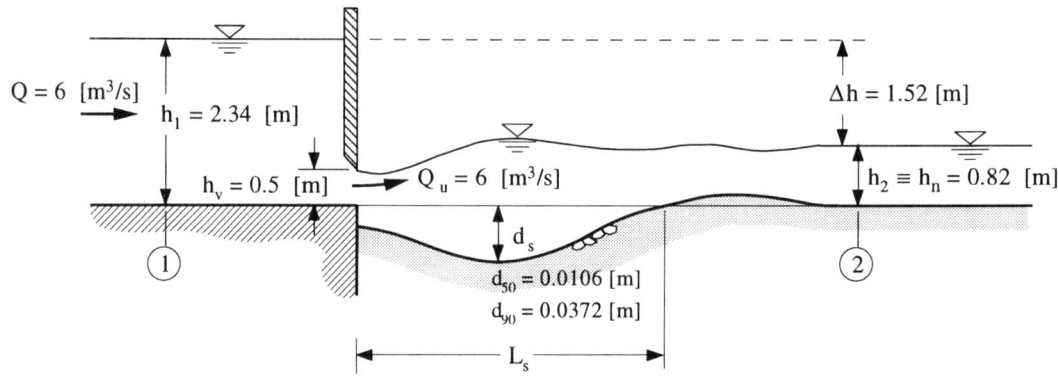

LOCAL SCOUR

The scour depth, d_s, shall be calculated using two different formulae.

a) Scour depth according to the relation of *Eggenberger* et *Müller* (1944) :

$$h + d_s = w \frac{\Delta h^{0.5} q^{0.6}}{d_{90}^{0.4}} \qquad (9.34)$$

where : $\Delta h = 2.34 - 0.82 = 1.52$ [m]
$q_u \equiv q = 6/4 = 1.5$ [m²/s]
$d_{90} = 37.2$ [mm]
$w = 10.35$ [s$^{0.6}$/m$^{0.3}$] ...... for a submerged jet, $h_v < h_2$;

consequently one obtains :

$$d_s = 10.35 \left(\frac{1.52^{0.5} \times 1.5^{0.6}}{37.2^{0.4}} \right) - 0.82 = \mathbf{3.01} \ [m]$$

The length of the scour hole is :

$$L_s \approx 6 (h + d_s) \approx 23 \ [m]$$

b) Scour depth according to the relation of *Breusers* (1991) :

$$\frac{d_s}{h_v} = 0.008 \left(\frac{U_u}{u_{*cr}} \right)^2 \qquad (9.36)$$

where : $h_v = 0.5$ [m]
$U_u = q/h_v = 3.0$ [m/s]
u_{*cr} : for $d_{50} = 10.6$ [mm] :
with $d_* = d_{50} \left(\frac{\rho_s - \rho}{\rho} \cdot \frac{g}{v^2} \right)^{1/3} = 268$ [-]

$\rho_s = 2650$ [kg/m³], $\rho = 1000$ [kg/m³], $v = 1.0 \times 10^{-6}$ [m²/s]
using Fig. 3.13 $\Rightarrow$ $\tau_{*cr} = 0.055$ [–]

$\tau_{*cr} = \frac{\tau_{ocr}}{(\gamma_s - \gamma) d_{50}} \Rightarrow \tau_{ocr} = 9.44$ [N/m²]

$u_{*cr} = \left(\frac{\tau_{ocr}}{\rho} \right)^{1/2} = \left(\frac{9.44}{1000} \right)^{1/2} = 0.097$ [m/s] ;

consequently one obtains :

$$d_s = 0.5 \times 0.008 \left(\frac{3.0}{0.097}\right)^2 = \textbf{3.8 [m]}$$

being valid for $U_u / u_{*cr} = (3.0/0.097) = 30.88 < 100$

The length of the scour hole is :

$$L_s \approx (5 \text{ to } 7)\, d_s = 19 \text{ to } 27 \text{ [m]}$$

iia)   Flow over the structure

The following schema shows the problem with the known parameters.

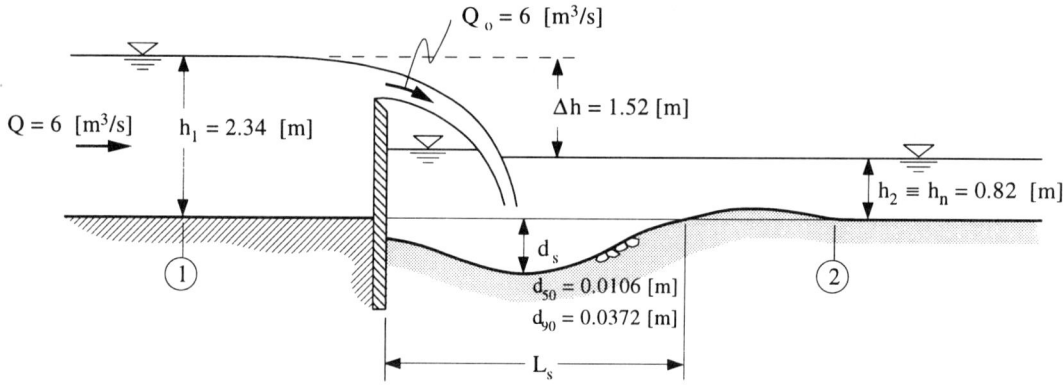

The scour depth, d_s, shall be calculated using three different formulae.

a)   Scour depth according to the relation of *Eggenberger* et *Müller* (1944) :

$$h + d_s = w \frac{\Delta h^{0.5}\, q^{0.6}}{d_{90}^{0.4}} \qquad (9.28)$$

where : $\Delta h = 1.52$ [m]
$q_o \equiv q = 1.5$ [m²/s]
$d_{90} = 37.2$ [mm]
$w = 22.88$ [s$^{0.6}$/m$^{0.3}$] ;

consequently one obtains :

$$d_s = 22.88 \left(\frac{1.52^{0.5} \times 1.5^{0.6}}{37.2^{0.4}}\right) - 0.82 = \textbf{7.65 [m]}$$

LOCAL SCOUR 659

b) Scour depth according to the relation of *Kotoulas* (1967):

$$h + d_s = 1.9 \frac{1}{g^{0.35}} \left(\frac{\Delta h^{0.35} q^{0.7}}{d_{95}^{0.4}} \right) \qquad (9.31)$$

where: $d_{95} \approx d_{90} = 0.0372$ [m];

consequently one obtains:

$$d_s = 1.9 \frac{1}{9.81^{0.35}} \left(\frac{1.52^{0.35} \times 1.5^{0.7}}{0.0372^{0.4}} \right) - 0.82 = \mathbf{4.08 \; [m]}$$

c) Scour depth according to the relation of *Mason* et *Arumugam* (1985):

$$h + d_s = K_M \frac{1}{g^{0.3}} \left(\frac{\Delta h^y q^x}{d_m^{0.1}} \right) h^{0.15} \qquad (9.33)$$

where: $\Delta h = 1.52$ [m]
$d_m \approx d_{50} = 0.0106$ [m]
$K_M = 6.42 - 3.10 \, \Delta h^{0.1} = 3.19$
$y = 0.15 + \Delta h/200 = 0.16$
$x = 0.6 - \Delta h/300 = 0.59$;

consequently one obtains:

$$d_s = 3.19 \frac{1}{9.81^{0.3}} \left(\frac{1.52^{0.16} \times 1.5^{0.59}}{0.0106^{0.1}} \right) 0.82^{0.15} - 0.82 = \mathbf{2.52 \; [m]}$$

iib) Flow over and under the structure

The following schema shows the problem with the known parameters.

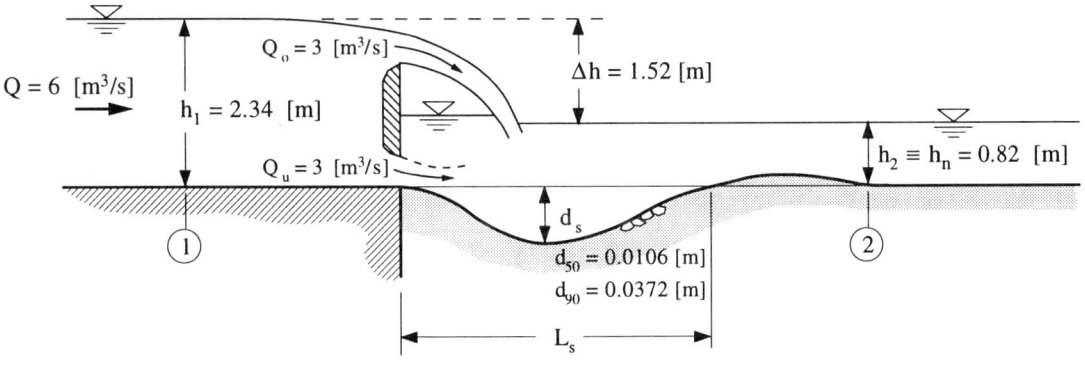

The scour depth, d_s, shall be calculated according to the relation of *Eggenberger* et *Müller* (1944):

$$h + d_s = w \frac{\Delta h^{0.5} q^{0.6}}{d_{90}} \tag{9.38}$$

where: $\Delta h = 1.52$ [m]
$q = q_o + q_u = 0.75 + 0.75 = 1.5$ [m²/s]

$$w = 22.88 - \frac{1}{0.0049 R_q^3 - 0.0063 R_q^2 + 0.029 R_q + 0.064} \tag{9.39}$$

with $R_q = q_o/q_u = 1 \Rightarrow w = 11.96$ [s$^{0.6}$/m$^{0.3}$];

consequently one obtains:

$$d_s = 11.96 \left(\frac{1.52^{0.5} \times 1.5^{0.6}}{37.2^{0.4}} \right) - 0.82 = \mathbf{3.60} \text{ [m]}$$

Formula	overflow	—	$Q_o = 6$ [m³/s]	$Q_o = 3$ [m³/s]
	underflow	$Q_u = 6$ [m³/s]	—	$Q_u = 3$ [m³/s]
Eggenberger et *Müller*		$d_s = 3.0$ [m] $L_s = 23$ [m]	$d_s = 7.6$ [m]	$d_s = 3.6$ [m]
Breusers		$d_s = 3.8$ [m] $L_s \approx 24$ [m]	—	—
Kotoulas		—	$d_s = 4.1$ [m]	—
Mason et *Arumugam*		—	$d_s = 2.5$ [m]	—

The results of this study are summarized in the above table.

– Calculation of scour at underflow, q_u, obtained with two relations, give comparable values.

– Calculation of scour at overflow, q_o, obtained with three relations do not give conclusive results. A value of $d_s = 4.1$ [m] should be taken, since the relation of *Eggenberger* et *Müller* (1944) is known as rather overestimating.

– Obviously the scour depth for the overflow is considerably larger than for the other two scenarios. The smallest scour depth is therefore obtained if the gate is operated as sluice gate, allowing only an underflow.

LOCAL SCOUR   661

9.6.2 Problems, unsolved

Ex. 9.1

The discharge of a river is estimated at $Q = 410$ [m³/s]. The wetted section was measured as being $A = 470$ [m²] and the average depth as being $h = 3.0$ [m]. The channel slope is $S_f = 0.00016$ [-], made up of sediments having a diameter of $d_{50} = 0.78$ [mm] with $\sigma_g = 2.2$.
The bridge crossing this river will have a cylindrical, elongated pier, having the dimensions of $D_p = 2.4$ [m] and $L_p = 8.9$ [m]. The pier is set at $\alpha = 10°$ to the flow direction.
Calculate the scour depth, d_s, at the pier.

Ex. 9.2

Consider the same river, as was (??) described in Ex. 9.1.
Instead of one single cylindrical, elongated pier, there will be installed two piers being cylindrical, round of $D_p = 2.4$ [m]. The distance between these piers shall both be $a = 10$ [m].
Calculate the scour depth for the front and the rear pier.

Ex. 9.3

The river, described in Ex. 9.1, shall in the future be regulated. It will maintain its average depth of $h = 3.0$ [m], but the average flow velocity will be reduced by 20 %.
Determine the scour depth, d_s, at the pier.

Ex. 9.4

The data are the ones of Ex. 9.C.
Develop different scenarios, by variation of the angle of approach, θ, and of the abutment length, L_A, so that the scour depth does not exceed a value of $d_s = 5.0$ [m].

Ex. 9.5

The data of Ex. 9.C are to be used, but the abutment is a vertical wall, having a length of $L_A = 15$ [m].
Determine the resulting scour depth.

Ex. 9.6

A channel of a rectangular, wide cross section has a width of $B = 90$ [m]. The bed slope was measured as $S_f = 0.0005$ [-] and the granulometry is taken as $d_{50} = 0.32$ [mm].

The design discharge of the channel is Q = 55.8 [m³/s] ; the resulting flow depth was previously calculated, being h = 0.89 [m], but this value should be controlled.
A long constriction is foreseen, resulting in a width reduction of $B_1/B_2 = 2$. The resulting constricted section should be free of erosion.
What will be the scour depth, d_s ?

Ex. 9.7

A hydraulic model study was performed to investigate the overfall scour. The following data are available :

q   =   0.011 [m²/s]
Δh   =   0.19   [m]
h   =   0.04   [m]
d_{90} =   1.5   [mm]

Determine the scour depth to be expected. The model test gave a final scour depth of $(h+d_s) = 0.25$; is this a reasonable value ?

Ex. 9.8

Determine the equilibrium scour depth, d_s , which will take place downstream of a sluice gate. A flow of q = 2.5 [m²/s] passes when sluice-gate opening is set at $h_v = 0.4$ [m]. The upstream and downstream flow depths are $h_1 = 8.0$ [m] and $h_2 = 3.5$ [m]. The particle size of the sediments in the downstream channel is $d_{90} = 6$ [mm].

Ex. 9.9

Repeat Ex. 9.8, but sluice gate functions as an overflow weir. Subsequently determine the upstream flow depth, h_1 , for the special case where almost no scour is allowed.

LIST OF SYMBOLS

SYMBOL	UNITS (SI)	MEANING
A	m²	cross section, area
a_h, (a_{ht})	cm²/s	thermal diffusivity (turbulent)
B	m	width of channel, at water surface
	m³/s³	reduced sediment flux per unit width : $B = g'hU$
B_s, B_r	-	constants of integration (velocity distribution)
b	m	width of channel, at bed
C	m$^{1/2}$/s	coefficient of Chézy
C (c)	-	average (local) concentration by volume or by mass of a substance
	kg/m³	average (local) densiometric concentration of a substance
C_s, (c_s)	-	concentration (local) of granular material
$C_{(BOD)}$	mg/l	concentration of biochemical oxygen demand
$C_{(OD)}$	mg/l	concentration of oxygen dissolved in water
$\overline{C}$	-	average concentration of air
C^+, C^-	-	characteristics (positive, negative)
c	m/s	celerity of gravity wave (perturbation)
c_b	-	local concentration, close to the bed
c_f, $\vec{c_f}$	-	extensive property
c_k'	m/s	celerity of a diffusive wave
c_k	m/s	celerity of a kinematic wave
c_t	m/s	celerity of a translatory wave
c_w	m/s	absolute celerity of (gravity) wave
D_h	m	hydraulic depth
D_p	m	diameter of cylinder (pier)
$D_{(OD)}$	mg/l	deficit of oxygen dissolved in water

d	m	diameter of grain (granulate)
d_s	m	(local) scour depth
d_x	m	diameter of grain, equal to x % in the granulometric curve
d_*	-	dimensionless diameter of grain
E_s	-	entrainment coefficient of sediments from the bed
E_w	-	entrainment coefficient of ambient fluid (turbidity current)
Fr	-	Froude number : $Fr = U/\sqrt{gh}$
Fr_D	-	densimetric Froude number
f	-	friction coefficient of Weisbach-Darcy
f'	-	friction coefficient due to grain roughness
f''	-	friction coefficient due to bed forms
f_{CT}	-	friction coefficient of a turbidity current
G_s	kg/s;(N/s)	total solid discharge in mass (in weight)
g	m/s²	gravitational acceleration
g'	m/s²	reduced gravity : $g' = g\, C_s\, (\rho_s-\rho)/\rho$
H	m	total (energy) head
	m	wave height
	m	water depth at dam
H_D	m	height of weir (spillway)
H_s	m	specific energy
h	m	flow (water) depth
	m	flow depth of integral scale (turbidity current)
h	m	position (elevation) of water surface : $\mathit{h} = h + z'$
h_c	m	critical (water) depth
h_n	m	normal (uniform) water depth
h_p	m	water depth at plunge point (turbidity current)
h_r	m	head loss
h_t	m	height, where the velocity, u , is zero (turbidity current)
i_b	-	fraction of granulometric curve of bed material

LIST OF SYMBOLS

Symbol	Units	Description
i_{sb}	-	fraction of granulometric curve of solid discharge, q_{sb}
K	m^2/s	coefficient in parabolic model of solid transport (see eq. 6.11)
$K(h)$	m^3/s	conveyance of channel
K_D	-	coefficient of discharge (weir)
K_s	$m^{1/3}/s$	coefficient of Strickler
K_v	-	coefficient of discharge (gate)
K_x, K'_x	m^2/s (cm^2/s)	dispersivity (longitudinal, turbulent)
k_a	$1/s$	coefficient of reaeration
k_r	$1/s$	coefficient of reaction
k_s	mm	standard (uniform, equivalent) roughness
$k_t, (k)$	m^2/s (cm^2/s)	(coefficient of) turbulent diffusivity (molecular)
L	m	length
	m	wave length
L_A	m	length of abutment
L_y	m	distance, where transversal diffusion extends over entire channel width
L_z	m	distance, where vertical diffusion extends over entire flow depth
l	m	mixing length
M_0	g	total mass of a substance
M	-	hydraulic exponent (second)
m	-	side slope (bank)
N	-	hydraulic exponent
n	$s/m^{1/3}$	coefficient of Manning
P	m	wetted perimeter
p	Pa	effective pressure
	-	porosity of granulate
p_e	-	probability of erosion after Einstein
p_f	Pa	pressure at bed
p_t	Pa	total pressure (charge)
p^*	Pa	driving (piezometric) pressure

Q	m³/s	(liquid) discharge
Q_s	m³/s	total solid discharge
q	m²/s	unit discharge
q_l	m²/s	supplementary discharge per unit length
q_s	m²/s	unit total solid discharge
q_{sb}	m²/s	solid discharge as bed load per unit width
q_{ss}	m²/s	solid discharge as suspended load per unit width
q_{sw}	m²/s	solid discharge as washload per unit width
R	-	specific density of submerged granular material : $R = (\rho_s - \rho_a)/\rho_a$
R_h	m	hydraulic radius : $R_h = A/P$
R_h'	m	hydraulic radius due to grain roughness
R_h''	m	hydraulic radius due to bed forms
R_{hb}	m	hydraulic radius of the channel bed
Re	-	Reynolds number : $Re = 4UR_h/\nu$ or $4Uh/\nu$
Re'	-	Reynolds number : $Re' = Uh/\nu$
Re_*	-	Reynolds number of particles : $Re_* = u_* d/\nu$
Ri	-	(global) Richardson number : $Ri = g'h \cos\alpha/U^2$
r	m	radius
	m	radius of a curve (bend)
S_c	-	critical slope
S_e	-	energy slope : $S_e = -(dH/dx)$
S_f	-	bed slope : $S_f = -(dz/dx)$
S_w	-	water (surface) slope : $S_w = -(dh/dx)$
s_s	-	relative density of particle
T	°C	temperature
t	s	time ; interval of time
U	m/s	average velocity in a cross section
	m/s	average velocity of integral scale (turbidity current)
U_c	m/s	maximum velocity in a section at $z' = \delta$
	m/s	average velocity at critical depth

LIST OF SYMBOLS

Symbol	Units	Description
$U_{cr} \equiv U_E$	m/s	velocity of erosion or critical velocity
U_D	m/s	velocity of sedimentation
U_f	m/s	velocity of front of turbidity current
$\bar{U}$	m/s	average velocity (turbidity current)
$\tilde{U}$	m/s	average velocity, produced by a wave
u, v, w	m/s	local (average) velocity, $u(x)$, $v(y)$, $w(z)$
u', v', w'	m/s	fluctuations components of the local velocity
u_*	m/s	friction velocity
$\vec{V}(u,v,w)$	m/s	vector of (local) velocity
v_{ss}	m/s	settling velocity of particle
$x, y, z(z')$	m	Cartesian coordinates
$z \equiv z_b, z'$	m	position (elevation) of bed
$\mathfrak{z}$	-	exponent of Rouse: $\mathfrak{z} = v_{ss}/\kappa u_*'$
α	°	slope of bed
	°	angle of a curve (bend)
α_e	-	correction coefficient for velocity distribution (energy)
β	°	front angle (oblique jump)
	-	equilibrium parameter of longitudinal pressure gradient
β_u	-	correction coefficient for velocity distribution (momentum)
$\gamma = \rho g$	N/m³	specific weight
Δ	m	height of dune
Δh	m	height of (translatory) wave: $\Delta h = h_2 - h_1$
	m	drop in water depth: $\Delta h = h_1 - h_2$
Δz	m	height of floor change
ε_s	m²/s	diffusivity of solid particles

Symbol	Units	Description
$\varepsilon_t, \vec{\varepsilon}_t, (\varepsilon_m)$	m²/s (cm²/s)	mass diffusivity, turbulent (molecular), of a substance in a mixture
$\vec{\varepsilon}_t \, (\varepsilon_{tx}, \varepsilon_{ty}, \varepsilon_{tz})$	m²/s (cm²/s)	turbulent diffusivity (longitudinal, transversal, vertical)
η	-	ratio of heights : $\eta = h/h_n$
θ	°	inclination of side walls
	°	angle of deflection
κ	-	Karman constant : $\kappa = 0.4$
λ	m	length of dune
$\nu, (\nu_t)$	m²/s	kinematic (turbulent) viscosity
ξ_g, ξ_s, ξ_α	-	coefficients (local scour) of sediment grading, of shape and of angle of approach
ξ_M	-	roughness parameter after Meyer-Peter
ζ_H	-	hiding coefficient after Einstein
ζ_p	-	lift-force correction coefficient after Einstein
Π	-	wake parameter of Coles
ρ	kg/m³	density (of fluid)
ρ_s	kg/m³	density of solid particle
ρ_t	kg/m³	average density of turbidity current
σ^2	m²	variance
τ_o	Pa	wall or bed-shear stress
τ_{ocr}	Pa	critical shear stress
τ_{zx}	Pa	tangential shear stress
τ_*	-	dimensionless shear stress : $\tau_* = \tau_o/[d(\gamma_s - \gamma)]$
φ	°	angle of repose
Φ	-	integral of Bakhmeteff
	-	intensity (dimensionless) of solid discharge
Ψ	-	intensity (dimensionless) of shear stress

BIBLIOGRAPHY

ABBOTT M.B. (1979) : *Computational Hydraulics*
    Pitman Publ. Ltd, London, GB

ABBOTT M.B. et D. BASCO (1989) : *Computational Fluid Dynamics*
    Longman Sci. & Techn., Harlow, GB

ACKERS P. et W.R. WHITE (1973) : *Sediment Transport: New Approach and Analysis*
    Proc., Am. Soc. Civil Engrs., Vol. 99, HY11, USA

AKIYAMA J. et H. STEFAN (1985) : *Turbidity Current with Erosion and Deposition*
    Proc., Am. Soc. Civil Engrs., Vol. 111, HY12, USA

ALBERTSON M.L. et al. (1960) : *Fluid Mechanics for Engineers*
    Prentice-Hall Inc., Englewood Cliffs, NJ, USA

ALTINAKAR M.S., W.H. GRAF et E.J. HOPFINGER (1990) : *Weakly Depositing Turbidity Current on Small Slopes*
    J. Hydr. Res., Vol. 28/1, NL

ALTINAKAR M.S., W.H. GRAF et E.J. HOPFINGER (1993) : *Water and Sediment Entrainment in Weakly Depositing Turbidity Currents on Small Slopes*
    Proc., XXV Congr., Int. Ass. Hydr. Res., Vol. 2, Tokyo, J

ALTINAKAR M.S., W.H. GRAF et E.J. HOPFINGER (1995) : *Flow Structure in Turbidity Currents*
    J. Hydr. Res., Vol. 34, No. 5, NL

APMANN R.P. (1973) : *Estimating Discharge from Superelevation in Bends*
    Proc., Am. Soc. Civil Engrs., Vol. 99, HY1, USA

ASHIDA K. et M. MICHIUE (1971) : *An Investigation of River Degradation Downstream of a Dam*
    Proc., XIV Congr., Int. Ass. Hydr. Res., Vol. 3, Paris, F

BAKHMETEFF B.A. (1932) : *Hydraulics of Open Channel Flow*
    McGraw-Hill, New York, USA

BARNES H.H. (1967) : *Roughness Characteristics of Natural Channels*
    Water Supply Paper No. 1849; U.S. Geological Survey, USA

BATHURST J.C., W.H. GRAF et H.H. CAO (1987) : *Bed Load Discharge Equations for Steep Mountain Rivers*
    in "Sediment Transport in Gravel-bed Rivers"; C. Thorne et al. (Ed.)
    J. Wiley & Sons, Ltd, Chichester, GB

BAUER S. et W.H. GRAF (1971) : *Free Overfall as Flow Measuring Device*
    Proc., Am. Soc. Civil Engrs., Vol. 97, IR1, USA

BIRD R., W. STEWART et E. LIGHTFOOT (1960) : *Transport Phenomena*
J. Wiley & Sons, Inc., New York, USA

BONNEFILLE R. , J. GODDET (1959) : *Etudes des courants de densité en canal*
Proc., VIII Congr., Int. Ass. Hydr. Res., Vol. 2, Montreal, C

BOUVARD M. (1984) : *Barrages mobiles et ouvrages de dérivation*
Eyrolles, Paris, F

BRADSHAW P. (Ed.) (1978) : *Turbulence; Topics in Applied Physics* -
Vol. 12; Springer Verlag, Berlin, D

BREUSERS H.N. et al. (1977) : *Local Scour around cylindrical Piers*
J. Hydr. Res., Vol. 15/3, NL

BREUSERS H.N. (1991) : *Scour by Jets*
ch. 6 in "Scouring" ; H. Breusers et A.J. Raudkivi (Ed.)
IAHR, Hydraulic Structures Design Manual N° 2 ;
A.A. Balkema, Rotterdam, NL

BRITTER R.E. et P. LINDEN (1980) : *The Motion of the Front of a Gravity Current*
J. Fluid Mechanics, Vol. 88, part 3, GB

BROWNLIE W. (1983) : *Flow Depth in Sand-Bed Channels*
Proc., Am. Soc. Civil Engrs., Vol. 109, HY7, USA

CARDOSO A.H., W.H. GRAF et G. GUST (1989) : *Uniform Flow in a Smooth Open Channel*
J. Hydr. Res., Vol. 27/5, NL

CARDOSO A.H., W.H. GRAF et G. GUST (1991) : *Steady Gradually Accelerating Flow in a Smooth Open Channel*
J. Hydr. Res., Vol. 29/4, NL

CARLIER M. (1972) : *Hydraulique générale et appliquée*
Eyrolles, Paris, F

CHEN Y.H., F.M. HOLLY, K. MAHMOOD et D.B. SIMONS (1975) : *Transport of Material by Unsteady Flow*
in "Unsteady Flow in Open Channels"; K. Mahmood et al. (Ed.)
Water Res. Publ., Fort Collins, CO, USA

CHOW V.T. (1959) : *Open Channel Hydraulics*
McGraw-Hill, New York, USA

CHOW V.T., D. MAIDMENT et L. MAYS (1988) : *Applied Hydrology*
McGraw-Hill, New York, USA

COLEMANN N.L. (1981) : *Velocity Profiles with suspended Sediment*
J. Hydr. Res., Vol. 19, No 3, NL

CORREIA L. et W.H. GRAF (1988) : *Grain-size Distribution and Armoring in Gravel-Bed Rivers : a case study*
Rapp. Annuel, Lab. Rech. Hydr, EPF-Lausanne, CH

CORREIA L., B. KRISHNAPPAN et W.H. GRAF (1992) : *Fully Coupled Unsteady Mobile Boundary Flow Model*
   Proc., Am. Soc. Civil Engrs., Vol. 118, JHE, No 3, USA

CRANK J. (1956, 1989) : *The Mathematics of Diffusion*
   Oxford University Press, London, GB

CRAUSSE E. (1951) : *Hydraulique des canaux découverts*
   Eyrolles, Paris, F

CSANADY G.T. (1973) : *Turbulent Diffusion in the Environment*
   Reidel Publ. Co., Dordrecht, NL

CUNGE J., F. HOLLY et A. VERWEY (1980) : *Practical Aspects of Computational River Hydraulics*
   Pitman Publ. Ltd, London, GB

DAILY J.W. et D.R. HARLEMAN (1966) : *Fluid Dynamics*
   Addison-Wesley Publ. Co., Reading, MA, USA

DINGMAN S. (1984) : *Fluvial Hydrology*
   W. Freeman et Co., New York, USA

EGASHIRA S. et K. ASHIDA (1980) : *Studies on the Structure of Density Stratified Flows*
   Bull., Disaster Prevention Res. Inst., Vol. 29, Kyoto, J

EGGENBERGER W. et R. MÜLLER (1944) : *Experimentelle und theoretische Untersuchungen über das Kolkproblem*
   Mitteil., Versuchsanstalt f. Wasserbau, N° 5, Zürich, CH

EINSTEIN H.A. (1950) : *The Bed-Load Function for Sediment Transportation in Open Channel Flows*
   US Dept. Agr., Soil Conserv. Service, T.B. No 1026, Washington, USA

EXNER F.M. (1925) : *Über die Wechselwirkung zwischen Wasser und Geschiebe in Flüssen*
   Sitzungsber., Akad. Wissenschaften, pt. IIa, Bd. 134, Wien, A

FAIR G.M., J. GEYER et D. OKUN (1968) : *Water and Wastewater Engineering*
   Vol. 2, J. Wiley & Sons, New York, USA

FAVRE H. (1935) : *Ondes de translation*
   Dunod, Paris, F

FISCHER H. (1966) : *A Note on the One-dimensional Dispersion Model*
   Air & Wat. Pollut. Int. J., Vol. 10, GB

FISCHER H. et al. (1979) : *Mixing in Inland and Coastal Waters*
   Acad. Press, New York, USA

FLAMANT A. (1923) : *Hydraulique*
   Librairie Polytechn., Béranger, Paris, F

FORCHHEIMER P. (1930) : *Hydraulik*
   Teubner-Verlag, Leipzig, D

FRENCH R. (1986) : *Open-Channel Hydraulics*
McGraw-Hill, New York, USA

GHETTI A. (1981) : *Idraulica*
Libreria Cortina, Padova, I

GILL M.A. (1987) : *Nonlinear Solution of Aggradation and Degradation in Channels*
J. Hydr. Res., Vol. 25, No 5, NL

GRAF W.H. (1966) : *On the Determination of the Roughness Coefficient*
Bull., Int. Ass. Sci. Hydrology, XIe année, No 1, B

GRAF W.H. (1971; 1984) : *Hydraulics of Sediment Transport*
McGraw-Hill, New York; Water Res. Publ., Littleton, CO, USA

GRAF W.H. (1983) : *The Behaviour of a Silt-Laden Current*
Water Power and Dam Construction, Vol. 35, No 9, GB

GRAF W.H. (1988) : *Mixing by Diffusion in Lakes*
Proc., Asian and Pac. Reg. Conf. - IAHR -, Spec. Lect. 3, Kyoto, J

GRAF W.H. (1991) : *Flow Resistance over a Gravel Bed*
Chap. A in "Fluvial Hydraulics in Mountain Regions"; A. Armanini (Ed.)
Springer Verlag, Berlin, D

GRAF W.H. et E.R. ACAROGLU (1968) : *A Physical Model for Sediment Transport in Conveyance Systems*
Bull., Int. Assoc. Scient. Hydrology, Vol. XIII, No 3, B

GRAF W.H. et al. (1987) : *Flow Resistance in Steep Channels*
Comm. du Lab. d'Hydraulique No 54; EPF-Lausanne, CH

GRAF W.H. & M.S. ALTINAKAR (1991; 1995) : *Hydrodynamique*
Eyrolles, Paris, F ; Presses polytechniques romandes, Lausanne, CH

GRAF W.H. et G. PAZIS (1977) : *Erosion et déposition*
Comm. du Lab. d'hydraulique No 37; EPF-Lausanne, CH

GRAF W.H. et T. SONG (1995) : *Sediment Transport in Unsteady Flow*
Proc., XXVI Congr., Int. Assoc. Hydr. Res., Vol. 1, London, GB

GRAF W.H. et L. SUSZKA (1987) : *Sediment Transport in Steep Channels*
J. Hydrosc. and Hydr. Eng'g. Vol. 5, No 1, J

GRAF W.H. et B. YULISTIYANTO (1996) : *Experiments on Flow Around a Cylinder ; the Velocity and Vorticity Fields*
Rapport annuel, Laboratoire de recherches hydrauliques, EPF, Lausanne, CH

GRISHANIN K.V. (1969) : *River Dynamics*
(en russe), Hydrometeorological Publisher, Leningrad, R

GRISHANIN K.V. (1990) : *Foundations of Alluvial Stream Dynamics*
(en russe), Transport, Moscow, R

HAYASHI T. (1953) : *Mathematical Theory and Experiment of Flood Waves*
Trans., Jap. Soc. Civ. Engrs., No 18, Sept., Tokyo, J

HENDERSON F.M. (1966) : *Open Channel Flow*
    Macmillan Comp., New York, USA

HINZE J.O. (1975) : *Turbulence*
    McGraw-Hill Book Comp., New York, USA

HOFFMANN C.J. (1977) : *Spillways*
    Chap. IX in "Design of Small Dams"
    U.S. Dept. of Interior, B. of Reclamation; Washington, USA

HOLLY F.M. et J.L. RAHUEL (1990) : *New numerical/physical Framework for mobile-bed Modelling; part 1 and 2*
    J. Hydr. Res., Vol. 28, No 4 et 5, NL

HUG M. (Ed.) (1975) : *Mécanique des Fluides appliquée*
    Eyrolles, Paris, F

IPPEN A. (1950) : *Channel Transitions and Controls*
    Ch. VIII in "Engineering Hydraulics"; H. Rouse (Ed.)
    J. Wiley & Sons, New York, USA

IWASA Y. et S. AYA (1991) : *Predicting Longitudinal Dispersion Coefficients in Open-channel Flow*
    in "Environmental Hydraulics"; J. Lee et al. (Ed.)
    A. Balkema, Rotterdam, NL

JAEGER C. (1954) : *Hydraulique technique*
    Dunod, Paris, F

JANSEN P. et al. (1979) : *Principles of River Engineering*
    Pitman Publ. Ltd, London, GB

JARAMILLO W.F. et S.C. JAIN (1984) : *Aggradation and Degradation of Alluvial-Channel Beds*
    Iowa Inst. Hydr. Res., Rep. No 274, Iowa, USA

JOBSON H.E. et N. YOTSUKURA (1973) : *Mechanics of Heat Transfer in non-stratified open Channel Flow*
    Ch. 8 in "Environmental Impact on Rivers"; H. Shen (Ed.)
    Shen-Publisher, Fort Collins, Colorado, USA

JOHNSON P.A. (1995) : *Comparison of Pier-Scour Equations using Field Data*
    Proc., Am. Soc. Civil Engrs., Vol. 121, JHE 8, USA

KEULEGAN G.H. (1938) : *Laws of Turbulent Flow in Open Channels*
    J. Res., Natl. Bur. of Standards, U.S. Dept. of Comm., Vol. 21, USA

KINSMAN B. (1965) : *Wind Waves*
    Prentice Hall, Englewood Cliffs, NJ, USA

KIRONOTO B. et W.H. GRAF (1993) : *Turbulence Characteristics in Rough Uniform Open-Channel Flow*
    Proc., Instn Civ. Engrs, Wat., Marit. et Energy, Vol. 106 (Dec.), GB

KIRONOTO B. et W.H. GRAF (1994) : *Turbulence Characteristics in Rough Non-Uniform Open-Channel Flow*
 Proc., Instn Civ. Engrs, Wat., Marit. et Energy, Vol. 112 (Dec.), GB

KOMURA S. (1971) : *River-bed Variations at long Constrictions*
 Proc., XIV Congr., Int. Assoc. Hydr. Res., Vol. 3 , Paris, F

KOTOULAS D. (1967) : *Das Kolkproblem im Rahmen der Wildbachverbauung*
 Mitteil., Schweizer Anstalt f. forstliches Versuchswesen, Vol. 43/1, Birmensdorf, CH

KOZENY J. (1953) : *Hydraulik*
 Springer-Verlag, Wien, A

KRISHNAPPAN B. (1981) : *Unsteady, Non Uniform, Mobile Boundary Flow Model - MOBED*
 Hydr. Div., Nat. Water Res. Inst., Burlington, Ontario, CA

LAMB H. (1945) : *Hydrodynamics*
 Dover Publ., New York, USA

LAURSEN E.M. (1963) : *An Analysis of Relief Bridge Scour*
 Proc., Am. Soc. Civil Engrs., Vol. 89, HY3, USA

LENAU C. et A. HJELMFELD (1992) : *River Bed Degradation due to abrupt Outfall Lowering*
 Proc., Am. Soc. Civil Engrs., Vol. 118, HY6, USA

LIGGETT J. (1975) : *Stability*
 in "Unsteady Flow in Open Channels"; K. Mahmood et al. (Ed.)
 Water Res. Publ., Fort Collins, CO, USA

LIGGETT J. et J. CUNGE (1975) : *Numerical Methods of Solution of the Unsteay Flow Equations*
 in "Unsteady Flow in Open Channels"; K. Mahmood et al. (Ed.)
 Water Res. Publ., Fort Collins, CO, USA

MARTIN H. (1989) : *Plötzlich veränderliche instationäre Strömungen*
 in "Techn. Hydromechanik-2"; G. Bollrich (Ed.)
 VEB-Verlag, Berlin, D

MASON P.J. and K. ARUMUGAM (1985) : *Free-Jet Scour below Dams and Flip Buckets*
 Proc., Am. Soc. Civil Engrs., Vol. 111, JHE 2, USA

MELVILLE B.W. (1992) : *Local Scour at Bridge Abutments*
 Proc., Am. Soc. Civil Engrs., Vol. 118, JHE 4, USA

METCALF & EDDY, Inc. (1972) : *Wastewater Engineering*
 McGraw-Hill Book Comp., New York, USA

MIDDLETON G.V. (1984) : *Mechanics of Sediment Movement; Short Course*
 Soc. Econ. Paleont. et Mineral., Providence, R.I., USA

MONIN A. et A. YAGLOM (1971) : *Statistical Fluid Mechanics*
 The MIT Press, Cambridge, MA, USA

NAUDASCHER E. (1987) : *Hydraulik der Gerinne und Gerinnebauwerke*
　　Springer-Verlag, Wien, A

NORDIN C. et al. (1989) : *Testing Design Methods for local Scour at Bridge Piers*
　　Personal communication, USA

PADET J. (1993) : *Fluides en écoulement*
　　Masson, Paris, F

PARKER G., M. GARCIA, Y. FUKUSHIMA et W. YU (1987) : *Experiments on Turbidity Currents over an Erodible Bed*
　　J. Hydr. Res., Vol. 25, No 1, NL

PAZIS G. et W.H. GRAF (1977) : *Weak Sediment Transport*
　　Proc., Am. Soc. Civil Engrs., Vol. 103 HY7, USA

RAJARATNAM N. et D. MURALIDHAR (1964) : *End Depth for Exponential Channels*
　　Proc., Am. Soc. Civil Engrs., Vol. 90, IR1, USA

RANGA-RAJU K.G. (1981) : *Flow through Open Channels*
　　Tata McGraw-Hill, New Dehli, IN

RAUDKIVI A.J. (1976, 1990) : *Loose Boundary Hydraulics*
　　Pergamon Press, Oxford, GB

RAUDKIVI A.J. (1991) : *Scour at Bridge Piers*
　　ch. 5 in "Scouring" ; H. Breusers et A.J. Raudkivi (Ed.)
　　IAHR, Hydraulic Structures Design Manual N° 2 ;
　　A.A. Balkema, Rotterdam, NL

RESCH F. et H. LEUTHEUSSER (1972) : *Le ressaut hydraulique : Mesure de turbulence*
　　La Houille blanche, Vol. 4/2, F

REYNOLDS A.J. (1974) : *Turbulent Flows in Engineering*
　　J. Wiley & Sons, London, GB

RIBBERINK J. et J. van der SANDE (1985) : *Aggradation in Rivers due to Overloading*
　　J. Hydr. Res., Vol. 23, No 3, NL

ROTTA J.C. (1972) : *Turbulente Strömungen*
　　B.C. Teubner, Stuttgart, D

ROUSE H. (1938) : *Fluid Mechanics for Hydraulic Engineers*
　　Dover Publ., New York, USA

RUTHERFORD J.C. (1994) : *River Mixing*
　　J. Wiley & Sons, Ltd., Chichester, GB

SAUVAGET P. (1985) : *Dispersion in Rivers and Coastal Waters*
　　Ch. 2 in "Developments in Hydraulic Engineering – 3"; P. Novak (Ed.)
　　Elsevier Appl. Sci. Publishers, London, GB

SCHLICHTING H. (1979) : *Boundary-Layer Theory*
　　McGraw-Hill Book Comp., New York, USA

SHEN H.W. (1971) : *River Mechanics, Vol. II, Chap. 23*
　　H.W. Shen publ., Fort Collins, Color., USA

SILBERMAN E. et al. (1963) : *Friction Factors in Open Channels*
　　Proc., Am. Soc. Civil Engrs., Vol. 89, HY2, USA

SILBER R. (1968) : *Etude et tracé des écoulements permanents
　　en canaux et rivières*
　　Dunod, Paris, F

SILVESTER R. (1964) : *Hydraulic Jump in all Shapes of Horizontal Channels*
　　Proc., Am. Soc. Civil Engrs., Vol. 90, HY1, USA

SINNIGER R. et W. HAGER (1989) : *Constructions Hydrauliques*
　　Presses polytechniques romandes, Lausanne, CH

SONI J.P., R.J. GRADE et K.R. RAJU (1980) : *Aggradation in Streams by Overloading*
　　Proc., Am. Soc. Civil Engrs., Vol. 106, HY1, USA

STOKER J.J.. (1975) : *Water Waves*
　　Interscience Publ. Inc., New York, USA

STREETER V. (1971) : *Fluid Mechanics*
　　McGraw-Hill, New York, USA

STRICKLER A. (1923) : *Beiträge zur Frage der Geschwindigkeitsformeln...*
　　Mitteilung No. 16; Amt f. Wasserwirtschaft, Bern, CH

SUGIO S. (1972) : *Variations in Bed Form and Mean Velocity
　　in Alluvial Channels and Streams*
　　Bull. Fac. Eng'g., Vol. 9/1,2, Tokushima Univ., J

TAYLOR G.I. (1954) : *The Dispersion of Matter in turbulent Flow through a Pipe*
　　Proc., Roy. Soc. London, Vol. (A) 223, GB

TENNEKES H. et J.L. LUMLEY (1972) : *A First Course in Turbulence*
　　The MIT Press, Boston, MA, USA

TU H. et W.H. GRAF (1992) : *Velocity Distribution
　　in Unsteady Open-Channel Flow over Gravel Bed*
　　J. Hydrosci. and Hyd. Eng'g, Vol 10/1, J

TU H. et W.H. GRAF (1992a) : *Vertical Distribution of Shear Stress
　　in Unsteady Open-Channel Flow*
　　Proc., Instn Civ. Engrs, Wat., Marit. et Energy, Vol. 92 (June), GB

TURNER J.S. (1973) : *Buoyancy Effects in Fluids*
　　Cambridge Univ. Press, Cambridge, GB

VIESSMAN W., T. HARBOUGH et J. KNAPP (1972) : *Introduction to Hydrology*
　　Intext Educ. Publ., New York, USA

VRIES M. de (1965) : *Considerations about non-steady Bedload Transport
　　in Open Channels*
　　Delft Hydr. Lab.; series 1 (18.66), NL

VRIES M. de (1973) : *River-Bed Variations - Aggradation and Degradation*
    Delft Hydr. Lab.; Publ. No 107, NL

VRIES M. de (1985) : *A Sensitivity Analysis Applied to Morphological Computations*
    Delft Techn. Univ, Comm. on Hydr.; Rep. 85-2, NL

VREUGDENHIL C.B. et M. de VRIES (1973) : *Analytical Approaches to non-Steady Bedload Transport;*
    Delft Hydr. Lab.; Rep. S-78/N, NL

WALLISCH S. (1990) : *Äquivalente Sandrauhigkeiten...*
    DVWK Schriften 92; P. Parey Verlag, Hamburg, D

WAN Z. et Z. WANG (1994) : *Hyperconcentrated Flow*
    IAHR-Monograph; Balkema Publ, Rotterdam, NL

WHITE F.M. (1974) : *Viscous Fluid Flow*
    McGraw-Hill Book Company, New York, USA

WHITE W.R., H. MILLI et A. CRABBE (1973) : *Sediment Transport: An Appraisal of available Methods*
    Hydr. Res. Stat., Rept. INT 119, Wallingford, GB

WOOD I. (1985) : *Air Water Flows*
    Proc., XXI Congr., Int. Ass. Hydr. Res., Vol. 6, Melbourne, AUS

YALIN M.S. (1972) : *Mechanics of Sediment Transport*
    Pergamon Press, Oxford, GB

YÜCEL O. et W.H. GRAF (1971) : *Bed Load Deposition in Reservoirs*
    Proc., XV Congr., Int. Ass. Hydr. Res., Vol. A, Istanbul, TR

INDEX

Aggradation, 365, 368, 408, 412

Armouring, 382, 454

Attenuation of wave, 279

Bakhmeteff, equation of —, 139, 156

Bed load (*see* : transport of sediments)

Bresse, equation of —, 140, 154

Channel
— artificial, 3, 109, 124, 194
— geometry of —, 4, 5, 214
— natural, 3, 120, 152, 214
— on adverse slope, 142, 148
— on critical slope, 140, 142, 147
— on horizontal slope, 142, 148
— on mild slope 141, 142, 143, 195, 206, 225
— on steep slope, 141, 142, 146, 225
— prismatic, 4, 135, 253
— type of —, 3
— with mobile bed, 358

Characteristics, 258, 261, 294, 322, 337, 343

Celerity
— absolute, 35, 259
— of translatory wave, 287
— of wave, 31, 35, 185, 259, 273, 361

Chézy, coefficient of —, 77, 78, 355

Chow, equation of —, 158, 196

Coefficient of diffusivity (*see* : diffusivity)

Coefficient of friction, 53, 57, 59, 73, 74, 113, 481
— mobile bed, 84, 113, 358

Coefficient of reaction, 563, 566

Colebrook-White, relation of —, 57, 75

Condition of Courant, 263, 265, 308

Continuity, equation of —, 19, 41, 47, 71, 135, 253, 358, 359

Convection, 519, 520
— with reaction, 566

Convection-diffusion, 520, 523, 524
— laminar regime, 520, 529
— — continuous source, 541, 576
— — instantaneous source, 529
— turbulent regime, 520, 527, 529, 542, 551, 598

Convection-dispersion (*see* : dispersion)

Conveyance, 88

Curve, flow in —, 100, 126

Curve of stage-discharge, 116, 256, 357, 455

Dam break, 292, 334

Degradation, 365, 367, 400

Delta, formation of —, 412, 433

Diffusion, 519, 520
— molecular, 519
— of heat, 521
— of mass, 519
— of momentum, 521
— pure, 525, 532, 573
— transversal, 551, 582
— vertical, 576

Diffusion of a source
— continuous, 537, 541
— instantaneous, 533, 529

Diffusivity, 519, 522, 529
— longitudinal, 549
— molecular, 522, 529
— of heat, 521
— of mass, 521
— of momentum, 521
— transversal, 548, 582
— turbulent, 522, 527, 530, 543
— vertical, 385, 547, 576

INDEX

Discharge, 20, 30, 88, 92, 163, 170, 172
— curve of —, 27
— lateral, 20, 191

Discharge, calculation of —
— fixed bed, 88, 90, 91, 113
— mobile bed, 92, 113, 120

Dispersion, 543
— continuous injection, 561, 562
— instantaneous injection, 560, 561
— longitudinal, 558, 561, 587, 590, 598
— time-limited injection, 560, 561
— with reaction, 563

Dispersivity, 554

Dunes (mini-dunes, antidunes), 82, 85

Energy
— equation of —, 21, 136, 180
— specific, 25, 180, 183, 235

Entrainment of air, 107, 125

Entrainment, coefficient of —
— of fluid, 471, 475, 482
— of sediments, 476, 484

Equation(s)
— hydrodynamic, 36
— hydrodynamic equation for turbidity current, 474
— of balance, 523
— of gradually varied flow, 137, 140
— of momentum, 174, 191
— of motion, 36, 40, 46, 72, 136, 254, 358
— of Reynolds, 37, 529
— of Saint-Venant, 24, 47, 71, 135, 253, 257
— of Saint-Venant - Exner, 358, 360, 363, 370, 422
— of water surface, 136
— of wave, 34

Exner, relation of —, 359

Exponent
— hydraulic, 156, 158
— of Rouse, 387

Flow
— critical, 9, 26, 28
— fluvial, 9, 27, 141
— gradually varied, 7, 71, 135, 412
— laminar, 9, 51, 59
— lateral, 20, 191
— mobile bed, 84, 358
— non-uniform, 6, 7, 24, 46, 69, 133, 412
— of a water-sediment mixture, 354
— rapidly varied, 7, 163
— regimes of —, 8, 10
— torrential, 9, 27, 103, 141, 185, 241
— turbulent, 9, 52, 53, 59, 526
— types of —, 5, 125, 257
— uniform, 6, 7, 24, 40, 69, 71, 363, 481
— unsteady, 7, 24, 251, 258, 282, 358
— with wave, 35

Friction, coefficient of — (*see* : coefficient of friction)

Froude, number of —, 8, 29, 82, 103, 106, 470

Gates (underflow), 172, 225, 328

Granulometry, 382

Head (energy) loss, 23, 24, 71, 73, 177

Hydraulic drop, 168

Hydrograph, 256, 277, 299, 312, 326

Instability at the surface, 30, 105, 147, 174

Jump
— hydraulic, 174, 225
— oblique, 185

Kutter, coefficient of —, 79, 80

Law of —
— Fick, 519
— Fourier, 521
— Newton, 521

Manning, coefficient of —, 79, 80, 82, 449

Method of characteristics, 260, 294

Method, explicit, 264, 300

Method, implicit, 266

Mixing (*see* : diffusion)

Moody-Stanton, diagram of —, 75

Oxygen demand, biochemical, 568

Oxygen, dissolved
— balance of —, 571, 604
— concentration of —, 567

Oxygenation, 567, 568

Plunge point, 473, 505

Point of control, 142, 149, 163, 196

Pressure
— distribution of —, 12, 39
— driving —, 37
— gradient of —, 44, 49, 50, 58, 60
— on bed, 13, 22

Property, extensive, 519, 523

Reoxygenation, 567, 570

Reservoir (emptying), 205, 294, 334

Reynolds, number of —, 8, 59, 75, 95, 372
— analogy of —, 386

Richardson, number of —, 470, 482

Roughness, 53, 76, 79, 448
— composite, 81

Scour (local), 613
— abutment, 628, 629, 632, 649
— constriction, 633, 635, 638, 652
— pier, 614, 618, 625, 645, 647
— prevention, 626, 651
— structures (hydraulic), 638, 656

Section of channel, 4
— of maximum discharge, 91
— of variable width, 188, 241
— stable, 98, 122

Sediments (granulat), 353, 372, 382, 455

Shear stress, 23, 72, 83, 96, 97, 118, 121, 372
— distribution of —, 42, 49, 97
— Reynolds —, 39, 43, 61, 63, 64, 527

Shields, relation of —, 96, 117, 121, 373

Slope
— critical, 140
— of channel bed, 5, 73, 124, 136
— of energy, 24, 71, 136, 254
— of water surface, 5, 7, 136

Spectrum of energy, 66

Spillways (*see* : weirs)

Stage-discharge relation
— for sediments, 357, 463
— for water, 88, 113, 357, 463

Strickler, coefficient of —, 79, 448

Substance
— active, 563
— passive, 519

Suspended load (*see* : transport of sediments)

Transitions (geometrical), 179
— of variable width, 183
— with variable bed floor, 180, 235

Transport with reaction, 566, 604

Transport of matter (*see* : diffusion and convection)

Transport of sediments, 355, 396, 412
— as bed load, 371, 375, 412, 434, 448, 457
— as suspended load, 384, 389, 459
— as total load, 393, 396, 455
— as wash load, 398

Turbidity currents, 354, 467, 491
— form of interface, 479
— front of —, 485
— hydrodynamic equations, 474
— structure of —, 489, 508

Turbulence, 61, 64, 66, 353, 526

Uniformity, notion of —, 71

Velocity
— average, 12, 59, 73
— critical (of erosion), 94, 120, 223
— distribution of —, 10, 42, 48, 50, 54, 55, 59
— of sedimentation, 91, 388

Water depth, 9
— conjugate, 174
— critical, 29, 138, 169, 171, 194
— normal, 89, 109, 138, 194

Water surface (in varied flow), 136
— calculation of —, 149, by :
  — direct integration, 153, 196
  — graphical integration, 161
  — successive approximations, 149, 200, 218, 227, 412
— forms of —, 141, 195, 217

Wave of translation
— meeting of —, 289
— negative, 282, 290, 330, 334
— positive, 282, 285
— simple, 257, 294

Waves
— cross, 104, 189
— diffusive, 257, 276
— dynamic, 254, 257, 300
— flood, 280, 322
— gravity, 31, 34, 185, 275
— kinematic, 257, 270, 275, 281, 322
— of bed forms, 82, 86, 361
— roll, 106
— translatory, 282

Weirs, 163
— broad crested, 170, 239

Weisbach-Darcy, coefficient of —, 73, 75

FLUVIAL HYDRAULICS Flow and Transport Processes in Channels of Simple Geometry **W.H. GRAF and M.S. ALTINAKAR** John Wiley and Sons, Ltd, 1998	**COMPUTER PROGRAMS** written by M. S. Altinakar

Six programs, written in FORTRAN 77 standard, are available for solving the exercises presented in the book :

EXPLIC : Program for solving the problems of unsteady flow in open channels of trapezoidal, rectangular or triangular cross section. It is based on an explicit finite-difference scheme.
EXPLIC has been used in Chap. 5 for solving the problem Ex. 5.B.

DELTA : Program for computing one-dimensional sediment transport as bed load, taking into account the modification of the bed profile due to erosion and deposition.
DELTA has been used in Chap. 6 for solving the problem Ex. 6.C.

GRAVIT: Program for solving the equations describing the one-dimensional, steady, gradually varied flow of gravity currents (conservative or non conservative) in supercritical regime (Ri<1) using the 4th-order Runge-Kutta method.
GRAVIT has been used in Chap. 7 for solving the problem Ex. 7.A.

DIFFTUR: Program written for solving turbulent convection-diffusion problems (in a vertical or horizontal plane) in a bounded space (for example in a channel of rectangular cross section) by using the method of imaginary sources.
DIFFTUR has been used in Chap. 8 for drawing Fig. 8.9 and for solving the problems Ex. 8.B and Ex. 8.C.

DIFDIS: Program written for solving the problems of convection-dispersion in a turbulent flow resulting from the continuous discharge of a source.
DIFDIS has been used in Chap. 8 for drawing Fig. 8.7 and for solving the problem Ex. 8.D.

DISPTUR: Program written for solving the problems of longitudinal convection-dispersion (one-dimensional), with and without reaction, in an open channel. The program can handle the problems with multiple injections in space or in time.
DISPTUR has been used in Chap. 8 for drawing Fig. 8.11 and for solving the problems Ex. 8.E and Ex. 8.F.

The program codes have a modular structure and are written in a pedagogical style. The names of the variables are chosen to recall the notation used in the book. Each source code contains an exhaustive list of variables together with variable types and short descriptions. The exercices to be solved using the programs are marked by a diskette sign.

The IBM-PC and Macintosh versions of these programs (source codes, the executable files and the sample data files) as well as an acrobat (pdf) file containing more than 50 pages of supplementary text, graphics and quicktime videos can be downloaded from our web site :

http://lrhwww.epfl.ch/books.html

together with
the terms of service agreement and license, including copyright, rules of use and other additional legal clauses.

The web site is designed to provide a continuous exchange between the authors and the readers. From time to time the new versions of the programs and other up-to-date information will be made available to the readers at this site.

Remarks and suggestions concerning the book and the programs are always welcome. Please contact the authors at :
M. S. Altinakar,   Laboratoire de Recherches Hydrauliques / DGC / EPFL, CH-1015, LAUSANNE
Phone   :   +41 21 693 23 75 (secretariat)   Fax   :   +41 21 693 67 67
               +41 21 693 32 87 (direct line)   E-mail   :   mustafa.altinakar@epfl.ch